1 in.	=	*0.0254 m (International foot)
1 ft	=	*0.3048 m (International foot)
1.15 miles	≅	1 minute (1′) of latitude ≅ 1 nautical mile
3.141 592 654	=	π
6 miles	=	length, width, of normal township
* 10 sq. chains (ch²)	=	(Gunter's) 1 acre
* 15° longitude	=	width of one time zone = 360°/24 hr
15°F		changes length of 100-ft steel tape by 0.01 ft
* 16½ ft	=	1 rod = 1 pole = 1 perch = ¼ ch (Gunter's)
* 20°C	=	standard temperature (Celsius) in taping = 68°F
23°26½′	=	maximum declination of sun at solstices
23^{h}56^{m}04.091^s	=	length of sidereal day in mean solar time, and 3^{m}55.909^s solar time short of mean solar day; also 3^{m}56.555^s sidereal time short of mean solar day
* 24 hr	=	360° of longitude
* 25.4 mm	=	1 in. (International foot)
* 36	=	number of sections in normal township
57°17′44.8″	=	1 radian (rad) = 57.295 779 51°
* 66 ft	=	length of Gunter's chain = 100 links (lk)
69.1 miles	≅	1° latitude
* 80 ch	=	(Gunter's) 1 mile
100	=	usual stadia interval factor
101 ft	≅	1 second (1″) of latitude
* 400 grads (gons)	=	360°
480 ch	=	width and length of normal township
490 lb/ft³	=	density of steel for tape weight computations
640 acres	=	one normal section of 1 mile²
6076.10 ft	=	1 nautical mile
4,046.9 m²	=	1 acre
* 6400 mils	=	360°
5,729.578 ft	=	radius of 1° curve, arc definition
5,729.651 ft	=	radius of 1° curve, chord definition
10,000 km	=	distance from equator to pole (basis for length of meter)
* 43,560 ft²	=	1 acre
206,264.806 sec	=	1 radian = cot 1 sec = 180°/π in sec
299,792.5 km/sec	=	speed of light, and other electromagnetic waves, in vacuum
6,356,752.3 m	=	earth's polar semi-axis (GRS80 ellipsoid)
6,378,137.0 m	=	earth's equatorial semi-axis (GRS80 ellipsoid)
6,356,583.8 m	=	earth's polar semi-axis (Clarke ellipsoid 1866)
6,378,206.4 m	=	earth's equatorial semi-axis (Clarke ellipsoid 1866)
6,371,000 m	=	mean radius of earth = 20,902,000 ft
29,000,000 lb/in.²	=	Young's modulus of elasticity for steel = 2,000,000 kg/cm²

*Denotes exact value. All others correct to figures shown.

ELEMENTARY
SURVEYING

ELEMENTARY
SURVEYING

NINTH EDITION

PAUL R. WOLF
Professor Emeritus, Civil and Environmental Engineering
University of Wisconsin at Madison

RUSSELL C. BRINKER
Adjunct Professor of Civil Engineering
New Mexico State University

Sponsoring Editor: John Lenchek
Development Editor: Nicholas Murray
Art Director/Design Supervisor: Jess Schaal
Cover Design: Roseanne Lufrano
Cover Illustration: Darryl Ligasan
Production Administrator: Randee Wire
Project Coordination and Text Design: Elm Street Publishing Services, Inc.
Compositor: Weimer Graphics, Inc.
Printer and Binder: R. R. Donnelley & Sons Company
Cover Printer: R. R. Donnelley & Sons Company

Elementary Surveying, Ninth Edition

Library of Congress Cataloging-in-Publication Data

Wolf, Paul R.
    Elementary surveying / Paul R. Wolf, Russell C. Brinker.—9th
ed.
        p.   cm.
    Includes index.
    ISBN 0–06–500399–3
    1. Surveying. I. Brinker, Russell C. (Russell Charles),
 . II. Title.
TA545.W77 1993   93–8582
526.9—dc20

    94 95 96 9 8 7 6 5 4 3 2

CONTENTS

3 SURVEYING FIELD NOTES ═══════ 46

4 DISTANCE MEASUREMENT; TAPING ═══ 59

5 ELECTRONIC DISTANCE MEASUREMENT   86

6 LEVELING—THEORY, METHODS, EQUIPMENT ══════════ 106

PART I   THEORY AND METHODS   106

PART II   EQUIPMENT FOR DIFFERENTIAL LEVELING   117

7 LEVELING—FIELD PROCEDURES AND COMPUTATIONS ══════════ 132

11 FIELD OPERATIONS WITH TRANSITS, THEODOLITES, AND TOTAL STATIONS 203

12 TRAVERSING ═══════════ 229

13 TRAVERSE COMPUTATIONS ══ 241

14 AREA ═══════════════ 269

15 STADIA ═══════════════ 286

16 TOPOGRAPHIC SURVEYS ═══════ 298

17 MAPPING ═══════ 325

18 ASTRONOMICAL OBSERVATIONS ═══ 346

19 CONTROL SURVEYS ══════════ 384

20 SATELLITE AND INERTIAL SURVEYING SYSTEMS ══════ 419

21   STATE PLANE COORDINATES ══════ 454

22   BOUNDARY SURVEYS ══════════ 477

23 SURVEYS OF THE PUBLIC LANDS ══ 498

24 CONSTRUCTION SURVEYS ═══ 520

25 HORIZONTAL CURVES ═══ 543

26   VERTICAL CURVES ═══════════ 573

27   VOLUMES ═══════════ 588

28 PHOTOGRAMMETRY ══════ 606

29 INTRODUCTION TO GEOGRAPHIC INFORMATION SYSTEMS ══════ 644

APPENDIX A   INSTRUMENT TESTING AND ADJUSTING ══════ 670

APPENDIX B   COORDINATE GEOMETRY IN SURVEYING CALCULATIONS ═══════ 681

APPENDIX C   PROPAGATION OF RANDOM ERRORS AND LEAST-SQUARES ADJUSTMENT   693

PART I   ERROR PROPAGATION   693

PART II   LEAST-SQUARES ADJUSTMENT   695

APPENDIX D   EXAMPLE NOTE FORMS ═══ 711

APPENDIX E   TRIGONOMETRIC FORMULAS FOR SOLVING TRIANGLES ═══════ 721

PREFACE

Surveying is currently in an age of spiraling technological change. Advances are occurring at an unparalleled rate, and are profoundly affecting virtually all areas of surveying field and office practice. New instruments that have revolutionized field activities include the GPS satellite surveying system, electronic digital theodolites, total station instruments with their automatic data collectors (the newest of which are operated robotically), automatic digital levels that read bar-coded rods and employ image processing techniques, new laser alignment devices, and others.

In the office, computer hardware and software improvements affect procedures used for processing information. Personal computers and workstations having greater capabilities and lower cost are continually being introduced. New instruments for plotting maps and scanning documents have been perfected, and software that performs all types of surveying computations, compiling maps, and automatically drawing contours has been introduced.

Perhaps the most significant recent development in surveying has been geographic information systems (GISs). These systems, consisting of both hardware and software, enable the user to store, integrate, manipulate, analyze, and display virtually any type of spatially related information about our environment. They are used at all levels of government, in business, and in private industry, and they are being applied in many diverse areas to aid in planning, design, management, and decision making. Geographic information systems are significantly affecting all areas of surveying.

To accommodate these changes, this ninth edition of *Elementary Surveying* has been substantially revised throughout. New material has been added to cover advancing technologies. In particular, coverage of total stations, automatic data collectors, and GPS has been updated and expanded. The chapter on state plane coordinates has been completely rewritten to accommodate changes brought about by NAD83. A new chapter has been added on the subject of geographic information systems, and a new section on least squares adjustment has been included in the appendixes. In addition to these major changes, numerous other improvements, including revisions to text material and figures, have been made throughout to update the book and improve its clarity of presentation.

Although many changes have been made, this ninth edition follows the same approach of previous editions: it provides a readable textbook of basic theory and practical material for both field and office work. Chapters are arranged in the order generally found most convenient for college courses. Earlier chapters present basic material suitable for a first course in surveying, whereas later chapters cover more advanced and specialized topics in enough depth to provide sufficient material for a second course. More than 900 end-of-chapter problems are included, and to assist students in self-study, answers to many are given in Appendix G.

As in previous editions, the text emphasizes the theory of errors and correlation of theory and practical field methods. To remind students that surveyors must constantly strive to reduce the sizes of errors and eliminate mistakes, lists of typical errors and mistakes are given at the ends of most chapters. Practical suggestions that stem from the authors' many years of experience are interjected throughout the text so that students can benefit from them.

As noted above, much new material has been added to cover recently developed equipment. However, traditional instruments such as tapes, levels, and theodolites continue to be used in large numbers for many surveying jobs. These traditional instruments continue to be covered in this book, but coverage of older outdated equipment and procedures has been reduced or eliminated. Taping, dumpy and wye levels, transits, plane table, stadia, and triangulation are not covered in the same depth as in earlier editions, and the number of tables at the end of the book has also been reduced.

Although automatic data collectors are gradually being used in larger numbers for recording field measurements, clear and complete handwritten field notes continue to be important in surveying. For this reason, the subject of noteforms is discussed in a separate chapter, and sample noteforms for many of the common types of surveys are collected in Appendix D. Noteforms for a few others are given within the text.

An **Instructor's Manual** is available to adopters of the book. It contains answers to all end-of-chapter problems and a diskette with several valuable computer programs. The diskette has its own documentation, and the menu-driven programs include traverse computations, with area calculation; astronomical azimuth reduction; two-dimensional coordinate transformation; horizontal and vertical curve computation; least squares adjustment; oblique triangle solution; and stadia reduction. At the instructor's discretion, copies of this diskette can be made available to students.

In addition to the changes noted above, numerous other improvements have been incorporated in this ninth edition. A summary of major changes follows:

1. A new chapter covering the subjects of geographic and land information systems has been added.
2. Discussion of total station instruments has been substantially expanded, and modifications in field and office procedures that result from their use are presented throughout the book.
3. Coverage of the global positioning system has been completely updated and expanded.
4. The material on automatic data collectors has been modernized and enlarged, and the use of these devices for various types of field surveys described.

5. The new NAD83 and NGVD88 horizontal and vertical datums have been described and discussed.
6. The chapter on errors and their analysis has been significantly revised.
7. A new section on least squares adjustment has been included in the appendixes. It features several practical examples illustrating applications of this important adjustment procedure.
8. A description of the new electronic digital level has been included.
9. Discussion of new laser equipment for construction layout has been updated.
10. Coverage of radial traversing with total station instruments has been expanded.
11. The chapter on topographic surveying has been completely revised and includes discussion of the new topographic detailing procedures of radial surveys by total station and the use of portable GPS units.
12. Digital elevation models (DEMs) and triangulated irregular networks (TINs) have been described, and the impacts they have on field surveys for topographic mapping discussed.
13. The discussion on computer-aided drafting (CAD) systems has been substantially updated and expanded.
14. The new GPS high-accuracy reference networks (HARNs) that are being installed in most states are described.
15. The chapter on state plane coordinates has been completely revised to include methods of computing in both NAD27 and NAD83.
16. Coverage of boundary surveying has been augmented. Discussion on single- and double-proportionate measurement has been increased, and examples given.
17. The chapter on construction surveys has been revised and new material added to reflect recent changes in equipment and procedures in this important area.
18. The method of staking a circular curve by coordinates with total station instruments has been described, and an example given.
19. The chapter on photogrammetry has been updated to introduce digital or "softcopy" photogrammetry, and the use of GPS in aircraft to reduce or eliminate the need for traditional photogrammetric ground control surveys.
20. The bibilographies at the end of each chapter have been updated and expanded.

Paul R. Wolf
Russell C. Brinker

ACKNOWLEDGMENTS

Past editions, as well as this one, have benefited from the ideas and reviews of numerous educators and practitioners. For their help, the authors are extremely grateful. In this edition professors who reviewed material or otherwise assisted include:

John W. Adcox, University of North Florida
Rajendra J. Aggarwala, University of Michigan
Don Andersen, North Dakota State University
R. B. Buckner, East Tennessee State University
Earl F. Burkholder, Oregon Institute of Technology
Robert Burtch, Ferris State University
James K. Crossfield, California State University, Fresno
Bon A. Dewitt, University of Florida
Charles D. Ghilani, Pennsylvania State University, Wilkes-Barre
Kandiah Jeyapalan, Iowa State University
Andrew C. Kellie, Murray State University
David R. Knowles, University of Arkansas
Francis McKelvey, Michigan State University
Gerald W. Mahun, Pennsylvania State University, Wilkes-Barre
David Mezera, University of Wisconsin—Madison
John P. Powers, Jr., University of Arkansas—Little Rock
James P. Reilly, New Mexico State University
Jud Rouch, University of Arkansas—Little Rock
Robert J. Schultz, Oregon State University
William C. Taylor, Michigan State University
Fred Thomack, Madison Area Technical College
David A. Tyler, University of Maine
Alan P. Vonderohe, University of Wisconsin—Madison

Special thanks are expressed to Professor Charles Ghilani of Pennsylvania State University, Wilkes-Barre, whose contributions have been extraordinarily valuable.

Practitioners who assisted include Ted Koch, Wisconsin State Cartographer; Gene ~~Heferman, Paul Hartzheim, James We~~ndels, and Ken Worden of the Wisconsin Depart- ... of Sokkia Corporation; David Scott of Pentax ...nowles and Senne, Inc.; and Nancy von Meyer ...aduate students who assisted include Rajendra ...i, Jenn-Taur Lee, and especially Tim Ruhren. ...owledge the contributions of charts, maps, and ...odetic Survey and the U.S. Geological Survey, ...ation received from the many instrument man- ...and described herein. ...ers who may have been inadvertently omitted, ...ally, acknowledgments would not be complete ...Mrs. Lynn Wolf and Mrs. Millie Brinker, who ...t and steadfast support. Millie Brinker passed ...as in preparation. She will be deeply missed. ...he authors will gratefully accept any construc- ...tions for improvements.

P. R. W.
R. C. B.

1 ROD = 1 PERCH = 1 POLE = 16 1/2 '

1 VARA = ~33"

1 GUNTER'S CHAIN (CH) = 66' = 100 LINKS (LK) = 4 RODS

1 MILE = 80 GUNTER'S CHAIN

NAUTICAL MILE = 6076.10'

= LENGTH OF 1 MINUTE OF LAT or LONG AT THE EQUATOR

ACRE = 10 SQ. CHAINS = 43560

400 GRADS = CIRCLE

100 CENTESIMAL MIN/GRAD

100 CENTESIMAL SEC/MIN

2π RADIANS = 360°

1

INTRODUCTION

1-1 DEFINITION OF SURVEYING

Surveying has traditionally been defined as the science, art, and technology of determining the relative positions of points above, on, or beneath the earth's surface, or of establishing such points. In a more general sense, however, surveying can be regarded as that discipline which encompasses all methods for measuring, processing, and disseminating information about the physical earth and our environment. Surveying has been important since the beginning of civilization. Its earliest applications were in measuring and marking boundaries of property ownership. Throughout the years its importance has steadily increased with the growing demand for a variety of maps and charts, and the need for establishing accurate line and grade to guide construction operations.

Today, the importance of measuring and monitoring our environment is becoming increasingly critical as our population expands, land values appreciate, our natural resources dwindle, and human activities continue to pollute our land, water, and air. Using modern ground, aerial, and satellite technologies, and computers for data processing, contemporary surveyors are now able to measure and monitor the earth and its natural resources on literally a global basis. Never before has so much survey data been available for assessing current conditions, making sound planning decisions, and formulating policy in a host of land-use, resource development, and environmental preservation applications. Still the major areas of surveying practice continue to be in boundary surveying, mapping, and in making the measurements necessary to provide accurate line and grade for construction.

Recognizing the increasing breadth and importance of the practice of surveying, the *International Federation of Surveyors* (see Section 1-9) recently adopted the following definition:

> Practice of the surveyor's profession may involve one or more of the following activities that may occur either on, above, or below the surface of the land or the sea, and may be carried out in association with other professionals.
>
> **1.** Determination of the shape of the Earth and measurement of all facts needed to determine the size, position, shape, and contour of any part of the Earth's surface, and the provisions of plans, maps, files and charts recording these facts.

2. Positioning of objects in space, and positioning of physical features, structures, and engineering works on, above, or below the surface of the Earth.
3. Determination of the positions of boundaries of public or private land, including national and international boundaries, and registration of those lands with appropriate authorities.
4. Design, establishment, and administration of land and geographic information systems, collection and storage of data within those systems, and analysis and manipulation of that data to produce maps, files, charts and reports for use in the planning and design processes.
5. Planning of the use, development, and re-development of property, and management of that property, whether urban or rural, and whether land or buildings, including determination of values, estimation of costs, and the economic application of resources such as money, labor, and materials taking into account relevant legal, economic, environmental and social factors.
6. Study of the natural and social environment, measurement of land and marine resources, and the use of this data in planning and development in urban, rural and regional areas.

The breadth and diversity of the practice of surveying, as well as its importance in modern civilization, are readily apparent from this definition. The thrust of this book is to describe the equipment and procedures used in pursuing these areas of practice.

1-2 HISTORY OF SURVEYING

The oldest historical records in existence today which bear directly on the subject of surveying state that this science had its beginning in Egypt. Herodotus recorded that Sesostris (about 1400 B.C.) divided the land of Egypt into plots for the purpose of taxation. Annual floods of the Nile River swept away portions of these plots, and surveyors were appointed to replace the bounds. These early surveyors were called *rope-stretchers,* since their measurements were made with ropes having markers at unit distances.

As a consequence of this work, early Greek thinkers developed the science of geometry. Their advance, however, was chiefly along the lines of pure science. Heron stands out prominently for applying science to surveying in about 120 B.C. He was the author of several important treatises of interest to surveyors, including *The Dioptra,* which related the methods of surveying a field, drawing a plan, and making calculations. It also described one of the first pieces of surveying equipment recorded, the *diopter* [Figure 1-1(a)]. For many years Heron's work was the most authoritative among Greek and Egyptian surveyors.

Significant development in the art of surveying came from the practical-minded Romans, whose best known writing on surveying was by Frontinus. Although the original manuscript disappeared, copied portions of Frontinus's work have been preserved. This noted Roman engineer and surveyor, who lived in the first century, was a pioneer in the field, and his essay remained the standard for many years.

The engineering ability of the Romans was demonstrated by their extensive construction work throughout the empire. Surveying necessary for this construction resulted in the organization of a surveyors' guild. Ingenious instruments were developed and used. Among these were the *groma* [Figure 1-1(b)], used for sighting; the *libella,* an A frame with a plumb bob, for leveling; and the *chorobates,* a horizontal straightedge about 20 ft long with supporting legs and a groove on top for water to serve as a level.

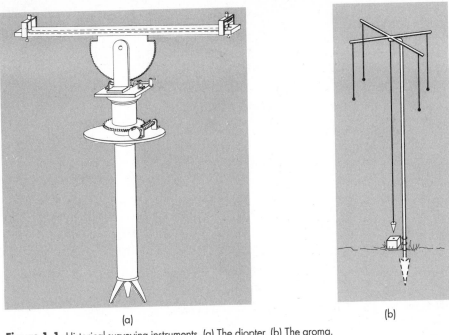

(a)   (b)

Figure 1-1 Historical surveying instruments. (a) The diopter. (b) The groma.

One of the oldest Latin manuscripts in existence is the *Codex Acerianus,* written in about the sixth century. It contains an account of surveying as practiced by the Romans and includes several pages from Frontinus's treatise. The manuscript was found in the tenth century by Gerbert and served as the basis for his text on geometry, which was largely devoted to surveying.

During the Middle Ages, Greek and Roman science was kept alive by the Arabs. Little progress was made in the art of surveying, and the only writings pertaining to it were called "practical geometry."

In the thirteenth century Von Piso wrote *Practica Geometria,* which contained instructions on surveying. He also authored *Liber Quadratorum,* dealing chiefly with the *quadrans,* a square brass frame having a 90° angle and other graduated scales. A movable pointer was used for sighting. Other instruments of the period were the *astro-labe,* a metal circle with a pointer hinged at its center and held by a ring at the top, and the *cross staff,* a wooden rod about 4 ft long with an adjustable cross arm at right angles to it. The known lengths of the arms of the cross staff permitted distances to be measured by proportion and angles.

Early civilizations assumed the earth to be a flat surface, but by noting the earth's circular shadow on the moon during lunar eclipses and watching ships gradually disappear as they sailed toward the horizon, it was slowly deduced that the planet actually curved in all directions.

Determining the true size and shape of the earth has intrigued humans for centuries. History records that a Greek named Eratosthenes was among the first to compute its

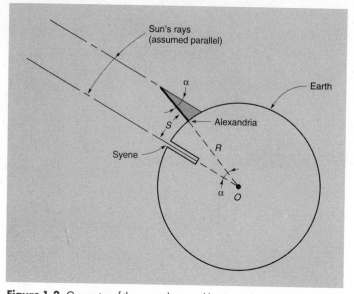

Figure 1-2 Geometry of the procedure used by Eratosthenes to determine the earth's circumference.

dimensions. His procedure, which occurred about 200 B.C., is illustrated in Figure 1-2. Eratosthenes had concluded that the Egyptian cities of Alexandria and Syene were located approximately on the same meridian, and he had also observed that at noon on the summer solstice, the sun was directly overhead at Syene. (This was apparent because on that day the image of the sun could be seen reflecting from the bottom of a deep vertical well there.) He reasoned that at that moment, the sun, Syene, and Alexandria were in a common meridian plane, and if he could measure the arc length between the two cities, and the angle it subtended at the earth's center, he could compute the earth's circumference. He determined the angle by measuring the length of the shadow cast at Alexandria from a tall vertical staff of known length. The arc length was found from multiplying the number of caravan days between Syene and Alexandria by the average daily distance traveled. From these measurements Eratosthenes calculated the earth's circumference to be about 25,000 mi. Subsequent precise geodetic measurements using better instruments but techniques similar geometrically to Erathosthenes's have shown his value, though slightly too large, to be amazingly close to the currently accepted one. (Actually, as explained in Chapter 19, the earth approximates an oblate spheroid having an equatorial radius about $13\frac{1}{2}$ mi longer than the polar radius.)

In the eighteenth and nineteenth centuries the art of surveying advanced more rapidly. The need for maps and location of national boundaries caused England and France to make extensive surveys requiring accurate triangulation; thus geodetic surveying began. The U.S. Coast and Geodetic Survey (now the National Geodetic Survey of the U.S. Department of Commerce) was established by an act of Congress in 1807. Initially its charge was to perform hydrographic surveys and prepare nautical charts. Later its activities were expanded to include establishment of control monuments throughout the country.

Increased land values and the importance of precise boundaries, along with the demand for public improvements in the canal, turnpike, and railroad eras, brought surveying into a prominent position. More recently, the large volume of general construction, numerous land subdivisions with better records required, and demands posed by the fields of exploration and ecology have entailed an augmented surveying program. Surveying is still the sign of progress in the development and use of the earth's resources.

In addition to meeting a host of growing civilian needs, surveying has always played an important role in our nation's defense activities. World Wars I and II, the Korean and Vietnam conflicts, and the recent Operation Desert Storm, each created staggering demands for precise measurements and accurate maps. These military operations also provided the stimulus for improving instruments and methods to meet these needs. Surveying also contributed to, and benefited from, the space program where new equipment and systems were needed to supply precise control for missile alignment and mapping of proposed moon landing sites. As a result of this evolution, surveyors now have an array of "high-tech" instruments available for use. These include electronic distance measuring (EDM) equipment, laser devices, north-seeking gyroscopes, improved aerial cameras, helicopters, inertial surveying systems, precise satellite positioning devices, remote sensors, and various-size computers and peripherals for automatic data processing and mapping.

Traditional instruments for ground surveys—the transit, dumpy level, and steel tape—have now been supplanted by modern theodolites, EDM devices, total station instruments, and automatic levels (Figure 1-3). In the mapping field, aerial methods called *photogrammetry* (see Chapter 28) have replaced ground surveys on all except small projects. Figure 1-4 shows a photogrammetric stereoplotter equipped with a modern computerized digital recording system being used to prepare a map from aerial photographs. Conventional ground methods, however, are still essential for establishing locations of horizontal and vertical control points, property corners, and construction layout.

1-3 GEODETIC AND PLANE SURVEYS

Two general classifications of surveys are *geodetic* and *plane.* They differ principally in the assumptions on which the computations are based, although field measurements for geodetic surveys are usually performed to a higher order of accuracy than those for plane surveys.

In geodetic surveying, the curved surface of the earth is considered by performing the computations on an *ellipsoid* (curved surface approximating the size and shape of the earth—see Chapter 19). It is now becoming common to do geodetic computations in a three-dimensional, earth-centered Cartesian coordinate system. The calculations involve solving equations derived from solid geometry and calculus. Geodetic methods are employed to determine relative positions of widely spaced monuments and to compute lengths and directions of the long lines between them. These monuments serve as the basis for referencing other subordinate surveys of lesser extent.

In the past, field measurements for geodetic surveys consisted primarily of angles observed using ground-based theodolites, and distances measured with tapes or electronic devices. Although these types of measurements are still used, the new Global Positioning System (GPS) can perform geodetic surveys with much greater accuracy,

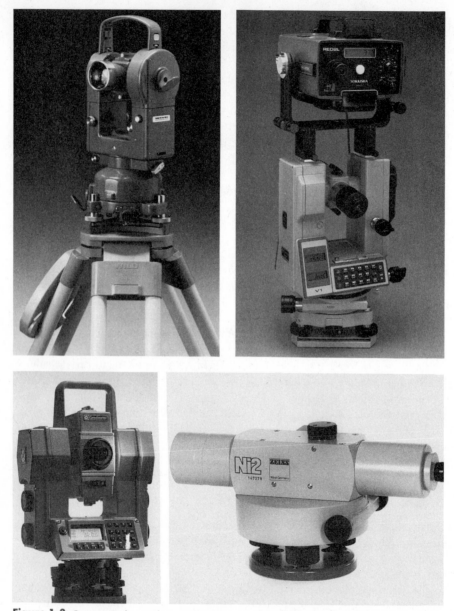

Figure 1-3 Conventional ground surveying instruments. (a) Wild T-2 theodolite. (b) Lietz Sokkia DT2 digital theodolite with RED2L electronic distance measuring instrument. (c) Geodimeter 500 total station (combined digital theodolite and electronic distance meter). (d) Zeiss Ni2 automatic level. (Courtesy Leica, Inc.; Sokkia Corporation; Geotronics of North America, Inc.; and Carl Zeiss, Inc.)

speed, and economy. GPS receivers (Figure 1-5) enable ground stations to be located precisely by electronically measuring distances to satellites operating in known orbits. Inertial surveying systems are another relatively new type of equipment available to surveyors. These devices are carried in helicopters or ground vehicles and employ gyroscopes and accelerometers to measure components of positional change in north-

Figure 1-4 Galileo Stereosimplex G6 stereoplotter with digital recording system. (Courtesy Wisconsin Department of Transportation.)

Figure 1-5 Trimble GPS field surveyor 4000SE with internal antenna. (Courtesy Trimble Navigation and Tony Grissim.)

ing, easting, and elevation. GPS and inertial systems are described in more detail in Chapter 20.

In plane surveying, except for leveling, the reference base for field work and computations is assumed to be a flat horizontal surface. The direction of a plumb line (and thus gravity) is considered parallel throughout the survey region, and all measured angles are presumed to be plane angles. For areas of limited size, the surface of our vast ellipsoid is actually very nearly flat. On a line 5 mi long, the ellipsoid arc and chord lengths differ by only about 0.02 ft. A plane surface tangent to the ellipsoid has departed only about 0.7 ft at 1 mi from the point of tangency. In a triangle having an area of 75 mi^2, the difference between the sum of the three ellipsoidal angles and three plane angles is about 1″. It is evident, therefore, that except in surveys covering extensive areas, the earth's surface can be approximated as a plane, thus simplifying computations and techniques. In general, algebra, plane and analytical geometry, and plane trigonometry are used in plane surveying calculations. Even for very large areas such as those involved in state plane coordinate systems described in Chapter 21, plane surveying can be used, with some allowances being made in computations. This book concentrates primarily on methods of plane surveying, an approach that satisfies the requirements of most projects.

1-4 IMPORTANCE OF SURVEYING

Surveying is one of the world's oldest and most important arts because, as noted previously, from the earliest times it has been necessary to mark boundaries and divide land. Surveying has now become indispensable to our modern way of life. The results of today's surveys are being used to **(1)** map the earth above and below sea level; **(2)** prepare navigational charts for use in the air, on land, and at sea; **(3)** establish property boundaries of private and public lands; **(4)** develop data banks of land-use and natural resource information which aid in managing our environment; **(5)** determine facts on the size, shape, gravity, and magnetic fields of the earth; and **(6)** prepare charts of our moon and planets.

Surveying continues to play an extremely important role in many branches of engineering. For example, surveys are required to plan, construct, and maintain highways, railroads, rapid-transit systems, buildings, bridges, missile ranges, launching sites, tracking stations, tunnels, canals, irrigation ditches, dams, drainage works, urban land subdivisions, water supply and sewage systems, pipelines, and mine shafts. Surveying or surveying methods are commonly employed in laying out assembly lines and jigs, fabricating and placing large equipment, providing control for aerial photography, and in many related tasks in agronomy, archeology, astronomy, forestry, geography, geology, geophysics, landscape architecture, meteorology, paleontology, and seismology, but particularly in military and civil engineering. Optical alignment is an application of surveying in shop practice (installation of machinery, fabrication of airplanes, ships, and so on).

All engineers must know the limits of accuracy possible in construction, plant design and layout, and manufacturing processes, even though someone else may do the actual surveying. In particular, surveyors and civil engineers who are called on to design and plan surveys must have a thorough understanding of the methods and instruments used, including their capabilities and limitations. This knowledge is best obtained by making measurements with the kinds of equipment used in practice to get a true concept of the theory of errors, and the small but recognizable differences that occur in observed quantities.

In addition to stressing the need for reasonable limits of accuracy, surveying emphasizes the value of significant figures. Surveyors and engineers must know when to work to hundredths of a foot instead of to tenths or thousandths, or perhaps the nearest foot, and what precision in field data is necessary to justify carrying out computations to the desired number of decimal places. With experience, they learn how available equipment and personnel govern procedures and results.

Neat sketches and computations are the mark of an orderly mind, which in turn is an index of sound engineering background and competence. Taking field notes under all sorts of conditions is excellent preparation for the kind of recording and sketching expected of all engineers. Performing later office computations based on the notes underscores their importance. Additional training that has a carryover value is obtained in arranging computations in an organized manner.

Engineers who design buildings, bridges, equipment, and so on are fortunate if their estimates of loads to be carried are correct within 5%. Then a factor of safety of 2 or more is applied. But except for topographic work, only exceedingly small errors can be tolerated in surveying, and there is no factor of safety. Traditionally, therefore, surveying stresses both manual and computational precision.

1-5 SPECIALIZED TYPES OF SURVEYS

Many types of surveys are so specialized that a person proficient in a particular discipline may have little contact with the other areas. Persons seeking careers in surveying and mapping, however, should be knowledgeable in every phase, since all are closely related in modern practice. Some important classifications are described briefly here.

Control surveys establish a network of horizontal and vertical monuments that serve as a reference framework for other surveys.

Topographic surveys determine locations of natural and artificial features and elevations used in map making.

Land, boundary, and *cadastral surveys* are (usually) closed surveys to establish property lines and corners. The term *cadastral* is now generally applied to surveys of the public lands systems. There are three major categories: *original surveys* to establish new section corners in unsurveyed areas that still exist in Alaska and several western states; *retracement surveys* to recover previously established boundary lines; and *subdivision surveys* to establish monuments and delineate new parcels of ownership. *Condominium surveys,* which provide a legal record of ownership, are a type of boundary survey.

Hydrographic surveys define shorelines and depths of lakes, streams, oceans, reservoirs, and other bodies of water. *Sea surveying* is associated with port and offshore industries and the marine environment, including measurements and marine investigations made by shipborne personnel.

Route surveys are made to plan, design, and construct highways, railroads, pipelines, and other linear projects. They normally begin at one control point and progress to another in the most direct manner permitted by field conditions.

Construction surveys provide line, grade, control elevations, horizontal positions, dimensions, and configurations for construction operations. They also secure essential data for computing construction pay quantities.

As-built surveys document exact final location and layout of engineering works, and record any design changes that may have been incorporated into the construction. These are particularly important when underground facilities are constructed, so their precise locations are known, and thus unexpected damage to them can be avoided during later installation of other underground utilities.

Mine surveys are performed above and below ground to guide tunneling and other operations associated with mining, including geophysical surveys for mineral and energy resource exploration.

Solar surveys map property boundaries, solar access easements, and position obstructions and collectors according to sun angles, and meet other requirements of zoning boards and title insurance companies.

Optical tooling (also referred to as *industrial surveying* or *optical alignment*) is a method of making extremely accurate measurements for manufacturing processes where small tolerances are required.

Except for control surveys, most other types described are usually performed using plane surveying procedures; but geodetic methods may be employed on the others if a survey covers an extensive area or requires extreme accuracy.

Ground, aerial, and *satellite surveys* are broad classifications sometimes used. Ground surveys utilize measurements made with ground-based equipment such as tapes, EDM devices, levels, theodolites, and total station instruments. Aerial surveys may be accomplished by either *photogrammetry* or *remote sensing.* Photogrammetry uses cameras that are usually carried in airplanes, whereas remote sensing employs cameras and other types of sensors that can be transported in either aircraft or satellites. Procedures for securing and reducing aerial data are described in Chapter 28. Aerial surveys have been used in all the specialized types of surveying listed, except for optical tooling, and in this area *terrestrial* (ground-based) photographs are often used. Satellite surveys include the determination of ground locations from measurements made to satellites using GPS receivers, or the use of satellite images for mapping and monitoring large regions of the earth.

1-6 LAND AND GEOGRAPHIC INFORMATION SYSTEMS

Land Information Systems (LISs) and *Geographic Information Systems* (GISs) are new areas of activity which have rapidly assumed positions of major prominence in surveying. These computer-based systems enable storing, integrating, manipulating, analyzing, and displaying virtually any type of spatially related information about our environment. LISs and GISs are being used at all levels of government, and by businesses, private industry, and public utilities to assist in management and decision making. Specific applications have occurred in many diverse areas and include natural resource management, facilities siting and management, land records modernization, demographic and market analysis, emergency response and fleet operations, infrastructure management, and regional, national, and global environmental monitoring.

Data stored within LISs and GISs may be both natural and cultural, and be derived from surveys, existing maps, charts, aerial and satellite photos, statistics, tabular data, and other documents. Specific types of information, or so-called layers, may include political boundaries, individual property ownership, population distribution, locations of natural resources, transportation networks, utilities, zoning, hydrography, soil types, land use, vegetation types, wetlands, and many many more.

An essential ingredient of all information entered into LIS and GIS databases is that it be spatially related, that is, located in a common geographic reference framework. Only then are the different layers of information physically relatable so they can be analyzed using computers to support decision making. This geographic positional requirement will place a heavy demand upon surveyors in the future, who will play key roles in designing, implementing, and managing these systems. Surveyors from virtually all of the specialized areas described in the preceding section will be involved in developing the needed databases. Their work will include establishing the required basic control frameworks; conducting boundary surveys and preparing legal descriptions of property ownership; performing topographic and hydrographic surveys by ground, aerial, and satellite methods; compiling and digitizing maps; and assembling a variety of other digital data files.

The last chapter of this book, Chapter 29, is devoted to the topic of land and geographic information systems. This subject seems appropriately covered at the end, after each of the other types of surveys needed to support these systems has been discussed.

1-7 FEDERAL SURVEYING AGENCIES

Several agencies of the U.S. government perform extensive surveying and mapping. Four of the major ones are:

1. The Coast and Geodetic Survey, now the National Geodetic Survey (NGS) and part of the National Ocean Survey (NOS), was originally organized to map the coast. Its activities include control surveys, preparation of nautical and aeronautical charts, photogrammetric surveys, tide and current studies, collection of magnetic data, gravimetric surveys, and worldwide control survey operations that involve satellites. The basic control points established by this organization are the foundation for all large-area surveying. The NGS also plays a major role in coordinating and assisting in activities related to the development of modern LISs and GISs at local, state, and national levels.

2. The Bureau of Land Management (BLM), originally established in 1812 as the General Land Office, directs the public lands surveys. Lines and corners have been set for most public lands in the conterminous United States, but much work remains in Alaska and is proceeding with "modern techniques."

3. The U.S. Geological Survey (USGS), established in 1879, has the responsibility for preparing maps of the entire country. Its standard $7\frac{1}{2}'$ quadrangle maps show topographic and cultural features, and are suitable for general use as well as a variety of engineering and scientific purposes. Nearly 10 million copies are distributed each year. Currently the USGS is engaged in a comprehensive program to develop a national digital cartographic database, which will consist of map data in a computer-readable format.

4. The Defense Mapping Agency (DMA) prepares maps and associated products, and provides services for the Department of Defense and all land combat forces. It is divided into the following military mapping groups: Aerospace Center, Defense Mapping School, Hydrographic Center, Inter-American Geodetic Survey, and Topographic Center. The DMA Topographic Center fulfills a key mission in an era when accurate mapping, charting, and geodesy products are essential to realize the complete potential of new weapons. Technological advances in weaponry demand corresponding improvements in mapping, charting, and geodesy to obtain accuracies that were just dreams only a few years ago.

In addition to these four agencies, units of the Corps of Engineers, U.S. Army, have made extensive surveys for emergency and military purposes. Some of these surveys provide data for engineering projects, such as those connected with flood control. Extensive surveys have also been conducted for special purposes by nearly 40 other federal agencies, including the Forest Service, National Park Service, International Boundary Commission, Bureau of Reclamation, Tennessee Valley Authority, Mississippi River Commission, U.S. Lake Survey, and Department of Transportation. Likewise, many cities, counties, and states have had extensive surveying programs, as have various utilities.

1-8 THE SURVEYING PROFESSION

Land or boundary surveying is classified as a learned profession because the modern practitioner needs a wide background of technical training and experience and must exercise a considerable amount of independent judgment. Registered (licensed) professional surveyors must have a thorough knowledge of mathematics—particularly geometry and trigonometry, but also calculus; competence with computers; a solid understanding of surveying theory, instruments, and methods in the areas of geodesy, photogrammetry, remote sensing, and cartography; some competence in economics (including office management), geography, geology, astronomy, and dendrology; and a familiarity with laws pertaining to land and boundaries. They should be knowledgeable in both field operations and office computations. Above all, they are governed by a professional code of ethics, and are expected to charge reasonable fees for their work.

The personal qualifications of surveyors are as important as their technical ability in dealing with the public. They must be patient and tactful with clients and their sometimes hostile neighbors. Few people are aware of the painstaking research of old records required before field work is started. Diligent, time-consuming effort may be needed to locate corners on nearby tracts for checking purposes as well as to find corners for the property in question.

Permission to trespass on private property or to cut obstructing tree branches and shrubbery must be obtained through a proper approach. Such privileges are not conveyed by a surveying license or by employment in a state highway department (but a court order can be secured if a landowner objects to necessary surveys).

All 50 states, Guam, and Puerto Rico have registration laws for professional surveyors and engineers (as do the provinces of Canada). Some states presently have separate licensing boards for surveyors. In general, a surveyor's license is required to make property surveys, but not for construction, topographic, or route work, unless boundary corners are set.

To qualify for registration as either a professional land surveyor (LS) or professional engineer (PE) it is necessary to have an appropriate college degree, although some states allow relevant experience in lieu of formal education. In addition, candidates must acquire two or more years of additional practical experience, and also pass a two-day comprehensive written examination. In most states a common national examination covering fundamentals and principles and practice of land surveying is now used. However, two hours of the exam are devoted to local legal customs and aspects. Thus transfer of registration from one state to another has become easier.

Some states also require continuing education units (CEUs) for registration renewal, and many more are considering legislation that would add this requirement. Typical state laws require that a licensed land surveyor sign all plats, assume responsibility for any liability claims, and take an *active part* in the field work.

1-9 PROFESSIONAL SURVEYING ORGANIZATIONS

There are many professional organizations in the United States and worldwide that serve the interests of surveying and mapping. Generally the objectives of these organizations are the advancement of knowledge in the field, encouragement of communication among surveyors, and upgrading of standards and ethics in surveying practice. *The American Congress on Surveying and Mapping* (ACSM) is the foremost professional surveying organization in the United States. Founded in 1941, ACSM regularly sponsors technical meetings at various locations throughout the country. These meetings bring together large numbers of surveyors for presentation of papers, discussion of new ideas and problems, and exhibitions of the latest in surveying equipment. ACSM publishes a quarterly journal, *Surveying and Land Information Systems,* and also regularly publishes its newsletter, *The ACSM Bulletin.*

As noted in the preceding section, all states require persons who perform boundary surveys to be licensed. Most states also have professional surveyor societies or organizations with membership open only to those within the state who are licensed. These state societies are often affiliated with ACSM, and offer benefits similar to those of ACSM, except that they concentrate on matters of state and local concern.

The American Society for Photogrammetry and Remote Sensing (ASPRS) is a sister organization of ACSM. ASPRS, like ACSM, is devoted to the advancement of the fields of measurement and mapping, although its major interests are directed toward the use of aerial and satellite imagery for achieving these goals. ASPRS cosponsors technical meetings with ACSM, and its monthly journal *Photogrammetric Engineering and Remote Sensing* regularly features surveying and mapping articles.

The *Surveying Engineering Division* of the *American Society of Civil Engineers* (ASCE) is also dedicated to professional matters related to surveying, and publishes quarterly the *Journal of Surveying Engineering.* Other organizations in the United States which support the profession of surveying and mapping include the *Urban and Regional Information Systems Association* (URISA) and *Automated Mapping and Facilities Management* (AM/FM).

The *Canadian Institute of Geomatics* (CIG) is the foremost professional organization in Canada concerned with surveying. Its objectives parallel those of ACSM. This organization, formerly the *Canadian Institute of Surveying and Mapping* (CISM), disseminates information to its members through its *CIG Journal*, formerly the *CISM Journal.*

The *International Federation of Surveyors* (FIG), founded in 1878, fosters the exchange of ideas and information among surveyors worldwide. The acronym *FIG* stems from its French name, *Fédération Internationale des Géométres.* FIG membership consists of professional surveying organizations from many countries throughout the world. ACSM has been a member since 1959. FIG is organized into nine technical commissions, each concerned with a specialized area of surveying. FIG sponsors international conferences, usually at three- or four-year intervals, and its commissions also hold periodic symposia where delegates gather for the presentation of papers on subjects of international interest.

1-10 FUTURE CHALLENGES IN SURVEYING

Surveying is currently in the midst of a revolution in the way data are measured, recorded, processed, stored, retrieved, and shared. This is due in large part to developments in satellite and computer technology. Concurrent with technological advancements, society continues to demand more data, with increasingly higher standards of accuracy, than ever before. Consequently in a few years the demands on surveyors will be very different from what they are now.

In the future, the National Geodetic Reference System, a network of horizontal and vertical control points, must be maintained and supplemented to meet requirements of increasingly higher order surveys. New topographic maps with larger scales as well as digital map products are necessary for better planning. Existing maps of our rapidly expanding urban areas need revision and updating to reflect changes, and more and better map products are needed of the older parts of our cities to support urban renewal programs and infrastructure maintenance and modernization. Large quantities of data will be needed to plan and design new rapid-transit systems to connect our major cities, and surveyors will face new challenges in meeting the precise standards required in staking alignments and grades for these systems.

In the future, assessment of environmental impacts of proposed construction projects will call for more and better maps and other data. GISs and LISs that contain a variety of land-related data such as ownership, location, acreage, soil types, land uses, and natural resources must be designed, developed, and maintained. Cadastral surveys of the yet unsurveyed public lands are essential. Monuments set years ago by the original surveyors have to be recovered and remonumented for preservation of property boundaries. Appropriate surveys with very demanding accuracies will be necessary to position drilling rigs as mineral and oil explorations press further offshore.

Other future challenges include making precise deformation surveys for monitoring existing structures such as dams, bridges, and skyscrapers to detect imperceptible movements that could be precursors to catastrophes caused by their failure. Timely measurements and maps of the effects of natural disasters such as earthquakes, floods, and hurricanes will be needed so that effective relief and assistance efforts can be planned and implemented. In the space program, the desire for maps of neighboring planets will continue. And we must increase our activities in measuring and monitoring natural and human-caused global changes (glacial growth and retreat, volcanic activity, large-scale deforestation, and so on) that can potentially affect our land, water, atmosphere, energy supply, and even our climate.

These and other opportunities offer professionally rewarding indoor or outdoor (or both) careers for numerous people with suitable training in the various branches of surveying.

PROBLEMS

NOTE: Answers for some of these problems, and some in later chapters, can be obtained by consulting a dictionary, the bibliographies, later chapters, and professional surveyors.

1-1 Explain the difference between geodetic and plane surveys.

1-2 Describe some surveying applications in:
   (a) Agriculture   **(b)** Forestry   **(c)** Mining

1-3 Why is it necessary to make accurate surveys of underground utilities?

1-4 Why is accurate surveying important in fabricating huge ships and airplanes?

1-5 How did the early Romans control elevations for the long, large aqueducts built to bring water to their cities?

1-6 Why should a surveyor have a working knowledge of dendrology?

1-7 Why should an individual demand a survey before agreeing to purchase a farm, city lot, or home?

1-8 Do laws in your state or community specify the accuracy required for surveys made to lay out a subdivision? If so, what limits are set?

1-9 What organizations in your state will furnish maps and surveying reference data to surveyors and engineers?

1-10 List the legal requirements for registration as a land surveyor in your state.

1-11 Explain how aerial photographs can be valuable in surveying.

1-12 How does every surveyor benefit from the U.S. satellite program?

BIBLIOGRAPHY

American Society of Civil Engineers and American Congress on Surveying and Mapping. 1978. *Definitions of Surveying and Mapping Terms.* Manual (No. 34).

Bedini, S. A. 1993. "Thomas Jefferson, Statesman Surveyor." *Professional Surveyor* 13 (No. 2):32.

Brandenberger, A. J., and S. K. Ghosh. 1991. "Status of the World's Topographic Mapping, Geodetic Networks, and National Mapping Agencies." *Surveying and Land Information Systems* 51 (No. 3):178.

Brinker, R. C., and R. Minnick. 1987. *The Surveying Handbook.* New York: Van Nostrand Reinhold.

Dix, W. 1991. *Recollections of the American Congress on Surveying and Mapping 1941–1991.* Washington, D.C.: ACSM.

Dix, W. S. 1978. "Surveying and Mapping—50 Years of Progress—1928–1978." *Surveying and Mapping* 38 (No. 4):301.

Frodge, S. L. 1992. "Deformation Surveys." *ACSM Bulletin* (No. 138):32.

Funk, T. J., and M. Lafler. 1991. "Desert Storm Surveying." *Point of Beginning* 17 (No. 1):10.

Fusco, N. 1990. "Safe Surveying on Hazardous Sites." *Professional Surveyor* 10 (No. 3):24.

Gastaldi, A. 1991. "Surveying Alaska's Iditarod Trail." *Professional Surveyor* 11 (No. 6):4.

Grisham, D. 1993. "The Surveyor and the Information Highway in the Real World." *Professional Surveyor* 13 (No. 2):14.

Haney, T. 1993. "Mapping in the Civil War." *ACSM Bulletin* (No. 142):24.

Kor, J. S., and D. Scapuzzi. 1991. "Performing the Master Surveys for the Superconducting Super Collider." *Point of Beginning* 16 (No. 6):12.

Kreisle, W. F. 1988. "History of Engineering Surveying." *ASCE Journal of Surveying Engineering* 114 (No. 3):102.

McLaughlin, J. D., and S. E. Nichols. 1987. "Parcel-Based Land Information Systems." *Surveying and Mapping* 47 (No. 1):155.

Minnick, R. 1985. *Plotters and Patterns of American Land Surveying.* Rancho Cordova, Calif.: Landmark.

National Research Council. 1980. *Need for Multipurpose Cadastre.* Washington, D.C.: National Academy Press.

Onsrud, H. J. 1987. "Challenge to the Profession: A Formal Legal Education for Surveyors." *Surveying and Mapping* 47 (No. 1):31.

Pallamary, M. 1991. "Surveying the San Francisco Earthquake." *Professional Surveyor* 10 (No. 1):4.

Petersohn, F., et al. 1986. "How Can Information Be Used." *Professional Surveyor* 6 (No. 4):21.

Plasker, J., et al. 1990. "U.S. National Report to FIG." *Surveying and Land Information Systems* 50 (No. 2):61.

Quinn, A. O. 1981. "A Report of the Activities of the Private Sector in the Surveying and Mapping Profession." *Surveying and Mapping* 41 (No. 1):55.

Radcliffe, E. 1991. "Breakthrough Beneath the Sea—Surveying the English Channel Tunnel." *Point of Beginning* 16 (No. 5):10.

Ridgeway, H. H. 1982. "Surveying: The Profession and Its Requirements." *ASCE Journal of the Surveying and Mapping Division* 108 (No. SU1):18.

Thorpe, C. 1984. "The Babylonian Marker." *Professional Surveyor* 4 (No. 2):7.

Vonderohe, A. P. 1987. "Solar Access Surveys." *ASCE Journal of the Surveying Engineering Division* 113 (No. 1):2.

2
THEORY OF MEASUREMENTS AND ERRORS

2-1 INTRODUCTION

Making measurements, and subsequent computations and analyses using them, are fundamental tasks of surveyors. The process requires a combination of human skill and mechanical equipment applied with the utmost judgment. No matter how carefully made, however, measurements are never exact and will always contain errors.

Surveyors, whose work must be performed to exacting standards, should therefore thoroughly understand the different kinds of errors, their sources and expected magnitudes under varying conditions, and their manner of propagation. Only then can they select instruments and procedures necessary to reduce error sizes to within tolerable limits.

Of equal importance, surveyors must be capable of assessing the magnitudes of errors in their measurements so that either their acceptability can be verified or, if necessary, new ones taken. The design of measurement programs, comparable to other engineering design, is now practiced. Matrix algebra and digital computers are tools now commonly used by surveyors to plan measurement projects, and to investigate and distribute errors after results have been obtained.

2-2 TYPES OF MEASUREMENTS IN SURVEYING

Five kinds of measurements illustrated in Figure 2-1 form the basis of traditional plane surveying: (1) horizontal angles, (2) horizontal distances, (3) vertical (or zenith) angles, (4) vertical distances, and (5) slope distances. Horizontal angles, as angle *AOB*, and horizontal distances, *OA* and *OB*, are measured in horizontal planes; vertical angles, such as *AOC*, are measured in vertical planes. Zenith angles, as *EOC*, are also measured in vertical planes. Vertical lines, such as *AC* and *BD*, are measured in the direction of gravity; slope distances, as *OC*, are determined along inclined planes. By using combi-

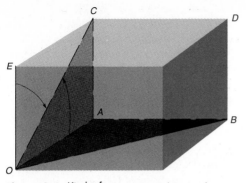

Figure 2-1 Kinds of measurements in surveying.

nations of these basic measurements it is possible to compute relative positions between any points. Equipment and procedures for making each of these basic kinds of measurements are described in later chapters of this book.

2-3 UNITS OF MEASUREMENT

Magnitudes of measurements (or of values derived from measurements) must be given in terms of specific units. In surveying, the most commonly employed units are for *length, area, volume,* and *angle.* Two different systems are in use for specifying units of measured quantities, the *English* and *metric* systems. The metric system is currently used in all except three countries of the world: the United States, Burma, and Liberia. Because of its widespread adoption, the metric system is called the *International System of Units* and abbreviated *SI.*

The basic unit employed for length measurements in the English system is the *foot,* whereas the *meter* is used in the metric system. In the past, two different definitions have been used to relate the foot and meter. Although they differ slightly, their distinction must be made clear in surveying. In 1893, the United States officially adopted a standard in which 39.37 in. was exactly equivalent to 1 m. Under this standard, the foot was equivalent to 0.3048006+ m. In 1959 a new standard was officially adopted in which the inch was equal to exactly 2.54 cm. Under this standard, one ft equals exactly 0.3048 m. This current unit, known as the *international foot,* differs from the previous one by about 1 part in 500,000, or approximately 1 foot per 100 miles. This small difference is thus only important for very precise surveys conducted over long distances. Because of the vast number of surveys performed prior to 1959, it would have been extremely difficult and confusing to change all related documents and maps that already existed. Thus the old standard, now called the *U.S. survey foot,* is still used. Individual states have the option of officially adopting either standard. The National Geodetic Survey uses the meter in its distance measurements, thus it is unnecessary to specify the foot unit. Those making conversions from metric units, however, must know the appropriate standard for their state.

Because the English system has long been the officially adopted standard for measurements in the United States, except for geodetic surveys, the linear units of feet and *decimals* of a foot are most commonly used by surveyors. In construction, feet and inches are often used. Because surveyors perform all types of surveys including geodetic, and they also provide measurements for developing construction plans and guiding building operations, they must understand all the various systems of units and be capable of making conversions between them. *Caution must always be exercised to ensure that measurements are recorded in their proper units, and conversions correctly made.*

A summary of the length units used in past and present surveys in the United States includes the following:

$$
\begin{array}{rl}
1 \text{ foot} =& 12 \text{ inches} \\
1 \text{ yard} =& 3 \text{ feet} \\
1 \text{ inch} =& 2.54 \text{ centimeters (basis of international foot)} \\
1 \text{ meter} =& 39.37 \text{ inches} = 3.2808 + \text{ feet (basis of U.S. survey foot)} \\
1 \text{ rod} =& 1 \text{ pole} = 1 \text{ perch} = 16\tfrac{1}{2} \text{ feet} \\
1 \text{ vara} =& \text{approximately 33 inches (old Spanish unit often encountered} \\
& \text{in the southwestern United States)} \\
1 \text{ Gunter's chain (ch)} =& 66 \text{ feet} = 100 \text{ links (lk)} = 4 \text{ rods} \\
1 \text{ mile} =& 5280 \text{ feet} = 80 \text{ Gunter's chains} \\
1 \text{ nautical mile} =& 6076.10 \text{ feet (nominal length of a minute of latitude, or of} \\
& \text{longitude at the equator)} \\
1 \text{ fathom} =& 6 \text{ feet}
\end{array}
$$

In the English system, areas are given in *square feet* or *square yards.* The most common unit for large areas is the *acre.* Ten square chains (Gunter's) equal 1 acre. Thus an acre contains 43,560 ft^2, which is the product of 10 and 66^2. The *arpent* (equal approximately to 0.85 acre, but varying somewhat in different states) was used in land grants of the French crown. When employed as a linear term, it refers to the length of a side of 1 square arpent.

Volumes in the English system can be given in *cubic feet* or *cubic yards.* For very large volumes, for example, the quantity of water in a reservoir, the *acre-foot* unit is used. It is equivalent to the area of an acre having a depth of 1 ft, and thus is 43,560 ft^3.

The unit of angle used in surveying is the *degree,* defined as $\frac{1}{360}$ of a circle. One degree (1°) equals 60 min, and 1 min equals 60 sec. Divisions of seconds are given in tenths, hundredths, and thousandths. Other methods are also used to subdivide a circle, for example, 400 *grads* (with 100 *centesimal min*/grad and 100 *centesimal sec*/min. Another term, *gons,* is now used interchangeably with grads. The military services use *mils* to subdivide a circle into 6400 units.

A *radian* is the angle subtended by an arc of a circle having a length equal to the circle's radius. Therefore, 2π rad $= 360°$, 1 rad $= 57°17'44.8'' = 57.2958°$, and 0.01745 rad $= 1°$.

2-4 INTERNATIONAL SYSTEM OF UNITS (SI)

As previously noted, the meter[1] is the basic unit for length in the metric or SI system. Subdivisions of the meter (m) are the *millimeter* (mm), *centimeter* (cm), and *decimeter*

---

[1]Originally the meter was defined as $\frac{1}{10,000,000}$ of the earth's meridional quadrant. When the metric system was legalized for use in the United States in 1866, a meter was defined as the interval under certain physical conditions between lines on an international prototype bar made

(dm), equal to 0.001 m, 0.01 m, and 0.1 m, respectively. A *kilometer* (km) equals 1000 m, or approximately five-eighths of a mile.

Areas in the metric system are specified using the *square meter* (m^2). Large areas, for example tracts of land, are given in *hectares* (ha), where one hectare is equivalent to a square having sides of 100 meters. Thus there are 10,000 m^2, or approximately 2.471 acres per hectare. The *cubic meter* (m^3) is used for volumes in the SI system. Either degrees, minutes, and seconds, or the radian, are accepted SI units for angles.

During the 1960s and 1970s, a significant effort was made in the United States towards adoption of SI. However, costs and frustrations associated with making the change generated substantial resistance, and the effort was temporarily stalled. Now Congress has recognized the need for the United States to officially adopt the SI system if we are to compete in the rapidly developing global economy. Accordingly in 1988 Congress enacted the *Omnibus Trade and Competitiveness Act* to designate the metric system as the preferred system of weights and measures for U.S. trade and commerce. That Act, together with a subsequent *Executive Order* issued in July of 1991, required each federal agency to develop a definite metric conversion plan by November 30, 1991, and directed those agencies to use the SI standard in their procurements, grants, and other business-related activities to the extent economically feasible. As an example of one agency's response, the Federal Highway Administration adopted a plan calling for **(1)** use of metric units in all publications and correspondence after September 30, 1992, and **(2)** use of metric units on all plans and contracts for federal highways after September 30, 1996. Although the Act and Executive Order do not mandate states, counties, cities or industries to convert to metric, strong incentives are provided, e.g., these entities cannot obtain certain federal matching funds unless they comply in using SI. In light of these developments, it appears that the metric system will indeed soon become the official system for use in the U.S. The change should be welcomed by surveyors who are presently burdened with unit conversions and awkward computations involving yard, foot, and inch units.

This book uses both English and SI units in discussion and example problems.

2-5 SIGNIFICANT FIGURES

In recording measurements, an indication of the accuracy attained is the number of digits (significant figures) recorded. By definition, the number of significant figures in any measured value includes the positive (certain) digits plus one (*only one*) digit that

---

of 90% platinum and 10% iridium, and accepted as equal to 39.37 in. A copy of the bar is held by the U.S. Bureau of Standards and compared periodically with the international standard stored in France.

In 1960, at the General Conference on Weights and Measures (CGPM), the United States and 35 other nations agreed to redefine the meter as the length of 1,650,763.73 waves of the orange-red light produced by burning the element krypton (Kr-86). That definition permitted industries to make more accurate measurements and to check their own instruments without recourse to the standard meter-bar in Washington. The wavelength of this light is a true constant, whereas there is a risk of instability in the metal meter-bar. The CGPM met again in 1983 and established the current definition of the meter as the length of the path traveled by light in a vacuum during a time interval of 1/299,792,458 sec. Obviously, with this definition, the speed of light in a vacuum becomes exactly 299,792,458 m/sec. The purported advantage of this latest standard is that the meter is more accurately defined, since it is in terms of time, the most accurate of our basic measurements.

is estimated or rounded off, and therefore questionable. For example, a distance measured with a tape whose smallest graduations are 0.01 ft, and recorded as 73.52 ft, is said to have four significant figures; in this case the first three digits are certain and the last is rounded off and therefore questionable.

Any properly recorded measurement can be presumed to have a maximum uncertainty of plus or minus half its last digit. Thus for the example of 73.52, its uncertainty is ±0.005. This conclusion can be derived intuitively, for if a value as high as 73.5251 had been observed, it would have presumably been rounded up and recorded as 73.53; and if it had been as low as 73.5149, it would have been rounded down and recorded as 73.51.

To be consistent with the theory of errors, it is essential that data be recorded with the correct number of significant figures. If a significant figure is dropped off in recording a value, the time spent in acquiring certain precision has been wasted. On the other hand, if data are recorded with more figures than those that are significant, false precision will be implied and time may be wasted in making computations.

The number of significant figures is often confused with the number of decimal places. Decimal places may have to be used to maintain the correct number of significant figures, but in themselves they do not indicate significant figures. Some examples follow:

Two significant figures: 24, 2.4, 0.24, 0.0024, 0.020
Three significant figures: 364, 36.4, 0.000364, 0.0240
Four significant figures: 7621, 76.21, 0.0007621, 24.00

Zeros at the end of an integral value may cause difficulty because they may or may not indicate significant figures. In the value 2400 it is not known how many figures are significant; there may be two, three, or four. The preferred method of eliminating this uncertainty is to express the value in terms of powers of 10. The significant figures in the measurement are then written as a number between 1 and 10, including the correct number of zeros at the end, and the decimal point is placed by annexing a power of 10. As an example, 2400 becomes $2.400 \times (10)^3$ if both zeros are significant, $2.40 \times (10)^3$ if one is, and $2.4 \times (10)^3$ if there are only two significant figures. Alternatively, a bar may be placed over the last significant figure, as $240\overline{0}$, $24\overline{0}0$, and $2\overline{4}00$ for 4, 3, and 2 significant figures, respectively.

In addition, subtraction, multiplication, and division it is imperative that the number of significant figures given in answers be consistent with the measured values. The following three steps will achieve this for addition or subtraction: (1) identify the column containing the rightmost significant digit in each number being added or subtracted; (2) perform the addition or subtraction; and (3) round the answer so that its rightmost significant digit occurs in the leftmost column identified in step (1). Two examples illustrate the procedure.

(a)	(b)
46.7518	378.
+ 1.02	− 0.1
+375.0	377.9 (ans. 378.)
422.7718 (ans. 422.8)	

In **(a)**, the digits 8, 2, and 0 are the rightmost significant ones in the numbers 46.7518, 1.02, and 375.0, respectively. Of these, the 0 in 375.0 is leftmost with respect to the decimal. Thus the answer 422.7718 obtained in adding the numbers is rounded to 422.8, with its rightmost significant digit occurring in the same column as the 0 in 375.0. In **(b)**, the digits 8 and 1 are rightmost, and of these the 8 is leftmost. Thus the answer 377.9 is rounded to 378.

In multiplication, the number of significant figures in the answer is equal to the least number of significant figures in any of the factors. For example, $362.56 \times 2.13 = 772.2528$ when multiplied out, but the answer is correctly given as 772. Its three significant figures are governed by the three significant digits in 2.13. Likewise, in division the quotient should be rounded off to contain only as many significant figures as the least number of significant figures in either the divisor or the dividend. These rules for significant figures in computations stem from error propagation theory, and are discussed further in Section 2-22.

In surveying, four specific types of problems relating to significant figures are encountered and must be understood.

1. Field measurements are given to some specific number of significant figures, thus dictating the number of significant figures in answers derived by computing using them. In an intermediate calculation it is common practice to carry at least one more digit than required and then round off the answer to the correct number of significant figures.

2. There may be an implied number of significant figures. For instance, the length of a football field might be specified as 100 yards. But in laying out the field, such a distance would probably be measured to the nearest hundredth of a foot, not the nearest half-yard.

3. Each factor may not cause an equal variation. For example, if a steel tape 100.00 ft long is to be corrected for a change in temperature of 15°F, one of these numbers has five significant figures while the other has only two. A 15° variation in temperature changes the tape length by only 0.01 ft, however. Therefore an adjusted tape length to five significant figures is warranted for this type of data. Another example is the computation of a slope distance from horizontal and vertical distances, as in Figure 2-2. The vertical distance V is given to two significant figures, and the horizontal distance H is measured to five significant figures. From these data the slope distance S can be computed to five significant figures. For small angles of slope, a considerable change in the vertical distance produces a relatively small change in the difference between slope and horizontal distances.

4. Measurements are recorded in one system of units but may have to be converted to another. A good rule to follow in making these conversions is to retain in the answer

Figure 2-2 Slope correction.

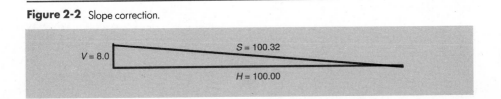

a number of significant figures equal to those in the measured value. As an example, to convert 178 ft $6\frac{3}{8}$ in. to meters, the number of significant figures in the measured value would first be determined by expressing it in its smallest units of $\frac{1}{8}$ in., or $(178 \times 12 \times 8) + (6 \times 8) + 3 = 17{,}139$. Thus the measurement contains five significant figures, and the answer $17{,}139 \div (8 \times 39.37$ in./m$) = 54.416$ m is properly expressed with five significant figures. (Note that 39.37 used in the conversion is an exact constant and does not limit the number of significant figures.)

2-6 ROUNDING OFF NUMBERS

Rounding off a number is the process of dropping one or more digits so the answer contains only those digits that are significant or necessary in subsequent computations. In rounding off numbers to any required degree of precision in this text, the following procedures will be observed:

1. When the digit to be dropped is lower than 5, the number is written *without* the digit. Thus 78.374 becomes 78.37. Also 78.3749 rounded to four figures becomes 78.37.
2. When the digit to be dropped is exactly 5, the nearest *even* number is used for the preceding digit. Thus 78.375 becomes 78.38, and 78.385 is also rounded to 78.38.
3. When the digit to be dropped is greater than 5, the number is written with the preceding digit *increased* by 1. Thus 78.376 becomes 78.38.

Procedures 1 and 3 are standard practice. When rounding the value 78.375 in 2, however, some people always take the next higher hundredth, whereas others invariably use the next lower hundredth. Using the nearest even digit, however, establishes a uniform procedure and produces better balanced results in a series of computations. It is improper procedure to perform two-stage rounding where, for example, in rounding 78.3749 to four digits it would be first rounded to five figures, yielding 78.375, and then rounded again to 78.38. The correct answer in rounding 78.3749 to four figures is 78.37.

2-7 DIRECT AND INDIRECT MEASUREMENTS

Measurements may be made directly or indirectly. Examples of *direct measurements* are applying a tape to a line, fitting a protractor to an angle, or turning an angle with a theodolite.

An *indirect measurement* is secured when it is not possible to apply a measuring instrument directly to the quantity to be measured. The answer is therefore determined by its relationship to some other measured value or values. As an example, the distance across a river can be found by measuring the length of a line on one side, the angle at each end of this line to a point on the other side, and then computing the distance by one of the standard trigonometric formulas. Many indirect measurements are made in surveying, and since all contain errors, it is inevitable that quantities computed from them will also contain errors. The manner by which errors in measurements combine to produce erroneous computed answers is called *error propagation*. This topic is discussed further in Section 2-22.

2-8 ERRORS IN MEASUREMENTS

By definition an error is the difference between a measured value for a quantity and its true value, or

$$E = X - \overline{X} \tag{2-1}$$

where E is the error in a measurement, X the measured value, and $\overline{X}$ its true value.

It can be unconditionally stated that **(1)** *no measurement is exact,* **(2)** *every measurement contains errors,* **(3)** *the true value of a measurement is never known, and, therefore,* **(4)** *the exact error present is always unknown.* These facts are demonstrated by the following. When a distance is measured with a scale divided into tenths of an inch, the distance can be read only to hundredths (by interpolation). If a better scale graduated in hundredths of an inch were available and read under magnification, however, the same distance might be estimated to thousandths of an inch. And with a scale graduated in thousandths of an inch, a reading to ten-thousandths might be possible. Obviously, accuracy of measurements depends on the scale's division size, reliability of equipment used, and human limitations in estimating closer than about one-tenth of a scale division. As better equipment is developed, measurements more closely approach their true values, but they can never be exact. Note that *measurements,* not *counts* (of cars, pennies, marbles, or other objects), are under consideration here.

2-9 MISTAKES

These are observer blunders and are usually caused by a misunderstanding of the problem, carelessness, fatigue, missed communication, or poor judgment. Examples include transposition of numbers such as recording 73.96 instead of the correct value of 79.36; reading an angle counterclockwise, but indicating it as a clockwise angle in the field notes; sighting the wrong target; or recording a taped distance as 762.38 instead of 862.38 because of failure to include a full tape length.

Large mistakes such as these are not considered in the succeeding discussion of errors. They must be detected by systematic checking of all work, and eliminated by redoing part of the job or even all of it. It is very difficult to detect small mistakes because they merge with errors. When not exposed, these little mistakes will therefore be incorrectly treated as errors.

2-10 SOURCES OF ERRORS IN MAKING MEASUREMENTS

Errors in measurements stem from three sources and are classified accordingly.

Natural errors are caused by variations in wind, temperature, humidity, atmospheric pressure, atmospheric refraction, gravity, and magnetic declination. An example is a steel tape whose length varies with changes in temperature.

Instrumental errors result from any imperfection in the construction or adjustment of instruments and from the movement of individual parts. For example, the graduations on a scale may not be perfectly spaced, or the scale may be warped. The effect of many instrumental errors can be reduced, or even eliminated, by adopting proper surveying procedures or applying computed corrections.

Personal errors arise principally from limitations of the human senses of sight and touch. As an example, a small error occurs in the measured value of a horizontal angle if the vertical cross hair in a theodolite is not aligned perfectly on the target, or if the target is the top of a rod which is being held slightly out of plumb.

2-11 TYPES OF ERRORS

Errors in measurements are of two types: *systematic* and *random*.

Systematic errors result from factors which comprise the "measuring system" and include the environment, instrument, and observer. So long as system conditions remain constant, the systematic errors will likewise remain constant. If conditions change, the magnitudes of systematic errors also change. Because systematic errors tend to accumulate, they are often called *cumulative errors.*

Conditions producing systematic errors conform to physical laws which can be modeled mathematically. Thus if the conditions are known to exist and can be measured, a correction can be computed and applied to observed values. An example of a constant systematic error is the use of a 100-ft steel tape that has been calibrated and found to be 0.02 ft too long. It introduces a 0.02-ft error each time it is used, but the error is readily eliminated by applying a correction. An example of a variable systematic error is the change in length of a steel tape resulting from temperature differentials that occur during the period of the tape's use. If the temperature changes are measured, length corrections can be computed by a simple formula, as explained in Chapter 4.

Random errors are the errors that remain after mistakes and systematic errors have been eliminated. They are caused by factors beyond the control of the observer, obey the laws of probability, and are sometimes called *accidental errors.* They are present in all surveying measurements.

The magnitudes and algebraic signs of random errors are matters of chance. There is no absolute way to compute or eliminate them, but they can be estimated using adjustment procedures known as *least squares* (see Section 2-26 and Appendix C). Random errors are also known as *compensating errors,* since they tend to partially cancel themselves in a series of measurements. For example, a person interpolating to hundredths of a foot on a tape graduated only to tenths, or reading a level rod marked in hundredths, will presumably estimate too high on some values and too low on others. Individual personal characteristics may nullify such partial compensation, however, since some people are inclined to interpolate high, others interpolate low, and many favor certain digits—for example, 7 instead of 6 or 8, 3 instead of 2 or 4, and particularly 0 instead of 9 or 1.

2-12 PRECISION AND ACCURACY

A *discrepancy* is the difference between two measured values of the same quantity. A small discrepancy indicates there are probably no mistakes and random errors are small. Small discrepancies do not preclude the presence of systematic errors, however.

Precision refers to the degree of refinement or consistency of a group of measurements, and is evaluated on the basis of discrepancy size. If multiple measurements are made of the same quantity and small discrepancies result, this indicates high precision. The degree of precision attainable is dependent on equipment sensitivity and observer skill.

Accuracy denotes the absolute nearness of measured quantities to their true values. The difference between precision and accuracy is perhaps best illustrated with refer-

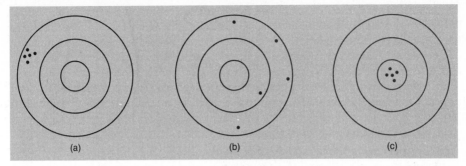

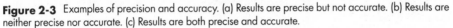

Figure 2-3 Examples of precision and accuracy. (a) Results are precise but not accurate. (b) Results are neither precise nor accurate. (c) Results are both precise and accurate.

ence to target shooting. In Figure 2-3(a), for example, all five shots exist within a close grouping, indicating a precise operation; that is, the shooter was able to repeat the procedure with a high degree of consistency. However, the shots are far from the bull's-eye and therefore not accurate. This probably results from misaligned rifle sights. Figure 2-3(b) shows randomly scattered shots that are neither precise nor accurate. In Figure 2-3(c) the closely spaced grouping, in the bull's-eye, represents both precision and accuracy. The shooter who obtained the results in (a) was perhaps able to produce the shots of (c) after aligning the rifle sights. In surveying, this would be equivalent to the calibration of measuring instruments.

As with the shooting example, a survey can be precise without being accurate. To illustrate, if refined methods are employed and readings taken carefully, say to 0.001 ft, but there are instrumental errors in the measuring device and corrections are not made for them, the survey will not be accurate. As a numerical example, two measurements of a distance with a tape assumed to be 100.000 ft long that is actually 100.050 ft might give results of 453.270 and 453.272 ft. These values are precise, but they are not accurate, since there is a systematic error of approximately 0.23 ft in each. The *precision* obtained would be expressed as $0.002/453.271 = 1/220,000$, which is excellent, but accuracy of the distance is only $0.23/453.271 = 1$ part in 2000. Also, a survey may appear to be accurate when rough measurements have been taken. For example, the angles of a traverse may be read with a compass to only the nearest $\frac{1}{4}°$, and yet produce a zero misclosure error. On good surveys, precision and accuracy are consistent throughout.

2-13 ELIMINATING MISTAKES AND SYSTEMATIC ERRORS

All field operations and office computations are governed by a constant effort to eliminate mistakes and systematic errors. It would of course be preferable if mistakes never occurred, but because humans are fallible, this is not possible. In the field mistakes can be minimized by experienced observers who alertly perform their measurements using standardized repetitive procedures. Mistakes that do occur can be corrected only if discovered. Comparing several measurements of the same quantity is one of the best ways to identify mistakes. Making a commonsense estimate and analysis is another. Assume that five measurements of a line are recorded as follows: 567.91, 576.95,

567.88, 567.90, and 567.93. The second value disagrees with the others, apparently because of a transposition of figures in reading or recording. This mistake can be eradicated by **(a)** repeating the measurement or **(b)** casting out the doubtful value.

When a mistake is detected, it is usually best to repeat the measurement. If, however, a sufficient number of other measurements of the quantity is available and in agreement, as in the foregoing example, the widely divergent result may be discarded. Serious consideration must be given to the effect on an average before discarding a value. It is seldom safe to change a recorded number, even though there appears to be a simple transposition in figures. Tampering with physical data is always bad practice and will certainly cause trouble, even if done infrequently.

Systematic errors can be calculated and proper corrections applied to the measurements. Procedures for making these corrections to all basic surveying measurements are described in the chapters that follow. In some instances it may be possible to adopt a field procedure that automatically eliminates systematic errors. For example, a leveling instrument out of adjustment causes incorrect readings, but if all backsights and foresights are made the same length, the errors cancel in differential leveling (see Chapters 6 and 7).

2-14 PROBABILITY

At one time or another, everyone has had an experience with games of chance, such as coin flipping, card games, or dice, which involve probability. In basic mathematics courses, laws of combinations and permutations are introduced. It is shown that things which happen randomly or by chance are governed by mathematical principles referred to as probability.

Probability may be defined as the ratio of the number of times a result should occur to its total number of possibilities. In the toss of a fair die, for example, there is a one-sixth probability that a 2 will come up. This simply means that there are six possibilities, and only one of them is a 2. In general if a result may occur in m ways and fail to occur in n ways, then the probability of its occurrence is $m/(m + n)$. The probability that any result will occur is a fraction between 0 and 1, 0 indicating impossibility and 1 denoting absolute certainty. Since any result must either occur or fail, the sum of the probabilities of occurrence and failure is 1. Thus if $\frac{1}{6}$ is the probability of throwing a 2 with one toss of a die, then $(1 - \frac{1}{6})$, or $\frac{5}{6}$, is the probability that a 2 will not come up.

The theory of probability is applicable in many sociological and scientific measurements. In Section 2-11 it was pointed out that random errors exist in all surveying work. This can perhaps be better appreciated by considering the measuring process, which generally involves execution of several elementary tasks. Besides instrument selection and calibration, these tasks may include setup, centering, aligning, or pointing the equipment; setting, matching, or comparing index marks; and reading or estimating values from graduated scales, dials, or gages. Because of equipment and observer imperfections, measurements cannot be made exactly, so they will always contain random errors. The magnitudes of these errors, and the frequency with which errors of a given size occur, follow the laws of probability.

For convenience, the term *error* will be used to mean only random error for the remainder of this chapter. It will be assumed that all mistakes and systematic errors have been eliminated before random errors are considered.

2-15 MOST PROBABLE VALUE

It has been stated earlier that in physical measurements the true value of any quantity is never known. Its *most probable value* can be calculated, however, if redundant measurements have been made. Redundant measurements are observations in excess of the minimum needed to determine a quantity. For a single unknown, such as a line length that has been directly and independently measured a number of times using the same equipment and procedures,[2] the first measurement establishes a value for the quantity and all additional observations are redundant. The most probable value in this case is simply the arithmetic mean, or

$$\overline{M} = \frac{\sum M}{n} \tag{2-2}$$

where $\overline{M}$ is the most probable value of the quantity, $\sum M$ the sum of the individual measurements M, and n the total number of observations. Equation (2-2) can be derived using the principle of least squares, which is based on the theory of probability.

In more complicated problems, where the observations are not made with the same instruments and procedures, or if several interrelated quantities are being determined through indirect measurements, most probable values are calculated by employing least-squares methods, as discussed in Section 2-26 and Appendix C. The treatment here relates to multiple direct observations of the same quantity using the same equipment and procedures.

2-16 RESIDUALS

Having determined the most probable value of a quantity, it is possible to calculate *residuals*. A residual is simply the difference between any measured value of a quantity and its most probable value. So

$$v = M - \overline{M} \tag{2-3}$$

where v is the residual in any measurement M, and $\overline{M}$ is the most probable value for the quantity. Residuals are theoretically identical to errors, with the exception that residuals can be calculated whereas errors cannot because true values are never known. Thus residuals rather than errors are the values actually dealt with in the analysis and adjustment of survey data. Because they are so similar, however, in practice and in this text, the terms error and residual are frequently used interchangeably.

2-17 OCCURRENCE OF RANDOM ERRORS

To analyze the manner in which random errors occur, consider the data of Table 2-1, which represents 100 repetitions of an angle measurement made with a precise theodolite (described in Chapter 10). Assume these measurements are free from mistakes and systematic errors. For convenience in analyzing the data, only the seconds portions of the measurements are tabulated. The data have been rearranged in column (1) so that

---

[2]The significance of using the same equipment and procedures is that the measurements are of equal reliability or *weight*. The subject of unequal weights is discussed in Section 2-25.

TABLE 2-1 ANGLE MEASUREMENTS FROM PRECISE THEODOLITE

MEASURED VALUE (1)	NO. (2)	RESIDUAL (SEC) (3)	MEASURED VALUE (1-CONT.)	NO. (2-CONT.)	RESIDUAL (SEC) (3-CONT.)
27°43′19.5″	1	−5.5	27°43′25.1″	3	0.1
20.0	1	−5.0	25.2	1	0.2
20.5	1	−4.5	25.4	1	0.4
20.8	1	−4.2	25.5	2	0.5
21.2	1	−3.8	25.7	3	0.7
21.3	1	−3.7	25.8	4	0.8
21.5	1	−3.5	25.9	2	0.9
22.1	2	−2.9	26.1	1	1.1
22.3	1	−2.7	26.2	2	1.2
22.4	1	−2.6	26.3	2	1.3
22.5	2	−2.5	26.4	1	1.4
22.6	1	−2.4	26.5	1	1.5
22.8	2	−2.2	26.6	3	1.6
23.0	1	−2.0	26.7	1	1.7
23.1	2	−1.9	26.8	2	1.8
23.2	2	−1.8	26.9	1	1.9
23.3	3	−1.7	27.0	1	2.0
23.6	1	−1.4	27.1	3	2.1
23.7	1	−1.3	27.4	1	2.4
23.8	2	−1.2	27.5	2	2.5
23.9	3	−1.1	27.6	1	2.6
24.0	5	−1.0	27.7	2	2.7
24.1	3	−0.9	28.0	1	3.0
24.3	1	−0.7	28.6	2	3.6
24.5	2	−0.5	28.7	1	3.7
24.7	3	−0.3	29.0	1	4.0
24.8	3	−0.2	29.4	1	4.4
24.9	2	−0.1	29.7	1	4.7
25.0	2	0.0	30.8	1	5.8
			$\Sigma = 2499.4$	$\Sigma = 100$	

Mean = 2499.4/100 = 25.0″
Most Probable Value = 27°43′25.0″

entries begin with the smallest measured value and are listed in increasing size. If a certain value was obtained more than once, the number of times it occurred, or its *frequency,* is tabulated in column (2).

From Table 2-1 it can be seen that the *dispersion* (range in measurements from smallest to largest) is 11.3 sec. Beyond assessing the dispersion, and noticing a general trend for measurements toward the middle of the range to occur with greater frequency, however, it is difficult to analyze the distribution pattern of the measurements by simply scanning the tabular values. To assist in studying the data, a *histogram* can be prepared. This is simply a bar graph showing the sizes of the measurements (or their residuals) versus their frequency of occurrence. It gives an immediate visual impression of the distribution pattern of the measurements (or their residuals).

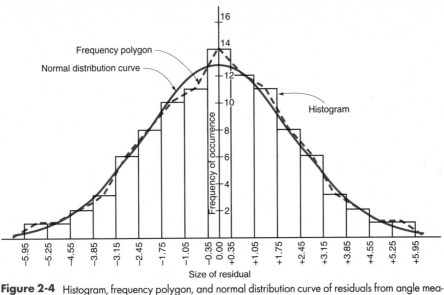

Figure 2-4 Histogram, frequency polygon, and normal distribution curve of residuals from angle measurements made with theodolite.

For the data of Table 2-1 a histogram showing the frequency of occurrence of the residuals has been developed and is plotted in Figure 2-4. To plot a histogram of residuals, it is first necessary to compute the most probable value for the measured angle. This has been done using Eq. (2-2). As shown at the bottom of Table 2-1, its value is 27° 43'25.0". Then using Eq. (2-3), residuals for all measured values are computed. These are tabulated in column (3) of Table 2-1. The residuals vary from −5.5" to +5.8". (The absolute value of the sum of these two extremes is the dispersion, or 11.3".)

To obtain a histogram with an appropriate number of bars for portraying the distribution of residuals adequately, the interval of residuals represented by each bar, or the *class interval,* was chosen as 0.7". This produced 17 bars on the graph. The range of residuals covered by each interval, and the number of residuals that occur within each interval, are listed in Table 2-2. By plotting class intervals on the abscissa against the number (frequency of occurrence) of residuals in each interval on the ordinate, the histogram of Figure 2-4 was obtained.

If the adjacent top center points of the histogram bars are connected with straight lines, the so-called *frequency polygon* is obtained. The frequency polygon for the data of Table 2-1 is superimposed as a heavy dashed blue line in Figure 2-4. It graphically displays essentially the same information as the histogram.

If the number of measurements being considered in this analysis were increased progressively, and accordingly the histogram's class interval taken smaller and smaller, ultimately the frequency polygon would approach a smooth continuous curve, symmetrical about its center, like the one shown with the heavy solid blue line in Figure 2-4. For clarity this curve is shown separately in Figure 2-5. The curve's "bell shape" is characteristic of a *normally distributed* group of errors, and thus it is often referred to as the *normal distribution curve.* Statisticians frequently call it the *normal density curve,* since it shows the densities of errors of various sizes. In surveying, normal or

TABLE 2-2 RANGES OF CLASS INTERVALS AND NUMBER OF RESIDUALS IN EACH INTERVAL

HISTOGRAM INTERVAL (SEC.)	NUMBER OF RESIDUALS IN INTERVAL
− 5.95 to − 5.25	1
− 5.25 to − 4.55	1
− 4.55 to − 3.85	2
− 3.85 to − 3.15	3
− 3.15 to − 2.45	6
− 2.45 to − 1.75	8
− 1.75 to − 1.05	10
− 1.05 to − 0.35	11
− 0.35 to + 0.35	14
+ 0.35 to + 1.05	12
+ 1.05 to + 1.75	11
+ 1.75 to + 2.45	8
+ 2.45 to + 3.15	6
+ 3.15 to + 3.85	3
+ 3.85 to + 4.55	2
+ 4.55 to + 5.25	1
+ 5.25 to + 5.95	1
	$\Sigma = 100$

Figure 2-5 Normal distribution curve.

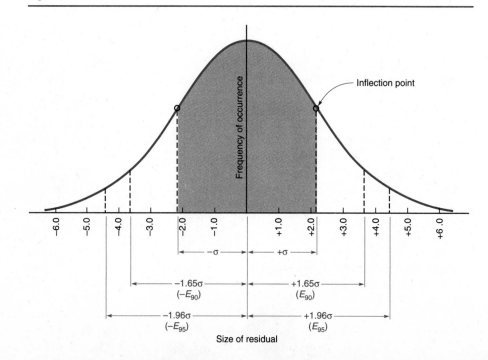

very nearly normal error distributions almost always occur, and henceforth in this book that condition is assumed.

In practice, histograms and frequency polygons are seldom used to represent error distributions. Instead, normal distribution curves which approximate them are preferred. (Note how closely the normal distribution curve superimposed on Figure 2-4 agrees with the histogram and the frequency polygon.)

As demonstrated with the data of Table 2-1, the histogram for a set of observations shows the probability of occurrence of an error of a given size graphically by bar areas. For example, 14 of the 100 residuals (errors) in Figure 2-4 are between −0.35″ and +0.35″. This represents 14% of the errors, and the center histogram bar, which corresponds to this interval, is 14% of the total area of all bars. Likewise, the area between ordinates constructed at any two abscissas of a normal distribution curve represents the percent probability that an error of that size exists. Since the area sum of all bars of a histogram represents all errors, it therefore represents all probabilities, and thus its sum equals 1. Likewise, the total area beneath a normal distribution curve is also 1.

If the same measurements of the preceding example had been taken using better equipment and more caution, smaller errors would be expected and the normal distribution curve would be similar to that in Figure 2-6(a). Compared to Figure 2-5, this curve is taller and narrower, showing that a greater percentage of values has smaller errors, and fewer measurements contain big ones. Thus the measurements are more precise. For readings taken less precisely, the opposite effect is produced, as illustrated in Figure 2-6(b), which shows a shorter and wider curve. In all three cases, however, the curve maintained its characteristic symmetric bell shape.

From these examples it is seen that relative precisions of groups of measurements become readily apparent by comparing their normal distribution curves. The normal distribution curve for a set of observations is computed using parameters derived from the residuals, but the procedure is beyond the scope of this text.

2-18 GENERAL LAWS OF PROBABILITY

From an analysis of the data in the preceding section, and the curves in Figures 2-4 through 2-6, some general laws of probability can be stated:

1. Small residuals (errors) occur more often than large ones; that is, they are more probable.
2. Large errors happen infrequently and are therefore less probable; for normally distributed errors, unusually large ones may be *mistakes* rather than random errors.
3. Positive and negative errors of the same size happen with equal frequency; that is, they are equally probable. (This enables an intuitive deduction of Eq. (2-2) to be made; that is, the most probable value for a group of repeated measurements, made with the same equipment and procedures, is the mean.)

2-19 MEASURES OF PRECISION

As shown in Figures 2-5 and 2-6, although the curves have similar shapes, there are significant differences in their dispersions; that is, their abscissa widths differ. The magnitude of dispersion is an indication of the relative precisions of the measurements.

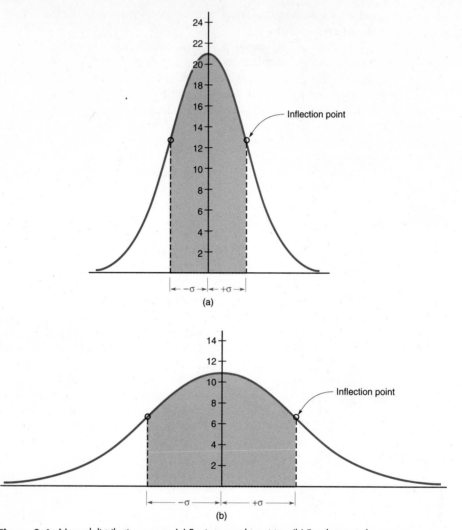

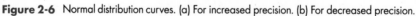

Figure 2-6 Normal distribution curves. (a) For increased precision. (b) For decreased precision.

Other statistical terms more commonly used to express precisions of groups of measurements are *standard deviation* (also often called *standard error*) and *variance*. The equation for standard deviation is

$$\sigma = \pm \sqrt{\frac{\Sigma v^2}{n - 1}} \tag{2-4}$$

where σ is the standard deviation of a group of measurements of the same quantity, v the residual of an individual observation, Σv^2 the sum of squares of the individual residuals, and n the number of observations. Variance is equal to σ^2, *the square of the standard deviation*.

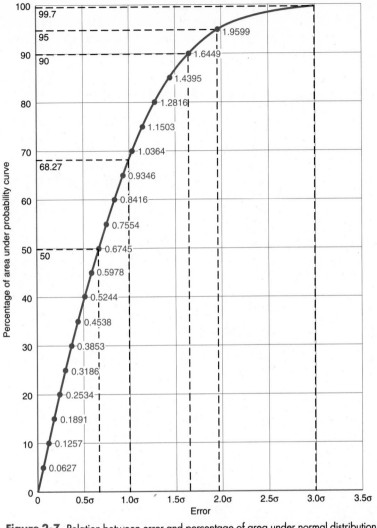

Figure 2-7 Relation between error and percentage of area under normal distribution curve.

Note that in Eq. (2-4) the standard deviation has both plus and minus values. On the normal distribution curve, the numerical value of the standard deviation is the abscissa at the *inflection points* (locations where the curvature changes from concave downward to concave upward). In Figures 2-5 and 2-6 these inflection points are shown. Note the closer spacing between them for the more precise measurements of Figure 2-6(a) as compared to Figure 2-6(b).

Figure 2-7 is a graph showing the percentage of the total area under a normal distribution curve that exists between ranges of residuals (errors) having equal positive and negative values. The abscissa scale is shown in multiples of the standard deviation. From this curve, the area between residuals of $+\sigma$ and $-\sigma$ equals 68.27% of the total area under the normal distribution curve. Hence it gives the range of residuals which can be expected to occur 68.27% of the time. This relation is shown more clearly on the curves in Figures 2-5 and 2-6, where the areas between $\pm\sigma$ are shown shaded. The percentages shown in Figure 2-7 apply to all normal distributions; regardless of curve shape or the numerical value of the standard deviation.

2-20 INTERPRETATION OF STANDARD DEVIATION

It has been shown that the standard deviation establishes the limits within which measurements are expected to fall 68.27% of the time. In other words, if a measurement is repeated ten times, it will be expected that approximately seven of the results will fall within the limits established by the standard deviation, and conversely about three of them will fall anywhere outside these limits. Another interpretation is that one additional measurement will have a 68.27% chance of falling within the limits set by the standard deviation. A third deduction is that the true value has a 68.27% probability of falling within the standard deviation limits.

When Eq. (2-4) is applied to the data of Table 2-1, a standard deviation of ±2.19 sec is obtained. In examining the residuals in the table, 70 of the 100 values, or 70 percent, are actually smaller than 2.19 sec. This illustrates that the theory of probability closely approximates reality.

2-21 THE 50, 90, and 95% ERRORS

From the data given in Figure 2-7, the probability of an error of any percentage likelihood can be determined. The general equation is

$$E_P = C_P\sigma \tag{2-5}$$

where E_P is a certain percentage error and C_P the corresponding numerical factor taken from Figure 2-7.

By Eq. (2-5), after extracting appropriate multipliers from Figure 2-7, the following are expressions for errors that have a 50, 90, and 95% chance of occurring:

$$E_{50} = 0.6745\sigma \tag{2-6}$$

$$E_{90} = 1.6449\sigma \tag{2-7}$$

$$E_{95} = 1.9599\sigma \tag{2-8}$$

The 50% error, or E_{50}, is the so-called *probable error*. It establishes limits within which the measurements should fall 50% of the time. In other words, a measurement has the same chance of coming within these limits as it has of falling outside of them. In the past, probable error was often used in discussing random errors, but now it is seldom used.

The 90 and 95% errors are commonly used to specify precisions required on surveying projects. Of these, the 95% error, also frequently called the *two-sigma (2σ) error*, is most often specified. As an example, a particular project may call for the 95% error to be less than or equal to a certain value for the work to be acceptable. For the data of Table 2-1, applying Eqs. (2-7) and (2-8), the 90 and 95% errors are ±3.60 sec and ±4.29 sec, respectively. The 90 and 95% errors are shown graphically in Figure 2-5.

Surveyors also often use the so-called *three-sigma* (3σ) *error* as a criterion for rejecting individual measurements. From Figure 2-7, there is a 99.7% probability that an error will be less than this amount. Thus within a group of observations, any value whose residual exceeds 3σ is considered to be a mistake, and either a new measurement must be taken or the computations based on one less value.

The x axis is an asymptote of the normal distribution curve, so the 100% error cannot be evaluated. This means that no matter what size error is found, a larger one is theoretically possible.

To clarify definitions and use the equations given in Sections 2-17 through 2-21, suppose that a line has been measured ten times using the same equipment and procedures. The results are shown in column (1) of the following table. It is assumed that no mistakes exist, and that the measurements have already been corrected for all systematic errors. Compute the most probable value for the line length, its standard deviation, and errors having 50, 90, and 95% probability.

EXAMPLE 2-1

LENGTH (ft) (1)	RESIDUAL v (ft) (2)	v^2 (3)
538.57	+0.12	0.0144
538.39	−0.06	0.0036
538.37	−0.08	0.0064
538.39	−0.06	0.0036
538.48	+0.03	0.0009
538.49	+0.04	0.0016
538.33	−0.12	0.0144
538.46	+0.01	0.0001
538.47	+0.02	0.0004
538.55	+0.10	0.0100
$\Sigma = 5384.50$	$\Sigma = -0.00$	$\Sigma v^2 = 0.0554$

SOLUTION

By Eq. (2-2), $\overline{M} = \dfrac{5384.50}{10} = 538.45$ ft.

By Eq. (2-3), the residuals are calculated. These are tabulated in column (2) and their squares listed in column (3). Note that in column (2) the algebraic sum of residuals is zero. For measurements of equal reliability, except for roundoff, this column should always total zero and thus provide a computational check.

By Eq. (2-4), $\sigma = \pm\sqrt{\dfrac{\Sigma v^2}{n-1}} = \pm\sqrt{\dfrac{0.0554}{9}} = \pm 0.078$ ft.

By Eq. (2-6), $E_{50} = \pm 0.6745\sigma = \pm 0.6745(0.078) = \pm 0.05$ ft.
By Eq. (2-7), $E_{90} = \pm 1.6449(0.078) = \pm 0.13$ ft.
By Eq. (2-8), $E_{95} = \pm 1.9599(0.078) = \pm 0.15$ ft.

The following conclusions can be drawn concerning this example.

1. The most probable line length is 538.45 ft.
2. The standard deviation of a single measurement is ±0.08 ft (rounded from 0.078). Accordingly, the normal expectation is that 68% of the time a recorded length will lie between 538.37 and 538.53 ft; that is, about seven values will lie within these limits. (Actually seven of them do.)
3. The probable error (E_{50}) is ±0.05 ft. Therefore it can be anticipated that half, or five, of the measurements will fall in the interval 538.40 to 538.50 ft. (Four values do.)
4. The 90% error is ±0.13 ft, and thus nine of the measured values can be expected to be within the range of 538.32 and 538.58 ft.
5. The 95% error is ±0.15 ft, so the length can be expected to lie between 538.30 and 538.60 95% of the time. (Note that all measurements indeed are within the limits of both the 90 and 95% errors.)

2-22 ERROR PROPAGATION

It was stated earlier that because all measurements contain errors, any quantities computed from them will likewise contain errors. The process of evaluating errors in quantities computed from measured values which contain errors is called *error propagation*. In the subsections that follow, simple specific cases of error propagation common in surveying are discussed, and examples are presented. All formulas can be derived from the general law of error propagation, given in Section C-1 of Appendix C.

2-22.1 Error of a Sum

The following formula gives the propagated error in the sum of quantities, each of which contains a different random error:

$$E_{\text{sum}} = \pm\sqrt{E_a^2 + E_b^2 + E_c^2 + \cdots} \qquad (2\text{-}9)$$

where E represents any specified percentage error (such as σ, E_{50}, E_{90}, or E_{95}), and a, b, and c are the separate measurements.

EXAMPLE 2-2   Assume that a line is measured in three sections, with the individual parts equal to (753.81, ±0.12), (1238.40, ±0.28), and (1062.95, ±0.20) ft, respectively. Determine the line's total length and its anticipated standard deviation.

SOLUTION
Total length = 753.81 + 1238.40 + 1062.95 = 3055.16 ft.

By Eq. (2-9); $E_{\text{sum}} = \pm\sqrt{0.012^2 + 0.028^2 + 0.020^2} = \pm 0.036$ ft.

2-22.2 Error of a Series

Sometimes a series of similar quantities, such as the angles within a closed polygon, are read with each measurement being in error by about the same amount. The total error in the sum of all measured quantities of such a series is called the *error of the series,* designated as E_{series}. If the same error E in each measurement is assumed and Eq. (2-9) applied, the series error is

$$E_{series} = \pm\sqrt{E^2 + E^2 + E^2 + \cdots} = \pm\sqrt{nE^2} = \pm E\sqrt{n} \qquad (2\text{-}10)$$

where E represents the error in each individual measurement and n the number of measurements.

This equation shows that when the same operation is repeated, random errors tend to balance out and the resulting error of a series is proportional to the square root of the number of observations. This equation has extensive use—for instance, to determine the allowable misclosure error for angles of a traverse, as discussed in Chapter 12.

Assume that any distance of 100 ft can be taped with an error of ± 0.02 ft if certain techniques are employed. Determine the error in taping 5000 ft using these skills.   EXAMPLE 2-3

SOLUTION
Since the number of 100-ft lengths in 5000 ft is 50 then by Eq. (2-10)

$$E_{series} = \pm E\sqrt{n} = \pm 0.02\sqrt{50} = \pm 0.14 \text{ ft}$$

A distance of 1000 ft is to be taped with an error of not more than ± 0.10 ft. Determine how accurately each 100-ft length must be measured to ensure that the error will not exceed the permissible limit.   EXAMPLE 2-4

SOLUTION
Since by Eq. (2-10), $E_{series} = E\sqrt{n}$ and $n = 10$, the allowable error E in 100 ft is

$$E = \frac{E_{series}}{\sqrt{n}} = \frac{0.10}{\sqrt{10}} = \pm 0.03 \text{ ft}$$

Suppose it is required to tape a length of 2500 ft with an error of not more than ± 0.10 ft. How accurately must each tape length be measured?   EXAMPLE 2-5

SOLUTION
Since 100 ft is again considered the unit length, $n = 25$, and by Eq. (2-10), the allowable error E in 100 ft is

$$E = \frac{0.10}{\sqrt{25}} = \pm 0.02 \text{ ft}$$

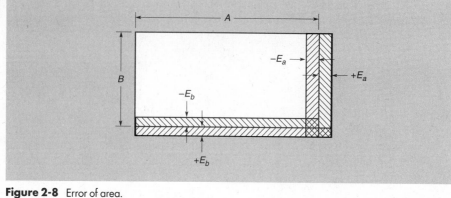

Figure 2-8 Error of area.

Analysis of Examples 2-4 and 2-5 shows that a larger number of possibilities provides a greater chance for errors to cancel out.

2-22.3 Error in a Product

The equation for propagated error in a product AB, where E_a and E_b are the respective errors in A and B, is

$$E_{prod} = \pm\sqrt{A^2E_b^2 + B^2E_a^2} \tag{2-11}$$

The physical significance of the error propagation formula for a product is illustrated in Figure 2-8, where A and B are shown to be measured sides of a rectangular parcel of land with errors E_a and E_b, respectively. The product AB is the parcel area. In Eq. (2-11), $\sqrt{A^2E_b^2} = AE_b$ represents either of the longer (horizontal) crosshatched bars, and is the error due to either $-E_b$ or $+E_b$. The term $\sqrt{B^2E_a^2} = BE_a$ is represented by the shorter (vertical) crosshatched bars, which is the error resulting from either $-E_a$ or $+E_a$.

EXAMPLE 2-6

For the rectangular lot illustrated in Figure 2-8, measurements of sides A and B with their 95% errors are $(252.46, \pm0.053)$ and $(605.08, \pm0.072)$ ft, respectively. Calculate the parcel area and the expected 95% error in the area.

SOLUTION

Area = $252.46 \times 605.08 = 152{,}760$ ft^2.

By Eq. (2-11), $E_{95} = \pm\sqrt{(252.46)^2(0.053)^2 + (605.08)^2(0.072)^2} = \pm30.7$ ft^2.

Example 2-6 can also be used to demonstrate the validity of one of the rules of significant figures in computation. The computed area is actually 152,758.4968 ft^2. However, the rule for significant figures in multiplication (see Section 2-5) states that there cannot be more significant figures in the answer than in any of the individual factors used. Accordingly, the area should be rounded off to 152,760 (five significant figures). From Eq. (2-11), with an error of ± 30.7 ft, the answer could be $152{,}758.4968$ ± 30.7, or from $152{,}727.8$ to $152{,}789.2$ ft^2. Thus the fifth digit in the answer is seen to be questionable, and hence the number of significant figures specified by the rule is verified.

2-22.4 Error of the Mean

Equation (2-2) stated that the most probable value of a group of repeated measurements is the arithmetic mean. Since the mean is computed from individual measured values, each of which contains an error, the mean is also subject to error. By applying Eq. (2-10) it is possible to find the error for the sum of a series of measurements where each one has the same error. Since the sum divided by the number of measurements gives the mean, the error in the mean may be found by the relation

$$E_m = \frac{E_{sum}}{n}$$

Substituting Eq. (2-10) for E_{sum},

$$E_m = \frac{E\sqrt{n}}{n} = \frac{E}{\sqrt{n}} \tag{2-12}$$

where E is the specified percentage error of a single measurement, E_m the corresponding percentage error of the mean, and n the number of observations.

The error of the mean at any percentage probability can be determined and applied to all criteria that have been developed. For example, the standard deviation of the mean, $(E_{68})_m$ or σ_m, is

$$(E_{68})_m = \sigma_m = \frac{\sigma}{\sqrt{n}} = \pm \sqrt{\frac{\Sigma v^2}{n(n-1)}} \tag{2-13a}$$

and the 90 and 95% errors of the mean are

$$(E_{90})_m = \frac{E_{90}}{\sqrt{n}} = \pm 1.6449 \sqrt{\frac{\Sigma v^2}{n(n-1)}} \tag{2-13b}$$

$$(E_{95})_m = \frac{E_{95}}{\sqrt{n}} = \pm 1.9599 \sqrt{\frac{\Sigma v^2}{n(n-1)}} \tag{2-13c}$$

These equations show that *the error of the mean varies inversely as the square root of the number of repetitions.* Thus to double the accuracy—that is, to reduce the error by one-half—four times as many measurements must be made.

Calculate the standard deviation of the mean and the 90% error of the mean for the measurements of Example 2-1.   EXAMPLE 2-7

SOLUTION

By Eq. (2-13a), $\sigma_m = \dfrac{\sigma}{\sqrt{n}} = \pm \dfrac{0.078}{\sqrt{10}} = \pm 0.025$ ft.

Also, by Eq. (2-13b), $(E_{90})_m = \pm 1.6449\,(0.025) = \pm 0.041$ ft.

These values show the error limits of 68 and 90% probability for the line's length. It can be said that the true line length has a 68% chance of being within ± 0.025 of the mean, and a 90% likelihood of falling not farther than ± 0.041 ft from the mean.

2-23 APPLICATIONS

The example problems show that the equations of error probability are applied in two ways:

1. To analyze measurements that have already been made, for comparison with other results or with specification requirements.
2. To establish procedures and specifications in order that the required results will be obtained.

The application of the various error probability equations must be tempered with judgment and caution. Recall that they are based on the assumption that the errors conform to a smooth and continuous normal distribution curve, which in turn is based on the assumption of a large number of measurements. Frequently in surveying only a few observations—often from four to eight—are taken. If these conform to a normal distribution, then the answer obtained using probability equations will be reliable; if they do not, the conclusions could be misleading. In the absence of knowledge to the contrary, however, an assumption that the errors are normally distributed is still the best available.

2-24 CONDITIONAL ADJUSTMENT OF MEASUREMENTS

In Section 2-8 it was emphasized that the true value of any measured quantity is never known. In some types of problems, however, the sum of several measurements must equal a fixed value; for example, the sum of the three angles in a plane triangle has to total 180°. In practice, therefore, the measured angles are adjusted to make them add to the required amount. Correspondingly, distances—either horizontal or vertical—must also often be adjusted to meet certain conditional requirements. The methods used will be explained in later chapters, where the operations are taken up in detail.

2-25 WEIGHTS OF MEASUREMENTS

It is evident that some measurements are more precise than others because of better equipment, improved techniques, and superior field conditions. In making adjustments, it is consequently desirable to assign *relative weights* to individual observations. It can logically be concluded that if an observation is very precise, it will have a small standard deviation or variance, and thus should be weighted more heavily (held closer to its measured value) in an adjustment than an observation of lower precision. From this reasoning it is deduced that weights of observations should bear an inverse relationship to precision. In fact, it can be shown that relative weights are inversely proportional to variances, or

$$W_a \propto \frac{1}{\sigma_a^2} \tag{2-14}$$

where W_a is the weight of an observation a, which has a variance σ_a^2. Thus the higher the precision (the smaller the variance), the larger should be the relative weight of the measured value being adjusted. In some cases variances are unknown originally, and weights must be assigned to measured values based on estimates of their relative precision. If a quantity is measured repeatedly and the individual observations have varying weights, the weighted mean can be computed from the expression

$$\overline{M}_w = \frac{\Sigma WM}{\Sigma W} \tag{2-15}$$

where $\overline{M}_w$ is the weighted mean, ΣWM the sum of the individual weights times their corresponding observations, and ΣW the sum of the weights.

Suppose four measurements of a distance are recorded as 482.16, 482.17, 482.20, and 482.18 and given weights of 1, 2, 2, and 4, respectively by the survey-party chief. Determine the weighted mean.   EXAMPLE 2-8

SOLUTION
By Eq. (2-15),

$$\overline{M}_w = \frac{482.16 + 482.17(2) + 482.20(2) + 482.18(4)}{1 + 2 + 2 + 4} = 482.18 \text{ ft}$$

In computing adjustments involving unequally weighted measurements, corrections applied to observed values should be made inversely proportional to the relative weights.

Assume the measured angles of a certain triangle and their relative weights are $A = 49°51'15''$, $W_a = 1$; $B = 60°32'08''$, $W_b = 2$; and $C = 69°36'33''$, $W_c = 3$. Perform a weighted adjustment of the angles.   EXAMPLE 2-9

SOLUTION
Angle adjustments are made inversely proportional to their weights, as in the accompanying tabulation. Angle C with the greatest weight (3) gets the smallest correction, $2x$; B receives $3x$; and A, $6x$.

	MEASURED ANGLE	WT	CORRECTION	NUMERICAL CORR.	ROUNDED CORR.	ADJUSTED ANGLE
A	49°51'15"	1	$6x$	+2.18"	+2"	49°51'17"
B	60°32'08"	2	$3x$	+1.09"	+1"	60°32'09"
C	69°36'33"	3	$2x$	+0.73"	+1"	69°36'34"
Sum	179°59'56"	$\Sigma = 6$	$11x$	+4.00"	+4"	180°00'00"

$$11x = 4'' \text{ and } x = +0.36''$$

It must be emphasized again that adjustment computations based on the theory of probability are valid only if systematic errors and mistakes have been eliminated by employing proper procedures, equipment, and calculations.

2-26 LEAST-SQUARES ADJUSTMENT

As explained in Section 2-24, most surveying measurements must conform to certain geometrical conditions. The amounts by which they fail to meet these conditions are called *misclosures,* and they indicate the presence of random errors. Various procedures are used to distribute these errors and to produce mathematically perfect geometrical conditions. Some simply apply corrections of the same size to all measured values, where each correction equals the total misclosure divided by the number of measurements. Others introduce corrections in proportion to assigned weights. Still others employ rules of thumb, for example, the "compass rule" described in Chapter 13 for adjusting closed traverses.

Because random errors in surveying conform to the mathematical laws of probability and are "normally distributed," the most appropriate adjustment procedure should be based upon these laws. Least squares is such a method. It is not a new procedure, having been applied by the German mathematician Karl Gauss as early as the latter part of the eighteenth century. Until the advent of computers, however, it was only used sparingly because of the lengthy calculations involved.

Least squares is suitable for adjusting any of the basic types of surveying measurements described in Section 2-2, and is applicable to all of the commonly used surveying procedures. The method enforces the condition that *the sum of the weights of the measurements times their corresponding squared residuals is minimized.* This fundamental condition, which is developed from the equation for the normal error distribution curve, provides most probable values for the adjusted quantities. In addition, it also (1) determines precisions of the adjusted values, (2) reveals the presence of large errors and mistakes so steps can be taken to eliminate them, and (3) makes possible the optimum design of survey procedures in the office before going into the field to take measurements.

The basic assumptions that underlie least-squares theory are as follows: (1) Mistakes and systematic errors have been eliminated so only random errors remain; (2) the number of observations being adjusted is large; and (3) the frequency distribution of errors is normal. Although these assumptions are not always met, least-squares adjustment still provides the most rigorous error treatment available, and hence it has become very popular and important in modern surveying. A more detailed discussion of the subject is presented in Appendix C.

2-1 Convert the following distances given in meters to feet:
    (a) 2914.57 m   **(b)** 738.29 m   ***(c)** 1257.30 m

2-2 Convert the following distances given in feet to meters:
    (a) 537.52 ft   **(b)** 10,028.75 ft   ***(c)** 1328.3 ft

2-3 Compute the lengths in feet corresponding to the following distances measured with a Gunter's chain:
    (a) 15 ch 37 lk   **(b)** 75 ch 41 lk   ***(c)** 19 ch 12 lk

2-4 Express 327,740 ft^2 in:
    (a) acres   **(b)** hectares   ***(c)** square Gunter's chains

2-5 Convert 16.786 hectares to:
    (a) acres   **(b)** square Gunter's chains

2-6 What are the lengths in feet and decimals for the following distances shown on a building blueprint:
    (a) 82 ft $10\frac{3}{8}$ in.   ***(b)** 50 ft $2\frac{3}{4}$ in.

2-7 What is the area in acres of a rectangular parcel of land measured with a Gunter's chain if the recorded sides are as follows:
    (a) 79 ch 17 lk and 65 ch 53 lk   **(b)** 67 ch 23 lk and 41 ch 28 lk

2-8 Compute the area in acres of triangular lots shown on a plat having the following recorded right-angle sides:
    (a) 325.72 ft and 438.05 ft   ***(b)** 10 ch 61 lk and 6 ch 95 lk

2-9 A distance is expressed as 215,256.91 U.S. survey feet. What is the length in international feet?

2-10 What are the radian and degree-minute-second equivalents for the following angles given in grads:
    (a) 48.00 grads   **(b)** 72.37 grads   **(c)** 206.15 grads

2-11 Give answers to the following problems in the correct number of significant figures:
    (a) sum of 91.72, 0.00154, 156, and 9.7
    (b) sum of 0.052, 1130, 6.4374, and 348.6
    (c) product of 2185.29 and 3.1
    ***(d)** quotient of 4930.27 divided by 3.59

2-12 Express the value or answer in powers of 10 to the correct number of significant figures:
    (a) 73,862   **(b)** 8360   ***(c)** square of 18,763
    (d) sum of (21.158 + 0.3 + 146.1) divided by 7.2

2-13 Convert the adjusted angles of a triangle to radians and show a computational check:
    (a) 48°30′16″, 81°10′46″, and 50°18′58″
    ***(b)** 28°57′13″, 64°33′00″, and 86°29′47″

2-14 Explain the difference between systematic and random errors. Give an example of each.

2-15 Discuss the difference between precision and accuracy.

A distance *AB* is measured repeatedly using the same equipment and procedures, and the results, in feet, are listed in Problems 2-16 through 2-20. Calculate **(a)** the line's most probable length, **(b)** the standard deviation of a single measurement, and **(c)** the standard deviation of the mean for each set of results.

***2-16** 728.56, 728.59, 728.58, 728.54, 728.57, 728.62, 728.71, 728.53, 728.59, and 728.46

2-17 Same as Problem 2-16, but discard one measurement, 728.71.

2-18 Same as Problem 2-16, but discard two measurements, 728.71 and 728.46.

2-19 Same as Problem 2-16, but include two additional measurements, 728.53 and 728.55.

2-20 Same as Problem 2-16, but include five additional measurements, 728.57, 728.64, 728.40, 728.49, and 728.61.

In Problems 2-21 through 2-25, determine the range within which measurements should fall **(a)** 50% of the time and **(b)** 90% of the time. List the percentage of values that actually fit within these ranges.

***2-21** For the data of Problem 2-16.

2-22 For the data of Problem 2-17.

*Asterisks indicate problems that have answers given in Appendix G.

2-23 For the data of Problem 2-18.

2-24 For the data of Problem 2-19.

2-25 For the data of Problem 2-20.

In Problems 2-26 through 2-28, an angle is measured repeatedly using the same equipment and procedures. Calculate **(a)** the angle's most probable value, **(b)** the standard deviation of a single measurement, and **(c)** the standard deviation of the mean.

*2-26 46°30′02″, 46°29′52″, 46°30′22″, 46°29′42″, 46°30′12″, and 46°30′02″

2-27 Same as Problem 2-26, but with two additional measurements, 46°30′02″ and 46°30′12″.

2-28 Same as Problem 2-26, but with four additional measurements, 46°30′02″, 46°30′12″, 46°30′12″, and 46°30′02″.

2-29 A field party is capable of making taping measurements with a standard deviation of ±0.010 ft per 100-ft tape length. What standard deviation would be expected in a distance of 3600 ft taped by this party?

2-30 Repeat Problem 2-29, except that the standard deviation per 30-m tape length is ±0.010 m and a distance of 1500 m is taped. What is the expected 90% error in 1500 m?

*2-31 A distance of 2500 ft must be taped in a manner to ensure a standard deviation smaller than ±0.05 ft. What must be the standard deviation per 100-ft tape length to achieve the desired precision?

2-32 Lines of levels were run requiring n instrument setups. If the rod reading for each backsight and foresight has a standard deviation σ, what are the standard deviations in each of the following level lines?

    (a) $n = 40, \sigma = \pm0.005$ ft   **(b)** $n = 20, \sigma = \pm0.50$ mm

2-33 A line AC was measured in 2 sections AB and BC, with lengths and standard deviations listed below. What is the standard deviation in the total length AC?

    (a) $AB = 238.21, \pm0.06$ ft; $BC = 460.74, \pm0.10$ ft

    (b) $AB = 1720.416, \pm0.124$ m, $BC = 1061.702, \pm0.096$ m

2-34 Line AD is measured in three sections, AB, BC, and CD, with lengths and standard deviations as listed below. What is the standard deviation in the total length AD?

    *(a) $AB = 573.12 \pm0.06$ ft; $BC = 1274.83, \pm0.10$ ft; $CD = 942.78, \pm0.09$ ft

    (b) $AB = 382.965$ m, ±0.020 m; $BC = 495.030$ m, ±0.025 m; $CD = 526.520$ m, ±0.028 m

2-35 A distance AB was measured four times as 577.38, 577.36, 577.40, and 577.39 ft. The measurements were given weights of 2, 3, 1, and 2, respectively, by the observer. **(a)** Calculate the weighted mean for distance AB. **(b)** What difference results if later judgment revises the weights to 2, 1, 3, and 3?

2-36 Determine the weighted mean for the following angles and weights:

    (a) 98°24′51″, wt 1; 98°24′43″, wt 2; 98°24′54″, wt 3; 98°24′45″, wt 1

    (b) 45°27′09″, wt 2; 45°27′05″, wt 1; 45°27′12″, wt 4; 45°27′10″, wt 2

2-37 Specifications for measuring angles of an n-sided polygon limit the total angular misclosure to E. How accurately must each angle be measured for the following values of n and E?

    (a) $n = 16, E = 12″$ **(b)** $n = 12, E = 30″$

2-38 What is the area of a rectangular field and its expected error for the following recorded values:

    *(a) 658.10, ±0.12 ft; by 802.56, ±0.15 ft

    (b) 275.15, ±0.08 ft; by 469.55, ±0.10 ft

    (c) 128.52, ±0.015 m, by 180.40, ±0.020 m

2-39 Adjust the angles of triangle ABC for the following angular values and weights:

    (a) $A = 53°31′35″$, wt 2; $B = 59°40′16″$, wt 1; $C = 66°48′17″$, wt 3

    (b) $A = 70°12′55″$, wt 1; $B = 43°42′52″$, wt 4; $C = 66°04′03″$, wt 2

2-40 Determine relative weights and perform a weighted adjustment (to the nearest second) for angles A, B, and C of a plane triangle, given the following four observations for each angle:

*Asterisks indicate problems that have answers given in Appendix G.

Angle A	Angle B	Angle C
52°22'50"	72°20'24"	55°17'02"
52°23'02"	72°20'38"	55°17'06"
52°23'00"	72°20'42"	55°16'59"
52°22'56"	72°20'20"	55°17'01"

2-41 A line of levels was run from bench marks A to B, B to C, and C to D. The elevation differences obtained between bench marks, with their standard deviations, are listed below. What is the difference in elevation from bench mark A to D and the standard deviation of that elevation difference?

*(a) BM A to BM B = +35.87, ±0.10 ft; BM B to BM C = −53.05, ±0.15 ft; and BM C to BM D = −48.90, ±0.08 ft

(b) BM A to BM B = +17.82, ±0.05 ft; BM B to BM C = +39.20, ±0.08 ft; and BM C to BM D = +69.44 ±0.13 ft

(c) BM A to BM B = −56.218, ±0.012 m; BM B to BM C = +49.387, ±0.020 m; and BM C to BM D = +65.580, ±0.016 m

*Asterisks indicate problems that have answers given in Appendix G.

BIBLIOGRAPHY

American Congress on Surveying and Mapping. 1980. *Metric Practice Guide for Surveying and Mapping.* Washington, D.C.: ACSM.

Barry, B. A. 1978. *Errors in Practical Measurement in Science, Engineering, and Technology.* New York: John Wiley and Sons.

Braunstein, P. 1991. "Surveying and the International System of Units." *Point of Beginning* 16 (No. 6):10

Buckner, R. B. 1984. *Surveying Measurements and Their Analysis.* Rancho Cordova, Calif.: Landmark.

Carver, G. P. "Metrication—An Economic Wake-up Call for Surveyors and Mappers." *Point of Beginning* 19 (No. 1): 74.

Goldman, D. 1984. "Measuring Up to Metric." *Professional Surveyor* 4 (No. 3):12.

Lane, A. 1992. "Metric Mandate." *Professional Surveyor* 12 (No. 2):13.

Mikhail, E. 1976. *Observations and Least Squares.* New York: Harper & Row.

Mikhail, E. M., and G. Gracie. 1981. *Analysis and Adjustment of Survey Measurements.* New York: Van Nostrand.

Miller, M. 1986. "La Vara of New Mexico." *Professional Surveyor* 6 (No. 1):34.

Toscano, P. 1991. "The Gunter's Chain." *Surveying and Land Information Systems* 51 (No. 3):155.

Whitten, C. A., et al. 1980. "Planning for Metrication for Surveying and Mapping." *Bulletin, American Congress on Surveying and Mapping* (No. 70):37.

Whitten, C. A. 1984. "Which Foot to Use—International or U.S. Survey." *Point of Beginning* 10 (No. 3):62.

Wolf, P. R. 1980. *Adjustment Computations: Practical Least Squares for Surveyors,* 2nd ed. Rancho Cordova, Calif.: Landmark.

3

SURVEYING FIELD NOTES

3-1 INTRODUCTION

Field notes are the record of work done in the field. They typically contain measurements, sketches, descriptions, and many other items of miscellaneous information. In the past, field notes were prepared exclusively by hand lettering in field books or special note pads as the work progressed and data were gathered. Recently, however, automatic data collectors have been introduced which are interfaced with many different modern surveying instruments. As the work progresses, they create computer files containing a record of measured data. Automatic data collectors are rapidly gaining popularity, but when used, the numerical data they generate are often supplemented by manually prepared sketches and descriptions. Regardless of the manner or form in which the notes are taken, they are extremely important. That is why this topic is discussed in a separate chapter.

Whether prepared manually, created by an electronic data collector, or a combination of these forms, surveying field notes are the only permanent record of work done in the field. If the data are incomplete, incorrect, lost, or destroyed, much or all of the time and money invested in making the measurements and records has been wasted. Hence the job of data recording is frequently the most important and difficult one in a surveying party. Field books and computer files containing information gathered over a period of weeks are worth many thousands of dollars because of the high costs of maintaining a party of two, three, or more persons in the field.

Recorded field data are used in the office to perform computations, make drawings, or both. The office personnel using the data are usually not the same people who took the notes in the field. Accordingly, it is essential that notes be intelligible to anyone without verbal explanations.

Property surveys are subject to court review under some conditions, so field notes become an important factor in litigation. Also, because they may be used as references in land transactions for generations, it is necessary to index and preserve them properly. The saleable "goodwill" of a surveyor's business depends largely on the office library of field books. Cash receipts may be kept in an unlocked desk drawer, but field books are stored in a fireproof safe!

Topics pertinent to handwritten field notes and automatic data collectors are presented in this chapter in Parts I and II, respectively.

PART I
HANDWRITTEN FIELD NOTES

3-2 GENERAL REQUIREMENTS OF HANDWRITTEN FIELD NOTES

The following points are considered in appraising a set of field notes:

Accuracy. This is the most important quality in all surveying operations.

Integrity. A single omitted measurement or detail can nullify use of the notes for computing or plotting. If the project was far from the office, it is time consuming and expensive to return for a missing measurement. Notes should be checked carefully for completeness before leaving the survey site, and never "fudged" to improve closures.

Legibility. Notes can be used only if they are legible. A professional-looking set of notes is likely to be professional in quality.

Arrangement. Noteforms appropriate to the particular survey contribute to accuracy, integrity, and legibility.

Clarity. Advance planning and proper field procedures are necessary to ensure clarity of sketches and tabulations, and to minimize the possibility of mistakes and omissions. Avoid crowding notes; paper is relatively cheap. Costly mistakes in computing and drafting are the end results of ambiguous notes.

Appendix D contains examples of handwritten field notes for a variety of surveying operations. Each is identified by its plate number. These notes have been prepared keeping the above points in mind.

In addition to the items stressed in the foregoing, certain other guidelines must be followed to produce acceptable handwritten field notes. The notes should be lettered with a sharp pencil of at least 3-H hardness so that an indentation is made in the paper. Books so prepared will withstand damp weather in the field (or even a soaking) and still be legible, whereas graphite from a soft pencil, or ink from a pen or ballpoint, leaves an undecipherable smudge under such circumstances.

For handwritten notes, the Reinhardt system of slope lettering is generally employed for clarity and speed; it requires a minimum number of simple strokes to form any letter. This style of lettering is described in engineering drawing textbooks, and is used in the example noteforms of Appendix D.

Erasures of recorded data are not permitted in field books. If a number has been entered incorrectly, a line is run through it without destroying the number's legibility, and the proper value is noted above it (see Plate D-2, left page). If a partial or entire page is to be deleted, diagonal lines are drawn through opposite corners, and VOID is lettered prominently on the page, giving the reasons.

Field notes are presumed to be "original" unless marked otherwise. Original notes are those taken at the same time measurements are being made. If the original notes are copied, they must be so marked. Copied notes may not be accepted in court because they are open to question concerning possible mistakes, such as interchanging numbers, and omissions. The value of a distance or an angle placed in the field book from memory, 10 minutes after the observation, is definitely unreliable.

Students are tempted to scribble notes on scrap sheets of paper for later transfer in neater form to the field book. This practice may result in the loss of some or all of the original data and defeats one purpose of a surveying course—to provide experience in taking notes under actual field conditions. In a real job situation, a surveyor is not likely to spend any time at night transcribing scribbled notes. Certainly, an employer will not pay for this evidence of incompetence.

3-3 TYPES OF FIELD BOOKS

Since field books contain valuable data, suffer hard wear, and must be permanent in nature, it is good economy to use only the best for practical work. Various kinds of field books are available, but bound and loose-leaf types are the most common.

The bound book, a standard for many years, has a sewed binding, a hard cover of leatherite, polyethylene, or covered hardboard, and contains 80 leaves. Its use ensures maximum testimony acceptability for property survey records in courtrooms.

Bound duplicating books enable copies of the original notes to be made through carbon paper in the field. The alternate duplicate pages are perforated to enable their easy removal for advance shipment to the office.

Loose-leaf books have come into wide use because of many advantages, which include (1) assurance of a flat working surface, (2) simplicity of filing individual project notes, (3) ready transfer of partial sets of notes between field and office, (4) provision for holding pages of printed tables, diagrams, formulas, and sample forms, (5) the possibility of using different rulings in the same book, and (6) a saving in sheets and thus cost since none are wasted by filing partially filled books. A disadvantage is the possibility of losing sheets.

Stapled or spiral-bound books are not suitable for practical work. They may be satisfactory for abbreviated surveying courses that have only a few field periods, because of limited service required and low cost. Special column and page rulings provide for particular needs in leveling, angle measurement, topographic surveying, cross-sectioning, and so on.

A camera is a helpful notekeeping "instrument." A moderately priced, reliable, lightweight camera can be used to adequately document monuments set or found, and to provide records of other valuable information or admissible field evidence. Tape recorders can also be used in certain circumstances, particularly where lengthy written explanations would be needed to document conditions or provide detailed descriptions.

3-4 KINDS OF NOTES

Four types of notes are kept in practice: (1) sketches, (2) tabulations, (3) descriptions, and (4) combinations of these. The most common type is a combination form, but an experienced recorder selects the version best fitted to the job at hand. The noteforms in Appendix D illustrate some of these types, and apply to field problems described in this text. Other examples are included within the text at appropriate locations.

Sketches often greatly increase the efficiency with which notes can be taken. They are especially valuable to persons in the office who must interpret the notes without benefit of the notetaker's presence. The proverb about one picture being worth 10,000 words might well have been written for notekeepers.

For a simple survey, such as measuring the distances between stations on a series of lines, a sketch showing the lengths is sufficient. In measuring the length of a line forward and backward, a tabulation properly arranged in columns is adequate, as in Plate D-1 in Appendix D. The location of a reference point may be difficult to identify without a sketch, but often a few lines of description are enough. Bench marks usually are so described, as in Plate D-2.

In notekeeping this axiom is always pertinent: *When in doubt about the need for any information, include it and make a sketch. It is better to have too many data that not enough.*

3-5 ARRANGEMENTS OF NOTES

Note styles and arrangement depend on departmental standards and individual preference. Highway departments, mapping agencies, and other organizations engaged in surveying furnish their field personnel with sample noteforms, similar to those in Appendix D, to aid in preparing uniform and complete records that can be checked quickly.

It is desirable for students to have as guides predesigned sample sets of noteforms covering their first field work to set high standards and save time. The noteforms shown in Appendix D are a composite of several models. They stress the open style, especially helpful for beginners, in which some lines or spaces are skipped for clarity. Thus angles measured at a point *A* (see Plate D-6) are placed opposite *A* on the page, but distances measured between *A* and *B* on the ground are recorded on the line between *A* and *B* in the field book.

Left- and right-hand pages are practically always used in pairs and therefore carry the same page number. A complete title should be lettered across the top of the left page and may be extended over the right one. Titles may be abbreviated on succeeding pages for the same survey project. Location and type of work are placed beneath the title. Some surveyors prefer to confine the title on the left page and keep the top of the right one free for date, party, weather, and other items. This design is revised if the entire right page has to be reserved for sketches and bench-mark descriptions. Arrangements shown in Appendix D demonstrate the flexibility of noteforms.

The left page is generally ruled in six columns designed for tabulation only. Column headings are placed between the first two horizontal lines at the page top and follow from left to right in the anticipated order of reading and recording. The upper part of the left or right page must contain four items:

1. *Project name, location, date, time of day* (A.M. *or* P.M.), *and starting and finishing times.* These entries are necessary to document the notes and furnish a timetable as well as to correlate different surveys. Precision, troubles encountered, and other facts may be gleaned from the time required for a survey.

2. *Weather.* Wind velocity, temperature, and adverse weather conditions such as rain, snow, sunshine, and fog have a decided effect on accuracy in surveying operations. Surveyors are unlikely to do their best possible work at temperatures of 15°F or with rain pouring down their necks. Hence weather details are important in reviewing field notes, in applying corrections to tape lengths due to temperature variations, and for other purposes.

3. *Party.* The names and initials of party members and their duties are required for documentation and future reference. Jobs can be described by symbols, such as $\bar{\top}$ for instrument operator, ϕ for rodperson, and N for notekeeper. The party chief is frequently the notekeeper.

4. *Instrument type and number.* The type of instrument used (with its make and serial number) and its degree of adjustment affect the accuracy of a survey. Identification of the specific equipment employed may aid in isolating some errors—for example, a particular tape with an actual length that is later found to disagree with the distance recorded between its end graduations.

To permit ready location of desired data, each field book must have a table of contents that is kept current daily. In practice, surveyors cross-index their notes on days when field work is impossible.

3-6 SUGGESTIONS FOR RECORDING NOTES

Observing the suggestions given in preceding sections, together with those listed here, will eliminate some common mistakes in recording notes.

1. Letter the notebook owner's name and address on the cover and first inside page in India ink. Number all field books for record purposes.
2. Begin a new day's work on a new page. For property surveys having complicated sketches, this rule may be waived.
3. Employ any orderly, standard, familiar noteform type but, if necessary, design a special arrangement to fit the project.
4. Include explanatory statements, details, and additional measurements if they might clarify the notes for field and office personnel.
5. Record what is read without performing any mental arithmetic. Write down what you read!
6. Run notes down the page, except in route surveys, where they usually progress upward to conform with sketches made while looking in the forward direction. (See Plate D-10.)
7. Use sketches instead of tabulations when in doubt. Carry a straightedge for ruling lines and a small protractor to lay off angles.
8. Make drawings to general proportions rather than to exact scale or without plan, and recognize that the usual preliminary estimate of space required is too small. Letter parallel with or perpendicular to the appropriate features, showing clearly to what they apply. Dimension lines, as used in machine graphics, are seldom necessary.
9. Exaggerate details or sketches if clarity is thereby improved, or prepare separate diagrams.
10. Line up descriptions and drawings with corresponding numerical data. For example, a bench-mark description should be placed on the right-hand page opposite its elevation, as in Plate D-2.
11. Avoid crowding. If it is helpful to do so, use several right-hand pages of descriptions and sketches for a single left-hand sheet of tabulation. Similarly, use any

number of pages of tabulation for a single drawing. Paper is cheap compared with the value of time that might be wasted by office personnel in misinterpreting compressed field notes, or by requiring a party to return to the field for clarification.

12. Use explanatory notes when they are pertinent, always keeping in mind the purpose of the survey and needs of the office force. Put these notes in open spaces to avoid conflict with other parts of the sketch.

13. Employ conventional symbols and signs for compactness.

14. Have north at the top, or left side, of all sketches if possible. A meridian arrow is vital.

15. Keep tabulated figures inside of and off column rulings, with decimal points and digits in line vertically.

16. Make a mental estimate of all measurements before receiving and recording them in order to eliminate large mistakes.

17. Repeat aloud values given for recording. For example, before writing down a distance of 124.68, call out "one, two, four, point six, eight" for verification by the person who submitted the measurement.

18. Place a zero before the decimal point for numbers smaller than 1; that is, record 0.37 instead of .37.

19. Show the precision of measurements by means of significant figures. For example, record 3.80 instead of 3.8 only if the reading was actually determined to hundredths.

20. Do not superimpose one number over another or on lines of sketches, and do not try to change one figure to another, as a 3 to a 5.

21. Make all possible arithmetic checks on the notes and record them before leaving the field.

22. Compare all misclosures and error ratios while in the field. On large projects where daily assignments are made for several parties, completed work is shown by satisfactory closures.

23. Arrange essential computations made in the field so they can be checked later.

24. Title, index, and cross-reference each new job or continuation of a previous one by client's organization, property owner, and description.

25. Sign surname and initials in the lower right-hand corner of the right page on all original notes. This places responsibility just as signing a check does.

PART II
AUTOMATIC DATA COLLECTORS

3-7 INTRODUCTION TO AUTOMATIC DATA COLLECTORS

Advances in computer technology in recent years have led to the development of sophisticated automatic data collection systems for taking field notes. These devices are about the size of a pocket calculator and produced by a number of different manufacturers. They are available with a variety of features and capabilities.

Data collectors can be directly wired to (interfaced with) modern surveying instruments, and when operated in that mode they can automatically receive and store data in computer compatible files as measurements are taken. Control of the measurement and storage operations is maintained through the data collector's keyboard. For clarification of the notes, the operator inputs point identifiers and other descriptive information along with the measurements as they are being recorded automatically. When a job is completed, or at day's end, the files can be transferred directly to a computer for further processing.

In using automatic data collectors, the usual preliminary information such as date, party, weather, time, and instrument number is entered manually into the file through the keyboard. For a given type of survey, the data collector's internal microprocessor is programmed to follow a specific sequence of steps. The operator identifies the type of survey to be performed from a menu, or by means of a code, and then follows instructions that appear on the unit's screen. Step-by-step prompts will guide the operator to either **(a)** input "external" data (which may include station names, descriptions, or other information), or **(b)** press a key to initiate the automatic recording of measured values.

Data collectors store information in either binary or ASCII (American Standard Code for Information Interchange) format. Binary storage is faster and more compact, but usually the data must be translated to ASCII before they can be read or edited. Most data collectors enable an operator to scroll through stored data, displaying them on the screen for review and editing while still at the job site.

The organizational structures used by different data collectors in storing information vary considerably from one manufacturer to the next. They all follow specific rules, and once they are understood, the data can be readily interpreted by both field and office personnel. The disadvantage of having varied data structures from different manufacturers is that a new system must be learned with each instrument of different make. Efforts have been made toward standardizing the data structures. The *Survey Data Management System* (SDMS), for example, has been adopted by the American Association of State Highway and Transportation Officials (AASHTO) and is recommended for all surveys involving highway work. The example field notes for a radial survey given in Table 16-1 of Section 16-9 are in the SDMS format.

Most manufacturers of modern surveying equipment have developed data collectors specifically to be interfaced with their own instruments. The Sokkia SDR 33 automatic data collector shown in Figure 10-14 is an example. This unit, although made to be interfaced with Sokkia instruments, can also be adapted to certain others. In the figure it is shown connected to the Sokkia SET 2B total station, an instrument which can automatically measure and record both angles and distances (see Sections 5–7 and 10–12). In addition to serving as a data collector, the SDR 33 also doubles as a hand-held computer and is able to perform a variety of time-saving computations directly in the field.

Some automatic data collectors can also be operated as *electronic field books*. In the electronic field book mode, the data collector is not interfaced with a surveying instrument. Instead of handwriting the data in a field book, the notekeeper enters measurements manually into the data collector by means of keyboard strokes after readings are taken. This has the advantage of enabling field notes to be recorded directly in a computer format ready for further processing, even though the surveying instruments being used are older and cannot be directly interfaced with data collectors. Data collectors provide the utmost in efficiency, however, when they are interfaced with surveying instruments such as total stations having automatic readout capabilities, and are operated in the automatic data collection mode.

Figure 3-1 SDC71 automatic data collector. (Courtesy ABACUS, a Division of Calculus, Inc.)

The Hewlett-Packard/Abacus SDC71 data collector shown in Figure 3-1 is a so-called third-party unit; that is, it is made by an independent company to be interfaced with instruments manufactured by others. It can be either operated in the electronic field book mode or interfaced with a variety of instruments for automatic data collection. Besides being a data collector, it is also a programmable computer.

The Husky Hunter 16 shown in Figure 3-2 is another third-party data collector, also with considerable versatility. It is actually an MS-DOS based hand-held computer and therefore can accept and run programs developed on another computer without modification (except that its screen size is limited). It can be used as a stand-alone electronic field book for manual entry of data, or interfaced to a variety of different makes and models of surveying instruments for automatic data collection.

3-8 TRANSFER OF FILES FROM DATA COLLECTORS

At regular intervals, usually at lunch time and at the end of a day's work or when a survey has been completed, the information stored in files within a data collector is transferred to another device. This is a safety precaution to avoid accidental loss of substantial amounts of data. Ultimately, of course, the files will be downloaded to the host computer, which will perform computations or generate maps and plots from the data. Depending on the peripheral equipment available, different procedures for data transfer can be used. In one method that is particularly convenient when surveying in

Figure 3-2 Husky Hunter 16 data collector downloading to lap-top computer. (Courtesy Wisconsin Department of Transportation.)

remote locations, data can be returned to the home office via telephone lines using devices called *modems*. Thus office personnel can immediately begin using the data.

Another method of data transfer consists in downloading straight into a computer by direct hookup via an RS-232 cable. This can be done by bringing the data collector into the office, or it can be done in the field if a battery-powered lap-top computer is available. Figure 3-2 illustrates this process.

A third data transfer alternative, which can also be done in the field, consists in using a portable disk drive. As shown in Figure 3-3, this transfers data to diskettes, which in turn can be transported to the office for further processing.

It is also convenient to obtain hard-copy printouts of data files periodically. This not only facilitates review of the data for completeness before leaving the field, it also provides a back-up record of the work done. Figure 3-4 shows a battery-operated portable printer being used in the field to create a hard-copy listing from a file stored in the SDC71 data collector.

Figure 3-3 Downloading files from Husky Hunter 16 data collector to diskettes using portable disk drive. (Courtesy Wisconsin Department of Transportation.)

Figure 3-4 Using a portable printer to obtain listing of field notes stored in automatic data collector. (Courtesy Wisconsin Department of Transportation.)

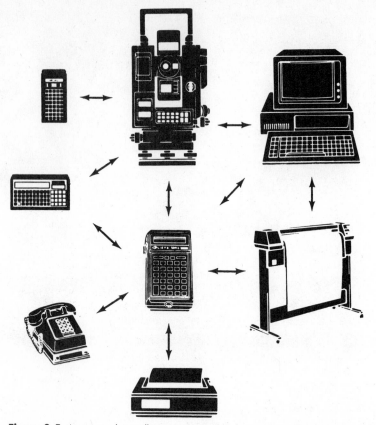

Figure 3-5 Automatic data collector—a central component in modern computerized surveying systems. (Courtesy Sokkia Corporation.)

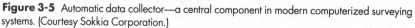

Some surveying instruments, for example, the Topcon ITS-1 total station shown in Figure 10-15, have the capability of storing data internally on memory cards. These cards can in turn be taken to the office, where the files can be downloaded using special card readers. Figure 10-16 shows this operation.

From the preceding discussion, and as illustrated in Figure 3-5, automatic data collectors are central components of modern computerized surveying systems. In these systems, data flow automatically from the field instrument through the collector to the printer, computer, plotter, and other units in the system. The term "field-to-finish systems" is often applied when this form of instrumentation is utilized in surveying.

3-9 ADVANTAGES AND DISADVANTAGES OF AUTOMATIC DATA COLLECTORS

The major advantages of automatic data collection systems are that (1) mistakes in reading and manually recording measurements in the field are precluded and (2) the time to process, display, and archive the field notes in the office is reduced significantly. Systems that incorporate computers can execute some programs in the field, which adds a significant advantage. As an example, the data for a survey can be corrected for systematic errors and misclosures computed, so verification that a survey meets closure requirements is made before the crew leaves a site.

Automatic data collectors are most useful when large quantities of information must be recorded, for example, in topographic surveys or cross-sectioning. In Section 16-9.1 their use in topographic surveying is described, and an example set of notes taken for that purpose is presented and discussed.

Although automatic data collectors have many advantages, they also present some dangers and problems. There is the slight chance, for example, that files could be accidentally erased through carelessness, or they could be lost due to malfunction or damage to the unit. Some difficulties are also created by the fact that sketches cannot be entered into the computer. This problem can be overcome, however, by supplementing the files with sketches made simultaneously with the measurements. Another unresolved question concerns the legal value of computer files, especially those containing boundary survey notes, for example. To aid in this regard, many data collection systems have a provision that protects files from any editing or tampering. In addition, hardcopy outputs can, and should, be obtained at the end of each day or job, and signed by the party chief.

Automatic data collectors are available from a large number of manufacturers. They must be capable of transferring data through various hardware in modern surveying systems such as that illustrated in Figure 3-5. Since equipment varies considerably, it is important when considering the purchase of a data collector to be certain it fits the equipment owned or perhaps needed in the future.

PROBLEMS

3-1 What information should normally be included in a good set of field notes?

3-2 Why should a pen not be used in field notekeeping?

3-3 Explain why data should always be entered directly into the field book at the time measurements are made, rather than on scrap paper for neat transfer to the field book later.

3-4 Why are erasures not permitted in a field book?

3-5 Why should a zero be placed before a decimal point in tabulating values, such as 0.73?

3-6 Explain the reason for item 20 in Section 3-6 when recording field notes.

3-7 Explain why ruled vertical and horizontal lines are necessary on field book pages.

3-8 When should sketches be made instead of just recording data?

3-9 Justify the requirement to list in a field book the makes and serial numbers of all instruments used on a survey.

3-10 Refer to Section 3-4 and list the type, or types, of field notes used in each of Plates D-1 through D-10.

3-11 Discuss the advantages of automatic data collection systems.

3-12 What disadvantage might occur if an automatic data recorder is used by a land surveyor?

BIBLIOGRAPHY

Brinker, R. C., B. A. Barry, and R. Minnick. 1981. *Noteforms for Surveying Measurements,* 2nd ed. Rancho Cordova, Calif.: Landmark.

Brinker, R. C., and R. Minnick. 1987. *The Surveying Handbook.* New York: Van Nostrand Reinhold, Chap. 1.

Hummel, J. 1992. "Florida's Electronic Field Book." *Point of Beginning* 17 (No. 4):44.

Koo, T. K., and M. K. Lim. 1990. "Entity Coded Neutral File Format for Integrating Data Collectors." *ASCE Journal of Surveying Engineering* 116 (No. 2):93.

Onsrud, H. J., and R. J. Hintz. 1991. "Evidentiary Admissibility and Reliability of Automated Field Recorder Data." *Surveying and Land Information Systems* 51 (No. 1):23.

Pafford, F. W. 1962. *Handbook of Survey Notekeeping.* New York: John Wiley and Sons.

Roth, A. W. 1992. "P.O.B. 1992 Data Collector Survey." *Point of Beginning* 17 (No. 6):30.

Shrestha, R. L. 1990. "Formats and Specifications for an Electronic Field Book." *Surveying and Land Information Systems* 50 (No. 3):215.

Turner, H. 1987. "A Model to Integrate Data Collectors." *Surveying and Mapping* 47 (No. 2):117.

Whalen, C. T. 1988. "Data Recording—One Man's Experience." *Point of Beginning* 13 (No. 4):88.

4

DISTANCE MEASUREMENT; TAPING

PART I
METHODS OF LINEAR MEASUREMENT

4-1 INTRODUCTION

Distance measurement is generally regarded as the most fundamental of all surveying observations. Even though many angles may be read in a survey, the length of at least one line must be measured to supplement the angles in locating points.

In plane surveying the distance between two points means the horizontal distance. If the points are at different elevations, the distance is the horizontal length between plumb lines at the points.

Lengths of lines may be specified in different units. The unit generally used in plane surveying in the United States is the foot, decimally divided. In architectural and machine work and on some construction projects, the unit is a foot divided into inches and fractions of an inch. Geodetic surveying usually employs the meter. Chains, varas, rods, and other units have been, and still are, utilized in some localities and for special purposes.

4-2 METHODS OF MAKING LINEAR MEASUREMENTS

In surveying, linear measurements have been obtained by many different methods. These include (1) pacing, (2) odometer readings, (3) optical rangefinders, (4) tacheometry (stadia), (5) subtense bars, (6) taping, (7) electronic distance measurement (EDM), (8) inertial methods, (9) satellite systems, and others. Of these, taping, EDM, and satellite systems are most commonly used by surveyors today. In particular, the satellite-supported *Global Positioning System* (GPS) is rapidly replacing all other systems due to many advantages, but most notably because of its range, accuracy, and efficiency. Methods (1) through (5) are discussed briefly in the following sections. Taping is described in detail in Part II of this chapter, EDM is covered in Chapter 5, and inertial and satellite systems are described in Chapter 20.

Triangulation is a method for determining positions of points from which horizontal distances can be computed (see Chapter 19). In this procedure, lengths of lines are computed trigonometrically from measured base lines and angles. *Photogrammetry* can also be used to obtain horizontal distances. This topic is covered in Chapter 28.

Besides these methods, distances can also be estimated, a technique useful in making field note sketches and checking measurements for mistakes. With practice, estimating can be done quite accurately.

4-3 PACING

Distances obtained by pacing are sufficiently accurate for many purposes in surveying, engineering, geology, agriculture, forestry, and military field sketching. Pacing is also used to detect blunders that may occur in making distance measurements by more accurate methods.

Pacing consists of counting the number of steps or paces in a required distance. The length of an individual's pace must first be determined. This is best done by walking with natural steps back and forth over a measured level course at least 300 ft long, and dividing the known distance by the average number of steps. For short distances the length of each pace is needed, but the number of steps taken per 100 ft is desirable for checking long lines.

It is possible to adjust one's pace to an even 3 ft, but a person of average height finds such a step tiring if maintained for very long. The length of an individual's pace varies when going uphill or downhill and changes with age. For long distances, a pocket instrument called a *pedometer* can be carried to register the number of paces, or a *passometer* attached to the body or leg counts the steps. Some surveyors prefer to count *strides,* a stride being two paces.

Pacing is one of the most valuable things learned in surveying, since it has practical applications for everybody and requires no equipment. Experienced pacers can measure distances of 100 ft or longer with an accuracy of $\frac{1}{50}$ to $\frac{1}{100}$ of the distance if the terrain is open and reasonably level.

4-4 ODOMETER READINGS

An odometer converts the number of revolutions of a wheel of known circumference to a distance. Lengths measured by an odometer on a vehicle are suitable for some preliminary surveys in route-location work. They also serve as a rough check on measurements made by other methods. Other types of measuring wheels are available and useful for determining short distances, particularly on curved lines. Odometers give surface distances which should be corrected to horizontal if the ground slopes severely (see Section 4-13). With odometers, an accuracy of approximately $\frac{1}{200}$ of the distance is reasonable.

4-5 OPTICAL RANGEFINDERS

These instruments operate on the same principle as rangefinders on single-lens reflex cameras. Basically, when focused, they solve for the object distance f_2 in Eq. (6-12),

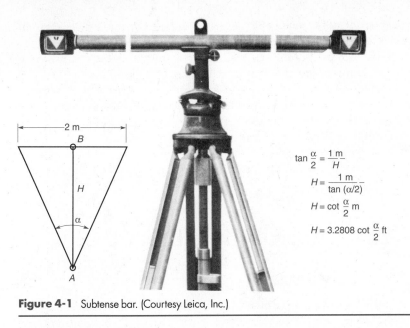

$$\tan \frac{\alpha}{2} = \frac{1\ m}{H}$$

$$H = \frac{1\ m}{\tan(\alpha/2)}$$

$$H = \cot \frac{\alpha}{2}\ m$$

$$H = 3.2808 \cot \frac{\alpha}{2}\ ft$$

Figure 4-1 Subtense bar. (Courtesy Leica, Inc.)

where focal length f and image distance f_1 are known. An operator looks through the lens and adjusts the focus until a distant object viewed is focused in coincidence, whereupon a distance reading is obtained. These instruments are capable of accuracies of 1 part in 50 at distances up to 150 ft, but accuracy diminishes as the length increases. They are suitable for reconnaissance, sketching, or checking more accurate measurements for mistakes.

4-6 TACHEOMETRY

Tacheometry (*stadia* is the more common term in the United States) is a surveying method used to quickly determine the horizontal distance to, and elevation of, a point. Stadia measurements are obtained by sighting through a telescope equipped with two or more horizontal cross hairs at a known spacing. The apparent intercepted length between the top and bottom hairs is read on a graduated rod held vertically at the desired point. The distance from telescope to rod is found by proportional relationships in similar triangles. An accuracy of $\frac{1}{500}$ of the distance is achieved with reasonable care. A detailed explanation of the method is given in Chapter 15.

4-7 SUBTENSE BAR

This indirect distance measuring procedure involves reading the angle subtended by two precisely spaced targets on a subtense bar. An Invar[1] subtense bar mounted on a tripod is shown in Figure 4-1 (along with a geometric diagram). The bar is set over one point (B in the figure), leveled by means of a level vial, and aligned perpendicular to the survey line by means of a sighting device on top of the bar. Fixed targets near the bar ends are precisely 2 m apart. The horizontal angle α between targets is measured with

---

[1]Invar is a metal that has a low coefficient of thermal expansion and thus maintains its true length precisely despite temperature variations (see Section 4-9.2).

a theodolite set over the other point (*A* in the figure) and the horizontal distance computed. Equations for computing horizontal distances are given in Figure 4-1.

Distances determined by the subtense method are always horizontal, even though inclined sights are taken, because α is measured in a horizontal plane. For sights of 500 ft (150 m) or shorter, and using a 1″ theodolite, an accuracy of 1 part in 3000 or better can be achieved. Accuracy diminishes with increased line length, but this can be offset somewhat by repeating the angles several times, or by taking readings from both ends of the line and averaging. The subtense method was often used in the past to obtain distances over inaccessible courses such as over bodies of water. EDM devices and GPS have now almost totally replaced this procedure, however.

PART II
DISTANCE MEASUREMENTS BY TAPING

4-8 INTRODUCTION TO TAPING

Measurement of a horizontal distance by taping consists of applying the known length of a graduated tape directly to a line a number of times. Two types of problems arise: **(1)** measuring an unknown distance between fixed points, such as two stakes in the ground, and **(2)** laying out a known or required distance with only the starting mark in place.

Taping is performed in six steps: **(1)** lining in, **(2)** applying tension, **(3)** plumbing, **(4)** marking tape lengths, **(5)** reading the tape, and **(6)** recording the distance. The application of these steps in taping on level and sloping ground is detailed in Sections 4-11 and 4-12.

4-9 TAPING EQUIPMENT

Various types of equipment used for taping in the United States, past and present, are described in this section.

4-9.1 Historical Equipment

Early surveyors struggled with braced timber panels and wood and metal poles. These devices resulted in the term *pole* as a unit of measure. Its length was $16\frac{1}{2}$ ft, the same as the *rod*.

A *Gunter's chain* was the best measuring device available to surveyors in the United States for many years and is referred to in old field notes and deeds. It was 66 ft (4 poles) long and had 100 links, each link equal to 0.66 ft or 7.92 in. The links were made of heavy wire, had a loop at each end, and were joined together by three rings (Figure 4-2). The outside ends of the handles fastened to the end links were the 0- and 66-ft marks. Successive tags had one, two, three, or four teeth to mark every tenth link from each end. The center tag was round. With 600 or 800 connecting link and ring surfaces subject to frictional wear, hard use elongated the chain, and its length had to be adjusted by means of bolts in the handles.

Distances measured with chains were recorded either in chains and links or in chains and decimals of chains—for example, 7 ch 94.5 lk or 7.945 ch. Decimal parts of links were estimated. The 66-ft length of the Gunter's chain was selected because of its

Figure 4-2 Gunter's chain.

relevance to the mile and the relationship of a square chain to an acre. Thus 1 ch = $\frac{1}{80}$ mile, and 10 ch^2 = 10 $\times$ 66^2 = 43,560 ft^2 = 1 acre.

An *engineer's chain* had the same construction as a Gunter's chain, but was 100 ft long and each of its 100 links had a length of 1 ft.

Chains are seldom, if ever, used today, although a steel tape graduated like a Gunter's chain is manufactured. Nevertheless, the many chain surveys on record oblige the modern practitioner to understand the limits of accuracy possible with this equipment, and the conversion of distances recorded in chains and links to feet. The U.S. Bureau of Land Management still converts its distance measurements to chains, although the data are obtained from modern EDM and total station instruments in feet. The term *chaining* continues to be used interchangeably with *taping,* even though tapes rather than chains are employed.

Before the thin flat steel now used in tapes could be produced efficiently, *wires* were utilized for measuring lengths. They still are practical in special cases—for example, hydrographic surveys.

4-9.2 Tapes in Current Use

Surveyor's and engineer's tapes are made of steel $\frac{1}{4}$ to $\frac{3}{8}$ in. wide and weigh 2 to 3 lb/ 100 ft. Those graduated in feet have lengths of 100, 200, 300, and 500 ft. They are marked in feet, tenths and hundredths. Metric tapes have standard lengths of 30, 60, 100, and 150 m. The 100-ft tape is by far the most common. All can either be wound on a reel [see Figure 4-3(a)] or done up in loops.

Invar tapes are made of a special nickel steel (35% nickel and 65% steel) to reduce length variations caused by differences in temperature. The thermal coefficient of

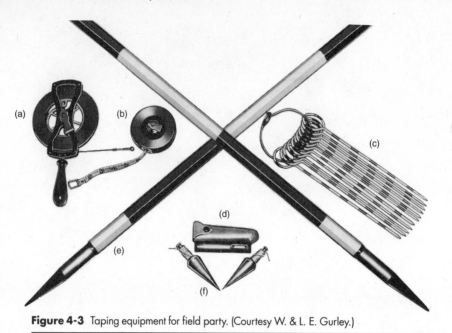

Figure 4-3 Taping equipment for field party. (Courtesy W. & L. E. Gurley.)

expansion and contraction is only about $\frac{1}{30}$ to $\frac{1}{60}$ that of an ordinary steel tape. The metal is soft and somewhat unstable. This weakness, along with their cost of perhaps ten times that of steel tapes, made them suitable only for precise geodetic work and as a standard for comparison with working tapes. A somewhat newer version, the *Lovar tape,* has properties and a cost between those of steel and Invar tapes.

Cloth (or *metallic*) *tapes* are actually made of high-grade linen, $\frac{5}{8}$ in. wide with fine copper wires running lengthwise to give additional strength and prevent excessive elongation. Metallic tapes commonly used are 50, 100, and 200 ft long and come on enclosed reels [see Figure 4-3(b)]. Although not suitable for precise work, metallic tapes are convenient and practical for many purposes.

Fiberglass tapes come in a variety of sizes and lengths, and are usually wound on a reel. They can be employed for the same types of work as metallic tapes.

4-9.3 Taping Accessories

Chaining pins or taping pins are used to mark tape lengths. Most taping pins are made of number 12 steel wire, sharply pointed at one end, have a round loop at the other end, and are painted with alternate red and white bands [see Figure 4-3(c)]. Sets of 11 pins carried on a steel ring are standard.

The *hand level*, described in Section 6-13, is a simple instrument used to keep the tape ends at equal elevations when measuring over rough terrain [see Figure 4-3(d)].

Tension handles facilitate the application of a desired standard or known tension. A complete unit consists of a wire handle, a clip to fit the ring end of the tape, and a spring balance reading up to 30 lb in $\frac{1}{2}$-lb graduations.

Clamp handles are used to apply tension by a positive, quick grip using a scissors-type action on any part of a steel tape without damage to the tape or injury to hands.

A *pocket thermometer* permits reading data for making temperature corrections. It is about 5 in. long, graduated from perhaps -30 to $+120°F$ in 1 or 2° divisions, and kept in a protective metal case.

Range poles (lining rods) made of wood, steel, or aluminum are about 1 in. thick and 6 to 10 ft long. They are round or hexagonal in cross section and marked with alternate 1-ft-long red and white bands which can be used for rough measurements [see Figure 4-3(e)]. A wooden range pole has a metal shoe at the base. The main utility of range poles is to mark the line being measured so that the tape's alignment can be maintained.

Plumb bobs for taping [see Figure 4-3(f)] should weigh a minimum of 8 oz and have a fine point. At least 6 ft of good-quality string or cord, free of knots, is necessary. Plumb-bob points are now standardized to simplify replacement.

4-10 CARE OF TAPING EQUIPMENT

The following points are pertinent in the care of tapes and range poles:

1. Considering the cross-sectional area of the average surveyor's steel tape and its permissible stress, a pull of 100 lb will do no damage. If the tape is kinked, however, a pull of less than 1 lb can break it. Therefore always check to be certain that any loops and kinks are eliminated before tension is applied.
2. If a tape gets wet, wipe it first with a dry cloth, then with an oily one.
3. Tapes should be either kept on a reel or "thrown" into circular loops, but not handled both ways.
4. Each tape should have an individual number or tag to identify it.
5. Broken tapes can be mended by riveting or applying a sleeve device, but a mended tape should not be used on important work.
6. Range poles are made with the metal shoe and point in line with the section above. This alignment may be lost if the pole is used improperly.

4-11 TAPING ON LEVEL GROUND

The subsections that follow describe six steps in taping on level ground using a 100-ft tape.

4-11.1 Lining In

Using range poles, the line to be measured should be marked at both ends, and at intermediate points where necessary, to ensure unobstructed sight lines. The forward tapeperson is lined in by the rear tapeperson (or by a transit or theodolite for greater accuracy). Directions are given by vocal or hand signals.

4-11.2 Applying Tension

The 100-ft end of a tape is held over the first (rear) point by the rear tapeperson, while the forward tapeperson, holding the zero end, is lined in. For accurate results the tape must be straight and the two ends held at the same elevation. A specified tension, generally between 10 and 25 lb, is applied. To maintain a steady pull, tapepersons wrap the leather thong at the tape's end around one hand, keep forearms against their bodies, and face at right angles to the line. In this position, they are off the line of sight. Also, the body need only be tilted to hold, decrease, or increase the pull. Sustaining a constant tension with *outstretched* arms is difficult, if not impossible, for a pull of 15 lb or more. Good communication between head and rear tapepersons will avoid jerking the tape, save time, and produce better results.

4-11.3 Plumbing

Weeds, brush, obstacles, and surface irregularities may make it undesirable to lay a tape on the ground. Instead, the tape is held above ground in a horizontal position. Each end point on the tape is marked by placing the plumb-bob string over the proper tape graduation and securing it with one thumb. The rear tapeperson continues to hold a plumb bob over the fixed point, while the forward tapeperson marks the length. In measuring a distance shorter than a full tape length, the forward tapeperson moves the plumb-bob string to a point on the tape over the ground mark.

4-11.4 Marking Tape Lengths

When the tape has been lined in properly, tension has been applied, and the rear tapeperson is over the point, "stick" is called out. The forward tapeperson then places a pin exactly opposite the zero mark of the tape and calls "stuck." The marked point is checked by repeating the measurement until certainty of its correct location is assured.

After checking the measurement, the forward tapeperson signals that the point is OK, the rear tapeperson pulls up the rear pin, and they move ahead. The forward tapeperson drags the tape, paces roughly 100 ft, and stops. Just before the 100-ft end reaches the pin that has been set, the rear tapeperson calls "tape" to notify the forward tapeperson that they have gone 100 ft. The process of measuring 100-ft lengths is repeated until a partial tape length is needed at the end of the line.

4-11.5 Reading the Tape

There are two common styles of graduations on 100-ft surveyor's tapes. *It is necessary to identify the type being used before starting work* to avoid making 1-ft mistakes repeatedly.

The more common type of tape has a total graduated length of 101 ft. It is marked from 0 to 100 by full feet in one direction, and has an additional foot preceding the zero mark graduated from 0 to 1 ft in tenths or in tenths and hundredths in the other direction. In measuring the last partial tape length of a line with this kind of tape, a full-foot graduation is held by the rear tapeperson at the last pin set [like the 87-ft mark in Figure 4-4(a)]. The graduations on the extra foot between zero and the tape end should straddle the closing point. The head tapeperson reads the additional length of 0.68 ft beyond the zero mark. To ensure correct recording, the rear tapeperson calls "87." The head tapeperson repeats and adds the partial foot reading, calling "87.68." Since part of a foot has been added, this type of tape is known as an *add tape*.

The other kind of tape found in practice has a total graduated length of 100 ft. It is marked from 0 to 100 with full-foot increments, and the first foot at each end (from 0 to 1 and from 99 to 100) is graduated in tenths or in tenths and hundredths. With this kind of tape, the last partial tape length is measured by holding a full-foot graduation at the last chaining pin set such that the graduated section of the tape between the zero mark and the 1-ft mark straddles the closing point. This is indicated in Figure 4-4(b), where the 88-ft mark is being held on the last chaining pin and the tack marking the end of the line is opposite 0.32 ft read from the zero end. The partial tape length is then $88.00 - 0.32 = 87.68$ ft. The quantity 0.32 ft is said to be *cut off* and hence this type of tape is called a *cut tape*. To ensure subtraction of a foot from the number at the full-foot graduation used, the following field procedure and calls are recommended: Rear tapeperson calls "88"; forward tapeperson says "cut point three-two"; rear tapeperson answers "eighty seven point six eight"; forward tapeperson replies "check."

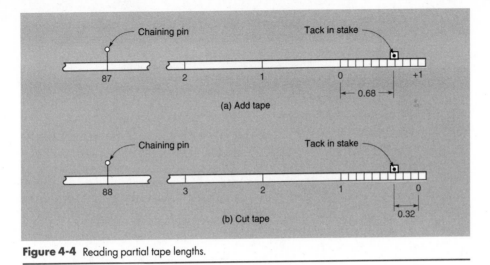

Figure 4-4 Reading partial tape lengths.

An advantage of the add tape is that it is easier to use because no subtraction is needed when measuring decimal parts of a foot. Its disadvantage is that careless tape persons will sometimes make measurements of 101.00 ft and record them as 100.00 ft. The cut tape practically eliminates this mistake.

The same routine should be used throughout all taping by a party, and the results tested in every possible way. A single mistake in subtracting the partial foot when using a cut tape will destroy the precision of a hundred other measurements. For this reason, the add tape is more nearly foolproof. The greatest danger for mistakes in taping arises when changing from one style of tape to the other.

4-11.6 Recording the Distance

Accurate field work may be canceled by careless recording. After the partial tape length is obtained at the end of a line, the rear tapeperson determines the number of full 100-ft tape lengths by counting the pins collected from the original set of 11. For distances longer than 1000 ft, a notation is made in the field book when the rear tapeperson has ten pins and one remains in the ground. This signifies a tally of ten full tape lengths and is traditionally called an "out." The forward tapeperson starts out again with ten pins and the process is repeated.

Although taping procedures may appear to be relatively simple, high precision is difficult to achieve, especially for beginners. Taping is a skill that can best be taught and learned by field demonstrations and practice.

4-12 HORIZONTAL MEASUREMENTS ON SLOPING GROUND

In taping on uneven or sloping ground, it is standard practice to hold the tape horizontally and use a plumb bob at one or perhaps both ends. It is difficult to keep the plumb line steady for heights above the chest. Wind exaggerates this problem and may make accurate work impossible.

Where a 100-ft length cannot be held horizontally without plumbing from above shoulder level, shorter distances are measured and accumulated to total a full tape length. This procedure, called *breaking tape,* is illustrated in Figure 4-5.

Figure 4-5 Breaking tape.

As an example of this operation, assume that when taping down slope, the 100-ft end of the tape is held at the rear point, and the forward tapeperson can advance only 30 ft without being forced to plumb from above the chest. A pin is therefore set beneath the 70-ft mark, as in Figure 4-6. The rear tapeperson moves ahead to this pin and holds the 70-ft graduation there while another pin is set at, say, the 25-ft mark. Then, with the 25-ft graduation over the second pin, the full 100-ft distance is marked at the zero point. To avoid kinking the tape, the full 100-ft length is pulled ahead by the forward tapeperson. In this way, the partial tape lengths are added mechanically to make a full 100 ft by holding the proper graduations, and no mental arithmetic is required. The rear tapeperson returns the pins set at the intermediate points to the forward tapeperson to keep the tally clear on the number of full tape lengths established. In all cases the tape is leveled by eye or hand level, with the tapepersons keeping in mind the natural tendency to have the downhill end of a tape too low. Practice will improve the knack of holding a tape horizontally by keeping it perpendicular to the vertical plumb-bob string.

Taping downhill is preferable to measuring uphill for two reasons. First, in taping downhill the rear point is held steady on a fixed object while the other end is plumbed. In taping uphill, the forward point must be set while the other end is wavering somewhat. Second, if breaking tape is necessary, the head tapeperson can more conveniently use the hand level to proceed downhill a distance which renders the tape horizontal when held comfortably at chest height.

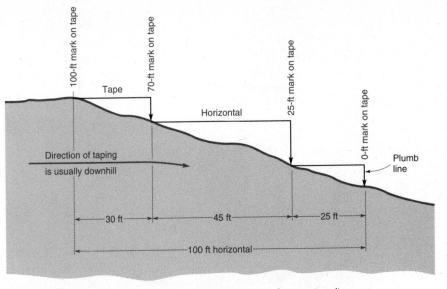

Figure 4-6 Procedure for breaking tape (when tape is not in box or on reel).

4-13 SLOPE MEASUREMENTS

In measuring the distance between two points on a steep slope, rather than break tape every few feet, it may be desirable to tape along the slope and compute the horizontal component. This requires measurement also of either the angle of inclination α or the difference in elevation d (Figure 4-7). Breaking tape is more time consuming and generally less accurate due to the accumulation of random errors from marking tape ends and keeping the tape level and aligned for many short sections.

In Figure 4-7, if angle α is determined, the horizontal distance between points A and B can be computed from the relation

$$H = L \cos \alpha \tag{4-1}$$

Figure 4-7 Slope measurement.

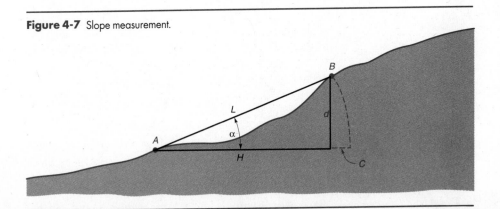

where H is the horizontal distance between points, L the slope length separating them, and α the vertical angle from horizontal, usually obtained with an Abney hand level and clinometer (see Figure 6-18), transit, or theodolite (see Chapter 10).

If the difference in elevation d between the ends of the tape is measured, which is done by leveling (see Chapters 6 and 7), the horizontal distance can be computed using the following expression derived from the Pythagorean theorem:

$$H = \sqrt{L^2 - d^2} \tag{4-2}$$

Another approximate formula, obtained from the first term of a binomial expansion of the Pythagorean theorem, may be used to reduce slope distances to horizontal:

$$H = L - \frac{d^2}{2L} \text{ (approx.)} \tag{4-3}$$

In Eq. (4-3) the term $d^2/2L$ equals C in Figure 4-7 and is a correction to be subtracted from the measured slope length to obtain the horizontal distance. The error in using the approximate formula for a 100-ft length grows with increasing slope, but for grades up to 10% (10-ft elevation difference per 100 ft), the answer is correct to the nearest 0.001 ft. The approximate formula is useful for making quick estimates, without a calculator, of error sizes produced for varying slope conditions.

EXAMPLE 4-1   A distance of 575.28 ft was measured along a smooth slope. The slope angle was measured and found to be 6°22′. What is the horizontal distance?

SOLUTION
By Eq. (4-1), $H = (575.28) \cos 6°22′ = 571.73$ ft.

EXAMPLE 4-2   A distance of 290.43 ft was measured along a smooth slope from A to B. The elevations of A and B were measured and found to be 865.2 and 891.4 ft, respectively. What is the horizontal distance from A to B?

SOLUTION
Elevation difference $d = 891.4 - 865.2 = 26.2$ ft.
By Eq. (4-2), $H = \sqrt{(290.43)^2 - (26.2)^2} = 289.25$ ft.
By Eq. (4-3), $H = 290.43 - \dfrac{(26.2)^2}{2 \times 290.43} = 290.43 - 1.18 = 289.25$ ft

4-14 SOURCES OF ERROR IN TAPING

There are three fundamental sources of error in taping.

1. *Instrumental errors.* A tape may differ in actual length from its nominal graduated length because of a defect in manufacture or repair, or as a result of kinks.

2. *Natural errors.* The horizontal distance between end graduations of a tape varies because of the effects of temperature, wind, and weight of the tape itself.
3. *Personal errors.* Tapepersons may be careless in setting pins, reading the tape, or manipulating the equipment.

The most common types of taping errors are discussed in the subsections that follow. They stem from instrumental, natural, and personal sources. Some types produce systematic errors, others random errors.

4-14.1 Incorrect Length of Tape

Incorrect length of a tape can be one of the most important errors. It is systematic. Tape manufacturers do not guarantee steel tapes to be exactly their graduated nominal length—for example, 100.00 ft—nor do they provide a standardization certificate unless requested and paid for as an extra. The true length is obtained by comparing it with a standard tape or distance. The National Institute of Standards and Technology (NIST)[2] will make such a comparison and certify the exact distance between end graduations under given conditions of temperature, tension, and manner of support.

A 100-ft steel tape usually is standardized for each of two sets of conditions—for example, 68°F, a 12-lb pull, with the tape fully supported throughout; and 68°F, a 20-lb pull, with the tape supported at the ends only. Schools and surveying offices normally have a precisely measured 100-ft line or at least one standardized tape that is used only to check other tapes subjected to wear.

An error due to incorrect length of a tape occurs each time the tape is used. If the true length, known by standardization, is not exactly equal to its nominal value of 100.00 ft recorded for every full length, the correction can be determined and applied from the formulas

$$C_l = \left(\frac{l - l'}{l'}\right)L \qquad (4\text{-}4)$$

and

$$\bar{L} = L + C_l \qquad (4\text{-}5)$$

where C_l is the correction to be applied to the measured (recorded) length of a line to obtain the true length, l the actual tape length, l' the nominal tape length, L the measured (recorded) length of line, and $\bar{L}$ the corrected length of line.

A 100-ft steel tape when compared with a standard is actually 100.02 ft long. What is the corrected length of the line measured with this tape and recorded to be 565.75 ft?   EXAMPLE 4-3

SOLUTION

By Eq. (4-4), $C_l = \left(\dfrac{100.02 - 100.00}{100.00}\right)565.75 = +0.11$ ft.

[2]Tapes can be sent to the National Institute of Standards and Technology, Building 220, Room A107, 1 Bureau Road, Gaithersburg, Md. 20899; telephone: (301) 975-4079.

By Eq. (4-5), $\bar{L} = 565.75 + 0.11 = 565.86$ ft.

The example illustrates that *in measuring unknown distances with a tape that is too long, a correction must be added. Conversely, if the tape is too short, the correction will be minus,* resulting in a decrease. (C_l is still added, but has a negative sign.)

An alternate method of making corrections for an incorrect tape length is to compute the amount by which a full tape length is too long or too short and then multiply it by the number of tape lengths in the line. Thus in Example 4-3 the tape is 0.02 ft too long, and the total correction is $5.6575 \times 0.02 = 0.11$ ft. This value is then added to the measured length of 565.75 to get 565.86 ft.

From a practical standpoint, the effect of any error is to make the tape length incorrect. Note that the true (actual) distance equals the measured distance plus a correction, and the proper algebraic sign for Eq. (4-5) is "built in." This is also true for the corrections discussed in succeeding sections. However, students should still try to reason whether a certain condition "makes" a tape too long or too short and apply the correction accordingly.

4-14.2 Temperature Other Than Standard

Steel tapes are standardized for 68°F (20°C) in the United States. A temperature higher or lower than this value causes a change in length that must be considered.

The coefficient of thermal expansion and contraction of steel used in ordinary tapes is approximately 0.00000645 per unit length per degree Fahrenheit, and 0.0000116 per unit length per degree Celsius. For any tape, the correction for temperature can be computed and applied using the formulas

$$C_t = k(T_1 - T)L \tag{4-6}$$

and

$$\bar{L} = L + C_t \tag{4-7}$$

where C_t is the correction in the length of a line due to nonstandard temperature, k the coefficient of thermal expansion and contraction of the tape, T_1 the tape temperature at time of measurement, T the tape temperature when it has standard length, L the measured (recorded) length of line, and $\bar{L}$ the corrected length of the line.

Errors due to temperature change may be practically eliminated by either **(a)** measuring temperature and making corrections according to Eqs. (4-6) and (4-7), or **(b)** using an Invar tape.

EXAMPLE 4-4   The recorded length of a line measured at 30.5°F with a steel tape that is 100.00 ft long at 68°F was 872.54 ft. What is the line's corrected length?

SOLUTION

By Eq. (4-6), $C_t = 0.00000645(30.5 - 68)872.54 = -0.21$ ft.
By Eq. (4-7), $\bar{L} = 872.54 - 0.21 = 872.33$ ft.

Errors due to temperature changes are systematic and have the same sign if the temperature is always above 68°F or always below that standard. When the temperature is above 68°F during part of the time occupied in measuring a long line, and below 68°F for the remainder of the time, the errors tend to partially balance each other, but corrections should still be computed and applied.

Temperature effects are difficult to assess in taping. The air temperature read from a thermometer may be quite different from that of the tape to which it is attached. Sunshine, shade, wind, evaporation from a wet tape, and other conditions make the tape temperature uncertain. Field experiments prove that temperatures on the ground or in the grass may be 10 to 25° higher or lower than those at shoulder height because of a 6-in. "layer of weather" (microclimate) on top of the ground. Since a temperature difference of 15°F produces a change of 0.01 ft per tape length, the importance of such large variations is obvious.

Shop measurements made with steel scales and other devices likewise are subject to temperature effects. The precision required in fabricating a large airplane or ship can be lost by this one cause alone.

4-14.3 Inconsistent Pull

When a steel tape is pulled with a tension greater than its standard, the tape will stretch and be longer than its standard length. Conversely, if less than standard pull is used, the tape will be shorter than its standard length. The *modulus of elasticity* of the tape regulates the amount that it stretches. The correction for pull can be computed and applied using the following formulas:

$$C_p = (P_1 - P)\frac{L}{AE} \qquad (4\text{-}8)$$

and

$$\bar{L} = L + C_p \qquad (4\text{-}9)$$

where C_p is the total elongation in tape length due to pull, in feet; P_1 the pull applied to the tape, in pounds; P the standard pull for the tape, in pounds; A the tape's cross-sectional area, in square inches; E the modulus of elasticity of steel, in pounds per square inch; L the measured (recorded) length of line, in feet; and $\bar{L}$ the corrected length, in feet. An average value of E is 29,000,000 lb/in.[2] for the kind of steel used in tapes. The cross-sectional area of a steel tape can be obtained from the manufacturer, by measuring its width and thickness with calipers, or by dividing the total tape weight by the product of its length (in feet) times the unit weight of steel (490 lb/ft³), and multiplying by 144 to convert square feet to square inches.

Eqs. (4-8) and (4-9) also apply when the metric system is used. In that case, to produce the correction C_p in meters, comparable units are P and P_1 in kilograms, L and $\bar{L}$ in meters, A in square centimeters, and E in kilograms per square centimeter. An average value of E for steel in these units is approximately 2,000,000 kg/cm².

Errors resulting from incorrect tension can be eliminated by (a) using a spring balance to measure and maintain the standard pull, or (b) applying a pull other than standard and making corrections for the deviation from standard according to Eqs. (4-8) and (4-9).

EXAMPLE 4-5   A steel tape that is 100.000 ft long under a pull of 12.0 lb when supported throughout, and has a cross-sectional area of 0.005 in.^{2}, is applied fully supported with a 20-lb pull to measure a line whose recorded length is 686.79 ft. What is the corrected length of the line?

SOLUTION

By Eq. (4-8), $C_p = \dfrac{(20 - 12)686.79}{0.005(29,000,000)} = +0.038$ ft.

By Eq. (4-9), $\overline{L} = 686.79 + 0.038 = 686.83$ ft.

Errors due to incorrect pull may be either systematic or random. The pull applied by even an experienced tapeperson is sometimes greater or less than the desired value. An inexperienced person, particularly one who has not used a spring balance on a tape, is likely to apply less than the standard tension consistently.

4-14.4 Sag

A steel tape not supported along its entire length sags in the form of a *catenary,* a good example being the cable of a suspension bridge. Sag shortens the horizontal (chord) distance between end graduations, because the tape length remains the same (Figure 4-8). Sag can be diminished (by greater tension) but not eliminated unless the tape is supported throughout.

The following formulas are used to compute the sag correction:

$$C_s = \frac{w^2 L_s^3}{24 P_1^2} \tag{4-10}$$

or

$$C_s = \frac{W^2 L_s}{24 P_1^2} \tag{4-11}$$

Figure 4-8 Effect of sag.

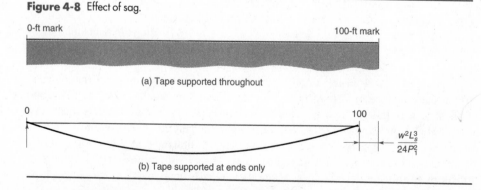

0-ft mark   100-ft mark

(a) Tape supported throughout

0   100

$\dfrac{w^2 L_s^3}{24 P_1^2}$

(b) Tape supported at ends only

where in the English system C_s is the correction for sag (difference between length of curve and straight line from one support to the next), in feet; L_s the unsupported length of the tape, in feet; w the weight of the tape per foot of length, in pounds; W the total weight of the tape between supports, in pounds (note that W equals wL_s); and P_1 the pull on the tape, in pounds. Metric system units for Eqs. (4-10) and (4-11) are kilograms for W, w, and P_1; and meters for C_s and L_s.

In measuring lines of unknown length, the sag correction is always negative. After a line has been measured in several segments, and a sag correction has been calculated for each segment, the corrected length is given by

$$\bar{L} = L + \sum C_s \qquad (4\text{-}12)$$

where $\bar{L}$ is the corrected length of the line, L the recorded length of the line, and $\sum C_s$ the sum of the individual sag corrections.

The effects of errors due to sag can be eliminated by (a) supporting the tape at short intervals or throughout, or (b) computing a sag correction for each unsupported segment and applying the total to the recorded length according to Eqs. (4-10) or (4-11), and (4-12).

A steel tape 100.000 ft long weighs 1.50 lb (0.0150 lb/ft) and is used supported at the ends only, as in Figure 4-8(b). A line is measured in three segments (the first two being 100 ft) using a 12-lb pull and recorded as 250.52 ft. What is the length of the line corrected for sag?   EXAMPLE 4-6

SOLUTION
By Eq. (4-11), for each 100-ft segment,

$$C_s = \frac{-(1.50)^2(100.00)}{24(12)^2} = -0.065 \text{ ft}$$

and by Eq. (4-10) for the 50.52-ft segment,

$$C_s = \frac{-(0.0150)^2(50.52)^3}{24(12)^2} = -0.008 \text{ ft}$$

The corrected distance by Eq. (4-12) is $\bar{L} = 250.52 + [-2(0.065) - 0.008] = 250.38$ ft.

Note in this example the dramatic decrease in sag correction to a practically negligible figure for the 50.52-ft tape length. Thus the effects of sag can be almost eliminated by supporting the tape at its midpoint when full tape lengths are measured.

As stated previously, when lines of unknown length are being measured, sag corrections are always negative, whereas positive corrections occur if the tension applied exceeds the standard pull. For any given tape, the so-called *normal tension* needed to offset these two factors can be obtained. Its value can either be computed by setting

Eqs. (4-8) and (4-11) equal to each other and solving for P_1, or it can be determined by experiment. Although applying the normal tension does eliminate the need to make corrections for both pull and sag, it is not commonly used because the required pull is often too great for convenient application.

4-14.5 Poor Alignment

If one end of a tape is off-line or the tape is snagged on an obstruction, a systematic error is introduced. The corrected length where an offset from alignment occurs can be calculated by Eq. (4-2) or (4-3), with both d and L in the horizontal plane; that is, d is the distance the tape is off-line and L the length of tape involved.

When a pin marking the end of a 100-ft length is set 1.4 ft off-line, by Eq. (4-3) the error in that measurement is $(1.4)^2/200 = -0.01$ ft. An equivalent error enters the next tape length when the succeeding pin is put correctly on-line. If the center of a 100-ft tape catches on brush and is 1.0 ft off-line, the error produced in the two 50-ft lengths is $2(1.0)^2/100 = -0.02$ ft.

Errors resulting from poor alignment are systematic in effect and always make the recorded length longer than the true distance. They may be reduced (but never eliminated) by care in setting pins, lining in properly, and keeping the tape straight. Snapping the tape while applying tension will straighten it. A moderate amount of field practice enables a rear tapeperson to sight the line's end mark and keep the forward tapeperson within much less than a foot off the correct course.

4-14.6 Tape Not Horizontal

The error caused by a tape being inclined in the vertical plane is the same as that resulting from it being off-line in the horizontal plane. Corrected lengths can also be determined by Eqs. (4-2) and (4-3), where d is the difference in elevation between the tape ends and L the tape length.

Errors due to the tape not being horizontal are systematic and always make the recorded length longer than the true length. They are reduced by using a hand level to keep elevations of the tape ends equal, or by running differential levels (see Section 7-4) over the taping points and applying corrections for elevation differences.

4-14.7 Improper Plumbing

Practice and steady nerves are necessary to hold a plumb bob still long enough to mark a point. The plumb bob will sway, even in calm weather. On very light slopes and on smooth surfaces such as pavements, inexperienced tapepersons obtain better results by laying the tape on the ground instead of plumbing. Experienced tapepersons plumb most measurements.

Errors due to improper plumbing are random, since they may make distances either too long or too short. The errors would be systematic, however, when taping directly against or in the direction of a strong wind.

Touching the plumb bob on the ground, or steadying it with one foot, decreases its swing. Practice in plumbing will reduce errors.

4-14.8 Faulty Marking

Chaining pins should be set perpendicular to the taped line but inclined 45° to the ground. This position permits plumbing to the point where the pin enters the ground without interference from the loop.

TABLE 4-1 SUMMARY OF ERRORS

ERROR TYPE	ERROR SOURCE*	SYSTEMATIC (S) OR RANDOM (R)	DEPARTURE FROM NORMAL TO PRODUCE 0.01-ft ERROR FOR 100-ft TAPE
Tape length	I	S	0.01 ft
Temperature	N	S or R	15°F
Pull	P	S or R	15 lb
Sag	N, P	S	0.6 ft at center for 100-ft tape standardized by support throughout
Alignment	P	S	1.4 ft at one end of 100-ft tape or 0.7 ft at midpoint
Tape not level	P	S	1.4 ft elevation difference between ends of 100-ft tape
Plumbing	P	R	0.01 ft
Marking	P	R	0.01 ft
Interpolation	P	R	0.01 ft

*I—instrumental; N—natural; P—personal.

Brush, stones, and grass or weeds deflect a chaining pin and may increase the effect of incorrect marking. Errors from these sources tend to be random and are kept small by carefully locating a point, then checking it.

When taping on solid surfaces such as pavement or sidewalks, pencil marks or scratches can be used to mark taped segments. Accuracy in taping on the ground can be increased by using tacks in stakes as markers rather than chaining pins.

4-14.9 Incorrect Reading or Interpolation

The process of reading to hundredths on tapes graduated only to tenths, or to thousandths on tapes graduated to hundredths, is called *interpolation*. Errors from this source are random over the length of a line. They can be reduced by care in reading, employing a magnifying glass, or using a small scale to determine the last figure.

4-14.10 Summary of Effects of Taping Errors

An error of 0.01 ft is significant in many surveying measurements. Table 4-1 lists the nine types of errors; classifies them as instrumental (I), natural (N), or personal (P), and systematic (S) or random (R); and gives the departure from normal that produces an error of 0.01 ft in a 100-ft length.

The accepted method of reducing errors on precise work is to make separate measurements of the same line with different tapes, at different times of day, and in opposite directions. An accuracy of $\frac{1}{10,000}$ can be obtained by careful attention to details.

4-15 TAPE PROBLEMS

All tape problems develop from the fact that a tape is either longer or shorter than its graduated "nominal" length because of manufacture, temperature changes, tension applied, or some other reason. There are only two basic types of taping tasks: An unknown distance between two fixed points can be *measured,* or a required distance can be *laid off* from one fixed point. Since the tape may be too long or too short for either task, there are four possible versions of taping problems, which are: (**1**) Measure with a tape that is too long. (**2**) Measure with a tape that is too short. (**3**) Lay off with a tape that is too long. (**4**) Lay off with a tape that is too short. The solution of a particular problem is always simplified and verified by drawing a sketch.

Assume that the fixed distance *AB* in Figure 4-9 is measured with a tape that is later found to be 100.03 ft long. Then (the conditions in the figure are greatly exaggerated) the first tape length would extend to point 1; the next, to point 2; and the third, to point 3. Since the distance remaining from 3 to *B* is less than the correct distance from the true 300-ft mark to *B*, the *recorded* length *AB* is too small and must be increased by a correction. If the tape had been too short, the *recorded* distance would be too large, and the correction must be subtracted.

In laying out a required distance from one fixed point, the reverse is true. The correction must be subtracted from the desired length for tapes longer than the nominal value and added for tapes that are shorter. A simple sketch like Figure 4-9 makes clear whether the correction should be added or subtracted for any of the four cases.

4-16 COMBINED CORRECTIONS IN A TAPING PROBLEM

In taping linear distances, several types of systematic errors often occur simultaneously. The following examples illustrate procedures for computing and applying corrections for the two basic types of problems, *measurement* and *layoff*.

EXAMPLE 4-7   A steel tape standardized at 68°F and supported throughout under a tension of 20 lb was found to be 100.012 ft long. The tape had a cross-sectional area of 0.0078 in.^{2} and a weight of 0.0266 lb/ft. This tape was held horizontal, supported at the ends only, with a constant tension of 15 lb, to measure a line from *A* to *B* in nine segments. The data given in the following table were recorded. Apply corrections for tape length, temperature, pull, and sag to determine the correct length of the line.

SECTION	MEASURED (RECORDED) DISTANCE (ft)	TEMPERATURE (°F)
A–1	100.000	58
1–2	100.000	58
2–3	100.000	59
3–4	100.000	59
4–5	100.000	59
5–6	100.000	60
6–7	100.000	60
7–8	100.000	60
8–B	70.564	61
	Σ 870.564	

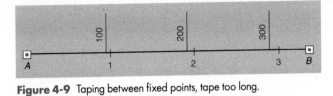

Figure 4-9 Taping between fixed points, tape too long.

SOLUTION

(a) The tape length correction by Eq. (4-4) is

$$C_l = \left(\frac{100.012 - 100.000}{100.000} \right) 870.564 = +0.104 \text{ ft}$$

(b) The temperature corrections by Eq. (4-6) are (*Note:* Separate corrections are required for distances measured at each different temperature.):

$$C_{t_1} = 0.00000645(58 - 68)200.000 = -0.013 \text{ ft}$$

$$C_{t_2} = 0.00000645(59 - 68)300.000 = -0.018 \text{ ft}$$

$$C_{t_3} = 0.00000645(60 - 68)300.000 = -0.016 \text{ ft}$$

$$C_{t_4} = 0.00000645(61 - 68)70.564 = -0.003 \text{ ft}$$

$$\sum C_t = -0.050 \text{ ft}$$

(c) The pull correction by Eq. (4-8) is

$$C_p = \frac{(15 - 20)870.564}{0.0078(29,000,000)} = -0.019 \text{ ft}$$

(d) The sag corrections by Eq. (4-10) are (*Note:* Separate corrections are required for the two different suspended lengths.):

$$C_{s_1} = -8 \left[\frac{(0.0266)^2(100.000)^3}{24(15)^2} \right] = -1.048 \text{ ft}$$

$$C_{s_2} = - \left[\frac{(0.0266)^2(70.564)^3}{24(15)^2} \right] = -0.046 \text{ ft}$$

$$\sum C_s = -1.094 \text{ ft}$$

(e) Finally, corrected distance AB is obtained by adding all corrections to the measured distance, or

$$AB = 870.564 + 0.104 - 0.050 - 0.019 - 1.094 = 869.505 \text{ ft}$$

EXAMPLE 4-8
The tape of Example 4-7 is to be used to lay off a horizontal distance CD of exactly 475.00 ft. The ground is on a smooth 3% grade; thus the tape will be used fully supported. Determine the correct slope distance to lay off if a pull of 15 lb is used and the temperature is 87°F.

SOLUTION

(a) The tape length correction, by Eq. (4-4), is

$$C_l = \left[\frac{100.012 - 100.000}{100.000} \right] 475.00 = +0.057 \text{ ft}$$

(b) The temperature correction, by Eq. (4-6), is

$$C_t = 0.00000645(87 - 68)475.00 = +0.058 \text{ ft}$$

(c) The pull correction, by Eq. (4-8), is

$$C_p = \frac{(15 - 20)475.00}{0.0078(29,000,000)} = -0.010 \text{ ft}$$

(d) Since this is a lay-off problem, all the corrections are subtracted. Thus the required horizontal distance to lay off, rounded to the nearest hundredth of a foot, is

$$CD_h = 475.00 - 0.057 - 0.058 + 0.010 = 474.88 \text{ ft}$$

(e) Finally, a rearranged form of Eq. (4-2) is used to solve for the slope distance (the difference in elevation d for use in this equation, for 475 ft on a 3% grade, is $4.75(3) = 14.25$ ft):

$$CD_s = \sqrt{(474.88)^2 + (14.25)^2} = 475.09 \text{ ft}$$

4-17 SPECIAL FIELD OPERATIONS USING A TAPE

Many problems arising in the field can be solved by taping. Some examples follow.

4-17.1 Laying out a Right Angle with a Tape

A right angle is readily laid out by the 3–4–5 method. In Figure 4-10(a), to erect a perpendicular to AD at A, measure 30 ft along AD and set point B. Then with the tape's zero graduation at B and the 100-ft mark at A, form a loop in the tape by bringing the 50- and 60-ft graduations together and pull each part of the tape taut to locate C. One

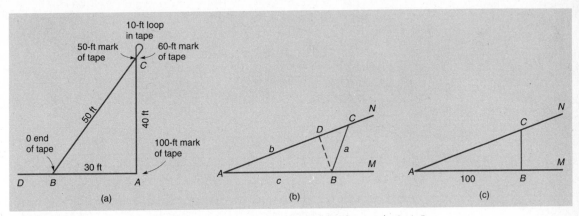

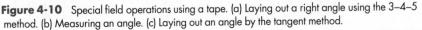

Figure 4-10 Special field operations using a tape. (a) Laying out a right angle using the 3–4–5 method. (b) Measuring an angle. (c) Laying out an angle by the tangent method.

person can make the layout alone by tying the tape thongs to stakes beyond A and B. Of course any distances in the proportions of 3, 4, and 5 could be used.

4-17.2 Measuring an Angle with a Tape by the Chord Method

If all three sides of a triangle are known, the angles can be computed. To find angle A of Figure 4-10(b), measure any definite lengths along AM and AN, such as AB and AC. Also measure BC. Then from the cosine law of trigonometry

$$\cos A = \frac{b^2 + c^2 - a^2}{2bc} \tag{4-13}$$

where a, b, and c are the sides of triangle ABC. For $b = 30.0$ ft, $c = 25.0$ ft, and $a = 12.5$ ft, angle A is calculated to be $24°09'$.

Alternatively, in Figure 4-10(b) an isosceles triangle may be formed by making AB equal to AC. Then

$$\sin \tfrac{1}{2}A = \frac{a}{2c} \tag{4-14}$$

Selecting a value of 50 ft for AB and AC simplifies the arithmetic. Thus if $AB = AC = 50.0$ ft and BC measures 20.92 ft, then $\sin \tfrac{1}{2}A = 0.2092$ and angle $A = 24°09'$.

4-17.3 Measuring an Angle with a Tape by the Tangent Method

If AD and a perpendicular BD are measured [see Figure 4-10(b)], $\tan A = BD/AD$. By making AD equal to 50 or 100 ft, the tangent is easily computed. To illustrate, if $AD = 100.00$ ft and BD measures 44.84 ft, then $\tan A = 0.4484$ and angle $A = 24°09'$. This procedure is not as convenient as the chord method because it requires establishing a perpendicular at D.

4-17.4 Laying off Angles

An angle can be laid off by reversing the tangent method just described. Along the initial side of the angle, a unit distance of 10, 20, 50, or 100 ft is laid off, like AB in Figure 4-10(c). A perpendicular BC is erected and if $AB = 100$, its length is made equal to 100 times the natural tangent of the desired angle. Points A and C are connected to give the required angle at A. This method is used by draftspersons as well as by surveyors in the field.

PROBLEMS

4-1 List six methods of measuring horizontal distances. Give an advantage and a disadvantage of each.

* **4-2** A student counted 188, 187, 186, 188, 186, and 187 paces in six trials of walking along a course of 500-ft known length on level ground. Then 211, 212, 210, and 212 paces were counted in walking four repetitions of an unknown distance AB. What is (a) the pace length and (b) the length of AB?

4-3 Similar to Problem 4-2, except 145, 146, 145, 144, and 145 paces were counted in five trials of walking a course of 400-ft known length. Then 406, 409, 408, and 405 paces were counted in walking four repetitions of a course AB of unknown length.

4-4 The following repeated readings were taken on a 2-m subtense bar with a 1″ theodolite. Compute the horizontal distance from theodolite to subtense bar.
(a) 0°20′18″, 0°20′17″, 0°20′21″, 0°20′24″
(b) 0°12′52″, 0°12′54″, 0°12′50″, 0°12′52″

4-5 List the nine common types of errors that occur in taping.

4-6 For the following data, compute the horizontal distance for a recorded slope distance AB.
(a) $AB = 327.28$ ft, slope angle $= 4°15′$
(b) $AB = 382.96$ m, difference in elevation A to $B = 18.3$ m
*(c) $AB = 651.54$ ft, grade $= 4.5\%$

4-7 Compute the horizontal distance between the ends of a 100-ft tape using approximate Eq. (4-3) and exact Eq. (4-2) for differences in elevation of 5, 10, 15, 20, and 25 ft. Carry out the computations far enough to show differences in results and tabulate your answers.

A 100-ft steel tape of cross-sectional area 0.0030 in.², weight 1.0 lb, and standardized at 68°F is 100.016 ft between end marks when supported throughout under a 12-lb pull. What is the true horizontal length of a recorded distance AB for the conditions given in Problems 4-8 through 4-13? (Assume horizontal taping and all full tape lengths except the last.)

	RECORDED DISTANCE AB (ft)	AVERAGE TEMPERATURE (°F)	MEANS OF SUPPORT	TENSION (lb)
*4-8	536.90	68	Throughout	12
4-9	845.72	68	Throughout	15
4-10	629.54	102	Throughout	16
4-11	420.31	68	Ends only	15
*4-12	966.35	22	Ends only	18
4-13	385.20	96	Ends only	20

*Asterisks indicate problems that have answers given in Appendix G.

For the tape of Problems 4-8 through 4-13, determine the true horizontal length of the recorded slope distance *BC* for the conditions shown in Problems 4-14 through 4-16. (Assume the tape was fully supported for all measurements.)

	RECORDED SLOPE DISTANCE *BC* (ft)	AVERAGE TEMPERATURE (°F)	TENSION (lb)	ELEVATION DIFFERENCE PER 100 ft (ft)
4-14	496.25	87	12	6.8
***4-15**	576.81	38	20	5.2
4-16	687.56	20	18	9.5

For the conditions given in Problems 4-17 through 4-21, determine the horizontal length of *CD* that must be laid out to achieve required true horizontal distance *CD*. Assume a 100-ft steel tape will be used, with cross-sectional area 0.0060 in.^{2}, weight 2.0 lb, and standardized at 68°F to be 100.014 ft between end marks when supported throughout with a 12-lb pull. (Assume horizontal taping and all full tape lengths except the last.)

	REQUIRED HORIZONTAL DISTANCE *CD* (ft)	AVERAGE TEMPERATURE (°F)	MEANS OF SUPPORT	TENSION (lb)
***4-17**	200.00	68	Throughout	12
4-18	378.68	37	Ends only	16
4-19	482.26	94	Throughout	15
***4-20**	97.00	46	Ends only	18
4-21	542.85	104	Ends only	20

For the tape of Problems 4-17 through 4-21, determine the slope length that must be laid out to achieve required true horizontal distance *DE* for the conditions of Problems 4-22 through 4-24. (Assume the tape will be fully supported for all measurements.)

	REQUIRED HORIZONTAL DISTANCE *DE* (ft)	AVERAGE TEMPERATURE (°F)	TENSION (lb)	SLOPE
***4-22**	200.00	17	12	3.5 ft/100 ft
4-23	325.00	88	15	5°30′ slope
4-24	618.42	55	18	6% grade

A 30-m steel tape measured 30.0150 m when standardized fully supported under a 7-kg pull at a temperature of 20°C. The tape weighed 0.90 kg and had a cross-sectional area of 0.028 cm^2. What is the true horizontal length of a recorded slope distance *AB* for the conditions given in Problems 4-25 through 4-27? (Assume the tape was fully supported for all measurements.)

*Asterisks indicate problems that have answers given in Appendix G.

	RECORDED SLOPE DISTANCE AB (m)	AVERAGE TEMPERATURE (°C)	TENSION (kg)	ELEVATION DIFFERENCE PER 100 m (m)
4-25	93.357	25	10	4.5
*4-26	382.408	6	8	2.7
4-27	241.950	12	9	1.8

A 30-m steel tape measured 29.9895 m when standardized fully supported under a 7-kg pull at a temperature of 20°C. The tape weighed 1.00 kg and had a cross-sectional area of 0.030 cm². Determine slope length BC that must be laid out using this tape to achieve the required true horizontal distance BC for the conditions given in Problems 4-28 and 4-29. (Assume a fully supported tape for all measurements.)

	REQUIRED HORIZONTAL DISTANCE BC (m)	AVERAGE TEMPERATURE (°C)	TENSION (kg)	SLOPE
4-28	80.000	8	10	6.5% grade
*4-29	265.275	35	9	4°20′ slope

4-30 What difference in temperature from standard, if neglected in the use of a steel tape, will cause an error of: **(a)** 1 part in 3000 **(b)** 1 part in 5000 **(c)** 1 part in 10,000?

4-31 When measuring a distance AB, the first taping pin was placed 1.5 ft to the right of line AB and the second pin was set 1.0 ft left of line AB. The recorded distance was 215.75 ft. Calculate the corrected distance. (Assume three taped segments, the first two 100 ft each.)

*4-32 In taping from A to B, a tree on-line necessitated setting an intermediate point C offset 4.5 ft to the side of line AB. Line AC was then measured as 368.92 ft along a uniform 4% slope. Line CB on horizontal ground was measured as 285.10 ft. Find the horizontal length of AB.

4-33 For a line measured with a 100-ft steel (add) tape having an extra graduated foot, chaining pins were set by mistake at the 101-ft mark. The length was recorded as exactly 800 ft. What is the correct distance?

4-34 The distance between two fixed points on a construction site was measured with the tape fully supported at a temperature of 102°F with a steel tape that was 100.000 ft long when fully supported at 68°F. It was recorded as 2680.60 ft. Compute the distance corrected for temperature. What distance would have been recorded if the temperature at the time of measurement had been 25°F?

4-35 What would the sag correction per tape length be if the tape of Problems 4-8 through 4-13 was used with a 20-lb pull and supported at: **(a)** the ends only **(b)** the midpoint and ends **(c)** the quarter points and ends?

4-36 A tape with a cross section of $\frac{5}{16} \times 0.025$ in. is 100.027 ft long when supported throughout at a tension of 15 lb. What is the length between end graduations if the same tension is applied, but the tape is supported at the ends and midpoint? If it is supported at the ends and quarter-points?

*4-37 A 100-ft steel tape having a cross-sectional area of 0.0048 in.² is exactly 100.000 ft long at 68°F when fully supported under a pull of 12 lb. By trial and error using Eqs. (4-8) and (4-11), determine the "normal tension" for this tape.

*Asterisks indicate problems that have answers given in Appendix G.

4-38 What error in a measured distance results for the following conditions:
   (a) One end of a 22-ft length of tape is off-line by 1.0 ft.
   (b) One end of a 46-ft length of tape is too high by 2.0 ft.
   (c) One end of a 100-ft tape is off-line by 1.6 ft and too low by 2.3 ft.
   *(d) One end of a 30-m tape is off-line by 25 cm and too high by 40 cm.

4-39 To determine the angle *AOB* between two intersecting fences without setting up a theodolite, convenient distances *OA* = 100.00 ft and *OB* = 80.00 ft are measured from the intersection along the fence lines. If distance *AB* is 78.37 ft, what is the intersection angle?

*4-40 Determine the most probable length of a line *AB*, the standard deviation, and the 90% error of a single measurement for the following series of taped measurements made under the same conditions: 486.25, 486.29, 486.23, 486.32, 486.28, and 486.25 ft.

4-41 The standard deviation of taping a 500-ft distance is ±0.08 ft. Using the same procedures, what should it be for a 2500-ft distance?

4-42 An irregular field is measured with a 100-ft steel tape that is actually 100.04 ft long, and the area is erroneously found to be 35.096 acres. What is the true area?

4-43 In measuring with a 100-ft standardized steel tape having a cross section of 0.25 × 0.025 in., if the tape is supported throughout, which of the following errors is most serious? **(a)** A constant temperature difference of +15°F from standard; **(b)** an alignment error of 0.5 ft on each tape length; **(c)** tape out of level 0.8 ft on each length; or **(d)** a tension variation of 5 lb from standard on each tape length.

*4-44 A line of 2-mi length must be laid off with a steel tape and have a standard deviation smaller than 0.50 ft. What standard deviation per tape length is permissible?

*Asterisks indicate problems that have answers given in Appendix G.

BIBLIOGRAPHY

Brinker, R. C., and R. Minnick. 1987. *The Surveying Handbook.* New York: Van Nostrand Reinhold, Chap. 3.
Golley, B. J., and J. Sneedon. 1974. "Investigation of Dynamic Taping." *ASCE Journal of the Surveying and Mapping Division* 100 (No. SU2):115.
Moffitt, F. H., and H. Bouchard. 1992. *Surveying.* New York: Harper Collins Publishers, Chap. 2.

5

ELECTRONIC DISTANCE MEASUREMENT

5-1 INTRODUCTION

A major advance in surveying instrumentation occurred approximately 40 years ago with the development of electronic distance measuring (EDM) instruments. These devices determine lengths by indirectly measuring the time it takes electromagnetic energy of known velocity to travel from one end of a line to the other and return. This indirect time measurement scheme consists in determining how many cycles of electromagnetic energy are required to travel the double path distance. The frequency (time required for each cycle) is precisely controlled by the EDM instrument and thus known, so the total travel time becomes known. Multiplying total time by velocity, and dividing by 2, yields the unknown distance.

The first EDM instrument was introduced in 1948 by the Swedish physicist Erik Bergstrand. His device, called the *geodimeter* (an acronym for geodetic distance meter), resulted from attempts to improve methods for measuring the velocity of light. The instrument transmitted visible light and was capable of accurately measuring distances up to about 25 mi (40 km) at night. In 1957 a second EDM apparatus, the *tellurometer,* was introduced. Designed in South Africa by T. L. Wadley, this instrument transmitted microwaves and was capable of measuring distances up to 50 mi (80 km) or more, day or night.

The potential value of these early EDM models to the surveying profession was immediately recognized. However, they were expensive and not readily portable for field operations. Furthermore, measuring procedures were lengthy, and mathematical reductions to obtain distances from observed values were difficult and time consuming. Continued research and development have overcome all of these deficiencies.

Prior to the introduction of EDM instruments, accurate distance measurements were made by taping. Although seemingly a relatively simple procedure, precise taping is one of the most difficult and painstaking of all surveying tasks. Now EDM instruments have made it possible to obtain accurate distance measurements rapidly and easily.

Given a line of sight, long or short lengths can be measured over bodies of water, busy freeways, or terrain that is inaccessible for taping.

The first generation of EDM instruments consisted of rather large stand-alone devices, which were mounted independently on tripods in a manner like that shown in Figure 5-1(a). Second-generation instruments were smaller and commonly mounted on theodolites, as shown in Figure 5-1(b) and (c). This arrangement was convenient because it enabled making distance and angle measurements from a single setup. Zenith (or vertical) angles read with the theodolite at the time of distance measurement could be used to compute horizontal and vertical distance components from the measured slope lengths.

In the current generation, EDM instruments have now been combined with *digital theodolites* and *microprocessors*. The resulting devices, called *total station instruments* (see Figure 5-8), can measure simultaneously and automatically both distances and angles. The microprocessor receives the measured slope length and zenith (or vertical) angle, calculates horizontal and vertical distance components, and displays them in real time. When equipped with *automatic data collectors* (see Section 3-7), they can record field notes electronically for transmission to computers, plotters, and other office equipment for processing. These so-called *field-to-finish* systems are gaining worldwide acceptance and changing the practice of surveying substantially.

5-2 CLASSIFICATION OF EDM INSTRUMENTS

The most common system for classifying EDM instruments is by the type of electromagnetic energy they transmit. Two categories are commonly employed in surveying: **(1)** *electro-optical* instruments, which transmit either laser or infrared light and **(2)** *microwave* equipment, which transmits invisible electromagnetic energy of very short wavelength.

In addition to their unique types of transmitted energy, other basic differences exist between instruments in these two categories. A major difference, for example, is that signals transmitted by electro-optical instruments are returned from the opposite end of the line by a passive prism reflector. Microwave systems, on the other hand, employ two identical units. One transmits the signal to the other located at the opposite end of the line. The second unit receives the signal and transmits it back to the original instrument. Other differences are noted in Sections 5-5 and 5-6, where these two types of EDM instruments are described in more detail.

5-3 PROPAGATION OF ELECTROMAGNETIC ENERGY

Electronic distance measurement is based on the rate and manner by which electromagnetic energy propagates through the atmosphere. The rate of propagation can be expressed with the following equation:

$$V = f\lambda \tag{5-1}$$

where V is the velocity of electromagnetic energy, in meters per second; f the modulated frequency of the energy, in *hertz*;[1] and λ the wavelength, in meters. The velocity of

[1]The hertz (Hz) is a unit of frequency equal to 1 cycle/sec. The kilohertz (KHz), megahertz (MHz), and gigahertz (GHz) are equal to 10^3 Hz, 10^6 Hz, and 10^9 Hz, respectively.

(a)

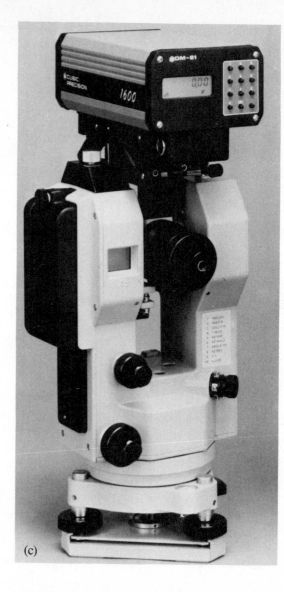

(c)

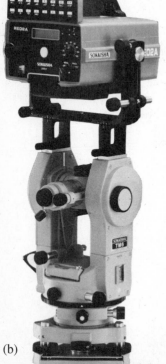

(b)

Figure 5-1 Electro-optical EDM instruments. (a) Rangemaster III laser instrument with independent tripod mount. (Courtesy Cubic Precision, Inc.) (b) RED2L infrared instrument attached to theodolite standards. (Courtesy Sokkia Corporation.) (c) DM-81 infrared instrument with theodolite telescope mount. (Courtesy Cubic Precision, Inc.)

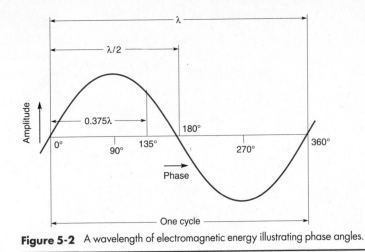

Figure 5-2 A wavelength of electromagnetic energy illustrating phase angles.

electromagnetic energy in a vacuum is approximately 299,792,458 m/sec. Its speed is slowed somewhat in the atmosphere according to the following equation:

$$V = c/n \tag{5-2}$$

where c is the velocity of electromagnetic energy in a vacuum, and n the atmospheric *index of refraction*. The value of n varies from about 1.0001 to 1.0005, depending on pressure and temperature, but is approximately equal to 1.0003. Thus, as will be discussed later, accurate electronic distance measurement requires that atmospheric pressure and temperature be measured so that the appropriate value of n is known.

The manner by which electromagnetic energy propagates through the atmosphere can be represented conceptually by the sinusoidal curve illustrated in Figure 5-2. This figure shows one wavelength, or *cycle*. Portions of wavelengths or the positions of points along the wavelength are given by *phase angles*. Thus in Figure 5-2 a 360° phase angle represents a full cycle, or a point at the end of a wavelength, while 180° is a half-wavelength, or the midpoint. An intermediate position along a wavelength having a phase angle of, say, 135° is $(\frac{135}{360})$, or 0.375 of a wavelength.

5-4 PRINCIPLES OF ELECTRONIC DISTANCE MEASUREMENT

In Section 5-1 it was stated that distances are measured electronically by determining the time it takes electromagnetic energy of known velocity to travel from one end of a line to the other and return. Indirectly this process involves determining the number of wavelengths in an unknown distance. Then, knowing the precise length of the wave, the distance can be determined. This is similar to relating an unknown distance to the calibrated length of a steel tape.

The procedure of measuring a distance electronically is depicted in Figure 5-3, where an EDM device has been centered over station A by means of a plumb bob or optical plummet. The instrument transmits a *carrier* signal of electromagnetic energy to station B. A reference frequency of precisely regulated wavelength has been superimposed or *modulated* onto the carrier. The signal is returned from B to the receiver,

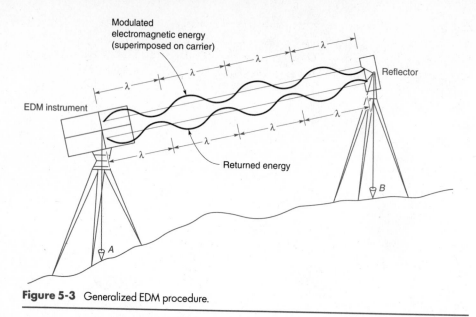

Figure 5-3 Generalized EDM procedure.

so its travel path is double the slope distance *AB*. In the figure the modulated electromagnetic energy is represented by a series of sine waves, each having wavelength λ. The unit at *A* determines the number of wavelengths in the double path, multiplies by the wavelength in feet or meters, and divides by 2 to obtain distance *AB*.

It would of course be highly unusual if a measured distance was exactly an integral number of wavelengths, as illustrated in Figure 5-3. Rather, some fractional part of a wavelength would in general be expected, for example, the partial value *p* shown in Figure 5-4. In that figure, distance *L* between the EDM instrument and reflector would be expressed as

$$L = \frac{n\lambda + p}{2} \tag{5-3}$$

where λ is the wavelength, *n* the number of full wavelengths, and *p* the length of the fractional part. The fractional length is determined by the EDM instrument from

Figure 5-4 Phase difference measurement principle.

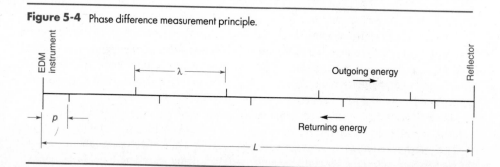

measurement of the *phase shift* (phase angle) of the returned signal. To illustrate, assume that the wavelength for the example of Figure 5-4 was precisely 20.000 m. Assume also that the phase angle of the returned signal was 115.7°, in which case length p would be $\left(\frac{115.7}{360}\right) \times 20.000 = 6.428$ m. Then from the figure, since $n = 9$, by Eq. (5-3), length L is

$$L = \frac{9(20.000) + 6.428}{2} = 93.214 \text{ m}$$

Considering the double path distance, the 20-m wavelength used in the example just given has an "effective wavelength" of 10 m. This is one of the fundamental wavelengths used in current EDM instruments. It is generated using a frequency of approximately 15 MHz.

EDM instruments cannot determine the number of full wavelengths in an unknown distance by transmitting only one frequency and wavelength. To resolve the ambiguity, they must transmit additional signals having longer wavelengths. This procedure is explained in the following section, which describes electro-optical EDM instruments.

5-5 ELECTRO-OPTICAL INSTRUMENTS

The majority of EDM instruments manufactured today are electro-optical and transmit infrared light as a carrier signal. This is primarily because its intensity can be modulated directly, considerably simplifying the equipment. Earlier models used tungsten or mercury lamps. They were bulky, required a large power source, and had relatively short operating ranges, especially during the day because of excessive atmospheric scatter. EDM instruments using coherent light produced by gas lasers followed. These were smaller and more portable, and were capable of making measurements of long distances in the daytime as well as at night.

Figure 5-5 is a generalized schematic diagram illustrating the basic method of operation of one particular type of electro-optical instrument. The transmitter uses a GaAs diode that emits *amplitude-modulated* (AM) infrared light. The frequency of modulation is precisely controlled by a crystal oscillator. The modulation process may be thought of as similar to passing light through a stove pipe in which a damper plate is spinning at a precisely controlled rate or frequency. When the damper is closed, no light passes. As it begins to open, light intensity increases to a maximum at a phase angle of 90° with the plate completely open. Intensity reduces to zero again with the damper closed at a phase angle of 180°, and so on. This intensity variation or amplitude modulation is properly represented by sine waves such as those shown in Figures 5-2 and 5-3.

As shown in Figure 5-5, a beam splitter divides the light emitted from the diode into two separate signals: an *external* measurement beam and an *internal* reference beam. The external one is carefully aimed at a retro-reflector that has been centered over the point at the line's other end. Final aiming is accomplished using slow-motion screws to change the instrument's direction slightly and maximize the returned signal's strength. Figure 5-6 shows a triple corner cube retro-reflector of the type used to return the external beam, coaxial, to the receiver.

The internal beam passes through a variable-density filter and is reduced in intensity to a level equal to that of the returned external signal, enabling a more accurate

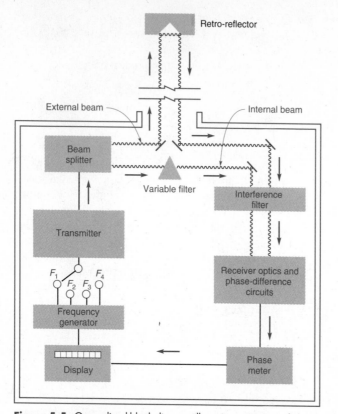

Figure 5-5 Generalized block diagram illustrating operation of electro-optical EDM instrument.

measurement to be made. Both internal and external signals go through an interference filter, which eliminates undesirable energy such as sunlight. The internal and external beams then pass through components to convert them into electric energy while preserving the phase shift relationship resulting from their different travel path lengths. A phase meter converts this phase difference into direct current having a magnitude proportional to the differential phase. This current is connected to a *null meter,* which is adjusted to null the current. The fractional wavelength is measured during the nulling process, converted to distance, and displayed.

To resolve the ambiguous number of full cycles a wave has undergone, EDM instruments transmit different modulation frequencies. The unit illustrated in the schematic of Figure 5-5 uses four frequencies: $F1$, $F2$, $F3$, and $F4$, as indicated. If modulation frequencies of 14.984 MHz, 1.4984 MHz, 149.84 KHz, and 14.984 KHz are used, and assuming the index of refraction is 1.0003, then their corresponding "effective" wavelengths are 10.000, 100.00, 1000.0, and 10,000 m, respectively. Assume that a distance of 3867.142 appears on the display as the result of measuring a line. The four rightmost digits, 7.142, are obtained from the phase shift measured while transmitting the 10.000-m wavelength at frequency $F1$. Frequency $F2$, having a 100.00-m wavelength, is then transmitted, yielding a fractional length of 67.14. This provides the digit 6 in the displayed distance. Frequency $F3$ gives a reading of 867.1, which provides the digit 8 in the answer, and finally, frequency $F4$ yields a reading of 3867, which supplies the

Figure 5-6 Triple retro-reflector.

digit 3, to complete the display. From this example it should be evident that the high resolution of a measurement (nearest 0.001m) is secured using the 10.000-m wavelength, and the others simply resolve the ambiguity of the number of these shorter wavelengths in the total distance.

With older instruments, changing of frequencies and nulling are done manually by setting dials and turning knobs. Now modern instruments incorporate microprocessors which control the entire measuring process. Once the instrument is aimed at the reflector and the measurement started, the final distance appears in the display almost instantaneously. Other changes in new instruments include improved electronics to control the amplitude modulation, and replacement of the null meter by an electronic phase detector. These changes have significantly improved the accuracy with which phase shifts can be determined, which in turn has reduced the number of different frequencies that need to be transmitted. Consequently, as few as two frequencies are now used on some instruments: one which produces a short wavelength to provide the high-resolution digits, and one with a long wavelength to provide the coarse numbers. To illustrate how this is possible, consider again the example measurement just described, which used four frequencies. Recall that a reading of 7.142 was obtained with the 10.000-m wavelength, and that 3867 was read with the 10,000-m wavelength. Note the overlap of the common digit 7 in the two readings. Assuming that both phase shift measurements are reliably made to four significant figures, the leftmost digit of the first reading should indeed be the same as the rightmost one of the second reading. If these digits are the same in the measurement, this provides a check on the instrument's operation. Modern instruments compare these overlapping digits, and will display an error message if they do not agree. If they do check, the displayed distance will take all four digits from the first (short-wavelength) reading, and the first three digits from the second reading.

Approximately 25 different electro-optical EDM models are currently being manufactured for exclusive measurement of distances. Many others are embodied in total station instruments. Maximum ranges of the infrared electro-optical instruments vary from about 1 km for the smallest instruments up to approximately 20 km for the larger ones. These ranges are restricted by power limitations of the diodes used to produce the infrared light, but for most surveying these distance capabilities are adequate. Laser electro-optical instruments and those using microwaves are available for longer distances, if needed.

The RED2L instrument of Figure 5-1(b) and the DM-81 of Figure 5-1(c) have maximum ranges of approximately 5 and $1\frac{1}{2}$ km, respectively, with triple prisms. They are useful for a variety of general survey projects. Both have purported accuracies in the range of about $\pm(5$ mm $+$ 5 ppm);[2] will automatically compute and display horizontal and vertical components of a slope distance after manual entry of the vertical angle via the control panel; and can be operated in the "tracking mode." In tracking, sometimes also called "stakeout," a required distance (horizontal, vertical, or slope) is entered by means of the control panel, and the difference between the desired distance and that to the reflector is rapidly updated and displayed. This feature, extremely useful in construction stakeout, is described further in Section 24-9. The Rangemaster III of Figure 5-1(a) has a range of more than 60 km and an accuracy of $\pm(5$ mm $+$ 1 ppm). Its most common use has been for control surveys.

5-6 MICROWAVE INSTRUMENTS

The measurement signal used by microwave devices consists of frequency modulation (FM) superimposed on the carrier wave. Like electro-optical instruments, microwave equipment operates on the phase shift principle and uses varying frequencies to resolve ambiguities in the unknown number of full wavelengths in a distance. The range of microwave devices is comparatively long, and they can operate in fog or light rain. Measurements made in such adverse weather conditions are somewhat limited in range, however.

A complete microwave EDM system consists of two portable identical units. Each includes all components necessary to make measurements—transmitter, receiver, antenna, circuitry—and built-in communication arrangement. Units are centered by plumb bobs or optical plummets over the terminal points of a course, with one instrument functioning in the "master," or transmitter, mode and the other in the "remote," or receiver, mode. Either may be operated as master or remote by simply changing a switch position.

Measurement with microwave devices requires an operator at each end of the line to take a set of readings while using the instrument in the master mode. Since both units contain temperature-stabilized wavelength calibration, this procedure gives two inde-

---

[2]ppm = parts per million. One ppm equals 1 mm/km. In a distance 5000 ft long, a 5-ppm error equals $5000 \times 5 \times 10^{-6} = 0.025$ ft.

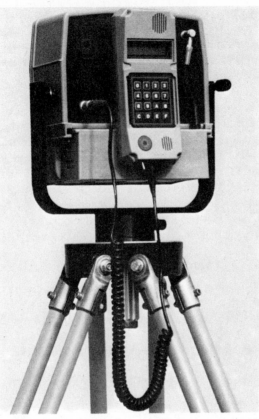

Figure 5-7 Microfix 100C microwave EDM instrument. (Courtesy Navigation Electronics.)

pendent measurements of the distance and a valuable check. The operators, who may not be in sight of each other because of fog, darkness, or distance, coordinate their work procedures by communicating on the built-in radio telephone. As with electro-optical instruments, temperature and pressure affect the energy's propagation velocity and must be measured. With microwave instruments, relative humidity must also be measured. This can be done with a wet and dry bulb thermometer called a *psychrometer.*

The Microfix 100C shown in Figure 5-7 is a lightweight and compact microwave EDM instrument. It can be adapted to fit standard theodolites and measures distances up to 60 km to an accuracy of $\pm(15 \text{ mm} + 3 \text{ ppm})$. The measurement system is fully automatic after pointing has been accomplished. This device can also operate in a "tracking" mode.

5-7 TOTAL STATION INSTRUMENTS

Total station instruments (also called electronic tacheometers) combine an EDM instrument, an electronic digital theodolite, and a computer in one unit. The electronic digital theodolite, described in more detail in Section 10-10, automatically measures and displays horizontal and zenith (or vertical) angles. Total station instruments measure distance and direction simultaneously and transmit the results automatically to a built-in computer. The horizontal and zenith (or vertical) angle and slope distance can be displayed; then upon keyboard commands, horizontal and vertical distance components are instantaneously computed and displayed. If the coordinates of the occupied station and a reference azimuth are input to the system, the coordinates of the sighted point are immediately obtained. This information can all be directly stored in an automatic data collector, thereby eliminating manual recording. These instruments are of tremendous value in all types of surveying, as will be discussed in later portions of this text.

The Pentax PTS III total station instrument shown in Figure 5-8 has a distance range of approximately 2 km (using a single prism) with a purported accuracy of ±(3 mm + 2 ppm) and reads angles to the nearest 5″. The Geodimeter 500 shown in Figure 1-3(c) has a similar distance range and accuracy, but reads angles to 1″. Total station instruments are discussed in detail in Chapter 10, and others are pictured and described. The Sokkia SET 2B of Figure 10-14 (shown with the SDR 33 data collector) also has a distance range of approximately 2 km (with a single prism) and reads angles to the nearest 2″.

5-8 ERRORS IN ELECTRONIC DISTANCE MEASUREMENT

As noted earlier, accuracies of EDM instruments are quoted in two parts: a constant error, and a value proportional to the distance measured. Specified errors vary for different instruments, but the constant portion is usually about ±3 mm, and the proportion is generally about ±3 ppm. The constant error is most significant on short distances; for example, with an instrument having a constant error of ±3 mm, a measurement of 30 m is good to only $\frac{3}{30,000} = \frac{1}{10,000}$, or 100 ppm. For a long distance, say 30 km, the constant error becomes negligible and the proportional part more important.

From the foregoing it is clear that except for very short distances, the order of accuracy possible with EDM instruments is very high. Errors can seriously degrade the measurements, however, and thus caution must always be exercised to minimize their effects. Sources of error in EDM work may be personal, instrumental, or natural. The subsections that follow identify and describe errors from each of these sources.

5-8.1 Personal Errors

Personal errors include inaccurate setups of EDM instruments and reflectors over stations, faulty measurements of instrument and reflector heights [needed for computing horizontal lengths (see Section 5-9)], and errors in determining atmospheric pressures and temperatures (and humidity if microwave instruments are used). These errors are largely random. They can be minimized by exercising utmost care and by using good-quality barometers, thermometers, and psychrometers.

Mistakes (not errors) in manually reading and recording displayed distances are common and costly. They can be eliminated with some instruments by obtaining the readings in both feet and meters and comparing them. Of course, automatic data collectors (see Section 3-7) also circumvent this problem.

5-8.2 Instrumental Errors

If EDM equipment is carefully adjusted and precisely calibrated, instrumental errors should be extremely small. To assure their accuracy and reliability, EDM instruments should be checked against a first-order base line at regular time intervals. For this

Figure 5-8 Pentax PTS III total station instrument. (Courtesy Pentax Corporation.)

purpose the National Geodetic Survey has established a number of accurate base lines in each state.[3] These are approximately a mile long and placed in relatively flat areas. Monuments are set at the ends and at intermediate points along the base line.

[3]For locations of base lines in your area, contact the National Geodetic Information Center, NOAA, National Geodetic Survey, N/CG17, SSMC3 Station 09535, 1315 East West Highway, Silver Spring, Md. 20910; telephone: (301) 713-3242.

Although most EDM instruments are quite stable, occasionally they become maladjusted and generate erroneous frequencies. This results in faulty wavelengths that degrade distance measurements in a manner similar to using a tape of incorrect length. Periodic checking of the equipment against a calibrated base line will detect the existence of measurement errors. It is especially important to make these checks if high-order surveys are being conducted.

The corner cube reflectors used with EDM instruments are another source of instrumental error. With these the "effective center" is actually behind the prism due to the slower rate at which light is able to travel through the glass cube, as compared to its travel through air. This often results in the effective center not coinciding with the plummet, a condition that produces an offset known as the "reflector constant." The constant can be as large as 70 mm and will vary from one reflector to another. If its value is known, it can be subtracted from distance measurements. Alternatively, the "electrical center" of the transmission unit can be shifted forward to compensate for the reflector constant. However, if an EDM instrument is being regularly used with several reflectors, this shift is impractical and the constant for each reflector should be known and subtracted from measured distances. With newer EDM instruments that use microprocessors, the constant can be entered via the keyboard and included in the computed corrections.

By comparing a base-line length to a distance measured with an EDM instrument, a so-called *system measurement constant* is ascertained. A correction for this systematic error can then be applied to all subsequent measurements. The constant thus determined combines the amount by which the electrical center of the instrument and the effective center of the reflector are offset forward or back from their plummets.

Although calibration using a first-order base line is preferred, if one is not available, the constant can still be obtained. In this procedure, three stations, A, B, and C, are established on a straight line, with distance AC roughly 1 mi and B approximately midway between A and C. The total length AC and the two parts AB and BC should be measured. For these measurements, the following equation can be written:

$$AC + K = (AB + K) + (BC + K)$$

from which

$$K = AC - AB - BC \tag{5-4}$$

where K is the system measurement constant to be added to measured distances. The procedure, including centering of the EDM instrument and reflector, should be repeated several times very carefully and the average value of K adopted. Since different reflectors have varying offsets, the test should be run with each one to be used, and the results marked on them to avoid confusion. For the most precise calibration, lengths AB and BC should preferably be carefully laid out as even multiples of the EDM instrument's shortest measurement wavelength. Otherwise the value of K obtained can be incorrect due to the instrument's cyclic error (ppm portion).

5-8.3 Natural Errors

Natural errors in EDM operations stem primarily from atmospheric variations in temperature, pressure, and humidity which affect the index of refraction and modify the wavelength of electromagnetic energy. These variables are measured and accounted for in accurate distance determination. Humidity can be neglected when using electro-

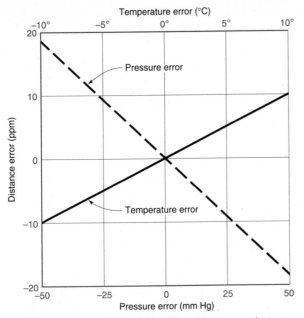

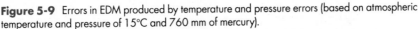

Figure 5-9 Errors in EDM produced by temperature and pressure errors (based on atmospheric temperature and pressure of 15°C and 760 mm of mercury).

optical instruments, but this variable is important when microwave instruments are employed.

Newer EDM instruments having built-in microprocessors use atmospheric variables, input through the keyboard, to compute corrected distances after making measurements but before displaying them. For older instruments, corrections are made by varying the transmission frequency, or they can be computed manually after the measurement. Equipment manufacturers provide tables and charts that assist in this process.

The magnitude of error in electronic distance measurement due to errors in measuring atmospheric pressure and temperature is indicated in Figure 5-9. Note that a 10°C temperature error, or a pressure difference of 25 mm (1 in.) of mercury, each produce a distance error of about 10 ppm. For microwave instruments, humidity is determined with a psychrometer that gives wet and dry bulb temperatures. An error of 1.5°C in the difference of the two bulbs is equivalent to approximately 10 ppm in distances measured with these types of instruments.

Microwaves transmit in a much wider pattern than infrared energy. Thus multiple reflections can occur from a ground or water surface and cause a condition called *ground swing,* which affects the accuracy of readings taken with microwave instruments. Errors from this source can be reduced by elevating the master and remote units as high above ground as possible, and by averaging the results of multiple measurements taken from both ends of the line using varying carrier frequencies.

5-9 COMPUTING HORIZONTAL LENGTHS FROM SLOPE DISTANCES

All EDM equipment measures the slope distance between stations. Many instruments can reduce these distances to their horizontal components automatically if the zenith

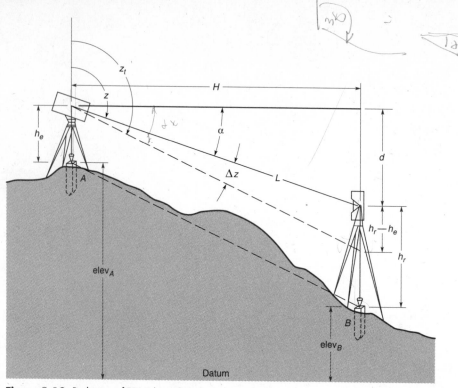

Figure 5-10 Reduction of EDM slope distance to horizontal.

(or vertical) angle is input. With some early models this cannot be done, so the reduction must be carried out manually. The procedures used, whether performed internally by the microprocessor or done manually, follow those outlined in this section. It is presumed, of course, that slope distances are first corrected for instrumental and atmospheric conditions.

Reduction of slope distances to horizontal can be based on elevation differences, or on zenith (or vertical) angle. Special conditions arise as a result of variations in field procedures and type of instrument mount. Long lines must be treated differently in reduction than short ones. These topics are considered in the subsections that follow.

5-9.1 Reduction of Short Lines by Elevation Differences

If difference in elevation is used to reduce slope distances to horizontal, during field operations heights h_e of the EDM instrument and h_r of the reflector above their respective stations are measured and recorded (see Figure 5-10). If elevations of stations A and B in the figure are known, either Eq. (4-2) or Eq. (4-3) will reduce the slope distance to horizontal, with the value of d (difference in elevation between EDM instrument and reflector) computed as follows:

$$d = (\text{elev}_A + h_e) - (\text{elev}_B + h_r) \qquad (5\text{-}5)$$

EXAMPLE 5-1   A slope distance of 165.360 m (corrected for meteorological conditions) was measured from A to B, whose elevations were 447.401 and 445.389 m above datum, respectively.

Find the horizontal length of line AB if the heights of the EDM instrument and reflector were 1.417 and 1.615 m above their respective stations.

SOLUTION

By Eq. (5-5), $d = (447.401 + 1.417) - (445.389 + 1.615) = 1.813$ m.

By Eq. (4-2), $H = \sqrt{(165.360)^2 - (1.813)^2} = 165.350$ m.

5-9.2 Reduction of Short Lines by Zenith or Vertical Angle

If zenith angle z (angle measured downward from the upward direction of the plumb line) is measured to the inclined path of the transmitted energy when measuring slope distance L (see Figure 5-10), then the following equation is applicable to reduce the slope length to its horizontal component:

$$\boxed{H} = L \sin z \qquad (5\text{-}6)$$

If vertical angle α (angle between horizontal and the inclined energy path) is measured (see Figure 5-10), then Eq. (4-1) is applicable for the reduction.

For most precise work, especially on longer lines, the zenith (or vertical) angle should be measured directly, reversed, and averaged (see Section 11-10). Also, the mean obtained from both ends of the line will compensate for earth curvature and refraction.

In some cases, rather than measuring the required zenith angle z, the true zenith angle z_t of Figure 5-10 between points A and B may be measured in a separate operation using a theodolite and a graduated rod. The true zenith angle is obtained by sighting the rod held on B at a height equal to the theodolite's hi (height of instrument above the station). If the true zenith angle has been measured, and a significant height difference exists between the EDM instrument and reflector when the slope distance is measured, as shown in Figure 5-10, angle z for use in Eq. (5-6) can be obtained by adding Δz algebraically to z_t, where Δz, in seconds, is calculated by

$$\Delta z = \frac{(h_e - h_r) \sin z_t}{L} \times 206{,}265''/\text{rad} \qquad (5\text{-}7)$$

On a downhill sight from point X to point Y, a slope distance of 230.75 ft was measured. The heights of EDM instrument and reflector were 5.8 and 4.8 ft, respectively. In a separate operation the true zenith angle from X to Y was measured with a theodolite to be $100°28'00''$. Find the corrected horizontal distance.

EXAMPLE 5-2

SOLUTION

By Eq. (5-7),

$$\Delta z = \frac{(5.8 - 4.8) \sin (100°28')}{230.75} \times 206.265 = 879'' = 14'39''$$

$$z_t = 100°28'00'' + 14'39'' = 100°42'39''$$

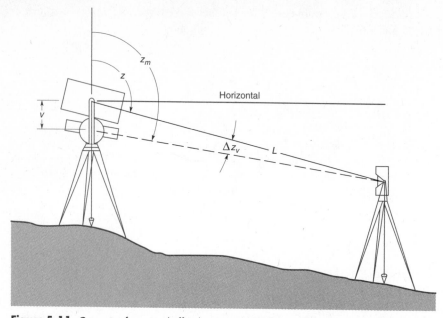

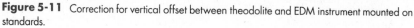

Figure 5-11 Correction for vertical offset between theodolite and EDM instrument mounted on standards.

By Eq. (5-6), $H = 230.75 \sin (100°42'39'') = 226.73$ ft.

In Example 5-2, if the angle correction Δz had not been made, H would have been calculated as $230.75 \times \sin (100°28') = 226.91$, and a mistake of 0.18 ft would have resulted.

Angle z can be measured directly with a theodolite, thus avoiding computation of Δz (even if a height difference exists between EDM instrument and reflector), by sighting on the graduated rod to a reading that compensates for the height difference. Assume, for example, that h_e and h_r at the time of distance measurement were 5.3 and 6.2 ft, respectively. For zenith angle measurement, if the theodolite has an hi of 5.6 ft, a reading of 6.5 ft sighted on the rod will produce required angle z. Alternatively, if the EDM instrument and theodolite fit the same tribrachs, the theodolite can be placed on the tripod when distance measurement is completed, but before the tripod is moved. Then, by sighting the reflector, required zenith angle z is obtained in spite of height differences. *Computations or special field procedures to account for height differences are best avoided, and this can usually be done by setting the EDM instrument and reflector in the same vertical position.*

5-9.3 Considerations for Different Theodolite Mounts

If the EDM instrument is mounted on a theodolite as shown in Figures 5-1(b) and (c), the energy will be transmitted from a point offset vertically above the theodolite's horizontal axis. Zenith angle z_m, measured to the center of the reflector, will normally be obtained at the time of distance measurement. If the EDM instrument is mounted on the telescope so it tilts with the theodolite's line of sight, correction for the vertical offset is not needed, and the horizontal distance is obtained using L and z_m in Eq. (5-6). If the mount is on the standards, however, as shown in Figure 5-11, an angular correction, Δz_v, will be needed to account for the vertical offset v. This angle, in seconds, is

$$\Delta z_v = \frac{v \sin z_m}{L} \times 206{,}265''/\text{rad} \qquad (5\text{-}8)$$

The correction must be added to z_m to get the correct angle for use in Eq. (5-6). Again, this correction can be avoided by sighting below the reflector a distance v when measuring the zenith angle.

For long lines, errors caused by ignoring height differences between EDM instrument and reflector, or vertical offsets of the transmission optics above the theodolite, are insignificant. They become important on short lengths, however.

A zenith angle of 97°25′ (downhill) was measured to the center of a reflector, and a slope distance of 153.72 ft was obtained. The EDM instrument was mounted on the standards and offset vertically 0.66 ft. What is the corrected horizontal distance? (Assume the theodolite and reflector heights are equal.)

EXAMPLE 5-3

SOLUTION
By Eq. (5-8),

$$\Delta z_v = \frac{0.66 \sin 97°25'}{153.72} \times 206{,}265 = 878'' = 14.6'$$

$$z = 97°25' + 0°14.6' = 97°39.6'$$

By Eq. (5-6), $L = 153.72 \sin (97°39.6') = 152.35$ ft.

Note that if the vertical offset had been ignored, the reduced horizontal distance would have been 152.43 ft, in error by 0.08 ft. Errors caused by ignoring height differences between instrument and reflector, or vertical offsets of the transmission optics above the theodolite, increase with steepness of slope.

5-9.4 Reduction of Long Lines to Horizontal

Although GPS is rapidly replacing other instrumentation for geodetic control surveys, because of the long-range capabilities of some EDM instruments, they are still sometimes used for this type of work. Computations involving long distances in geodetic surveys require that the measurements be reduced first to horizontal and then to either the *ellipsoid* or mean sea level. (The ellipsoid, as described in Section 19-1, is a uniform mathematical surface used to approximate the mean sea level figure of the earth.) If the lines are sufficiently long and measured on steep slopes, Eqs. (4-1) through (4-3) may not be suitable for making the reduction to horizontal. Reduction of long lines to horizontal, and to either the ellipsoid or the mean sea level, is discussed in Section 19-14. For lines up to nearly 2 mi in length, on slopes as steep as 10°, the error introduced by using Eqs. (4-1) through (4-3) does not exceed 1 part in 20,000.

PROBLEMS

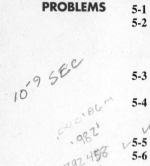

5-1 List the advantages of electronic distance measurement as compared to taping.

5-2 The speed of electromagnetic energy through the atmosphere at standard barometric pressure of 29.92 in. of mercury is accepted as 299,792.5 km/sec for measurements with an EDM instrument. What time lag in the equipment will produce an error of 100 ft in a measured distance?

5-3 If electromagnetic energy travels approximately 186,000 mi/sec under given conditions, what unit of distance corresponds to each nanosecond of time?

5-4 What "effective" wavelength results from transmitting electromagnetic energy through an atmosphere having an index of refraction of 1.0003, if the frequency is:
*(a) 7.4927 MHz   **(b)** 4.8703 MHz?

5-5 Explain briefly how a distance can be measured by the method of phase comparison.

5-6 How are variations in the propagation of electromagnetic energy due to atmospheric conditions accounted for in measuring distances with EDM instruments?

5-7 Which frequency, highest or lowest, is used to obtain the precision in electronic distance measurements? Explain.

*5-8 If an EDM instrument has a purported accuracy capability of $\pm(5 \text{ mm} + 5 \text{ ppm})$, what error can be expected in a measured distance of **(a)** 2500 ft **(b)** 1000 m **(c)** 10 mi?

*5-9 If a certain EDM instrument has an accuracy capability of $\pm(5 \text{ mm} + 3 \text{ ppm})$, what is the precision of measurements, in terms of $1/x$, for line lengths of: **(a)** 100 ft **(b)** 1000 ft **(c)** 10,000 ft?

5-10 To calibrate an EDM instrument, distances AC, AB, and BC along a straight line were measured as 4023.186 m, 1990.697 m, and 2032.535 m, respectively. What is the system measurement constant for this equipment? Compute the length of each segment corrected for the constant.

5-11 Same as Problem 5-10, except the distances are 4871.70 ft, 2410.55 ft, and 2461.21 ft, respectively.

5-12 Discuss the various systematic errors that affect electronic distance measurements.

5-13 Which causes a greater error in a line measured with an EDM instrument?
(a) A disregarded 5°C temperature variation from standard or **(b)** a neglected atmospheric pressure difference from standard of 5 mm of mercury?

*5-14 Same as Problem 5-13, except for **(a)** 10°F and **(b)** $\frac{1}{2}$ in. of mercury.

5-15 If the actual temperature and pressure at measurement time are assumed to be 15°C and 760 mm Hg, what temperature difference, in degrees Fahrenheit, will produce an error of 0.02 ft on a line determined with an EDM instrument to be: **(a)** 2500 ft **(b)** 2000 m?

*5-16 If the temperature and pressure at measurement time are 15°C and 760 mm Hg, what pressure difference, in inches of mercury, will produce an error of 0.01 ft on a line determined with an EDM instrument to be: **(a)** 1500 ft **(b)** 800 m?

5-17 If the temperature and pressure at measurement time are 15°C and 760 mm Hg, what will be the error in electronic measurement of a line 10 mi long if the temperature at the time of observing is recorded 5°C too low? Will the observed distance be too long or too short?

*5-18 For actual temperature and pressure of 15°C and 760 mm Hg, what neglected temperature difference from standard causes an EDM instrument measurement error of **(a)** 1 part in 250,000 **(b)** 1 part in 100,000?

*5-19 In Figure 5-10, h_e, h_r, elev$_A$, elev$_B$, and the measured slope length L were 5.35, 4.40, 825.75, 987.35, and 1438.29 ft, respectively. Calculate the horizontal length between A and B.

5-20 Similar to Problem 5-19, except that the values were 1.530, 1.675, 534.21, 686.29, and 1702.46 m, respectively.

5-21 In Figure 5-10, h_e, h_r, z_t, and the measured slope length L were 5.45 ft, 4.10 ft, 100°52′15″, and 875.13 ft, respectively. Calculate the horizontal length between A and B.

*5-22 Similar to Problem 5-21, except that the values were 1.40 m, 1.75 m, 80°35′24″, and 524.06 m, respectively.

*Asterisks indicate problems that have answers given in Appendix G.

✓ **5-23** In Figure 5-11, a zenith angle of 76°14'14" was measured. The EDM instrument was standard mounted and offset a distance of 0.75 ft vertically above the theodolite axis. If the theodolite and reflector heights were equal, what is the corrected horizontal distance for a recorded slope distance of 195.27 ft?

*5-24 Similar to Problem 5-23, except that the recorded zenith angle, vertical EDM instrument offset, and slope distance were 99°29'03", 20.0 cm, and 62.54 m, respectively.

5-25 Discuss the differences in theodolite mounting systems for EDM instruments and their effects on distance measurements and reduction procedures.

*Asterisks indicate problems that have answers given in Appendix G.

BIBLIOGRAPHY

Bell, T. P. 1978, "A Practical Approach to Electronic Distance Measurement." *Surveying and Mapping* 38 (No. 4):335.

Crisp, R. 1979. "Electronic Surveying and On-Site Recording of Geodetic Data." *Bulletin, American Congress on Surveying and Mapping* (No. 67):15.

Dracup, J. F., et al. 1979. *Surveying Instrumentation and Coordinate Computation Workshop Lecture Notes.* Falls Church, Va.: American Congress on Surveying and Mapping.

Greene, J. R. 1977. "Accuracy Evaluation in Electro-Optical Distance Measuring Instrument." *Surveying and Mapping* 37 (No. 3):247.

Kesler, J. M. 1973. "EDM Slope Reduction and Trigonometric Leveling." *Surveying and Mapping* 33 (No. 1):61.

Kivioja, L. A. 1978. "The EDM Corner Reflector Constant Is Not Constant." *Surveying and Mapping* 38 (No. 2):143.

Kivioja, L. A. 1983. "Reflector Constant and Refraction Index for EDM Corner Reflectors." *Surveying and Mapping* 43 (No. 4):399.

Reilly, J. P. 1990. "POB EDMI Survey 1990." *Point of Beginning* 16 (No. 2):34.

Romaniello, C. G. 1977. "EDM 1976." *Surveying and Mapping* 37 (No. 1):25.

Rueger, J. M. 1986. "Measurement and Computation of Short Periodic Errors of EDM Instruments." *Surveying and Mapping* 46 (No. 3):231.

Rueger, J. M. 1990. *Electronic Distance Measurement: An Introduction*, 3rd ed. New York: Springer-Verlag.

Saxena, N. K. 1975. "Electro-Optical Short Range Surveying Instruments." *ASCE, Journal of the Surveying and Mapping Division* 101 (No. SU1):17.

Stoughton, H. W. 1993. "Development and Application of Refractivity Correction: Formula for Optical and Infrared EDM Observations." *Surveying and Land Information Systems,* 53 (No. 2): 79.

Vogel, S. A. 1983. "National Geodetic Survey Calibration Base Line Program." *Bulletin, American Congress on Surveying and Mapping* (No. 86):19.

Witte, B. U., and W. Schwarz. 1982. "Calibration of Electro-Optical Rangefinders—Experience Gained and General Remarks Relative to Calibration." *Surveying and Mapping* 42 (No. 2):151.

6

LEVELING—THEORY, METHODS, EQUIPMENT

PART I
THEORY AND METHODS

6-1 INTRODUCTION

Leveling is the general term applied to any of the various processes by which elevations of points or differences in elevation are determined. It is a vital operation in producing necessary data for mapping, engineering design, and construction. Leveling results are used to (1) design highways, railroads, canals, sewers, water supply systems, and other facilities having grade lines that best conform to existing topography; (2) lay out construction projects according to planned elevations; (3) calculate volumes of earthwork and other materials; (4) investigate drainage characteristics of an area; (5) develop maps showing general ground configurations; and (6) study earth subsidence and crustal motion.

6-2 DEFINITIONS

Basic terms in leveling are defined in this section, some of which are illustrated in Figure 6-1.

  Vertical line. A line that follows the direction of gravity as indicated by a plumb line.
  Level surface. A curved surface that at every point is perpendicular to the local plumb line (the direction in which gravity acts). Level surfaces are approximately spheroidal in shape. A body of still water is the best example. Within local areas, level surfaces at different heights are considered to be concentric.[1]

---

[1]Due to flattening of the earth in the polar direction, level surfaces at different elevations are not truly concentric. This condition requires an *orthometric correction* for long north-south level

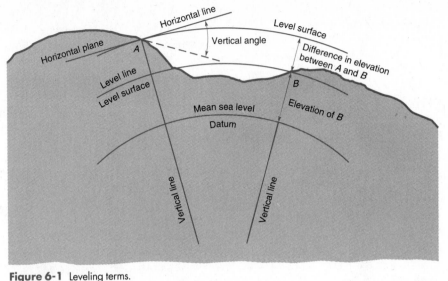

Figure 6-1 Leveling terms.

Level line. A line in a level surface—therefore, a curved line.

Horizontal plane. A plane perpendicular to the direction of gravity. In plane surveying, a plane perpendicular to the plumb line.

Horizontal line. A line in a horizontal plane. In plane surveying, a line perpendicular to the vertical.

Vertical datum. Any level surface to which elevations are referred (for example, mean sea level). Also called *vertical datum plane,* though not actually a plane.

Elevation. The vertical distance from a vertical datum, usually mean sea level, to a point or object. If the elevation of point *A* is 802.46 ft, *A* is 802.46 ft above some datum.

Mean sea level (MSL). The average height of the sea's surface for all stages of the tide over a 19-year period. It was arrived at from readings, usually taken at hourly intervals, at 26 gaging stations along the Atlantic and Pacific oceans and the Gulf of Mexico. The elevation of the sea differs from station to station depending on local influences of the tide; for example, at two points $\frac{1}{2}$ mi apart on opposite sides of an island in the Florida Keys, it varies by 0.3 ft. Therefore, to provide a common reference for elevations throughout North America, it was necessary to adopt a *mean sea level* (see Section 6-3). Scientists at the National Oceanic and Atmospheric Administration report that shrinking polar icecaps may have been causing the earth's sea level to rise at a rate slightly more than 0.1 in. a year since 1940, triple the rate of the preceding 50 years. If this phenomenon continues, low-lying coastal areas will be affected in future years.

Tidal datums. The vertical datums used in coastal areas for establishing property boundaries of lands bordering waters subject to tides. Tidal datums also provide

circuits in precise work. Its value, which is relatively small, is a function of the latitude and elevation of the level circuit. On a line of levels run from Seattle to Los Angeles, a corrrection of approximately 2 ft would be required.

the basis for locating fishing and oil drilling rights in tidal waters, and the limits of swamp and overflowed lands. Various definitions of tidal datums have been used in different areas, but the one most commonly employed is the *mean high water* (MHW) line. Others applied include *mean higher high water* (MHHW), *mean low water* (MLW), and *mean lower low water* (MLLW). Interpretations of tidal datums, and the methods by which they are determined, have been, and continue to be, the subject of numerous court cases.

Bench mark (BM). A relatively permanent object, natural or artificial, having a marked point whose elevation above or below an adopted datum is known or assumed. Common examples are metal disks set in concrete, reference marks chiseled on large rocks, nonmovable parts of fire hydrants, curbs, and so on.

Leveling. The process of finding elevations of points, or their difference in elevation.

Vertical control. A series of bench marks or other points of known elevation established throughout an area, also termed *basic control* or *level control*. The basic vertical control for the United States was derived from first- and second-order leveling. Less precise third-order leveling has been used to fill gaps between second-order bench marks, as well as for many other specific projects (see Chapter 19).

6-3 NORTH AMERICAN VERTICAL DATUM

Precise leveling operations to establish a distributed system of reference bench marks throughout the United States began in the 1850s. This work was initially concentrated along the eastern seaboard, but in 1887 the U.S. Coast and Geodetic Survey (USC&GS) began its first transcontinental leveling across the country's midsection. That project was completed in the early 1900s. By 1929 thousands of bench marks had been set. In that year the USC&GS began a general least-squares adjustment of all leveling thus far completed in the United States and Canada. The adjustment involved over 100,000 km of leveling and incorporated long-term data from the 26 tidal gaging stations; hence it was related to mean sea level. In fact, that network of bench marks with their resulting adjusted elevations defined the mean sea level datum. It was called the *National Geodetic Vertical Datum of 1929 (NGVD29)*.

Through the years after 1929, the NGVD29 deteriorated somewhat due to changes in sea level and shifting of the earth's crust. Also more than 625,000 km of additional leveling has been completed. To account for these changes and to incorporate the additional leveling, the National Geodetic Survey (NGS) has performed a new general readjustment. Work on this adjustment, which included more than 1.3 million observed elevation differences, began in 1978. Although not finished until 1991, its planned completion date was 1988, and thus it has been named the *North American Vertical Datum of 1988 (NAVD88)*. Besides the United States and Canada, Mexico was also included in this general readjustment. Bench mark elevations that were defined by the NGVD29 datum have changed by relatively small, but nevertheless significant amounts in the eastern half of the continental United States. The changes are much greater in the western part of the country, however, and reach 1.5 m in the Rocky Mountain region. It is therefore imperative that surveyors positively identify the datum to which their elevations are referred. Listings of the new elevations are available from the NGS.[2]

---

[2]Locations and NAVD88 elevations of bench marks can be obtained from the National Geodetic Information Center, NOAA, National Geodetic Survey, N/CG17, SSMC3 Station 09535, 1315 East West Highway, Silver Spring, MD 20910; telephone: (301) 713-3242.

6-4 CURVATURE AND REFRACTION

From the definitions of a level surface and a horizontal line, it is evident that the horizontal departs from a level surface because of *curvature* of the earth. In Figure 6-2 the deviation *DB* from a horizontal line through point *A* is expressed approximately by the formula

$$C_f = 0.667M^2 = 0.0239F^2 \tag{6-1a}$$

or

$$C_m = 0.0785K^2 \tag{6-1b}$$

where the departure of a level surface from a horizontal line is C_f in feet or C_m in meters, M is the distance AB in miles, F the distance in thousands of feet, and K the distance in kilometers.

Since points *A* and *B* are on a level line, they have the same elevation. If the line of sight were horizontal, the earth's curvature would cause a rod held at *B* to be read too high by length *BD*.

Light rays passing through the earth's atmosphere are bent or refracted toward the earth's surface, as shown in Figure 6-3. Thus a theoretically horizontal line of sight, like *AH* in Figure 6-2, is bent to the curved form *AR*. Hence the reading on a rod held at *R* is diminished by length *RH*.

The effect of refraction in making objects appear higher than they really are (and therefore rod readings too small) can be remembered by noting what happens when the sun is on the horizon, as in Figure 6-3. At the moment when the sun has just passed *below* the horizon, it is seen just *above* the horizon. The sun's diameter of approximately 32 min is roughly equal to the average refraction on a horizontal sight.

Displacement resulting from refraction is variable. It depends on atmospheric conditions, length of line, and the angle a sight line makes with the vertical. For a horizontal sight, refraction R_f in feet or R_m in meters is expressed approximately by the formula

$$R_f = 0.093M^2 = 0.0033F^2 \tag{6-2a}$$

or

$$R_m = 0.011K^2 \tag{6-2b}$$

Figure 6-2 Curvature and refraction.

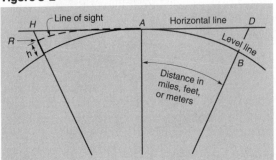

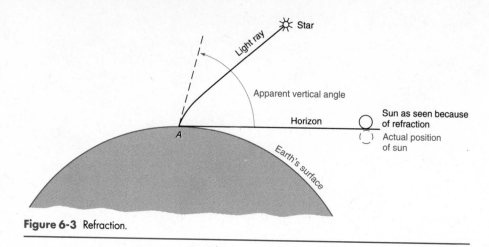

Figure 6-3 Refraction.

This is about one-seventh the effect of curvature of the earth, but in the opposite direction.

The combined effect of curvature and refraction, h in Figure 6-2, is approximately

$$h_f = 0.574M^2 = 0.0206F^2 \tag{6-3a}$$

or

$$h_m = 0.0675K^2 \tag{6-3b}$$

where h_f is in feet and h_m in meters.

For sights of 100, 200, and 300 ft, h_f = 0.00021 ft, 0.00082 ft, and 0.0019 ft, respectively, or 0.00067 m for a 100-m length. It will be explained in Section 7-4 that, although the combined effects of curvature and refraction produce rod readings that are slightly too large, proper field procedures can practically eliminate the error due to these causes.

6-5 METHODS TO DETERMINE DIFFERENCES IN ELEVATION

Differences in elevation have traditionally been determined by taping, differential leveling, barometric leveling, and indirectly by trigonometric leveling. Brief descriptions of these methods follow. Newer techniques, described in Chapter 20, utilize satellite and inertial systems. Elevation differences can also be determined using photogrammetry, as discussed in Chapter 28.

6-5.1 Taping Method

Application of a tape to a vertical line between two points is sometimes possible. This method is used to measure depths of mine shafts, to determine floor elevations in condominium surveys, and in the layout and construction of multistory buildings, pipelines, and so forth. When a water or sewer line is being laid, a graduated pole or rod may replace the tape (see Section 24-4).

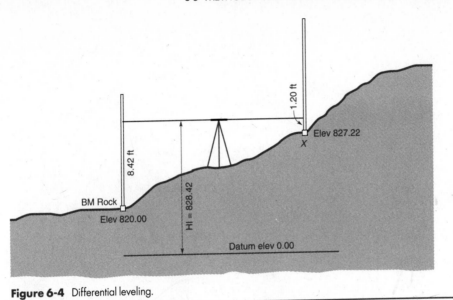

Figure 6-4 Differential leveling.

6-5.2 Differential Leveling

In this most commonly employed method, a telescope with suitable magnification is used to read graduated rods held on fixed points. A horizontal line of sight within the telescope is established by means of a level vial or automatic compensator.

The basic procedure is illustrated in Figure 6-4. An instrument is set up approximately halfway between BM Rock and point X. Assume the elevation of BM Rock is known to be 820.00 ft. After leveling the instrument, a plus sight taken on a rod held on the BM gives a reading of 8.42 ft. A *plus sight* ($+S$), also termed *backsight* (BS), is the reading on a rod held on a point of known or assumed elevation. This reading is used to compute the *height of instrument* (HI), defined as the vertical distance from datum to the instrument line of sight. Direction of the sight—whether forward, backward, or sideways—is not important. The term *plus sight* is preferable to *backsight,* but both are used. Adding the plus sight 8.42 ft to the elevation of BM Rock, 820.00, gives an HI of 828.42 ft.

If the telescope is then turned to bring into view a rod held on point X, a *minus sight* ($-S$), also called *foresight* (FS), is obtained. In this example it is 1.20 ft. A minus sight is defined as the rod reading on a point whose elevation is desired. The term *minus sight* is preferable to *foresight*. Subtracting the minus sight, 1.20 ft, from the HI, 828.42, gives the elevation of point X as 827.22 ft.

Differential leveling theory and applications can thus be expressed by two equations, which are repeated over and over,

$$\text{HI} = \text{elev} + \text{BS} \tag{6-4}$$

and

$$\text{elev} = \text{HI} - \text{FS} \tag{6-5}$$

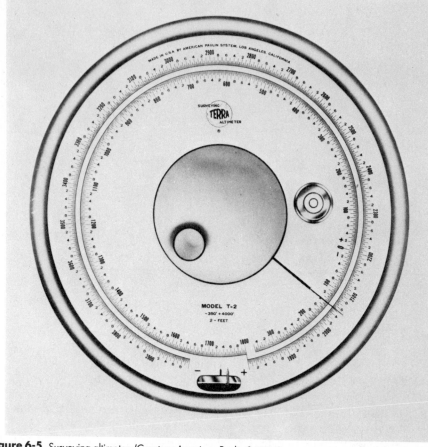

Figure 6-5 Surveying altimeter. (Courtesy American Paulin System.)

Since differential leveling is by far the most commonly used method to determine differences in elevation, it will be discussed in detail in Chapter 7.

6-5.3 Barometric Leveling

The barometer, an instrument that measures air pressure, can be used to find relative elevations of points on the earth's surface. Figure 6-5 shows a surveying altimeter. Calibration of the scale on different models is in multiples of 1 or 2 ft, $\frac{1}{2}$ or 1 m.

Air pressures are affected by circumstances other than difference in elevation, such as sudden shifts in temperature and changing weather conditions due to storms. Also, during each day a normal variation in barometric pressure amounting to perhaps a 100-ft difference in elevation occurs. This variation is known as the *diurnal range*.

In barometric leveling, various techniques can be used to obtain correct elevation differences in spite of pressure changes that result from weather variations. In one of these, a *control* barometer remains on a bench mark (base) while a *roving* instrument is taken to points whose elevations are desired. Readings are made on the base at stated intervals of time, perhaps every 10 min, and the elevations recorded along with temperature and time. Elevation, temperature, and time readings with the roving barometer are

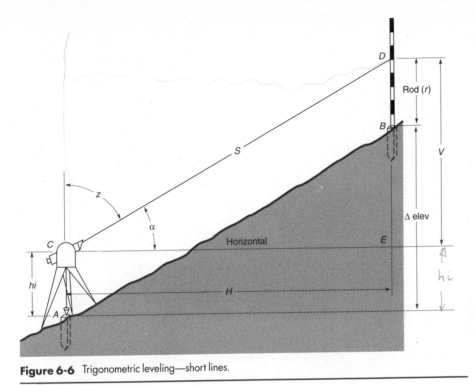

Figure 6-6 Trigonometric leveling—short lines.

taken at critical points and adjusted later in accordance with changes observed at the control points. Methods of making field surveys by barometer have been developed in which one, two, or three bases may be used. Other methods employ leapfrog or semi-leapfrog techniques. In stable weather conditions, and by using several barometers, elevations correct to within ±2 to 3 ft are possible.

Barometers have been used in the past for work in rough country where extensive areas had to be covered but a high order of accuracy was not required. They are seldom used today, however, having given way to other more modern and accurate equipment.

6-5.4 Trigonometric Leveling

The difference in elevation between two points can be determined by measuring (**1**) the inclined or horizontal distance between them and (**2**) the *zenith angle* or the *vertical angle* to one point from the other. (Zenith and vertical angles, described in more detail in Section 11-10, are measured in a vertical plane. Zenith angles are measured downward from vertical, and vertical angles are measured up or down from horizontal.) Thus in Figure 6-6 if slope distance S and zenith angle z or vertical angle α between C and D are measured, then V, the elevation difference between C and D, is

$$V = S \cos z \qquad (6\text{-}6)$$

or

$$V = S \sin \alpha \qquad (6\text{-}7)$$

Alternatively, if horizontal distance H between C and D is measured, then V is

$$V = H \cot z \tag{6-8}$$

or

$$V = H \tan \alpha \tag{6-9}$$

The difference in elevation (Δ elev) between points A and B in Figure 6-6 is given by

$$\Delta \text{ elev} = hi + V - r \tag{6-10}$$

where hi is the height of the instrument above point A and r is the reading on the rod held at B when zenith angle z or vertical angle α is read. If r is made equal to hi, then these two values cancel in Eq. (6-10) and simplify the computations.

Note the distinction in this text between HI and hi. Although both are called *height of instrument*, as described in Section 6-5.2, *HI is the elevation of the instrument above datum*, while *hi is the height of the instrument above an occupied point.*

For short lines (up to about 1000 ft in length) elevation differences obtained in trigonometric leveling are appropriately depicted by Figure 6-6 and properly computed using Eqs. (6-6) through (6-10). For longer lines, however, earth curvature and refraction become factors that must be considered. Figure 6-7 illustrates the situation. Here an instrument is set up at C over point A. Sight D is made on a rod held at point B, and zenith angle z_m or vertical angle α_m is measured. The true difference in elevation (Δ elev) between A and B is vertical distance HB between level lines through A and B, which is equal to $HG + GF + V - ED - r$. Since HG is the instrument height hi, GF is earth curvature C [see Eqs. (6-1)], and ED is refraction R [see Eqs. (6-2)], the elevation difference can be written as

$$\Delta \text{ elev} = hi + V + (C - R) - r \tag{6-11}$$

The value of V in Eq. (6-11) is obtained using one of Eqs. (6-6) through (6-9), depending on what quantities are measured. Again if r is made equal to hi, these values cancel. Also, the term $(C - R)$ is given by Eqs. (6-3). Thus, except for the addition of the curvature and refraction correction, long and short sights may be treated the same in trigonometric leveling computations. Note that in developing Eq. (6-11), angle F in triangle CFE was assumed to be 90°. Of course as lines become extremely long, this assumption does not hold. However, for lengths within the practical range, errors caused by this assumption are negligible.

The hi used in Eq. (6-11) can be obtained by simply measuring the vertical distance from the occupied point up to the instrument's horizontal axis (axis about which the telescope rotates) using a graduated rule or rod. An alternate method can be used to determine the elevation of a point that produces accurate results and does not require measurement of the hi. In this procedure, the instrument is set up at a location where it is approximately equidistant from a point of known elevation (bench mark) and the one whose elevation is to be determined. The slope distance and zenith (or vertical) angle are measured to each point. Because the distances from the two points are approxi-

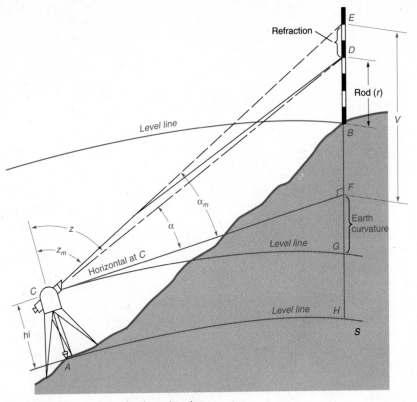

Figure 6-7 Trigonometric leveling—long lines.

mately equal, curvature and refraction errors cancel. Also, since the same instrument setup applies to both readings, the hi values cancel, and if the same rod reading r is sighted when making both angle readings, they cancel. Thus the elevation of the unknown point is simply the bench mark elevation, minus V calculated for the bench mark, plus V computed for the unknown point, where the V values are obtained using either Eq. (6-6) or (6-7).

EXAMPLE 6-1

The slope distance and zenith angle between points A and B were measured with a total station instrument as 9585.26 ft and 81°42′20″, respectively. The hi and rod reading r were equal. If the elevation of A is 1238.42 ft above mean sea level, compute the elevation of B.

SOLUTION
By Eq. (6-3a), the curvature and refraction correction is

$$h_f = 0.0206 \left[\frac{9585.26 \times \sin 81°42′20″}{1000} \right]^2 = 1.85 \text{ ft}$$

(Theoretically the horizontal distance should be used in computing curvature and refraction. In practice it is approximated by multiplying the slope distance by the sine of the zenith angle.)

By Eqs. (6-6) and (6-11), the elevation difference is (note that hi and r cancel)

$$V = 9585.26 \cos 81°42'20'' = 1382.77 \text{ ft}$$

$$\Delta \text{ elev} = 1382.77 + 1.85 = 1384.62 \text{ ft}$$

Finally, the elevation of B is

$$\text{elev}_B = 1238.42 + 1384.62 = 2623.04 \text{ ft above mean sea level}$$

In this calculation note that if curvature and refraction had been ignored, an error of 1.85 ft would have resulted in the elevation for B. Although Eq. (6-11) was derived for an uphill sight, it is also applicable to downhill sights. In that case the algebraic sign of V obtained in Eqs. (6-6) through (6-9) will be negative, however, because α will be negative or z greater than 90°.

For uphill sights curvature and refraction is added to a positive V to increase the elevation difference. For downhill sights, it is again added, but to a negative V, which decreases the elevation difference. Therefore, if "reciprocal" zenith (or vertical) angles are read (measuring the angles from both ends of a line), and V is computed for each and averaged, the effects of curvature and refraction cancel. Alternatively, the curvature and refraction correction can be completely ignored if one calculation of V is made using the average of the reciprocal angles. This assumes atmospheric conditions remain constant, so that refraction is equal for both angles. Hence they should be measured within as short a time period as possible. This method is preferred to reading the zenith (or vertical) angle from one end of the line and correcting for curvature and refraction, as in Example 6-1. The reason is that Eqs. (6-3) assume a standard atmosphere, which may not actually exist at the time of measurement.

EXAMPLE 6-2   For Example 6-1, assume that at B the slope distance was measured again as 9585.26 ft and the zenith angle was read as 98°19'06''. The instrument height and r were equal. Compute **(a)** the elevation difference from this end of the line, and **(b)** the elevation difference using the mean of reciprocal angles.

SOLUTION

(a) By Eq. (6-3a), $h_f = 1.85$ (the same as for Example 6-1).
By Eqs. (6-6) and (6-11) (note that hi and r cancel),

$$\Delta \text{ elev} = 9585.26 \cos 98°19'06'' + 1.85 = -1384.88 \text{ ft}$$

Note that this disagrees with the value of Example 6-1 by 0.26 ft. (The sight from B to A was downhill, hence the negative sign.) The difference of 0.26 ft is most probably due partly to measurement errors and partly to refraction changes that occurred during the time interval between vertical angle measurements. The average elevation difference for the two ends is 1384.75 ft.

(b) The average zenith angle is $\dfrac{81°42'20'' + 180° - 98°19'06''}{2} = 81°41'37''$.

By Eq. (6-10), Δ elev $= 9585.26 \cos 81°41'37'' = 1384.75$ ft.

Note that this checks the average value obtained using the curvature and refraction correction.

With the advent of EDM equipment, and more recently total station instruments, trigonometric leveling has become an increasingly common method for rapid and convenient measurement of elevation differences. It is used for topographic mapping, construction stakeout, control surveys, and other tasks. It is particularly valuable in rugged terrain. In trigonometric leveling, accurate zenith (or vertical) angle measurements are critical. For precise work a 1″ to 3″ theodolite is recommended and angles should be read direct and reversed from both ends of a line. Also, errors due to uncertainties in refraction are mitigated if sight lengths are limited to about 1000 ft.

PART II
EQUIPMENT FOR DIFFERENTIAL LEVELING

6-6 CATEGORIES OF LEVELS

Instruments used for differential leveling can be classified into three categories: *dumpy levels, tilting levels,* and *automatic levels.* Although each differs somewhat in design, all have two common components: **(1)** a *telescope* to create a line of sight and enable an operator to view a graduated rod superimposed on a reticle, and **(2)** a system to orient the line of sight in a horizontal plane. Dumpy and tilting levels use *level vials* to orient the line of sight; automatic levels employ so-called *automatic compensators.*

Instruments in each of these three categories are described in this chapter. *Hand levels,* although not commonly used for differential leveling, have many special uses where rough elevation differences over short distances are needed. They are also discussed in this chapter. Transits and theodolites can also be used for differential leveling. These instruments are described in Chapter 10, and their use as levels is explained in Section 11-17.

Electronic levels that transmit beams of either visible laser or invisible infrared light are another category of leveling instruments. They are not commonly employed in differential leveling, but are used extensively for establishing elevations on construction projects. They are described in Chapter 24.

6-7 TELESCOPES

The telescopes of leveling instruments define the line of sight and magnify the view of a graduated rod against a reference reticle, thereby enabling accurate readings to be obtained. The dumpy level telescope illustrated in Figure 6-8 is representative of most leveling instruments. Contained within the metal cylindrical tube are four main parts: objective lens, negative lens, reticle, and eyepiece.

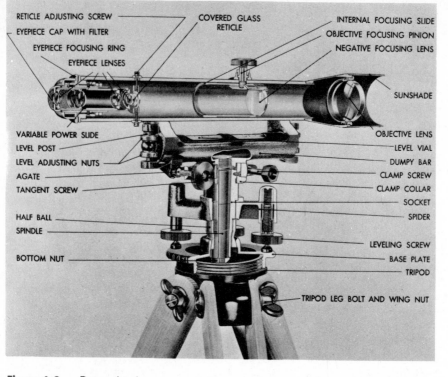

Figure 6-8   Dumpy level.

Objective Lens. This compound lens, securely mounted in the main tube's object end, has its optical axis reasonably concentric with the tube axis. Its main function is to gather incoming light rays and direct them toward the negative focusing lens.

Negative Lens. The negative lens is located between the objective lens and reticle. It should be mounted in a sliding tube so its optical axis coincides with that of the objective lens. Its function is to focus rays of light that pass through the objective lens onto the reticle plane.

Reticle. The reticle consists in a pair of perpendicular reference lines (usually called cross hairs) mounted at the principal focus of the objective optical system. The point of intersection of the cross hairs, together with the optical center of the objective system, forms the so-called *line of sight,* also sometimes called the *line of collimation.* The reticle is held in place in the main cylindrical tube by two pairs of opposing capstan screws located at right angles to each other, one pair horizontal and the other vertical, to facilitate adjusting the line of sight (see Appendix A). Cross hairs are fine lines etched on a thin round glass plate. Additional lines parallel to and equidistant from the primary lines are added to the reticle for special purposes such as for stadia (see Chapter 15). The reticle is mounted within the main telescope tube to place the lines in a horizontal-vertical orientation.

Eyepiece. The eyepiece is a microscope (usually with magnification from about 25 to 40×) for viewing the image focused by the objective lens system at the plane of the reticle.

Focusing is an important function to be performed in using a telescope. The process is governed by the fundamental principle of lenses stated in the following formula:

$$\frac{1}{f_1} + \frac{1}{f_2} = \frac{1}{f} \tag{6-12}$$

where f_1 is the distance from lens to image at the reticle plane, f_2 the distance from lens to object, and f the lens focal length. The focal length of any lens is a function of the radii of the ground spherical surfaces of the lens, and of the index of refraction of the glass from which it is ground. It is a constant for any particular single or compound lens. To focus for each varying f_2 distance, f_1 must be changed to maintain the equality of Eq. (6-12).

Focusing the telescope of a level is a two-stage process. First the eyepiece lens must be focused. Since the reticle's position in the telescope tube remains fixed, the distance between it and the eyepiece lens must be adjusted to suit the eye of an individual observer. This is done by bringing the cross hairs to a clear focus, that is, making them appear as black as possible when sighting at the sky or a distant, light-colored object. Once this has been accomplished, the adjustment need not be changed for the same observer, regardless of sight length, unless the eyes tire.

The second stage of focusing occurs after the eyepiece has been adjusted. Objects at varying distances from the telescope are brought to sharp focus at the plane of the cross hairs by turning the focusing knob. This moves the negative focusing lens to change f_1 and create the equality in Eq. (6-12) for varying f_2 distances.

After focusing, if the cross hairs appear to travel over the object sighted when the eye is shifted slightly in any direction, *parallax* exists. Either the objective lens, the eyepiece, or both must be refocused to eliminate this effect if accurate work is to be done.

6-8 LEVEL VIALS

Level vials are used to orient many different surveying instruments with respect to the direction of gravity. There are two basic types: the *tube* vial and the *circular* or so-called "bull's-eye" version. On dumpy and tilting levels, tube vials are used to precisely orient the line of sight horizontal prior to making rod readings. Bull's-eye vials are used on tilting and automatic levels for quick rough leveling. The principles of both types of vials are identical.

A tube level is a glass tube manufactured so that its upper inside surface precisely conforms to an arc of a given radius (see Figure 6-9). The tube is sealed at both ends, and except for a small air bubble, it is filled with a sensitive liquid. The liquid must be nonfreezing, quick-acting, and maintain a bubble of relatively stable length for normal temperature variations. Purified synthetic alcohol is generally used. As the tube is tilted, the bubble moves, always to the highest point in the tube because air is lighter than the liquid. Uniformly spaced graduations etched on the tube's exterior surface, and spaced 2 mm apart, locate the bubble's relative position.

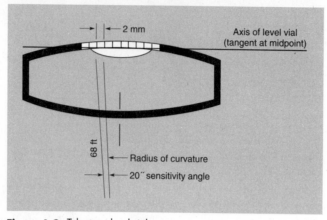

Figure 6-9 Tube-type level vial.

The *axis* of a level vial is an imaginary longitudinal line tangent to the upper inside surface at its midpoint. When the bubble is centered in its run, the axis should be a horizontal line, as in Figure 6-9. For a leveling instrument that uses a level vial, if it is in proper adjustment, its line of sight is parallel to its level vial axis. Thus by centering the bubble, the line of sight is made horizontal.

The *sensitivity* of a level vial is determined by its radius of curvature established in grinding. The larger the radius, the more sensitive a bubble. A highly sensitive bubble, necessary for precise work, may be a handicap in rough surveys because more time is required to center it.

A properly designed level has a vial sensitivity correlated with the *resolving power* (resolution) of its telescope. A slight movement of the bubble should be accompanied by a small but discernible change in the observed rod reading at a distance of about 200 ft. Sensitivity of a level vial is expressed in two ways: by **(1)** the angle, in seconds, subtended by one division on the scale, and **(2)** the radius of the tube's curvature. If one division subtends an angle of 20″ at the center, it is called a 20″ bubble. A 20″ bubble on a vial with 2-mm division spacings has a radius of approximately 68 ft.[3] The sensi-

[3]The relationship between sensitivity and radius is readily determined. In radian measure, an angle θ subtended by an arc whose radius and length are R and S, respectively, is given as

$$\theta = \frac{S}{R}$$

Thus for a 20″ bubble with 2-mm vial divisions, by substitution,

$$\frac{20''}{206,265''/\text{rad}} = \frac{2 \text{ mm}}{R}$$

Solving for R,

$$R = \frac{2 \text{ mm} \times 206265''/\text{rad}}{20'' \times 304.8 \text{ mm/ft}} = 67.7 \text{ ft}$$

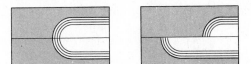

Figure 6-10 Coincidence-type level vial correctly set in left view; twice the deviation of the bubble shown in right view.

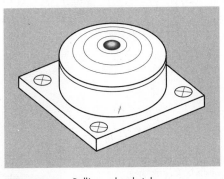

Figure 6-11 Bull's-eye level vial.

tivity of level vials on most current dumpy and tilting levels ranges from approximately 20 to 40″.

Figure 6-10 illustrates the *coincidence*-type tube level vial used on precise equipment. The bubble is centered by bringing the two ends together to form a smooth curve. A prism splits the image of the bubble and makes the two ends visible simultaneously. This arrangement enables bubble centering to be done more accurately.

Bull's-eye level vials are spherical in shape (see Figure 6-11), the inside surface of the sphere being precisely manufactured to a specific radius. Like the tube version, except for an air bubble, bull's-eye vials are filled with liquid. The vial is graduated with concentric circles having 2-mm spacings. Its axis is actually a plane tangent to the radius point of the graduated concentric circles. When the bubble is centered in the smallest circle, the axis should be horizontal.

Besides their use for rough leveling of tilting and automatic levels, bull's-eye vials are also used on EDM instruments, total stations, tribrachs, rod levels, prism poles, and many other surveying instruments. Their sensitivity is much lower than that of tube vials—generally in the range of from 2 to 25′ per 2-mm division.

6-9 DUMPY LEVEL

As shown in Figure 6-8, the telescope of a dumpy level is rigidly fastened to the *dumpy bar*, also called the *level bar*.[4] The bar is centered on an accurately machined vertical

---

[4]In an older type level similar to the dumpy, the telescope rested in supports on the level bar called *wyes*. The telescope could be removed, which facilitated adjusting the instrument. This type of level, called a *wye level*, is no longer made, but a few are still in use.

spindle that sits in a conical *socket* of the *leveling head*. The spindle makes the dumpy bar revolve in a horizontal plane when the instrument is properly adjusted.

A dumpy level has its level vial (tube type) set in the dumpy bar (see Figure 6-8) and is thereby somewhat protected. The vial always remains in the same vertical plane as the telescope, but screws at each end permit vertical adjustment (see Appendix A) or replacement of the vial.

The conical socket into which the vertical spindle of the dumpy bar fits is carried by four large *leveling screws*. They rest on a plate fastened to the tripod top and are in two pairs at right angles to each other. The leveling screws are used to center the bubble in leveling the instrument. (This leveling process is described in Section 7-2.) For an instrument in proper adjustment, after it has been leveled, the line of sight generates a horizontal plane as the telescope is revolved about its vertical axis.

6-10 TILTING LEVEL

Tilting levels are used for the most precise work, and are also widely employed for general purposes. With these instruments, quick approximate leveling is achieved using a bull's-eye bubble and the leveling screws. On some tilting levels, a ball-and-socket arrangement (with no leveling screws) permits the head to be tilted and quickly locked nearly level. Precise level in preparation for readings is then obtained by carefully centering a telescope bubble. This is done for each sight, after aiming at the rod, by tilting or rotating the telescope slightly in a vertical plane about a fulcrum at the vertical axis of the instrument. A micrometer screw under the eyepiece controls this movement.

The tilting feature saves time and increases accuracy, since only one screw need be manipulated to keep the line of sight horizontal as the telescope is turned about a vertical axis. The telescope bubble is viewed through a system of prisms from the observer's normal position behind the eyepiece. A prism arrangement splits the bubble image into two parts. Centering the bubble is accomplished by making the images of the two ends coincide, as in Figure 6-10.

The tilting level shown in Figure 6-12 has a four-screw leveling head, 30× magnification, and sensitivity of the level vial equal to 20″/2 mm. Figure 6-13 shows a tilting level with a ball-and-socket leveling head. It is characterized by a short telescope, streamlined construction, small size, and light weight. Its telescope magnification is also 30×, and its level vial sensitivity 18″/2 mm.

6-11 AUTOMATIC LEVELS

Automatic levels of the type pictured in Figure 6-14 incorporate a self-leveling feature. With most of these instruments, rough leveling using a three-screw leveling head approximately centers a bull's-eye bubble, although some models have a ball-and-socket arrangement. After the bull's-eye bubble is manually centered, an *automatic compensator* takes over, levels the line of sight, and keeps it level.

The operating principle of the automatic compensator used in the instrument of Figure 6-14 is shown schematically in Figure 6-15. The system consists of prisms suspended from wires to create a pendulum. The wire lengths, support locations, and nature of the prisms are such that only horizontal rays reach the intersection of cross hairs. Thus a horizontal line of sight is achieved even though the telescope itself may be slightly tilted away from horizontal. Damping devices shorten the time for the pendulum to come to rest, so the operator does not have to wait.

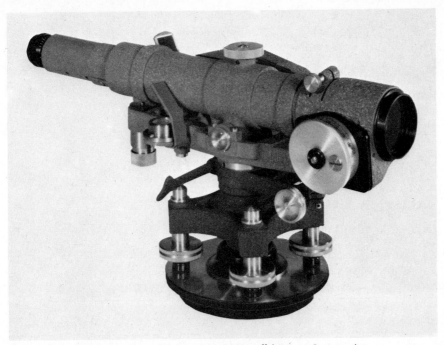

Figure 6-12 Tilting level with micrometer. (Courtesy Keuffel & Esser Company.)

Figure 6-13 Kern GK 23-C tilting level. (Courtesy Leica, Inc.)

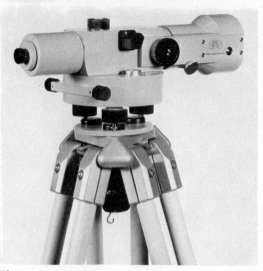

Figure 6-14 Zeiss automatic Ni 2 level with micrometer. (Courtesy Carl Zeiss, Inc.)

Figure 6-15 Compensator of self-leveling level. (Courtesy Keuffel & Esser Company.)

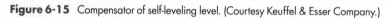

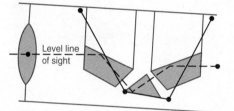

When telescope tilts up, compensator swings backward.

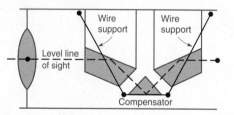

Telescope horizontal

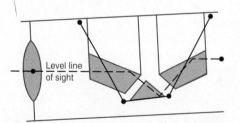

When telescope tilts down, compensator swings forward.

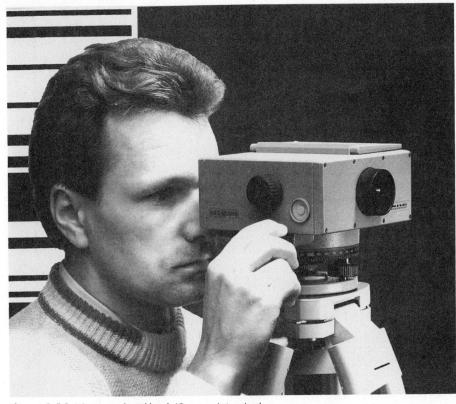

Figure 6-16 Electronic digital level. (Courtesy Leica, Inc.)

Automatic levels have become popular for general use because of the ease and rapidity of their operation. Some are precise enough for second-order and even first-order work if a *parallel-plate micrometer* is attached to the telescope front as an accessory. When the micrometer plate is tilted, the line of sight is displaced parallel to itself, and decimal parts of rod graduations can be measured by means of a graduated dial.

Under certain internal and environmental conditions, the damping devices of an automatic level compensator can stick. To check, with the instrument leveled and focused, it is necessary to read the target, tap the tripod, and after it vibrates, determine whether the same reading is obtained. Also, some unique compensator problems, such as residual stresses in the flexible links, can introduce systematic errors if not corrected by an appropriate observational routine on first-order work. Another recently discovered problem is that some automatic compensators are affected by magnetic fields which result in systematic errors in rod readings. The sizes of the errors are azimuth dependent, maximum for lines run north and south, and can exceed 1 mm/km. Thus it is of concern only for high-order control leveling.

The newest type of automatic level, the *electronic digital level,* is pictured in Figure 6-16. It is classified in the automatic category because it uses a pendulum compensator to level itself, after an operator accomplishes rough leveling with a bull's-eye bubble. With its telescope and cross hairs, the instrument could be used to obtain readings manually, just like any of the automatic levels. This instrument, however, is designed to operate by employing electronic digital image processing. After leveling the instrument, its telescope is turned toward a special bar-coded rod (also shown in Figure 6-16) and focused. At the press of a button, the image of bar codes in the telescope's field of view is captured and processed. This processing consists of an on-board computer comparing the captured image to the rod's entire pattern, which is stored in memory. When a match is found, which takes about 4 sec, the rod reading is displayed digitally. It can be recorded manually or automatically stored in the instrument's data collector.

The length of rod appearing within the telescope's field of view is a function of the distance from the rod. Thus as a part of its image processing, the instrument is also able to automatically compute the sight length, a feature convenient for balancing backsights and foresights. The instrument's maximum range is approximately 100 m, and its accuracy in rod readings is ± 0.5 mm. The bar-coded rod has either foot or meter graduations on the other side. Thus an operator can manually check any digitally displayed rod reading by turning the rod.

6-12 TRIPODS

Leveling instruments, whether dumpy, tilting, or automatic, are all mounted on tripods. A sturdy tripod in good condition is essential to obtain accurate results. Several types are available. The legs are made of wood or metal, may be fixed or adjustable in length, and solid or split. All models are shod with metallic conical points and hinged at the top, where they connect to a metal head. An adjustable-leg tripod is advantageous for setups in rough terrain or in a shop, but the type with a fixed-length leg may be slightly more rigid. The split-leg model is lighter than the solid type, but less rugged.

6-13 HAND LEVEL

The hand level (Figure 6-17) is a hand-held instrument used on low-precision work, or to obtain quick checks on more precise work. It consists of a brass tube approximately 6 in. long, having a plain glass objective and peep-sight eyepiece. A small level vial mounted above a slot in the tube is viewed through the eyepiece by means of a prism or 45° angle mirror. A horizontal line extends across the tube's center.

Figure 6-17 Hand level.

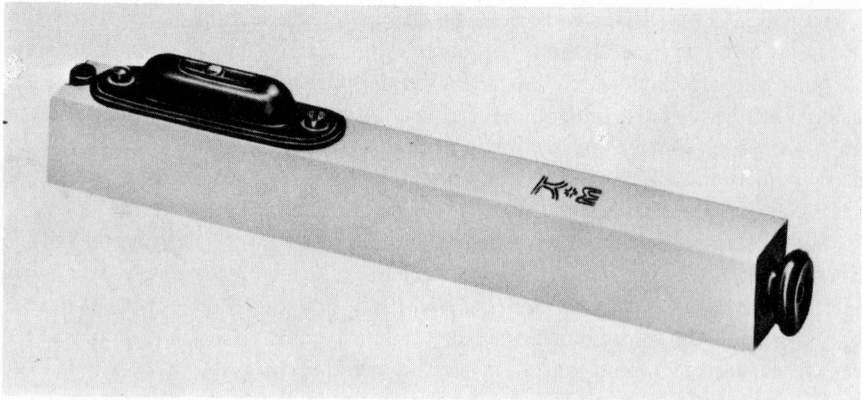

Figure 6-18 Abney hand level and clinometer.

The prism or mirror occupies only half the inside of the tube, and the other part is open to provide a clear sight through the objective. Thus the rod being sighted, and the reflected image of the bubble, are visible beside each other with the horizontal cross line superimposed.

The instrument is held in one hand and leveled by raising or lowering the objective end until the cross line bisects the bubble. Resting the level against a rod or staff provides stability and increases accuracy. The Abney hand level and clinometer, shown in Figure 6-18, functions as a hand level and also permits rough measurements of vertical angles and slopes.

6-14 LEVEL RODS

A variety of level rods are available, some of which are shown in Figure 6-19. They are made of wood, fiberglass, or metal and have graduations in feet and decimals, or meters and decimals.

A wide choice of patterns, colors, and graduations on single-piece, two-piece, and three-section leveling rods is available. The various types, usually named for cities or states, include the Philadelphia, New York, Boston, Troy, Chicago, San Francisco, and Florida rods.

The Philadelphia rod, shown in Figure 6-19(a) and (b), is the type most commonly used in college surveying classes. It consists of two sliding sections graduated in hundredths of a foot and joined by brass sleeves *a* and *b*. The rear section can be locked in position by a clamp screw *c* to provide any length from a *short rod* for readings of 7 ft or less to a *long rod (high rod)* for readings up to 13 ft. *When the high rod is needed, it must be fully extended, otherwise a serious mistake will result in the reading.* Graduations on the front faces of the two sections read continuously from zero at the base to 13 ft at the top for the high-rod setting.

Rod graduations are accurately painted, alternate black and white spaces 0.01 ft wide. The 0.1- and 0.05-ft marks are emphasized by spurs extending the black painting. Tenths are designated by black figures, and foot marks by red numbers, all straddling the proper graduation. Rodpersons should keep their hands off the painted markings, particularly in the 3- to 5-ft section, where a worn face will make the rod unfit for use. A Philadelphia rod can be read accurately with a level at distances up to about 250 ft.

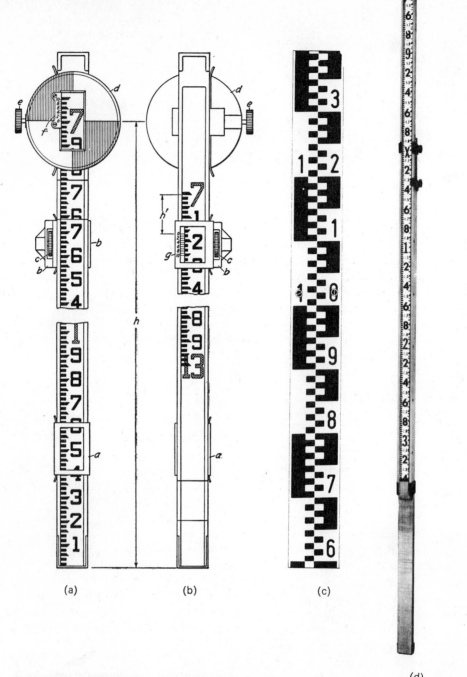

Figure 6-19 (a) Philadelphia rod (front). (b) Philadelphia rod (rear). (c) Leveling rod with metric graduations. (Courtesy Leica, Inc.) (d) Lenker direct-reading rod. (Courtesy Lenker Manufacturing Company.)

Philadelphia rods can be equipped with targets [d in Figure 6-19(a) and (b)] for use on long sights. When employed, the rodperson sets the target at the instrument's line-of-sight height according to hand signals from the instrument operator. It is fixed using clamp e, then read and recorded by the rodperson. The *vernier* at f (see Section 10-4), can be used to obtain readings to the nearest 0.001 ft if desired.

The Chicago rod, consisting of independent sections (usually three) which fit together but can be disassembled, is widely used on construction surveys. The San Francisco model has separate sections that slide past each other to extend or compress its length, and is generally employed on control, land, and other surveys. Both are conveniently transported in vehicles.

The direct-reading Lenker level rod [Figure 6-19(d)] has numbers in reverse order on an endless graduated steel-band strip that can be revolved on the rod's end rollers. Figures run down the rod and can be brought to a desired reading—for example, the elevation of a bench mark. Rod readings are preset for the backsight, and then, due to the reverse order of numbers, foresight readings give elevations directly without calculating HIs and subtracting FSs.

A rod consisting of a wooden frame and an Invar strip to eliminate the effects of humidity and temperature changes is used on precise work. The Invar strip, attached at its ends only, is free to slide in grooves on each side of the wooden frame. Rods for precise work are often graduated in meters on one side and in feet on the other. Readings of both sides are compared to eliminate mistakes.

Safety in traffic and near heavy equipment is an important consideration. The Quadpod, an adjustable stand that clamps to any leveling rod, reduces traffic hazards and labor costs.

PROBLEMS

6-1 How far will a horizontal line depart from the earth's curved surface in $\frac{1}{2}$, 1, 5, and 10 mi?

6-2 Why is it important for a bench mark to be a stable, relatively permanent object?

*6-3 Compute and tabulate the combined effect of curvature and refraction on level sights of 50, 100, 200, 500, and 1000 ft.

6-4 Similar to Problem 6-3, except for sights of 25, 50, 100, 150, 200, and 500 m.

6-5 On a large lake without waves, how far from shore is a sailboat when the top of its 40-ft mast disappears from the view of a person lying at the water's edge?

*6-6 Similar to Problem 6-5, except for a 15-m mast and a person whose eye height is 1.6 m above the water's edge.

6-7 Readings on a line of differential levels are taken to the nearest 0.01 ft. For what maximum distance can the earth's curvature and refraction be neglected?

*6-8 Similar to Problem 6-7, except readings are to the nearest millimeter.

Successive plus and minus sights taken on a downhill line of levels are listed in Problems 6-9 and 6-10. What error results from curvature and refraction?

6-9 125, 200; 100, 210; 90, 195; 100, 250 ft

6-10 35, 60; 25, 55; 30, 75; 20, 60 m

*6-11 What error results if the curvature and refraction correction is neglected in trigonometric leveling for sights: (a) 2500 ft long (b) 5000 ft long (c) $1\frac{1}{2}$ mi long?

*Asterisks indicate problems that have answers given in Appendix G.

6-12 The slope distance and zenith angle measured from point P to point Q were 13,608.21 ft and 96°17'18", respectively. The instrument and rod target heights were equal. If the elevation of point P is 2574.86 ft above mean sea level, what is the elevation of point Q?

6-13 The slope distance and vertical angle measured from point X to point Y were 16,789.50 ft and +7°28'50". The instrument and rod target heights were equal. If the elevation of point X is 3208.57 ft above mean sea level, what is the elevation of point Y?

6-14 Similar to Problem 6-12, except the slope distance was 6,976.32 m, the zenith angle was 85°19'52", and the elevation of point P was 975.572 m above mean sea level.

*6-15 In trigonometric leveling from point A to point B, the slope distance and zenith angle measured at A were 23,051.82 ft and 83°41'16". At B these measurements were 23,051.85 ft and 96°21'31", respectively. If the instrument and rod target heights were equal, calculate the difference in elevation from A to B.

6-16 In trigonometric leveling from point C to point D, the slope distance and vertical angle measured at C were 13,347.88 ft and −6°39'24". At D these measurements were 13,347.86 ft and +6°37'53", respectively. If the instrument and rod target heights were equal, and the elevation of point C was 3629.05 ft above mean sea level, what is the elevation of point D?

6-17 What is the sensitivity of a level vial with 2-mm divisions for: **(a)** a radius of 13.8 m **(b)** a radius of 20.6 m?

6-18 What is the sensitivity of a level vial with 0.01-ft graduations for: **(a)** a 206-ft radius of curvature **(b)** a 52-ft radius of curvature?

6-19 With the bubble centered, a 300-ft-length sight gives a reading of 5.132 ft. After moving the bubble four divisions off center, the reading is 5.250 ft. For 2-mm vial divisions, what is: **(a)** the vial radius of curvature in feet **(b)** the angle in seconds subtended by one division?

6-20 Similar to Problem 6-19, except the sight length was 400 ft, the initial reading was 4.253 ft, and the final reading was 4.481 ft.

*6-21 An observer fails to check the bubble, and it is off 2 divisions on a 250-ft sight. What error results for a 20" bubble?

6-22 Similar to Problem 6-21, except a 30" bubble is off two divisions on a 40-m sight.

*6-23 Sunshine on the forward end of a 20"/2-mm level vial bubble draws it off one division, giving a 6.14-ft plus sight on a 175-ft shot. Compute the correct reading.

6-24 List in tabular form, for comparison, the advantages and disadvantages of a tilting level versus an automatic level.

6-25 A slope distance of 80 ft and 15% slope are read with an Abney level. Compute the horizontal distance and elevation difference.

*6-26 If a BS of 4.37 ft is taken on BM A, elevation 972.14 ft, and an FS of 7.86 ft is read on point X, calculate the HI and the elevation of point X.

6-27 Similar to Problem 6-26, except a BS of 9.62 ft is taken on BM A and an FS of 1.48 ft read on point X.

6-28 State the causes of parallax when using surveying equipment.

6-29 List four conditions that can cause properly set BM monuments to settle or move.

6-30 Compare the advantages of a three-screw leveling head and a four-screw arrangement.

6-31 Describe the manner by which rod readings are obtained with the new electronic digital level.

*Asterisks indicate problems that have answers given in Appendix G.

═══════════

BIBLIOGRAPHY   Berry, R. M. 1977. "Observational Techniques for Use with Compensator Leveling Instruments for First-Order Levels." *Surveying and Mapping* 37 (No. 1):15.

Briscoe, J. 1983. "The Use of Tidal Datums in the Law." *Surveying and Mapping* 43 (No. 2):115.

Caspary, W. F., and H. Heister. 1989. "A Partially Automated Microprocessor-Controlled Leveling System." *Surveying and Land Information Systems* 49 (No. 2):73.

Hintz, R. J., and D. W. Gibson. 1986. "Sine Curve Fitting of Tidal Data in Estimation of Mean High Water." *Surveying and Mapping* 46 (No. 3):201.

Holdahl, S. R. 1983. "The Correction for Leveling Refraction and Its Impact on Definition of the North American Vertical Datum." *Surveying and Mapping* 43 (No. 2):123.

Holdahl, S. R. 1986. "Correcting for Magnetic Error." *Bulletin, American Congress on Surveying and Mapping* (No. 103):31.

Holdahl, S. R. 1992. "A Dynamic Vertical Reference System." *Surveying and Land Information Systems* 52 (No. 2):92.

Huff, L. C. 1987. "Rapid Precision Leveling System." *Bulletin, American Congress on Surveying and Mapping* (No. 107):17.

Kaula, W. M. 1987. "The Need for Vertical Control." *Surveying and Mapping* 47 (No. 1):57.

Kim, R. R. 1984. "Accurate Trigonometric Elevations Are Easily Attainable." *Point of Beginning* 10 (No. 1):38.

Maddux, W. S. 1982. "Datum Extrapolation by Simultaneous Comparison of Partial Tidal Cycles." *Surveying and Mapping* 42 (No. 2):139.

Reilly, J. P. 1991. "1991 Level Survey." *Point of Beginning* 17 (No. 2):44.

Sliva, L. 1987. "Some Aspects of Bench Mark Stability." *Surveying and Mapping* 47 (No. 2):155.

Stoughton, H. W. 1984. "Historical Overview of Vertical Datums." *Professional Surveyor* 4 (No. 6):32.

Teskey, W. 1992. "Trigonometric Leveling in Precise Engineering Surveys." *Surveying and Land Information Systems* 52 (No. 1):46.

Vanicek, P. 1991. "Vertical Datum and NAVD88." *Surveying and Land Information Systems* 51 (No. 2):83.

Zilkoski, D., et al. 1992. "Results of the General Adjustment of the North American Vertical Datum of 1988." *Surveying and Land Information Systems* 52 (No. 3):133.

7
LEVELING—FIELD PROCEDURES AND COMPUTATIONS

7-1 INTRODUCTION

Chapter 6 covered the basic theory of leveling, briefly described the different procedures used in determining elevations, and showed examples of most types of leveling equipment. This chapter concentrates on differential leveling, and discusses handling the equipment, running and adjusting simple leveling loops, and performing some project surveys to obtain data for field and office use. Some special variations of differential leveling, useful or necessary in certain situations, are presented. *Profile leveling,* to determine the configuration of the ground surface along some established reference line, is discussed in Section 7-9. Finally errors in leveling are discussed. Leveling procedures for construction and other surveys, along with those of higher order to establish the nationwide vertical control network, will be covered in later chapters.

7-2 CARRYING AND SETTING UP A LEVEL

The safest way to transport a leveling instrument in a vehicle is to leave it in the container. The case closes properly only when the instrument is set correctly in the padded supports. A level should be removed from its container by lifting the level bar or base, *not* by grasping the telescope. The head must be screwed snugly on the tripod. If the head is too loose, the instrument is unstable; if too tight, it may "freeze."

The legs of a tripod must be tightened correctly. If each leg falls slowly of its own weight after being placed in a horizontal position, it is properly adjusted. Clamping them too tightly strains the plate and screws. If the legs are loose, a wobbly setup results.

It is generally unnecessary to set a level over any particular point. Therefore it is inexcusable to have the base plate badly out of level before using the leveling screws.

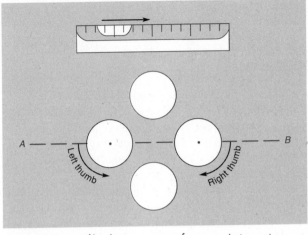

Figure 7-1 Use of leveling screws on a four-screw instrument.

On side-hill setups, placing one leg on the uphill side and two on the downhill slope eases the problem. On very steep slopes, some instrument operators prefer two legs uphill and one downhill for stability. The most convenient height of setup is one that enables the observer to sight through the telescope without stooping or stretching.

In leveling a four-screw head, the telescope is rotated until it is over two opposite screws as in the direction *AB* of Figure 7-1. The bubble is approximately centered by using the thumb and first finger of each hand to adjust the opposite screws simultaneously. This procedure is repeated with the telescope over the other two leveling screws. Time is wasted by centering the bubble exactly on the first try, since it will be thrown off during the cross leveling. Working with each pair of screws about three times should complete the job.

Leveling screws are turned in opposite directions at the same speed by both hands, unless the intention is to tighten or loosen the leveling head. A simple but useful rule in centering a bubble, illustrated in Figure 7-1, is: *A bubble follows the left thumb.*

If one hand turns faster than the other, the screws loosen, the head rocks on two screws, or they bind. Final precise adjustment can be made with one screw only. Leveling screws should be snug, not wrench tight, to save time and avoid damage to threads and base plate. A good observer senses the proper setting of all leveling screws to permit ready movement without jamming the threads. Instruments should be leveled on the base plate before being replaced in the case.

For automatic and tilting levels that have a three-screw head and bull's-eye bubble, the telescope is aligned over one screw and thus made perpendicular to a line through the other two. The bull's-eye bubble is centered by alternately turning one screw and then the other two. The telescope need not be rotated during the process.

Inexperienced instrument operators working on a steep hillside are likely to find, after completing the leveling process, that the telescope is too low for sighting the upper turning point or bench mark. To avoid this, a hand level can be used to check for proper height of the setup before precisely leveling the instrument. As another alternative, the instrument can be quickly set up without attempting to level it carefully. Then the rod is sighted making sure the bubble is somewhat back of center. If it is visible for this placement, it obviously will also be seen when the instrument is leveled.

7-3 DUTIES OF A RODPERSON

The duties of a rodperson are relatively simple. Like a tape handler, however, a rodperson can nullify the best efforts of an observer by failing to follow a few basic rules.

A level rod must be held plumb to give the correct reading. In Figure 7-2, point *A* is below the line of sight by vertical distance *AB*. If the rod is tilted to position *AD*, an erroneous reading *AE* is obtained. It can be seen that the smallest reading possible, *AB*, is the correct one and is secured only when the rod is plumb.

Waving the rod is a procedure used to ensure that the rod is plumb when a reading is taken. The process consists of *slowly* tilting the rod top, first toward the instrument and then away from it. The observer watches the readings alternately increase and decrease, and then selects the minimum value, the correct one. Beginners tend to swing the rod too fast and through too long an arc. Small errors can be introduced in the process if the bottom of the rod is resting on a flat surface. A rounded-top monument, steel spike, or thin edge makes an excellent benchmark or turning point.

On still days the rod can be plumbed by letting it balance of its own weight while lightly supported by the fingertips. An observer makes certain the rod is plumb in the lateral direction by checking its coincidence with the vertical wire and signals for any adjustment necessary. The rodperson can save time by sighting along the side of the rod to line it up with a telephone pole, tree, or side of a building. Plumbing along the line toward the instrument is more difficult, but holding the rod against the toes, stomach, and nose will bring it close to a plumb position.

A rod level of the type shown in Figure 7-3 ensures *fast* and *correct* rod plumbing. Its L shape is designed to fit the rear and side faces of a rod, while the bull's-eye bubble is centered to plumb the rod in *both directions*. A plumb bob suspended alongside the rod can also be used, and in this procedure the rod is adjusted in position until its edge is parallel with the string.

Figure 7-2 Plumbing a level rod.

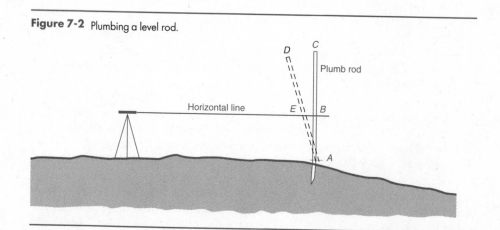

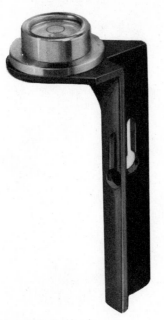

Figure 7-3 Rod level.

EXAMPLE 7-1

In Figure 7-2, if $AB = 10$ ft and $EB = 6$ in., what error results?

SOLUTION

Using the rightmost term of Eq. (4-3) in Section 4-13, the error e is

$$e = \frac{d^2}{2L} = \frac{0.5^2}{2 \times 10} = 0.012 \text{ ft, or } 0.01 \text{ ft}$$

Errors of this magnitude are serious, whether the results are carried out to hundredths or thousandths. They make careful plumbing necessary, particularly for high-rod readings.

7-4 DIFFERENTIAL LEVELING

Figure 7-4 illustrates the procedure followed in differential leveling. In running a circuit of differential levels, the first reading is always a plus sight taken on a bench mark. From it the *HI* can be computed using Eq. (6-4). Then a minus sight is taken on the first turning point, TP_1 in the figure, and by Eq. (6-5) its elevation is obtained. The process of taking a plus sight, followed by a minus sight, is repeated over and over until the circuit is completed.

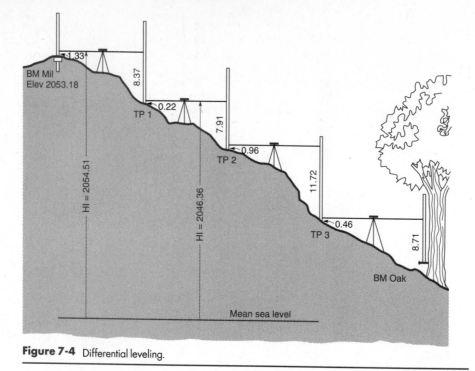

Figure 7-4 Differential leveling.

As shown in the example of Figure 7-4, four instrument setups were required to complete the run from BM Mil to BM Oak. A tabular form of field notes is used for differential leveling, and the addition and subtraction to compute HIs and elevations done directly in the notes. Field notes for the example of Figure 7-4 are given in Plate D-2 of Appendix D. These notes also show the data for the return run from BM Oak back to BM Mil to complete the circuit.

The intermediate points upon which the rod is held in running a line of differential levels are called *turning points* (TPs). Two rod readings are taken on each, a minus sight followed by a plus sight. Turning points should be solid objects with a definite high point.

In differential leveling, horizontal lengths for the plus and minus sights should be made approximately equal. This can be done by pacing, by stadia measurements (see Chapter 15), by counting rail lengths or pavement joints if working along a track or roadway, or by any other convenient method. Balancing plus and minus sights will eliminate errors due to instrument maladjustment (most important) and the combined effects of the earth's curvature and refraction, as shown in Figure 7-5. Here e_1 and e_2 are the combined curvature and refraction errors for the plus and minus sights, respectively. If D_1 and D_2 are made equal, e_1 and e_2 are also equal. In calculations e_1 is added and e_2 subtracted; thus they cancel each other.

Figure 7-5 can also be used to illustrate the importance of balancing sight lengths if a collimation error exists in the instrument's line of sight. This condition exists, if after leveling the instrument, its line of sight is not horizontal. Suppose in Figure 7-5 for example, that because the line of sight is systematically directed below horizontal, an error e_1 results in the plus sight. But if D_1 and D_2 are made equal, an error e_2 (equal to

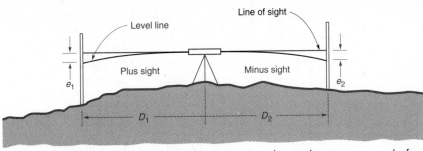

Figure 7-5 Balancing plus and minus sight distances to cancel errors due to curvature and refraction.

e_1) will result on the minus sight and the two will cancel, thus eliminating the effect of the instrumental error.

On slopes it may be somewhat difficult to balance lengths of plus and minus sights, but usually it can be done by following a zigzag path.

A bench mark is described in a field book the first time used, and thereafter it is referred to by noting the page number on which it was recorded. Descriptions begin with the general location and must include enough details to enable a person unfamiliar with the area to find the mark readily (see Appendix D, Plates D-2 and D-3). A bench mark is usually named for some prominent object it is on or near, to aid in describing its location; one word is preferable. Examples are BM River, BM Tower, BM Corner, and BM Bridge. On extensive surveys, bench marks are often numbered consecutively. Although an advantage in identifying relative positions along a line, this method is more subject to mistakes in field marking or recording.

Turning points are also numbered consecutively but not described in detail, since they are merely a means to an end and usually will not have to be relocated. If possible, however, it is advisable to select turning points that can be relocated, so if reruns on long lines are necessary because of blunders, field work can be reduced.

Before a party leaves the field, all possible note checks must be made to detect any mistakes in arithmetic and verify achievement of an acceptable closure. *The algebraic sum of the plus and minus sights applied to the first elevation should give the last elevation.* This computation checks the addition and subtraction for all HIs and turning points unless compensating mistakes have been made. When carried out for each left-hand page of tabulations, it is termed the *page check*. On Plate D-2, for example, note that the page check is secured by adding the sum of backsights, 40.24, to the starting elevation 2053.18, and then subtracting the sum of foresights, 40.21, to obtain 2053.21, which checks the last elevation.

Leveling should always be checked by running closed circuits or *loops*. This can be done either by returning to the starting bench mark, as demonstrated in Plate D-2, or by ending the circuit at another bench mark of equal or higher reliability. If a closure is made by returning to the initial bench mark, then the final elevation should agree with the starting elevation. The amount by which they differ is the *loop misclosure*. Note that in Plate D-2 a loop misclosure of 0.03 ft was obtained.

If closure is made to another bench mark, the *section misclosure* is the difference between the closing bench mark's given elevation and its elevation obtained after leveling through the circuit. Specifications, or purpose of the survey, fix permissible misclosures (see Section 7-5). If the allowable misclosure is exceeded, one or more additional runs must be made. When acceptable misclosure is achieved, final elevations are obtained by making an adjustment (see Section 7-6).

Note that in running a level circuit between bench marks, *a new instrument setup has to be made before starting the return run to get a complete check.* In Plate D-2, for example, a minus sight of 8.71 was read on BM Oak to finish the run out, and a plus sight of 11.95 was recorded to start back, showing that a new setup had been made. Otherwise an error in reading the final minus sight would be accepted for the first plus sight on the run back. An even better check is secured by tying the run to a different bench mark.

If the elevation above mean sea level is available for the starting bench mark, elevations then determined for all intermediate points along the circuit will also be referenced to mean sea level. However, if the starting bench mark's elevation above mean sea level is not known, an assumed value may be used and all elevations converted to mean sea level later by applying a constant.

A lake or pond undisturbed by wind, inflow, or outflow, or even a slow-moving stream can serve as an extended turning point. Stakes driven flush with the lake or stream surface, or rocks whose high points are at this level should be used.

Double-rodded lines of levels are sometimes used on important work. In this procedure, plus and minus sights are taken on two turning points, using two rods from each setup, and the readings carried in separate noteform columns. A check on each instrument setup is obtained if the HI agrees for both lines.

7-5 PRECISION

Precision in leveling is increased by repeating measurements, making frequent ties to established bench marks, using high-quality equipment, keeping it in good adjustment, and performing the measurements carefully. No matter how carefully the work is executed, however, errors will exist and will be evident in the form of misclosures, as discussed in Section 7-4. To determine whether or not work is acceptable, misclosures are compared with permissible values on the basis of either number of setups or distance covered. Various organizations set precision standards based on their project requirements. For example, on a simple construction survey, an allowable misclosure of $C = 0.02$ ft $\sqrt{n}$ might be used, where n is the number of setups. Note that this criterion was applied for the level circuit of Plate D-2.

The Federal Geodetic Control Subcommittee (FGCS) recommends the following formula to compute allowable misclosures:[1]

$$C = m\sqrt{K} \tag{7-1}$$

where C is the allowable *loop* or *section* misclosure, in millimeters; m a constant; and K the total length leveled, in kilometers. For "loops" (circuits that begin and end on the same bench mark) K is the total perimeter distance, and the FGCS specifies constants of 4, 5, 6, 8, and 12 mm for the five classes of leveling, designated, respectively, as **(1)**

---

[1]The FGCS was formerly the FGCC (Federal Geodetic Control Committee). Their complete specifications for leveling are available in a booklet entitled "Standards and Specifications for Geodetic Control Networks" (September 1984), which can be obtained from the National Geodetic Information Center, NOAA, National Geodetic Survey, N/CG17, SSMC3 Station 09535, 1315 East West Highway, Silver Spring, MD 20910; telephone: (301) 713-3242.

first-order class I, (2) first-order class II, (3) second-order class I, (4) second-order class II, and (5) third-order. For "sections" (leveling that begins on one bench mark and closes on another), the constants are the same, except that 3 mm applies for first-order class I. The particular order of accuracy recommended for a given type of project is discussed in Section 19-6.

A differential leveling loop is run from an established BM A to a point 2 mi away and back, with a misclosure of 0.056 ft. What order leveling does this represent?

EXAMPLE 7-2

SOLUTION

$$C = \frac{0.056 \text{ ft}}{0.00328 \text{ ft/mm}} = 17 \text{ mm}$$

$$K = (2 \text{ mi} + 2 \text{ mi}) \times 1.61 \text{ km/mi} = 6.4 \text{ km}$$

By a rearranged form of Eq. (7-1), $m = \dfrac{C}{\sqrt{K}} = \dfrac{17}{\sqrt{6.4}} = 6.7$.

This leveling meets the allowable 8-mm tolerance level for second-order class II work, but does not quite meet the 6-mm level for second-order class I, and if that standard had been specified, the work would have to be repeated.

Since distance leveled is proportional to number of instrument setups, the misclosure criteria can be specified using that variable. As an example, if sights of 200 ft are taken, thereby spacing instrument setups at 400 ft, approximately 8.2 setups/km will be made. For second-order class II leveling, the allowable misclosure will then be, again by Eq. (7-1),

$$C = \frac{8}{\sqrt{8.2}} \sqrt{n} = 2.8 \sqrt{n}$$

where C is the allowable misclosure, in millimeters; and n the number of times the instrument is set up.

It is important to point out that meeting FGCS misclosure criteria alone does not guarantee that a certain order of accuracy has been met. Because of compensating errors, it is possible, for example, that crude instruments and low-order techniques can produce small misclosures, yet intermediate elevations along the circuit may contain large errors. To help ensure that a given level of accuracy has indeed been met, besides stating allowable misclosures, the FGCS also specifies equipment and procedures that must be used to achieve a given order of accuracy. These specifications identify calibration requirements for leveling instruments (including rods), and they also outline required field procedures that must be used. Then if the misclosure specified for a given order of accuracy has been met, while employing appropriate instruments and procedures, it can be reasonably expected that all intermediate elevations along the circuit are established to that order.

Field procedures specified by the FGCS include minimum ground clearances for the line of sight, allowable differences between backsight and foresight distances, and maximum sight lengths. As examples, sight lengths of not more than 50 m are permitted for first-order class I, while lengths up to 90 m are allowed for third order. As noted in Section 7-8, stadia is a convenient method for measuring the lengths of backsights and foresights to verify their acceptance.

7-6 ADJUSTMENTS OF SIMPLE LEVEL CIRCUITS

Since permissible misclosures are based on lengths of lines or number of setups, it is logical to adjust elevations in proportion to these values. Elevation differences d and lengths of lines l are shown for a circuit in Figure 7-6. The misclosure found by algebraic summation of the elevation differences is $+0.24$ ft. Adding lengths of the lines yields a total circuit length of 3.0 mi. Elevation adjustments are then (0.24 ft/3.0) times the corresponding lengths, giving -0.08, -0.06, -0.06, and -0.04 ft (shown in the figure). The adjusted elevation differences (shown in black) are used to get the final elevations of bench marks (also shown in black in the figure). Any misclosure which fails to meet tolerances may require reruns instead of adjustment.

In Plate D-2 adjustment for misclosure was made based on the number of instrument setups. Thus after verifying that the misclosure of 0.03 ft was within tolerance, the correction per setup was $\frac{0.03}{7} = 0.004$ ft. Since errors in leveling accumulate, the first point receives a correction of 1×0.004, the second 2×0.004, and so on. The corrected elevations are rounded off to the nearest hundredth of a foot, however.

Level circuits with different lengths and routes are sometimes run from scattered reference points to obtain the elevation of a given bench mark. The most probable value for a bench mark can then be computed from a weighted mean of the observations, the weights varying inversely with line lengths.

In running level circuits, especially long ones, it is recommended that some turning points or bench marks used in the first part of the circuit be included again on the return run. This creates a multiloop circuit, and if a blunder or large error exists, its location can be isolated to one of the smaller loops. This saves time because only the smaller loop containing the blunder or error needs to be rerun.

Figure 7-6 Adjustment of level circuit based on lengths of lines.

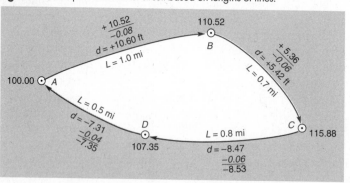

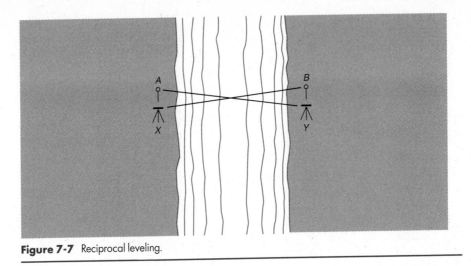

Figure 7-7 Reciprocal leveling.

Although least squares is the best method for adjusting circuits that contain two or more loops (see Section 2-26 and Appendix C), an approximate procedure can also be employed. In this method each loop is adjusted separately, beginning with the one farthest from the closing bench mark.

7-7 RECIPROCAL LEVELING

Sometimes in leveling across topographic features such as rivers, lakes, and canyons, it is difficult or impossible to keep plus and minus sights short and equal. Reciprocal leveling may be utilized at such locations.

As shown in Figure 7-7, a level is set up on one side of a river at X, near A, and rod readings are taken on points A and B. Since XB is very long, several readings are taken for averaging. This is done by reading, turning the leveling screws to throw the instrument out of level, releveling, and reading again. The process is repeated two, three, four, or more times. Then the instrument is moved close to Y and the same procedure followed.

The two differences in elevation between A and B, determined with an instrument first at X and then at Y, normally will not agree because of curvature, refraction, and personal and instrumental errors. In the procedure just outlined, however, the long foresight from X to B is balanced by the long backsight from Y to A. Thus the average of the two elevation differences cancels the effects of curvature and refraction, so the result is accepted as the correct value if the precision of the two differences appears satisfactory. Delays at X and Y should be minimized because refraction can change with changing atmospheric conditions.

7-8 THREE-WIRE LEVELING

As implied by its name, three-wire leveling consists in making rod readings on the upper, middle, and lower cross hairs. Formerly it was used mainly for precise work, but now it is common on projects requiring only ordinary precision. The method has the advantages of **(1)** providing checks against rod reading blunders, **(2)** producing greater

Sta.	Sight	Stadia	Sight	Stadia	Elev.
THREE-WIRE LEVELING					
TAYLOR LAKE ROAD					
BM A					103.8432
	0.718		1.131		
	0.633	8.5	1.051	8.0	+0.6337
	0.550	8.3	.0972	7.9	104.4769
3	1.901	16.8 3	3.154	15.9	−1.0513
	+0.6337		−1.0513		
TP 1					103.4256
	1.151		1.041		
	1.082	6.9	0.969	7.2	+1.0820
	1.013	6.9	0.897	7.2	104.5076
3	3.246	13.8 3	2.907	14.4	−0.9690
	+1.0820		−0.9690		
TP 2					103.5386
	1.908		1.264		
	1.841	6.7	1.194	7.0	+1.8410
	1.774	6.7	1.123	7.1	105.3796
3	5.523	13.4 3	3.581	14.1	−1.1937
	+1.8410		−1.1937		
BM B					104.1859
Σ	+3.5567		Σ	−3.2140	*Check*
Page Check:					
	103.8432	+3.5567	−3.2140	= 104.1	859

Figure 7-8 Sample field notes for three-wire leveling.

accuracy because averages of three readings are available, and **(3)** furnishing stadia measurements of sight lengths to assist in balancing backsight and foresight distances (stadia is discussed in Chapter 15). In the three-wire procedure the difference between the upper and middle readings is compared with that between the middle and lower values. They must agree within one or two of the smallest units being recorded (usually 0.1 or 0.2 of the least count of the rod graduations); otherwise the readings are repeated. An average of the three readings is actually used, but as a computational check it must be very close to the middle wire figure. The difference between the upper and lower wire readings multiplied by the instrument *stadia interval factor* (see Section 15-2) gives the sight distances.

A sample set of field notes for the three-wire method is presented in Figure 7-8. Backsight readings on BM *A* of 0.718, 0.633, and 0.550 m taken on the upper, middle, and lower wires, respectively, give upper and lower differences (multiplied by 100) of

8.5 and 8.3 m, which agree within acceptable tolerance. Stadia measurement of the backsight length (the sum of the upper and lower differences) is 16.8 m. The average of the three backsight readings on BM *A*, 0.6337 m, agrees within 0.0007 m of the middle reading. The stadia foresight length of 15.9 m at this setup is within 0.9 m of the backsight length and is satisfactory. The HI (104.4769 m) for the first setup is found by adding the average backsight reading to the elevation of BM *A*. Subtracting the average foresight reading on TP 1 gives its elevation (103.4256 m). This process is repeated for each setup.

7-9 PROFILE LEVELING

Before engineers can properly design linear facilities such as highways, railroads, transmission lines, aqueducts, canals, sewers, and water mains, they need accurate information about the topography along the proposed routes. Profile leveling, which yields elevations at definite points along a reference line, provides the needed data. The subsections that follow discuss topics pertinent to profile leveling and include staking and stationing the reference line, field procedures for profile leveling, and drawing and using the profile.

7-9.1 Staking and Stationing the Reference Line

Depending on the particular project, the reference line may be a single straight segment, as in the case of a short sewer line; a series of connected straight segments which change direction at angle points, as with transmission lines; or straight segments joined by curves, which occur with highways and railroads. The required alignment for any proposed facility will normally have been selected as the result of a preliminary design, which is usually based on a study of existing maps and aerial photos. The reference alignment will most often be the proposed construction centerline, although frequently offset reference lines are used.

To stake the proposed reference line, key points such as the starting and ending points and angle points will be set first. Then intermediate stakes will be placed on line, usually at 100-ft intervals, but sometimes at closer spacing. Distances for staking can be taped, or measured using an EDM instrument or total station operating in its tracking mode (see Sections 10-12 and 24-9).

In route surveying, a system called *stationing* is used to specify the relative position of any point along the reference line. The starting point is usually designated with some arbitrary value, for example, $10+00$ or $100+00$, although $0+00$ can be used. If the beginning point was $10+00$, the first stake 100 ft along the line from it would be designated $11+00$, the one 200 ft along the line $12+00$, and so on. The term *full station* is applied to each of these points set at 100-ft increments. A point located between two full stations, say 84.9 ft beyond station $17+00$, would be designated $17+84.9$. Thus locations of intermediate points are specified by their nearest preceding full station and their *plus*. In the designation of station $17+84.9$, the plus is 84.9.

In rugged areas and in urban situations, *half-stations* with a plus of 50 ft are often also staked. Even *quarter-stations* (at 25-ft increments) are sometimes placed in these situations.

Stationing not only provides a convenient unambiguous method for specifying positions of points along the reference line, it also gives the distances between points. For example, stations $24+18.3$ and $17+84.9$ are $(2418.3 - 1784.9)$, or 633.4 ft, apart.

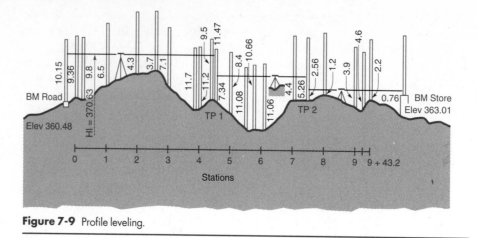

Figure 7-9 Profile leveling.

7-9.2 Field Procedures for Profile Leveling

Profile leveling consists simply of differential leveling with the addition of *intermediate foresights* (minus sights) taken at required points along the reference line. Figure 7-9 illustrates an example of the field procedure, and the notes in Plate D-3 relate to this example. As shown in the figure, the leveling instrument is initially set up at a convenient location and a plus sight of 10.15 ft taken on the bench mark. Adding this to the bench mark elevation yields an HI of 370.63 ft. Then intermediate foresights are taken on points along the profile, such as $0+00$, $0+20$, and $1+00$. (If the reference line's beginning is far removed from the bench mark, differential levels running through several turning points may be necessary to get the instrument into position to begin taking intermediate foresights on the profile line.) Notice that the noteform for profile leveling contains all the same column headings as differential leveling, but is modified to include another column labeled "Intermediate Sight."

When distances to intermediate sights become too long, or if terrain variations or vegetation obstruct rod readings ahead, the leveling instrument must be moved. This is done by establishing a turning point, as TP 1 in Figure 7-9. After reading a minus sight on the turning point, the instrument is moved ahead, leveled, and a plus sight taken on the same turning point. This establishes a new HI, with the instrument then in position for further intermediate sights. This procedure is repeated until the profile is completed.

Intermediate sights are taken at all full stations, and if specifications require, sometimes at half- or even quarter-stations. They are also taken at high and low points along the alignment, as well as at changes in slope. Intermediate sights are also taken on "critical" points such as railroad tracks, highway centerlines, gutters, and drainage ditches. As presented in Plate D-3, rod readings are normally only taken to the nearest 0.1 ft where the rod is held on the ground, but on critical points, and for all plus and minus sights taken on turning points and bench marks, the readings are recorded to the nearest hundredth of a foot.

In profile leveling, lengths of intermediate foresights vary, and in general they will not equal the backsight length. Thus errors due to an inclined line of sight and to curvature and refraction will occur. Because errors from these sources increase with increasing sight lengths, on important work the instrument's condition of adjustment should be checked (see Appendix A), and excessively long intermediate foresight distances should be avoided.

Instrument heights (HIs) and elevations of all turning points are computed immediately after each backsight and foresight. However, elevations for intermediate foresights are not computed until after the circuit is closed on either the initial bench mark or another. Then the circuit misclosure is computed, and if acceptable, an adjustment is made and elevations of intermediate points are calculated. The procedure is described in the following subsection.

As in differential leveling, the page check should be made for each left-hand sheet. In profile leveling, however, intermediate foresights play no part in this computation. As illustrated in Plate D-3, the page check is made by adding the algebraic sum of the column of plus sights and the column of minus sights to the beginning elevation. This should equal the last elevation tabulated on the page for either a turning point or the ending bench mark if that is the case, as it is in the example of Plate D-3.

7-9.3 Drawing and Using the Profile

Prior to drawing the profile, it is first necessary to compute elevations along the reference line from the field notes. This cannot be done, however, until an adjustment has been made to distribute any misclosure in the level circuit. In the adjustment process, HIs are adjusted, because they will affect computed profile elevations. The adjustment is made progressively in proportion to the total number of HIs in the circuit. The procedure is illustrated in Plate D-3, where the misclosure was 0.03 ft. Since there were three HIs, the correction applied to each is $\frac{-0.03}{3} = -0.01$ ft per HI. Thus a correction of -0.01 was applied to the first HI, -0.02 ft to the second, and -0.03 ft to the third. Adjusted HIs are shown in Plate D-3 in parentheses above their unadjusted values. It is unnecessary to correct turning point elevations since they are of no consequence.

After adjusting the HIs, profile elevations are computed by subtracting intermediate foresights from their corresponding adjusted HIs. The profile is then drawn by plotting elevations on the ordinate versus their corresponding stations on the abscissa. By connecting adjacent plotted points, the profile is realized.

Until recently, profiles were manually plotted, usually on special paper like the type shown in Figure 7-10. Now with computer-aided drafting and design (CADD) systems (see Section 17-17) it is only necessary to enter the stations and elevations into the computer and this special software will plot and display the profile on the screen. Hard copies, if desired, may be obtained from automatic plotters interfaced with the computer such as that shown in Figure 17-9.

In drawing profiles, the vertical scale is generally exaggerated with respect to the horizontal scale to make differences in elevation more pronounced. A ratio of 10:1 is frequently used, but flatness or roughness of the terrain determines the desirable proportions. Thus for a horizontal scale of 1 in. = 100 ft, the vertical scale might be 1 in. = 10 ft. The scale actually employed should be plainly marked.

Plotted profiles are used for many purposes, such as (**1**) determining depth of cut or fill on proposed highways, railroads, and airports; (**2**) studying grade-crossing problems; and (**3**) investigating and selecting the most economical grade, location, and depth for sewers, pipelines, tunnels, irrigation ditches, and other projects.

The *rate of grade* (or *gradient* or *percent grade*) is the rise or fall in feet per 100 ft, or in meters per 100 m. Thus a grade of 2.5% means a $2\frac{1}{2}$-ft difference in elevation per

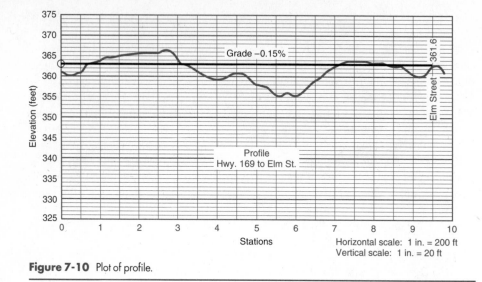

Figure 7-10 Plot of profile.

100 ft horizontally. Ascending grades are plus; descending grades, minus. A grade line of -0.15%, chosen to give somewhat equal cuts and fills, is shown in Figure 7-10. Along this grade line, elevations drop at the rate of 0.15 ft per 100 ft. The process of staking grades is described in Chapter 24.

The term *grade* is also used to denote the elevation of the finished surface on an engineering project.

7-10 GRID, CROSS-SECTION, OR BORROW-PIT LEVELING

Grid leveling is a method for locating contours (see Section 16-9.3). It is accomplished by staking an area in squares of 10, 20, 50, 100, or more feet (or comparable meter lengths) and determining the corner elevations by differential leveling. Rectangular blocks, say 50 by 100 ft, that have the longer sides roughly parallel with the direction of most contour lines may be preferable on steep slopes. The grid size chosen depends on the project extent, ground roughness, and accuracy required.

The same process, termed *borrow-pit-leveling,* is employed on construction jobs to ascertain quantities of earth, gravel, rock, or other material to be excavated or filled. The procedure is covered in Section 27-10 and Plate D-4.

7-11 USE OF THE HAND LEVEL

A hand level can be used for some types of leveling when a low order of accuracy is sufficient. The instrument operator takes a plus and minus sight while standing in one position and then moves ahead to repeat the process. A hand level is useful, for example, in cross-sectioning to obtain a few additional rod readings on sloping terrain where a turning point would otherwise be required.

7-12 SOURCES OF ERROR IN LEVELING

All leveling measurements are subject to three sources of error: (**1**) instrumental, (**2**) natural, and (**3**) personal. These are summarized in the subsections that follow.

7-12.1 Instrumental Errors

Line of Sight. A dumpy or tilting level in proper adjustment should have its line of sight and level vial axis parallel. Then, with the bubble centered, a horizontal plane, rather than a conical surface, is generated as the telescope is revolved. Also, if the compensators of automatic levels are operating properly, they should always produce a truly horizontal line of sight. If these conditions are not met, a so-called *line of sight* (or *collimation*) error exists, and serious errors in rod readings can result. These errors are systematic, but they are canceled in differential leveling if the horizontal lengths of plus and minus sights are kept equal. The error may be serious in going up or down a steep hill where all plus sights are longer or shorter than all minus sights, unless care is taken to run a zigzag line. The size of the collimation error can be determined in a simple field procedure (see Section 19-16), and if backsights and foresights cannot be balanced, a collimation correction can be made.

Cross Hair Not Exactly Horizontal. Reading the rod near the center of the horizontal cross hair will eliminate or minimize this potential error.

Rod Not Correct Length. Inaccurate divisions on a rod cause errors in measured elevation differences similar to those resulting from incorrect markings on a tape. Uniform wearing of the rod bottom makes HI values too large, but the effect is canceled when included in both plus and minus sights. Rod graduations should be checked by comparing them with those on a standardized tape.

Tripod Legs Loose. Tripod leg bolts that are too loose or too tight allow movement or strain that affects the instrument head. Loose metal tripod shoes cause unstable setups.

7-12.2 Natural Errors

Curvature of the Earth. As noted in Section 6-4, a level surface curves away from a horizontal plane at the rate of $0.667M^2$, or about 0.7 ft. per mile. The effect of curvature of the earth is to increase the rod reading. Equalizing lengths of plus and minus sights in differential leveling cancels the error due to this cause.

Refraction. Light rays coming from an object to the telescope are bent, making the line of sight a curve concave to the earth's surface, which thereby decreases rod readings. Balancing the lengths of plus and minus sights usually eliminates errors due to refraction. Large and sudden changes in atmospheric refraction may be important in precise work, however. Errors due to refraction tend to be random over a long period of time, but could be systematic on one day's run.

Temperature Variations. Heat causes leveling rods to expand, but the effect is not important in ordinary leveling. If the level vial of a dumpy or tilting level is heated, the liquid expands and the bubble shortens. This does not produce an error (although it may be inconvenient), unless one end of the tube is warmed more than the other and the bubble therefore moves toward the heated end. Other parts of the instrument warp because of uneven heating, and this distortion affects the adjustment. Shading the level by means of a cover when carrying it, and by an umbrella when it is set up, will reduce or eliminate heat effects. These precautions are followed in precise leveling.

Air boiling or heat waves near the ground surface or adjacent to heated objects make the rod appear to wave and prevent accurate sighting. Raising the line of sight by high tripod setups, taking shorter sights, avoiding any that pass close to heat sources (such as buildings and stacks), and using the lower magnification of a variable-power eyepiece reduce the effect.

Wind. Strong wind causes the instrument to vibrate and makes the rod unsteady. Precise leveling is not attempted on windy days.

Settlement of the Instrument. Settlement of the instrument after a plus sight has been taken makes the minus sight too small and therefore the recorded elevation of the next point too large. The error is cumulative in a series of setups on soft material. Unusual care is required in setting up a level on spongy ground, blacktop, or ice. Readings must be taken in quick order, perhaps using two rods and two observers to preclude walking around the instrument. Alternating the order of taking plus and minus sights also helps somewhat to reduce this error.

Settlement of a Turning Point. This condition causes an error similar to that resulting from settlement of the instrument. It can be avoided by selecting firm, solid turning points or, if none are available, using a steel *turning pin.*

7-12.3 Personal Errors

Bubble Not Centered. In working with dumpy or tilting levels, errors caused by the bubble not being exactly centered at the time of sighting are the most important of any, particularly on long sights. If the bubble runs between the plus and minus sights, *it must be recentered before the minus sight is taken.* Experienced observers develop the habit of checking the bubble before and after each sight, a procedure simplified with some instruments which have a mirror-prism arrangement permitting a simultaneous view of the level vial and rod.

Parallax. Parallax caused by improper focusing of the objective or eyepiece lens results in incorrect rod readings. Careful focusing eliminates this problem.

Faulty Rod Readings. Incorrect rod readings result from parallax, poor weather conditions, long sights, improper target settings, and other causes, including mistakes such as those due to careless interpolation and transposition of figures. Short sights selected to fit weather and instrument conditions reduce the magnitude of reading errors. If a target is used, the rodperson should read the rod, and the observer should check it independently.

Rod Handling. Serious errors caused by improper plumbing of the rod are eliminated by using a rod level that is in adjustment, or holding the rod parallel to a plumb-bob string. Banging the rod on a turning point for the second (plus) sight may change the elevation of a point.

Target Setting. If a target is used, it may not be clamped at the exact place signaled by the observer because of slippage. A check sight should always be taken after the target is clamped.

7-13 MISTAKES

A few common mistakes in leveling are listed here.

Improper Use of a Long Rod. If the vernier reading on the back of a damaged Philadelphia rod is not exactly 6.500 or 7.000 for the short rod, the target must be set to read the same value before extending the rod.

Holding the Rod in Different Places for the Plus and Minus Sights on a Turning Point. The rodperson can avoid such mistakes by using a well-defined point or by outlining the rod base with lumber crayon, keel, or chalk.

Reading a Foot Too High. This mistake usually occurs because the incorrect foot mark is in the telescope's field of view near the cross line; for example, an observer may read 5.98 instead of 4.98. Noting the foot marks above and below the horizontal cross line will prevent this mistake.

Waving a Flat-Bottom Rod While Holding It on a Flat Surface. This action produces an incorrect rod reading because rotation is about the rod edges instead of the center or front face. In precise work, plumbing with a rod level or other means is preferable to waving. This procedure also saves time.

Recording Notes. Mistakes in recording, such as transposing figures, entering values in the wrong column, and making arithmetic mistakes, can be minimized by having the notekeeper repeat the value called out by an observer, and by making the standard field-book checks on rod sums and elevations.

Touching Tripod or Instrument During Reading Process. Beginners may center the bubble, put one hand on the tripod or instrument while reading a rod, and then remove the hand while checking the bubble, which has now returned to center but was off during the observation.

7-14 REDUCING ERRORS AND ELIMINATING MISTAKES

Errors in leveling are reduced (but never eliminated) by carefully adjusting and manipulating both instrument and rod (see Appendix A for procedures) and establishing standard field methods and routines. The following routines prevent most large errors or quickly disclose mistakes: **(1)** checking the bubble before and after each reading (if an automatic level is not being used), **(2)** using a rod level, **(3)** keeping the horizontal lengths of plus and minus sights equal, **(4)** running lines forward and backward, and **(5)** making the usual field book arithmetic checks.

7-1 What errors in leveling are eliminated by keeping the lengths of plus and minus sights equal?

7-2 Describe three different methods that can be used to ensure that the rod is held plumb while taking rod readings in differential leveling.

7-3 How can errors due to settlement of the instrument and rod be reduced or eliminated?

7-4 Explain how errors due to lack of instrument adjustment can be practically eliminated in running a line of differential levels.

7-5 Why is it advisable to set up a level with all three tripod legs on, or in, the same material (concrete, gravel, soil) if possible?

*7-6 Compute the distance a rod extended for a 10-ft reading must be out of plumb to introduce an error of 0.01 ft.

7-7 Similar to Problem 7-6, except for a 3.0-m reading and an error of 1 mm.

*7-8 What error results on a 200-ft sight with a level if the rod reading is 10.50 but the top of the 12-ft rod is 9 in. out of plumb?

7-9 Prepare a set of level notes for the data listed. Perform a check and adjust the misclosure. Elevation of BM 7 is 652.54 ft. If the total loop length is 3000 ft, what order of leveling is represented?

POINT	+S (BS)	−S (FS)
BM 7	6.39	
TP 1	4.91	5.54
BM 8	5.68	9.81
TP 2	2.74	4.98
TP 3	9.46	6.24
BM 7		2.58

7-10 Similar to Problem 7-9, except the elevation of BM 7 is 895.62 ft and the loop length 2000 ft.

*7-11 A differential leveling loop began and closed on BM Bridge (elevation 1237.28 ft). The BS and FS distances were kept approximately equal. Readings taken in order are 8.59 on BM Bridge, 6.54 and 4.87 on TP 1, 7.50 and 6.08 on BM X, 7.23 and 2.80 on TP 2, and 1.11 on BM Bridge. Prepare, check, and adjust the notes.

7-12 A differential leveling circuit began on BM Hydrant (elevation 860.71 ft) and closed on BM Rock (elevation 864.93 ft). The BS and FS distances were kept approximately equal. Readings taken in order are 3.78 on BM Hydrant, 5.43 and 6.05 on TP 1, 5.29 and 4.86 on BM 1, 7.40 and 8.81 on BM 2, 2.47 and 6.77 on TP 2, and 5.42 on BM Rock. Prepare, check, and adjust the notes.

7-13 A differential leveling loop started and closed on BM Rivet, elevation 946.02 ft. The BS and FS distances were kept approximately equal. Readings taken in order are 2.65, 7.23, 3.51, 8.40, 5.06, 6.11, 7.93, 5.82, 9.05, and 0.68. Prepare, check, and adjust the notes.

*7-14 A level setup midway between X and Y reads 5.18 on X and 6.80 on Y. When moved within a few feet of X, readings of 4.74 on X and 6.32 on Y are recorded. What is the true elevation difference, and the reading required on Y to adjust the instrument?

7-15 A level is set up near C (elev 193.436 m) and then near D. Rod readings taken in order are C = 1.788 m, D = 1.321 m, D = 1.296 m, and C = 1.669 m. Compute the elevation of D, and the reading required on C to adjust the instrument.

7-16 The peg adjustment test shows that a level's line of sight is inclined downward 0.0045 ft/100. What is the allowable difference between BS and FS distances at each setup (neglecting curvature and refraction) to keep elevations correct within 0.001 ft?

*Asterisks indicate problems that have answers given in Appendix G.

*7-17 Reciprocal leveling gives the following readings in feet from a setup near *A*: on *A*, 2.071; on *B*, 8.254, 8.259, and 8.257. At the setup near *B*: on *B*, 9.112; on *A*, 2.926, 2.930, and 2.927. The elevation of *B* is 1099.600. Compute the misclosure and elevation of *A*.

7-18 Reciprocal leveling across a canyon provides the data listed. The correct elevation of *Y* is 549.732 ft. The elevation of *X* is required. Instrument at *X*: $+S = 5.476$, $-S = 9.039$, 9.035, and 9.037. Instrument at *Y*: $+S = 8.362$; $-S = 4.798$, 4.799, and 4.799.

*7-19 Prepare a set of three-wire leveling notes for the data given and make the page check. The elevation of BM *X* is 230.054 m. Rod readings (in meters) are (*H* denotes upper cross-wire readings, *M* middle wire, and *L* lower wire): BS on BM *X*: $H = 1.683$, $M = 1.453$, $L = 1.224$; FS on TP 1: $H = 2.959$, $M = 2.707$, $L = 2.454$; BS on TP 1: $H = 2.254$, $M = 2.054$, $L = 1.854$; FS on BM *Y*: $H = 1.013$, $M = 0.817$, $L = 0.620$.

7-20 Similar to Problem 7-19, except the elevation of BM *X* is 985.730 ft, and rod readings (in feet) are: BS on BM *X*: $H = 6.573$, $M = 6.321$, $L = 6.070$; FS on TP 1: $H = 5.949$, $M = 5.653$, $L = 5.356$; BS on TP 1: $H = 5.470$, $M = 5.195$, $L = 4.921$; FS on BM *Y*: $H = 5.674$, $M = 5.453$, $L = 5.231$.

*7-21 Prepare a set of profile leveling notes for the data listed and show the page check. The elevation of BM *A* is 1275.39 ft, and the elevation of BM *B* is 1264.78 ft. Rod readings are: BS on BM *A*, 5.68; intermediate foresight (IFS) on 1+00, 4.3; FS on TP 1, 9.56; BS on TP 1, 10.02; intermediate foresight on 2+00, 11.1; on 3+00, 6.1; FS on TP 2, 8.15; BS on TP 2, 3.28; intermediate foresight on 3+64, 1.51; on 4+00, 3.1; on 5+00, 6.4; FS on TP 3, 7.77; BS on TP 3, 3.16; FS on BM *B*, 7.23.

7-22 Same as Problem 7-21, except the elevation of BM *A* = 730.78 ft, the elevation of BM *B* = 722.75 ft, and the BS on BM *A* = 8.20 ft.

7-23 Plot the profile in Problem 7-21 and design a grade line between stations 0+00 and 5+00 that balances cut and fill areas.

*7-24 If the elevations on a certain project at stations 10+00 and 14+00 are 1232.47 and 1248.06, respectively, what is the percent grade connecting these two points?

7-25 Same as Problem 7-24, except for elevations of 671.85 and 653.92 at stations 17+50 and 21+75, respectively.

7-26 Differential leveling between BMs *A, B, C, D*, and *A* gives elevation differences in feet of -16.532, $+23.485$, $+38.421$, and -45.360, and distances in miles of 0.4, 0.6, 0.8, and 0.3, respectively. If the elevation of *A* is 5095.705 ft, compute the adjusted elevations of BMs *B, C*, and *D*, and the order of leveling.

7-27 Leveling from BM *X* to *W*, BM *Y* to *W*, and BM *Z* to *W* gives differences in elevation of -29.13, $+27.31$, and $+11.29$ ft, respectively. Distances in feet are $XW = 2500$, $YW = 3000$, and $ZW = 4000$. True elevations of the bench marks in feet are $X = 559.82$, $Y = 503.36$, and $Z = 519.42$. What is the adjusted elevation of *W*?

7-28 A 12-ft level rod was calibrated and its graduated scale was found to be uniformly expanded so that the distance between its 0 and 12.00 marks was actually 12,01 ft. How will this affect elevations determined with this rod for **(a)** circuits run on relatively flat ground **(b)** circuits run downhill **(c)** circuits run uphill?

7-29 After running a line of levels between BM Sign and BM Road, examination showed that the level rod had a repaired base plate on the bottom, thus making the rod too long. Is the elevation determined for BM Road correct? Explain.

7-30 A line of levels with 24 setups (48 rod readings) was run from BM Rock to BM Pond with readings taken to the nearest 0.01 ft; hence any one could have an error of ±0.005 ft. For reading errors only, what total error would be expected in the BM Pond elevation?

7-31 Same as Problem 7-30, except for 18 setups and readings to the nearest 1 mm with possible error of ±0.5 mm each.

*7-32 Compute the permissible misclosure for the following lines of levels: **(a)** an 8-mi loop of third-order levels **(b)** a 20-km section of second-order class I levels **(c)** a 40-km loop of first-order class I levels

*Asterisks indicate problems that have answers given in Appendix G.

7-33 What are the primary differences in running "ordinary" differential leveling and "precise" three-wire leveling?

7-34 Describe the advantage of running a double-rodded line of levels.

7-35 List four considerations that govern a rodperson's selection of turning points and bench marks.

BIBLIOGRAPHY

Berry, R. M. 1977. "Observational Techniques for Use with Compensator Leveling Instruments for First-Order Levels." *Surveying and Mapping* 37 (No. 1):15.

Cyran, E. J. 1983. "Aerial Profiling of Terrain System." *ASCE, Journal of Surveying Engineering* 109 (No. 2):136.

Federal Geodetic Control Committee. 1984. *Standards and Specifications for Geodetic Control Surveys.* Silver Spring, Md.: National Geodetic Information Branch, NOAA.

Holdahl, S. R., W. E. Strange, and R. J. Harris. 1987. "Empirical Calibration of Zeiss Ni-1 Instruments to Account for Magnetic Errors." *Manuscripta Geodectica* 12:28.

Kellie, A. C. 1987. "On the Level." *Professional Surveyor* 7 (No. 3):33.

Kulp, E. F. 1970. "High Precision Levels with Automatic Instruments." *ASCE, Journal of the Surveying and Mapping Division* 96 (No. SU2):121.

Selley, A. P. 1977. "A Trigonometric Leveling Crossing of the Strait of Belle Isle." *Canadian Surveyor* 31:249.

Whalen, C. T., and E. I. Balacz. 1977. "Test Results of First-Order Class III Leveling." *Surveying and Mapping* 37 (No. 1):45.

8

ANGLES, BEARINGS, AND AZIMUTHS

8-1 INTRODUCTION

Determining the locations of points and orientations of lines frequently depends on measurements of angles and directions. In surveying, directions are given by *bearings* and *azimuths* (see Sections 8-5 and 8-6).

As described in Section 2.2, angles measured in surveying are classified as *horizontal* or *vertical,* depending on the plane in which they are measured. Horizontal angles are the basic measurements needed for determining bearings and azimuths. Vertical (or zenith) angles are used in trigonometric leveling, stadia (see Chapter 15), and for reducing measured slope distances to horizontal.

Angles are most often *directly* measured in the field by total station, theodolite, or transit, but a compass or sextant can also be used. Angles can be constructed without measurement on a planetable sheet (see Chapter 16). The compass, transit, theodolite, and total station are discussed in succeeding chapters.

The sextant, shown in Figure 8-1, is a hand-held instrument that can be used to measure angles in any plane. It consists of a telescope and mirrors, which enable simultaneous viewing of two points. In measuring an angle with a sextant, the apex is at the eye of the observer, who sights through the telescope. Adjustments are made to the mirrors, one of which is half-silvered, until the two points being viewed coincide. In that position, the angle between them can be read from a graduated arc.

Sextants are rarely used today, having been replaced by other more modern positioning equipment. They were used in the past principally by navigators to find their positions by observing angles to the stars, but also by surveyors to locate their hydrographic survey boats by measuring angles to control points on shore. The natural ability of humans to stand erect in spite of boat pitch and roll made sextants ideal for these uses. In measuring angles between distant objects such as stars, accuracies to within ±1′ are possible, but they diminish for shorter sights.

An angle can be measured *indirectly* by the tape method described in Section 4-17, and its value computed from the relationships of known quantities in a triangle or other simple geometric figure.

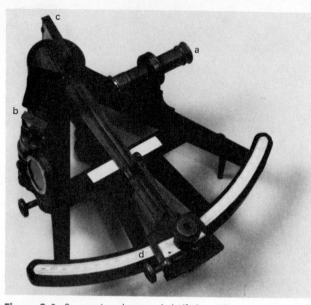

Figure 8-1 Sextant. (*a*, telescope; *b*, half-silvered horizon mirror; *c*, index mirror; *d*, graduated arc with vernier.)

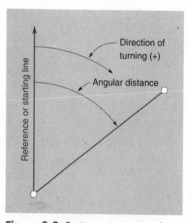

Figure 8-2 Basic requirements in determining an angle.

Three basic requirements determine an angle. As shown in Figure 8-2, they are **(1)** *reference* or *starting line*, **(2)** *direction of turning*, and **(3)** *angular distance* (value of the angle). Methods of computing bearings and azimuths described in this chapter are based on those three elements.

8-2 UNITS OF ANGLE MEASUREMENT

A purely arbitrary unit defines the value of an angle. The *sexagesimal* system used in the United States and many other countries is based on degrees, minutes, and seconds, with the last unit further divided decimally. In Europe the *grad* or *gon* is a standard unit (see Section 2-3). Radians may be more suitable in computations, and in fact are employed in digital computers, but the sexagesimal system will continue to be used in most U.S. surveys rather than decimal degrees, radians, or grads.

8-3 KINDS OF HORIZONTAL ANGLES

The kinds of horizontal angles most commonly measured in surveying are (1) *interior angles,* (2) *angles to the right,* and (3) *deflection angles.* Because they differ considerably, the kind used must be clearly indicated in field notes.

Interior angles, shown in Figure 8-3, are measured on the inside of a closed polygon. Normally the angle at each apex within a polygon is measured. Then, as discussed in Section 12-7, a check can be made on their values because the sum of all angles in any polygon must equal $(n - 2)180°$, where n is the number of angles. Polygons are commonly used for boundary surveys and many other types of work. Surveyors normally refer to them as *closed traverses* rather than polygons, as discussed in Chapters 12 and 13.

Exterior angles, located outside a closed polygon, are explements of interior angles. The advantage to be gained by measuring them is their use as another check, since the sum of the interior and exterior angles at any station must total 360°.

Angles to the right are measured *clockwise from the rear to the forward station.* Note: As a survey progresses, stations are commonly identified by consecutive alphabetical letters (as in Figure 8-3), or by increasing numbers. Thus, the interior angles of Figure 8-3(a) are also angles to the right. *Angles to the left,* turned counter-clockwise from the rear station, are illustrated in Figure 8-3(b). Note that the polygons of Figure 8-3 are "right" and "left"—that is, similar in shape but turned over like the right and left hands. Figure 8-3(b) is shown only to emphasize a serious mistake that occurs if clockwise and counterclockwise angles are mixed. *To avoid this confusion, it is recommended that a uniform procedure of always measuring angles clockwise be adopted,* and the direction of turning noted in the field book with a sketch.

Figure 8-3 Closed polygon. (a) Clockwise interior angles (angles to the right). (b) Counterclockwise interior angles (angles to the left).

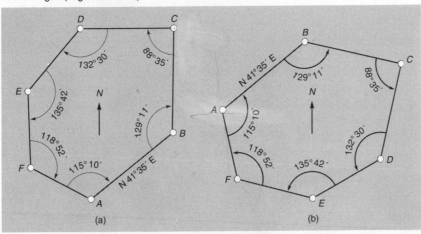

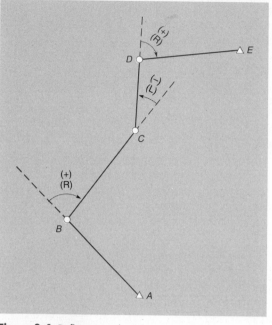

Figure 8-4 Deflection angles.

Deflection angles (Figure 8-4) are measured from an extension of the back line, to the forward station. They are used principally on the long linear alignments of route surveys. As illustrated in the figure, deflection angles may be measured to the right (clockwise) or to the left (counterclockwise) depending upon the direction of the route. Clockwise angles are considered plus, and counterclockwise ones minus, as shown in the figure. Deflection angles are always smaller than 180°, and the direction of turning is identified by appending an *R* or *L* to the numerical value. Thus the angle at B in Figure 8-4 is (R), and that at C is (L). Deflection angles are the only exception where counterclockwise measurement of angles should be made.

8-4 DIRECTION OF A LINE

The direction of a line is the horizontal angle between it and an arbitrarily chosen reference line called a *meridian*. Different meridians are used for specifying directions. The one most commonly employed is an *astronomic* (often called *geodetic* or *true*) meridian. At any point, it is the north-south reference line that passes through the earth's geographic poles. Astronomic meridians derive their name from the field operation to obtain them, which consists in making observations on the sun or stars, as described in Chapter 18.

A *magnetic* meridian is defined by a freely suspended magnetic needle that is influenced by the earth's magnetic field only. Magnetic meridians are discussed in detail in Chapter 9.

An *assumed* meridian can be established by merely assigning any arbitrary direction—for example, taking a certain street line to be true north. The directions of all other lines are then found in relation to it. Disadvantages of using an assumed meridian are the difficulty, or perhaps impossibility, of reestablishing it if the original points are lost, and its nonconformance with other surveys and maps.

Surveys based on a state or other plane coordinate system employ a *grid* meridian for reference. Grid north is the direction of true north for a selected central meridian, and held parallel to it over the entire area covered by a plane coordinate system (see Chapter 21). Obviously, the terms *true* or *due* north, if used in a survey, must be explained, since they may not specify a unique line.

8-5 BEARINGS

Bearings represent one system for designating directions of lines. The bearing angle of a line is defined as the acute horizontal angle between a reference meridian and the line. The angle is measured from either the north or south toward the east or west, to give a reading smaller than 90°. The proper quadrant is shown by the letter N or S preceding the angle, and E or W following it. Thus a properly expressed bearing includes quadrant letters and an angular value. An example is N80°E.

In Figure 8-5 all bearings in quadrant *NOE* are measured clockwise from the meridian. Thus the bearing of line *OA* is N70°E. All bearings in quadrant *SOE* are counterclockwise from the meridian, so *OB* is S35°E. Similarly, the bearing of *OC* is S55°W and that of *OD*, N30°W.

True bearings are measured from the local astronomic or true meridian, *magnetic* bearings from the local magnetic meridian, *assumed* bearings from any adopted meridian, and *grid* bearings from the appropriate grid meridian. Magnetic bearings can be obtained in the field by observing the magnetic needle of a compass, and used along with measured angles to get *computed* bearings.

In Figure 8-6 it is assumed that a compass is set up successively at points *A, B, C,* and *D* and bearings read on lines *AB, BA, BC, CB, CD,* and *DC*. Bearings *AB, BC,* and

Figure 8-5 Bearing angles.

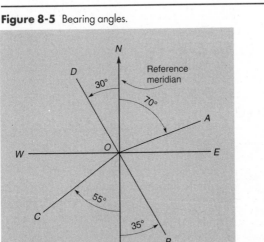

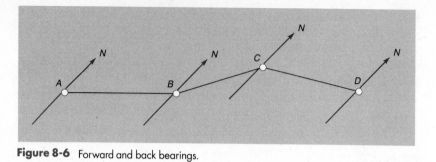

Figure 8-6 Forward and back bearings.

CD are called *forward* bearings; those of *BA, CB,* and *DC, back* bearings. Forward bearings have the same numerical value as back bearings but opposite letters. If bearing *AB* is N44°E, bearing *BA* is S44°W.

In boundary surveys, the term *record* bearing refers to that quoted in a previous survey, and *deed* bearing to one used in a property deed description.

8-6 AZIMUTHS

Azimuths are horizontal angles measured *clockwise* from any reference meridian. In plane surveying, azimuths are generally measured from north, but astronomers, the military, and the National Geodetic Survey have used south as the reference direction.

Examples of azimuths measured from north are shown in Figure 8-7. As illustrated, they can range from 0 to 360° in value, and they do not require letters to identify the quadrant. Thus the azimuth of *OA* is 70°; of *OB*, 145°; of *OC*, 235°; and of *OD*, 330°. Azimuths may be *true, magnetic, grid,* or *assumed,* depending on the reference meridian used. To avoid any confusion, it is necessary to state in the field notes, at the

Figure 8-7 Azimuths.

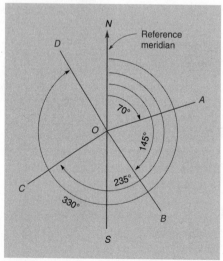

beginning of work, what reference meridian applies for azimuths, and whether they are measured from north or south.

A line's forward direction can be given by its *forward* azimuth, and its reverse direction by its *back* azimuth. Forward azimuths are converted to back azimuths, and vice versa, by adding or subtracting 180°. For example, if the azimuth of *OA* is 70°, the azimuth of *AO* is 70° + 180° = 250°. If the azimuth of *OC* is 235°, the azimuth of *CO* is 235° − 180° = 55°.

Azimuths can be read directly on the graduated circle of a total station instrument, repeating theodolite (or transit) after the instrument has been oriented properly. As explained in Section 12-2, this can be done by sighting along a line with its known azimuth on the circle and then turning to the desired course. Azimuths *(directions)* are used advantageously in some topographic, control, and other surveys as well as in computations.

8-7 COMPARISON OF BEARINGS AND AZIMUTHS

Because bearings and azimuths are encountered in so many surveying operations, a comparative summary of their properties given in Table 8-1 should be helpful. Bearings are readily computed from azimuths by noting the quadrant in which the azimuth falls, then converting as shown in the table.

8-8 CALCULATION OF BEARINGS

Most types of surveys, but especially those which employ *traverses,* require computation of bearings (or azimuths). A traverse, as described in Chapter 12, is a series of connected lines whose lengths, and angles at the junction points, have been measured. Figures 8-3 and 8-4 illustrate examples. Traverses have many uses. To survey the boundary lines of a piece of property, for example, a "closed-polygon" type traverse like those of Figure 8-3 would normally be used. A highway survey from one city to another would usually involve a traverse like that of Figure 8-4. Regardless of the type used, it is necessary to compute the directions of its lines.

TABLE 8-1 COMPARISON OF BEARINGS AND AZIMUTHS

BEARINGS	AZIMUTHS
Vary from 0 to 90°	Vary from 0 to 360°
Require two letters and a numerical value	Require only a numerical value
May be true, magnetic, grid, assumed, forward, or back	Same as bearings
Are measured clockwise and counterclockwise	Are measured clockwise only
Are measured from north and south	Are measured either from north only or from south only in any one survey

Example directions for lines in the four quadrants (azimuths from north)	
N54°E	54°
S68°E	112° (180° − 68°)
S51°W	231° (180° + 51°)
N15°W	345° (360° − 15°)

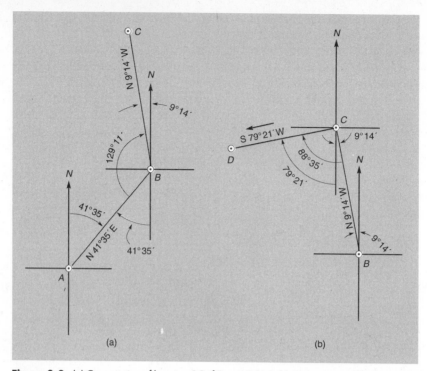

Figure 8-8 (a) Computation of bearing *BC* of Figure 8-3(a). (b) Computation of bearing *CD* of Figure 8-3(a).

Computations of bearings of lines are simplified by drawing sketches similar to those in Figure 8-8 showing all data. In Figure 8-8(a), the bearing of line *AB* from Figure 8-3(a) is N41°35′E, and the angle at *B* turned clockwise (to the right) from known line *BA* is 129°11′. Then the bearing angle of line *BC* is 180° − (41°35′ + 129°11′) = 9°14′, and from the sketch, the bearing of *BC* is N9°14′W.

In Figure 8-8(b) the clockwise angle at *C* from *B* to *D* was measured as 88°35′. The bearing of *CD* is 88°35′ − 9°14′ = S79°21′W. Continuing this technique, the bearings in Table 8-2 have been determined for all lines in Figure 8-3(a).

TABLE 8-2 BEARINGS OF LINES IN FIGURE 8-3(a)

COURSE	BEARING
AB	N41°35′E
BC	N9°14′W
CD	S79°21′W
DE	S31°51′W
EF	S12°27′E
FA	S73°35′E
AB	N41°35′E (Check)

The bearing of any starting course *should always* be recomputed as a check using the last angle. Any discrepancy shows that **(a)** an arithmetic error was made or **(b)** the angles were not properly adjusted prior to computing bearings. In Table 8-2 note that the bearing of *AB* in Figure 8-3(a), obtained by employing the 115°10′ angle measured at *A*, yields a bearing of N41°35′E, which agrees with the starting bearing. Students should compute each bearing of Figure 8-3(a) to verify values given in Table 8-2.

Traverse angles must be adjusted to the proper geometric total before bearings are computed. As noted earlier, in a closed-polygon traverse, the sum of interior angles equals $(n - 2)180°$, where n is the number of angles or sides. If the traverse angles fail to close by, say, 2′ and are not adjusted prior to computing bearings, the original and computed check bearing of *AB* will differ by the same 2′, assuming there are no other calculating errors.

8-9 COMPUTING AZIMUTHS

Many surveyors prefer azimuths over bearings for directions of lines because they are easier to work with, especially when computing traverses with computers. Also sines and cosines of azimuth angles provide correct algebraic signs for latitudes and departures as discussed in Chapter 13.

Azimuth calculations, like those for bearings, are best made with the aid of a sketch. Figure 8-9 illustrates computations for azimuth *BC* in Figure 8-3(a). Azimuth *BA* is found by adding 180° to azimuth *AB*: 180° + 41°35′ = 221°35′. Then clockwise angle *B*, 129°11′, is added to azimuth *BA* to get azimuth *BC* = 221°35′ + 129°11′ = 350°46′. This general process of adding (or subtracting) 180° to obtain the back azi-

Figure 8-9 Computation of azimuth *BC* of Figure 8-3(a).

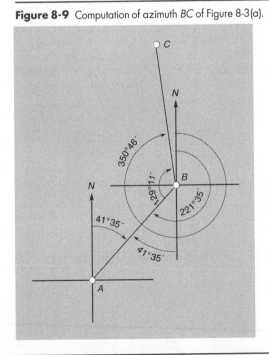

TABLE 8-3 COMPUTATION OF AZIMUTHS (FROM NORTH)

ANGLES TO THE RIGHT [FIGURE 8-3(a)]	
41°35′ = AB + 180°00′	31°51′ = ED + 135°42′
221°35′ = BA + 129°11′	167°33′ = EF + 180°00′
350°46′ = BC − 180°00′	347°33′ = FE + 118°52′
170°46′ = CB + 88°35′	466°25′ − *360°00′
259°21′ = CD − 180°00′	106°25′ = FA + 180°00′
79°21′ = DC + 132°30′	286°25′ = AF + 115°10′
211°51′ = DE − 180°00′	401°35′ − *360°00′
31°51′ = ED	41°35′ = AB (Check)

*When a computed azimuth exceeds 360°, the correct azimuth is obtained by merely subtracting 360°.

muth and then adding the clockwise angle is repeated for each line until the azimuth of the starting line is recomputed. If a computed back azimuth or azimuth exceeds 360°, then 360° is subtracted from it and the computations are continued. These calculations are conveniently handled in tabular form, as illustrated in Table 8-3. This table lists the calculations for all azimuths of Figure 8-3(a). Note that a check was secured by recomputing the beginning azimuth using the last angle. The procedures illustrated in Table 8-3 for computing azimuths are systematic and readily programmed for computer solution.

8-10 MISTAKES

Some mistakes made in using bearings and azimuths are:

1. Confusing magnetic and true bearings.
2. Mixing clockwise and counterclockwise angles.
3. Interchanging bearings for azimuths.
4. Failing to change bearing letters when using the back bearing of a line.
5. Using an angle at the wrong end of a line in computing bearings—that is, using angle A instead of angle B when starting with line AB as a reference.
6. Not including the last angle to recompute the starting bearing or azimuth as a check—for example, angle A in traverse ABCDEA.
7. Subtracting 360°00′ as though it were 359°100′ instead of 359°60′, or using 90° instead of 180° in bearing computations.
8. Adopting an assumed reference line that is difficult to reproduce.

9. Reading degrees and decimals from a calculator as though they were degrees, minutes, and seconds.

10. Failing to adjust traverse angles before computing bearings or azimuths if there is a misclosure.

8-1 Why is it important to adopt a standard angle-measuring procedure, such as always turning angles clockwise?

8-2 Why are deflection angles often used in route surveys?

8-3 How are back bearings and back azimuths used?

*8-4 Convert: **(a)** $37°28'$ to radians **(b)** 58.263 grads to degrees, minutes, and seconds **(c)** $28°17'32''$ to decimal degrees

8-5 Convert: **(a)** $167°22'05''$ to radians **(b)** $73°50'27''$ to grads **(c)** 338.1474 degrees to degrees, minutes, and seconds

In Problems 8-6 through 8-8, convert the azimuths from north to bearings.

*8-6 $32°05'$, $129°17'25''$, $237°48'$, and $295°$

8-7 $29°$, $123°48'$, $243°12'$, and $346°23'40''$

8-8 $79°12'30''$, $94°$, $193°52'$, and $279°16'28''$

Convert the bearings in Problems 8-9 through 8-11 to azimuths from north and compute the angle, smaller than $180°$, between successive bearings.

*8-9 N37°15'E, S51°32'E, S24°31'W, and N56°14'W

8-10 N63°50'E, S45°28'E, S9°54'W, and N90°00'W

8-11 N0°00'00''E, N57°28'16''E, S2°17'46''W, and S88°29'00''W

Compute the azimuth from north of line CD in Problems 8-12 through 8-14. (Azimuths of AB are also from north.)

*8-12 Azimuth $AB = 86°20'$; clockwise angles $ABC = 76°53'$, $BCD = 257°10'$.

8-13 Bearing $AB = $ S41°22'W; clockwise angles $ABC = 138°47'$, $BCD = 185°50'$.

8-14 Azimuth $AB = 291°40'$; clockwise angles $ABC = 153°10'$, $BCD = 337°46'$.

8-15 Compute the interior angles of lot 15 in Figure 22-2.

*8-16 Calculate the interior angles of lot 16 in Figure 22-2.

8-17 For a bearing $DE = $ S47°31'W and angles to the right, compute the bearing of FG if angle $DEF = 147°35'$ and $EFG = 201°52'$.

8-18 Similar to Problem 8-17, except the azimuth of DE is $197°28'42''$, and angles to the right DEF and EFG are $39°28'50''$ and $275°10'08''$, respectively.

Course AB of a five-sided traverse runs due north. Station C is westerly from B. From the given balanced interior angles, compute and tabulate the bearings and azimuths from north for each side of the traverses in Problems 8-19 through 8-21.

8-19 $A = 80°12'$, $B = 120°05'$, $C = 60°46'$, $D = 82°24'$, $E = 196°33'$

*8-20 $A = 77°18'$, $B = 121°08'$, $C = 103°45'$, $D = 99°29'$, $E = 138°20'$

8-21 $A = 59°17'40''$, $B = 108°06'15''$, $C = 96°27'54''$, $D = 130°24'12''$, $E = 145°43'59''$

In Problems 8-22 and 8-23, compute and tabulate the bearings of the sides of a regular hexagon (polygon with six equal angles), given the starting direction of side AB.

8-22 Bearing of $AB = $ N35°18'E (Station C is easterly from B.)

8-23 Azimuth of $AB = 205°20'38''$ (Station C is westerly from B.)

*8-24 Similar to Problem 8-22, except for a regular pentagon.

*Asterisks indicate problems that have answers given in Appendix G.

8-25 Similar to Problem 8-23, except for a regular pentagon.

8-26 Calculate the course bearings for the following deflection-angle open (route survey) traverse. Bearing $0 + 00$ to $7 + 53.2 = S86°03'40''E$; then $8°45'32''R$ to $11 + 77.8$; $12°08'16''L$ to $14 + 29.3$; $14°53'46''R$ to $20 + 06.6$; and $12°34'10''L$ to $25 + 48.0$. Sketch the traverse.

*8-27 Similar to Problem 8-26, except calculate azimuths given the first course azimuth of $23°17'25''$.

Compute bearings of all lines for a closed traverse *ABCDEFGHIJA* that has the following balanced clockwise interior angles, using the fixed bearings listed in Problems 8-28 and 8-29. $JAB = 178°07'$, $ABC = 119°53'$, $BCD = 177°36'$, $CDE = 99°36'$, $DEF = 147°18'$, $EFG = 144°35'$, $FGH = 163°30'$, $GHI = 120°50'$, $HIJ = 101°17'$, $IJA = 187°18'$. (Note: Station C is westerly from Station B.)

8-28 Bearing $AB = N17°48'W$

8-29 Bearing $DE = S52°27'E$

*8-30 Similar to Problem 8-28, except that azimuths from north are required, and fixed azimuth $AB = 102°38'$ (from north).

8-31 Similar to Problem 8-28, except that azimuths from north are required, and fixed azimuth $CD = 146°42'$ (from north).

8-32 Given balanced deflection angles of the following closed traverse, compute the bearings. The bearing of AB is $N75°21'W$. The deflection angles are $B = 83°12'L$, $C = 96°48'L$, $D = 49°59'L$, $E = 45°33'L$, $F = 38°22'R$, and $A = 122°50'L$.

*8-33 Similar to Problem 8-32, except compute azimuths from north if the azimuth from north of line AB is $216°32'$.

8-34 How can you quickly verify the formula $(n - 2)180°$ for the sum of interior angles of a polygon?

8-35 An angle *APB* is measured at different times using various instruments and procedures. The results, which are assigned certain weights, are as follows: $62°10'28''$, wt 3; $62°10'30''$, wt 1; and $62°10'27''$, wt 4. What is the most probable value of the angle?

8-36 Similar to Problem 8-35, but with an additional measurement of $62°10'29''$, wt 2.

*Asterisks indicate problems that have answers given in Appendix G.

BIBLIOGRAPHY Brinker, R. C., and R. Minnick. 1987. *The Surveying Handbook*. New York: Van Nostrand Reinhold, Chap. 5.

Moffitt, F. H., and H. Bouchard. 1992. *Surveying*, 9th ed. New York: Harper & Row, Chap. 4.

9
THE COMPASS

9-1 INTRODUCTION

The compass has long been an important instrument for surveying and in many other applications. It has been used by navigators and others for many centuries to determine directions. Prior to the invention of the transit, a compass furnished surveyors with the most practical way to measure horizontal angles and directions of lines.

Compasses were used extensively for early boundary surveys in the United States. The resulting property descriptions, many of which are still in effect today, were given in terms of magnetic directions. Thus an understanding of compasses is necessary to check and retrace the original land lines on which they were used.

Even though modern technology has led to instruments that are more convenient and far more accurate for measuring angles and directions, compasses are still important to surveyors in many special situations. Besides their already stated use for retracing boundary lines whose descriptions are given by magnetic directions, other uses include (**1**) quickly making blunder checks of angle measurements made with more accurate equipment; (**2**) performing rough reconnaissance or preliminary surveys; (**3**) maintaining parallel range lines during hydrographic surveys (see Chapter 16); and (**4**) locating Polaris for azimuth observations at dusk before the star is visible to the naked eye (see Chapter 18). An important recent application has been their use in Alaska to set final property corners of the U.S. public land system from nearby "approximate" corners established using inertial surveying systems. A new application is their use in orienting the antennas of GPS receivers (see Chapter 20).

9-2 THEORY OF THE COMPASS

A compass consists of a magnetized steel needle mounted on a pivot at the center of a graduated circle. Unless disturbed by a local attraction (see Section 9-5), the needle aligns itself with the earth's magnetic field and points toward magnetic north (in the northern hemisphere). In 1990 the north and south magnetic poles were located at approximately 77° north latitude, 102° west longitude, and 65° south latitude, 140° east longitude, respectively. They move constantly.

The earth's magnetic forces align the needle and pull or dip one end of it below the horizontal position. The *angle of dip* varies from 0° near the equator to 90° at the magnetic poles. In the northern hemisphere, the south end of the needle is weighted with a very small coil of wire to balance the dip effect and keep it horizontal. The position of the coil can be adjusted to conform to the latitude in which the compass is used. Weights on transit compasses are set for an average latitude of 40°N and usually do not have to be changed for any location in the United States except Alaska.

As the compass box is turned, the needle continues to point toward magnetic north and gives a reading that is dependent on the graduated circle position.

9-3 MAGNETIC DECLINATION

Magnetic declination is the horizontal angle from the true geographic meridian to the magnetic meridian. Navigators call this angle *variation* of the compass; the armed forces use the term *deviation*. An east declination exists if the magnetic meridian is east of true north; a west declination occurs if it is west of true north.

Because the magnetic pole positions are constantly changing, magnetic declinations at all locations also undergo continual changes. For any given time, the declination at any location can be obtained (if there is no local attraction) by establishing a true meridian from astronomical observations, and then reading a compass while sighting along the true meridian.

A chart showing magnetic declinations for a specific epoch of time is called an *isogonic map.* Lines on such maps connecting points that have the same declination are called *isogonic lines.* The line made up of points that have zero declination is termed the *agonic line.* On it the magnetic needle defines true north as well as magnetic north.

Figure 9-1 is an isogonic map covering the contiguous 48 states of the United States for the year 1990.[1] The agonic line (heavy solid line) cuts diagonally across the country through Wisconsin, Illinois, Kentucky, Tennessee, Alabama, and northwest Florida. It is gradually moving westward. Points to the west of the agonic line have east declinations; points to the east have west declinations. As a memory aid, the needle can be thought of as pointing *toward* the agonic line. Note there is a 40° difference in declination between the states of Maine and Washington. This is a huge change if a pilot flies by compass between the two states! In Alaska, 1990 declinations varied from about 2°E near the island of Attu in the western Aleutian Islands to about 34°E in the far northeast. The declination at Anchorage was about 24°E, and 27°E at Fairbanks and Juneau. In the Hawaiian Islands, the declination was approximately 11°E in 1990.

The *annual change* in declination shown by dashed lines on larger and more detailed isogonic maps, and on Figure 9-1, aids in estimating the declination a few years before and after the chart date. As an example, Figure 9-1 shows that Cleveland, Ohio, had a magnetic declination of about 7°W in 1990, and an average annual change of approximately 5.5′ westerly. Thus Cleveland's estimated declination in 1995 is 7°W + 5 × 5.5′ = 7°28′W.

Secular change (see Section 9-4) for longer intervals should be computed from available tables which extend back to the earliest times likely to be significant in such problems. The best way to determine the declination at a given location on any date is to make an astronomical observation or use existing control lines. If this is not possible, an approximate declination can be obtained from an isogonic map.

---

[1]The U.S. Geological Survey produces large detailed isogonic charts of the contiguous United States, Alaska, and Hawaii every 5 years. Copies can be obtained from U.S. Geological Survey Map Sales, Box 25286, Denver, Colo. 80225.

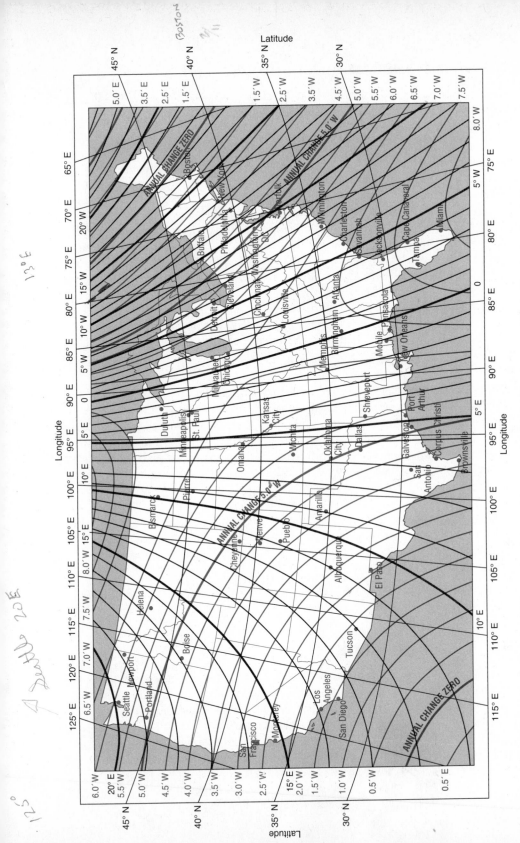

Figure 9-1 Isogonic chart showing distribution of magnetic declination (solid lines) and annual change (dashed lines) in the United States for 1990. (Courtesy U.S. Geological Survey.)

9-4 VARIATIONS IN MAGNETIC DECLINATION

It has been stated that magnetic declinations at any point vary over time. These variations can be categorized as secular, daily, annual, and irregular, and are summarized as follows.

Secular Variation. Because of its magnitude, this is the most important of the variations. Unfortunately, no general law or mathematical formula has been found to enable precise long-term predictions of secular variation, and its past behavior can be described only by means of detailed tables and charts derived from observations. Records which have been kept at London for four centuries show a range in magnetic declination from 11°E in 1580, to 24°W in 1820, back to 5°W in 1985. Secular variation changed the magnetic declination at Baltimore, Maryland, from 5°11′W in 1640 to 0°35′W in 1800, 5°19′W in 1900, 7°25′W in 1950, 8°43′W in 1975, and 10°20′W in 1990 (with an annual change of 4.7′W).

In retracing old property lines run by compass or based on the magnetic meridian, it is necessary to allow for the difference in magnetic declination at the time of the original survey and at the present date. The difference is generally due mostly to secular variation.

Daily Variation. Daily variation of the magnetic needle's declination causes it to swing through an arc averaging approximately 8′ for the United States. The needle reaches its extreme easterly position at about 8 A.M. and its most westerly position at about 1:30 P.M. Mean declination occurs at around 10:30 A.M. and 8 P.M. These hours and the daily swing range change with latitude and season of the year, but neglect of the daily variation is normally well within the range of error expected in compass readings.

Annual Variation. This periodic swing is less than 1 min of arc and can be neglected. It must not be confused with the annual change (the amount of secular-variation change in one year) shown on some isogonic maps.

Irregular Variations. Unpredictable magnetic disturbances and storms can cause short-term irregular variations of a degree or more.

9-5 LOCAL ATTRACTION

The magnetic field is affected by metallic objects and direct-current electricity, both of which cause a local attraction. As an example, when set up beside an old-time streetcar with overhead power lines, the compass needle would swing toward the car as it approached, then follow it until it was out of effective range.

If the source of an artificial disturbance is fixed, all bearings from a given station will be in error by the same amount. Angles calculated from bearings taken at the station will be correct, however.

Local attraction is present if the forward and back bearings of a line differ by more than the normal observation errors. Consider the following compass bearings read on a series of lines:

AB .... N24°15'W	CD .... N60°00'E
BA .... S24°10'E	DC .... S61°15'W
BC .... N76°40'W	DE .... N88°35'E
CB .... S76°40'E	ED .... S87°25'W

Forward bearing *AB* and back bearing *BA* agree reasonably well, indicating that little or no local attraction exists at *A* or *B*. The same is true for point *C*. However, the bearings at *D* differ from corresponding bearings taken at *C* and *E* by roughly 1°15' to the west of north. Local attraction therefore exists at point *D* and deflects the compass needle approximately 1°15' to the west of north.

It is evident that to detect local attraction, successive stations on a compass traverse have to be occupied and forward and back bearings read, even though the directions of all lines could be determined by setting up an instrument only on alternate stations.

9-6 THE SURVEYOR'S COMPASS

A surveyor's compass is shown in Figures 9-2 and 9-3. George Washington and thousands of surveyors before and after him used this type of instrument to run land lines which still determine property holdings and therefore must be retraced. So it is important to understand them.

The compass circle is graduated in degrees or half-degrees but can be read to perhaps 5 or 10' by estimation. The instrument consists of a metal base plate *A* (see Figure

Figure 9-2 Surveyor's compass. (Courtesy W. & L. E. Gurley.)

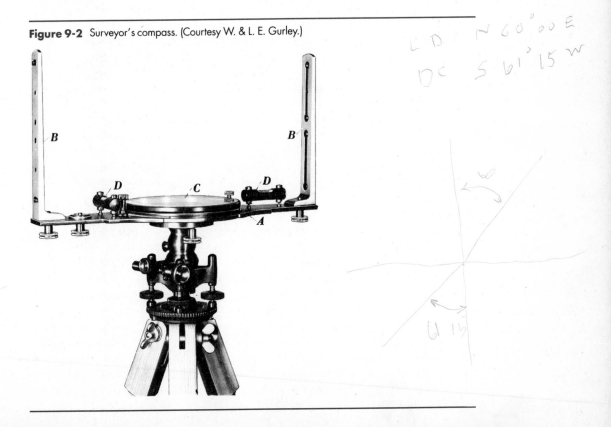

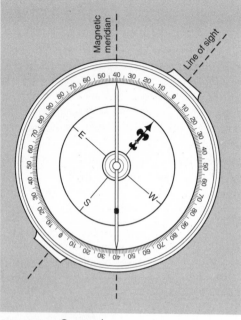

Figure 9-3 Compass box.

9-2) with two vertical sight vanes *B* at the ends. The sight vanes are strips of metal with vertical slits to define the line of sight. The compass box is at *C*, and two small level vials *D* are mounted on the plate perpendicular to each other. When the bubbles in these vials are centered, the plate and compass are horizontal and ready for use.

The compass box (see Figure 9-3) has a conical point at its center to support the needle and a glass cover to protect it. A circular scale at the outer rim of the box is graduated in degrees and half-degrees. Zero marks are at the north and south points in line with the sight-vane slits. Graduations are numbered in multiples of 10°, clockwise and counterclockwise from 0° at the north and south, to 90° at the east and west. As the sight vanes and compass box are revolved, the needle establishes the bearing of the line observed.

Note that letters E and W on the compass box are reversed from their normal positions to give *direct readings* of bearings. Thus in Figure 9-3, the sight-line bearing through the vanes is N40°E.

Accuracy of a compass depends on the sensitivity of its needle. A sensitive needle is readily attracted toward a small piece of iron held nearby, but settles back in the original position each time the stimulus is removed. Sensitivity itself results from the needle having (**1**) proper shape and balance, (**2**) strong magnetism, (**3**) a sharp conical pivot point, and (**4**) a smooth cup that bears on the pivot. Tapping the glass cover releases a needle that does not swing freely. Touching the cover with a moistened finger removes static electricity that may otherwise affect the needle.

Remagnetizing the needle is relatively easy, but reshaping the pivot is difficult. *To retain a conical shape of the pivot point and prevent blunting to a spherical or flat form that produces sluggishness, the needle should be lifted from the pivot when not in use.* A lever arm is provided to raise the needle from the pivot and press it against the glass cover when the instrument is being moved or boxed.

Early compasses were supported by a single leg called a *Jacob staff.* A ball-and-socket joint and clamp were used to level the instrument and set the plate in a horizontal position. Later versions of compasses were mounted on a base with a four-screw leveling head, as shown in Figure 9-2.

The compass box of a transit is similar in construction to the surveyor's compass. Zero marks at the north and south points are on a line parallel with and beneath the telescope's line of sight. A special adjustment on some compasses swings the graduated circle through an arc to compensate for a given declination, thus permitting "true" bearings to be read from the needle. Because of different local attractions at succeeding stations, its use may not be very practical.

9-7 THE FORESTER'S AND GEOLOGIST'S COMPASS

Figure 9-4 shows a type of compass employed by geologists and the U.S. Forest Service. It can be used as a hand-held instrument or supported on a staff or tripod.

The instrument is made of aluminum, has brass sights, and a declination adjustment for the raised (upper) graduated compass ring. The beveled (lower) ring is used to turn right angles or to measure vertical angles by placing an edge of the base on a level surface.

9-8 BRUNTON COMPASS

Figure 9-5 shows a Brunton compass (also called a pocket transit), which combines the main features of a sighting compass, prismatic compass, hand level, and clinometer. The instrument is convenient and sufficiently accurate for topographic and preliminary

Figure 9-4 Forester's and geologist's compass.

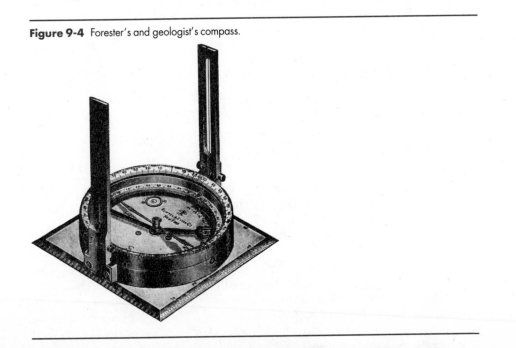

Figure 9-5 Brunton pocket transit.

surveys of many kinds. It can be hand-held or mounted on a Jacob staff or tripod. The device is widely used by geologists.

A Brunton compass consists of a brass case hinged on two sides. The cover, shown on the left in Figure 9-5, has a fine mirror and centerline on the inside face. A hinged sighting point at the extreme left and the sighting point at the far right are folded outward when the instrument is in use. The bearing of a line is read from the compass needle position, while the point observed is reflected through the sight vane on a mirror. A declination adjustment can be made by revolving the raised compass ring.

The *clinometer* (vertical-angle) arc, inside the compass ring, is graduated to degrees and read to the nearest 5′ by a vernier on the clinometer arm. To read vertical angles or grade percentages, the compass is held vertically instead of horizontally. Another arc gives grade percentages for both elevation and depression. Another small and convenient instrument is the liquid-filled Suunto compass. Graduated in degrees, readings can be estimated to 10′.

9-9 TYPICAL COMPASS PROBLEMS

Typical problems in compass surveys require the conversion of true bearings to magnetic bearings, magnetic bearings to true bearings, and magnetic bearings to magnetic bearings for the declinations existing at different dates.

EXAMPLE 9-1   Assume the magnetic bearing of a property line was recorded as S43°30′E in 1862. The magnetic declination at the survey location was 3°15′W. The true bearing is needed for a subdivision property plan.

SOLUTION
A sketch similar to Figure 9-6 makes the relationship clear and should be used by beginners to avoid mistakes. True north is designated by a full-headed long arrow and magnetic north by a half-headed shorter arrow. The true bearing is seen to be S43°30′E + 3°15′ = S46°45′E. Using different colored pencils to show the direction of true north, magnetic north, and lines on the ground helps clarify the sketch.

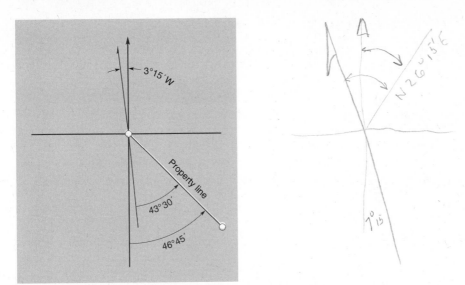

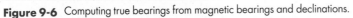

Figure 9-6 Computing true bearings from magnetic bearings and declinations.

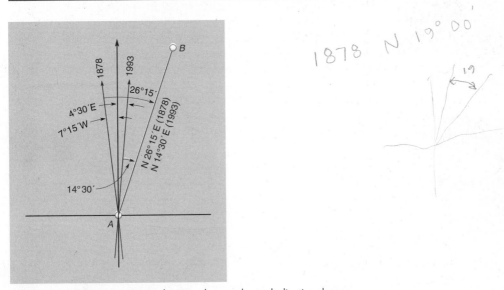

Figure 9-7 Computing magnetic bearing changes due to declination changes.

Assume the magnetic bearing of line *AB* read in 1878 was N26°15′E. The declination at the time and place was 7°15′W. In 1993 the declination was 4°30′E. The magnetic bearing in 1993 is needed.

EXAMPLE 9-2

SOLUTION

The declination angles are shown in Figure 9-7. The magnetic bearing of line *AB* is equal to the earlier date bearing minus the sum of the declination angles, or

$$N26°15′E − (7°15′ + 4°30′) = N14°30′E$$

9-10 SOURCES OF ERROR IN COMPASS WORK

Some sources of errors in using the compass are:

1. Compass out of level
2. Pivot not sharp or off the graduated circle center
3. Needle or sight vanes bent
4. Magnetism of needle weak
5. Magnetic variations
6. Local attraction caused by overhead power lines, underground ore deposits, chaining pins, metal range poles, loose-leaf field books, parked car nearby, and so on

9-11 MISTAKES

Some fairly typical mistakes in compass work are:

1. Reading the wrong end of the needle
2. Setting off the declination on the wrong side of north
3. Setting incorrect declination when reading magnetic bearings
4. Parallax (reading while looking from the side of the needle instead of over and
5. Failing to check the forward and back bearings when possible
6. Not making a sketch showing known and desired data

PROBLEMS

9-1 Determine from Figure 9-1 the declinations for 1990 and the expected declinations in 1995 at Boston, Miami, Kansas City, Los Angeles, and Portland, Oregon.

9-2 What was the total difference in magnetic declination between New York City and Seattle in 1990?

9-3 At the present rate of declination change (see Figure 9-1), approximately how fast, in miles per year, and in what direction is the agonic line moving across the United States?

9-4 Assuming a constant declination annual rate of change in Figure 9-1 (it has not been), compute the approximate declination at Atlanta for 1975. Check your answer by referring to the 1975 isogonic chart in the 6th edition of this book.

9-5 Would a vernier be useful on a surveyor's compass? Explain.

*9-6 The magnetic declination at a certain place is 6°35′E. What is the magnetic bearing there: (a) of true north (b) of true south (c) of true west?

9-7 Same as Problem 9-6, except the magnetic declination at the place is 16°35′W.

9-8 Explain why the letters E and W on a compass (see Figure 9-3) are reversed from their normal positions.

9-9 The magnetic bearing of an old survey line originally recorded as N7°15′W is now N2°15′E. What has been the magnetic declination change and its direction?

*9-10 The magnetic bearing of line XY in 1925 was N46°28′E when the declination was 3°30′W. The declination now is 1°45′E. What magnetic bearing should be used to retrace line XY?

9-11 How could you determine the magnetic declination at a point?

*Asterisks indicate problems that have answers given in Appendix G.

For Problems 9-12 through 9-14 the observed magnetic bearing of line *AB* and its true magnetic bearing are given. Compute the amount and direction of local attraction at point *A*.

	OBSERVED MAGNETIC BEARING	TRUE MAGNETIC BEARING
*9-12	N82°15E	N79°45'E
9-13	S3°00'E	S0°15'W
9-14	N1°30'W	N4°00'E

9-15 Can local attraction be defined by measuring with a compass at one station? Explain.

9-16 Classify the type of error resulting from local attraction.

9-17 Does local attraction at a point affect the size of an angle computed from the magnetic bearings read there? Explain.

9-18 After finding no location attraction at station *A*, bearings read were *AB* = N35°30'W, *BA* = S38°30'E, and *BC* = N62°00'E. What is the acute angle at *B* and the correct bearing of *BC*?

9-19 Readings on a true north line using a surveyor's compass to check local declination were N10°50'E, N10°45'E, N10°40'E, N10°45'E, N10°45'E, and N10°45'E. Compute the declination's most probable value and its standard error.

In Problems 9-20 and 9-21 convert the given magnetic bearings to true bearings.

*9-20 *AB* = N60°25'W, *BC* = S47°30'W, declination 6°30'E

9-21 *AB* = S10°45'E, *BC* = N17°00'E, declination 13°30'W

What magnetic bearing is needed to retrace a line for the conditions stated in Problems 9-22 through 9-25?

	1875 MAGNETIC BEARING	1875 DECLINATION	PRESENT DECLINATION
9-22	N65°35'E	3°30'W	2°30'E
9-23	S73°10'W	2°20'E	1°15'W
*9-24	N89°55'W	0°30'W	4°15'E
9-25	S45°30'E	7°15'E	5°20'W

In Problems 9-26 through 9-29 calculate the magnetic declination in 1870 based on the following data from an old survey record.

	1870 MAGNETIC BEARING	PRESENT MAGNETIC BEARING	PRESENT MAGNETIC DECLINATION
9-26	S00°15'E	S4°45'E	5°30'E
*9-27	S50°30'W	S62°15'W	15°40'E
9-28	N02°30'W	N02°15'E	3°15'E
9-29	N24°30'E	N21°10'E	4°45'E

*Asterisks indicate problems that have answers given in Appendix G.

Compute any location attractions and list the corrected bearings of *BC* and *CD* for Problems 9-30 and 9-31. The accepted bearing of *AB* is N36°30′E.

9-30 *BA* = S33°00′W, *BC* = S68°30′E, *CB* = N70°15′W, *CD* = S0°00′E
*9-31 *BA* = S35°15′W, *BC* = N82°10′E, *CB* = S86°30′W, *CD* = N75°45′W

The observed compass bearings of a five-sided traverse *ABCDEA* are given in Problems 9-32 and 9-33. Calculate the interior angles and explain the closure.

	AB	*BC*	*CD*	*DE*	*EA*
9-32	N23°30′E	N61°15′E	N87°45′W	S41°40′W	N68°20′E
*9-33	S64°30′E	S12°45′E	S60°00′W	N15°45′E	N43°15′E

*Asterisks indicate problems that have answers given in Appendix G.

BIBLIOGRAPHY

Boyum, B. H. 1982. "The Compass That Changed Surveying." *Professional Surveyor* 2:28.
Brinker, R. C., and R. Minnick. 1987. *The Surveying Handbook*. New York: Van Nostrand Reinhold, Chap. 20.
Easa, S. M. 1989. "Analytical Solution of Magnetic Declination Problem." *ASCE, Journal of Surveying Engineering* 115 (No. 3):324.
Kratz, K. E. 1990. "Compass Surveying with a Total Station." *Point of Beginning* 16 (No. 1):30.
Sipe, F. H. 1980. *Compass Land Surveying*. Rancho Cordova, Calif.: Landmark.
Sipe, H. F. 1990. "A Clinic on the Open-Sight Compass." *Surveying and Land Information Systems* 50 (No. 3):229.

10

THE TRANSIT, THEODOLITE, AND TOTAL STATION

10-1 INTRODUCTION

Transits, theodolites, and especially total stations are perhaps the most widely used surveying instruments today. The transit and theodolite are fundamentally equivalent and can accomplish basically the same tasks. Their most important application is measuring horizontal and vertical (or zenith) angles (see Chapter 11), but they can also be used to obtain horizontal distances and determine elevations of points by stadia (see Chapter 15), accomplish low-order differential leveling (see Section 11-17), and establish alignments—in particular to prolong straight lines (see Section 11-13).

Total station instruments incorporate an electronic, or digital, theodolite, an electronic distance measuring (EDM) instrument, and a microprocessor in the same unit. Thus they can accomplish all of the tasks of transits and theodolites, and in addition they can also measure distances accurately. Furthermore, they can make computations with these measurements and display the results in real time.

The main components of a transit, a theodolite, and the theodolite portion of a total station include a sighting telescope and two graduated circles mounted in mutually perpendicular planes. Prior to measuring angles, the *horizontal* circle is oriented into a horizontal plane, which automatically puts the other circle in a vertical plane. Horizontal and vertical (or zenith) angles can then be measured directly in their respective planes of reference. Until recently, level vials were the usual means of orienting the circles, but most newer instruments now use automatic compensators or electronic tilt-sensing mechanisms.

In chronological order, the transit was the first of these three instruments, with the theodolite and total station following in order. The first American transit was produced by William Young of Philadelphia in 1831. The term *transit* was adopted for the instrument because its telescope could be *transited,* or reversed in direction, by rotating it about a "horizontal axis." In Europe the name *transiting theodolite* was adopted for this type of angle-measuring instrument. Europeans eventually dropped the adjective and retained the name *theodolite.*

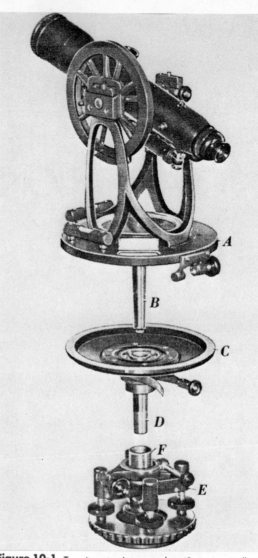

Figure 10-1 Transit parts. *A*, upper plate; *B*, inner spindle; *C*, lower plate; *D*, outer spindle; *E*, leveling head; *F*, socket. (Courtesy W. & L. E. Gurley.)

There is no internationally accepted understanding among surveyors on the exact difference denoted by the terms *transit* and *theodolite*. Currently, however, the most commonly used criterion is their general design, particularly their graduated circles and systems for reading them. Transits have an "open-circle" design (see Figures 10-1 and 10-2), in which their graduated circles, made of metal, are visible to an operator, and are read with the aid of verniers (see Section 10-4). Theodolites, on the other hand, feature an enclosed design (see Figures 10-7 through 10-10 and 10-12). Older theodolites have graduated circles made of glass. They are not directly visible to an operator and must be read by means of an internal microscopic optical system. These instruments are called *optical-reading theodolites*. Newer instruments, called *digital theod-*

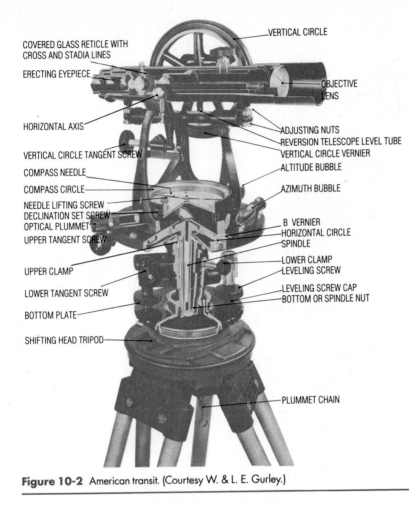

COVERED GLASS RETICLE WITH
CROSS AND STADIA LINES

ERECTING EYEPIECE

HORIZONTAL AXIS

VERTICAL CIRCLE TANGENT SCREW

COMPASS NEEDLE

COMPASS CIRCLE

NEEDLE LIFTING SCREW
DECLINATION SET SCREW
OPTICAL PLUMMET

UPPER TANGENT SCREW

UPPER CLAMP

LOWER TANGENT SCREW

BOTTOM PLATE

SHIFTING HEAD TRIPOD

VERTICAL CIRCLE

OBJECTIVE
LENS

ADJUSTING NUTS
REVERSION TELESCOPE LEVEL TUBE
VERTICAL CIRCLE VERNIER
ALTITUDE BUBBLE

AZIMUTH BUBBLE

B VERNIER
HORIZONTAL CIRCLE
SPINDLE

LOWER CLAMP
LEVELING SCREW

LEVELING SCREW CAP
BOTTOM OR SPINDLE NUT

PLUMMET CHAIN

Figure 10-2 American transit. (Courtesy W. & L. E. Gurley.)

olites, automatically resolve the circle readings using electronic systems, and display them in digital form. Other distinctions between transits and theodolites are described subsequently in this chapter. A few instruments, such as that in Figure 10-3, which combine some design features of both the transit and theodolite, are called *optical-reading transits.* They have glass circles and are read from glass verniers viewed through magnifying glasses.

In general theodolites are capable of greater precision and accuracy in angle measurements than transits. Because of this and other advantages, during the period from the 1960s through the 1980s, theodolites were purchased much more frequently than transits in the United States. Now in the 1990s, total station instruments, with their automatic angle and distance readout capabilities and built-in microprocessors for real-time data processing, are rapidly replacing both types of instruments. Some transits and many theodolites are still in use, however.

Transits, theodolites, and total stations are described in order in Parts I, II, and III which follow in this chapter. In spite of differences between these instruments, many of their parts, and their fundamental relationships, are the same.

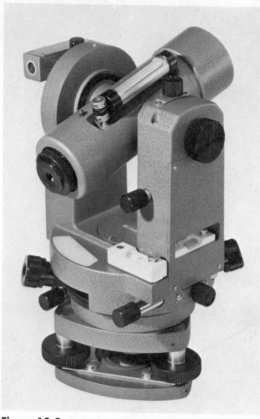

Figure 10-3 Optical reading transit.

PART I
THE TRANSIT

10-2 PARTS OF A TRANSIT

Transits are manufactured for general and special uses, but all have three main parts:
(**1**) alidade and upper plate, (**2**) lower plate, and (**3**) leveling head. These are shown in
their relative positions in Figure 10-1 and assembled in Figure 10-2. Reference to these
figures will lead to a better understanding of the descriptions given in the following
sections.

10-2.1 Upper Plate

The alidade containing an upper plate (*A* in Figure 10-1) is a horizontal circular plate
combined with a vertical *spindle B,* which enables the plate to revolve about a vertical
axis. The tapered design of American transit spindles assures that despite wear, unless
damaged by dirt or an accident, they will still seat and center properly. Attached to the
plate are two level vials, one parallel with the telescope (*altitude bubble*) and the other

at right angles to it *(azimuth bubble)* (see Figure 10-2), and two *verniers,* referred to as *A* and *B,* set 180° apart. Provisions are made for adjusting verniers and level vials.

Two vertical *standards* are cast as an integral part of the upper plate to support the horizontal *cross arms* of the telescope in bearings. The telescope revolves in a vertical plane about the centerline through the arms, called the *horizontal axis.*

The telescope, similar to that of a dumpy level (Section 6-7), contains an eyepiece, a reticle, and an objective lens system. A sensitive telescope level vial is attached to the telescope tube so the transit can be used as a leveling instrument on work where lower magnification and lesser sensitivity of the telescope vial are satisfactory.

The telescope is said to be in the *normal* or *direct* position when the telescope level vial is below it. Turning the telescope on its horizontal axis puts the level vial above, and the instrument is said to be in a *plunged, inverted,* or *reversed* mode. To permit use of the telescope for leveling in either the normal or inverted position, a *reversion vial* (curved and graduated on both its top and bottom so it is usable in both positions) is desirable.

A *vertical circle* supported by a cross arm turns with the telescope as it is revolved. The circle normally is divided into $\frac{1}{2}^\circ$ spaces with readings to the nearest minute obtained from a vernier having 30 divisions. The vernier is mounted on one standard with provisions for adjustment. If set properly, it should read zero when the telescope bubble is centered. If out of adjustment, a constant *index error* is read from the circle with the bubble centered and must be applied to all vertical angles, with appropriate sign, to get correct values.

The upper plate also contains the *compass box* and holds the *upper tangent screw.*

A *vertical-circle clamp* (for the horizontal axis) is tightened to hold the telescope horizontal or at any desired inclination. After the clamp is set, a limited range of vertical movement can be obtained by manipulating the *vertical-circle tangent screw* (also called the *vertical slow-motion screw*).

10-2.2 Lower Plate

The lower plate (*C* in Figure 10-1) is a horizontal circular plate graduated on its upper face. Its underside is attached to a vertical, hollow, tapered spindle *D* into which the upper plate's spindle fits precisely. The upper plate completely covers the lower plate, except for two openings where the verniers exactly meet the graduated circle.

The *upper clamp* (see Figure 10-2) fastens the upper and lower plates together. A small range of movement is possible after clamping by using the *upper tangent screw* (located on the upper plate).

10-2.3 Leveling Head

The leveling head (*E* in Figure 10-1) consists of a bottom horizontal plate and a "spider," with four *leveling screws.* The leveling screws, set in cups to prevent scoring the bottom plate, are partly or completely enclosed for protection against dirt and damage. The bottom plate has a collar threaded to fit on the tripod head.

A socket (*F* in Figure 10-1) of the leveling head includes a *lower clamp* (see Figure 10-2) to fasten the lower plate. The *lower tangent screw* is used to make precise settings after the lower clamp is tightened. The base of the socket is fitted into a ball-and-socket joint resting on the bottom plate of the leveling head, on which it slides horizontally. A *plummet chain* attached to the center of the spindle holds a plumb-bob string. An *optical plummet,* which is a telescope through the vertical center (spindle), is available on some transits. It points vertically when the plates are level, and is viewed at right angles (horizontally) by means of a prism for ease of observation.

A recapitulation of the use of the various clamps and tangent screws may be helpful to the beginner. The vertical-circle clamp and tangent screw on one standard control movement of the telescope in the vertical plane. The upper clamp fastens the upper and lower plates together, and an upper tangent screw permits a small differential movement between them. A lower clamp fastens the lower plate to the socket, after which a lower tangent screw turns the plate through a small angle. If the upper and lower plates are clamped together, they will, of course, move freely as a unit until the lower clamp is tightened.

Tripods for transits and levels, either fixed- or adjustable-leg types, may be used interchangeably.

10-3 CIRCLE SCALES

The *horizontal circle* of the lower plate may be divided in various ways, but generally it is graduated into 30 or 20' spaces. For convenience in measuring angles to the right or left, graduations are numbered from 0 to 360° both clockwise and counterclockwise. Figure 10-4 shows these arrangements.

Vertical circles of most transits are graduated into 30' spaces. They are usually numbered from zero at the bottom (for a horizontal sight) to 90° in both directions (for

Figure 10-4 Transit verniers. (Least count of (a) = 1'; least count of (b) = $\frac{1}{2}'$.)

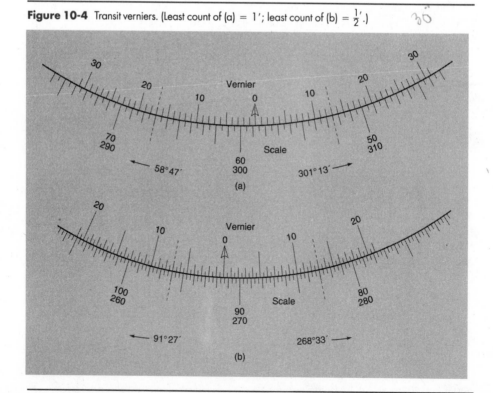

vertical sights), and then back to zero at the top. This facilitates reading both elevation and depression angles (see Section 11-10) with the telescope in either a direct or an inverted position.

10-4 TRANSIT VERNIERS

A vernier is a short auxiliary scale set parallel to and beside a primary scale. It enables reading fractional parts of the smallest main-scale divisions without interpolation. Verniers are used on transits for making angle readings.

Figures 10-4(a) and (b) show two different vernier and angle scale combinations commonly used on transits. Both are *double* verniers; that is, they can be read clockwise or counterclockwise from the index (zero) mark. Considering the clockwise vernier scale of Figure 10-4(a), it is constructed so that 30 of its divisions cover 29 half-degree (30′) divisions on the graduated circle. Thus each vernier division equals $\frac{29}{30}$ × 30′ = 29′. The difference between the length of one main-scale division and one vernier division is therefore 01′. This is the so-called *least count* of this vernier.

In general, the least count of a vernier is given by

$$\text{least count} = d/n \qquad (10\text{-}1)$$

where d is the value of the smallest main-scale division, and n the number of vernier divisions that span $(n - 1)$ main-scale units. By Eq. (10-1), the least count of the vernier of Figure 10-4(a) is $\frac{30'}{30}$ = 01′. This verifies the intuitive determination given above. An observer cannot make readings using a vernier without first determining its least count.

For the vernier of Figure 10-4(a), if the index mark were perfectly aligned with a certain main-scale division, say the 58°30′ mark (reading clockwise), and then the vernier advanced 01′ clockwise, the first vernier division would then be aligned with the first main-scale division to the left of the index mark and the reading would be 58°31′. If the circle were moved another 01′, the second vernier division would then be aligned with the second main-scale division left of the index mark and the reading would be 58°02′, and so on. In Figure 10-4(a) the seventeenth vernier division aligns with a main-scale division (shown by a dashed line), thus the circle has been advanced 17′ beyond 58°30′ and the clockwise reading is 58°47′.

With the double verniers shown in Figure 10-4(a) and (b), an observer can measure angles both clockwise and counterclockwise. This is particularly convenient for measuring deflection angles, which can be right or left. With double verniers there will be two matching lines, one for the clockwise angle, and the other for the counterclockwise angle. The counterclockwise circle reading in Figure 10-4(a) is 301°00′ + 13′ = 301°13′. Note that the sum of the clockwise and counterclockwise angles should be 360°00′.

In Figure 10-4(b) the smallest circle graduation is 20′, and 40 vernier divisions span 39 on the circle. Thus by Eq. (10-1), the least count is $\frac{20'}{40}$ = $\frac{1}{2}'$. The reading of the clockwise angle is 91°20′ + 07′ = 91°27′; for the counterclockwise circle it is 268°20′ + 13′ = 268°33′.

An understanding of verniers is best obtained by practice in reading various types, and by calculating and sketching the least count for different combinations of scale and vernier divisions. Typical mistakes in reading minutes and seconds from verniers result from the following:

1. Not using a magnifying glass
2. Reading in the wrong direction from zero, or on the wrong side of a double vernier
3. Failing to determine the least count correctly
4. Omitting the partial degree graduations of say 20 or 30′ when the index is beyond those marks

10-5 PROPERTIES OF THE TRANSIT

Transits are designed to have a proper balance between magnification and resolution of the telescope, least count of the vernier, and sensitivity of the plate and telescope bubbles. An average length of sight of about 300 ft is assumed in design. Thus a standard 1′ instrument has the following properties:

Magnification, 18 to 30×
Field of view, 1° to 1°30′
Resolution, 3 to 5″
Minimum focus, about 3 to 8 ft
Sensitivity of plate levels per 2-mm division, 60 to 100″
Sensitivity of telescope vial per 2-mm division, 30 to 60″
Weight of instrument head without tripod, 11 to 18 lb

Transit reticles usually include vertical and horizontal center hairs and additional stadia hairs, as shown in Figure 10-5(b), (c), and (d). Short stadia lines, used on glass reticles [Figure 10-5(b) and (d)], avoid confusion between the center and stadia hairs. A quarter hair, located halfway between the upper and middle hairs [Figure 10-5(c)], is sometimes used to increase the range of stadia readings, as noted in Chapter 15. The arrangement shown in Figure 10-5(d) avoids covering the object sighted and aids in centering. Other types of reticles designed specifically for making sun observations are described in Section 18-16.

A transit is a *repeating instrument* because horizontal angles can be measured by repetition any number of times and their total added on the plates. Advantages of the

Figure 10-5 Arrangement of cross hairs.

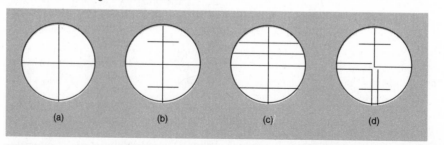

(a)   (b)   (c)   (d)

repeating procedure are (**1**) better accuracy obtained through averaging, and (**2**) disclosure of mistakes and errors by comparing values of the single and multiple readings.

10-6 HANDLING AND SETTING UP A TRANSIT

A transit is taken from its box by holding the leveling head, the underside of the lower plate, or the standards (*not* lifting by the telescope). It should be screwed securely on the tripod. A transit carried indoors should be balanced in a horizontal position under one arm, with the instrument head forward. The same method is suitable in areas covered with brush. In open terrain the instrument may be balanced on a shoulder. When a transit is carried, the telescope should be clamped *lightly* in a position perpendicular to the plates. The plate clamps should also be set *lightly* to prevent swinging, while still permitting ready movement if the instrument is bumped.

The wing nuts on the tripod legs must be tight to prevent slippage and rotation of the head. They are correctly adjusted if each tripod leg falls slowly of its own weight when placed in a horizontal position. If the wing nuts are overly tight, or if pressure is applied to the legs crosswise (which can break them) instead of lengthwise to fix them in the ground, the tripod is in a strained position. The result may be an unnoticed movement of the instrument head after observations have begun. Tripod legs should be well spread to furnish stability and place the telescope at a height convenient for the observer. Tripod shoes must be tight. Proper field procedures can eliminate most instrument maladjustments, but there is no way to take care of a poor tripod with dried-out wooden legs, except to discard or repair it.

A plumb-bob string hung on a hook at the bottom of the spindle using a slipknot permits raising or lowering the bob without retying, and avoids knots. A small slide attachment is also useful in accomplishing this purpose. The plumb bob must be brought directly over a definite point such as a tack in a wooden stake, and the plates leveled. The tripod legs can be moved in, out, or sideways to approximately level the plates before the leveling screws are used. Shifting the legs affects the position of the plumb bob, however, which makes it more difficult to set up a transit than a level.

Two methods are used to bring the plumb bob within about $\frac{1}{4}$ in. of the proper point. In the first method, the transit is set over the mark and one or more legs are moved to bring the plumb bob into position. One leg may be moved circumferentially to level the plates without greatly disturbing the plummet. Beginners sometimes have difficulty with this method because at the start the transit center is too far off the point, or the plates are badly out of level. Several movements of the tripod legs may then fail to both level the plates and center the plumb bob while maintaining a convenient height of instrument. If an adjustable-leg tripod is used, one or two legs can be lengthened or shortened to bring the bob directly over the point.

In the second method, which is particularly suited to level on uniformly sloping ground, the transit is set up near the point and the plates are approximately leveled by moving the tripod legs as necessary. Then, with one tripod leg held in the left hand, another under the left armpit, and the third supported in the right hand, the transit is lifted and placed over the mark. A slight shifting of one leg should bring the plumb bob within perhaps $\frac{1}{4}$ in. of the proper position and leave the plates practically level.

The plummet is precisely centered by loosening all four leveling screws and sliding them on the bottom plate using the ball-and-socket shifting-head device, which permits a limited movement. To assure mobility in any direction, the shifting head should be approximately centered on the bottom plate before setting up the instrument and when boxing it.

A transit is accurately leveled by means of the four leveling screws in somewhat the same manner described for a level. However, each level vial on the upper plate is first set over a pair of opposite screws, and because there are two vials available, the tele-scope position need not be changed in the initial leveling process. After both bubbles are carefully centered, the telescope is rotated 180°. As explained in Section 11-18.1, and illustrated in Figure 11-17, if the bubbles run, they will indicate *double* the dislev-elment. They should therefore be brought back halfway, and the telescope rotated back to its original direction. If the bubbles remain in their halfway positions, even though not centered, the instrument is level; if not, the process is repeated until the bubbles stay in the same location no matter what direction the telescope is pointed. If after leveling the instrument, the bubbles are very far off center, it may be desirable to adjust the vials, as described in Appendix A.

If the plumb bob is still over the mark after leveling, the instrument is ready for use. But if the plates are badly out of level, or the leveling screws not uniformly set, the plummet moves off the mark during leveling. The screws must then be loosened and shifted again, and the transit releveled. It is evident that time can be saved by starting with the plates reasonably level to avoid excessive manipulation and possible binding of the screws.

PART II
THE THEODOLITE

10-7 CHARACTERISTICS OF OPTICAL-READING THEODOLITES

In Section 10-1 it was noted that there are two different types of theodolites, *optical-reading* theodolites, which are read by means of internal microscopes, and *digital* theodolites, which electronically resolve angle readings and display them in numerical form. In this section, and in Sections 10-8 and 10-9, optical-reading theodolites are discussed. Digital theodolites are described in Section 10-10.

Theodolites differ from transits in appearance (they are generally more compact, lightweight, and "streamlined") and in design by a number of features, the more important of which are as follows:

1. The *telescopes* are short, have reticles etched on glass, and are equipped with rifle sights or collimators for rough pointing.

2. The *horizontal* and *vertical circles* are made of glass with graduation lines and numerals etched on the circles' surfaces. The lines are very thin and more sharply defined than can be achieved by scribing them on the metal circles used on transits. Precisely graduated circles with small diameters can be obtained, and this is one reason the instruments are so compact. Circles are divided into conventional sexagesimal degrees and fractions (360°), or into centesimal *grads* or *gons* (full circle divided into 400^g).

Figure 10-6 Standard tribrachs for most theodolites are designed for interchanging various accessories. Here tribrachs are shown to be compatible with a sighting target (left) and an EDM reflector (right). (Courtesy Leica, Inc.)

3. The *vertical circle* of most theodolites is precisely indexed with respect to the direction of gravity in one of two ways: **(a)** by an *automatic compensator* or **(b)** by a *collimation level* or *index level,* usually the coincidence type connected to the reading system of the vertical circle. Both provide a more accurate plane of reference for measuring vertical angles than the plate levels used on transits. Vertical circle readings are *zenith* angles, that is, 0° occurs with the telescope pointing vertically, and either 90° or 270° is read when it is horizontal.

4. The circle *reading systems* consist basically of a microscope with the optics inside the instrument. A reading eyepiece is generally adjacent to the telescope eyepiece or located on one of the standards. Some instruments have optical micrometers for fractional reading of circle intervals (micrometer scale visible through reading microscope); others are direct-reading. With most theodolites a mirror located on one standard can be adjusted to reflect light into the instrument and brighten the circles for daytime use. They can be equipped with a battery-operated internal lighting system for night and underground operation. Some theodolites can also use the battery-operated system in lieu of mirrors for daytime work.

5. Rotation about the *vertical axis* occurs within a steel cylinder or on precision ball bearings, or a combination of both.

6. The *leveling head* consists of three screws or *cams.*

7. The *bases* or *tribrachs* of theodolites are often designed to permit interchange of the instrument with sighting targets, prisms, and EDM instruments without disturbing centering over the survey point. Figure 10-6, for example, shows the placement of a sighting target and an EDM reflector on a theodolite tribrach.

8. An *optical plummet,* built into the base or alidade of most theodolites, replaces the plumb bob and permits centering with great accuracy.

9. A compass can be attached to a theodolite as an accessory, but it is not an integral part of the instrument, as it is with transits.

10. The tripods used with theodolites are the wide-frame type, and most have adjustable legs. Some are all metal and feature devices for preliminary leveling of the tripod head and mechanical centering ("plumbing") to eliminate the need for a plumb bob or optical plummet.

10-8 OPTICAL-READING *REPEATING* THEODOLITES

Optical-reading theodolites are divided into two basic categories: the *repeating* type and the *directional* model. Repeating theodolites are equipped with a double vertical axis, usually cylindrical in shape, or a repetition clamp. The double vertical axis is similar to the double spindle arrangement used on transits. This design enables horizontal angles to be repeated any number of times and added directly on the instrument's circle.

Figures 10-7 and 10-8 show examples of repeating type theodolites. The optical reading system of each instrument is shown in the small inset figures. Each of these theodolites reads directly to the nearest minute, with estimation possible to 0.1′. Both instruments have vertical-circle automatic compensators, telescopes with standard eyepieces of 30× magnification, optical plummets, and plate bubble sensitivity of 30″/2-mm division. These two instruments are representative of many others of this type.

The reading system of the Lietz TS6 theodolite in Figure 10-7 consists of a graduated glass scale having a span of 1° which appears superimposed on the degree divisions of the main circle. This scale is read directly by means of a microscope whose

Figure 10-7 Lietz TS6 optical-reading repeating theodolite. (Courtesy Sokkia Corporation.)

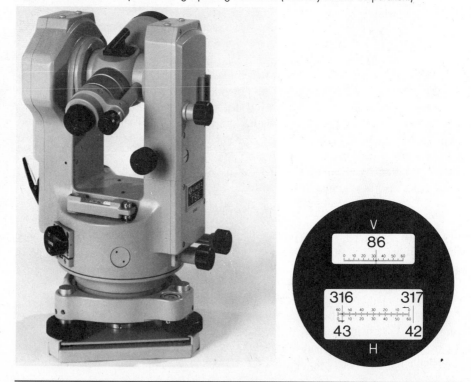

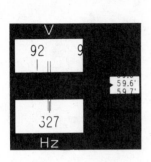

Figure 10-8 T-1 optical-reading repeating theodolite. (Courtesy Leica, Inc.)

small eyepiece can be seen beside the main telescope in the figure. To take a reading, it is simply necessary to observe which degree number lies within the 1° span of the glass scale and select the minute indicated by the index mark. The vertical and clockwise horizontal-circle readings indicated for the TS6 in Figure 10-7 are 86°32.5′ and 316°56.5′, respectively. (The counterclockwise horizontal circle reading is 43°03.5′.) Thus on this instrument the horizontal and vertical circles can be viewed and read simultaneously through the reading microscope.

The reading system of the T-1 in Figure 10-8 consists of an optical micrometer, which is also viewed with a microscope through a small eyepiece next to the main telescope. To make a reading, the operator must first center the reference mark between the two bifilar lines of a degree mark by turning the micrometer knob. The micrometer spans 1° of the main circle, and by centering the reference mark, the minutes portion of the angle can be read in the micrometer window on the right side of the reading microscope's field of view. The horizontal angle indicated in Figure 10-8 is 327°59.6′. The micrometer must be set again to read the vertical angle.

10-9 OPTICAL-READING *DIRECTIONAL* THEODOLITES

A directional theodolite is a nonrepeating type of instrument that has no lower slow motion. *Directions* rather than angles are read. After a sight has been taken on a point, the line's direction is read on the circle. An observation on the next mark gives a new direction, so the angle between the lines can be found by subtracting the first direction from the second.

Directional theodolites have a single vertical axis and therefore cannot measure angles by the repetition method. They do, however, have a *circle-orienting drive* to make a rough setting of the horizontal circle at any desired position.

On all directional theodolites each reading represents the *mean* of two diametrically opposed sides of the circle, made possible because the operator simultaneously views

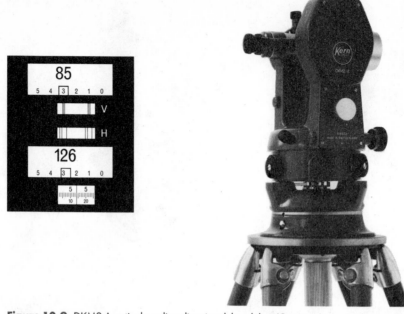

Figure 10-9 DKM2-A optical-reading directional theodolite. (Courtesy Leica, Inc.)

both sides of it through internal optics. This reading procedure, equivalent to averaging readings of the A and B verniers of a transit, automatically compensates for eccentricity errors (see Section 11-18.1).

Typical directional theodolites are shown in Figures 10-9 and 10-10. Each has a micrometer that permits reading the horizontal and vertical circles directly to 1″, with estimation possible to the nearest 0.1″. Both have automatic compensators for orienting the vertical circle, optical plummets, and plate bubbles with 20″/2-mm division sensitivity. These instruments are representative of many other similar ones in the directional category.

An inset for each figure illustrates the circle reading system of the instrument. The vertical and horizontal circles of the DKM2-A (see Figure 10-9) carry two concentric scales, one with single-line graduations, the other with bifilar divisions. The circle-reading microscope shows a portion of one scale superimposed on the diametrically opposed portion of the other. In reading an angle, the operator turns the optical micrometer knob to shift the two scales until the single-line graduations appear centered within the bifilar lines. As the micrometer is revolved, there is a simultaneous movement of a *field stop*, which frames the number of 10′ divisions in the reading. Centering with the microscope must be done separately for the horizontal and vertical circles.

In Figure 10-9 the micrometer has been set to read the vertical circle so the single graduations are centered within the bifilar lines of the window labeled V. (Note they are not centered in the H window.) The reading is 85° (see directly in the upper window) plus 3 × 10 or 30′ (the 3 being taken from within the field stop frame of the same window), plus 5′14.0″ (from the lower window). Thus the vertical circle reading is

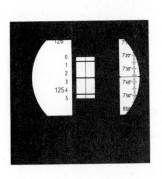

Figure 10-10 Th-2 optical-reading directional theodolite. (Courtesy Carl Zeiss, Inc.)

85°35′14.0″. The micrometer setting should be repeated before deciding to record 14″ instead of 13 or 15″.

The reading system of the Th-2 shown in Figure 10-10 is similar to that of the DKM2-A. A selector knob permits viewing either the horizontal or the vertical circle through the microscope; both circles cannot be seen simultaneously. The center smaller rectangular window of Figure 10-10 shows the graduations on diametrically opposed parts of the circle. The micrometer has already been adjusted for reading by making opposite graduations coincide, and in that position the number corresponding to a multiple of 10′ in the reading is indicated in the left window directly beside the number of degrees. Minutes and seconds portions of the reading are taken from the rightmost window. The angle thus indicated in the Th-2 is 125°, plus 4 × 10′ (both from the left window), plus 7′36″ (in the right window), giving a final reading of 125°47′36″.

The reading system of a widely used, older version T-2 directional theodolite is illustrated in Figure 10-11. In this arrangement the circles are divided into 20′ spaces and the whole-degree marks numbered.

A reading is obtained by turning the micrometer knob to make the division lines (which move in opposite directions) coincident. Since an average reading of the two sides of the circle is secured, each graduation is counted as 10′ to eliminate a later division by 2. The micrometer range is limited to 10′. The set of numbers seen upside down is on the circle's opposite side.

In Figure 10-11 a direction of 265°40′ is read by counting the number of divisions (4) between 265° and its diametrically opposite graduation, 85°. A micrometer scale gives the additional minutes and seconds—in this example, 7′23.6″. The vertical circle, read in the same manner, furnishes the zenith angle.

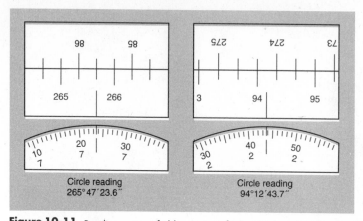

Figure 10-11 Reading system of older version of T-2 optical-reading directional theodolite.

An additional extremely precise directional instrument used for first-order triangulation, the DKM-3, is shown in Figure 19-13.

10-10 ELECTRONIC DIGITAL THEODOLITES

Electronic digital theodolites can automatically read and display horizontal and vertical angles, thus eliminating the need for manual reading of scales on graduated circles through microscopes. The basic design and appearance of these instruments (see Figure 10-12) are the same as those of the standard theodolites described in preceding sections. The fundamental difference is the manner in which they resolve angular values and display them externally in digital form.

The systems utilized to measure and display angular values automatically are similar to those now used in modern checkout machines in grocery and department stores. They operate by passing a beam of light through unique patterns of black bars with varying thicknesses and spacing. The Topcon DT-10 of Figure 10-12 is representative of the way electronic digital theodolites operate, and is briefly described here. Two glass circles are mounted parallel, one on top of the other, with a slight spacing between them. The *rotor* (lower circle) contains a pattern of equally divided alternate dark lines and light spaces. The *stator* (upper circle) contains a slit-shaped pattern which has the same pitch as that of the rotor circle. A *light-emitting diode* (LED) directs collimated light through the circles from below toward a photo detector cell above.

When an angle is turned with the DT-10, the rotor moves with respect to the stator, creating alternating variations of light intensity. The photo detector senses these variations, converts them into electrical pulses, and passes them to a microprocessor for conversion into digital values. The digits are displayed using a *liquid crystal diode* (LCD). Some employ LEDs, but LCDs use less power. They do however, require illumination for night work. The Topcon DT-10 can resolve horizontal and vertical angles to an accuracy of 3″.

Two separate systems like that just described are used in an electronic digital theodolite: one for measuring horizontal angles and another for vertical (or zenith) angles. The displayed angles can be read and recorded manually in field books, or the instruments can be equipped with automatic data collectors to eliminate recording.

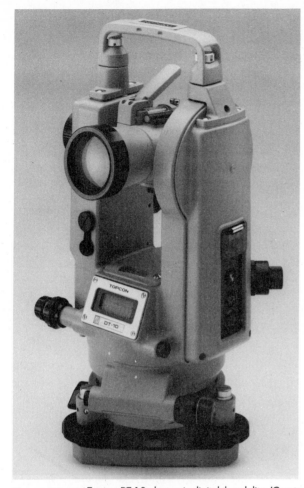

Figure 10-12 Topcon DT-10 electronic digital theodolite. (Courtesy Topcon America Corporation.)

Approximately 20 different models are currently manufactured. Their angle display resolutions range from 0.1 to 30″. Most employ automatic compensators for vertical-circle orientation. These devices are similar to those used on automatic levels (see Section 6-11), and automatically align the vertical circle so that 0° is oriented precisely upward (opposite the direction of gravity).

The technologies embodied in these new microprocessor-controlled electronic digital theodolites provide several significant advantages to surveyors. As examples, **(1)** the circles can be instantaneously zeroed by simply pressing a button, or they can be initialized to any value by entry through the keyboard (valuable for setting the reference azimuth for a backsight); **(2)** angles can be measured with increasing values either left or right; and **(3)** angles measured by repetition can be added to provide the total, even though 360° may have been passed one or more times. Other advantages are that mistakes in reading angles are greatly reduced, the speed of operation is increased, and the cost to produce these instruments is lower.

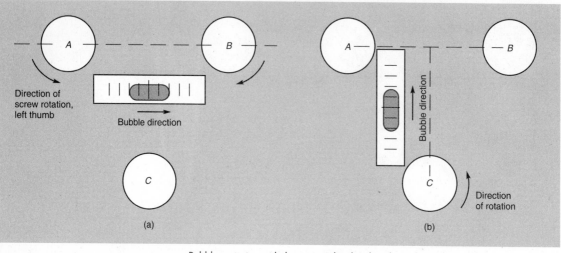

Figure 10-13 Bubble centering with three-screw leveling head.

10-11 HANDLING AND SETTING UP A THEODOLITE

Theodolites should be carefully lifted from their carrying cases by grasping the standards (some newer instruments are equipped with handles for this purpose), and the instrument securely fastened to the tripod by means of a tribrach. The tripod with the instrument is placed over the ground point in the manner described for transit setups (see Section 10-6). Beginners can use a plumb bob to approximate the required setup position. Precise centering over the point is done by means of an optical plummet, which provides a line of sight directed downward collinear with the theodolite's vertical axis. The instrument must be level for the optical plummet to define a vertical line. Most theodolite tribrachs have a relatively insensitive bull's-eye bubble to facilitate rough preliminary leveling before beginning final leveling with the plate bubble. Some tribrachs also contain an optical plummet.

The setup process using an instrument with an optical plummet, tribrach mount with bull's-eye bubble, and adjustable-leg tripod is most easily accomplished in the following steps: (**1**) adjust the position of the tripod legs by lifting and moving the instrument as a whole until the point is near the optical plummet's line of sight; (**2**) plant the legs and center the bull's-eye bubble by adjusting the tripod leg lengths (the point will still be nearly on the optical plummet's line of sight); (**3**) level the instrument using the plate bubble and leveling screws; and (**4**) loosen the tribrach screw and translate the instrument (do not rotate it) to carefully center the plummet cross hair on the point. Repeat steps (3) and (4) until precise leveling and centering are accomplished. Before starting, the instrument should be centered on the tripod head to permit maximum translation [step (4)] in any direction.

It was noted earlier that theodolites have a three-screw leveling head and a single plate bubble. To level this type of instrument, the plate bubble is placed parallel to the line through any two foot screws, as the line through *A* and *B* in Figure 10-13(a). The bubble is centered by turning these two screws, then rotated 90°, as shown in Figure 10-13(b), and centered again using the third screw (*C*) only. This process is repeated and carefully checked to ensure that the bubble remains centered. As with the transit and level, and as illustrated in Figure 10-13, *the bubble moves in the direction of the left thumb when the foot screws are turned.* A solid tripod setup is essential for theodolites that have very sensitive bubbles, and the instrument must be shaded if set up in bright sunlight. Otherwise the bubble will expand and run toward the warmer end as the liquid is heated.

PART III
TOTAL STATION INSTRUMENTS

10-12 CHARACTERISTICS OF TOTAL STATION INSTRUMENTS

Total station instruments combine three basic components—an EDM instrument, an electronic digital theodolite, and a computer or microprocessor—into one integral unit. These devices, also called *electronic tacheometers,* can automatically measure horizontal and vertical angles as well as slope distances from a single setup. From these data they can instantaneously compute horizontal and vertical distance components, elevations, and coordinates, and display the results on an LCD. They can also store the data, either on board or in external data collectors. A Lietz SET 2B total station with handheld data collector is shown in Figure 10-14.

Figure 10-14 Lietz SET 2B total station with SDR 33 data collector. (Courtesy Sokkia Corporation.)

Total stations offer many advantages for almost all types of surveying. They are used for topographic, hydrographic, cadastral, and construction surveys. With approximately 40 different models available from which to choose, they are currently the dominant instruments in surveying. In fact, a 1992 survey revealed that approximately 70% of firms engaged in surveying were using total station instruments.

The two measuring components of a total station, the EDM instrument and digital theodolite, have already been described in Chapter 5 and in Section 10-10, respectively. The EDM instrument component installed in a total station is relatively small, and as shown in the instrument of Figure 10-14, it is mounted with the telescope between the standards of the digital theodolite. Consequently, most instruments are not much larger than a theodolite. Among the available instruments, their weights vary from about 10 to 20 lb. Although the EDM instruments are small, they still have distance ranges adequate for most work. Lengths up to about 1 to 2 km can be measured with a single prism, and up to about 3 or 4 km with a triple prism.

The angle resolution of available total stations varies from as low as a half-second for precise instruments suitable for control surveys, up to 20″ for instruments made specifically for construction stakeout. Formats used for displaying angles also vary with different instruments. For example, the displays of some actually show the degree, minute, and second symbols, but others use only a decimal point to separate the number of degrees from the minutes and seconds. Thus 315.1743 is actually 315°17′43″.

Some instruments allow a choice of units, such as the display of angular measurements in degrees, minutes, and seconds, or in grads (gons). Distances may be shown in either feet or meters. Also, certain instruments enable the choice of displaying either zenith angles or vertical angles. These choices are entered through the keyboard, and the microprocessor performs the conversions accordingly. The keyboard for instrument control and data entry is located just above the leveling head, as shown in Figure 10-14. For convenience when operating in both the direct and reverse modes, an identical keyboard exists on both sides of most instruments.

The time required to make and display angle and distance measurements is generally in the range of about 3 to 7 sec when operated in the normal mode, and less than 0.5 sec when operated in the tracking mode. In the normal mode higher precision results because multiple measurements are made and the average taken. In the tracking mode, used primarily for stationing alignments and staking out construction projects, a prism is held on line near the anticipated final location of a stake. A measurement is quickly taken to the prism, the distance that it must be moved forward or back is instantly computed and displayed, the prism moved, another check of distance made, and if correct the stake set. This procedure is discussed in more detail in Chapter 24. The Lietz SET 2B of Figure 10-14 has a distance range of about $2\frac{1}{2}$ km with one prism, and up to about 3 km with a triple prism. Its horizontal and vertical angle resolution is 2″.

10-13 FUNCTIONS PERFORMED BY TOTAL STATIONS

Total stations, with their microprocessors, can perform a variety of functions and computations, depending on how they are programmed. Most are capable of assisting an operator, step by step, through several different types of basic surveying operations. After selecting the type of survey from a menu, prompts will automatically appear on the display to guide the operator through each step. An example illustrating a topographic survey conducted using this procedure is given in Section 16-9.1.

In addition to providing guidance to the operator, microprocessors of total stations can perform many different types of computations. The capabilities vary with different instruments, but some standard computations include (**1**) averaging of multiple angle and distance measurements; (**2**) correcting electronically measured distances for prism constants, atmospheric pressure, and temperature; (**3**) making curvature and refraction corrections to elevations determined by trigonometric leveling; (**4**) reducing slope distances to their horizontal and vertical components; (**5**) calculating point elevations from the vertical distance components (supplemented with keyboard input of instrument and reflector heights); and (**6**) computing coordinates of surveyed points from horizontal angle and horizontal distance components (supplemented with keyboard input of coordinates for the occupied station, and a reference azimuth. The subject of coordinate computations is covered in Chapter 13.

Many total stations, but not all, are also capable of making corrections to measured horizontal and vertical angles for various instrumental errors. For example, by going through a simple calibration process, the index error of the vertical circle can be determined, stored in the microprocessor, and then a correction applied automatically each time a zenith angle is measured. A similar calibration and correction procedure applies to errors that exist in horizontal angles due to imperfections in the instrument. Some total stations are also able to correct for personal errors, such as imperfect leveling of the instrument. By means of tilt-sensing mechanisms, they automatically measure the amount and direction of dislevelment, and then make corrections to measured horizontal and vertical angles for this condition.

10-14 DATA COLLECTION OPTIONS

Measurements can be stored "on board" with some total stations, and with others they are transferred to an external hand-held data collector. Two options are available in the on-board category. In the first option, the data can be stored directly in the memory of the microcomputer, and later downloaded to an external storage device via an RS-232 connection. This can occur in the field if an external memory unit or lap-top computer is available, or the instrument can be taken directly to the office for downloading. The second option features on-board storage systems like that shown with the Topcon ITS-1 total station in Figure 10-15, which stores information in removable memory cards. The card for the system shown has 256 kbytes of memory. When one card is filled, it can be removed and another quickly installed. With this system, memory is virtually unlimited, and the instrument never has to leave the field. A system for loading data from the memory card of Figure 10-15 into a computer is shown in Figure 10-16.

Several different versions of hand-held data collectors are available. Often they can do much more than store measurements. Some are actually computers which extend the system's overall computing capabilities well beyond those available only through the total station's microprocessor. Others are able to provide the control for operating the total station through commands entered on the data collector's keyboard. Because the hand-held units can be larger, they can have larger keyboards with more and larger keys. This can speed operations by increasing menu use, which reduces keystrokes.

10-15 SETUP AND OPERATION OF TOTAL STATIONS

Because they contain delicate electronic components, total stations are not as rugged as transits and theodolites. They must be packed and transported carefully, handled gently,

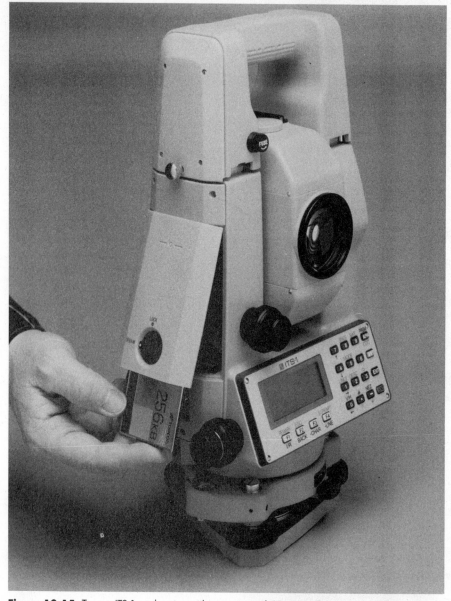

Figure 10-15 Topcon ITS-1 total station with memory card. (Courtesy Topcon America Corporation.)

and carefully removed from their cases. Total stations have three-screw leveling heads, and if equipped with a precise level vial, they are leveled using the same procedure as theodolites, described in Section 10-11. Some instruments, such as the Geodimeter 500 shown in Figure 1-3(c), do not have traditional level vials. That instrument is equipped with an electronic dual-axis leveling system in which four probes sense a liquid (horizontal) surface. After preliminary leveling is done by means of the tribrach bull's-eye bubble, signals from the probes are processed to form an image on the LCD display

Figure 10-16 Downloading files from external memory cards to a PC using a card reader. (Courtesy Topcon America Corporation.)

which guides an operator in performing rough leveling. The three leveling screws are used, but the instrument need not be turned about its vertical axis in the leveling process. After rough leveling, the amount and direction of any residual dislevelment is automatically and continuously received by the microprocessor, which corrects measured horizontal and vertical angles accordingly in real time.

As noted earlier, total stations are controlled with entries made either through their built-in keyboards or through the keyboards of hand-held data collectors. Details for operating each individual total station vary somewhat, and therefore are not described here. They are covered in the owner's manual provided with the purchase of an instrument. The use of total stations for specific types of surveys is described in later chapters.

The accuracy achieved with total stations is not merely a function of their ability to accurately resolve angles and distances. It is also related to operator procedure and to the condition of peripheral equipment. Operator procedure pertains to matters such as careful centering and leveling of the instrument, accurate pointing at targets, and taking averages of multiple angle measurements made in both direct and reversed positions. Peripheral equipment that can affect accuracy includes tribrachs, optical plummets, prisms, and prism poles. Tribrachs must provide a snug fit without slippage. Optical plummets that are out of adjustment cause instruments to be set up erroneously over the measurement point. Crooked prism poles or poles with maladjusted bull's-eye bubbles also cause errors in placement of the prism over the point being measured. (Normally optical plummets and bull's-eye bubbles can be adjusted easily; see Appendix A.) Prisms should be checked frequently to determine their constants (see Section 5-8.2) and their values stored for use in correcting distance measurements.

10-16 SERVO-DRIVEN AND REMOTELY OPERATED TOTAL STATIONS

The total station shown in Figure 1-3(c) can be equipped with a servo drive mechanism which enables it to aim itself automatically at a point to be set. It is only necessary to identify the point's number with a keyboard entry. The computer retrieves the direction to the point from storage or computes it and activates a servomotor to turn the telescope to that direction within a few seconds. This feature is particularly useful for construction stakeout, but it is also convenient in control surveying when multiple pointings are made in measuring angles by repetition. In this instance, final precise pointing is done manually.

The remote positioning unit (RPU 500) shown in Figure 10-17 enables control of a total station instrument from a distance. The remote positioning unit, which is attached to a prism pole, has a built-in telemetry link for communication with the total station. It can be operated with the Geodimeter 500 shown in Figure 1-3(c). A person is needed at the total station only for making the pointings. Otherwise all control is exercised through the keyboard of the remote positioning unit. Although this system still requires two people to conduct a survey, it enables the most experienced person in the party to control the measurements and also to make decisions about what to survey and where to hold the prism. This speeds the work, but perhaps more importantly, it eliminates mistakes in identifying points that can occur when the prism is far from the total station and cannot be clearly seen.

Geotronics recently introduced its Geodimeter model 4000, a new robotic total station that results in a one-person surveying party. The instrument can be totally controlled using a remote device similar to the RPU 500. The 4000 is equipped with an automatic search and aim function, as well as a link for communication with the remote positioning unit. It has servomotors for automatic aiming at the prism both horizontally and vertically. To operate the system, the total station must first be set up and oriented. This consists in entering the coordinates of the point where the total station is located, and taking a backsight along a line of known azimuth. Once oriented, the total station is sighted using the telescope of the remote positioning unit and the vertical angle of sight is telemetered to the total station. The total station's vertical servomotor automatically sets its telescope at the required vertical angle. Its horizontal servomotor activates and swings around until it finds the prism. Once the total station has found the remote positioning unit (which only takes a few seconds) and locks onto it, it will automatically follow its further movements. If lock is lost, the search routine is simply repeated. The remote positioning unit not only serves as the control unit for the system, but it also operates as a data collector.

Figure 10-17 Geodimeter remote positioning unit RPU 500. (Courtesy Geotronics of North America, Inc.)

√**10-1** List the fundamental differences between a transit and a theodolite.

√**10-2** Explain the differences between optical theodolites and elecronic digital theodolites.

10-3 Why should a transit, theodolite, or total station not be lifted by holding the telescope?

10-4 Why are two plate level vials used on a transit instead of just one?

10-5 What is the purpose of loosening the wing nuts of a transit or theodolite tripod before setting up?

10-6 Explain the importance of the upper clamp, upper tangent screw, lower clamp, and lower tangent screw in measuring an angle with a repeating instrument.

10-7 Determine the least count for the following transit main scale and vernier combinations:

    (a) Sixty divisions on a vernier span fifty-nine 10′ divisions of the main scale.

    (b) Forty divisions on a vernier cover thirty-nine 20′ divisions of the main scale.

    ***(c)** Thirty divisions on a vernier span twenty-nine 30′ divisions of the main scale.

√**10-8** Discuss the basic differences between repeating and directional theodolites.

√**10-9** Explain the procedure of leveling a theodolite that has three leveling screws and a single plate bubble.

*Asterisks indicate problems that have answers given in Appendix G.

10-10 List several advantages and disadvantages of an optical plummet compared with a plumb bob.

10-11 Explain "run of the micrometer" on the directional theodolite.

10-12 Can the repetition method be used for measuring vertical angles with a transit, theodolite, or total station?

10-13 Explain why a directional theodolite cannot be used to measure horizontal angles by repetition.

10-14 Name and describe an instrument used to accurately position a theodolite on top of a tower, exactly above its ground station.

10-15 What is a gyroscopic theodolite, and where can it be used most advantageously?

10-16 Explain how an electronic digital theodolie is able to automatically measure angles.

10-17 List the advantages of electronic digital theodolites.

10-18 Discuss the advantages of total station instruments as compared to transits and theodolites.

10-19 Name the three basic components that comprise a total station instrument.

10-20 List the standard types of computations that the microprocessors of most total station instruments are capable of performing.

10-21 Discuss the automatic data collection options that are available with modern total station instruments.

BIBLIOGRAPHY

Ali, A. E. 1991. "Electronic Theodolites: Comparison Test." *ASCE, Journal of Surveying Engineering* 117 (No. 1):3.

Beeks, D. 1987. "Evolution of the Transit." *Professional Surveyor* 7 (No. 5):10.

Brinker, R. C., and R. Minnick. 1987. *The Surveying Handbook.* New York: Van Nostrand Reinhold, Chap. 5.

Kivioja, L. A., and J. E. Pettey. 1973. "Wobbles of the Horizontal Axis of a Theodolite." *Surveying and Mapping* 33 (No. 4):481.

Moffitt, F. H., and H. Bouchard. 1992. *Surveying,* 9th ed. New York: Harper-Collins, Chap. 4.

Reilly, J. P. 1992. "P.O.B. 1992 Total Station Survey." *Point of Beginning* 17 (No. 4):24.

Reilly, J. P. 1992. "P.O.B. 1992 Transit/Theodolite Survey." *Point of Beginning* 18 (No. 2):78.

Steeves, P. A., and C. Horsfall. 1985. "Optical Plummet Adjustment." *Surveying and Mapping* 45 (No. 4):343.

Thousand, F. R. 1993. "Development and Use of the Total Station." *Wisconsin Professional Surveyor* (No. 154): 14.

11

FIELD OPERATIONS WITH TRANSITS, THEODOLITES, AND TOTAL STATIONS

11-1 INTRODUCTION

As mentioned in Section 10-1, transits and theodolites are employed principally for measuring horizontal and vertical angles. In some cases, unknown angular values must be determined so that positions of points can be calculated; in others, known angles are laid off to establish points at fixed locations given on construction plans. Transits and theodolites are also sometimes used for low-order differential leveling (see Section 11-17) and for measuring horizontal and vertical distances by stadia (see Chapter 15). Further uses include prolonging straight lines and staking construction alignments.

Total stations can also accomplish all of these tasks. For measuring or laying out angles, methods employed with total stations are basically the same as those used with transits and theodolites. However, with their built-in EDM instruments, total stations can measure longer distances with much greater accuracy than transits and theodolites can using the stadia procedure. This capability significantly expands their utility.

Descriptions of the field procedures used to accomplish basic measurements with transits, theodolites, and total stations are the subjects of this chapter. Uses of these instruments on specific types of surveys are covered in later chapters.

11-2 RELATIONSHIP OF ANGLES AND DISTANCES

Determining the relative positions of points often involves measurements of both angles and distances. The best quality surveys result when there is compatibility between the accuracies of these two different kinds of measurements. To select instruments and survey procedures necessary for achieving consistency, and to evaluate the effects of errors due to various sources, it is helpful to remember the relationships between angles and distances given here and illustrated in Figure 11-1.

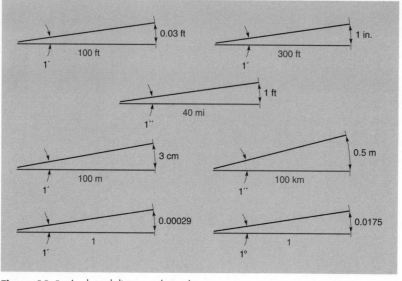

Figure 11-1 Angle and distance relationships.

$$1' \text{ of arc} = 0.03 \text{ ft at } 100 \text{ ft, or } 3 \text{ cm at } 100 \text{ m (approx.)}$$

$$1' \text{ of arc} = 1 \text{ in. at } 300 \text{ ft (approx.; actually } 340 \text{ ft)}$$

$$1'' \text{ of arc} = 1 \text{ ft at } 40 \text{ mi, or } 0.5 \text{ m at } 100 \text{ km (approx.)}$$

$$\sin 1' = \tan 1' = 0.00029 \text{ (approx.)}$$

$$\sin 1° = \tan 1° = 0.0175 = 0.01\tfrac{3}{4} \text{ (approx.)}$$

In accordance with the relationships listed, approximately a $1'$ error results in a measured angle if the line of sight is misdirected by 1 in. over a distance of 300 ft. This illustrates the importance of precisely setting the instrument and targets over their respective points, especially where short sights are involved. If an angle is expected to be accurate to within $\pm\tfrac{1}{2}'$ for sights of 500 ft, then the distance must be correct to within $500 \times (0.00029/2) = \pm0.07$ ft for compatibility. To appreciate the precision capabilities of a high-quality theodolite, an instrument reading to the nearest $0.1''$ is theoretically capable of measuring the angle between two points approximately 1 in. apart and 40 mi away! (As discussed in Section 11-4, errors in centering the instrument, sighting the point, and reading the circle make it difficult, if not impossible, to actually accomplish this accuracy, however.)

11-3 MEASURING A HORIZONTAL ANGLE WITH THE TRANSIT

Horizontal angles are measured with a transit by operating the upper clamp, lower clamp, and tangent screws. The vertical-circle clamp and tangent screw are used to bring the object sighted to the center of the field of view.

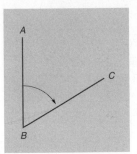

Figure 11-2 Measuring an angle.

Beginners may find it helpful to remember the following rules covering the use of the upper and lower clamps:

1. The lower clamp is used for backsighting only.
2. The upper clamp is used for setting the plates to zero, or any desired angle, and for foresighting.

The upper clamp and tangent screw are used to set 0°00′ (or any desired value) on the horizontal circle before sighting along the reference line, and to obtain a differential movement between the upper and lower plates when foresighting. The lower clamp and tangent screw are used to bring the line of sight along a reference line from which an angle is to be measured. The step-by-step procedure for direct measurement of angle *ABC* in Figure 11-2 is outlined to illustrate operation of the upper and lower motions.

1. Set up the instrument over point *B* and level it. Loosen both motions. Estimate the size of the angle as a check on the value to be obtained.
2. Set the circle to read approximately zero by holding the upper plate while turning the lower by tangential pressure on its underside. Tighten the upper clamp (snug but not wrench-tight). The upper and lower plates are now locked together.
3. Bring the vernier index mark precisely opposite the zero reading by means of the upper tangent screw. Always use a positive (clockwise direction) turn for the final setting for any tangent screw. If the zero mark is run beyond the point, back off and always finish with a positive motion. This prevents *backlash* (release of the spring tension, which can change the plate position).
4. Sight point *A* through the telescope. Set the vertical cross hair on, or almost on, the centerline of the range pole or other object marking *A* by turning the instrument with both hands on the plate edge or on the standards (*not* on the telescope).
5. Tighten the lower clamp. The lower plate is now fastened to the socket.
6. Carefully set the cross hair on the mark by means of the lower tangent screw, finishing with a positive motion. Both motions are now clamped together and thus to the socket, the horizontal circle reads zero, and the telescope is pointing to *A*. The transit is therefore *oriented,* since the line of sight is in a known direction with the proper value (0°00′) on the circle. Read the compass bearing for line *BA*.

7. Loosen the upper clamp and turn the upper plate until the vertical hair is on, or almost on, point *C*. The lower plate containing the graduated circle is still clamped to the socket. Tighten the upper clamp.

8. Carefully set the vertical hair on mark *C* by means of the upper tangent screw.

9. Read the angle on the horizontal circle, using the vernier *ahead* of the zero mark (*in the same clockwise direction as the angle was turned*). Read the compass bearing for line *BC*. Check the angle by comparing the measured value with the angle computed from compass bearings.

Since instrument setup, line clearing, objects to be sighted, and so on, are ready, little extra time is required to make repeated measurements.

11-4 MEASURING ANGLES BY REPETITION WITH A REPEATING INSTRUMENT

Measurements of horizontal angles should be repeated two or more times and the average taken. This procedure increases precision, eliminates certain instrumental errors, and prevents mistakes from going undetected. If an angle is to be measured by repetition, the method just described is followed for the first reading. Then, with the reading for the first angle left on the circle, a backsight is taken on *A,* as before, *using only the lower clamp and tangent screw to retain the angle setting.* The instrument is now oriented in the starting position, but the single-angle value is on the circle, instead of 0°00′.

The upper clamp is loosened, point *C* sighted again, the upper clamp tightened, and the cross hair brought precisely on the mark with the upper tangent screw. The sum of the first two turnings of the angle is now on the circle. This process can be continued for the number of repetitions desired. An even number of repetitions should be taken, half with the telescope *normal* (direct) and half with the telescope *plunged* (reversed). This eliminates by "reversion" the effects of some possible maladjustments of the instrument described in Section 11-18 and Appendix A. (Alternative terms for direct and reversed positions of the telescope are *face left* and *face right,* which refer to whether the vertical circle is left or right of the telescope.)

In measuring horizontal angles by repetition with a transit, theodolite, or total station, the instrument can be releveled if necessary between individual angle measurements. *The leveling screws should never be used to relevel the instrument between backsight and foresight, however.* To do so would reference the backsight in one plane and the foresight in another, causing an erroneous measurement.

After making multiple horizontal angle measurements, the total angle accumulated on the circle is divided by the number of repetitions to obtain the average value. With transits and optical-reading theodolites, the sum of the repetitions for an angle may be larger than 360°, making it necessary to add a multiple of 360° to the final reading before dividing.[1] It is necessary, therefore, to record the single angle after the first foresight so that the correct multiple of 360° can be determined.

If the instrument is in adjustment, carefully centered over a station, and leveled, then the two sources of error that remain in measuring a horizontal angle are pointing the

---

[1]This is unnecessary with electronic digital theodolites and total stations, because the absolute sum of all angles, regardless of the number of times 360° has been passed, is stored by the microprocessor.

telescope and reading the circle. A general formula for computing the maximum random error resulting from pointing and reading when measuring angles by repetition, developed from Eq. (2-10), is

$$E = \frac{1}{N}\sqrt{E_0^2 + 2NE_p^2 + E_R^2} \qquad (11\text{-}1)$$

where N is the number of repetitions of the angle, E_R the reading error, E_0 the error in making the initial setting to zero (it equals E_R), and E_p the pointing error (there are two for each angle). Reading errors are generally considered equal to half the least count of the vernier for a transit, or half the smallest graduations for a theodolite. Experience shows that with good equipment, an average observer can point the instrument (align the vertical cross hair) within about 2 to 5″.

A theodolite with smallest circle graduations of 1′ is used and an angle measured by repetition twice direct and twice reversed (2D, 2R). Assuming pointing errors of ±3″, compute the expected maximum error in the angle. Compare this with the error expected if the angle were measured four times independently and averaged.

EXAMPLE 11-1

SOLUTION

If small circle graduation errors are neglected, there remain two reading errors—initial setting and final reading—but eight pointings. Then, taking E_0 and E_R to be ±30″, by Eq. (11-1) the error in the angle obtained by repetition is

$$\tfrac{1}{4}\sqrt{(30)^2 + 8(3)^2 + (30)^2} = \tfrac{1}{4}(43.3) = 10.8''$$

If the same angle were measured four times independently and the results averaged, the maximum random error, from Eq. (2-12), would be

$$E = \frac{1}{\sqrt{4}}\sqrt{(30)^2 + 2(3)^2 + (30)^2} = 21.3''$$

The advantage of the repetition method is evident!

The procedure just outlined is applicable for all repeating instruments, including transits, theodolites (whether optical-reading or electronic digital), and total station instruments. *Direct angles* measured by repetition are commonly used in all types of surveys.

11-5 CLOSING THE HORIZON

Closing the horizon is the process of measuring the angles around a point to obtain a check on their sum, which should equal 360°00′00″. For example, if in Figure 11-3 only angles x and y are needed, it is desirable also to turn angle z to close the horizon at A. The method provides an easy way for a beginner to test pointings and readings.

Figure 11-4 shows the left-hand page of notes covering the measurements of the angles in Figure 11-3. The angles are first turned around the horizon by making a

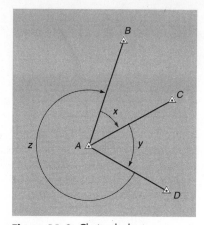

Figure 11-3 Closing the horizon.

Figure 11-4 Field notes for closing the horizon.

CLOSING THE HORIZON

Point Sighted	DIRECT Circle Rdg	Angle	REVERSED Circle Rdg	Angle	Avg Angle
B	0°00'00"		180°00'04"		
		42°12'14"		42°12'12"	42°12'13"
C	42°12'14"		222°12'16"		
		59°56'16"		59°56'14"	59°56'15"
D	102°08'30"		282°08'30"		
		257°51'34"		257°51'34"	257°51'34"
A	0°00'04"		180°00'04"		
				Σ =	360°00'02"
				−	360°00'00"
				Misclosure =	02"

pointing and reading at each station with the instrument in the direct mode. A final pointing on the initial backsight station provides a check, because it should equal the initial reading, allowing for reasonable random errors. Each individual angle is computed by subtracting the previous reading. Because of this procedure, therefore, setting the circle to exactly 0°00'00" for the initial pointing, as indicated in Figure 11-3, is unnecessary. With some precise instruments perfect zero settings are difficult, or even impossible, to make and time should not be wasted trying.

After completing the direct-mode readings, the instrument is plunged, or reversed, and the process repeated. As stated in Section 11-4, this procedure eliminates certain possible instrumental errors. Also, assuming the instrumental errors are small, the inversion process should produce readings for all pointings that differ by very nearly 180° from the first set—thus additional checks on the readings occur as the inverted work progresses. Angles computed from the inverted set are averaged with the direct ones, and the sum of these average angles calculated. The amount by which the sum differs from 360° (2" in this example) is the horizon misclosure. Comparison of the misclosure with permissible values will determine whether the work is acceptable or must be repeated.

11-6 LAYING OUT AN ANGLE WITH A REPEATING INSTRUMENT

To lay out an angle BAC equal to 25°30' with an instrument at point A (Figure 11-5), the circle is set to zero and point B is sighted using the lower motion. The upper clamp is loosened, the telescope turned until the circle reads 25°30', and the upper clamp again tightened. The line of sight establishes AC at the proper angle with AB.

To layout out an angle BAC equal to 25°30'40" by repetition with a 1' instrument, an angle BAC' of 25°30' is laid out as previously described and point C' marked. The angle BAC' is then measured by repetition as many times as the desired precision requires. The difference between angle BAC' and 25°30'40" can be marked off by measuring distance AC' and locating C by the following relation: distance $C'C = AC'$ tan $C'AC$. The angle BAC can then be turned by repetition as a check.

In Figure 11-5, if angle BAC' is found by repetition to be 25°30'20", then $C'AC =$ 20". If distance AC' is 300 ft, then $C'C = 300$ tan 20" = 300 (0.00029/3) = 0.029 ft. As discussed in Section 11-4, to eliminate the possible effects of instrumental errors, equal numbers of direct and reversed angles should also be measured and averaged in this lay-out procedure.

Figure 11-5 Laying out an angle by repetition.

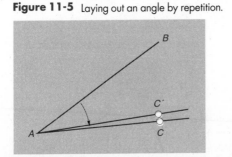

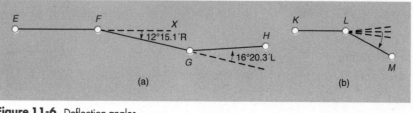

Figure 11-6 Deflection angles.

11-7 DEFLECTION ANGLES

A deflection angle (described in Section 8-3) is a horizontal angle measured from the prolongation of the preceding line, right or left, to the following line. In Figure 11-6(a) the deflection angle at F is 12°15.1′ to the right (12°15.1′R). At G the deflection angle is 16°20.3′L.

A straight line between terminal points is theoretically the most economical route to build and maintain for highways, railroads, pipelines, canals, and transmission lines. Practically, obstacles and conditions of terrain and land use require bends in the route, but deviations from a straight line are kept as small as possible. Use of deflection angles is more convenient for alignment layout, note taking, and computations.

If an instrument is in *perfect* adjustment (which is unlikely), the deflection angle at F [see Figure 11-6(a)] is measured by setting the circle to zero and backsighting on point E with the telescope plunged, then plunging again. The line of sight is now on EF extended and directed toward X. The upper clamp is loosened for the foresight, point G sighted, the upper clamp tightened, the vertical cross hair carefully set on the mark by means of the upper tangent screw, and the angle read.

Deflection angles are subject to serious errors if the instrument is not in adjustment, and may be read as larger or smaller than their correct values, depending on whether the line of sight after plunging is to the right or left of the true prolongation [see Figure 11-6(b)].

To eliminate errors from this cause, angles are usually doubled or quadrupled by the following procedure: The first backsight is taken with the circle set at zero and the telescope in the direct position. After plunging, the angle is measured and kept on the circle. A second backsight is taken using the lower motion, retaining the first angle, and keeping the telescope reversed. The telescope is plunged back to the normal position for the foresight, and the angle remeasured. Dividing the total angle by 2 gives an average angle from which instrumental errors have been eliminated by cancellation. In outline fashion, the method is as follows:

1. Backsight with the telescope normal. Plunge and measure the angle.
2. Backsight with the telescope plunged. Plunge again and measure the angle.
3. Read the total angle and divide by 2 for an average.

Of course, the precision in deflection angle measurement can be increased by making four, six, or eight repetitions and averaging.

Figure 11-7 shows the left-hand page of field notes for measuring the deflection angles at stations F and G of Figure 11-6. The procedure just outlined was followed. Four repetitions of each angle were taken with the instrument alternated from direct to reversed with each repetition. Readings were recorded only after the first, second, and

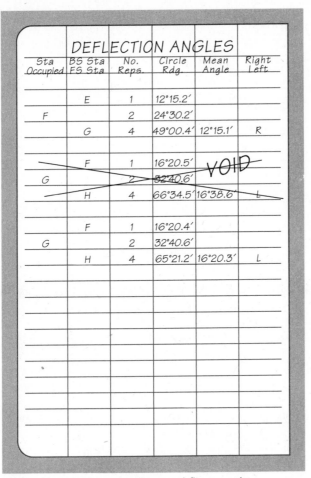

		DEFLECTION ANGLES			
Sta Occupied	BS Sta FS Sta	No. Reps.	Circle Rdg.	Mean Angle	Right Left
	E	1	12°15.2′		
F		2	24°30.2′		
	G	4	49°00.4′	12°15.1′	R
	F	1	16°20.5′	~~VOID~~	
G		~~2~~	~~32°40.6′~~		
	H	4	66°34.5′	16°38.6″	~~L~~
	F	1	16°20.4′		
G		2	32°40.6′		
	H	4	65°21.2′	16°20.3′	L

Figure 11-7 Field notes for measuring deflection angles.

fourth repetitions. *The final angle is the mean obtained by dividing the last recorded value by the number of repetitions*—four in this case. The purpose of the first two values is only to provide checks; that is, the second reading should be twice the first, and the final mean angle should be equal to the first, allowing of course for expected errors.

If a mistake should occur, as in the first set of angles measured at station *G* of Figure 11-7, lines should be drawn through the incorrect data, the word "VOID" written beside them, and the measurements repeated. In this voided data set, half the second recorded value (16°20.3′) agrees reasonably well with the first value (16°20.5′), but the final mean (16°38.6′) does not agree with the first recorded value. The most probable cause for this mistake is use of the upper tangent screw instead of the lower one for the third or fourth backsights.

11-8 AZIMUTHS

Azimuths are measured from a reference direction which itself must be determined from **(a)** a previous survey, **(b)** the magnetic needle, **(c)** a solar or star observation,

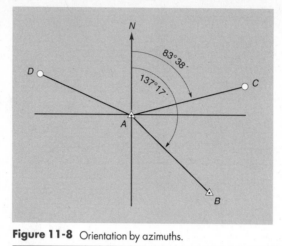

Figure 11-8 Orientation by azimuths.

(d) GPS observations, (e) a north-seeking gyro, or (f) assumption. Suppose that in Figure 11-8 the azimuth of line *AB* is known to be 137°17′ from true north. With a repeating instrument to find the azimuth of any other line from *A*, such as *AC*, first set 137°17′ on the circle numbered in the clockwise direction and backsight on point *B*. The instrument is now "oriented," since the line of sight is in a known direction with the corresponding angle on the circle. The next steps are to loosen the upper motion, turn the telescope clockwise to *C*, and read the clockwise angle. In this case the reading is 83°38′, which is the true azimuth of *AC*.

Note that after the lower clamp and tangent screw are used to backsight on point *B*, they are not disturbed regardless of the number of angles read from point *A*. When the circle reads zero, the telescope is pointing true north or in the direction of the reference meridian. As a check, if a transit equipped with a compass is being used, the needle can be lowered and read. If the telescope is pointing true north, the needle should read the declination of the place (provided there is no local attraction at point *A*).

In Figure 11-8, if the instrument is set up at point *B* instead of *A*, the azimuth of *BA* (317°17′) or the back azimuth of *AB* is put on the circle and point *A* sighted. The upper motion is loosened, and sights are taken on points whose azimuths from *B* are desired. Again, if the instrument is turned until the circle reads zero, the telescope points true north (or along the reference meridian).

The procedures for measuring angles described in Sections 11-4 through 11-8 are applicable with any type of repeating instrument, including transits, theodolites, and total stations.

11-9 MEASURING WITH A DIRECTIONAL THEODOLITE

As noted in Section 10-9, directional theodolites can be used for determining horizontal angles, but field procedures consist of measuring "directions," which are simply horizontal circle readings taken to successive stations sighted around the horizon. The difference in directions between any two stations is the angle.

Figure 11-9 shows a set of field notes for directions measured at station *A* of Figure 11-3. The notes actually are the results of four *positions*, each representing a reading at

Position No.	Station Sighted	Reading Direct	Reading Reverse	Mean	Reduced Direction
(1)	(2)	(3)	(4)	(5)	(6)
		o ′ ″	o ′ ″	″	o ′ ″
1	B	0 00 05	180 00 04	04	0 00 00
	C	42 12 15	222 12 16	16	42 12 12
	D	102 08 33	282 08 28	30	102 08 26
2	B	45 00 03	225 00 06	04	0 00 00
	C	87 12 14 (12)	267 12 20	16	42 12 12
	D	147 08 35	327 08 28	32	102 08 28
3	B	90 00 08	270 00 05	04	0 00 00
	C	132 12 20	312 12 22	21	42 12 15
	D	192 08 28	12 08 31	30	102 08 24
4	B	135 00 07	315 00 05	06	0 00 00
	C	177 12 15	357 12 19	17	42 12 11
	D	237 08 30	57 08 34	32	102 08 26

DIRECTIONS OBSERVED FROM STATION A

Figure 11-9 Field notes for measuring directions.

every station around the horizon with the instrument in both the direct and reversed modes.

Although there is no lower motion on a directional theodolite, the horizontal circle can be roughly indexed to selected values. To distribute readings around the entire circle, and hence minimize possible circle graduation errors, the initial sighting in the direct mode for the first position is set near 0°00′ and then advanced approximately 180°/n for the first pointing of each successive position, where n is the number of positions being measured. For the field notes of Figure 11-9, the initial readings for the four positions in the direct mode were indexed near 0°, 45°, 90°, and 135°, and the beginning readings with the telescope plunged therefore were near 180°, 225°, 270°, and 315°, thus providing a uniform distribution of readings around the circle.

In the field notes of Figure 11-9, the position number is in column (1); the station sighted in column (2); readings taken in the direct and reversed modes are in columns (3) and (4), respectively; mean values of the seconds portion for the direct and reversed readings are in column (5); and the reduced direction (obtained by subtracting the mean value for station B from all other mean directions) is in column (6). Note that at position number 2, column (6), values of 0°00′00″, 42°12′12″, and 102°08′28″ were obtained by subtracting the initial mean reading of 45°00'04" from the other values in column (5) for that position.

The sets of values in column (6) should be compared for agreement and acceptance criteria before leaving the station occupied so that additional positions can be measured if necessary. Angles can be calculated from the directions, but in control survey computations, directions are often preferable.

11-10 MEASURING VERTICAL (OR ZENITH) ANGLES

A vertical angle is the difference in direction between two intersecting lines measured in a vertical plane. As commonly used in surveying, it is the angle above or below a horizontal plane through the point of observation. Angles above the horizontal plane are called *plus angles,* or *angles of elevation.* Those below it are *minus angles,* or *angles of depression.* Transits measure vertical angles.

Most theodolites are designed so that vertical-circle readings give *zenith angles.* A zenith angle is measured in a vertical plane from the zenith (point directly overhead) to another point. In equation form, the relationship between vertical angles and zenith angles is

$$z = 90° - \alpha \qquad\qquad (11\text{-}2)$$

where z and α are the zenith and vertical angles, respectively.

With a theodolite, therefore, a reading of 0° corresponds to the telescope pointing vertically upward. In face left, with the telescope horizontal, the reading is 90°, and if the telescope is elevated 30° above horizontal, the reading is 60°. In face right, the horizontal reading is 270°, and with the telescope raised 30° above the horizon it is 300°.

Vertical angles and zenith angles are measured in trigonometric leveling, in EDM work for reduction of measured slope distances to horizontal, and in stadia.

To measure a vertical angle with a transit, the instrument is set up over a point and carefully leveled. The bubble in the telescope level vial should remain centered when the telescope is clamped in a horizontal position and rotated 360° about its vertical axis. If the vernier on the vertical arc does not read 0°00' when the bubble is centered, there is an *index error* that must be added to, or subtracted from, all readings. Confusion of signs is eliminated by placing in the field notes a statement such as "Index error is minus 2' to be subtracted from angles of depression and added to angles of elevation."

The horizontal cross hair is set approximately on the point to which a vertical angle is being measured, and the telescope clamped. Careful pointing is made by using the vertical-circle tangent screw. The vertical circle is read, and any index error applied to get the true angle above or below the horizon.

To eliminate the index error resulting from displacement of the vernier on the vertical arc, and lack of parallelism of the line of sight and telescope level vial, an average of two readings should be taken. One is secured with the telescope normal, the second with it inverted. This method requires a transit equipped with a complete vertical circle and reversion level vial.

Measurement of zenith angles with a theodolite follows the same general procedure just described, except that the vertical circle is oriented by either an automatic compensator or an index level vial. If the latter is used, serious errors result if the index level bubble is not centered prior to reading angles. As with the transit, instrumental errors are compensated for by computing the mean from an equal number of direct and reversed observations. With zenith angles, the mean is computed from:

$$\bar{z}_D = \frac{\Sigma z_D}{n} + \frac{n(360°) - (\Sigma z_D + \Sigma z_R)}{2n} \qquad (11\text{-}3)$$

where $\bar{z}_D$ is the mean value of the zenith angle [expressed according to its direct mode (face left) value], Σz_D the sum of direct zenith angles, Σz_R the sum of reversed (face right) angles, and n the number of z_D and z_R pairs of zenith angles read.

A zenith angle was read twice direct giving values of 70°00'10" and 70°00'12", and twice reversed yielding readings of 289°59'44" and 289°59'42". What is the mean zenith angle? **EXAMPLE 11-2**

SOLUTION

Two pairs of zenith angles are read, so $n = 2$. The sum of direct angles is 140°00'22", and that of reversed values if 579°59'26". Then by Eq. (11-3);

$$\bar{z}_D = \frac{140°00'22"}{2} + \frac{2(360°) - (140°00'22" + 579°59'26")}{2 \times 2}$$

$$= 70°00'11" + 0°00'03" = 70°00'14"$$

11-11 MEASURING ANGLES WITH DIGITAL THEODOLITES AND TOTAL STATIONS

Except for their method of automatically resolving angles, the mechanical operation of digital theodolites and total station instruments is similar to that for transits and optical-reading theodolites. Their design includes a vertical axis about which the instrument turns in azimuth, a horizontal axis for transiting the telescope, and a clamp and tangent screw for pointing.

To measure a horizontal angle, a backsight is taken using the clamp and tangent screw, and an initial value is entered in the display. Zero can be set if direct angles are being measured, but any required value may be entered if orienting on a line of known azimuth. The angle is then turned and a precise pointing for the foresight made again, using the clamp and tangent screw. The angular value is automatically displayed in the instrument. To eliminate instrumental errors and increase precision, angles can be repeated any number of times in both direct and reversed modes and the average taken. Built-in computers will automatically perform the averaging and display the results.

The procedure for measuring zenith angles with digital theodolites and total stations is the same as that outlined in Section 11-10 for optical-reading theodolites. When the instrument is leveled, 90° will automatically be displayed if the telescope is horizontal and face left. When a point is sighted, the zenith angle to it will automatically appear in the display.

Special capabilities are available with many electronic digital theodolites and total station instruments to enhance their accuracy and expedite operation. The Geodimeter

500 of Figure 1-3(c), for example, has a dual-axis automatic compensator which senses any misorientation of the circles. This information is relayed to the built-in computer which corrects for any index error in the vertical circle, and any dislevelment of the horizontal circle before displaying angular values. This real-time tilt sensing and correction feature makes it necessary to perform only rough leveling of the instrument, thus reducing setup time. In addition, the Geodimeter 500 measures angles by integration of electronic signals over the entire circle simultaneously; thus errors due to graduations and eccentricities (see Section 11-18.1) are eliminated. Furthermore, the computer also corrects horizontal angles for instrumental errors if the axis of sight is not perpendicular to the horizontal axis, or if the horizontal axis is not perpendicular to the vertical axis. This feature eliminates the need for averaging equal numbers of direct and reversed readings. With these advantages, and more, it is obvious why these instruments are rapidly replacing the older types.

11-12 SIGHTS AND MARKS

Objects commonly used for sights when making angle measurements with transits and theodolites on plane surveys include range poles, chaining pins, pencils, plumb-bob string, and tripod-mounted targets (see Figure 10-6). For short sights, string is preferred to a range pole because the small diameter permits more accurate centering. Small red and white targets of thin plastic or cardboard placed on the string extend the length of observation possible.

An error is introduced if the range pole sighted is not plumb. The observer must sight as low as possible on the pole when the mark itself is not visible, and the rodperson has to take special precautions in plumbing the pole, perhaps using a rod level or plumb bob.

When total station instruments are used for angle measurements only, these same types of targets may be used. If distance is also being measured, however, prism poles like those shown in Figure 11-10 are used. The pole shown on the left is hand held and kept vertical by means of a bull's-eye bubble. [The bubble should be regularly checked for adjustment, and adjusted if necessary (see Section A-9)]. This pole has graduations for easy determination of the prism's height. The tripod mount shown on the right in Figure 11-10 is centered over the point using the tribrach's optical plummet, or a plumb bob. With either type of pole, for angle measurements the vertical cross hair should bisect the pole just below the prism. Errors can result if the prism itself is sighted, especially on short lines.

In layout work on construction and in topographic mapping, *permanent* backsights and foresights may be established. These can be marks on structures such as walls, steeples, water tanks, and bridges, or they can be fixed artificial targets. They provide definite points on which the instrument operator can check orientation without the help of a rodperson.

11-13 PROLONGING A STRAIGHT LINE

On route surveys, straight lines may be continued from one point through several others. To prolong a straight line from a backsight, the vertical cross hair is aligned on the back point by means of the lower motion, the telescope plunged, and a point, or points, set ahead on line.

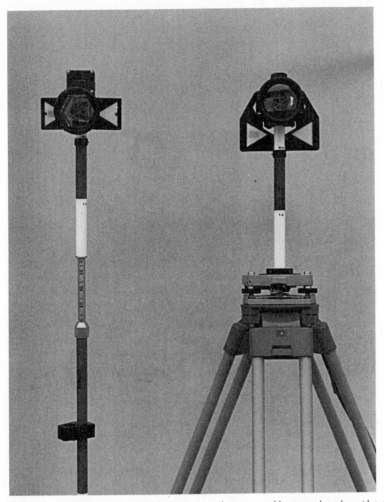

Figure 11-10 Prism poles used for measuring distances and horizontal angles with total station instruments.

To eliminate the effects of instrument maladjustment, the same procedure used in making a number of adjustments, known as the *principle of reversion,* is employed. The method applied, actually *double reversion,* is termed *double centering.* Figure 11-11 shows a simple use of the principle in drawing a right angle with a defective triangle. Lines *OX* and *OY* are drawn with the triangle in "normal" and "reversed" positions. Angle *XOY* represents twice the error in the triangle at the 90° corner, and its bisector establishes a line perpendicular to *AB.*

In practice, instruments should always be kept in good adjustment but used as though they might not be.

To prolong line *AB* of Figure 11-12 by double centering, the instrument is set up at *B,* a backsight taken on *A,* and the first point *C'* located with the telescope plunged. The lower motion is released and a second backsight taken on point *A,* this time with

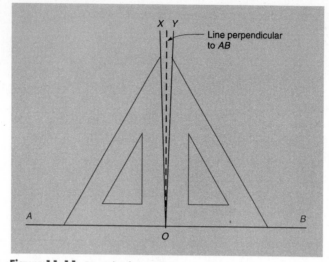

Figure 11-11 Principle of reversion.

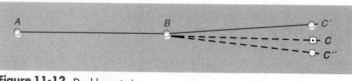

Figure 11-12 Double centering.

the telescope still plunged. The telescope is transited again to its normal position and point C'' marked. Distance $C'C''$ is bisected to get point C, on the line AB prolonged.

In outline form, the procedure is as follows:

1. Backsight on point A with the telescope direct. Plunge to the reversed position and set point C'.
2. Backsight on point A with the telescope reversed. Plunge to a direct position and set point C''.
3. Split the distance $C'C''$ to locate point C. Note that $C'C''$ represents *twice* the plunging error (and, as described in Appendix A, *four times* the adjustment error).

11-14 PROLONGING A LINE PAST AN OBSTACLE

Buildings, trees, telephone poles, and other objects may block survey lines. Four of the various methods used to extend lines past an obstacle are the **(1)** equilateral-triangle method, **(2)** right-angle-offset method, **(3)** measured-offset method, and **(4)** equal-angle method. Short backsights can introduce and accumulate errors, so procedures using distant points should be followed whenever possible.

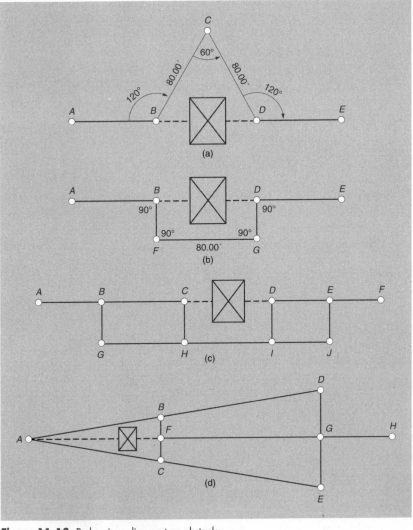

Figure 11-13 Prolonging a line past an obstacle.

11-14.1 Equilateral-Triangle Method

To prolong line *AB* in Figure 11-13(a), at *B* a 120° angle is turned off from a backsight on *A*, and a distance *BC* of 80.00 ft (or any distance necessary) is measured to locate point *C*. The instrument is then moved to *C*, a backsight taken on *B*, an angle of 60° measured, and a distance *CD* = *BC* = 80.00 ft laid off to mark point *D*. The instrument is moved to *D*, backsighted on *C*, and an angle of 120° turned. The line of sight *DE* is now along *AB* prolonged.

11-14.2 Right-Angle-Offset Method

To prolong *AB* of Figure 11-13(b), with instrument setups at points *B, F, G,* and *D*, 90° angles are turned. Distances *FG* and *BF* = *GD* need be only large enough to clear the obstruction, but longer lengths provide more accuracy in angle measurements.

The lengths shown in Figure 11-13(a) and (b) permit students to check their taping and instrument manipulation by combining the two methods.

11-14.3 Measured-Offset Method

To avoid the four 90° angles with short sights and consequently possible large errors, measured offsets obtained by swinging arcs with the tape can be used [see Figure 11-13(c)]. A long base is established for checkpoints on *GHIJ* if desired. This method is preferred to obtain the accuracy necessary for most survey work.

11-14.4 Equal-Angle Method

This method is excellent when field conditions are suitable. Equal angles just large enough to clear the obstacles are turned from the line at point *A*, and equal distances *AB = AC* and *AD = AE* measured in Figure 11-13(d). The line through points *F* and *G* at the midpoints of *BC* and *DE*, respectively, provides an extension of *AH* through the obstacle. Very little additional clearing is necessary using this method to bypass a large tree on line in wooded or brushy areas.

11-15 BALANCING IN

Occasionally it is necessary to set up on a line between two points already established but not intervisible—for example, *A* and *B* in Figure 11-14. This process is called *balancing in* or *wiggling in*.

Location of a trial point *C′* on line is estimated and the instrument set over it. A sight is taken on point *A* from point *C′* and the telescope plunged. If the line of sight does not pass through *B*, the instrument is moved laterally a distance *CC′* estimated from the proportion $CC' = BB' \times \frac{AC}{AB}$, and the process repeated. Several trials may be required to locate point *C* exactly, or close enough for the purpose at hand. The shifting head of the instrument is used to make the final small adjustment.

A method for getting a close first approximation of required point *C* takes two persons, *X* able to see point *A* and *Y* having point *B* visible, as shown in Figure 11-14. Each lines the other in with the visible point in a series of adjustments, and two range poles are placed at least 20 ft apart on the course established. An instrument set at point *C* in line with the poles should be within a few tenths of a foot of the required location.

Figure 11-14 Balancing in.

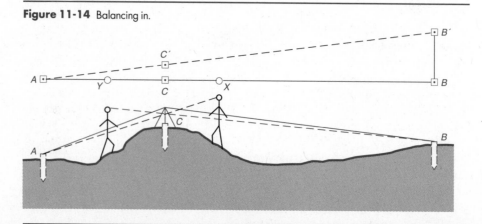

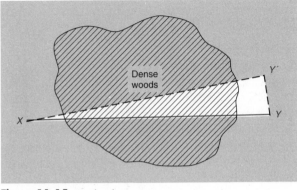

Figure 11-15 Random line.

11-16 RANDOM LINE

On many surveys it is necessary to run a line between two established points which are not intervisible because of obstructions. This situation arises repeatedly in property surveys. To solve the problem, a trial line, or *random line,* may be run from one point in the approximate direction of the other, and then corrected to the required orientation. In Figure 11-15, for example, it is necessary to run line XY. On the basis of a compass bearing, or information from maps and other sources, random line XY' is run. Lengths XY' and YY' (distance perpendicular to XY'), by which the random line misses point Y, are measured, and the angle YXY' is found from its calculated tangent. The correct line from X to Y can then be run by turning off the computed angle $Y'XY$ from line XY', or points on XY may be set by computed right-angle offsets from XY'.

11-17 TRANSITS AND THEODOLITES AS LEVELS

As noted in Section 6-6, transits and theodolites can be used as levels, although it is less convenient and generally yields results that are somewhat less accurate than those obtained with instruments designed especially for leveling. To use a transit as a level, after centering the plate bubbles of the instrument with the foot screws, the telescope bubble is carefully centered using the vertical-circle slow-motion screw. If the instrument is in proper adjustment, this will create a horizontal sight line so rod readings can be made, and the transit is operated essentially as a tilting level.

To create a horizontal line of sight with theodolites, after centering the plate bubble using the foot screws, and centering the vertical circle bubble (if the instrument is equipped with one), the vertical-circle reading (zenith angle) is set to 90°00.0′ with the instrument in the face left mode (270°00.0′ if the telescope is inverted or face right), whence the line of sight is horizontal so rod readings can be taken. In the case of micrometer theodolites, this requires setting the micrometer on zero first and then fixing the circle reading with the slow-motion screw.

11-18 SOURCES OF ERROR IN TRANSIT, THEODOLITE, AND TOTAL STATION WORK

Errors in using transits, theodolites, and total stations result from instrumental, natural, and personal sources. These are described in the subsections that follow.

11-18.1 Instrumental Errors

Figure 11-16 shows the fundamental reference axes of a theodolite. These same axes also apply to transits and total stations. (Transits have one additional reference line, the axis of the telescope bubble.) For a properly adjusted instrument, the four axes of Figure 11-16 must bear specific relationships to each other. These are: (1) the vertical axis should be perpendicular to the axis of the plate bubble, (2) the horizontal axis should be perpendicular to the vertical axis, and (3) the axis of sight should be perpendicular to the horizontal axis. For a transit, an additional required condition (not shown in Figure 11-16) is that the axis of the telescope bubble must be parallel to the axis of sight. If these relationships become maladjusted, errors result in measured angles unless proper field procedures are observed. A discussion of errors caused by a maladjustment of these axes, and of other sources of instrumental errors follows.

1. *Plate bubble out of adjustment.* If the axis of the plate bubble is not perpendicular to the vertical axis, the latter will not be truly vertical when the plate bubble is centered. *This condition causes errors in measured horizontal and vertical angles that cannot be eliminated by averaging direct and reversed readings.* The plate bubble is out of adjustment if, after centering, it runs when the instrument is rotated 180° in azimuth. The situation is illustrated in Figure 11-17. With the telescope initially pointing to the right and the bubble centered, the axis of the bubble tube is horizontal, as indicated by the solid line labeled *ABT*-1. Because the bubble is out of adjustment, it is not perpendicular to the vertical axis of the instrument, but instead makes an angle of $90° - \alpha$ with it. After turning the telescope 180°, it points left and the axis of the bubble tube is in the position indicated by the dashed line labeled *ABT*-2. The angle between the axis of the bubble tube and vertical axis is still $90° - \alpha$, but as shown in the figure, the indicated bubble run is E. From the figure's geometry, $E = 2\alpha$, which is double the bubble's maladjustment. The vertical axis can be made truly vertical by bringing the bubble back *half of the bubble run,* using the foot screws. Then, even though it is not centered, the bubble should stay in the same position as the instrument is rotated in azimuth, and accurate angles can be measured.

Although instruments can be used to obtain accurate results with their plate bubbles maladjusted, it is inconvenient and time consuming, so the required adjustment should be made. Procedures for making this and other instrument adjustments are described in Appendix A. Some digital theodolites and total stations are equipped with *dual-axis compensators,* which are able to sense the amount and direction of vertical axis tilt automatically. They can make corrections computationally in real time to both horizontal and vertical angles for this condition. The instruments equipped with *single-axis compensators* can only correct vertical angles.

2. *Horizontal axis not perpendicular to vertical axis.* This situation causes the axis of sight to define an inclined plane as the telescope is plunged and, therefore, if the backsight and foresight have differing angles of inclination, incorrect horizontal angles will result. Errors from this origin can be canceled by averaging an equal number of direct and reversed readings, or by double centering if prolonging a straight line.

3. *Axis of sight not perpendicular to horizontal axis.* If this condition exists, as the telescope is plunged, the axis of sight generates a cone whose axis coincides with the horizontal axis of the instrument. The greatest error from this source occurs when plunging the telescope, as in prolonging a straight line or measuring deflection angles. Also, when the angle of inclination of the backsight is not equal to that of the foresight,

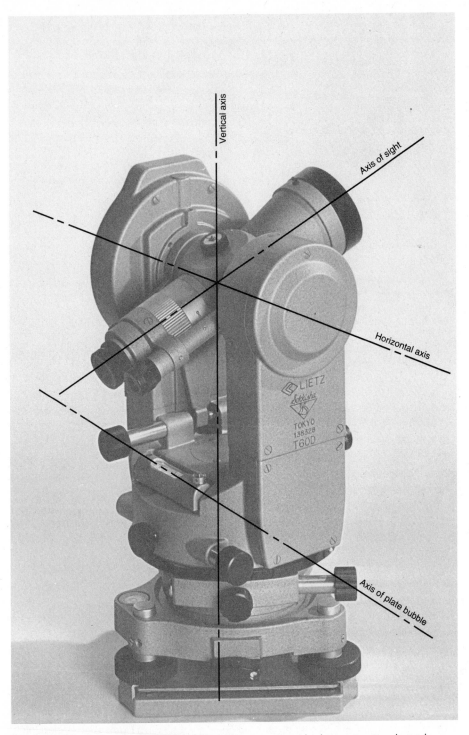

Figure 11-16 Reference axes of a theodolite. (The same axes apply also to a transit and a total station instrument.)

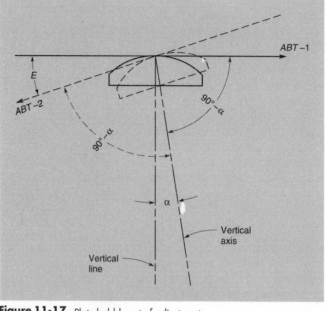

Figure 11-17 Plate bubble out of adjustment.

measured horizontal angles will be incorrect. These errors are eliminated by double centering and by averaging equal numbers of direct and reversed readings.

4. *Vertical-circle index error.* When the axis of sight is horizontal, a vertical angle of zero, or a zenith angle of either 90° or 270°, should be read; otherwise an index error exists. The error can be eliminated by computing the mean from equal numbers of vertical (or zenith) angles read in the direct and reversed modes. With most digital theodolites and total stations, the index error can be determined by carefully reading the same zenith angle both direct and reversed. The value is then computed, stored, and automatically applied to all measured zenith angles.

5. *Eccentricity of centers.* This condition exists if the geometric center of the graduated horizontal (or vertical) circle does not coincide with its center of rotation. Errors from this source are usually small. With transits they are eliminated in horizontal angles by averaging readings taken with both the *A* and *B* verniers (which are diametrically opposed or spaced 180° apart). With transits vertical-circle readings cannot be corrected for these errors. Many optical reading theodolites permit viewing diametrically opposed parts of both circles and making an average reading which eliminates these errors. Digital theodolites and total stations can also be equipped with systems that automatically average readings taken on opposite sides of the circles.

6. *Circle graduation errors.* If graduations around the circumference of a horizontal or vertical circle are nonuniform, errors in measured angles will result. These errors are generally very small. They can be minimized in horizontal angles measured with a transit or theodolite by averaging repeated angle measurements, but after each repetition the circle is advanced so that readings are uniformly spaced around the entire arc. This cannot be done for vertical angles with transits or theodolites. Some digital theodolites and total stations cannot space readings around either circle. Others always use readings taken from many locations around the circles for each measured horizontal and vertical angle, thus providing an elegant system for eliminating these errors.

7. *Axis of transit telescope bubble not parallel to axis of sight.* If this occurs for a transit, the axis of sight is inclined upward or downward when the telescope bubble is centered. It causes an error in vertical angles, and rod readings when the transit is used as a level. The effect is eliminated in vertical angles by averaging equal numbers of direct and reversed readings or applying an index error correction. In leveling its effect is canceled by balancing backsight and foresight distances.

8. *Eccentricity of transit verniers.* When the A and B transit vernier readings differ by exactly 180° for all positions, the circles are concentric and the verniers correctly set. If the readings disagree by a uniform number other than 180°, the verniers are offset, and it is best to use only the A vernier or take the mean of both verniers.

9. *Errors due to peripheral equipment.* Additional instrumental errors can result from worn tribrachs, optical plummets that are out of adjustment, unsteady tripods, and sighting poles with maladjusted bull's-eye bubbles. This equipment should be regularly checked and kept in good condition or adjustment. Procedures for adjusting instruments are outlined in Appendix A.

11-18.2 Natural Errors

1. *Wind.* Wind vibrates a transit and deflects the plumb bob. Shielding the instrument, or even suspending observations on precise work, may be necessary on windy days. An optical plummet is helpful in this situation.

2. *Temperature effects.* Temperature differentials cause unequal expansion of various parts of transits, theodolites, and total stations. This causes bubbles to run, which can produce erroneous observations. Temperature effects are reduced by shielding instruments from sources of heat or cold.

3. *Refraction.* Unequal refraction bends the line of sight and may cause an apparent shimmering of the observed object. It is desirable to keep lines of sight well above the ground and avoid sights close to buildings, smokestacks, and even large individual bushes in generally open spaces. In some cases, observations may have to be postponed until atmospheric conditions have improved.

4. *Settling of tripod.* The weight of an instrument may cause the tripod to settle, particularly when set up on soft ground or blacktop pavement. When a job involves crossing swampy terrain, stakes should be driven to support the tripod legs and work at a given station completed as quickly as possible. Stepping near a tripod leg, or touching one while looking through the telescope, will demonstrate the effect of settlement on the position of the bubble and cross hairs.

11-18.3 Personal Errors

1. *Instrument not set up exactly over point.* The plumb bob or optical plummet position should be checked at intervals during the time a station is occupied, to be certain it remains centered and the instrument is over the point.

2. *Bubbles not centered perfectly.* The bubbles must be checked frequently, but NEVER releveled between a backsight and a foresight—only *before* starting and *after* finishing an angle measurement.

3. *Improper use of clamps and tangent screws.* An observer must form good operational habits and be able to identify the various clamps and tangent screws by their touch without looking at them. Final setting of tangent screws is always made with a positive motion to avoid backlash. Clamps should be tightened just once and not checked again to be certain they are secure.

4. *Poor focusing.* Correct focusing of the eyepiece on the cross hairs, and of the objective lens on the target, is necessary to prevent parallax. Objects sighted should be placed as near the center of the field of view as possible. Focusing affects pointing, which is an important source of error.

5. *Overly careful sights.* Checking and double checking the position of the cross-hair setting on a target wastes time and actually produces poorer results than one fast observation. The cross hair should be aligned quickly and the next operation begun promptly.

6. *Careless plumbing and placement of rod.* One of the most common errors results from careless plumbing of a rod when only the top can be seen by the instrument operator because of brush or other obstacles in the way. Another is due to planting a pole off-line behind a point to be sighted.

7. *Transit vernier misinterpolated.* When a transit is used, using a magnifying glass and exercising caution will help reduce the size of these errors. Also, the number of minutes on the scale passed over by a vernier index should be estimated to check a reading.

11-19 MISTAKES

Some common mistakes in transit, theodolite, and total station work are

1. Sighting on, or setting up over, the wrong point
2. Calling out or recording an incorrect value
3. Reading the wrong circle, that is, counterclockwise instead of clockwise
4. Turning the wrong tangent screw
5. Reading a transit vernier in the wrong direction
6. Leaning on the tripod, or placing a hand on the instrument when pointing or taking readings.

PROBLEMS

11-1 Determine the angles subtended for the following conditions:
   (a) a 1-in.-diameter pipe sighted by theodolite from 400 ft away
   (b) a 2-in.-stake sighted by total station from $\frac{1}{4}$ mi away
   *(c) a 5-mm-diameter chaining pin seen by theodolite from 40 m away
11-2 What is the error in a measured direction for the situations noted?
   (a) setting a theodolite $\frac{1}{2}$ in. to the side of a tack on a 200-ft sight
   (b) lining in the edge (instead of center) of an 8-mm-diameter pencil at 30 m
   (c) sighting the edge (instead of center) of a $1\frac{1}{4}$-in.-diameter range pole 125 ft away
   *(d) sighting a range pole that is 2 in. off-line on a 500-ft sight
11-3 Intervening terrain obstructs the line of sight so only the top of an 8-ft-long range pole can be seen on an 600-ft sight. If the range pole is out of plumb and leaning sidewise $\frac{1}{2}$ in. per vertical foot, what maximum angular error results?

*Asterisks indicate problems that have answers given in Appendix G.

11-4 What are the primary sources of error in measuring angles with transits, theodolites, and total station instruments, assuming they are properly leveled and centered?

11-5 Describe a method of reading angles with a transit, theodolite, or total station that will eliminate most instrumental errors caused by improper construction and poor adjustment.

11-6 Cite the advantages of measuring angles by repetition.

11-7 In measuring angles with a transit, theodolite, or total station instrument, what errors are eliminated by averaging an equal number of measurements made with the telescope normal (face left) and plunged (face right)?

11-8 How are errors in circle graduations of transits and theodolites minimized in measuring horizontal angles? How are they minimized with total station instruments?

11-9 What are the advantages of closing the horizon around a point where only one angle is actually required?

*11-10 In measuring an angle by repetition, the reading after the first turning in a direct position was 64°13.6′. The reading after the fourth turning in the reversed position was 256°54.8′. Determine the angle.

11-11 Similar to Problem 11-10, except first and fourth readings were 94°10.0′ and 16°39.6′, respectively.

11-12 Similar to Problem 11-10, except six repetitions were made, and the first and sixth readings were 251°30.3′ and 69°01.2′, respectively.

*11-13 An interior angle x and its explement y were turned to close the horizon. Each angle was measured once direct and once reversed using the repetition method. Starting with an initial backsight setting of 0°00.0′ for each angle, the readings after the first and second turnings of angle x were 49°36.4′ and 99°13.0′, and the readings after the first and second turnings of angle y were 310°22.2′ and 260°46.0′. Calculate each angle and the horizon misclosure.

11-14 Similar to Problem 11-13, except the readings for angle x were 140°57.7′ and 281°55.3′ and those for angle y, 219°02.6′ and 78°05.0′.

In Figure 11-3, normal and plunged observed directions with a theodolite from A to points B, C, and D are listed in Problems 11-15 and 11-16. Find the values of the three angles.

*11-15 Normal: 28°31′23″, 91°56′43″, 255°08′26″
Plunged: 208°31′19″, 271°56′45″, 75°08′24″

11-16 Normal: 110°56′10″, 189°36′41″, 286°41′26″
Plunged: 290°56′09″, 9°36′42″, 106°41′28″

11-17 The angles at point X were measured with a 10″ theodolite. Based on 12 readings, the standard deviation of the angle was ± 5.0″. If the same procedure is used in measuring each angle within a four-sided polygon, what is the expected standard deviation of closure?

*11-18 Similar to Problem 11-17, except six readings were taken with a standard deviation of ± 8.0″.

11-19 An angle $ABC = 42°16′13″$ must be laid off with a 0.1′ theodolite. After a 400-ft backsight on point A, point C is marked 500 ft away with an angle of 42°16.2′ set on the circle. Angle ABC, measured six times by repetition, gives a final reading of 253°37.0′. What offset at C will produce the required angle?

*11-20 Similar to Problem 11-19, except an angle of 30°43′53″ is required to a point 800 ft away and the reading after repeating the angle six times is 184°23.8′.

11-21 Approximately how close to the true value would a horizontal angle be when read by repetition eight times with a 0.1′ theodolite?

*11-22 The line of sight of a theodolite is out of adjustment by 5″. **(a)** In prolonging a line by plunging the telescope between backsight and foresight, but not double centering, what angular error is introduced? **(b)** What off-line linear error results on a foresight of 1500 ft?

*Asterisks indicate problems that have answers given in Appendix G.

11-23 A line PQ is prolonged to point R by double centering. Two foresight points R' and R'' are set. What angular error would be introduced in a single plunging based on the following lengths of QR and $R'R''$, respectively?
  (a) 560.50 ft and 0.32 ft
  (b) 321.60 m and 4.25 cm
  (c) 895.00 ft and $3\frac{7}{8}$ in.

11-24 Describe four methods of prolonging a straight line past an obstacle on line.

11-25 Explain the term ''balancing in'' and describe a situation where it is necessary.

11-26 What are ''random lines'' and when are they useful?

11-27 In Figure 11-15 random line XY' of length 2638.05 ft is run, and falling distance $Y'Y$ is measured as 23.17 ft. What angle $Y'XY$ must be laid off from line XY' to establish the correct direction of XY?

11-28 What is the index error, and how can its value be obtained and the error eliminated?

*11-29 With a transit telescope normal, a vertical angle to point A was $+10°35'$, and with the telescope plunged, it was $+10°41'$. What is the index error? Compute the correct vertical angle to point B if the reading to it is $-7°14'$ with the telescope normal.

11-30 Similar to Problem 11-29, except for a theodolite with zenith angle readings of $95°20.3'$ in face left and $264°38.9'$ in face right, and $103°51.5'$ on point B in face left.

11-31 Similar to Problem 11-29, except for a theodolite with zenith angle readings of $86°17'21''$ and $86°17'19''$ face left, $273°42'31''$ and $273°42'33''$ face right, and $91°08'48''$ on point B in face left.

11-32 Give the functional relationship between vertical angles and zenith angles.

11-33 If pointings can be made with an accuracy of $\pm 3''$, what is the maximum error expected in measuring a horizontal angle by repetition for the following situations:
  (a) 4D and 4R with transit reading and setting errors of $\pm 30''$ each
  (b) 3D and 3R with theodolite reading and setting errors of $\pm 0.1'$ each
  *(c) 6D and 6R with theodolite reading and setting errors of $\pm 5''$ each

11-34 What number of repetitions is required to measure a horizontal angle to an accuracy of $2''$ with an instrument having a reading and setting error of $\pm 6''$? Assume reasonable pointing errors.

11-35 To reduce the effects of circle graduation errors in precision direction measurements with a directional theodolite, if eight positions are being measured, what should be the approximate circle readings for initial sightings at each position with the telescope direct?

11-36 What error in horizontal angles is consistent with the following linear precisions?
  (a) $\frac{1}{200}, \frac{1}{500}, \frac{1}{2000}, \frac{1}{5000},$ and $\frac{1}{10,000}$
  (b) $\frac{1}{300}, \frac{1}{800}, \frac{1}{1500}, \frac{1}{3000},$ and $\frac{1}{15,000}$
  *(c) $\frac{1}{400}, \frac{1}{1000}, \frac{1}{2500}, \frac{1}{4000},$ and $\frac{1}{20,000}$

11-37 List the reference lines or axes of a theodolite. What relationships should these axes bear to one another for an instrument in perfect adjustment?

*Asterisks indicate problems that have answers given in Appendix G.

BIBLIOGRAPHY   Brinker, R. C., and R. Minnick. 1987. *The Surveying Handbook*. New York: Van Nostrand Reinhold, Chap. 5.

Clark, M. M., and R. B. Buckner. 1992. ''A Comparison of Precision in Pointing to Various Targets at Different Distances.'' *Surveying and Land Information Systems* 52 (No. 1): 41.

Dracup, J. F., et al. 1979. *Surveying Instrumentation and Coordinate Computation Workshop Lecture Notes*. Falls Church, Va.: American Congress on Surveying and Mapping.

Madkour, M. F. 1969. ''Round Versus Positions.'' *ASCE, Journal of the Surveying and Mapping Division* 95 (No. SU1): 151.

Moffitt, F. H., and H. Bouchard. 1992. *Surveying*. 9th ed. New York: Harper-Collins, Chap. 6.

12

TRAVERSING

12-1 INTRODUCTION

A traverse is a series of consecutive lines whose lengths and directions have been determined from field measurements. *Traversing,* the act of establishing traverse stations and making the necessary measurements, is one of the most basic and widely practiced means of determining the relative locations of points.

There are two basic kinds of traverses: *closed* and *open*. Two categories of closed traverses exist: *polygon* and *link*. With the polygon type, as shown in Figure 12-1(a) the lines return to the starting point, thus forming a closed figure (geometrically and mathematically closed). Link traverses finish upon another station that has a positional accuracy equal to or greater than that of the starting point. The link type (geometrically open, mathematically closed), as illustrated in Figure 12-1(b), must have a closing

Figure 12-1 Examples of closed traverses.

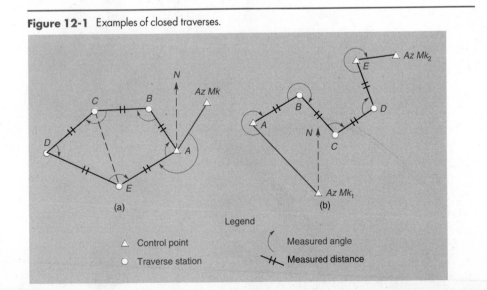

(a)

(b)

Legend

△ Control point   Measured angle

○ Traverse station   ⊢⊢ Measured distance

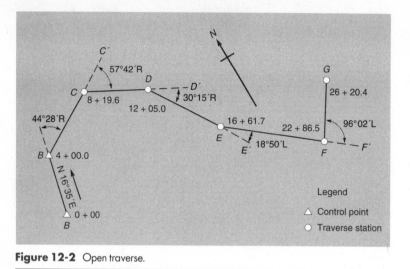

Figure 12-2 Open traverse.

reference direction—for example, line *E–Az Mk₂*. Closed traverses provide checks on the measured angles and distances, an extremely important consideration. They are used extensively in control, construction, property, and topographic surveys.

An open traverse (geometrically and mathematically open) (Figure 12-2) consists of a series of lines that are connected but do not return to the starting point or close upon a point of equal or greater order accuracy. Open traverses should be avoided because they offer no means of checking for errors and mistakes. If they are used, measurements *must* be repeated to guard against mistakes.

Hubs (wooden stakes with tacks to mark the points), steel stakes, or pipes are set at each traverse station *A, B, C,* and so on, in Figures 12-1 and 12-2, where a change in direction occurs. Spikes, "P-K" nails, and scratched crosses are used on blacktop pavement. Chiseled or painted marks are made on portland cement concrete. Traverse stations are sometimes interchangeably called *angle points* because an angle is measured at each one.

12-2 METHODS OF MEASURING TRAVERSE ANGLES OR DIRECTIONS

The methods used in measuring angles or directions of traverse lines vary and include **(1)** compass bearings, **(2)** interior angles, **(3)** angles to the right, **(4)** deflection angles, and **(5)** azimuths. These are described in the following subsections.

12-2.1 Traversing by Compass Bearings

The surveyor's compass, described in Chapter 9, was designed for use as a traversing instrument. Bearings are read directly on the compass as sights are taken along the lines (courses) of the traverse. Engineers' transits equipped with compasses can also be used to read bearings directly. The accuracy of bearings read by this method is limited by the compass to about ± 15′. Nowadays compasses are seldom used for traversing.

12-2.2 Traversing by Interior Angles

Interior-angle traverses are used for many types of work, but they are especially convenient for property surveys. Although interior angles could be measured either clockwise or counterclockwise, *to reduce mistakes in reading, recording, and computing, they should always be turned clockwise* from the backsight station to the foresight station. The procedure is illustrated in Figure 12-1(a). In this text, except for left deflection angles, clockwise turning will always be assumed. Furthermore, when angles are designated by three station letters or numbers in this text, the backsight station will be given first, the occupied station second, and the foresight station third. Thus angle *EAB* of Figure 12-1(a) was measured at station *A,* with the backsight on station *E* and the foresight at station *B.*

Interior angles may be improved by averaging equal numbers of direct and reversed readings. As a check, exterior angles may also be measured to close the horizon (see Section 11-5). In the traverse of Figure 12-1(a), a reference line *A–Az MK* of known direction exists. Thus the clockwise angle at *A* from *Az Mk* to *E* must also be measured to enable determining the directions of all other lines. This would not be necessary if the traverse contained a line of known direction, like *AB* of Figure 8-3, for example.

12-2.3 Traversing by Angles to the Right

Angles measured clockwise from a backsight on the "rearward" traverse station to a foresight on the "forward" traverse station [see Figure 12-1(b)] are called *angles to the right.* According to this definition, to avoid ambiguity in angle-to-the-right designations, the "sense" of the forward traverse direction must be established. This is normally done by consecutive numbering or lettering of traverse stations so that they increase in the forward direction. Thus in Figure 12-1(b), for example, the direction from *A* to *B, B* to *C, C* to *D,* and so on, is forward. Measured angles to the right can also be checked (and improved) by averaging equal numbers of direct and reversed readings.

12-2.4 Traversing by Deflection Angles

Route surveys are commonly run by deflection angles measured to the right or left from the lines extended, as indicated in Figure 12-2. A deflection angle is not complete without a designation R or L, and, of course, it cannot exceed 180°. Each angle should be doubled or quadrupled to reduce instrumental errors, and an average value determined. Deflection angles can be obtained by subtracting 180° from angles to the right. Positive values so obtained denote right deflection angles; negative ones are left.

12-2.5 Traversing by Azimuths

If a repeating instrument is being used, traverses can be run by azimuths. The process permits reading azimuths of all lines directly, thus eliminating the need to calculate them. In Figure 12-3 azimuths are measured clockwise from the north end of the meridian through the angle points. The instrument is oriented at each setup by sighting on the previous station with either the back azimuth on the circle (if angles to the right are turned) or the azimuth (if deflection angles are turned), as described in Section 11-8. Then after opening the upper motion, the forward station is sighted. The resulting angle on the horizontal circle will be the forward line's azimuth.

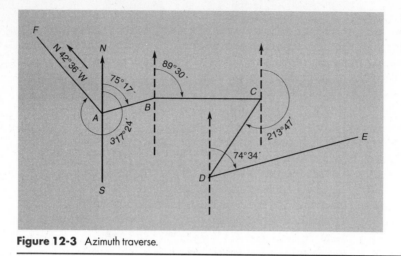

Figure 12-3 Azimuth traverse.

12-3 MEASUREMENT OF TRAVERSE LENGTHS

The length of each traverse line must be measured, and this is usually done by the simplest and most economical method capable of satisfying the required precision of a given project. Due to their speed, convenience, and accuracy, EDM instruments and total stations are most often used, although taping can be employed. Averages of distances measured both forward and back by stadia (see Chapter 15) give results suitable for various types of work, such as establishing low-precision control for topographic mapping. For some geological and agricultural work, pacing may be accurate enough.

The precision specified for a traverse to locate boundaries is governed by state statutes. On construction work, allowable limits of closure depend on the use and extent of the traverse and project type. Bridge location, for example, demands a high degree of precision.

In closed traverses, each line is measured and recorded as a separate distance. On long, open traverses for highways and railroads, distances are carried along continuously from the starting point. In Figure 12-2, for example, beginning with station $0 + 00$ at point A, 100-ft stations ($1 + 00$, $2 + 00$, $3 + 00$, and $4 + 00$) are marked until hub B at station $4 + 00.0$ is reached. Then stations $5 + 00$, $6 + 00$, $7 + 00$, $8 + 00$, and $8 + 19.6$ are set along course BC to C, and so on. The length of an open-traverse line is the difference between stations at its ends; thus the length of line BC is $819.6 - 400.0 = 419.6$.

12-4 SELECTION OF TRAVERSE STATIONS

Positions selected for setting traverse stations vary with the type of survey. In general, guidelines to consider in choosing them include accuracy, utility, and efficiency. Of course, intervisibility between adjacent stations, forward and back, must be maintained for angle and distance measurements. The stations should also ideally be set in convenient locations that allow for easy access. Ordinarily stations are placed to create lines that are as long as possible. This not only increases efficiency by reducing the number of instrument setups, but it also increases accuracy in angle measurements. Utility may override using very long lines, however, because intermediate hubs, or stations at strategic locations, may be needed to complete the survey's objectives.

Each different type of survey will have its unique requirements concerning traverse station placement. On property surveys, for example, traverse stations are placed at each corner if the actual boundary lines are not obstructed and can be occupied. If offset lines are necessary, a stake is located near each corner to simplify the measurements and computations. Long lines and rolling terrain may necessitate extra stations.

On route surveys, stations are set at each angle point, and at other locations where necessary to obtain topographic data or extend the survey. Usually the centerline is run before construction begins, but it will likely be destroyed and need replacement one or more times during various phases of the project. An offset traverse can be used to avoid this problem.

A traverse run to provide control for topographic mapping serves as a framework to which map details such as roads, buildings, streams, and hills are referenced. Station locations must be selected to permit complete coverage of the area to be mapped. *Spurs* consisting of one or more lines may branch off as open *(stub)* traverses to reach vantage points. Their use should be discouraged, however, because a check on their positions cannot be made.

12-5 REFERENCING TRAVERSE STATIONS

Traverse stations, like bench marks, may be lost if not properly described and preserved. Referencing these stations, or creating *ties* to them, is done to aid in finding them or to relocate them if they are destroyed.

Figure 12-4 and Plate D-1 in Appendix D present typical traverse ties. As illustrated, these ties consist of distance measurements made to nearby fixed objects. Short lengths (less than 100 ft) are convenient if a steel tape is being used, but, of course, the distance to definite and unique points is a controlling factor. Two ties, preferably at about right angles to each other, are sufficient, but three should be used to allow for the possibility

Figure 12-4 Referencing a point.

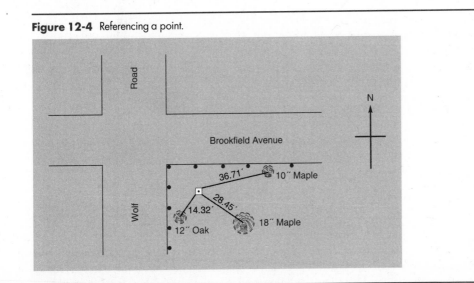

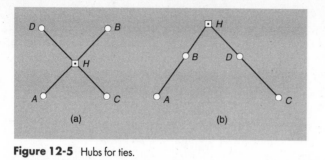

Figure 12-5 Hubs for ties.

that one reference mark may be destroyed. Ties to trees can be measured in hundredths of a foot if nails are driven into them. (Permission should be obtained, however, before driving nails into trees.)

If natural or existing features such as trees, utility poles, or corners of buildings are not available, stakes may be driven and used as ties. Figure 12-5(a) shows an arrangement of *straddle hubs* well suited to tying in a point such as *H* on a highway centerline or elsewhere. Reference points *A* and *B* are carefully set on the line through *H*, as are *C* and *D*. Lines *AB* and *CD* should be roughly perpendicular, and the four points should be placed in safe locations, outside of areas likely to be disturbed. It is recommended that a third point be placed on each line to serve as an alternate in the event one point is destroyed. The intersection of the lines of sight of two theodolites set up at *A* and *C*, and simultaneously aimed at *B* and *D*, respectively, will recover the point. The traverse hub *H* may also be found by intersecting strings stretched between diagonally opposite ties if the lengths are not too long. Hubs in the position illustrated by Figure 12-5(b) are sometimes used, but are not as desirable as straddle hubs for stringing.

12-6 TRAVERSE FIELD NOTES

The importance of notekeeping was discussed in Chapter 3. Since a traverse is itself the end on a property survey and the basis for all other data in mapping, a single mistake or omission in recording is one too many. All possible field and office checks must therefore be made. A partial set of field notes for an interior-angle traverse run using a total station instrument is shown in Plate D-5 of Appendix D. Plate D-6 shows the notes for a traverse run with theodolite and tape.

12-7 ANGLE MISCLOSURE

The angular misclosure for an interior-angle traverse is the difference between the sum of the measured angles and the geometrically correct total for the polygon. The sum, Σ, of the interior angles of a closed polygon is

$$\Sigma = (n - 2)180°$$

(12-1)

where *n* is the number of sides, or angles, in the polygon. This formula is easily derived from known facts. The sum of the angles in a triangle is 180°; in a rectangle, 360°; and in a pentagon, 540°. Thus each side added to the three required for a triangle increases the sum of the angles by 180°.

Figure 12-1(a) shows a five-sided figure in which, if the sum of the measured interior angles equals 540°00.8', the angular misclosure is 0.8'. Misclosures result from the accumulation of random errors in the angle measurements. Permissible misclosure can be computed by the formula

$$c = K \sqrt{N} \qquad (12\text{-}2)$$

where N is the number of angles, and K a constant that depends on the level of accuracy specified for the survey. The Federal Geodetic Control Subcommittee (FGCS) recommends constants for five different orders of traverse accuracy: *first order, second order class II, second order class II, third order class I, and third order class II*. Values of K for these orders, from highest to lowest, are 1.7″, 3″, 4.5″, 10″, and 12″, respectively. Thus if the traverse of Figure 12-1(a) were being executed to third-order class II standards, its allowable misclosure error would be 12″ × $\sqrt{5}$ = 27″.

The algebraic sum of the deflection angles in a closed-polygon traverse equals 360°, clockwise (right) deflections being considered plus and counterclockwise (left) deflections, minus. This rule applies if lines do not crisscross, or if they cross an even number of times. When lines in a traverse cross an odd number of times, the sum of right deflections equals the sum of left deflections.

A closed-polygon azimuth traverse is checked by setting up on the starting point a second time after having occupied the successive stations around the traverse and orienting by back azimuths. The azimuth of the first side should be the same as the original value. Any difference is the misclosure. If the first point is not reoccupied, the interior angles computed from the azimuths will automatically check the proper geometric total, even though one or more of the azimuths may be incorrect.

Although angular misclosures cannot be directly computed for link traverses, the angles can still be checked. The true direction or bearing of the first line may be determined from two intervisible stations with a known azimuth between them, or from a sun or Polaris observation, as described in Chapter 18. Measured angles are then applied to calculate the bearings of all traverse lines. The last line's computed bearing is compared with its known value, or the result obtained from another sun or Polaris observation. On long traverses, intermediate lines can be similarly checked. In using sun or Polaris observations to check angles on traverses of long east-west extent, allowance must be made for *convergence of meridians*. This topic is discussed in Section 19-12.

12-8 LOCATING BLUNDERS IN TRAVERSE MEASUREMENTS

A numerical or graphic analysis can often be used to determine the location of a mistake, and thereby save considerable field time in making necessary remeasurements. For example, if the sum of the interior angles of a five-sided traverse gives a bad misclosure—say, 10°01'—it is likely that one mistake of 10° and several small errors accumulating to 1' have been made. A method of graphically locating the station at which the mistake occurred, so only one point need be reoccupied, will be illustrated. The procedure, shown for a five-sided traverse, can be used for traverses having any number of sides.

In Figure 12-6(a), the traverse has been plotted by using the measured lengths and angles and shows a linear misclosure error AA'. The perpendicular bisector of line AA'

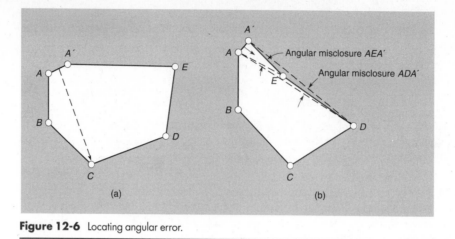

Figure 12-6 Locating angular error.

points to the mistaken angle—in this case, *C*. A correction applied to this angle will swing the traverse through an arc to eliminate the linear error *AA'*.

If, as in Figure 12-6(b), a second angle point lies near the perpendicular bisector of *AA'*, then the station at which the mistake has been made is the one for which the angular misclosure, when plotted at that station, is subtended by *AA'*. Stated differently, if the misclosure angle equals *AEA'*, *E* is the mistaken station; if it equals *ADA'*, *D* is the station.

For a mistake made in the distance measurement of one traverse line, suspect the side parallel with the direction of the closing line.

A *cutoff line,* such as *CE* shown dashed in Figure 12-1(a), run between two stations on a traverse, produces smaller closed figures to aid in checking and isolating blunders. The additional measurements increase redundancy and hence precision.

12-9 TRAVERSING WITH TOTAL STATION INSTRUMENTS

Total station instruments, with their combined electronic digital theodolite and EDM instrument, speed the process of traversing significantly because they can measure angles and distances from a single setup. The measuring process is further aided because, as described in Section 10-12, angles and distances are resolved automatically and displayed. Furthermore, the microprocessors of total stations can perform traverse computations to reduce slope distances to their horizontal and vertical components, and instantaneously calculate and store station coordinates and elevations.

To illustrate a method of traversing with a total station instrument, refer to the traverse of Figure 12-1(b). With the instrument set up and leveled at station *A,* a backsight is carefully taken on *Az MK₁*. The azimuth of line *A–Az Mk₁* is initialized on the horizontal circle by entering it in the unit using the keyboard. The coordinates and elevation of station *A* are also entered in memory. Next a foresight is made on station *B.* The azimuth of line *AB* will now appear on the display, and upon keyboard command, can be stored in the microprocessor's memory. Slope distance *AB* is then measured and reduced to its horizontal and vertical components by the microprocessor. Then the line's departure and latitude are computed and added to the coordinates of station *A* to yield the coordinates of station B. (Departures, latitudes, and coordinates are described in Chapter 13.)

The procedure outlined for station A is repeated at station B, except that the back azimuth BA and coordinates of station B need not be entered; rather, they are recalled from the instrument's memory. From the setup at B, azimuth BC and coordinates of C are determined and stored. This procedure is continued until a station of known coordinates is reached, as E in Figure 12-1(b). Here the known coordinates of E are entered in the unit's computer and compared to those obtained for E through the traverse measurements. The misclosure is computed, displayed, and, if within allowable limits, distributed by the microprocessor to produce final coordinates of intermediate stations.

If desired, traverse station elevations can also be determined as a part of the procedure (usually the case for topographic surveys). Then entries h_e (height of instrument) and h_r (height of reflector) must be input (see Section 5-9). The microprocessor computes the slope distance vertical component, and includes a correction for curvature and refraction (see Section 6-5.4). The elevation difference is added to the occupied station's elevation to produce the next point's elevation. At the final station, any misclosure is determined by comparing the computed elevation with its known value, and if within tolerance, it is adjusted to produce elevations of intermediate traverse stations.

All data from traversing with a total station instrument can be stored in an automatic data collector for later printing and transfer to the office for computing and plotting (see Sections 3-7 through 3-9). Alternatively, the traverse notes can be recorded manually as illustrated with Plate D-5 in Appendix D.

12-10 RADIAL TRAVERSING

In certain situations it may be most convenient to determine the relative positions of points by *radial traversing*. In this procedure, as illustrated in Figure 12-7(a), some point O, whose position is assumed known, is selected from which all points to be located can be seen. If a point such as O does not exist, it can be established. It is also assumed that a nearby azimuth mark, like Z in Figure 12-7(a), is available, and that reference azimuth OZ is known. With a theodolite, or preferably a total station, at point O, after backsighting on Z, horizontal angles to all stations A through F are measured. Azimuths of all radial lines from O (as OA, OB, OC, and so on) can then be calculated. Horizontal lengths of all radiating lines are also measured. By using the measured lengths and azimuths, coordinates for each point can be computed. (The subject of coordinate computations is discussed in Chapter 13.)

It should be clear that in the procedure just described, each point A through F has been surveyed independently of all others, and that no checks on their computed positions exist. To provide a check, lengths AB, BC, CD, and so on, could be computed from the coordinates of points, and then these same lengths measured. This results in many extra setups and substantial additional field work, however, and thus defeats one of the major benefits of radial traversing. To solve the problem of gaining checks with a minimum of extra field work, the method presented in Figure 12-7(b) is recommended. Here a second hub O' is selected from which all other points can also be seen. The position of O' is determined by observations of horizontal angle and distance from station O. This second hub O' is then occupied, and horizontal angles and distances to all stations A through F are measured as before. With the coordinates of both O and O' known, and by using the two independent sets of angles and distances, two sets of

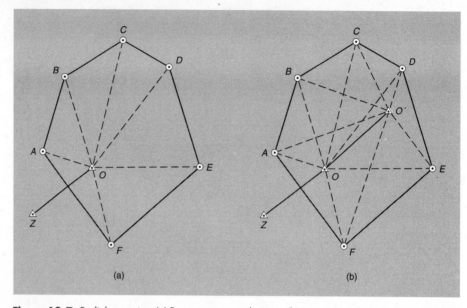

Figure 12-7 Radial traversing. (a) From one occupied station. (b) From two occupied points.

coordinates can be computed for each station, thus obtaining the check. If the two sets agree within a reasonable tolerance, the average can be taken. A better adjustment is obtained, however, using the method of least squares (see Section 2-26 and Appendix C). Although radial traversing can provide coordinates of many points in an area rapidly, the method is not as rigorous as running closed traverses.

Radial traversing is ideal for quickly establishing a large number of points in an area, and is especially efficient if a total station instrument is available. These devices not only enable both angle and distance measurements to be made from a single setup, but they also perform the calculations for azimuth, horizontal distance, and station coordinates in real time. Radial methods are also very convenient for laying out planned construction projects with a total station instrument. In this application, required coordinates of points to be staked are determined from the design, and the angles and distances that must be measured from a selected station of known position are computed. These are then laid out with a total station to set the stakes. The procedures are discussed in detail in Section 24-9.

12-11 SOURCES OF ERROR IN TRAVERSING

Some sources of error in running a traverse are:

1. Poor selection of stations, resulting in bad sighting conditions due to **(a)** alternate sun and shadow, **(b)** visibility of only the rod's top, **(c)** line of sight passing too close to the ground, **(d)** lines that are too long or too short, and **(e)** sighting into the sun
2. Errors in measurement of angles and distances
3. Failure to measure angles an equal number of times direct and reversed

12-12 MISTAKES

Some mistakes in traversing are:

1. Occupying or sighting on the wrong station
2. Incorrect orientation
3. Confusing angles to the right and left
4. Mistakes in notetaking

12-1 Explain the difference between closed and open traverses. Comment on the advisability of using open traverses.

12-2 Discuss guidelines to consider when selecting positions for the stations of a control traverse for topographic mapping.

12-3 Prepare a set of typical theodolite and tape field notes for the closed interior-angle traverse of Figure 12-1(a). (Refer to the example shown in Plate D-6.) Assume the bearing of AB is N39°20′W. Scale the lengths of lines, and measure the angles with a protractor. Include the angle at A from AZ Mk to E.

12-4 Similar to Problem 12-3, except for angles to the right in the traverse of Figure 12-1(b).

12-5 Similar to Problem 12-3, except for deflection angles in the traverse of Figure 12-1(b).

12-6 Prepare a set of typical theodolite and tape field notes for the closed traverse of Figure 8-3(a). (Refer to the example shown in Plate D-6.) Scale the lengths of lines.

*12-7 Compute and tabulate bearings for the traverse of Figure 12-2.

12-8 Cite the advantages of running a traverse using a total station instrument.

12-9 What should be the sum of the interior angles for a closed-polygon traverse that has:
(a) 6 sides **(b)** 8 sides **(c)** 12 sides

*12-10 Five interior angles of a six-sided closed-polygon traverse were measured as: $A = 89°10′$, $B = 143°46′$, $C = 86°23′$, $D = 162°51′$, and $E = 102°51′$. The angle at F was not measured. If all measured angles are assumed to be correct, what is the value of angle F?

12-11 Similar to Problem 12-10, except the traverse had seven sides with measured angles of $A = 139°28′12″$, $B = 103°17′40″$, $C = 86°48′19″$, $D = 140°24′51″$, $E = 169°22′37″$, and $F = 206°10′14″$. Compute the angle at G, which was not measured.

12-12 What is the correct algebraic sum of the deflection angles (5R and 3L) in a closed-polygon traverse, assuming: **(a)** no lines crisscross **(b)** lines crisscross once

*12-13 Four deflection angles of a closed five-sided traverse, none of whose lines crisscrosses, were measured as $A = 135°43′R$, $B = 92°14′R$, $C = 86°07′R$, and $D = 132°26′R$. The angle at E was not measured. If all measured angles are assumed to be correct, what is the value of angle E?

12-14 Similar to Problem 12-13, except the traverse was four-sided with measured deflection angles of $A = 124°17.3′L$, $B = 139°42.0′L$, and $C = 160°06.5′L$. Compute the angle at D which was not measured.

12-15 How can an angular check be obtained on **(a)** a closed-azimuth traverse **(b)** a closed-azimuth traverse with a cutoff line

12-16 Explain why it is important to carefully reference traverse stations once they have been placed.

12-17 Discuss criteria to consider when making reference ties to traverse stations.

*12-18 If the standard error for each measurement of a traverse angle is ±10″, what is the expected standard error of the misclosure in the sum of the angles for a six-sided traverse?

12-19 If the angles of a traverse are turned so that the 90% error of any angle is ±06″, prove that the 90% error of closure for a 10-sided traverse is equal to ±06″ $\sqrt{10}$.

*Asterisks indicate problems that have answers given in Appendix G.

12-20 In running an azimuth traverse, if the plate bubble of the instrument runs off-center between backsight and foresight readings, should it be recentered before taking the foresight reading? Explain.

12-21 The true azimuth from north of line *AB,* obtained from coordinates of monuments at *A* and *B,* is 111°28′. After a transit at station *A* has been oriented on this azimuth, point *C* is sighted and its azimuth found to be 156°23′. An open deflection angle traverse *ABCDEFGHI* is then run, and the following angles are read: *C* = 4°32′L, *D* = 10°38′L, *E* = 7°40′R, *F* = 3°10′R, *G* = 6°55′R, and *H* (to *I*) = 5°09′L. The true bearing of *IH* determined by a Polaris observation is N26°09′W. Are the traverse angles acceptable for ordinary work? Explain.

12-22 If point *A* in Problem 12-21 is at station 10 + 00 and hub *I* at station 58 + 75, what precision of linear measurements will be consistent with that of the angles?

12-23 The interior angles in a five-sided closed-polygon traverse were measured as *A* = 139°10′11″, *B* = 126°17′43″, *C* = 94°28′30″, *D* = 71°04′59″, and *E* = 108°58′31″. Compute the angular misclosure. For what FGCS order and class is this survey adequate?

12-24 Similar to Problem 12-23, except for a seven-sided traverse with measured angles of *A* = 164°12′12″, *B* = 86°21′33″, *C* = 204°19′06″, *D* = 57°28′53″, *E* = 70°16′36″, *F* = 174°56′08″, and *G* = 142°25′52″.

The recorded bearings and lengths for traverses are listed in Problems 12-25 through 12-28. If the lengths are assumed to be correct, which angle is most likely to be wrong?

*12-25 *AB* = S20°55′E, 288 ft; *BC* = S28°05′W, 455 ft; *CD* = N17°49′W, 416 ft; and *DA* = N55°11′E, 359 ft

12-26 *AB* = N42°58′E, 300 ft; *BC* = S42°02′E, 400 ft; *CD* = S59°20′W, 594 ft; and *DA* = N6°58′E, 282 ft

12-27 *AB* = N21°03′W, 696.8 ft; *BC* = S71°27′W, 478.4 ft; *CD* = S19°43′E, 438.4 ft; *DE* = S71°47′W, 268.0 ft; *EF* = S20°13′E, 451.2 ft; and *FA* = N54°32′E, 609.8 ft

12-28 *AB* = S18°30′E, 236 ft; *BC* = S37°30′W, 244 ft; *CD* = N70°30′W, 324 ft; *DE* = N18°30′W, 163 ft; and *EA* = N69°30′E, 437 ft

*12-29 In a radial traverse like that of Figure 12-7(a), the following measurements were obtained. Distances: *OA* = 185.52, *OB* = 325.80, *OC* = 450.63, *OD* = 451.01, *OE* = 388.66, and *OF* = 281.94 ft. Angles: *AOB* = 57°38′, *BOC* = 32°40′, *COD* = 24°30′, *DOE* = 50°51′, *EOF* = 77°08′, and *FOA* = 117°07′. Compute the horizon misclosure and adjust the angles. Using two sides and the included angle of each triangle, calculate lengths *AB, BC, CD, DE, EF,* and *FA.* Carefully plot the radial traverse using a scale of 100 ft per inch, and scale lengths *AB, BC,* and so on, from your plot. Compare them with your computed values.

12-30 Explain why it is advisable to use two instrument stations, as *O* and *O′* in Figure 12-7(b), when running radial traverses.

*Asterisks indicate problems that have answers given in Appendix G.

BIBLIOGRAPHY

Brinker, R. C., and R. Minnick. 1987. *The Surveying Handbook.* New York: Van Nostrand Reinhold, Chap. 8.

Dracup, J. F., et al. 1979. *Surveying Instrumentation and Coordinate Computation Workshop Lecture Notes.* Falls Church, Va.: American Congress on Surveying and Mapping.

Mattson, D. F. 1973. "Angular Reference Ties." *Surveying and Mapping* 33 (No. 1): 33.

Moffitt, F. H., and H. Bouchard. 1992. *Surveying* 9th ed. New York: Harper-Collins, Chap. 8.

13

TRAVERSE COMPUTATIONS

13-1 INTRODUCTION

Measured angles or directions of closed traverses are readily investigated before leaving the field. Linear measurements, even though repeated, are a more likely source of error and must also be checked. Although the calculations are more lengthy than angle checks, with today's programmable calculators and portable computers they can also be done in the field to determine, before leaving, whether a traverse meets the required precision. If specifications have been satisfied, the traverse is then adjusted to create perfect "closure" or geometric consistency among angles and lengths; if not, field measurements must be repeated until adequate results are obtained.

Investigation of precision, and acceptance or rejection of the field data, are extremely important in surveying. Adjustment for geometric closure is also crucial. In land surveying, for example, the law generally requires property descriptions to have exact geometric agreement.

Different procedures may be used for computing and adjusting traverses. These vary from elementary methods, to more advanced techniques based on the method of least squares. This text concentrates on elementary procedures. The usual steps followed in making elementary traverse computations are (1) adjusting angles or directions to fixed geometric conditions, (2) determining preliminary azimuths (or bearings) of the traverse lines, (3) calculating departures and latitudes and adjusting them for misclosure, (4) computing rectangular coordinates of the traverse stations, and (5) calculating the lengths and azimuths (or bearings) of the traverse lines after adjustment. These procedures are all discussed in this chapter, and are illustrated with several examples. The worked examples illustrate that instruments of different precision may be used to obtain the original data, but the basic computational procedures remain the same. Thus Examples 13-1 through 13-5, 13-8, and 13-11 utilize data obtained with an optical reading theodolite of moderate precision like those of Figures 10-7 and 10-8, where the angles were read directly to the nearest minute, with estimation to tenths of minutes. Other examples (13-6, 13-7, 13-9, and 13-10) are based on angles measured to the nearest second using a more precise instrument.

13-2 BALANCING ANGLES

The first step in traverse calculations is to balance (adjust) the angles to the proper geometric total. An exception occurs if magnetic bearings have been read directly with a compass for all courses, as described in Section 12-2.1, in which case no adjustment is possible. For closed traverses, angle balancing is done readily since the total error is known (see Section 12-7), although its exact distribution is not.

Angles of a closed traverse can be adjusted to the correct geometric total by applying one of two methods:

1. Applying an average correction to each angle where observing conditions were approximately the same at all stations. The correction is found by dividing the total angular misclosure by the number of angles.
2. Making larger corrections to angles where poor observing conditions were present.

Of these two methods, the first is almost always applied.

EXAMPLE 13-1   For the traverse of Figure 13-1 the measured interior angles are given in Table 13-1. Compute the adjusted angles using methods 1 and 2.

Figure 13-1 Traverse.

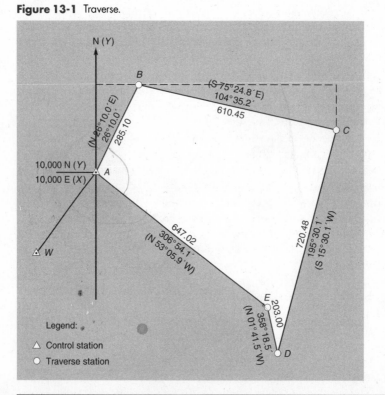

TABLE 13-1 ADJUSTMENT OF ANGLES

METHOD 1

POINT (1)	MEASURED ANGLE (2)	MULTIPLES OF AVERAGE CORRECTION (3)	CORRECTION ROUNDED TO 0.1′ (4)	SUCCESSIVE DIFFERENCES (5)	ADJUSTED ANGLE (6)
A	100°44.3′	0.24′	0.2′	0.2′	100°44.1′
B	101°35.1′	0.48′	0.5′	0.3′	101°34.8′
C	89°05.3′	0.72′	0.7′	0.2′	89°05.1′
D	17°11.9′	0.96′	1.0′	0.3′	17°11.6′
E	231°24.6′	1.20′	1.2′	0.2′	231°24.4′
Σ =	540°01.2′			Σ = 1.2′	Σ = 540°00.0′

METHOD 2

POINT (1)	MEASURED ANGLE (2)	ADJUSTMENT (7)	ADJUSTED ANGLE (8)
A	100°44.3′	0.2′	100°44.1′
B	101°35.1′	0.2′	101°34.9′
C	89°05.3′	0.0′	89°05.3′
D	17°11.9′	0.4′	17°11.5′
E	231°24.6′	0.4′	231°24.2′
Σ =	540°01.2′	Σ = 1.2′	Σ = 540°00.0′

SOLUTION

The computations are best arranged as shown in Table 13-1. The first part of the adjustment consists of summing the interior angles and determining the misclosure according to Eq. (12-1), which in this instance is $+01.2'$. The remaining calculations are tabulated, and the rationale for the procedures follows.

For work of ordinary precision it is reasonable to adopt corrections that are even multiples of the smallest recorded digit or decimal place for the angle readings. Thus in this example, corrections to the nearest 0.1′ will be made.

Method 1 consists of subtracting $01.2'/5 = 0.24'$ from each of the five angles. However, since the angles were read in multiples of 0.1′, applying corrections to the nearest hundredth of a minute would give a false impression of their precision. Therefore it is desirable to establish a pattern of corrections to the nearest 0.1′, as shown in Table 13-1. First multiples of the average correction of 0.24′ are tabulated in column (3). In column (4), each of these multiples has been rounded off to the nearest 0.1′. Then successive differences (adjustments for each angle) are found by subtracting the preceding value in column (4) from the one being considered. These are tabulated in column (5). The adjusted angles obtained by applying these corrections are listed in column (6). They must total exactly the true geometric value of $(n - 2)180°$.

In method 2, judgment is required because corrections are made to the angles expected to contain the largest errors. In this example, 0.4′ is subtracted from the angles at D and E, since they have the shortest sights (along line DE), and 0.2′ is subtracted from the angles at *A* and *B,* because they have the next shortest sights (along line *AB*). No correction was applied to angle *C* because of its long open sights. The sum of the corrections must of course equal the total misclosure. The adjustment made in this manner is shown in columns (7) and (8) of Table 13-1.

It should be noted that, although the adjusted angles by both methods satisfy the geometric condition of a closed figure, they may be no nearer to the true values than before adjustment. Unlike corrections for linear measurements, *the adjustments applied to angles are independent of the size of an angle.*

13-3 COMPUTATION OF PRELIMINARY AZIMUTHS OR BEARINGS

After balancing the angles, the next step in traverse computation is calculation of either preliminary azimuths or preliminary bearings. This requires the direction of at least one course within the traverse to be either known or assumed. For some computational purposes an assumed direction is sufficient, and in that case the usual procedure is to simply assign north as the direction of one of the traverse lines. On certain traverse surveys, the magnetic bearing of one line can be determined and used as a reference for determining the other directions. In most instances, however, as in boundary surveys, true directions are needed. This requirement can be met by **(1)** incorporating within the traverse a line whose true direction was established through a previous survey; **(2)** including one end of a line of known direction as a station in the traverse [for example, station *A* of line *A–Az Mk* of Figure 12-1(a)], and then measuring an angle from that reference line to a traverse line; or **(3)** determining the true direction of one traverse line by astronomical observations (see Chapter 18) or by other means.

If a line of known direction exists within the traverse, computation of preliminary azimuths (or bearings) proceeds as discussed in Chapter 8. Angles adjusted to the proper geometric total must be used; otherwise the azimuth or bearing of the first line, when recomputed after using all angles and progressing around the traverse, will differ from its fixed value by the angular misclosure. Azimuths or bearings at this stage are called "preliminary" because they will change after the traverse is adjusted, as explained in Section 13-11.

EXAMPLE 13-2   Compute preliminary azimuths and bearings for the traverse courses of Figure 13-1, based on a fixed azimuth of 234°17.6' for line *AW,* a measured angle to the right of 151°52.4' for *WAB,* and the angle adjustment by method 1 of Table 13-1.

SOLUTION

Preliminary azimuths of all lines are computed using the same tabular procedures as demonstrated in Section 8-9. These computations together with the results are given in Table 13-2. Note that the azimuth of *AB* was recalculated using the interior angle at *A* to obtain a computational check. Preliminary bearings (computed from the preliminary azimuths) are also listed.

TABLE 13-2 COMPUTATION OF AZIMUTHS AND BEARINGS

LINE	COMPUTATION	PRELIMINARY AZIMUTH	PRELIMINARY BEARING
AB	234°17.6′ + 151°52.4′ − 360°	26°10.0′	N26°10.0′E
BC	26°10.0′ + (180° − 101°34.8′)	104°35.2′	S75°24.8′E
CD	104°35.2′ + (180° − 89°05.1′)	195°30.1′	S15°30.1′W
DE	195°30.1′ + (180° − 17°11.6′)	358°18.5′	N01°41.5′W
EA	358°18.5′ + (180° − 231°24.4′)	306°54.1′	N53°05.9′W
AB	306°54.1′ + (180° − 100°44.1′) − 360°	26°10.0′ (check!)	N26°10.0′E

13-4 DEPARTURES AND LATITUDES

After balancing the angles and calculating preliminary azimuths (or bearings), traverse closure is checked by computing the *departure* and *latitude* of each line (course). As illustrated in Figure 13-2, the departure of a course is its orthographic projection on the east-west axis of the survey, and is equal to the length of the course multiplied by the sine of its azimuth (or bearing). Departures are sometimes called *eastings* or *westings*.

Also as shown in Figure 13-2, the latitude of a course is its orthographic projection on the north-south axis of the survey, and is equal to the course length multiplied by the cosine of its azimuth (or bearing). Latitude is also called *northing* or *southing*.

Figure 13-2 Departure and latitude of a line.

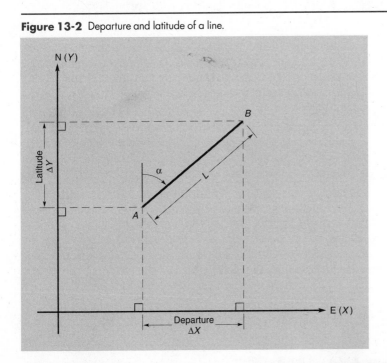

In equation form, the departure and latitude of a line are

$$\text{departure} = L \sin \alpha \tag{13-1}$$

$$\text{latitude} = L \cos \alpha \tag{13-2}$$

where L is the horizontal length and α the azimuth (or bearing) of the course.

Departures and latitudes are merely X and Y components of a line in a rectangular grid system, sometimes referred to as ΔX and ΔY. In traverse calculations, east departures and north latitudes are considered plus; west departures and south latitudes, minus. Azimuths (from north) used in computing departures and latitudes range from 0 to 360°, and the algebraic signs of sines and cosines automatically produce the proper algebraic signs of the departures and latitudes. Thus a line with an azimuth of 137°30′ has a positive departure and a negative latitude (its sine is plus and its cosine minus); a course of 323°18′ azimuth has a negative departure and a positive latitude.

In using bearings for computing departures and latitudes, the angles are always between 0 and 90°; hence their sines and cosines are invariably positive. Proper algebraic signs of departures and latitudes must therefore be assigned on the basis of the bearing angle directions, so a NE bearing has a plus departure and latitude, a SW bearing gets a minus departure and latitude, and so on. Because computers and hand calculators automatically affix correct algebraic signs to departures and latitudes through those of azimuth angle sines and cosines, it is more convenient to use azimuths than bearings for traverse computations.

13-5 DEPARTURE AND LATITUDE CLOSURE CONDITIONS

For a closed-polygon traverse like that of Figure 13-1, it can be reasoned that if all angles and distances were measured perfectly, the algebraic sum of the departures of all courses in the traverse should equal zero. Likewise, the algebraic sum of all latitudes should equal zero. And for closed link-type traverses like that of Figure 12-1(b), the algebraic sum of departures should equal the total difference in departure between the starting and ending control points. The same condition applies to latitudes in a link traverse. Because the measurements are not perfect, and errors exist in the angles and distances, the conditions just stated rarely occur. The amounts by which they fail to be met are termed *departure misclosure* and *latitude misclosure*. Their values are computed by algebraically summing the departures and latitudes, and comparing the totals to the required conditions.

The magnitudes of the departure and latitude misclosures for closed-polygon-type traverses give an "indication" of the precision that exists in the measured angles and distances. Large misclosures certainly indicate that either significant errors or even mistakes exist. Small misclosures usually mean the measured data are precise and free of mistakes, but *it is not a guarantee that systematic or compensating errors do not exist.*

13-6 TRAVERSE LINEAR MISCLOSURE AND RELATIVE PRECISION

Due to errors in the measured traverse angles and distances, if one were to begin at point A of a closed-polygon traverse like that of Figure 13-1, and progressively follow each course for its measured distance along its preliminary bearing or azimuth, one would finally return not to point A, but to some other nearby point A'. Point A' would be removed from A in an east-west direction by the departure misclosure, and in a north-south direction by the latitude misclosure. The distance between A and A' is termed the *linear misclosure* of the traverse. It is calculated from the following formula:

$$\text{linear misclosure} = \sqrt{(\text{departure misclosure})^2 + (\text{latitude misclosure})^2} \quad (13\text{-}3)$$

The *relative precision* of a traverse is expressed by a fraction that has the linear misclosure as its numerator and the traverse perimeter or total length as its denominator, or

$$\text{relative precision} = \frac{\text{linear misclosure}}{\text{traverse length}} \quad (13\text{-}4)$$

The fraction that results from Eq. (13-4) is then reduced to reciprocal form, and the denominator rounded to the same number of significant figures as the numerator. (This is illustrated in Example 13-3.)

Various projects require different relative precisions, and some surveys must meet rigid specifications if the work is to be accepted. Precision requirements for control survey traverses are listed in Chapter 19. Required relative precisions for property surveys may be set by state law (for example, a minimum of $\frac{1}{7500}$ in Minnesota and $\frac{1}{3000}$ in Wisconsin) and by cities and counties. A small city might decree $\frac{1}{5000}$; a large metropolitan area, $\frac{1}{10,000}$.

Based on the preliminary bearings from Table 13-2 and lengths shown in Figure 13-1, calculate the departures and latitudes, linear misclosure, and relative precision of the traverse.

EXAMPLE 13-3

SOLUTION

In computing departures and latitudes, the data and results are usually listed in a standard tabular form, such as that shown in Table 13-3. The column headings and rulings save time and simplify checking.

In Table 13-3, taking the algebraic sum of east (+) and west (−) departures gives the misclosure, 0.54 ft. Also, summing north (+) and south (−) latitudes gives the misclosure in latitude, − 0.72 ft. *Linear misclosure* is the hypotenuse of a small triangle with sides of 0.54 ft and − 0.72 ft, and in this example its value is, by Eq. (13-3),

$$\text{linear misclosure} = \sqrt{(0.54)^2 + (-0.72)^2} = 0.90 \text{ ft}$$

The *relative precision* for this traverse, by Eq. (13-4), is

$$\text{relative precision} = \frac{0.90}{2466.05} = \frac{1}{2700}$$

TABLE 13-3 COMPUTATION OF DEPARTURES AND LATITUDES

STATION	PRELIMINARY BEARING	LENGTH	DEPARTURE	LATITUDE
A				
	N26°10.0'E	285.10	+ 125.72	+ 255.88
B				
	S75°24.8'E	610.45	+ 590.77	− 153.74
C				
	S15°30.1'W	720.48	− 192.56	− 694.27
D				
	N01°41.5'W	203.00	− 5.99	+ 202.91
E				
	N53°05.9'W	647.02	− 517.40	+ 388.50
A				
		Σ = 2466.05	Σ = +0.54	Σ = −0.72

13-7 TRAVERSE ADJUSTMENT

For any closed traverse the linear misclosure must be adjusted (or distributed) throughout the traverse to close or balance the figure. This is true even though the misclosure is negligible in plotting the traverse at map scale. There are several elementary methods available for traverse adjustment, but the one most commonly used is the *compass rule* (Bowditch method). As noted earlier, adjustment by least squares is an advanced technique that can also be used. These two methods are discussed in the subsections that follow.

13-7.1 Compass (Bowditch) Rule

The compass, or Bowditch, rule adjusts the departures and latitudes of traverse courses in proportion to their lengths. Although not as rigorous as the least-squares method, it does result in a logical distribution of misclosures. Corrections by this method are made according to the following rules:

$$\frac{\text{correction in departure}}{\text{for } AB} = \frac{-(\text{total departure misclosure})}{\text{traverse perimeter}} \times \text{length of } AB \quad (13\text{-}5)$$

$$\frac{\text{correction in latitude}}{\text{for } AB} = \frac{-(\text{total latitude misclosure})}{\text{traverse perimeter}} \times \text{length of } AB \quad (13\text{-}6)$$

Note that the algebraic signs of the corrections are opposite those of the respective misclosures.

EXAMPLE 13-4   Using the preliminary azimuths from Table 13-2 and lengths from Figure 13-1, compute departures and latitudes, linear misclosure, and relative precision. Balance the departures and latitudes using the compass rule.

SOLUTION

A tabular solution for computing departures and latitudes is again used (see Table 13-4). Note that in this tabulation separate columns are used for east ($+$) and west ($-$) departures, and also for north ($+$) and south ($-$) latitudes. Note also that the results are exactly the same as those obtained in Example 13-3 using bearings.

To compute departure and latitude corrections by the compass rule, Eqs. (13-5) and (13-6) are used as demonstrated. By Eq. (13-5) the correction in departure for AB is

$$\frac{-0.54}{2466} \times 610 = -0.13 \text{ ft}$$

And by Eq. (13-6) the correction for the latitude of BC is

$$\frac{+0.72}{2466} \times 285 = +0.08 \text{ ft}$$

The other corrections are likewise found by multiplying a constant—the ratio of misclosure in departure, and latitude to the perimeter—by the successive course lengths.

In Table 13-4 the departure and latitude corrections are shown in parentheses above their unadjusted values. These corrections are added algebraically to their respective unadjusted values, and the corrected quantities tabulated in the "balanced" departure and latitude columns. A check is made of the computational process by algebraically summing the departure and latitude columns to verify that each is zero. In these columns, if rounding off causes an excess or a deficiency of 0.01 ft, this is eliminated by revising one of the corrections.

13-7.2 Least-Squares Method

As noted in Section 2-26, the method of least squares is based on the theory of probability, which models the occurrence of random errors. This results in adjusted values having the highest probability. Thus least squares provides the best and most rigorous traverse adjustment, but until recently the method has not been widely used because of the lengthy computations required. The availability of electronic computers has now made these calculations routine, and consequently the least-squares method has gained popularity.

In applying the least-squares method to traverses, angle and distance measurements are adjusted simultaneously. Thus no preliminary angle adjustment is made, as is done when using the compass rule. The least-squares method is valid for any type of traverse, and has the advantage that observations of varying precisions can be weighted appropriately in the computations. Examples illustrating some elementary least-squares adjustments are discussed in Appendix C. Adjusting traverses by this method is more advanced, however, and is beyond the scope of this text. An explanation of the procedures can be found in references cited in the bibliography at the end of Chapter 2.

13-8 RECTANGULAR COORDINATES

Rectangular X and Y coordinates of any point give its position with respect to an arbitrarily selected pair of mutually perpendicular reference axes. The X coordinate is

TABLE 13-4 BALANCING DEPARTURES AND LATITUDES BY THE COMPASS (BOWDITCH) RULE

STATION	LENGTH (FT)	PRELIMINARY AZIMUTH	DEPARTURE EAST +	DEPARTURE WEST −	LATITUDE NORTH +	LATITUDE SOUTH −	BALANCED DEPARTURE	BALANCED LATITUDE	COORDINATES X (EASTING)	COORDINATES Y (NORTHING)
A									10,000.00	10,000.00
	285.10	26°10.0′	(−0.06) 125.72		(+0.08) 255.88		+125.66	+255.96		
B									10,125.66	10,255.96
	610.45	104°35.2′	(−0.13) 590.77			(+0.18) 153.74	+590.64	−153.56		
C									10,716.30	10,102.40
	720.48	195°30.1′		(−0.16) 192.56		(+0.21) 694.27	−192.72	−694.06		
D									10,523.58	9408.34
	203.00	358°18.5′		(−0.05) 5.99	(+0.06) 202.91		−6.04	+202.97		
E									10,517.54	9611.31
	647.02	306°54.1′		(−0.14) 517.40	(+0.19) 388.50		−517.54	+388.69		
A									10,000.00	10,000.00
	2466.05		+716.49 −715.95 +0.54	−715.95	+847.29	−848.01 +847.29 −0.72	0.00	0.00		
		misclosures								

linear misclosure $= \sqrt{(0.54)^2 + (-0.72)^2} = 0.90$ ft

relative precision $= \dfrac{0.90}{2466} = \dfrac{0.90}{2466} = \dfrac{1}{2700}$

the perpendicular distance, in feet or meters, from the point to the Y axis; the Y coordinate is the perpendicular distance to the X axis. Although the reference axes are discretionary in position, in surveying they are normally oriented so that the Y axis points north-south, with north the positive Y direction. The X axis runs east-west, with positive X being east. Given the rectangular coordinates of a number of points, their relative positions are uniquely defined.

Coordinates are useful in a variety of computations, including **(1)** determining lengths and directions of lines (see Section 13-10); **(2)** calculating areas of land parcels (see Section 14-5); **(3)** making certain curve calculations (see Section 25-12); and **(4)** locating inaccessible points. Coordinates are also advantageous for plotting maps (see Section 17-7).

In practice, *state plane coordinate* systems, as described in Chapter 21, are most frequently used as the basis for rectangular coordinates in plane surveys. For many calculations, however, any arbitrary system may be used. As an example, coordinates may be arbitrarily assigned to one traverse station. To avoid negative values of X and Y, an origin is assumed south and west of the traverse such that one hub has, for example, coordinates $X = 1000$, $Y = 1000$, or any other suitable values. In a closed traverse, assigning $Y = 0$ to the most southerly point and $X = 0$ to the most westerly station saves time in calculations.

Given the X and Y coordinates of any starting point A, the X coordinate of the next point B is obtained by adding the adjusted departure of line AB to X_A. Likewise, the Y coordinate of B is the adjusted latitude of AB added to Y_A. In equation form this is

$$X_B = X_A + \text{departure } AB$$

$$Y_B = Y_A + \text{latitude } AB \qquad (13\text{-}7)$$

The process is continued around the traverse, successively adding departures and latitudes until the coordinates of starting point A are recalculated. If these recalculated coordinates agree exactly with the starting ones, a check on the coordinates of all intermediate points is obtained (unless compensating mistakes have been made).

Using the balanced departures and latitudes obtained in Example 13-4 (see Table 13-4) and starting coordinates $X_A = Y_A = 10,000.00$, calculate coordinates of the other traverse points.

EXAMPLE 13-5

SOLUTION

The process of successively adding balanced departures and latitudes to obtain coordinates is carried out in the two rightmost columns of Table 13-4. Note that the starting coordinates $X_A = 10,000.00$ and $Y_A = 10,000.00$ are recomputed at the end to provide a check.

13-9 ALTERNATIVE METHODS FOR MAKING TRAVERSE COMPUTATIONS

Procedures for making traverse computations that vary somewhat from those described in preceding sections can be adopted. One alternative is to adjust azimuths or bearings rather than angles. Another is to apply compass rule corrections directly to coordinates. These procedures are described in the subsections that follow.

13-9.1 Balancing Angles by Adjusting Azimuths or Bearings

In this method, "unadjusted" azimuths or bearings are computed based on the measured angles. These azimuths or bearings are then adjusted to secure a geometric closure, and to obtain preliminary values for use in computing departures and latitudes. The method is equally applicable to closed-polygon traverses, like that of Figure 13-1, or to closed traverses, as shown in Figure 12-1(b), which begin on one control station and end on another. The procedure of making the adjustment for angular misclosure in this manner will be explained by an example.

EXAMPLE 13-6   Table 13-5 lists measured angles to the right for the traverse of Figure 12-1(b). The azimuths of lines $A–Az\ Mk_1$ and $E–Az\ Mk_2$ have known values of 139°05′45″ and 86°20′47″, respectively. Compute unadjusted azimuths and balance them to obtain geometric closure.

SOLUTION

From the measured angles of column (2) in Table 13-5, unadjusted azimuths have been calculated and are listed in column (3). Because of angular errors, the unadjusted azimuth of the final line $E–Az\ Mk_2$ disagrees with its fixed value by 0°00′40″. This

TABLE 13-5 BALANCING TRAVERSE AZIMUTHS

STATION (1)	MEASURED ANGLE (2)	UNADJUSTED AZIMUTH (3)	AZIMUTH CORRECTION (4)	PRELIMINARY AZIMUTH (5)
$AZ\ MK_1$				
		319°05′45″		319°05′45″
A	283°50′10″			
		62°55′55″	− 8″	62°55′47″
B	256°17′18″			
		139°13′13″	− 16″	139°12′57″
C	98°12′41″			
		57°25′54″	− 24″	57°25′30″
D	103°30′34″			
		340°56′28″	− 32″	340°55′56″
E	285°24′59″			
		86°21′27″	− 40″	86°20′47″
$Az\ Mk_2$				

$$86°21′27″$$
$$-\ 86°20′47″$$
$$\text{misclosure} = \overline{\ \ 0°00′40″}$$
$$\text{correction per angle} = -40″/5 = -8″$$

represents the angular misclosure, which is divided by 5, the number of measured angles, to yield a correction of 8″ per angle. The corrections to azimuths, which accumulate and increase by 8″ for each angle, are listed in column (4). Thus line AB, which is based on one measured angle, receives an 8″ correction; line BC, which uses two measured angles, gets a 16″ correction; and so on. The final azimuth, $E-Az\ Mk_2$, receives a 40″ correction because all five measured angles have been included in its calculation. The corrected preliminary azimuths are listed in column 5.

13-9.2 Balancing Departures and Latitudes by Adjusting Coordinates

In this procedure, commencing with the known coordinates of a beginning station, unadjusted departures and latitudes for each course are successively added to obtain "preliminary" coordinates for all stations. For closed-polygon traverses, after progressing around the traverse, preliminary coordinates are recomputed for the beginning station. The difference between the computed preliminary X coordinate at this station and its known X coordinate is the departure misclosure. Similarly, the disagreement between the computed preliminary Y coordinate for the beginning station and its known value is the latitude misclosure. Corrections for these misclosures can be calculated using compass-rule Equations (13-5) and (13-6) and applied directly to the preliminary coordinates to obtain adjusted coordinates. The result is exactly the same as if departures and latitudes were first adjusted and coordinates computed from them.

Closed traverses like that of Figure 12-1(b) can be similarly adjusted. For this type of traverse, unadjusted departures and latitudes are also successively added to the beginning station's coordinates to obtain preliminary coordinates for all points, including the final closing station. Differences in preliminary X and Y coordinates, and the corresponding known values for the closing station, represent the departure and latitude misclosures, respectively. These misclosures are distributed directly to preliminary coordinates using the compass rule to obtain final adjusted coordinates. The procedure will be demonstrated by an example.

Table 13-6 lists the preliminary azimuths (from Table 13-5) and measured lengths (in feet) for the traverse of Figure 12-1(b). The known coordinates of stations A and E are $X_A = 12,765.48$, $Y_A = 43,280.21$, $X_E = 14,797.12$, and $Y_E = 44,384.51$ ft. Adjust this traverse for departure and latitude misclosures by making corrections to preliminary coordinates.

EXAMPLE 13-7

SOLUTION

From the lengths and azimuths listed in columns (2) and (3) of Table 13-6, departures and latitudes are computed and tabulated in columns (4) and (5). These unadjusted values are progressively added to the known coordinates of station A to obtain preliminary coordinates for all stations, including E, and are listed in columns (6) and (7). Comparing the preliminary X and Y coordinates of station E with its known values yields departure and latitude misclosures of $+0.18$ and -0.22 ft, respectively. From these values, the linear misclosure of 0.28 ft and relative precision of $\frac{1}{14,000}$ are computed (see Table 13-6).

TABLE 13-6 TRAVERSE ADJUSTMENT BY COORDINATES

STATION (1)	LENGTH (FT) (2)	PRELIMINARY AZIMUTH (3)	DEPARTURE (4)	LATITUDE (5)	PRELIMINARY COORDINATES		CORRECTIONS		ADJUSTED COORDINATES	
					X (6)	Y (7)	X (8)	Y (9)	X (FT) (10)	Y (FT) (11)
A	1045.50	62°55'47"	930.96	475.79	12,765.48	43,280.21			12,765.48	43,280.21
B	1007.38	139°12'57"	658.03	−762.76	13,696.44	43,756.00	−0.05 (−0.05)	0.06 (0.06)	13,696.39	43,756.06
C	897.81	57°25'30"	756.57	483.38	14,354.47	42,993.24	−0.05 (−0.10)	0.06 (0.12)	14,354.37	42,993.36
D	960.36	340°55'56"	−313.74	907.67	15,111.04	43,476.62	−0.04 (−0.14)	0.05 (0.17)	15,110.90	43,476.79
E					14,797.30	44,384.29	−0.04 (−0.18)	0.05 (9.22)	14,797.12	44,384.51

$\Sigma = 3911.05$ ft

$$\begin{array}{cc} 14,797.12 & 44,384.51 \\ +0.18 & -0.22 \end{array}$$

$$\text{misclosures} = \frac{14,797.30}{} \quad \frac{44,384.29}{}$$

$$\text{linear misclosure} = \sqrt{(0.18)^2 + (-0.22)^2} = 0.28 \text{ ft}$$

$$\text{relative precision} = \frac{0.28}{3911} = \frac{1}{14,000}$$

Compass-rule corrections for each course are computed and listed in columns (8) and (9). Their cumulative values obtained by progressively adding the corrections are given in parentheses in columns (8) and (9). Finally, by applying the cumulative corrections to the preliminary coordinates of columns 6 and 7, final adjusted coordinates listed in columns (10) and (11) are obtained.

13-10 LENGTHS AND AZIMUTHS OR BEARINGS FROM DEPARTURES AND LATITUDES, OR COORDINATES (INVERSION)

If the departure and latitude of a line are known, its length and azimuth or bearing are readily obtained from the following relationships:

$$\text{tan azimuth (or bearing)} = \frac{\text{departure}}{\text{latitude}} \qquad (13\text{-}8)$$

$$\text{length} = \frac{\text{departure}}{\text{sin azimuth (or bearing)}} \qquad (13\text{-}9)$$

$$= \frac{\text{latitude}}{\text{cos azimuth (or bearing)}} \qquad (13\text{-}10)$$

$$= \sqrt{(\text{departure})^2 + (\text{latitude})^2} \qquad (13\text{-}11)$$

Equations (13-7) can be written to express departures and latitudes in terms of coordinate differences ΔX and ΔY as follows:

$$\text{departure}_{AB} = X_B - X_A = \Delta X$$

$$\text{latitude}_{AB} = Y_B - Y_A = \Delta Y \qquad (13\text{-}12)$$

Substituting Eqs. (13-12) into Eqs. (13-8) through (13-11),

$$\text{tan azimuth (or bearing)} \, AB = \frac{X_B - X_A}{Y_B - Y_A} = \frac{\Delta X}{\Delta Y} \qquad (13\text{-}13)$$

$$\text{length} \, AB = \frac{X_B - X_A \, (\text{or} \, \Delta X)}{\text{sin azimuth (or bearing)} \, AB} \qquad (13\text{-}14)$$

$$= \frac{Y_B - Y_A \, (\text{or} \, \Delta Y)}{\text{cos azimuth (or bearing)} \, AB} \qquad (13\text{-}15)$$

$$= \sqrt{(X_B - X_A)^2 + (Y_B - Y_A)^2} \qquad (13\text{-}16\text{a})$$

$$= \sqrt{(\Delta X)^2 + (\Delta Y)^2} \qquad (13\text{-}16\text{b})$$

Formulas (13-8) through (13-16b) can be applied to any line whose coordinates are known, whether or not it was actually measured in the survey. Note that X_B and Y_B must be listed first in Eqs. (13-12) and (13-13), so that ΔX and ΔY will have the correct algebraic signs. Computing lengths and directions of lines from departures and latitudes, or from coordinates, is called *inversing*.

13-11 COMPUTING FINAL ADJUSTED TRAVERSE LENGTHS AND DIRECTIONS

In traverse adjustments, as illustrated in Examples 13-4 and 13-7, corrections are applied to the computed departures and latitudes to obtain adjusted values. These in turn are used to calculate X and Y coordinates of the traverse stations. By changing departures and latitudes in the adjustment process, their lengths and azimuths (or bearings) also change. In many types of surveys it is necessary to compute the changed, or "final adjusted," lengths and directions. If, for example, the purpose of the traverse was to describe the boundaries of a parcel of land, the final adjusted lengths and directions would be used in the recorded deed.

The equations developed in the preceding section permit computation of final values for lengths and directions of traverse lines based either on their adjusted departures and latitudes or on their coordinates.

EXAMPLE 13-8   Calculate the final adjusted lengths and azimuths of the traverse of Example 13-4 from the adjusted departures and latitudes listed in Table 13-4.

SOLUTION

Equations (13-8) and (13-11) are applied to calculate the adjusted length and azimuth of line *AB*. All others were computed in the same manner. The results are listed in Table 13-7.

By Eq. (13-8), $\tan \text{azimuth}_{AB} = \frac{125.66}{255.96} = 0.490936$; $\text{azimuth}_{AB} = 26°08.9'$.

By Eq. (13-11), $\text{length}_{AB} = \sqrt{(125.66)^2 + (255.96)^2} = 285.14$ ft.

Comparing the measured lengths of Figure 13-1 and Table 13-4 to the final adjusted values in Table 13-7, it can be seen that, as expected, all values have undergone change, some increasing and others decreasing. Notice that varying changes have also occurred in the azimuths.

EXAMPLE 13-9   Using coordinates, calculate adjusted lengths and azimuths for the traverse of Example 13-7.

TABLE 13-7 FINAL ADJUSTED LENGTHS AND DIRECTIONS FOR TRAVERSE OF EXAMPLE 13-4

LINE	BALANCED DEPARTURE	BALANCED LATITUDE	BALANCED LENGTH (FT)	BALANCED AZIMUTH
AB	125.66	255.96	285.14	26°08.9'
BC	590.64	−153.56	610.28	104°34.4'
CD	−192.72	−694.06	720.32	195°31.1'
DE	−6.04	202.97	203.06	358°17.6'
EA	−517.54	388.69	647.25	306°54.5'

TABLE 13-8 FINAL ADJUSTED LENGTHS AND AZIMUTHS FOR TRAVERSE OF EXAMPLE 13-7

| LINE | ADJUSTED | | ADJUSTED | |
	ΔX	ΔY	LENGTH (FT)	AZIMUTH
AB	930.91	475.85	1045.48	62°55′31″
BC	657.98	−762.70	1007.30	139°12′56″
CD	756.33	483.43	897.80	57°25′16″
DE	−313.78	907.72	960.42	340°55′51″

SOLUTION

Equations (13-13) and (13-16a) are used to demonstrate calculation of the adjusted length and azimuth of line AB. All others were computed in the same way. The results are listed in Table 13-8.

$$X_B - X_A = 13,696.39 - 12,765.48 = 930.91 = \Delta X$$

$$Y_B - Y_A = 43,756.06 - 43,280.21 = 475.85 = \Delta Y$$

By Eq. (13-13), tan azimuth$_{AB}$ = $\frac{930.91}{475.85}$ = 1.95630976; azimuth$_{AB}$ = 62°55′31″.
By Eq. (13-16a), length$_{AB}$ = $\sqrt{(930.91)^2 + (475.85)^2}$ = 1045.48 ft.

Comparing the adjusted lengths and azimuths of Table 13-8 with their unadjusted values of Table 13-6 reveals that all values have undergone changes of varying amounts.

13-12 USE OF OPEN TRAVERSES

Although open traverses are not generally recommended, there are situations where it is extremely helpful to run one and then compute the length and direction of the "closing line." In Figure 13-3, for example, suppose that improved horizontal alignment is planned for Taylor Lake and Atkins Roads, and a new construction line AE must be laid out. Due to dense forest, visibility between points A and E is not possible. A random line (see Section 11-16) could be run from A toward E and then corrected to the desired line, but that would be very difficult and time consuming due to tree density. One solution to this problem is to run open traverse ABCDE, which can be done quite easily along the cleared right-of-way of existing roads.

For this problem an assumed azimuth (for example, due north) can be taken for line UA, and assumed coordinates (for example, 10,000.00 and 10,000.00) can be assigned to station A. From measured lengths and angles, departures and latitudes of all lines, and coordinates of all points can be computed. From the resulting coordinates of stations A and E, the length and azimuth of closing line AE can be calculated. Finally the deflection angle α needed to reach E from A can be computed and laid off.

In running open traverses, extreme caution must be exercised in all measurements, because there is no check, and any errors or mistakes will result in an erroneous length and direction for the closing line. Utmost care must also be exercised in the calculations, although a rough check on them can be secured by carefully plotting the traverse and scaling the length of the closing line and the deflection angle.

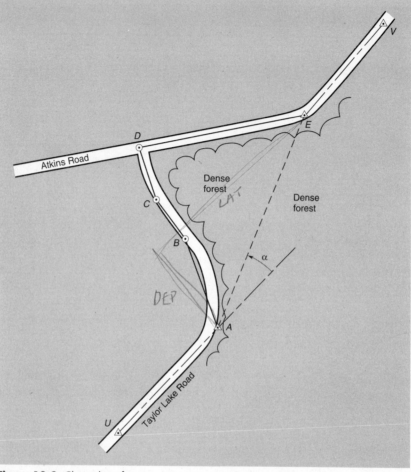

Figure 13-3 Closing line of an open traverse.

EXAMPLE 13-10   Compute the length and azimuth of closing line AE and deflection angle α of Figure 13-3, given the following measured data:

POINT	LENGTH (FT)	ANGLE TO THE RIGHT
A		115°18′25″
	3305.78	
B		161°24′11″
	1862.40	
C		204°50′09″
	1910.22	
D		273°46′37″
	6001.83	
E		

TABLE 13-9 COMPUTATIONS FOR CLOSING LINE

POINT	AZIMUTH	DEPARTURE	LATITUDE	X (FT)	Y (FT)
U					
	North (assumed)				
A				10,000.00	10,000.00
	295°18′25″	−2988.53	1413.11		
B				7,011.47	11,413.11
	276°42′36″	−1849.64	217.61		
C				5,161.83	11,630.72
	301°32′45″	−1627.93	999.39		
D				3,533.90	12,630.11
	35°19′22″	3470.15	4896.94		
E				7,004.05	17,527.05

SOLUTION

Table 13-9 presents a tabular solution for computing azimuths, departures and latitudes, and coordinates.

From the coordinates of points A and E, the ΔX and ΔY values of line AE are

$$\Delta X = 7,004.05 - 10,000.00 = -2,995.95 \text{ ft}$$

$$\Delta Y = 17,527.05 - 10,000.00 = 7,527.05 \text{ ft}$$

By Eq. (13-16b), the length of closing line AE is

$$\text{length}_{AE} = \sqrt{(-2995.95)^2 + (7527.05)^2} = 8101.37 \text{ ft}$$

By Eq. (13-13), the azimuth of closing line AE is

$$\tan \text{azimuth}_{AE} = \frac{-2995.95}{7527.05} = -0.39802446; \text{ azimuth}_{AE} = 338°17′46″$$

(Note that with a negative ΔX and positive ΔY the bearing of AE is northwest, hence the azimuth is 338°17′46″.)

Finally, the deflection angle α is the difference between the azimuths of lines AE and UA, or

$$\alpha = 338°17′46″ - 360° = -21°42′14″ \text{ (left)}$$

13-13 COORDINATE COMPUTATIONS IN BOUNDARY MEASUREMENTS

The computation of a bearing from the known coordinates of two points on a line is a common problem in boundary measurements. If the lengths and directions of lines from traverse points to the corners of a field are known, the coordinates of the corners can be determined and the lengths and bearings of all sides calculated.

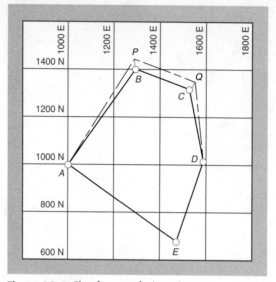

Figure 13-4 Plot of traverse for boundary survey.

EXAMPLE 13-11   In Figure 13-4, *APQDEA* is a parcel of land that must be surveyed, but due to obstructions, traverse stations cannot be set at *P* and *Q*. Therefore offset stations *B* and *C* are set nearby, and closed traverse *ABCDEA* run. Lengths and azimuths of lines *BP* and *CQ* are measured as 42.50 ft, 354°50.0′, and 34.62 ft, 26°39.9′, respectively. Traverse *ABCDEA* was computed and adjusted, and coordinates were determined for all stations. They are

POINT	*X* (ft)	*Y* (ft)
A	1000.00	1000.00
B	1290.65	1407.48
C	1527.36	1322.10
D	1585.70	1017.22
E	1464.01	688.25

Compute the length and bearing of property line *PQ*.

SOLUTION

1. Using Eqs. (13-1) and (13-2), the departures and latitudes of lines *BP* and *CQ* are:

$$\text{Dep}_{BP} = 42.50 \sin(345°50.0′) = -3.83 \text{ ft}$$
$$\text{Dep}_{CQ} = 34.62 \sin(26°39.9′) = 15.54 \text{ ft}$$

$$\text{Lat}_{BP} = 42.50 \cos(354°50.0') = 42.33 \text{ ft}$$
$$\text{Lat}_{CQ} = 34.62 \cos(26°39.9') = 30.94 \text{ ft}$$

2. From the coordinates of stations B and C and the departures and latitudes just calculated, the following tabular solution yields X and Y coordinates for points P and Q:

	X	Y		X	Y
B	1290.65	1407.48	C	1527.36	1322.10
BP	−3.83	+42.33	CQ	+15.54	+30.94
P	1286.82	1449.81	Q	1542.90	1353.04

3. From the coordinates of P and Q, the length and bearing of line PQ are found in the following manner:

	X	Y
Q	1542.90	1353.04
P	−1286.88	−1449.81
PQ	$\Delta X = 256.02$	$\Delta Y = -96.77$

By Eq. (13-13), tan bearing $PQ = \dfrac{256.02}{-96.77} = -2.64565$; $PQ = $ S69°17.7'E.

By Eq. (13-16b), length $PQ = \sqrt{(96.77)^2 + (256.02)^2} = 273.79 \text{ ft}$

By using Eqs. (13-13) and (13-16b), lengths and bearings of lines AP and QD can also be determined. As stated earlier, extreme caution must be used when employing this procedure, since no checks are obtained on the length and azimuth measurements of lines BP and CQ, nor are there any computational checks on the calculated lengths and bearings.

13-14 STATE PLANE COORDINATE SYSTEMS

Under ordinary circumstances, rectangular coordinate systems for plane surveys would be limited in size due to earth curvature. However, the National Geodetic Survey (NGS) developed statewide coordinate systems for each state in the United States, which retain an accuracy of 1 part in 10,000 or better while fitting curved earth distances to plane grid lengths.

State plane coordinates are related to latitude and longitude, so control survey stations set by the NGS and others can be tied to the systems. As additional stations are set and their coordinates determined, they too become usable reference points. These monumented control stations serve as starting points for local surveys and permit accurate restoration of obliterated or destroyed marks having known coordinates. If state plane coordinates of two intervisible stations are known, like A and Az Mk of Figure 12-1(a), the direction of line A–Az Mk can be computed and used to orient the theodolite or total station instrument at A. In this way, azimuths and bearings of traverse lines are obtained without the necessity of making astronomical observations or resorting to other means.

In the past some cities and counties have used their own coordinate systems for locating street, sewer, property, and other lines. Because of their limited extent and the resultant discontinuity at city or county lines, such local systems are less desirable than a statewide grid. Military grids are used to pinpoint the locations of objects by coordinates for fire control and other purposes.

A more extended discussion of state plane coordinates is given in Chapter 21.

13-15 TRAVERSE COMPUTATIONS USING COMPUTERS

Computers of various types and sizes are now widely used in surveying, and they are particularly convenient for making traverse computations. Small programmable hand-held units and battery-operated lap-top computers are commonly taken into the field and used to verify data for acceptable misclosures before returning to the office. In the office, personal computers of the type shown in Figure 13-5 are widely used.

A variety of software is available for use by surveyors. Some manufacturers supply standard programs, which include traverse computations, with the purchase of their equipment. Various software is also available for purchase from a number of suppliers. Spreadsheet software such as LOTUS 1-2-3, QUATTRO-PRO, EXCEL, and others can also be conveniently used with personal computers to calculate and adjust traverses. Of course, programs are also frequently written by surveying and engineering firms specifically for their own use. Standard programming languages employed include Fortran, Pascal, BASIC, and others.

A simple traverse computation program written in BASIC is given in Appendix F. It computes departures and latitudes, linear misclosure, and relative precision, and it performs adjustments by the compass (Bowditch) rule. In addition, the program calculates coordinates of the traverse points and the area within polygon traverses using the coordinate method (discussed in Section 14-5). In Section F-2 of Appendix F, Examples 13-4 and 13-7 are solved to demonstrate the use of the program, and to illustrate how its output is arranged in tabular form similar to Tables 13-4 and 13-6.

Besides performing routine computations such as traverse solutions, personal computers have several other valuable applications in surveying and engineering offices. As examples, they are being used with *computer-aided drafting* (CAD) software for plotting maps and drawing contours (see Section 17-17), and with increasing frequency they are also being employed to operate *geographic information system* (GIS) software (see Chapter 29).

13-16 MISTAKES IN TRAVERSE COMPUTATIONS

Some of the more common mistakes made in traverse computations are:

1. Failing to adjust the angles before computing azimuths or bearings
2. Applying angle adjustments in the wrong direction and failing to check the angle sum for proper geometric total
3. Interchanging departures and latitudes, or their signs
4. Confusing the signs of coordinates
5. Carrying out corrections beyond the number of decimal places in the original measurements

Figure 13-5 Personal computer being used to perform surveying computing and drafting. (Courtesy Wisconsin Department of Transportation.)

PROBLEMS

13-1 In adjusting measured traverse angles, why aren't adjustments made in proportion to angle sizes?

*13-2 The sum of eight interior angles of a closed-polygon traverse, each read to the nearest minute, is 1080°05′. What is the misclosure, and what correction would be applied to each angle in balancing them by method 1?

13-3 Similar to Problem 13-2, except the angles were read to the nearest 0.1′, and their sum was 719°58.6′ for a six-sided polygon traverse.

13-4 Similar to Problem 13-2, except the angles were read to the nearest second and their sum in a nine-sided polygon traverse was 1260°00′16″.

13-5 The algebraic sum of the deflection angles of a 10-sided traverse *ABCDEFGHIJA* with six left deflections, four right deflections, and no crisscrossing lines is 359°52′. Balance the angles by method 1.

13-6 Balance the angles in the following azimuth (from north) traverse by method 1, and compute and tabulate the bearings, assuming azimuth *AB* is correct: *AB* = 186°42′; *BC* = 48°35′; *CD* = 78°28′; *DE* = 188°44′; *EF* = 109°39′; *FA* = 300°12′; *AB* = 186°36′.

*13-7 Balance the following interior angles of a six-sided closed-polygon traverse using method 1. If the bearing of side *AB* is fixed at N86°18′E, calculate the bearings of the remaining sides. *A* = 89°16′; *B* = 91°25′; *C* = 101°29′; *D* = 156°13′; *E* = 196°44′; *F* = 84°56′. (*Note:* Line *BC* bears SE.)

*13-8 Compute departures and latitudes, linear misclosure, and relative precision for the traverse of Problem 13-7 if the lengths of the sides (in feet) are as follows: *AB* = 1743.09; *BC* = 578.95; *CD* = 438.98; *DE* = 774.41; *EF* = 563.92; and *FA* = 586.90.

*†13-9 Using the compass (Bowditch) rule, adjust the departures and latitudes of the traverse in Problem 13-8. If the coordinates of station *A* are *X* = 10,000.00 and *Y* = 10,000.00, calculate coordinates for the other stations and then the lengths and bearings of lines *AD* and *EB*.

13-10 Balance the following interior angles of a closed-polygon traverse to the nearest 0.1′ using method 1. Compute the azimuths assuming a correct azimuth of 293°30.2′ for line *AB*. *A* = 122°39.4′; *B* = 92°22.7′; *C* = 184°35.8′; *D* = 57°42.4′; *E* = 316°33.5′; *F* = 48°37.8′; *G* = 254°32.2′; *H* = 4°34.1′; *I* = 178°23.0′. (*Note:* Line *BC* bears NE.)

13-11 Determine departures and latitudes, linear misclosure, and relative precision for the traverse of Problem 13-10 if lengths of the sides (in feet) are as follows: *AB* = 812.05; *BC* = 649.68; *CD* = 742.89; *DE* = 986.25; *EF* = 2401.18; *FG* = 2374.02; *GH* = 1774.07; *HI* = 2960.10; and *IA* = 1534.24.

†13-12 Using the compass (Bowditch) rule, adjust the departures and latitudes of the traverse in Problem 13-11. If the coordinates of station *A* are *X* = 270,565.88 and *Y* = 59,341.35, calculate coordinates for the other stations and, from them, the lengths and bearings of lines *EA* and *BG*.

13-13 Same as Problem 13-10, except assume line *AB* has a fixed azimuth of 132°17.6′ and line *BC* bears NE.

13-14 Using the lengths from Problem 13-11 and azimuths from Problem 13-13, calculate departures and latitudes, linear misclosure, and relative precision of the traverse.

13-15 Adjust the departures and latitudes of Problem 13-14 using the compass (Bowditch) rule, and compute coordinates of all stations if the coordinates of station *A* are *X* = 16,385.28 ft and *Y* = 29,364.16 ft. Compute the length and azimuth of line *CG*.

13-16 Compute and tabulate for the following closed-polygon traverse: **(a)** bearings **(b)** departures and latitudes **(c)** linear misclosure and **(d)** relative precision. (Note: line *BC* bears NW.)

*Asterisks indicate problems that have solutions given in Appendix G.
†Daggers indicate problems that have solutions used in Chapter 14 problems.

LINE	BEARING	LENGTH (ft)	INTERIOR ANGLE (RIGHT)
AB	N32°21′E	701.82	$A = 128°58′$
BC		1009.91	$B = 138°04′$
CD		1109.48	$C = 110°24′$
DE		1320.51	$D = 125°57′$
EF		1447.66	$E = 85°57′$
FA		785.74	$F = 130°46′$

13-17 In Problem 13-16, if one side and/or angle is responsible for most of the error of closure, which is it likely to be?

†**13-18** Adjust the traverse of Problem 13-16 using the compass (Bowditch) rule. If the coordinates of point *A* are 5000.00E and 5000.00N, determine the coordinates of all other points. Find the length and bearing of line *AE*.

For the closed-polygon traverses given in Problem 13-19 through 13-21 (lengths in feet), compute and tabulate: **(a)** unbalanced departures and latitudes **(b)** linear misclosure **(c)** relative precision and **(d)** preliminary coordinates if $X_A = 1,000.00$ and $Y_A = 1,000.00$. Balance the traverses by coordinates using the compass (Bowditch) rule.

	COURSE	AB	BC	CD	DA
*†**13-19**	Bearing	N8°17′E	N87°02′E	S14°47′W	N68°43′W
	Length	404.03	622.13	653.16	550.94
†**13-20**	Bearing	N31°50′E	S58°10′E	S76°50′W	N58°10′W
	Length	292.80	878.40	413.14	585.60
†**13-21**	Bearing	S68°42′E	N25°03′E	N47°17′W	S25°05′W
	Length	385.94	1016.48	403.20	1164.40

†**13-22** Compute the linear misclosure, relative precision, and adjusted lengths and bearings for the sides after the departures and latitudes are balanced by the compass (Bowditch) rule in the following closed-polygon traverse.

LINE	LENGTH (m)	DEPARTURE (m)	LATITUDE (m)
AB	645.96	W613.60	N201.88
BC	1420.46	W504.75	S1327.75
CA	1587.74	E1118.86	N1126.53

+0.51

*†**13-23** Balance by the compass (Bowditch) rule the departures and latitudes listed in the following closed-polygon traverse. Calculate the linear misclosure, relative precision, and adjusted lengths and bearings.

LINE	AB	BC	CD	DA
Departure (ft)	E237.11	W667.68	W737.07	E1168.34
Latitude (ft)	N469.97	S314.03	S1041.39	N884.76

*Asterisks indicate problems that have solutions given in Appendix G.
†Daggers indicate problems that have solutions used in Chapter 14 problems.

234502

*13-24 The following data apply to a closed "link" traverse [like that of Figure 12-1(b)]. Compute preliminary azimuths, adjust them, and calculate departures and latitudes, misclosures in departure and latitude, and traverse relative precision. Balance the departures and latitudes using the compass (Bowditch) rule, and calculate coordinates of points B, C, and D. Compute the final lengths and azimuths of lines AB, BC, CD, and DE. Calculate the length and azimuth of the line from B to D.

STATION	MEASURED ANGLE*	ADJUSTED AZIMUTH	MEASURED LENGTH (ft)	X (ft)	Y (ft)
Az Mk₁					
		310°17′20″			
A	272°40′00″			35,287.13	26,482.48
			1432.07		
B	267°27′28″				
			1379.85		
C	87°02′31″				
			1229.77		
D	109°35′39″				
			1315.44		
E	270°29′59″			37,345.68	28,716.32
		57°32′52″	5357.13		
Az Mk₂					

*Angle to the right.

13-25 Similar to Problem 13-24, except use the following data:

STATION	MEASURED ANGLE*	ADJUSTED AZIMUTH	MEASURED LENGTH (ft)	X (ft)	Y (ft)
Az Mk₁					
		329°09′51″			
A	282°17′18″		2316.91	493,290.81	89,609.42
B	266°48′13″		2646.05		
C	89°16′53″		1984.73		
D	96°36′05″		2520.44		
E	291°27′38″			497,801.50	90,299.22
		105°36′08″			
Az Mk₂					

*Angle to the right.

†13-26 Two measurements that presented unusual difficulty in the field were omitted in the survey of a closed boundary, as shown in the following field notes. Compute the missing data. (Distances are in feet.)

*Asterisks indicate problems that have solutions given in Appendix G.
†Daggers indicate problems that have solutions used in Chapter 14 problems.

LINE	DISTANCE	BEARING
CD	1057.39	N85°20′E
DE	936.95	S82°32′E
EF	440.67	S22°20′W
FC	—	—

†**13-27** Similar to Problem 13-26, but determine the missing field data indicated in the following field notes. (Distances are in feet; *CD* is approximately 460 ft long.)

LINE	DISTANCE	BEARING
AB	610.67	N89°46′E
BC	397.17	S87°22′E
CD	—	S27°33′W
DE	351.90	—
EA	708.03	N40°23′W

The azimuths (from north) of a closed-polygon traverse are $AB = 46°27′$, $BC = 148°29′$, $CD = 201°44′$, and $DA = 248°06′$. If one measured distance contains a mistake, which course is most likely responsible for the closure conditions given in Problems 13-28 through 13-30? Is the course too long or too short?

13-28 Algebraic sum of departures $= -92.82$ ft, latitudes $= -37.05$ ft.

***13-29** Algebraic sum of departures $= -5.37$ ft, latitudes $= +8.43$ ft.

13-30 Algebraic sum of departures $= +34.44$ ft, latitudes $= +36.25$ ft.

†**13-31** Determine the lengths and bearings of the sides of a lot whose corners have the following *X* and *Y* coordinates (in feet): $A(0,0)$; $B(+370.23, +246.00)$; $C(-129.84, +390.45)$; $D(-232.84, +78.40)$.

*†**13-32** Compute the lengths and azimuths of the sides of a closed-polygon traverse whose corners have the following *X* and *Y* coordinates (in feet): $A(1000.00, 1000.00)$; $B(944.82, 1746.14)$; $C(1402.82, 1999.09)$; $D(1702.21, 1494.55)$.

***13-33** In searching for a record of the length and true bearing of a certain boundary line which is straight between *A* and *B,* the following notes of an old random traverse were found (survey by compass and Gunter's chain, declination 4°45′W). Compute the true bearing and length (in feet) of *BA.*

LINE	A–1	1–2	2–3	3–B
Magnetic bearing	North	N20°E	East	S46°30′E
Distance (ch)	8.33	25.06	16.88	8.90

13-34 The interior angles of a four-sided closed traverse *ABCDA* were all measured by the same observer using the same equipment and procedures, except the angles at *A* and *B* were turned six times (3D, 3R) and those at *C* and *D* eight times (4D, 4R). If the mean values at *A, B, C,* and *D* were 54°33.3′, 79°15.1′, 87°48.6′, and 138°20.6′, respectively, adjust the angles for misclosure.

***13-35** Similar to Problem 13-34, except for a five-sided closed traverse with *A* and *C* turned eight times (4D, 4R) and the others repeated four times (2D, 2R), giving mean values

*Asterisks indicate problems that have solutions given in Appendix G.

†Daggers indicate problems that have solutions used in Chapter 14 problems.

at *A, B, C, D,* and *E* of 32°29′02″, 116°50′18″, 80°01′49″, 94°17′19″, and 216°20′57″, respectively.

13-36 What uses are made of the relative precision for a closed traverse?

13-37 After adjustment of departures and latitudes, does a traverse actually close? Explain.

13-38 Why shouldn't traverse relative precision be carried out to a reciprocal such as 1/9376.85?

BIBLIOGRAPHY

Burkholder, E. F. 1986. "Coordinates, Calculators, and Intersections." *Surveying and Mapping* 46 (No. 1): 29.

Dracup, J. F. 1993. "Accuracy Estimates from the Least Squares Adjustment of Traverses by Conditions," *Surveying and Land Information Systems* 53 (No. 1) 11.

Dracup, J. F., et al. 1979. *Surveying Instrumentation and Coordinate Computation Workshop Lecture Notes.* Bethesda, Md.: American Congress on Surveying and Mapping.

Moffitt, F. H. and H. Bouchard. 1992. *Surveying,* 9th ed. New York: Harper-Collins, Chap. 8.

Navyasky, M. 1974. "Adjusting Observed Directions by Angles Using Equitable Weights." *Surveying and Mapping* 34 (No. 4): 355.

Sprinsky, W. H. 1987. "Traverse Adjustment for Observations from Modern Survey Instruments." *ASCE, Journal of the Surveying Engineering Division* 113 (No. 1): 70.

Stoughton, H. W. 1974. "The First Method to Adjust a Traverse Based on Statistical Considerations." *Surveying and Mapping* 34 (No. 2): 145.

Stoughton, H. W. 1975. "Computing Missing Elements of a Polygon." *Surveying and Mapping* 35 (No. 3): 217.

Valentine, W. 1984. "Practical Traverse Analysis." *ASCE, Journal of the Surveying Engineering Division* 110 (No. 1): 58.

Wolf, P. R. 1980. *Adjustment Computations: Practical Least Squares for Surveyors,* 2nd ed. Rancho Cordova, Calif.: Landmark.

14

AREA

14-1 INTRODUCTION

There are a number of important reasons for determining areas of tracts of land. One is to include the acreage of a parcel of land in the deed describing the property. Other purposes are to determine the acreages of fields, lakes, and so on, or the number of square yards to be surfaced, paved, seeded, or sodded. A special application is determining end areas for earthwork volume calculation (see Chapter 27).

In plane surveying, area is considered to be the orthogonal projection of the surface onto a horizontal plane. As noted in Chapter 2, the most common unit of area for large tracts is the acre: 1 acre $= 43,560 \, \text{ft}^2 = 10 \, \text{ch}^2$ (Gunter's). An acre lot, if square, would thus be $208.71 +$ ft on a side. In the metric system, area is given in square meters or *hectares:* 1 hectare $= 10,000 \, \text{m}^2 = 2.471 +$ acres.

14-2 METHODS OF MEASURING AREA

Both field and map measurements are used to determine area. *Field measurement* methods are the more accurate and include **(1)** division of the tract into simple figures (triangles, rectangles, and trapezoids), **(2)** offsets from a straight line, **(3)** coordinates, and **(4)** double-meridian distances. Each of these methods is described in sections that follow.

Methods of determining area from *map measurements* include **(1)** counting coordinate squares, **(2)** dividing the area into triangles, rectangles, or other regular geometric shapes, **(3)** digitizing coordinates, and **(4)** running a planimeter over the enclosing lines. These processes are described and illustrated in Section 14-8. Because maps themselves are derived from field measurements, methods of area determination invariably depend on this basic source of data.

14-3 AREA BY DIVISION INTO SIMPLE FIGURES

A tract can usually be divided into simple geometric figures such as triangles, rectangles, or trapezoids. The sides and angles of these figures can be measured in the field and their individual areas calculated and totaled. An example of a parcel subdivided into triangles is shown in Figure 14-1.

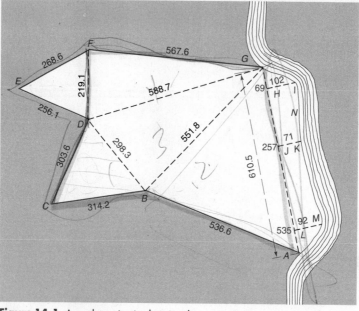

Figure 14-1 Area determination by triangles.

Formulas for computing areas of rectangles and trapezoids are well known. The area of a triangle whose sides are known can be computed by the formula

$$\text{area} = \sqrt{s(s-a)(s-b)(s-c)} \qquad (14\text{-}1)$$

where a, b, and c are sides of the triangle and $s = \frac{1}{2}(a+b+c)$. Another formula for the area of a triangle is

$$\text{area} = \tfrac{1}{2}ab \sin C \qquad (14\text{-}2)$$

where C is the angle included between sides a and b.

The choice of whether to use Eq. (14-1) or (14-2) will depend on which triangle parts are most conveniently measured—a decision ordinarily dictated by the nature of the area and the type of equipment available.

14-4 AREA BY OFFSETS FROM STRAIGHT LINES

Irregular tracts can be reduced to a series of trapezoids by measuring right-angle offsets from points along a measured reference line. The reference line is usually stationed, and positions where offsets are measured are given by their stations and pluses. The spacing between offsets may either be *regular* or *irregular*, depending on conditions. These two cases are discussed in the subsections that follow.

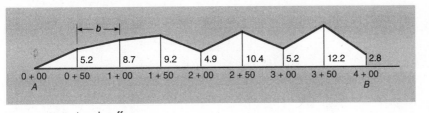

Figure 14-2 Area by offsets.

14-4.1 Regularly Spaced Offsets

Offsets at regularly spaced intervals are shown in Figure 14-2. For this case, the area is found by the formula

$$\text{area} = b\left(\frac{h_0}{2} + h_1 + h_2 + \cdots + \frac{h_n}{2}\right) \qquad (14\text{-}3)$$

where b is the length of a common interval between offsets, and $h_0, h_1, \ldots, h_n$ are the offsets. The regular interval for the example of Figure 14-2 is a half-station, or 50 ft.

Compute the area of the tract shown in Figure 14-2. EXAMPLE 14-1

SOLUTION
By Eq. (14-3),

$$\text{area} = 50\left(0 + 5.2 + 8.7 + 9.2 + 4.9 + 10.4 + 5.2 + 12.2 + \frac{2.8}{2}\right)$$

$$= 2860 \text{ ft}^2$$

In this example, a summation of offsets (terms within parentheses) can be secured by the *paper-strip method,* in which the area is plotted to scale and the midordinate of each trapezoid is successively added by placing tick marks on a long strip of paper. The area is then obtained by making a single measurement between the first and last tick marks and multiplying by width b.

14-4.2 Irregularly Spaced Offsets

For irregularly curved boundaries like that in Figure 14-3, the spacing of offsets along the reference line varies. Spacings should be selected so that the curved boundary is accurately defined when adjacent offset points on it are connected by straight lines. A formula for calculating area for this case is

$$\text{area} = \tfrac{1}{2}[a(h_0 + h_1) + b(h_1 + h_2) + c(h_2 + h_3) + \cdots] \qquad (14\text{-}4)$$

where $a, b, c, \ldots$ are the varying offset spaces, and $h_0, h_1, h_2, \ldots$ are the measured offsets.

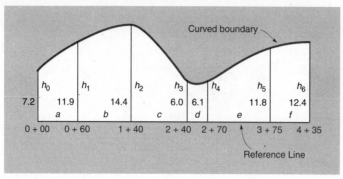

Figure 14-3 Area by offsets for a tract with curved boundary.

EXAMPLE 14-2   Compute the area of the tract shown in Figure 14-3.

SOLUTION
By Eq. (14-4),

$$\text{area} = \tfrac{1}{2}[60(7.2 + 11.9) + 80(11.9 + 14.4) + 100(14.4 + 6.0)$$

$$+ 30(6.0 + 6.1) + 105(6.1 + 11.8) + 60(11.8 + 12.4)]$$

$$= 4490 \text{ ft}^2$$

14-5 AREA BY COORDINATES

Computation of areas within closed polygons is most frequently done using the coordinate method. In this procedure, coordinates of each angle point in the figure must be known. They are normally obtained by traversing, although any method that yields the coordinates of these points is appropriate. If traversing is used, coordinates of the stations are computed after adjustment of the departures and latitudes, as discussed and illustrated in Chapter 13. The coordinate method is also applicable and convenient for computing areas of figures whose coordinates have been digitized using an instrument like that shown in Figure 29-8. The coordinate method is easily visualized; it reduces to one simple equation that applies to all geometric figures and is readily programmed for computer solution.

The procedure for computing areas by coordinates can be developed with reference to Figure 14-4. As shown in that figure, it is convenient (but not necessary) to adopt a reference coordinate system with the X and Y axes passing through the most southerly and the most westerly traverse stations, respectively. Lines BB', CC', DD', and EE' in the figure are constructed perpendicular to the Y axis. These lines create a series of trapezoids and triangles (shown by different color shadings). The area enclosed with traverse $ABCDEA$ can be expressed in terms of the areas of these individual trapezoids and triangles as

$$\text{area}_{ABCDEA} = B'BCC'B' + C'CDD'C' - AB'BA - DD'E'ED - AEE'A \quad (14\text{-}5)$$

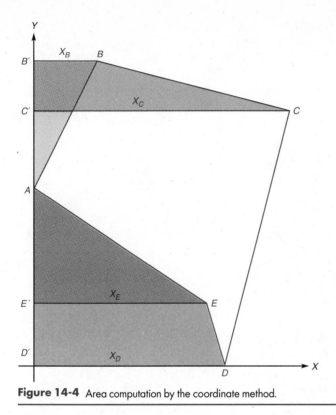

Figure 14-4 Area computation by the coordinate method.

The area of each trapezoid, for example $B'BCC'B'$ can be expressed in terms of lengths as

$$\text{area}_{B'BCC'B} = \frac{B'B + C'C}{2} \times B'C'$$

In terms of coordinate values, this same area $B'BCC'B'$ is

$$\text{area}_{B'BCC'B'} = \frac{X_B + X_C}{2}(Y_B - Y_C)$$

Each of the trapezoids and triangles of Eq. (14-5) can be expressed by coordinates in a similar manner. Substituting these coordinate expressions into Eq. (14-5), multiplying by 2 to clear fractions, and rearranging,

$$2(\text{area}) = X_B Y_A + X_C Y_B + X_D Y_C + X_E Y_D + X_A Y_E - X_A Y_B$$
$$- X_B Y_C - X_C Y_D - X_D Y_E - X_E Y_A \tag{14-6}$$

Equation (14-6) can be reduced to an easily remembered form by listing the X and Y coordinates of each point in succession in two columns, as shown in Eq. (14-7), *with coordinates of the starting point repeated at the end.* The products noted by diagonal arrows are ascertained, with dashed arrows considered plus and solid ones minus. *The algebraic summation of all products is computed and its absolute value divided by 2 to get area.*

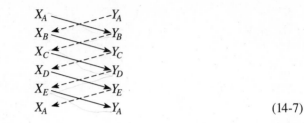

$$(14\text{-}7)$$

The procedure indicated in Eq. (14-7) is applicable to calculating any size traverse. The following formula, easily derived from Eq. (14-6), is a variation that can also be used,

$$\text{area} = \tfrac{1}{2}[X_A(Y_E - Y_B) + X_B(Y_A - Y_C) + X_C(Y_B - Y_D)$$
$$+ X_D(Y_C - Y_E) + X_E(Y_D - Y_A)] \tag{14-8}$$

It was noted earlier that for convenience, an axis system can be adopted in which $X = 0$ for the most westerly traverse point, and $Y = 0$ for the most southerly station. Magnitudes of coordinates and products are thereby reduced, and the amount of work lessened, since four products will be zero. Selection of a special origin like that just described is of little consequence, however, if the problem has been programmed for computer solution. Then the coordinates obtained from traverse adjustment can be used directly in the solution. A word of caution applies, however, if coordinate values are extremely large, as they would normally be—for example, if state plane values are used (see Chapter 21). In those cases, to ensure sufficient precision and prevent serious round-off errors, double precision should be used. Or, as an alternative, the decimal place in each coordinate can arbitrarily be moved n places to the left, the area calculated, and then multiplied by 10^{2n}.

Either Eq. (14-6) or Eq. (14-8) can be readily programmed for solution by computer. A program written in BASIC for computing traverses, including area by coordinates, is given in Appendix F.

Calculations for purposes of partitioning land—that is, cutting off a portion of a tract for title transfer—can be aided significantly by using coordinates. This subject is discussed in Chapter 22.

EXAMPLE 14-3   Figure 14-5 illustrates the same traverse as Figure 14-4. The computations in Table 13-4 apply to this traverse. Coordinate values shown in Figure 14-5, however, result from shifting the axes so that $X_A = 0.00$ (*A* is the most westerly station) and $Y_D = 0.00$ (*D* is the most southerly station). This was accomplished by subtracting 10,000.00 (the value of X_A) from all *X* coordinates, and subtracting 9408.34 (the value of Y_A) from all *Y* coordinates. Compute the traverse area by the coordinate method.

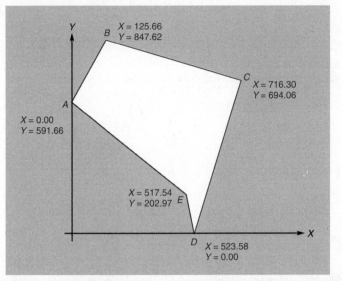

Figure 14-5 Traverse for computation of area by coordinates.

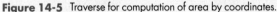

TABLE 14-1 COMPUTATION OF AREA BY COORDINATES

			DOUBLE AREA	
POINT	X	Y	PLUS	MINUS
A	0.00	591.66		
B	125.66	847.62	74,348	0
C	716.30	694.06	607,150	87,216
D	523.58	0.00	363,396	0
E	517.54	202.97	0	106,271
A	0.00	591.66	0	306,208
			1,044,894	$-499,695$
			$-499,695$	
			2) 545,199	
			272,600 ft^2 = 6.258 acres	

SOLUTION

These computations are best organized for tabular solution. Table 14-1 shows the procedure and results.

14-6 AREA BY DOUBLE-MERIDIAN DISTANCE METHOD

The area within a closed figure can also be computed by the double-meridian distance (DMD) method. This procedure requires balanced departures and latitudes of the tract's boundary lines, which are normally obtained by traversing. The DMD method is not as commonly used as the coordinate method because it is not as convenient, but

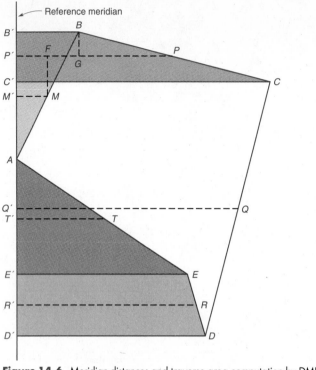

Figure 14-6 Meridian distances and traverse area computation by DMD method.

given the data from an adjusted traverse, it will yield the same answer. The DMD method is useful for checking answers obtained by the coordinate method.

By definition, *the meridian distance of a traverse course is the perpendicular distance from the midpoint of the course to the reference meridian.* To ease the problem of signs, a reference meridian usually is placed through the most westerly traverse station.

In Figure 14-6 the meridian distances of courses *AB, BC, CD, DE,* and *EA* are *MM'*, *PP'*, *QQ'*, *RR'*, and *TT'*, respectively. To express *PP'* in terms of convenient distances, *MF* and *BG* are drawn perpendicular to *PP'*. Then

$$PP' = P'F + FG + GP$$

$$= \text{meridian distance of } AB + \tfrac{1}{2} \text{ departure of } AB + \tfrac{1}{2} \text{ departure of } BC$$

Thus the meridian distance of any course of a traverse equals the meridian distance of the preceding course plus one-half the departure of the preceding course plus one-half the departure of the course itself. It is simpler to employ full departures of courses. Therefore DMDs equal to twice the meridian distances are used, and a single division by 2 is made at the end of the computation.

Based on the considerations described, the following general rule can be applied in calculating DMDs: *The DMD for any traverse course is equal to the DMD of the preceding course plus the departure of the preceding course plus the departure of the course itself.* Signs of the departures, east plus and west minus, must be considered.

When the reference meridian is taken through the most westerly station of a closed traverse and calculations of the DMDs are started with a course through that station, *the DMD of the first course is its departure.* Applying these rules, for the traverse in Figure 14-6,

DMD of AB = departure of AB

DMD of BC = DMD of AB + departure of AB + departure of BC

A check on all computations is obtained if the DMD of the last course, after computing around the traverse, is also equal to its departure but has the opposite sign. If there is a difference, the departures were not correctly adjusted before starting, or a mistake was made in the computations.

With reference to Figure 14-6, the area enclosed by traverse *ABCDEA* may be expressed in terms of trapezoid areas (shown by different color shadings) as

$$\text{area} = B'BCC'B' + C'CDD'C' - (AB'BA + DD'E'ED + AEE'A) \quad (14\text{-}9)$$

The area of each figure equals the meridian distance of a course times its balanced latitude. For example, the area of trapezoid $C'CDD'C' = Q'Q \times C'D'$, where $Q'Q$ and $C'D'$ are the meridian distance and latitude, respectively, of line *CD.* The DMD of a course multiplied by its latitude equals double the area. *Algebraic summation* of all double areas gives *twice the area* inside the entire traverse.

Signs of the products of DMDs and latitudes must be considered. If the reference line is passed through the most westerly station, all DMDs are positive. The products of DMDs and north latitudes are therefore plus, and those of DMDs and south latitudes are minus.

Using the balanced departures and latitudes listed in Table 13-4 for the traverse of Figure 14-6, compute the DMDs of all courses. $\quad$ EXAMPLE 14-4

SOLUTION

The calculations, done in tabular form following the general rule, are illustrated in Table 14-2.

Using the DMDs determined in Example 14-4, calculate the area within the traverse. $\quad$ EXAMPLE 14-5

SOLUTION

Computations for area by DMDs are generally arranged as in Table 14-3, although a combined form may be substituted. Sums of positive and negative double areas are obtained, and the absolute value of the smaller subtracted from that of the larger. The result is divided by 2 to get the area (272,600 ft²), and by 43,560 to obtain the number of acres (6.258). Note that the answer agrees with the one obtained using the coordinate method.

TABLE 14-2 COMPUTATION OF DMDs

Departure of AB = + 125.66 = DMD of AB
Departure of AB = + 125.66
Departure of BC = + 590.64
⎯⎯⎯⎯⎯
+ 841.96 = DMD of BC
Departure of BC = + 590.64
Departure of CD = − 192.72
⎯⎯⎯⎯⎯
+ 1239.88 = DMD of CD
Departure of CD = − 192.72
Departure of DE = − 6.04
⎯⎯⎯⎯⎯
+ 1041.12 = DMD of DE
Departure of DE = − 6.04
Departure of EA = − 517.54
⎯⎯⎯⎯⎯
+ 517.54 = − DMD of EA (Check)

TABLE 14-3 COMPUTATION OF AREA BY DMDs AND DPDs

COURSE	BALANCED DEPARTURE	BALANCED LATITUDE	DMD	DOUBLE AREAS		DPD	DOUBLE AREAS	
				PLUS	MINUS		PLUS	MINUS
AB	⊣ E125.66	⋋N255.96	+ 125.66	32,164		+ 255.96	32,164	
BC	⊣ E 590.64	− S153.56	+ 841.96		129,293	+ 358.36	211,662	
CD	− W192.72	− S694.06	+ 1239.88		860,551	− 489.26	94,290	
DE	⌐ W 6.04	⊣ N202.97	+ 1041.12	211,316		− 980.35	5,921	
EA	− W517.54	⋋N388.69	+ 517.54	201,163		− 388.69	201,163	
Total	0.00	0.00		444,643	989,844		545,200	00
					− 444,643			
					2)545,201			
					272,600 ft^2 = 6.258 acres			

If the total of minus double areas is larger than the plus value, it signifies only that DMDs were computed by going around the traverse in a clockwise direction. If the route had been from A through $E, D, C,$ and B and back to $A,$ the total plus double area would have been the greater. Areas carried out beyond the nearest square foot, or 0.0001 acre, cannot normally be justified for a traverse on which distances were measured to the nearest 0.01 ft and angles read to 0.1 or $\frac{1}{2}'$.

As a check, the area can be computed by *double parallel distances* (DPDs). *The DPD for any traverse course is equal to the DPD of the preceding course plus the latitude of the preceding course plus the latitude of the course itself.* In the DPD method, double areas of trapezoids are computed by multiplying the DPD of each course by its departure. Due regard must be given to algebraic signs of latitudes in DPD calculation, and of departures in multiplication, to get correct trapezoid areas. Algebraic summation of all trapezoid areas, and division of the absolute value by 2, gives the area. The last three columns in Table 14-3 show the area computation by DPDs for the traverse of Figure 14-6.

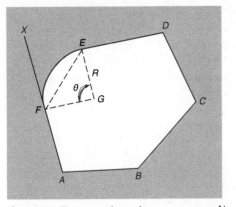

Figure 14-7 Area with circular curve as part of boundary.

In modern surveying and engineering offices, area calculations are seldom done by hand; rather, they are programmed for computer solution. If an area is computed by hand, however, it should be checked by using different methods, or by two persons who employ the same system. As an example, an individual working alone in an office could calculate areas by coordinates and check by DMDs. Experienced surveyors and engineers have learned that a half-hour spent checking computations in the field and office can eliminate lengthy frustrations at a later time.

14-7 AREA OF PARCELS WITH CIRCULAR BOUNDARIES

The area of a tract that has a circular curve for one boundary, as in Figure 14-7, can be found by dividing the figure into two parts: polygon *ABCDEGFA* and sector *EGF*. The radius $R = EG = FG$ and either central angle $\theta = EGF$ or length *EF* must be known or computed to permit calculation of sector area *EGF*. If R and central angle θ are known, then the area of sector $EGF = \pi R^2 \times (\theta°/360°)$. If chord length *EF* is known, angle $\theta = 2 \sin^{-1} \left(\frac{EF}{2R}\right)$, and the preceding equation is used to calculate the sector area. To obtain the tract's total area, the sector area is added to area *ABCDEGFA* found by the DMD or coordinate method.

14-8 AREA BY MEASUREMENTS FROM MAPS

To determine the area of a tract of land from map measurements, its boundaries must first be identified on an existing map or a plot of the parcel drawn from survey data. Then one of several available methods can be used to determine its area.

Accuracy in making area determinations from map measurements is directly related to the accuracy of the maps being used. Accuracy of maps, in turn, depends on the quality of the survey data from which they were produced, and also on the precision of the drafting process. Therefore if existing maps are being used to determine areas, their quality should first be verified.

Even with good-quality maps, areas measured from them will not normally be as accurate as those computed directly from survey data. Map scale and the device used to extract map measurements are major factors affecting the resulting area accuracy. If a map is plotted to a scale of 1000 ft/in., for example, and an engineer's scale is used which produces measurements good to ± 0.02 in., distances or coordinates scaled from this map can be no better than about $(\pm 0.02 \times 1000) = \pm 20$ ft. This uncertainty can produce substantial errors in areas. Differential shrinkage or expansion of the material upon which maps are drafted is another source of error in determining areas from map measurements. Changes in dimension of 2 to 3% are not uncommon for certain types of paper. (The subjects of maps and mapping are discussed in more detail in Chapters 16 and 17.)

Aerial photos can also be used as map substitutes to determine *approximate* areas if the parcel boundaries can be identified. The areas are approximate because the scale of an aerial photo is not uniform throughout, as explained in Chapter 28. Aerial photos are particularly useful for determining areas of irregularly shaped tracts, such as lakes.

Different procedures for determining areas from maps are described in the subsections that follow.

14-8.1 Area by Counting Coordinate Squares

A simple method for determining areas consists in overlaying the mapped parcel with a transparency having a superimposed grid. The number of grid squares included within the tract is then counted, with partial squares estimated and added to the total. Area is the product of the total number of squares times the area represented by each square. As an example, if the grids are 0.20 in. on a side, and a map at a scale of 200 ft/in. is overlaid, each square is equivalent to $(0.20 \times 200)^2 = 1600 \text{ ft}^2$.

14-8.2 Area by Scaled Lengths

If the boundaries of a parcel are identified on a map, the tract can be divided into triangles, rectangles, or other regular figures, the sides measured, and the areas computed using standard formulas and totaled.

14-8.3 Area by Digitizing Coordinates

A mapped parcel can be placed on a digitizing table which is interfaced with a computer, and the coordinates of its corner points quickly and conveniently recorded. From the file of coordinates, the area can be computed using one of Eqs. (14-6) through (14-8). It must be remembered, however, that even though coordinates may be digitized to the nearest 0.001 in., their actual accuracy can be no better than the map from which the data were extracted. Area determination by digitizing existing maps is now being practiced extensively in creating databases of geographic information systems.

14-8.4 Area by Planimeter

A planimeter measures the area contained within any closed figure that is circumscribed by its tracer. There are two types of planimeters: mechanical and electronic. The major parts of the mechanical type are a scale bar, graduated drum and disk, vernier, tracing point and guard, and anchor arm, weight, and point. The scale bar may be fixed or adjustable, as in Figure 14-8. For the standard fixed-arm planimeter, one revolution of the disk (dial) represents 100 in.^{2} and one turn of the drum (wheel) represents 10 in.^{2} The adjustable type can be set to read units of area directly for any particular map scale. The instrument touches the map at only three places: anchor point, drum, and tracing-point guard.

An electronic planimeter (Figure 14-9) operates similarly to the mechanical type, except that the results are given in digital form on a display console. Areas can be

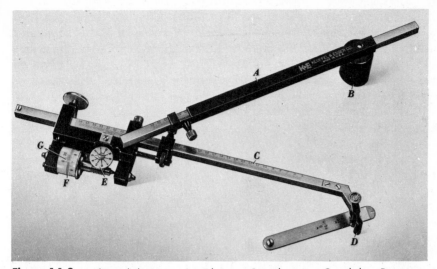

Figure 14-8 Mechanical planimeter: *A,* anchor arm; *B,* anchor point; *C,* scale bar; *D,* tracing point; *E,* disk; *F,* drum; *G,* vernier.

Figure 14-9 Electronic planimeter. (Courtesy Numonics Corporation.)

measured in units of square inches or square centimeters, and by setting an appropriate scale factor, they can be obtained directly in acres or hectares. Some instruments feature multipliers which can automatically compute and display volumes.

As an example of using an adjustable type of mechanical planimeter, assume the area within the traverse of Figure 14-5 will be measured. The anchor point beneath the weight is set in a position outside the traverse (if inside, a polar constant must be added), and the tracing point brought over corner *A*. An initial reading of 7231 is taken, the 7 coming from the disk, 23 from the drum, and 1 from the vernier. The tracing point is moved along the traverse lines from *A* to *B, C, D,* and *E,* and back to *A*. The point may be guided by a triangle or a straightedge, but normally it is steered freehand. A final reading of 8596 is made. The difference between the initial and final readings, 1365, is multiplied by the planimeter constant to obtain the area.

To determine the planimeter constant, a square area is carefully laid out, 5 in. on a side, with diagonals of 7.07 in., and its perimeter traced with the planimeter. If the difference between initial and final readings for the 5-in. square is, for example, 1250, then

$$5 \text{ in.} \times 5 \text{ in.} = 25 \text{ in.}^2 = 1250 \text{ units}$$

Thus the planimeter constant is

$$1 \text{ unit} = \frac{25}{1250} = 0.0200 \text{ in.}^2$$

Finally, the area within the traverse is

$$\text{area} = 1365 \text{ units} \times 0.0200 = 27.30 \text{ in.}^2$$

If the traverse is plotted at a map scale of 1 in. = 100 ft, then $1 \text{ in.}^2 = 10,000 \text{ ft}^2$ and the area measured is $273,000 \text{ ft}^2$.

As a check on planimeter operation, the outline may be traced in the opposite direction. The initial and final readings at point *A* should agree within a limit of perhaps two to five units.

The precision obtained in using a planimeter depends on operator skill, accuracy of the plotted map, type of paper, and other factors. Results correct to within $\frac{1}{2}$ to 1% can be obtained by careful work.

A planimeter is most useful for irregular areas, such as that in Figure 14-3, and has applications in many branches of engineering. The planimeter has been widely used in highway offices for determining areas of cross sections, and is also convenient for determining areas of drainage basins and lakes from measurements on aerial photos, checking computed areas in property surveys, and so forth.

14-9 SOURCES OF ERROR IN DETERMINING AREAS

Some sources of error in area computations are:

1. Errors in the field data from which coordinates or maps are derived
2. Making a poor selection of intervals and offsets to fit a given irregular boundary properly
3. Making errors in scaling from maps
4. Shrinkage and expansion of maps
5. Using coordinate squares that are too large and therefore make estimation of areas of partial blocks difficult

6. Making an incorrect setting of the planimeter scale bar
7. Running off and on the edge of the map sheet with the planimeter drum
8. Using different types of paper for the map and planimeter calibration sheet

14-10 MISTAKES

In computing areas, common mistakes include:

1. Forgetting to divide by 2 in the coordinate and DMD methods
2. Confusing signs of coordinates, departures, latitudes, and DMDs
3. Failing to check an area computation by a different method
4. Not drawing a sketch to scale or general proportion for a visual check
5. Not verifying the planimeter scale constant by tracing a known area

*14-1 Compute the area enclosed within polygon *BCDEFGB* of Figure 14-1 using triangles. (See Table E-2 of Appendix E for oblique triangle formulas.)

14-2 Similar to Problem 14-1, except for polygon *BCDFGAB* of Figure 14-1.

14-3 Compute the area enclosed between line *GHJLA* and the shoreline of Figure 14-1 using the offset method.

14-4 Determine the area between the traverse line from station 5 + 42 to station 7 + 00 and Black Creek of Figure 16-9 by the offset method.

*14-5 Compute the area between a lake and a straight line *AG*, from which offsets are taken at irregular intervals as follows (all distances in feet):

OFFSET POINT	A	B	C	D	E	F	G
Distance from A	0.00	75.90	198.55	248.60	311.85	437.25	506.00
Offset	35.2	91.8	148.0	248.1	356.9	200.2	38.5

14-6 Calculate by coordinates the area within the traverse of Problem 13-19.
14-7 Compute by coordinates the area enclosed by the traverse of Problem 13-20.
*14-8 Calculate by coordinates the area within the traverse of Problem 13-21.
*14-9 Compute the area enclosed in the traverse of Problem 13-19 using DMDs.
14-10 Determine the area within the traverse of Problem 13-20 using DMDs.
14-11 By the DMD method, find the area enclosed by the traverse of Problem 13-21.
14-12 Compute the area within the traverse of Problem 13-26 using the coordinate method. Check by DMDs.
14-13 Calculate the area inside the traverse of Problem 13-27 by DMDs and check by DPDs.
14-14 Calculate the area within the traverse of Problem 13-9 using the coordinate method.
*14-15 Determine the area inside the traverse of Problem 13-12 using the coordinate method.
*14-16 Compute the area enclosed by the traverse of Problem 13-18 using the DMD method. Check by coordinates.
14-17 Find the area within the traverse of Problem 13-22 using the DMD method. Check by coordinates.
14-18 Ascertain the area of the lot in Problem 13-31.
14-19 Calculate the area inside the traverse of Problem 13-32.
*14-20 Determine the area within the closed traverse of Problem 13-23 by the DMD method. Check by DPDs.
14-21 Plot the traverse of Problem 13-19 to a scale of 1 in. = 100 ft. Determine its surrounded area using a planimeter.

*Asterisks indicate problems that have answers given in Appendix G.

14-22 Similar to Problem 14-21, except for the traverse of Problem 13-20.

14-23 Plot the traverse of Problem 13-21 to a scale of 1 in. = 200 ft and find its enclosed area using a planimeter.

*14-24 The departures and latitudes (in 100-ft tape lengths) for a closed-polygon traverse *ABCDEFGA* follow. *AB:* E dep. = 6, N lat. = 1; *BC:* W dep. = 3, S lat. = 6; *CD:* E dep. = 5, S lat. = 4; *DE:* W dep. = 8, S lat. = 7; *EF:* W dep. = 4, N lat. = 9; *FG:* W dep. = 6, N lat. = 5. Calculate: **(a)** the departure and latitude of line *GA* **(b)** the length of line *GA* **(c)** the area within the traverse in acres by both coordinates and DMDs.

14-25 A closed-polygon traverse *ABCDEFGHA* has courses with departures and latitudes (in hundreds of feet) as specified. *AB:* E dep. = 3, N lat. = 7; *BC:* W dep. = 4, N lat. = 6; *CD:* W dep. = 8, S lat. = 2; *DE:* W dep. = 3, S lat. = 5; *EF:* E dep. = 7, S lat. = 4; *FG:* W dep. = 6, S lat. = 5; *GH:* E dep. = 8, S lat. = 5; *HA:* E dep. = 3, N. lat. = 8. **(a)** Which is the most westerly station? **(b)** Which is the most southerly station? **(c)** What are the length and bearing of the line connecting stations *H* and *E*? **(d)** Compute by DMDs the area enclosed in acres. **(e)** Check by coordinates.

✓**14-26** Compute by coordinates the area in hectares within a closed-polygon traverse *ABCDEFA* by placing the axis through the most westerly station. Departures and latitudes (in meters) follow. *AB:* E dep. = 40, N lat. = 50; *BC:* E dep. = 60, N lat. = 0; *CD:* E dep. = 40, S lat. = 60; *DE:* W dep. = 60, S lat. = 40; *EF:* W dep. = 100, S lat. = 20; *FA:* E dep. = 20, N lat. = 70.

Compute the unknown elements and areas enclosed by the traverses of Problems 14-27 and 14-28 by the coordinate method, selecting an axis in which *X* passes through the most southerly station and *Y* passes through the most westerly station.

14-27 COURSE	*AB*	*BC*	*CD*	*DA*
Bearing	Due east	S26°30'W	Due west	Unknown
Length (ft)	382.95	164.50	135.64	Unknown

*14-28 COURSE	*AB*	*BC*	*CD*	*DA*
Bearing	N55°15'W	S37°42'W	N62°30'E	Unknown
Length (ft)	459.90	402.48	201.29	Unknown

14-29 Calculate the area of a piece of property bounded by a traverse and circular arc described as follows: *AB:* S42°30'W, 540.00 ft; *BC:* S77°30'E, 540.00 ft; *CD:* N32°30'W, 270.00 ft; *DA:* a circular arc tangent to *CD* at point *D*.

*14-30 Similar to Problem 14-29, except *CD* is 249.75 ft.

14-31 Divide the area of the lot in Problem 14-29 into two equal parts by a line through point *B*. List in order the lengths and bearings of all sides for each parcel.

14-32 Partition the lot of Problem 14-30 into two equal areas by means of a line parallel to *BC*. Tabulate in clockwise consecutive order the lengths and bearings of all sides.

*14-33 Lot *ABCD* between two parallel street lines is 320.00 ft deep and has a 210-ft frontage (*AB*) on one street and a 240-ft frontage (*CD*) on the other. Interior angles at *A* and *B* are equal, as are those at *C* and *D*. What distances *AE* and *BF* should be laid off by a surveyor to divide the lot into two equal areas by means of a line *EF* parallel to *AB*?

14-34 The area of a field plotted on a map to a scale of 1 in. = 200 ft is measured with a

---

*Asterisks indicate problems that have answers given in Appendix G.

14-34 The area of a field plotted on a map to a scale of 1 in. = 200 ft is measured with a fixed-arm planimeter having a constant of 10, and 6.382 revolutions of the wheel are recorded. What is the area of the field in square feet and in acres?

*14-35 After an initial reading of 2327, a planimeter tracing point is run clockwise around the sides of a 5-in. square. When returned to the starting point, a reading of 2452 is made. If the beginning reading on point *A* of a plotted traverse is 1322, and 1796 when guided around the closed traverse back to *A,* compute the enclosed area in square feet if the plot is at a scale of 100 ft/in.

14-36 In Problem 14-24, if a tape assumed to be 100.00 ft long is found after standardization to be actually 100.04 ft long, how large an error in area is produced?

14-37 Write a computer program for calculating areas within closed polygons by the coordinate method.

14-38 Write a computer program for calculating areas within closed polygons by the DMD method.

*Asterisks indicate problems that have answers given in Appendix G.

Ahmed, F. A. 1983. "Area Computation Using Salient Boundary Points." *ASCE, Journal of the Surveying Engineering Division* 109 (No. 1): 54.

Chrisman, N. R., and B. S. Yandell. 1988. "Effects of Point Error on Area Calculations: A Statistical Model." *Surveying and Land Information Systems* 48 (No. 4): 241.

Easa, S. M. 1988. "Area of Irregular Region with Unequal Intervals." *ASCE, Journal of Surveying Engineering* 114 (No. 2): 50.

El-Hassan, I. M. 1987. "Irregular Boundary Area Computation by Simpson's 3/8 Rule." *ASCE, Journal of the Surveying Engineering Division* 113 (No. 3): 127.

Moffitt, F. H., and H. Bouchard. 1992. *Surveying,* 9th ed. New York: Harper-Collins, Chap. 8

BIBLIOGRAPHY

15

STADIA

15-1 INTRODUCTION

The stadia method, referred to as *tacheometry* in Europe, is a rapid and efficient way of indirectly measuring distances and elevation differences. The accuracy attainable with stadia is suitable for lower order trigonometric leveling, locating topographic details for mapping, measuring lengths of backsights and foresights in differential leveling, and making quick checks of measurements made by higher order methods. Stadia measurements can be taken with theodolites, transits, plane-table alidades, and levels.

Total station instruments with their capability of instantaneously measuring and displaying point coordinates and elevations from simultaneous angle and distance measurements, and new portable GPS receivers designed for rapid and extremely accurate real-time positioning are gradually relegating stadia to the background. Nevertheless, stadia is still suitable for many applications, and its use will likely continue for some time into the future.

15-2 PRINCIPLE OF THE STADIA

In addition to the center horizontal cross hair, a theodolite or transit reticle for stadia work has two additional horizontal cross hairs spaced equidistant from the center one, as illustrated with several reticles of Figure 10-5. With the line of sight horizontal and directed toward a graduated rod held vertically at a point some distance away, the interval appearing between the two stadia hairs of most surveying instruments is precisely $\frac{1}{100}$ of the distance to the rod. Thus the intercept on a rod held 100 ft away would be 1.00 ft. If rod intercepts are read to the nearest ± 0.01 ft, which can be routinely accomplished with ordinary care on average-length sights, distances can be determined directly to the nearest foot. This is sufficiently accurate for locating topographic details such as rivers, roads, and buildings, which are to be plotted on a map having a scale smaller than 1 in. = 100 ft, and sometimes for scales even larger, such as 1 in. = 50 ft.

The stadia method is based on the principle that in similar triangles, corresponding sides are proportional. In Figure 15-1, depicting a telescope with a simple lens, light rays from points A and B passing through the lens center form a pair of similar triangles

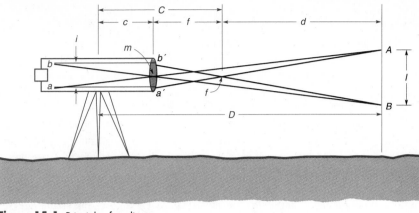

Figure 15-1 Principle of stadia.

AmB and *amb*. Here $AB = I$ is the rod intercept (*stadia interval*), and $ab = i$ is the spacing between stadia hairs.

Standard symbols used in stadia measurements and their definitions are as follows (refer to Figure 15-1):

f = *focal length* of lens (a constant for any particular compound objective lens)
i = spacing between stadia hairs (*ab* in Figure 15-1)
$\dfrac{f}{i}$ = *stadia interval factor*, usually 100 and denoted by K
I = rod intercept (*AB* in Figure 15-1), also called *stadia interval*
c = distance from instrument center (vertical axis) to objective lens center (varies slightly when focusing the objective lens for different sight lengths but is generally considered to be a constant)
C = *stadia constant* $= c + f$
d = distance from focal point F in front of telescope to face of rod
D = distance from instrument center to face $= C + d$

From similar triangles of Figure 15-1,

$$\frac{d}{f} = \frac{I}{i} \quad \text{or } d = \frac{f}{i}I = KI$$

Thus

$$D = KI + C \tag{15-1}$$

The geometry illustrated in Figure 15-1 pertains to a simplified type of *external focusing telescope*. It has been used because an uncomplicated drawing correctly shows the relationships and aids in deriving the stadia equation. These telescopes are now obsolete in surveying instruments. The objective lens of an *internal focusing telescope* (the type now used in surveying instruments) remains fixed in position, while a movable negative-focusing lens between the objective lens and the plane of the cross hairs changes direction of the light rays. As a result, the stadia constant, (C), is so small that it can be assumed equal to zero and drops out of Eq. (15-1). Thus the equation for distance on a horizontal stadia sight reduces to[1]

$$D = KI \qquad\qquad (15\text{-}2)$$

Fixed stadia hairs in theodolites, transits, levels, and alidades are carefully spaced by instrument manufacturers to make the stadia interval factor $\frac{f}{i} = K$ equal to 100. It should be determined the first time an instrument is used, although the manufacturer's exact value posted inside the carrying case will not change unless the cross hairs, reticle, or lenses are replaced or adjusted.

To determine the stadia interval factor K, rod intercept I for a horizontal sight of known distance D is read. Then in an alternate form of Eq. (15-2), the stadia interval factor is $K = \frac{D}{I}$. As an example, at a known (taped) distance of 300.0 ft, a rod interval of 3.01 was read. Then $K = 300.0/3.01 = 99.7$. Accuracy in determining K is increased by averaging values from several lines whose measured lengths vary from about 100 to 500 ft by 100-ft increments.

15-3 STADIA MEASUREMENTS ON INCLINED SIGHTS

Most stadia shots are inclined because of varying topography, but the intercept is read on a *plumbed rod* and the slope length "reduced" to horizontal and vertical distances.

In Figure 15-2 an instrument is set over point M and the rod held at O. With the middle cross hair set on point R to make RO equal to the height of the instrument $EM = hi$, the vertical angle (angle of inclination) is α. Note that in stadia work the height of instrument hi is again defined as the height of the line of sight above the point occupied (*not* HI, the height above datum as in leveling).

Let L represent the slope length ED, H the horizontal distance $EG = MN$, and V the vertical distance $RG = ON$. Then

$$H = L \cos \alpha \qquad\qquad (a)$$

$$V = L \sin \alpha \qquad\qquad (b)$$

If the rod could be held normal to the line of sight at point O, a reading $A'B'$, or I', would be obtained, making

$$L = KI' \qquad\qquad (c)$$

---

[1]A few older transits with external focusing telescopes are still in service. They can be recognized because the objective lens is seen to move as the telescope is focused for varying lengths of sights. To determine the constant C for one of these instruments, first focus for a long sight, say 1000 ft, and measure f, the distance from the objective lens to the reticle ring. Then focus for an average sight, about 200 ft, and measure c, the distance from the objective lens to the center of the instrument. The sum $f + c = C$, normally about 1 ft, is included in the stadia equations for these older external focusing instruments.

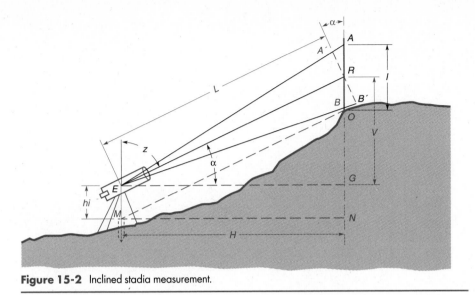

Figure 15-2 Inclined stadia measurement.

Since it is not practical to hold the rod at an inclination angle α, it is plumbed and reading AB, or I, taken. For the small angle at R on most sights, it is sufficiently accurate to consider angle $AA'R$ as a right angle. Therefore

$$I' = I \cos \alpha \qquad\qquad\text{(d)}$$

and substituting (d) into (c),

$$L = KI \cos \alpha \qquad\qquad\text{(e)}$$

Finally, substituting (e) into (a), the equation for horizontal distance on an inclined stadia sight is

$$H = KI \cos^2 \alpha \qquad\qquad\text{(15-3a)}$$

If zenith angles are read rather than vertical angles, then the horizontal distance is given by

$$H = KI \sin^2 z \qquad\qquad\text{(15-3b)}$$

where z is the zenith angle, equal to $90° - \alpha$.

The vertical distance is found by substituting (e) into (b), which gives

$$V = KI \sin \alpha \cos \alpha \qquad\qquad\text{(15-4a)}$$

or

$$V = KI \sin z \cos z \qquad\qquad\text{(15-4b)}$$

If the trigonometric identity $\frac{1}{2}\sin 2\alpha$ is substituted for $\sin\alpha\cos\alpha$, the formulas for vertical distance become

$$V = \frac{1}{2}KI\sin 2\alpha \qquad (15\text{-}5a)$$

or

$$V = \frac{1}{2}KI\sin 2z \qquad (15\text{-}5b)$$

In the final form generally used, K is assigned 100 and the formulas for the reduction of inclined sights to horizontal and vertical distances are

$$H = 100\,I\cos^2\alpha \qquad (15\text{-}6a)$$

or

$$H = 100\,I\sin^2 z \qquad (15\text{-}6b)$$

and

$$V = 50\,I\sin 2\alpha \qquad (15\text{-}7a)$$

or

$$V = 50\,I\sin 2z \qquad (15\text{-}7b)$$

From Figure 15-2, the elevation of point O is

$$\text{elev}_O = \text{elev}_M + hi + V - R \qquad (15\text{-}8)$$

From Eq. (15-8), the advantage of sighting an R value that is equal to the hi when reading the vertical (or zenith) angle is evident. Since the rod reading and hi are opposite in sign, if equal in magnitude they cancel each other and can be omitted from the elevation computation. If the hi cannot be seen because of obstructions, any rod reading can be sighted and Eq. (15-8) used. Setting the middle cross wire on a full-foot mark just above or below the hi simplifies the arithmetic.

EXAMPLE 15-1   Assume that in Figure 15-2, the elevation of M is 268.2 ft; $hi = EM = 5.6$ ft; rod intercept $AB = I = 3.28$ ft; zenith angle z to point D read at 5.6 ft on the rod is 85°44′; and $K = 100$. Compute horizontal distance H, elevation difference V, and the elevation of point O.

SOLUTION
From Eqs. (15-6b) and (15-7b)

$$H = 100(3.28) \sin^2(85°44.0') = 326 \text{ ft}$$

$$V = 50(3.28) \sin(171°28.0') = 24.3 \text{ ft}$$

By Eq. (15-8), $\text{elev}_O = 268.2 + 5.6 + 24.3 - 5.6 = 292.5$ ft.
(Note that since hi and R were equal, they could have been omitted in the elevation calculation.)

15-4 STADIA FIELD PROCEDURES

When the stadia method is employed, it is most often for locating details in topographic mapping. However, it can also be used to run control traverses for topographic mapping or for other lower order applications. Stadia is also adaptable for trigonometric leveling.

In topographic detailing by stadia, the theodolite or transit is set up on a control station and oriented in azimuth by backsighting along a line of known direction with the line's azimuth on the horizontal circle. Then azimuths can be read directly to all sighted points. Frequently, many shots may be taken from one instrument station; *thus orientation should be checked by sighting on the control line after every 10 or 20 topographic points, and before leaving the setup.*

As in all surveying operations, proper field techniques save time and reduce the number of mistakes. An order of field procedures well suited for topographic detailing is as follows:

1. Bisect the rod with the vertical cross hair.
2. With the middle cross hair approximately on the hi, set the lower hair at a full-foot mark (decimeter on metric rods).
3. Read the upper cross hair and mentally subtract the lower hair reading from it to get the rod intercept. Record (or call out to the notekeeper) only the rod intercept.
4. Move the middle cross hair to the hi by using the vertical tangent screw.
5. Release the rodperson for movement to the next point by giving the proper signal.
6. Read and record the horizontal angle (or azimuth).
7. Read and record the vertical (or zenith) angle.

This procedure enables an instrument operator to keep two or three rodpersons busy in open terrain where points to be located are widely separated. In step 3 of the above procedure, the rod interval is read after the lower cross hair has been moved to a full-foot mark, and the middle hair is not precisely on the hi or the graduation read for the vertical (or zenith) angle. This does not normally cause a significant error in computed horizontal or vertical distance to the point, however, except on long sights with steep vertical angles. Not having the rod plumb can cause appreciable errors, but use of a rod level eliminates this problem.

When stadia is used for traversing, rod intercepts and horizontal and vertical (or zenith) angles are measured both forward and back at each traverse point. Average values of horizontal distances and elevation differences are used in the subsequent computations. An elevation check should be secured by closing on the initial traverse point, or a nearby bench mark, and, if within tolerances, any misclosure distributed. The horizontal angles should also be checked for closure, and again, if within acceptable limits, the misclosure adjusted. Departures and latitudes can then be computed and the traverse precision checked by methods described in Chapter 13.

In employing stadia for trigonometric leveling, the HI (height of instrument above datum) is first determined. This can be done either by sighting on a station of known elevation, or by setting the instrument over such a point and measuring the height of the horizontal axis above it with a graduated rod. The elevation of any point can then be found by computation from its rod intercept and vertical (or zenith) angle.

15-5 STADIA FIELD NOTES

An example of stadia field notes for topographic detailing is shown in Plate D-7 of Appendix D. In this example, shots are also taken forward and back on adjacent traverse stations for checking purposes. Notice, for example, that with the instrument at station *B*, sights are taken on both stations *A* and *C*. Columns in the notes are arranged in the order of reading—point sighted, rod intercept, azimuth, and vertical angle. A sketch and pertinent data are placed on the right-hand page as usual. Numbering topographic points begins with 1 at the first point and continues successively to eliminate possible duplication. Reduction of the notes is generally done in the office, unless information is needed immediately.

15-6 REDUCTION OF STADIA NOTES

Prior to computers, stadia tables and special stadia slide rules were used to compute values of *H* and *V* for varying vertical (or zenith) angles and stadia intervals. Previous editions of this book describe these reduction procedures and include the tables. Handheld calculators and programmed computers are now most often used by surveyors to solve the stadia formulas. An example of a simple stadia computer program, written in BASIC, is given in Table 15-1. It solves Eqs. (15-6) through (15-8). Many statements in the program involve steps in data input, conversion of angles to radians (necessary with computers), and data output, which is printed to a file named by the user. The program will accept either vertical or zenith angles. A stadia interval factor *K* of 100 is assumed, and the rod reading *R* is assumed equal to *hi*.

In running the program, prompts will appear on the screen to guide the user. Data are input through the keyboard. Example 15-1 has been computed with the program to demonstrate its use, and Table 15-2 gives the output in a neatly arranged format. Note that a second stadia sight has been included, where *I* was 2.77 ft and the zenith angle 95°23′12″. (Zenith angles greater than 90° are equivalent to negative vertical angles, and the equations automatically affix a negative sign to computed elevation differences through the sine function.)

Computer solutions of stadia equations are particularly beneficial when large numbers of shots are involved, as in locating details for topographic mapping. Besides rapid and accurate processing, the data can be automatically stored in a file. If azimuths are also measured and input, coordinates as well as elevations can be determined for all points sighted. From the data file, a map can be automatically plotted using CAD software and a device such as that shown in Figure 17-9.

TABLE 15-1 PROGRAM IN BASIC FOR STADIA REDUCTION

```
10 REM   STADIA REDUCTION
20 PI = 4!*ATN(1!)
30 RAD = PI/180!
40 INPUT "ENTER FILENAME OF OUTPUT DATA";FI$
50 OPEN FI$ FOR OUTPUT AS #2
60 INPUT "ENTER NUMBER OF STADIA POINTS"; NPOINTS
70 INPUT "ENTER ELEVATION OF STATION OCCUPIED (Ft.)"; EL
80 INPUT "ENTER 1 FOR ZENITH OR 2 FOR VERTICAL ANGLES"; TYPE
90 PRINT #2,TAB(10); "STADIA REDUCTION"
100 PRINT #2,
110 PRINT #2, "POINT    ROD        ANGLE       HORIZ    ELEV    ELEVATION
120 PRINT #2, "  NO    INT   DEG MIN SEC    DIST     DIFF    OF POINT"
130 PRINT #2,"-----------------------------------------------------------"
140 FMT$ = " ###    ##.## ### ## ##.#    ###.#    ###.#    ####.#"
150 FOR INDEX = 1 TO NPOINTS
160     PRINT:PRINT "ENTER ROD INTERCEPT (Ft), AND ...."
170     PRINT "  VERTICAL/ZENITH ANGLE (DEG,MIN,SEC) FOR POINT #"; INDEX;
180     INPUT ROD, VDEG, VMIN, VSEC
190     SIGN = SGN(VDEG)
200     VMIN = ABS(VMIN)
210     VSEC = ABS(VSEC)
220     VRAD = SIGN*(ABS(VDEG)+VMIN/60+VSEC/3600)*RAD
230     IF TYPE = 1 THEN VRAD = PI/2 - VRAD
240     DIST = 100 * ROD * COS(VRAD)*COS(VRAD)
250     DELEV = 50 * ROD * SIN(2*VRAD)
260     ELEV = EL + DELEV
270     PRINT #2,USING FMT$; INDEX, ROD, VDEG, VMIN, VSEC, DIST, DELEV, ELEV
280     PRINT #2,
290 NEXT INDEX
300 CLOSE
```

TABLE 15-2 EXAMPLE OUTPUT FROM STADIA PROGRAM

POINT NO	ROD INT	ANGLE DEG	MIN	SEC	HORIZ DIST	ELEV DIFF	ELEVATION OF POINT
1	3.28	85	44	0.0	326.2	24.3	292.5
2	2.77	95	23	12.0	274.6	-25.9	242.3

So-called *self-reducing tacheometers* have been manufactured which enable direct readout of horizontal stadia distances and elevation differences. One system uses pairs of curved stadia lines which can be seen in the instrument's field of view. The spacing between these lines appears to move together or apart as the telescope is elevated or depressed. The spacing between them corresponds to variations of the trigonometric functions used in the reduction equations.

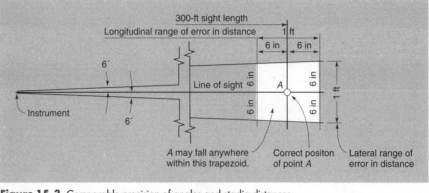

Figure 15-3 Comparable precision of angles and stadia distances.

15-7 STADIA PRECISION

A ratio of error of $\frac{1}{300}$ to $\frac{1}{500}$ can be obtained for a stadia traverse run with ordinary care and reading both foresights and backsights. Shorter sights, a long traverse, and special procedures can give better ratios. Errors in stadia work are usually the result of poor rod readings rather than incorrect angles. An error of 1′ in reading a vertical (or zenith) angle does not appreciably affect the horizontal distance. The same 1′ error develops a difference in elevation smaller than 0.1 ft on a 300-ft sight for usual-sized vertical (or zenith) angles.

Figure 15-3 shows that if stadia distances are determined to the nearest foot (the usual case), horizontal angles *to topographic points* need be read only to within about 6′ for comparable precision on 300-ft sights. A stadia distance given to the nearest foot is assumed to be correct to within about $\frac{1}{2}$ ft. If the same $\frac{1}{2}$-ft error is allowed laterally, the direction can be off by about 6′.

The accuracy of trigonometric leveling by stadia depends on the sight lengths and the sizes of vertical (or zenith) angles measured.

15-8 SOURCES OF ERROR IN STADIA WORK

Errors that occur in stadia work are both instrumental and personal, and include the following:

Instrumental Errors

1. Improper spacing of stadia wires
2. Index error in vertical or zenith angles
3. Incorrect length of rod graduations
4. Line of sight not established truly horizontal by level vials or automatic compensators

Personal Errors

1. Rod not held plumb (avoid by using a rod level)
2. Incorrect rod readings resulting from long sights
3. Careless leveling for vertical-arc readings

Most errors in stadia work can be eliminated by **(a)** properly manipulating the instrument, **(b)** limiting lengths of sights, **(c)** using a good rod and rod level, and **(d)** averaging readings in the forward and backward directions. Line-of-sight errors cannot easily be corrected by field procedures; the instrument should be adjusted.

15-9 MISTAKES

Some typical mistakes in stadia work are:

1. Mistakes in reading the rod intercept
2. Use of an incorrect stadia interval factor
3. Waving the rod (*The rod should always be held plumb.*)
4. Index error applied with wrong sign
5. Confusion of plus and minus vertical angles read with a transit (theodolites which read zenith angles eliminate this problem)

PROBLEMS

15-1 Explain why the rod must be held plumb for stadia readings.
15-2 What simplifying assumption is made in deriving the stadia formula for inclined sights?
15-3 Explain how the stadia interval factor K can be determined.
15-4 Why should the line of sight through the lower cross line be kept several feet above any intervening ground if possible?
15-5 Why is it important in stadia work to know the value of the instrument's index error?
15-6 Why should sights be taken both forward and backward on each course in running stadia traverses?
15-7 If a stadia interval factor of 100.0 is assumed, but is later found to be 98.8, what errors in horizontal distance result on a horizontal sight for a rod intercept of:
  (a) 1.40 ft
  (b) 3.65 ft
*15-8 Distances *AB, AC, AD,* and *AE* of 121.6, 213.1, 304.2, and 406.4 ft were laid out on a level course. With an internal focusing theodolite setup on *A,* rod intercepts of 1.21, 2.12, 3.03, and 4.04 ft were read with the rod held on *B, C, D,* and *E,* respectively. Determine the stadia interval factor.
15-9 Similar to Problem 5-8, except horizontal distances of 105.7, 214.4, 319.7, and 412.2 ft were laid out, and rod intercepts of 1.06, 2.15, 3.21, and 4.13 ft, respectively, were read on each point in order.
15-10 Compute the error in distance if the top of a 12-ft stadia rod is inclined 1.5 ft toward the observer and the rod intercept is 5.00 ft on a horizontal sight.
15-11 An instrument operator on a stadia sight cannot set the center hair on the *hi* or see the upper hair because of overhanging branches. How can the distance and elevation difference be found?

Compute the horizontal distances and elevation differences for the data in Problems 15-12 and 15-13. (An internal focusing instrument was used.)

15-12 Rod intercept = 4.07 ft, $\alpha = +6°12'$ (on *hi*), $K = 100.5$
*15-13 Rod intercept = 3.45 ft, $\alpha = -4°51'$ (on *hi*), $K = 99.0$

Calculate the horizontal distances and elevation differences for the notes in Problems 15-14 and 15-15. (Assume an internal focusing instrument.)

15-14 Rod intercept = 2.50 ft, $z = 95°30'$ (on *hi*), $K = 98.5$

*Asterisks indicate problems that have answers given in Appendix G.

*15-15 Rod intercept = 1.85 ft, z = 83°28′ (on hi), K = 101.0

15-16 Specifications for a stadia traverse require an accuracy limit of $\frac{1}{500}$. Within what range can zenith angles be ignored in reducing distances to horizontal?

At station X a sight is taken with an internal focusing instrument on BM Bridge, elevation 987.5 ft. Compute the elevation of point X in Problems 15-17 and 15-18 (K = 100).

*15-17 Rod intercept = 3.54 ft, α = +1°55′ to 7.0 ft on rod, hi = 4.9 ft

15-18 Rod intercept = 2.86 ft, z = 92°24′ to 8.5 ft on rod, hi = 5.2 ft

Calculate horizontal distance AB and the elevation of B for stadia readings taken at A, elevation = 455.6 ft, hi = 5.4 ft, for Problems 15-19 through 15-21 (rod intercepts I are in ft and K = 100).

15-19 I = 2.88, z = 93°32′ read to 9.7 ft on the rod

*15-20 I = 3.25, z = 86°52′ read to 6.8 ft on the rod

15-21 I = 4.01, z = 91°48′ read to 2.5 ft on the rod

How close must angles be read for consistent precision of angles and stadia distances in Problems 15-22 and 15-23?

*15-22 A sight of 300 ft

15-23 A sight of 110 m

At station E, elevation 980.7, a stadia foresight is taken on hub F; the instrument is moved to F, and a stadia backsight taken on E. Compute distance EF and the elevation of point F in Problems 15-24 and 15-25 (rod intercepts I are in ft and K = 100).

*15-24 At E, I = 2.98, z = 88°08′ to 5.4 on rod, hi = 5.4 ft
At F, I = 2.99, z = 92°08′ to 3.$\overset{\circ}{3}$ on rod, hi = 4.9 ft

15-25 At E, I = 3.25, α = −3°19′ to 9.2 on rod, hi = 5.1 ft
At F, I = 3.23, α = 3°42′ to 3.5 on rod, hi = 5.5 ft

15-26 An internal focusing theodolite having a stadia interval factor of 99.5 was used to run a stadia survey of a closed-polygon traverse $ABCDEA$. The following field notes were recorded:

STATION OCCUPIED	STATION SIGHTED	AZIMUTH	ROD INTERCEPT	ZENITH ANGLE
A	E	151°20.0′	2.48	91°39.3′
	B	45°00.0′	2.54	85°56.1′
B	A	225°00.0′	2.55	94°01.5′
	C	98°05.4′	2.82	86°14.6′
˙C	B	278°05.7′	2.81	93°44.8′
	D	180°00.0′	2.40	96°17.3′
D	C	0°00.0′	2.39	83°40.2′
	E	250°35.8′	3.62	92°52.0′
E	D	70°35.8′	3.60	87°06.1′
	A	331°20.0′	2.48	88°22.4′

Calculate averages for forward and back observations of both horizontal distances and elevation differences of each traverse course. Using average horizontal distances, compute departures and latitudes, the traverse misclosure, and relative precision. Balance the departures and latitudes using the compass (Bowditch) rule. If the coordinates of

*Asterisks indicate problems that have answers given in Appendix G.

station A are $X = 1000.0$ and $Y = 1000.0$, calculate coordinates for all other traverse stations. Assuming the elevation of station A is 1044.5 ft, use the average elevation differences to calculate elevations of all other traverse stations. Adjust the elevations.

BIBLIOGRAPHY

Brinker, R. C., and R. Minnick. 1987. *The Surveying Handbook.* New York: Van Nostrand Reinhold, Chaps. 3 and 21.

Concord, J. E. 1971. "Tacheometry in Surveying Engineering." *ASCE, Journal of the Surveying and Mapping Division* 97 (No. SU1): 39.

Easa, S. M. 1990. "Model of Stadia Surveying with Incomplete Intercepts." *ASCE, Journal of Surveying Engineering* 116 (No. 3): 139.

Easa, S. M. 1991. "Maximizing Accuracy in Stadia Surveying." *Surveying and Land Information Systems* 51 (No. 3): 149.

Miller, C. H., and J. K. Odum. 1983. "Calculator Program for Reducing Alidade or Transit Stadia Traverse Data." *Surveying and Mapping* 43 (No. 4): 393.

Moffitt, F. H., and H. Bouchard. 1992. *Surveying,* 9th ed. New York: Harper-Collins, Chap. 14.

Wolf, P. R., et al. 1978. "An Evaluation of Accuracies and Applications of Tacheometry." *Surveying and Mapping* 38 (No. 3): 231.

16

TOPOGRAPHIC SURVEYS

16-1 INTRODUCTION

Topographic surveys are made to determine the configuration (*relief*) of the earth's surface and to locate *natural* and *cultural* features on it. From the survey data, using various types of lines and conventional symbols, *topographic maps* that depict these natural and cultural features are produced. Topographic maps are simply graphic representations of portions of the earth's surface. Natural features normally shown on them include relief (hills, valleys, and other surface irregularities), *hydrography* (rivers, lakes, oceans, etc.), and vegetation. Cultural (artificial) features are the products of people, and include roads, railroads, trails, buildings, bridges, canals, and boundary lines. Names and legends on maps identify the features.

Topographic maps are made and used by engineers and planners to determine the most desirable and economical locations of highways, railroads, canals, pipelines, transmission lines, reservoirs, and other facilities; by geologists to investigate mineral, oil, water, and other resources; by foresters to locate access or haul roads, fire-control routes, and towers; by architects in housing and landscape design; by agriculturists in soil conservation work; and by archeologists, geographers, and scientists in numerous fields.

A *planimetric map* depicts only natural and cultural features in their plan positions and does not show relief. Objects shown in plan view are called *planimetric features*. *Hypsometric* maps show relief by using various conventions and procedures. *Contours* are most commonly used, and are preferred by surveyors and engineers. *Digital elevation models* (DEMs) and *three-dimensional perspective models* are newer methods for depicting relief, made possible by computers. *Color, hachures, shading,* and *tinting* can also be used, but these methods are not quantitative enough and are generally unsuitable for surveying and engineering work. Contours, digital elevation models, and three-dimensional perspective models are discussed in later sections of this chapter and in Chapter 17. Topographic maps combine the properties of both planimetric and hypsometric maps.

16-2 METHODS FOR TOPOGRAPHIC SURVEYING

Topographic surveys are conducted by either aerial (photogrammetric) or ground (field) methods, and are often a combination of both. Refined equipment and procedures available today have made photogrammetry accurate and economical. Hence almost all topographic mapping projects covering large areas now employ this method. Ground surveys are still frequently used, however, especially for preparing large-scale maps of small areas. Even when photogrammetry is utilized, ground surveys are necessary to establish control and to field-check mapped features for accuracy. This chapter concentrates on ground methods. Several field procedures for locating topographic features, both horizontally and vertically, will be described. Photogrammetry is discussed in Chapter 28.

16-3 CONTROL FOR TOPOGRAPHIC SURVEYS

The first requirement of any topographic survey is good control, whether the survey is done by ground or aerial methods. Control, discussed in Chapter 19, is classified as either horizontal or vertical.

Horizontal control for a topographic survey is provided by two or more points on the ground, permanently or semipermanently monumented, and precisely fixed in position horizontally by distance and direction, or coordinates. It is the basis for map scale and locating topographic features. Horizontal control is usually established by traversing, triangulation, trilateration, or inertial and satellite methods, and can be filled in photogrammetrically for large areas. For small areas, horizontal control for topographic work is generally established by a traverse, although a single line may suffice in some cases.

Until recently, triangulation and trilateration were the most economical procedures to establish basic control for surveys extending over large areas such as a state or the entire United States. These techniques are now giving way to satellite surveying systems (see Chapter 20). Monuments of the state plane coordinate systems, having been established by control surveys, are used to initiate surveys of all types, but unfortunately more are needed in most areas.

Vertical control is provided by bench marks in or near the tract to be surveyed, and becomes the foundation for correctly portraying relief on a map. A vertical control net is usually established by lines of levels starting from and closing on bench marks. Elevations are generally determined for all traverse hubs, with provision in some cases for additional marks to be set in nearby strategic locations. A lake surface is a continuous bench mark and may sometimes be used.

With the rapidly growing popularity of total station instruments, trigonometric leveling has now become practical (see Section 6-5.4). It is frequently used to establish vertical control for mapping, especially in rugged areas. Improved methods in satellite GPS surveys (see Chapter 20) have also made that procedure suitable for establishing vertical control for modern topographic surveys.

Specified maximum allowable closure errors for both horizontal and vertical control should be determined in advance of the field work, then used to guide it.

Topographic details are usually built on a framework of control points whose positions and elevations have been established. Any errors in their positions or elevations are reflected in the location of topography. It is advisable, therefore, to run, check, and adjust the horizontal and vertical control surveys before a topographic detail survey is begun, rather than carry on both processes simultaneously.

The methods selected to establish control and locate topographic details govern the speed, cost, and efficiency of a topographic survey. In later sections of this chapter, the different basic field procedures and the varying equipment that can be used to accomplish these tasks are described.

16-4 CONTOURS

As stated earlier, contours are most often used by surveyors and engineers to depict relief. The reason is that they provide an accurate quantitative representation of the terrain. Because planimetric features and contours are located simultaneously in most field topographic surveys, it is important to understand them and their characteristics before discussing the various field procedures used to position them.

A contour is a line connecting points of equal elevation. The shoreline of a lake is a visible contour, but in general, contours cannot be seen in nature. On maps contours represent the planimetric locations of the traces of level surfaces of different elevations (see the plan view of Figure 16-1). Contours are drawn by interpolating between points whose positions and elevations have been measured and plotted. Nowadays, with increasing frequency, computerized automated contouring systems are being used to interpolate and draw contours from digitized terrain data.

The vertical distance between level surfaces forming the contours is called the *contour interval.* For topographic quadrangles at 1:24,000 scale, the U.S. Geological Survey uses one of the following contour intervals: 5, 10, 20, 40, or 80 ft. Contour intervals of 1, 2, or 5 ft are commonly used on large-scale maps for engineering design.

Figure 16-1 Contour lines. (a) Plan view. (b), (c) Profile views.

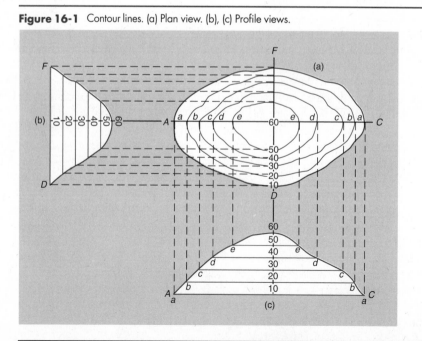

Metric units are sometimes used in place of English units for contours, in which case a contour interval of 0.5, 1, 2, 5, or 10 m is generally selected.

The contour interval selected depends on a map's purpose and scale, and the diversity of relief in the area. Reducing the interval requires more costly and precise field work. In regions where both flat coastal areas and mountainous terrain are included in a map, supplementary contours at one-half or one-fourth the basic interval are often drawn (and shown in dashed lines).

Figure 16-2 is a topographic map showing 10-ft contours. *Spot elevations* are given for critical points such as peaks, sags, streams, and highway crossings. Topographic mapping convention calls for drawing only those contours that are evenly divisible by the contour interval. Thus for the 10-ft contour interval of the map in Figure 16-2, contours such as the 1100, 1110, 1120, and 1130 are shown. Elevations are shown in breaks in the contour lines, and to avoid confusion, at least every fifth contour is labeled. To aid in reading topographic maps, every fifth contour (each that is evenly divisible by five times the contour interval) is drawn using a heavier line. Thus in Figure 16-2 the 1100, 1150, 1200, etc. contours are drawn more heavily.

16-5 CHARACTERISTICS OF CONTOURS

Although each contour line in nature has a unique shape, all contours adhere to a set of general characteristics. Important ones, fundamental to their proper field location and correct plotting, are listed.

1. Contour lines must close on themselves, either on or off a map. They *cannot* dead-end.
2. Contours are perpendicular to the direction of maximum slope.
3. The slope between adjacent contour lines is assumed to be uniform. (Thus it is necessary that breaks in grade be located in topographic surveys.)
4. The distance between contours indicates the steepness of a slope. Wide separation denotes gentle slopes; close spacing, steep slopes; even and parallel spacing, uniform slope.
5. Irregular contours signify rough, rugged country. Smooth lines imply gradual slopes and changes.
6. Concentric closed contours that increase in elevation represent hills. A contour forming a closed loop around lower ground is called a *depression contour*. (Hachures inside the lowest contour and pointing to the bottom of a hole or sink with no outlet make map reading easier.)
7. Contours of different elevations never meet except on a vertical surface such as a wall, cliff, or natural bridge. They cross only in the rare case of a cave or overhanging shelf. Knife-edge conditions are seldom, if ever, found in natural formations.
8. A contour cannot branch or wye into two contours of the same elevation.
9. Contour lines crossing a stream point upstream and form V's; they point down the ridge and form U's when crossing a ridge crest.
10. Contour lines go in pairs up valleys and along the sides of ridge tops.
11. A single contour of a given elevation cannot exist between two equal-height contours of higher or lower elevation. For example, an 820-ft contour cannot exist alone between two 810- or two 830-ft contours.

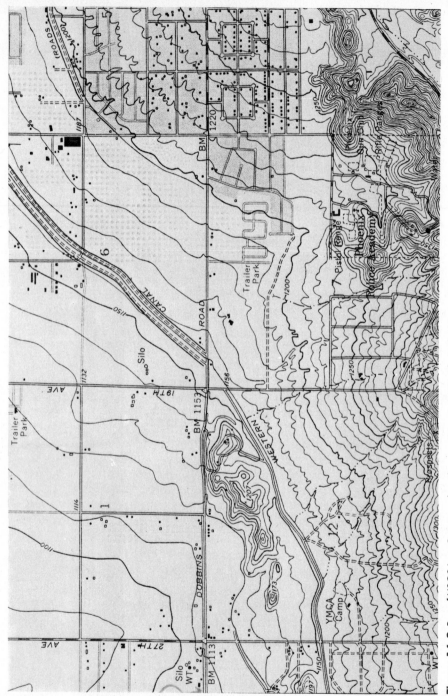

Figure 16-2 Part of USGS Lone Butte quadrangle map. (Courtesy U.S. Geological Survey.)

12. Cuts and fills for earth dams, levees, highways, railroads, canals, and so on, produce straight or geometrically curved contour lines with uniform, or uniformly graduated, spacing. Contours cross sloping or crowned streets in typical V- or U-shaped lines.

Keeping these characteristics in mind will (1) make it easier to visualize contours when looking at an area, (2) assist in selecting the best array of points to locate when conducting a topographic survey, and (3) prevent serious mistakes in sketching.

Numerous points may be needed to locate a contour in certain types of terrain. For example, in the unusual case of a level field that is at or near a contour elevation, the exact location of that contour would be time consuming or perhaps impossible to determine. In these situations, a uniform distribution of *spot elevations* can be used to portray the area's relief.

16-6 DIRECT AND INDIRECT METHODS OF LOCATING CONTOURS

Contours can be established by either the *direct method* (trace-contour method) or the *indirect method* (controlling-point method). The controlling-point method is generally more convenient and faster, and therefore it is most often selected. It is also the most frequent choice when data will be entered into a computer for automated contouring. These two methods are described in the subsections that follow.

16-6.1 Direct Method

This method is usually performed using a theodolite or transit. After the instrument is set up, the telescope oriented horizontally, and the HI established, the rod reading (foresight) that must be subtracted to give a specific contour elevation is determined. The rodperson selects trial points expected to give this minus sight, and is directed uphill or downhill by the instrument operator until the required reading is actually secured (within 0.1 to 0.5 ft, the allowable discrepancy depending on the terrain, contour interval, and specified accuracy).

In Figure 16-3 the instrument is set up at point *A,* elevation 674.3 ft, *hi* 4.9 ft, and HI 679.2 ft. If 5-ft contours are being located, a reading of 4.2 or 9.2 with the telescope level will place the rod on a contour point. For example, in Figure 16-3, the 9.2-ft rod reading means that point *X* lies on the 670-ft contour. After this point has been located

Figure 16-3 Direct method of locating contours.

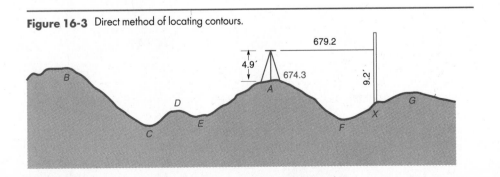

by trial, distance (usually by stadia) and azimuth are read and the process is repeated. Work is speeded by using a piece of red plastic flagging which can be moved up and down on the level rod to mark the required reading and eliminate searching for a number. The maximum distance between contour points in this method is determined by the terrain and accuracy required. Beginners have a tendency to take more sights than necessary in ordinary terrain. Contours are sketched by connecting located points having equal elevations. This is usually done as part of the office work, but they may also be drawn in the field book to clarify unusual conditions.

The direct method is suitable in gently rolling country, but generally not practical in undulating or rough, rugged terrain. Neither is it convenient for obtaining data to be used in a computer-driven automated contouring system.

16-6.2 Indirect Method

In the indirect method, the rod is set on "controlling points" which are critical to the proper definition of the topography. They include high and low points on the terrain, and locations where changes in ground slope occur, such as *B, C, D, E, F,* and *G* in Figure 16-3. Channels of drainage features and ridge lines must be included. Elevations are determined on these points using a theodolite or a total station, and employing trigonometric leveling (see Section 6-5), stadia (see Chapter 15), or from rod readings taken with the telescope level when possible. Azimuths and distances are also read to locate the points. The positions of controlling points are then plotted, and contours interpolated between elevations of adjacent points.

Figure 16-4(a) illustrates a set of controlling points labeled *A* through *N* that have been plotted according to their surveyed horizontal positions. Measured elevations (to the nearest foot) of the points are given in parentheses. Contours having a 10-ft interval have been sketched freehand between adjacent points by interpolation. It is improper to interpolate between points that cross controlling features such as gullies, streams, rivers, ridge lines, roads, and so on. Thus to properly draw the contours of Figure 16-4(a), with the stream located on the map, elevations were first interpolated along its thread between surveyed points *E, G, I,* and *J.* Interpolations were then made from the stream to points on either side. As an example, it would have been incorrect to interpolate between points *D* and *F.* Rather the elevation of the stream on the line between *D* and *F* (approximately 9 ft) was used to interpolate both ways from the stream to points *D* and *F.*

Note in Figure 16-4(a) that the gently curved contours tend to duplicate the naturally rolling topography of nature. Note also that contours crossing the stream form Vs pointing upstream.

16-7 DIGITAL ELEVATION MODELS AND AUTOMATED CONTOURING SYSTEMS

Data collected for use in automated contouring systems are an array of points whose horizontal positions are given by their *X* and *Y* coordinates, and whose elevations are given as *Z* coordinates. Such three-dimensional arrays provide a *digital* representation of the continuous variation of relief over an area, and are known as *digital elevation models* (DEMs). Alternatively, the term *digital terrain model* (DTM) is sometimes used.

Two basic geometric configurations are used in the field for collecting DEM data: the *grid method* and the *irregular method.* In the grid method, elevations are deter-

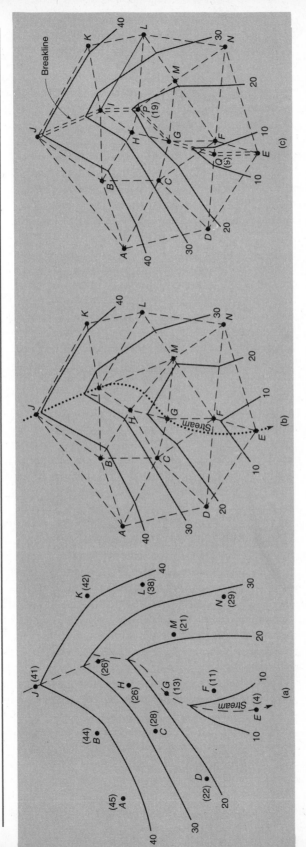

Figure 16-4 (a) Contours compiled by hand from controlling points A through N. (b) TIN model (dashed lines) constructed from data of (a), and contours derived from TIN model (solid lines). Stream is shown with dotted line. Note striking differences between the 10- and 20-ft contours of (a) and (b). (c) TIN model (dashed lines) constructed from data of (a) but with the addition of two points, P and Q, and the designation of lines EQ, QG, GP, PI, and IJ as breaklines. Contours shown with solid lines were derived from this TIN model. Note agreement of these contours with those of (a).

mined on points which conform to a regular square or rectangular grid. The procedure is described in Section 16-9.3, and sample field notes are given in Plate D-4 of Appendix D. From the array of grid data, the computer interpolates between points along the grid lines to locate contour points, and then draws the contour lines. The major disadvantage of this method is that critical high and low points and changes in slope do not generally occur at the grid intersections, and thus they are missed in the data collection process, which results in inaccurate relief portrayal.

The irregular method is simply the controlling-point method, but additional information (to be described later) is included. As previously noted, the controlling-point method involves determining the elevations of all high and low points, and points where slopes change. This of course produces a DEM with an irregularly spaced configuration of surveyed points.

The first step taken by computerized contouring systems that utilize irregularly spaced DEMs is to create a so-called *triangulated irregular network,* or TIN, model of the terrain from the DEM. It is very important to understand the TIN model concept to ensure that an appropriate array of controlling points is selected and surveyed in the field if an automated contouring system is to be used. A TIN model is constructed by connecting points in the array to create a network of adjoining triangles. The dashed lines of Figure 16-4(b) shows a TIN model created for the data in Figure 16-4(a). Various criteria can be used in the development of TIN models from an array of surveyed points, but one commonly used standard creates the "most equilateral network" of triangles.

In drawing contours, automated contouring systems generally make two assumptions concerning TIN models: (1) all triangle sides have a constant slope, and (2) the surface area of any triangle is a plane. Based on these assumptions, elevations of contour crossings are interpolated along triangle edges, and contours are constructed such that they change direction only at triangle boundaries. Contours derived in this manner from the TIN model of Figure 16-4(b) are shown in the figure as solid lines. Note the disparities between the hand-drawn contours of Figure 16-4(a) and those derived from the TIN model of Figure 16-4(b). Differences are particularly obvious between the 10- and 20-ft contours. These occur because (1) the computer did not interpret the curved thread of the stream [shown as a dotted line in Figure 16-4(b)], and (2) in creating the network of triangles, several sides were constructed which cross the stream, resulting in improper interpolation across the stream.

From this example it is apparent, as noted earlier, that additional information must be provided for computer-driven systems to depict contours accurately. That important added information is the identification of controlling features, also more often called *breaklines* in modern computer mapping terminology. (The term *discontinuity lines* is also sometimes used.) Breaklines are linear topographic features which have uniform slopes, and they must be identified to the computer in the input array. Then *the computer makes them sides of triangles in the TIN model and thus elevations are interpolated along them.* Streams, lake shores, roads, railroads, ditches, ridge lines, and so on, are examples of controlling features or breaklines. Curved breaklines such as streams must have enough data points so that when adjacent ones are connected with straight lines, they adequately define the feature's alignment.

The dashed lines of Figure 16-4(c) represent the TIN model constructed from the same data set as Figure 16-4(b), except that the stream (shown with a double dashed line) has now been identified as a breakline, and two additional points, *P* and *Q,* have been added to better approximate the curvature of the stream. In this figure, contours derived from the TIN model are shown. Note that these now very nearly duplicate the hand-drawn contours.

The important lesson of the foregoing is that if an automated contouring system is used, field points must be selected carefully, breaklines identified, and the data properly input to meet the system's assumptions. As indicated by this example, a few more controlling points may have to be surveyed, but the benefits of automated contouring systems make it worthwhile.

16-8 BASIC FIELD METHODS FOR LOCATING TOPOGRAPHIC DETAILS

Objects to be located in a topographic survey can range from single points or lines to meandering streams and complicated geological formations. The process of tying topographic details to the control net is called *detailing*. Regardless of their shape, all objects can be located by considering them as composed of a series of connected straight lines, with each line being determined by two points. Irregular or curved lines can be assumed straight between points sufficiently close together; thus detailing becomes a process of locating points.

When ground surveying instruments such as theodolites (or transits), tapes, EDM instruments and total stations are employed, any one of six different basic methods can be used to locate a point P in the field. These are illustrated in Figures 16-5(a) through (f). All rely on horizontal control. One line, *AB,* must be fixed in each of the first five methods. The positions of three points must be known or identifiable to apply the

---

Figure 16-5 Locating a point *P.*

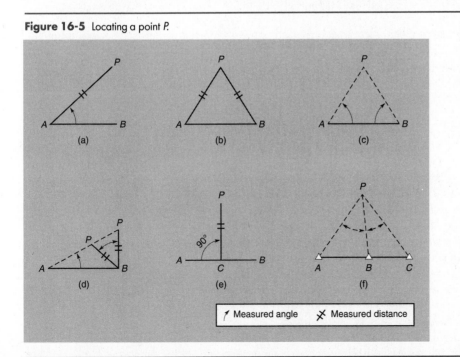

method of Figure 16-5(f), which is called *resection* or the *three-point problem.* Items to be measured in Figure 16-5(a) to (f) are, respectively:

(a) One angle and the adjacent distance, from either A or B
(b) Two distances, one each from A and B
(c) Two angles, one each from A and B
(d) One angle and the opposite distance (two possible points P)
(e) One distance AC along AB, and a right-angle offset distance
(f) Two angles from the point to be located to three control points

Method (a), known as *radiation,* is most often used and is the clear choice when total station instruments are employed. Methods (b) and (c) are *distance intersection* and *angle intersection,* respectively. Distance intersection is most practical when an EDM instrument is employed, and angle intersection when a theodolite is used. Angle intersection is particularly efficient to locate distant points over inaccessible terrain. Method (d) is seldom used since it requires the same basic measurements as method (a), with the disadvantage that it produces two possible locations for point P, as indicated in the figure. The perpendicular offset method (e) is frequently used on route surveys. The resection method (f) is occasionally convenient to locate an isolated point. In this method, for which the computational procedure is given in Section 19-11, with the theodolite at point P, angles are measured to three visible control stations.

From these methods for locating topographic details, an experienced party chief selects the one most appropriate for any given situation.

16-9 SPECIFIC FIELD METHODS FOR LOCATING TOPOGRAPHIC DETAILS

Location of planimetric features and contours can be accomplished by one of the following field procedures: **(1)** radiation by total station, **(2)** radiation by stadia, **(3)** coordinate squares or "grid" method, **(4)** offsets from a reference line, **(5)** plane-table, and **(6)** use of portable GPS units. An explanation of the uses, advantages, and disadvantages of each system follows.

16-9.1 Radiation by Total Station

In the radiation method, illustrated in Figure 16-5(a), a distance and a horizontal angle or azimuth are measured to each desired topographic detail point. The procedure is especially efficient if a total station instrument is used. These instruments simultaneously measure both distance and direction, compute horizontal distance and elevation difference components, and display coordinates and elevations of points in real time. Their utility on topographic surveys is further enhanced by using an automatic data collector (see Sections 3-7 through 3-9). This permits downloading the array of stored data directly into a computer for processing through an automated mapping system.

The field procedure of radiation with a total station can be made most efficient if the instrument is placed on a hill or ridge which overlooks a large part or all of the area to be surveyed. This permits more and longer radial lines and reduces the number of setups required.

An example illustrating the use of a total station with data collector for topographic mapping is given with reference to Figure 16-6 and the field notes of Table 16-1. In

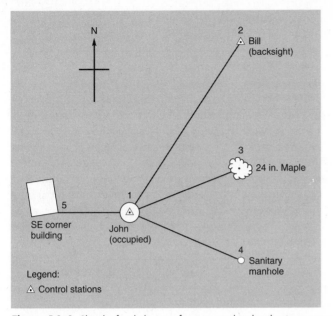

Figure 16-6 Sketch of radial survey for topographic details to accompany field notes of Table 16-1.

TABLE 16-1 EXCERPT OF AUTOMATIC DATA COLLECTOR FIELD NOTES OF A RADIAL SURVEY FOR TOPOGRAPHIC DETAILS

ENTRY	EXPLANATION
AC:SS	(Activity:Sideshot/keyboard entry by operator)
PN:3	(Point number: 3/keyboard entry by operator)
PD:24IN MAPLE	(Point description:24-in. maple/keyboard entry)
HZ:16.3744	(Horizontal angle:16°37′44″/by total station)
VT:90.2550	(Vertical "zenith" angle:90°25′50″/by total station)
DS:565.855	(Distance:565.855 ft/by total station)
AC:SS	(Activity:Sideshot/keyboard entry by operator)
PN:4	(Point number:4/keyboard entry by operator)
PD:SAN MH	(Point description:Sanitary manhole/keyboard)
HZ:70.3524	(Horizontal angle:70°35′24″/by total station)
VT:91.1548	(Vertical "zenith" angle:91°15′48″/by total station)
DS:436.472	(Distance:436.472 ft/by total station)
AC:SS	(Activity:Sideshot/keyboard entry by operator)
PN:5	(Point number:5/keyboard entry by operator)
PD:SE COR BLDG	(Point description:Southeast corner bldg/keyboard)
HZ:225.1422	(Horizontal angle:225°14′22″/by total station)
VT:88.3035	(Vertical "zenith" angle:88°30′35″/by total station)
DS:265.934	(Distance:265.934 ft/by total station)

Source: Courtesy ABACUS, A Division of Calculus, Inc.

Figure 16-6, a total station instrument was set up at station 1, John, and oriented in azimuth with a backsight on station 2, Bill. Measurements of azimuth, zenith angle, and distance, respectively, were then taken to points 3, 4, and 5 which are *sideshots* to topographic details.

When using an automatic data collector in this process, initial data for the setup are first entered in the unit via the keyboard. These include the X and Y coordinates of stations John and Bill, the elevation of John, and the heights of both the total station instrument and the reflector. The left-hand column of Table 16-1 illustrates an actual set of field notes recorded by a data collector during the process of taking sideshots on points 3, 4, and 5. Six entries were recorded per point. On each line, the entry to the left of the colon was automatically supplied by the computer and appeared on the data collector display at the time of measurement to prompt the operator. Entries to the right of the colon were supplied by the operator, either manually via the keyboard or by pressing the proper button on the total station instrument. Explanations to assist students in interpreting the data in Table 16-1 are given in parentheses.

16-9.2 Radiation by Stadia

Stadia radiation is the same as radiation by total station, except that distances are measured by stadia. This procedure is rapid and sufficiently accurate for most topographic surveys. Stadia distances, azimuths, and vertical or zenith angles are read for lines radiating from the theodolite, transit, or planetable to required points. The principles of stadia were discussed in detail in Chapter 15. Plate D-7 shows sample notes for a transit-stadia survey.

16-9.3 Coordinate Squares or "Grid" Method

The method of coordinate squares (grid method) is better adapted to locating contours than planimetric features, but can be used for both. The area to be surveyed is staked in squares 10, 20, 50, or 100 ft (5, 10, 20 or 40 m) on a side, the size depending on the terrain and accuracy required. A theodolite (or transit) can be used to lay out control lines at right angles to each other, such as *AD* and *D3* in Figure 16-7. Grid lengths are marked and the other corners staked and identified by the number and letter of intersecting lines.

Elevations of the corners can be obtained by differential or trigonometric leveling. Contours are interpolated between the corner elevations (along the sides of the blocks) by estimation or by calculated proportional distances. Elevations obtained by interpolation along the diagonals will generally not agree with those from interpolation along the four sides because the ground's surface is not a plane. Except for plotting contours, this is the same procedure as that used in the borrow-pit problem in Section 27-10.

In plotting contours by the grid method, a widely spaced grid can be used for gently sloping areas, but it must be made more dense for areas where the relief is rolling or rugged. A drawback of the method is that no matter how dense the grid, critical points (high and low spots and slope changes) will not generally occur at grid locations, thus degrading the accuracy of the resulting contour map.

16-9.4 Offsets from a Reference Line

This procedure is most often selected for mapping long linear areas, such as those necessary for route surveys. After the centerline or reference line has been staked and

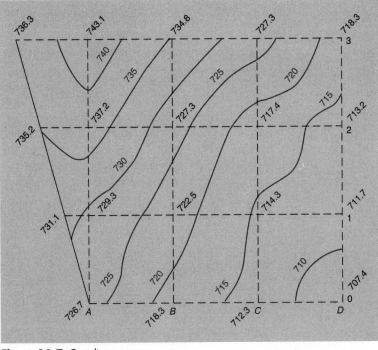

Figure 16-7 Coordinate squares.

stationed, planimetric details such as utilities, trees, fences, buildings, and so on, are located by offsets from it. Elevations for determining contour locations are also determined by offsets using the technique of *cross-sectioning,* as described in Section 27-3.

Figures 16-8 and 16-9 illustrate examples for locating planimetric features. In Figure 16-8, house *abcdea* is tied to reference line *AB* by perpendicular offsets. The building shape requires defining only two main corners, such as *a* and *b,* but corner *c* is also located as a check. All house sides are measured (usually by taping if the sides are short) and the lengths recorded on a sketch in the field book. Location of the barn can be done by making linear measurements from the house, as shown in the figure. It is good practice to make redundant measurements, such as two distances to each corner as shown, to provide a check.

Location of the second house *qrstq* in Figure 16-8 illustrates another practical method for mapping on route surveys using offsets from a reference line. Here the stations and pluses where "projections" of lines on linear objects intersect the reference line are noted. Distances out from the reference line, measured along the projected lines, are also recorded.

Location of a crooked creek by using the offset method is illustrated in Figure 16-9. At intervals along the reference line, offsets to the edge of the creek are measured. They can be taken at regular intervals, or spaced at distances that permit the curved line to be considered straight between successive offsets. For the examples of both Figures 16-8 and 16-9, all measurements may be shown on a sketch in the field book.

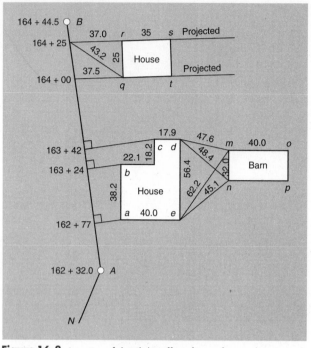

Figure 16-8 Location of details by offsets from reference line.

As noted, elevation data for contouring are located by taking cross sections. This consists in taking profiles of the ground surface at right angles to the reference line. The right angle can be measured using a theodolite, or a hand-held *pentagonal prism* of the type shown in Figure 16-10. Alternatively, the perpendicular can be approximately established by standing on the centerline, pointing the arms in opposite directions along the line, then bringing the palms of the hands together with arms outstretched in front of the body. (The eyes should be closed to prevent unconsciously aiming at a prominent object near the perpendicular line.) Rod readings are taken along the perpendicular line at all breaks in the ground surface and recorded with the respective distances out, left, and right. Plate D-8 is a sample set of crosssection notes.

Cross-section data can be used for compiling contours, but more commonly they are the basis for making earthwork volume computations. This application is described in detail in Chapter 27.

16-9.5 Planetable Method

In this method a map is compiled directly in the field on a drawing board, which is mounted on a tripod. An alidade, consisting of a telescope mounted on a pedestal rigidly attached to a straightedge base (Figure 16-11), is placed on the board. Using the alidade's telescope, a rod held at an object to be located is sighted, and a stadia rod interval and vertical angle read. From the stadia readings, the distance to the point and its elevation are computed. The direction to the object is drawn along the alidade straightedge, and the distance to it scaled to immediately locate the point on the map

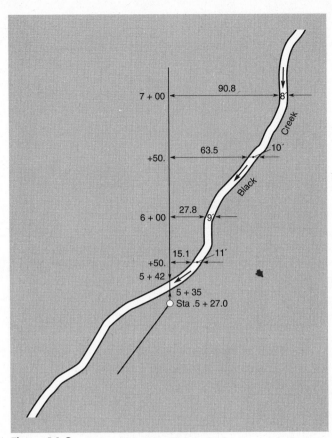

Figure 16-9 Location of creek by offset method.

Figure 16-10 Double pentagonal prism (on side). (Courtesy Leica, Inc.)

sheet. This is simply graphic radiation by stadia, as described in Section 16-9.2, except the equipment and procedure are different.

Planetable surveying eliminates the need to measure or record any horizontal angles. Elevations of points are written beside their plotted positions. The instrument operator

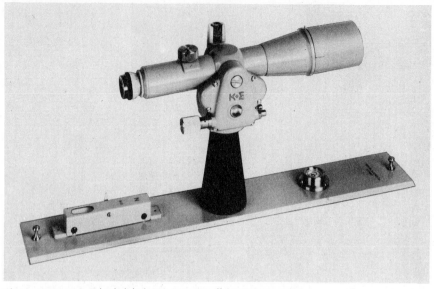

Figure 16-11 Standard alidade. (Courtesy Keuffel & Esser Company.)

sketches contours by either the direct or indirect method while looking at the area. Since the map is plotted in the field, coverage can be checked by observation.

The planetable is among the oldest of surveying instruments and is seldom used today, having given way to newer equipment such as total stations and GPS units. Thus only this brief description is retained in this current text. For a more detailed discussion of the planetable, an earlier edition of this book should be consulted.

16-9.6 Topographic Detailing with Portable GPS Units

The Global Positioning System (GPS), as explained in Chapter 20, is a new satellite-based surveying procedure that enables rapid determination of accurate positions anywhere on earth. Early versions of GPS receivers and postprocessing software required that the receivers remain stationary on a point for up to several hours to obtain precise ground positions. The equipment was not readily portable, and it was relatively expensive. Now recent research and development have led to smaller, less expensive hardware, and refined procedures known as *kinematic GPS surveying,* and *real-time kinematic GPS surveying* (see Section 20-6), which are ideal for topographic surveying.

With either of these kinematic GPS surveying procedures, point positions can be determined rapidly at any location where a receiver is placed. It is essential, however, that the receiver's antenna be open for clear visibility to satellites above. Thus GPS may not be suitable for direct location of individual large trees, tall buildings, or other objects which obscure satellites from view.

Receivers for topographic work are specially designed, small, and portable, and are interfaced with a keyboard for system control and entry of codes to identify features surveyed. The unit shown in Figure 16-12 performs real-time kinematic surveys. It can compute (in real-time) the coordinates of locations where the receiver is placed, and can store data for virtually an unlimited number of points in files. The files can then be directly downloaded to a computer for further processing, which could include auto-

Figure 16-12 Trimble "Site Surveyor" roving GPS receiver being operated in the real-time kinematic mode for topographic surveying. (Courtesy Trimble Navigation.)

matic map drafting. These systems make topographic data collection a simple and very fast one-person operation.

16-10 SELECTION OF FIELD METHOD

Selection of the field method to be used on any topographic survey depends on many considerations, including **(1)** purpose of the survey, **(2)** map use (accuracy required), **(3)** map scale, **(4)** contour interval, **(5)** size and type of area involved, **(6)** cost, **(7)** equipment and time available, and **(8)** experience of the survey personnel.

Items (1) to (5) are interdependent. The cost, of course, will be a minimum if the most suitable method is selected for a project. On large projects, personnel costs rather than equipment investment will usually govern. The method chosen by a private surveyor making a topographic survey of 50 or 100 acres, however, may be governed by the equipment owned.

16-11 HYDROGRAPHIC SURVEYS

Hydrographic surveys determine depths and surface configurations of the bottoms of water bodies. Usually the survey data are used to prepare hydrographic maps, although in navigation and dredging, they may be recorded in electrical formats for real-time analysis. Bodies of water surveyed include rivers, reservoirs, harbors, lakes, and oceans.

Hydrographic surveys and maps are used in a variety of ways. As examples, engineers employ them for planning and monitoring harbor and river dredging operations, and to ascertain reservoir capacities for flood control and water-supply systems; petroleum engineers use them to position offshore drilling facilities and locate underwater pipelines; navigators need them to chart safe passageways and avoid reefs, bars, and other underwater hazards; biologists and conservationists find them helpful in their study and management of aquatic life; and anglers use them to locate likely fish-holding structures.

Hydrographic maps are fundamentally topographic maps of the land areas beneath water surfaces. Thus field procedures for hydrographic surveys are similar to those for topographic work; hence the subject is discussed in this chapter. There are some basic differences, however, such as the fact that the land area being mapped cannot be seen, and the depth measurements must be made in water.

Two basic tasks involved in hydrographic surveys are **(1)** making *soundings* (measuring depths) from the water surface to bottom, and **(2)** locating the positions where soundings were made. Techniques used to perform these tasks vary depending on the water body's size, accuracy required, type of equipment to be used, and number of personnel available. The subsections that follow briefly describe procedures for mapping small to moderate-sized water bodies.

16-11.1 Equipment for Making Soundings

The size of a water body and its depth control the type of equipment used to measure depths. For shallow areas of limited size, a *sounding pole* can be used. This is usually a wooden or fiberglass staff resembling a level rod. It is perhaps 15 ft long, graduated in feet or tenths of feet, with a metal shoe on the bottom. Direct depth measurements are made by lowering the pole vertically into the water until it hits bottom, and then reading the graduation at the surface.

Lead lines can be used where depths are greater than can be reached with a sounding pole. These consist of a suitable length of stretch-resistant cord or other material, to which a heavy lead weight (perhaps 5 to 15 lb) is attached. The cord is marked with foot graduations, and these should be checked frequently against a steel tape for their accuracy. In use, the weight is lowered into the water, being careful to keep the cord vertical. The graduation at the surface is read when the weight hits bottom.

In deep water, or for hydrographic surveys of appreciable extent, electronically operated sonic depth recorders called *echo sounders* are used to measure depths. These devices, an example of which is shown in Figure 16-13, transmit an acoustic pulse vertically downward and measure the elapsed time for the signal to travel to the bottom,

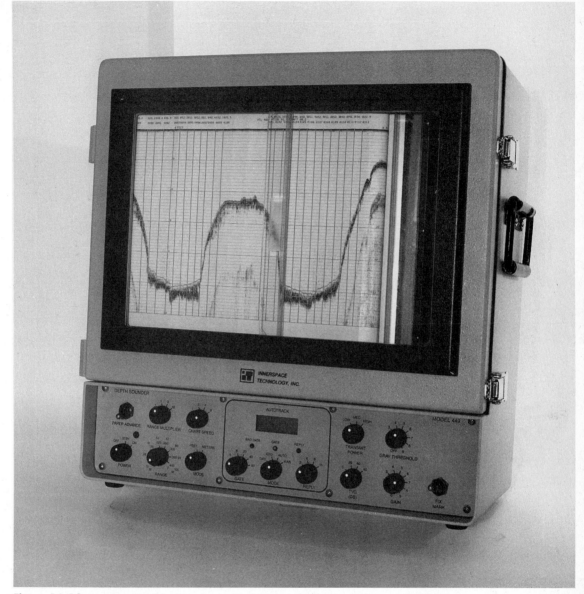

Figure 16-13 Innerspace model 449 hydrographic depth sounder. (Courtesy Innerspace Technology, Inc.)

be reflected, and return. The travel time is converted to depth, and displayed in either digital or graphic form. A graphic profile, such as that shown in Figure 16-14, is generally preferred, because it provides a visual record from which depths can be extracted. Also it can be referred to repeatedly for plotting and checking.

Sounding poles and lead lines yield spot depths and are restricted to use in relatively shallow water. Electronic depth sounders, however, provide continuous profiles of the

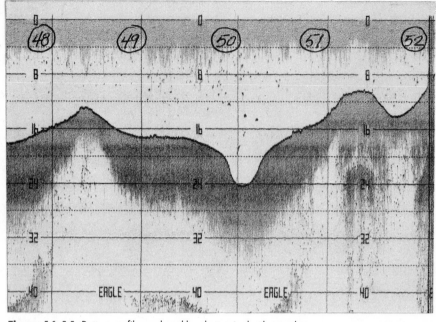

Figure 16-14 Bottom profile produced by electronic depth sounder.

surface beneath the boat's path and can be used in water of virtually any depth. The vertical scale of modern echo sounders can be changed by simply pressing buttons to accommodate varying ranges in depth. In the profile of Figure 16-14, for example, the chart's vertical range was set to 40 ft, and profile depths shown vary from 8 to 24 ft.

The reference plane from which depth soundings are measured is the water surface. Because of surface fluctuations, its elevation or *stage* at the time of survey must be determined with respect to a fixed datum—usually mean sea level. This can be done by running a level circuit to the water from a nearby bench mark. In situations where soundings are repeated at regular intervals, a graduated staff can be permanently installed in the water so that its stage, in feet above mean sea level, can be read directly each time soundings are repeated.

16-11.2 Locating Soundings

Any of the traditional ground surveying procedures illustrated in Figure 16-5 can be used to locate boat positions where soundings are taken. In addition to these techniques, other methods have also been applied in hydrographic surveys, for example, GPS. If ground surveying techniques are used, some horizontal control must first be established on shore. Ideal locations for control stations are on peninsulas or in open areas that afford a wide unobstructed view of the water body for tracking the sounding boat. The coordinate positions of the control points can be established by traverse, but triangulation and trilateration are also well suited for this work.

Among the various boat-positioning methods, *radiation* [Figure 16-5(a)] and *angle intersection* [Figure 16-5(c)] are frequently selected if total station instruments or theodolites are used. Radiation is particularly efficient, especially if a total station instru-

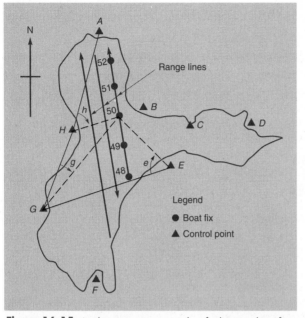

Figure 16-15 Angle intersection procedure for locating boat fixes along range lines.

ment is used, because only one person on shore is needed to track the boat. After setting up on one control station and backsighting another, an angle and a distance are measured to locate each boat position. Special total station instruments are manufactured for this work to facilitate sighting and measuring distances electronically to a moving target. From angles and distances, which are automatically read, the total station's computer determines the boat's coordinates. These can either be stored in an automatic data collector for later office use in mapping, or transmitted by radio to the boat if real-time positioning is required, as in controlling ongoing dredging.

Figure 16-15 illustrates the use of angle intersections in the hydrographic survey of a lake. Here the boat travels back and forth along *range lines* while the depth sounder continuously records bottom profiles. At regular intervals so-called *fixes* are taken by measuring angles to the boat from shore stations. Two angles establish the boat's position, but three or more provide redundancy and a check. In Figure 16-15, for example, angle measurements *e, g,* and *h* for fix number 50 (indicated by dashed lines) have been made from shore stations *E, G,* and *H,* respectively. Prior to measuring angles, the theodolites were oriented by backsighting on another visible control station, as at station *G* from *E.*

Flag or radio signals are given from the boat to coordinate fixes and ensure that angles from all shore stations are measured simultaneously. At the precise moment of any fix, the profile is also marked and the fix number noted. On Figure 16-15, for example, fixes 48 through 52 are identified, and on the profile of Figure 16-14, these fix locations are also marked. This correlates bottom depths with specific locations in the water body—a necessity for mapping.

If the boat is driven back and forth along parallel range lines to cover the area of interest, and then the area traversed again with perpendicular courses, a grid of profiles results from which contours can be drawn. In larger bodies of water a compass is valuable to assist in keeping the range lines parallel. Required accuracy dictates the spacing between range lines, with closer spacing yielding more accurate results.

Various other boat-positioning systems can be used, depending on circumstances. One that works well for hydrographic surveys of rivers or other relatively narrow water bodies consists in laying out uniformly spaced reference lines which cross the water. The lines are marked by placing tall painted stakes on the bank on either side. Then fixes can be taken as the sounding boat navigates along the reference lines. To position each fix along the lines, however, either a distance must be measured from one reference point, or an azimuth to the boat from an independent control point. When the boat is moving perpendicular to the marked lines, its passage across projections between stakes positions fixes, but again a distance or angle is needed to complete the fix location.

Kinematic GPS surveying (see Section 20-6) is ideal for establishing sounding locations for hydrographic surveys. With its many advantages, it is expected to replace most other hydrographic positioning techniques in the future.

16-11.3 Hydrographic Mapping

Procedures for preparing hydrographic maps do not differ appreciably from those used in topographic mapping discussed in Chapter 17. Basically, depths are plotted in their surveyed positions and contours drawn. If an echo sounder is used, depths are interpolated from the profiles and plotted between fix locations.

In addition to depth contours, the shoreline and other prominent features are also located on hydrographic maps. This is especially important for navigation and fishing maps, as the features are the means by which users line in and locate themselves on the water body. Planimetric features are most often located photogrammetrically (see Chapter 28), but the techniques for topographic mapping discussed in this chapter can also be used.

Modern hydrographic surveying systems utilize sophisticated electronic positioning and depth-recording devices. These, coupled with computers interfaced with plotters, enable rapid automated production of hydrographic maps in near real time. But the basic principles discussed here still apply.

16-12 ACCURACY SPECIFICATIONS FOR TOPOGRAPHIC SURVEYS

In the United States, *national map accuracy standards* (NMAS) give specifications for maximum errors permitted in horizontal positions and elevations shown on maps. They specify that for maps produced at scales larger than 1:20,000, not more than 10% of points tested shall be in error in horizontal position by more than $\frac{1}{30}$ in. (0.8 mm). To meet these specifications, on a map plotted to a scale of 1 in. = 100 ft, point positions would have to be correctly portrayed to within ± 3.3 ft. On smaller scale maps, the limit of error is $\frac{1}{50}$ in. (0.5 mm), or approximately 40 ft on the ground at a map scale of 1:24,000. These limits of accuracy apply to positions of well-defined points only, such as monuments, bench marks, highway intersections, and building corners.

The NMAS vertical accuracy requirements specify that not more than 10% of elevations tested shall be in error by more than one-half the contour interval, and none can exceed the interval. To meet this requirement, contours can be shifted by distances up to the horizontal positional tolerance (discussed above) if necessary.

Of course if these NMAS horizontal and vertical mapping criteria are to be met, topographic features must be surveyed to an accuracy higher than that called for in the mapping specifications. Published maps meeting these accuracy requirements usually have the following note in their legends: "This map complies with National Map Accuracy Standards."

The American Society for Photogrammetry and Remote Sensing (ASPRS) has adopted its own standards to govern photogrammetric production of large-scale maps. It specifies standards for three levels of accuracy, classes 1, 2, and 3. For a map to meet its class 1 standards, the root-mean-square (rms)[1] error in both X and Y coordinates of well-defined points must not exceed ± 0.01 in. at map scale. Thus for a map scale of 500 ft/in., the allowable rms error in X and Y coordinates is ± 5.0 ft. Vertical accuracy is specified in terms of the map's contour interval (CI). For class 1, the rms error of well-defined points must not exceed $\pm (CI/3)$. These horizontal and vertical standards are both relaxed by factors of 2 and 3 for class 2 and class 3 maps, respectively.

The American Society of Civil Engineers (ASCE) has also developed a set of standards for topographic mapping, which are aimed primarily at large-scale engineering maps. In addition to suggesting accuracies for various map scales, they also provide standards for contouring, map symbols, abbreviations, lettering, and other factors important in mapping.

The accuracy of any map may be tested by comparing the positions of points whose locations or elevations are shown on it with corresponding positions determined by surveys of a higher accuracy. Plotted horizontal positions of objects are checked by running an independent traverse or other survey to points selected by the person or organization for whom the map was made. To check vertical accuracy, elevations obtained from field profile surveys are compared with elevations taken from profiles made from plotted contours. These procedures provide a check on both field work and map drafting.

16-13 SOURCES OF ERROR IN TOPOGRAPHIC SURVEYS

Some sources of error in topographic surveys are:

1. Instrumental errors, especially an index error which affects vertical and zenith angles
2. Errors in reading instruments
3. Control not established, checked, and adjusted before topography is taken
4. Control points too far apart and poorly selected for proper coverage of an area
5. Sights taken on detail points that are too far away
6. Poor selection of points for contour delineation

---

[1] The rms error is defined as the square root of the average of squared discrepancies for points tested. Discrepancies are the differences between coordinates and elevations of points taken from the map and their values as determined by check surveys. The check surveys are governed by standards set by the Federal Geodetic Control Subcommittee (FGCS).

16-14 MISTAKES

Some typical mistakes in topographic surveys are:

1. Unsatisfactory equipment or field method for particular survey and terrain conditions
2. Mistakes in instrument reading and data recording
3. Failure to periodically check azimuth orientation when many detail points are located from one instrument station
4. Too few (or too many) contour points taken
5. Omission of some topograhic details

PROBLEMS

16-1 What is the difference between a planimetric map and a topographic map?

16-2 List five topographic details classified as ''cultural'' features not mentioned in Section 16-1.

16-3 Name the different conventions or procedures that can be used to portray relief.

16-4 What factors must be considered when selecting the contour interval to be used for a given topographic map.

*16-5 A contour map has an interval of 10 ft and a scale of 1:2400. If two adjacent contours are 0.75 in. apart, what is the average slope of the ground between the contours?

16-6 On a map whose scale is 1 in. = 500 ft, how far apart (in inches) would 5-ft contours be on a uniform slope (grade) of 6%?

*16-7 On a map drawn to a scale of 1 in. = 100 ft, contour lines are $\frac{1}{2}$ in. apart at a certain place. Contour interval is 5 ft. What is the ground slope, in percent, between adjacent contours?

16-8 Same as Problem 16-7, except for a 2-m interval, 40-mm spacing, and a map scale of 1:2000.

16-9 Sketch at a scale of 10 ft/in., the general shape of contours that cross a 40-ft-wide street having a +5.00% grade, a 12-in. parabolic crown, and a 6-in.-high curb.

16-10 Same as Problem 16-9, except a four-lane divided highway including a 30-ft-wide median strip, two 12-ft-wide lanes of pavement each side of the median having a +3.00% grade with a 1.00% side slope; and 10-ft-wide shoulders both sides with a 4% side slope. The median slopes 10.0% toward the center.

16-11 For a 10-ft contour interval, what is the greatest error in elevation expected of any definite point read from a map if it complies with National Map Accuracy Standards?

16-12 When should points located for contours be connected by straight lines? When by smooth curves?

The following table gives elevations at the corners of 50-ft coordinate squares, and they apply to Problems 16-13 through 16-15.

74	71	66	65	66	69
76	73	71	68	63	72
80	69	68	66	66	68

*16-13 About which number in the table can a 5-ft closed contour be drawn?

16-14 Draw 2-ft contours for the area.

16-15 Same as Problem 16-14, except at the bottom of the table add a fourth line of elevations: 83, 78, 72, 66, 61, and 65 (from left to right).

*Asterisks indicate problems that have answers given in Appendix G.

16-16 What are the disadvantages of locating contours in the field using the coordinate-squares method?

16-17 What is: **(a)** A digital elevation model (DEM)? **(b)** A triangulated irregular network (TIN) model?

16-18 Explain the importance of locating "breaklines" in the field if contours will be drawn using a computerized automated contouring system.

16-19 Why is the offsets-from-the-centerline method used more than any other ground system on route surveys to locate points?

16-20 Cite the advantages of locating topographic details by radial methods using a total station instrument with automatic data collector.

16-21 Discuss the advantages and disadvantages of performing topographic surveys using kinematic GPS surveying procedures.

16-22 What are the advantages of performing topographic surveys by planetable? The disadvantages?

16-23 Prepare a set of field notes to locate the topographic details in Figure 16-8. Scale additional distances and angles if necessary.

16-24 A rectangular lot running N–S and E–W is 150 × 100 ft. To locate contours it is divided into 50-ft square blocks, and the following horizontal rod readings are taken at the corners successively along the E–W lines, proceeding from west to east, and beginning with the northernmost line: 6.4, 5.8, 3.8; 6.7, 6.0, 5.2; 7.9, 6.8, 6.2; and 9.5, 8.2, 6.7. The HI is 180.8 ft. Sketch 1-ft contours.

16-25 List the various equipment used for making hydrographic depth soundings and discuss the limitations, advantages, and disadvantages of each.

*16-26 On a map having a scale of 200 ft/in. the distance between plotted fixes 49 and 50 of Figure 16-15 is 3.15 in. From measurements on the profile of Figure 16-14, determine how far from fix 50 the 20-ft contour (existing between fixes 49 and 50) should be plotted on the map.

16-27 Similar to Problem 16-26, except locate the 16-ft contour between fixes 50 and 51 if the corresponding map distance is 2.98 in.

16-28 Why is it important to show the shoreline and some planimetric features for navigation hydrographic maps?

*Asterisks indicate problems that have answers given in Appendix G.

BIBLIOGRAPHY

ASPRS. 1987. *Large Scale Mapping Guidelines.* Bethesda, Md.: American Society for Photogrammetry and Remote Sensing.

Baldwin, D., and C. Hempel. 1986. "Hydrographic Surveying." *Professional Surveyor* 6 (No. 5):29.

Brown, R. L. 1980. "Proposed Manual on Selection of Map Uses, Scales, and Accuracies for Engineering and Associated Purposes: Map Availability—Chapter VI." *ASCE, Journal of the Surveying and Mapping Division* 106 (No. SU1):149.

Carter, J. 1988. "Digital Representations of Topographic Surfaces." *Photogrammetric Engineering and Remote Sensing* 39 (No. 11):1577.

Crombie, B. W. 1977. "Contour Design and the Topographic Map User." *Canadian Surveyor* (No. 31):34.

Crosswell, P., and S. Clark. 1988. "Trends in Automated Mapping and Geographic Information System Hardware." *Photogrammetric Engineering and Remote Sensing* 39 (No. 11):1571.

Feldscher, C. B. 1980. "A New Manual on Map Uses, Scales and Accuracies." *ASCE Journal of the Surveying and Mapping Division* 106 (No. SU1):143.

Lee, J., et al. 1992. "Modelling the Effect of Data Errors on Feature Extraction from Digital Elevation Models." *Photogrammetric Engineering and Remote Sensing* 43 (No. 10):1461.

Odom, H. 1984. "Electronic Advancements in Hydrographic Surveying." *ASCE Journal of the Surveying Engineering Division* 110 (No. 1):21.

Ramivez, J. 1991. "Computer-Aided Mapping Systems: The Next Generation." *Photogrammetric Engineering and Remote Sensing* 42 (No. 1):85.

Roth, A. W. 1992. "Effectively Utilizing Your Contouring Program." *Point of Beginning* 17 (No. 2):32.

Shmutter, B., and Y. Doytsher. 1990. "Contouring-Simulation of Cartographic Procedure." *ASCE, Journal of Surveying Engineering* 116 (No. 4):193.

Skidmore, A., and B. Turner. 1992. "Map Accuracy Assessment Using Line Intersect Sampling." *Photogrammetric Engineering and Remote Sensing* 43 (No. 10):1453.

Wolf, P. R. 1986. "A Look Inside Wisconsin's Lakes." *Wisconsin Fins and Feathers* 15 (No. 7):55.

17

MAPPING

17-1 INTRODUCTION

Maps are visual expressions of portions of the earth's surface. Features are depicted on them using various combinations of points, lines, and standard symbols. Maps have traditionally been produced in *graphic,* or "hard-copy," form, that is, printed on paper or Mylar. Recently, however, their production in *digital,* or "soft-copy," form has been increasing. Maps in digital form are stored in a computer or on a disk, can be viewed on a screen, and, if desired, can be printed in hard-copy form.

Throughout the ages, maps have had a profound impact on human activities, and today the demand for them is perhaps greater than ever. They are important in engineering, resource management, urban and regional planning, management of the environment, construction, conservation, geology, agriculture, and many other fields. Maps show various features—for example, topography, property boundaries, transportation routes, soil types, vegetation, land ownership, and mineral and resource locations. Maps are especially important in engineering for planning project locations, designing facilities, and estimating contract quantities.

Maps comprise the very heart of modern land information systems (LISs) and geographic information systems (GISs). This rapidly emerging technology for spatial data analysis and management uses the computer to store, retrieve, manipulate, merge, analyze, and display information by means of digital maps (see Chapter 29). LISs and GISs have applications in virtually every field of endeavor. Spatial databases to support these systems are generally developed either by digitizing existing graphic maps, or by entering ground survey or photogrammetric data directly and generating new digital maps in the computer.

Maps of various types needed to create spatial databases for LISs and GISs include *topographic maps,* which show the natural and cultural features and relief in an area; *cadastral maps,* which give boundaries of land ownership; *natural resource maps,* which provide the location and distribution of forest and water resources, wetlands, soil types, and so on; *facilities maps,* which show existing transportation networks, water and sewer mains, and distribution lines for electric power; and *land-use maps,* which show the various activities of humans related to the land. Applications of LIS and GIS technology have been expanding at a phenomenal rate, and these activities will ensure a heavy demand for high-quality maps of various types and scales in the future.

The military services have always depended heavily on a steady flow of up-to-date maps and charts. During World War II, the Army Map Service, now the Topographic Center, Defense Mapping Agency (DMA), prepared more than 40,000 different maps covering approximately 400,000 square miles of the earth's surface; 500 million copies were printed. The Normandy invasion alone required 70 million copies of 3000 different maps. Because of the size and dispersion of military forces in Vietnam, the Army Map Service issued an estimated 500 million copies of maps to support that conflict. In Operation Desert Storm the high demand for maps and positional information continued. The latest technologies were employed to meet the needs, and included GPS units, aerial and satellite imagery, and GIS technology. In many instances, the maps were distributed in digital form and viewed on screens of portable computers.

Cartography, the term applied to the overall process of map production, includes map design, preparing or *compiling* manuscripts, final drafting, and reproduction. These processes all apply whether the maps are graphic or digital, and, except for reproduction, they are subjects discussed in this chapter.

17-2 MAP SCALE

Map scale is the ratio of a map distance to its corresponding true length on the ground. Map scales are given in three ways: **(1)** by *ratio* or *representative fraction,* such as 1:2000 or 1/2000; **(2)** by an *equivalence,* for example, 1 in. = 200 ft; and **(3)** *graphically* using either a bar scale or labeled grid lines spaced at uniform distances apart. Two bar scales, placed at right angles to each other and in diagonally opposite corners of a map sheet, or grid lines, permit accurate measurements to be made, even though the paper changes dimensions.

Choice of scale depends on the purpose, size, and required precision of the finished map. Dimensions of a standard sheet, type and number of topographic symbols, and accuracy requirements for scaling distances from it are some additional considerations. Map scales are usually selected to be compatible with one of the standard graduations on engineer's scales. These standard graduations have 10, 20, 30, 40, 50, or 60 units per inch. Thus scales of 1 in. = 100 ft and 1 in. = 1000 ft are compatible with the 10 scale; 1 in. = 200 ft and 1 in. = 2000 ft are consistent with the 20 scale, and so on.

Map scales are generally classified as *large, medium,* and *small.* Their respective scale ranges are as follows:

Large scale, 1 in. = 100 ft (1:1200) or larger
Medium scale, 1 in. = 100 to 1000 ft (1:1200 to 1:12,000)
Small scale, 1 in. = 1000 ft (1:12,000) or smaller

Maps in graphic form can have their scales enlarged or reduced photographically, by pantograph, or by opaque projector. Digital maps have virtually unlimited enlargement possibilities through the zoom capabilities of computers. Enlargements must be produced with caution, however, because any errors in the original maps are also magnified, and the enlarged product may not meet required accuracy standards.

17-3 MAPPING ORGANIZATIONS

Maps of various types are prepared by private surveying and engineering companies, industries, public utilities, cities, counties, states, and agencies of the federal government. Unfortunately, with so many organizations and agencies involved, in the past mapping activities were often not coordinated, and some duplication of effort occurred. Also, the existence of much valuable information was unknown and thus unavailable to many prospective users. Steps have been taken to improve this situation, however. At the federal level, the U.S. Geological Survey (USGS) now coordinates all mapping activities. Through their Earth Science Information Center (ESIC) they disseminate information nationwide on maps, digital cartographic products, and aerial photography available for any area.[1] Wisconsin and several other states have established state cartographers' offices. One of their functions is the dissemination of local information concerning maps and related products to surveyors, engineers, cartographers, and the general public.

17-4 NATIONAL MAPPING PROGRAM

The National Mapping Program has been established to provide maps and other cartographic products needed by citizens of the United States. It is the responsibility of the National Mapping Division of the USGS. The USGS began publishing topographic maps in 1886 as an aid to scientific studies. It now produces a variety of topographic maps at differing scales; however, its standard series has been at 1:24,000. In this series, individual sheets cover quadrangles of $7\frac{1}{2}'$ in both latitude and longitude. The entire United States (except Alaska) is covered at this scale, and over 57,000 maps are involved. On these maps, cultural features are shown in black, contours in brown, water features in blue, urban regions in red, and woodland areas in green. Topographic coverage of the United States is also available at scales of 1:50,000 (county maps), 1:62,500 (the older 15' quadrangles produced until about 1950), 1:100,000, and 1:250,000. Maps covering 15' quadrangles at a scale of 1:63,360 are standard for Alaska.

In addition to topographic maps, a variety of other special maps and related products are published as a part of the National Mapping Program. Maps are available showing geology, hydrology, land use, slope, and other categories of information. Orthophotomaps prepared from aerial photos (see Section 28-15) and satellite image maps are also produced. Index maps giving the status of topographic mapping in the United States and its territories, and information on how to order the maps and other products of the National Mapping Program, are available from the USGS.[2]

17-5 NATIONAL DIGITAL CARTOGRAPHIC DATABASE

As noted earlier, requirements for digital cartographic data are growing rapidly to support GISs. To meet these needs, the USGS is developing the National Digital Cartographic Database (NDCDB). This database, which is available to the public, contains two main types of digital data: (**1**) *digital line graphs* (DLGs) and (**2**) *digital elevation models* (DEMs). The database is being generated primarily by digitizing existing maps and other cartographic products.

Digital line graphs contain only the linear features or planimetry in an area. Included are political boundaries, hydrography, transportation networks, and the subdivision lines of the U.S. Public Land Survey System (see Chapter 23). The digital elevation

---

[1] Requests for information should be addressed to USGS-ESIC, 507 National Center, Reston, Va. 22092; telephone: (800)872-6277.
[2] Requests for information should be addressed to the U.S. Geological Survey, Map Distribution Center, Building 810, Box 25286, Denver, Colo. 80225; telephone: (303)236-7477.

models are arrays of elevation values, produced in a 30-m grid. The horizontal positions of points in the DEMs are X and Y coordinates referenced to the universal transverse Mercator coordinate system (see Chapter 21).

The NDCDB will provide the framework for referencing, overlaying, and merging other types of spatially related digital information in geographic information systems. USGS long-range plans contemplate the possibility of eventually filling all requests for maps with digital data rather than hard-copy maps. The data would be transmitted instantaneously. Users could view the maps on a screen, modify them as appropriate for their specific use, and then print their own hard copies as needed.

17-6 MANUAL AND COMPUTER-AIDED DRAFTING PROCEDURES

In general, all maps produced by surveyors and engineers fall into one of three categories: **(1)** topographic maps, which show the relief and natural and cultural features in an area (see Figures 16-2 and 17-13); **(2)** property and control maps resulting from boundary surveys or surveys of control networks (see Figures 17-14 and 22-6); and **(3)** construction maps and plans which provide the horizontal and vertical alignment information needed to guide construction work. Procedures for constructing these maps are described in the remaining sections of this chapter.

Maps may be drafted manually or produced with computer-aided drafting (CAD) systems. Manual procedures utilize standard drafting tools such as scales, protractors, compasses, triangles, and T squares. CAD systems employ computers programmed with special software and interfaced with electronic plotting devices (see Figure 17-9). With either approach, after deciding on scale and other factors that control overall map design, a manuscript is prepared. When completed, it is drafted or printed in hard-copy form.

In manual drafting, the manuscript is usually compiled in pencil. It should be prepared carefully to locate all features and contours as accurately as possible and be complete in every detail, including placement of symbols and letters. Lettering on the manuscript need not be done with extreme care, for its major purpose is to ensure good overall map design and proper placement. A well-prepared manuscript goes a long way toward achieving a good-quality final map.

The completed version of the manually compiled manuscript is drafted in ink, or *scribed*. If inked, the manuscript is placed on a light table, and features are traced on a stable-base transparent overlay material. Lettering is usually done first; then planimetric features and contours are traced.

Scribing is performed on sheets of transparent stable-base material coated with an opaque emulsion. Manuscript lines are transferred to the coating in a laboratory process. Lines representing features and contours are prepared by cutting and scraping to remove the coating. Special scribing tools are used to vary line weights and make standard symbols. Scribing is generally easier and faster than inking. Reproductions are made from the finished inked or scribed product.

In drafting maps with CAD, a soft-copy manuscript is compiled in the computer and displayed on its screen as work progresses. CAD software provides instructions to the computer which basically duplicate manual drafting functions. A file containing coordinates of points, as well as specific instructions on how to plot them, must be input in preparation for mapping. An operator interactively designs and compiles the map by entering commands into the computer's keyboard, or using a mouse to activate functions on a menu. Points, lines of various types, and a variety of symbols are available to the operator. Letters of differing sizes and styles can also be selected. When the manuscript is completely finished, the final map is drawn by simply activating the electronic plotter. (CAD systems are described in more detail in Section 17-17.)

Mapping with CAD systems has many advantages over manual methods, and therefore it is rapidly increasing in popularity. Nevertheless, many maps continue to be drawn manually. Furthermore, even if a computerized system is used, maps produced by it should be checked regularly for accuracy using manual techniques. For these reasons, and because CAD systems operate by basically duplicating manual procedures, it is still important to learn the basics of manual map drafting. These are covered in the sections that follow.

17-7 BASIC MANUAL PLOTTING PROCEDURES

Manual map drafting consists fundamentally of plotting individual points. Lines are then drawn from point to point to portray features. Though this process may seem simple in principle, accurate work requires skill, patience, and care.

Points may be plotted by coordinates, or by angles and distances. These procedures are discussed in the subsections that follow.

17-7.1 Plotting by Coordinates

To plot points by coordinates, the map sheet is first laid out precisely in a grid pattern with unit squares of appropriate size. Squares of 2, 4, or 5 in. are commonly used, and depending on map scale, they may represent 100, 200, 400, 500, or 1000 ft. The grids are constructed using a sharp, hard pencil, and are checked by carefully measuring diagonals. The grid lines are labeled with coordinate values, making sure that the range of coordinates covered on the map will accommodate the most extreme X and Y coordinates to be plotted.

Points are plotted by measuring their X and Y coordinates from the ruled grid lines. Mistakes in plotting can be detected by comparing scaled lengths (and directions) of lines with their values measured in the field or computed. Since each point is plotted independently, a mistake in one will not affect the others, and that point can simply be corrected. Plotting by this procedure is convenient for field measurements obtained by total stations or portable GPS units because those instruments provide coordinates directly. The coordinate method is employed by CAD systems.

Special instruments called *coordinatographs* (Figure 17-1) are also available for drawing the grid and plotting points by coordinates. Index marks set to coordinate values on perpendicular graduated X and Y rails permit rapid and accurate plotting to be performed. Some like that of Figure 17-1, are computer driven, which speeds the process and eliminates mistakes.

17-7.2 Plotting by Angles and Distances

Plotting a point by an angle and distance duplicates the radiation field procedure. An angle is laid off from a reference line to obtain the direction to the point. The required distance is then measured along the established direction to locate the point. Angles can be plotted by the *tangent* method, *chord* method, or with a *protractor*. Explanations of these procedures follow.

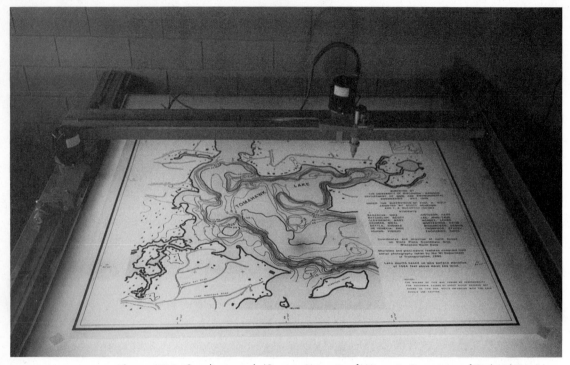

Figure 17-1 Coordinatograph. (Courtesy University of Wisconsin, Department of Civil and Environmental Engineering.)

To lay off an angle by the tangent method, a convenient distance is measured along a reference line to serve as a base. Thus in Figure 17-2, to plot a 12°14′ deflection angle at point A, length AB equal to 10 in. is first marked on the prolongation of the back line. A perpendicular with length equal to the distance AB times the natural tangent of 12°14′ (2.17 in.) is erected at B to locate point C. The line connecting A and C makes the desired angle with AB.

To lay off an angle by the chord method, as indicated in Figure 17-3, a convenient base length BD of, say, 10 units is first marked on side BA, locating point D. With the vertex B as the center and a radius of 10 units, an arc is swung. Then with D as the center and a radius equal to the chord for the desired angle, another arc is drawn. The intersection of the two arcs locates E. A line connecting points B and E forms the angle's other side. The chord for an angle is found by multiplying twice the distance BD by the sine of half the angle. Thus in Figure 17-3 the chord is $2 \times 10 \times \sin 16°14′ = 5.59$.

A protractor is a device made of paper, plastic, or metal, cut in a full circle or semicircle, with angle graduations along the circumference. A fine point identifies the circle center. The protractor is centered at the angle vertex with the zero line along one side and the proper angle point marked at its edge. Protractors are available in sizes having radii from 2 to 8 in. or more. A metal circle with a movable arm extending beyond the edge, as illustrated in Figure 17-4 is convenient. The arm A rotates about the center B and has a vernier C reading to minutes. Protractors do not generally yield accuracies equal to those of the tangent or chord methods.

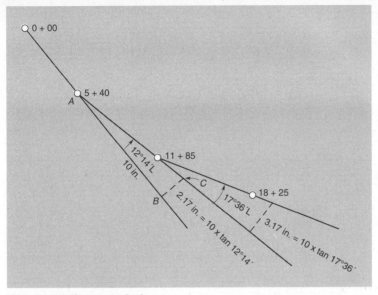

Figure 17-2 Tangent method.

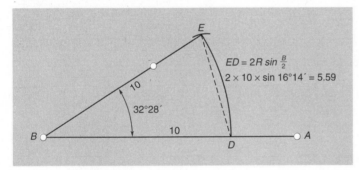

Figure 17-3 Chord method.

Figure 17-4 Protractor.

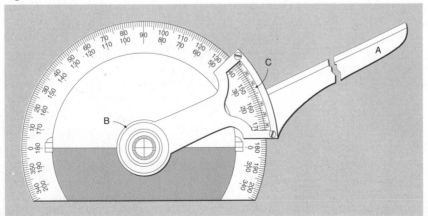

17-8 CONTOUR INTERVAL

As noted in Section 16-4, the choice of contour interval to be used on a topographic map depends on its intended use, required accuracy, type of terrain, and scale. If, according to National Map Accuracy Standards, elevations can be interpolated from a map to within one-half the contour interval, then for an accuracy of 1 ft, a 2-ft maximum interval is necessary. However, if only 10-ft accuracy is required, a 20-ft contour interval will suffice.

Terrain type and map scale combine to regulate the contour interval needed to produce a suitable density (spacing) of contours. Rugged terrain requires a larger contour interval than gently rolling country, and flat ground mandates a relatively small one to portray the surface adequately. Also if the map scale is reduced, the contour interval must be increased; otherwise lines are crowded, confuse the user, and possibly obscure other important details.

For average terrain, the following large and medium map scales and contour interval relationships generally provide suitable spacing:

ENGLISH SYSTEM		METRIC SYSTEM	
SCALE (ft/in)	CONTOUR INTERVAL (ft)	SCALE	CONTOUR INTERVAL (m)
50	1	1:500	0.5
100	2	1:1000	1
200	5	1:2000	2
500	10	1:5000	5
1000	20	1:10,000	10

17-9 PLOTTING CONTOURS

In manual drafting, points to be used in plotting contours are located using techniques described in Section 17-7. Contours found by the *direct method* are sketched through the points. Interpolation between plotted points is necessary for the *indirect method*.

Interpolation to find contour locations between points of known elevation can be done in several ways:

1. Estimating.
2. Scaling the distance between points of known elevation and locating the contour points by proportion.
3. Using special devices called *variable scales,* which contain a graduated spring. The spring may be stretched to make suitable marks fall on the known elevations.
4. Using a triangle and scale, as indicated in Figure 17-5. To interpolate for the 420-ft contour between point *A* at elevation 415.2 and point *B* at elevation 423.6, first set the 152 mark on any of the engineer's scales opposite *A*. Then, with one side of the triangle against the scale and the 90° corner at 236, the scale and triangle are pivoted together around *A* until the perpendicular edge of the triangle passes through point *B*. The triangle is then slid to the 200 mark and a dash drawn to intersect the line from *A* to *B*. This is the interpolated contour point *P.*

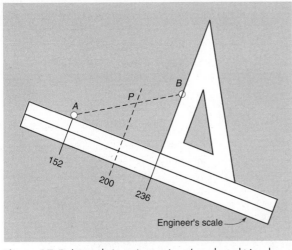

Figure 17-5 Interpolating using engineer's scale and triangle.

Contours are drawn only for elevations evenly divisible by the contour interval. Thus for a 20-ft interval, elevations of 800, 820, and 840 are shown, but 810, 830, and 850 are not. To improve legibility, every fifth line, evenly divisible by five times the contour interval, is made heavier. So for a 20-ft interval, the 800, 900, and 1000 lines would be heavier.

17-10 TOPOGRAPHIC SYMBOLS

Standard symbols are used to represent special topographic features, thereby making it possible to show many details on a single sheet. Figure 17-6 gives a few of the hundreds of symbols employed on topographic maps. Considerable practice is required to draw these symbols well at a suitable scale. They are, of course, routinely drawn by computer-driven automated plotting systems.

Before symbols are placed on a map, such things as buildings, roads, and boundary lines are plotted and inked. The symbols are then drawn, or cut from standard "stickup" sheets with an adhesive on the back and pasted on the map. A fully detailed map with coloring and shading is a work of art.

17-11 PLACING THE MAP ON A SHEET

To produce a map with good overall balance and ensure that all required information fits on the map, it is necessary to center the contents on the sheet. The example of Figure 17-7, a simple traverse and planimetric survey, will illustrate the procedure. Before any plotting is done, the proper scale for a sheet of given size must be selected. Assume in the example an 18- by 24-in. sheet with a 1-in. border on the left (for possible binding) and $\frac{1}{2}$-in. borders on the other three sides. A border line somewhat heavier than all other lines can be drawn to outline this area. If the most westerly station (A in the example) has been chosen as the origin of coordinates, then divide the total departure to the most easterly point C by the number of inches available for plotting in the east-west direction. The maximum scale possible in Figure 17-7 is

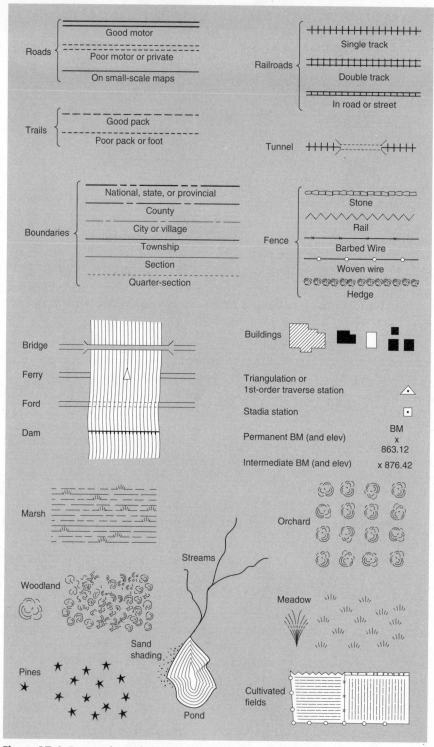

Figure 17-6 Topographic symbols.

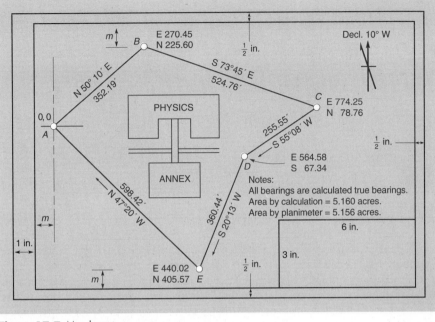

Figure 17-7 Map layout.

774.25 divided by 22.5, or 1 in. = 34 ft. The nearest standard scale that will fit is 1 in. = 40 ft.

This scale must be checked in the Y direction by dividing the total difference in Y coordinates, 225.60 + 405.57 = 631.17, by 40 ft, giving 15.8 in. required in the north-south direction. Since 17 in. are usable, a scale of 1 in. = 40 ft is satisfactory, although 1 in. = 50 ft provides a better border margin. If a scale of 1 in. = 40 ft is not suitable for the map's purpose, a sheet of different size should be selected, or perhaps more than one sheet must be employed to map the required area.

In Figure 17-7 the traverse is centered between the border lines in the Y direction by making each distance m equal to $\frac{1}{2}[17 - \frac{631.17}{40}]$, or 0.61 in. The same 0.61 in. can be used for the left side. "Weights" of the title, notes, and north arrow compensate for the traverse being to the left of the sheet center.

If topography is to be plotted outside the traverse, the maximum north-south and east-west distances to topographic features must be added to the traverse coordinates before computations are made for the scale and centering distances.

On some types of maps it is permissible to skew the meridian with respect to the borders of the sheet in order to better accommodate an area relatively long in the NE-SW or NW-SE direction, or to make street lines parallel with the borders. When a map is to be plotted by any means other than coordinates, it is advisable first to make a sketch showing controlling features. If this is done on tracing paper, orientation for the best fit and appearance is readily determined by rotating and shifting the tracing.

17-12 LETTERING

An important part of the contents of any map is the textual information. The title and all feature names, coordinate values, contour elevations, and other items must be

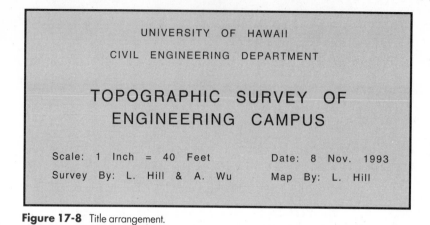

UNIVERSITY OF HAWAII

CIVIL ENGINEERING DEPARTMENT

TOPOGRAPHIC SURVEY OF ENGINEERING CAMPUS

Scale: 1 Inch = 40 Feet   Date: 8 Nov. 1993

Survey By: L. Hill & A. Wu   Map By: L. Hill

Figure 17-8 Title arrangement.

clearly identified. To produce a professional looking drawing, and one which clearly conveys the intended information, a suitable style of lettering must be selected. That style should be used consistently throughout the map, but the size varied in accordance with the importance of each particular item identified. Lettering that is too big or bold should not be used, but the letters must be large enough to be readable without difficulty.

Lettering should be carefully placed so that it is clearly associated with the item it identifies, and so that letters do not interfere with other features being portrayed. Normally best balance results if names are centered in the objects being identified. Also both appearance and clarity are generally improved by aligning letters parallel with linear objects that run obliquely, as has been done with the traverse lengths and bearings of Figure 17-7. For ease in map reading, letters should be placed so that the map is read from either the bottom or its right side.

Because of the importance of lettering to a map's overall appearance and utility, even when drafting is done manually the text is seldom hand-lettered. Rather, mechanical lettering devices that produce uniform sizes and styles of inked letters (the title of Figure 17-8 was produced with one), or special machines that print letters on adhesive-backed transparent tape are usually used. With the latter device, a variety of fonts and sizes is available. After the letters are printed, they are pasted onto the map, but can be lifted and moved later if necessary.

In map drafting with CAD systems, lettering is greatly simplified. A wide choice of fonts and sizes is available, and letters can be easily placed, aligned, rotated, and moved. To ensure a good-quality final product, however, the same rules stated above should be followed when using CAD.

17-13 MERIDIAN ARROW

Every map must display a meridian arrow for orientation purposes. It should preferably be near the top of the sheet, although it may be moved elsewhere for balance. An arrow should not be so large, elaborate, or heavily blacked in that it becomes the focal point of a sheet.

True, grid, or magnetic north (or all three) may be shown. A true-meridian arrow is identified by a full head and full feather; and a grid and/or magnetic arrow by a half-head and half-feather. The half-head and half-feather are put on the side away from the true north arrow to avoid touching it.

17-14 TITLE

The title may be placed wherever it will best balance the sheet, but is always kept outside the property lines on a boundary survey. Usually the title occupies the lower, right-hand corner, with any pertinent notes just above or to the left of it. Searching for a particular map in a bound set or loose pile of drawings is facilitated if all titles are in the same location. Since sheets are filed flat, bound on the left border, or hung from the top, the lower right-hand corner is the most convenient position.

The title should state the type of map, name of property or project and its owner or user, location or area, date completed, scale, contour interval, horizontal and vertical datums used, and name of the surveyor with his or her license number on property surveys. Additional data may be required on special-purpose maps. Lettering should be simple in style rather than ornate, and conform in size with the individual map sheet. Emphasis is placed on the most important parts of the title by increasing their height and using uppercase (capital) letters.

Perfect symmetry of outline about a vertical centerline is necessary, since the eye tends to exaggerate any defection. Also, an appearance of stability is obtained by having a full-width bottom line. An example title for the 18- by 24-in. sheet of Figure 17-7 is given in Figure 17-8.

No part of a map better portrays the artistic ability of the compiler than a neat, well-arranged title. Today many companies and government agencies use sheets with pre-printed title forms to be filled in with individual job data, or with CAD systems standard title blocks are stored, retrieved, and revised for each new project.

17-15 NOTES AND LEGENDS

Notes cover special features pertaining to the individual map, such as the following:

All bearings are true (or magnetic, or grid).
Coordinates are based on the NAD83.
Datum for elevations is the NAVD88.
Area by calculation is X acres (or hectares).

Notes must be in a prominent place where they are certain to be seen upon even a cursory examination of the map. The best location is just above or to the left of the title block in the lower right-hand corner of the map. A user then finds the desired map by title, and checks any special notes before examining the drawing.

Legends are used to explain the meaning of certain symbols used on maps, and to clarify the use of different line types. For example, triangles may represent traverse stations, squares could be houses, and crosses churches. Different line uses could be solid lines for paved roads, dashed lines for trails, and double lines for railroads.

17-16 DRAFTING MATERIALS

Polyester film and tracing and drawing papers are the materials commonly used for preparing maps in surveying and engineering offices. Polyesters such as Mylar are by far most frequently employed because they are dimensionally stable. They are also strong, durable, and waterproof, they take pencil, ink, and stickup items, and withstand erasing; so they are ideal for manual drafting. Tracing papers are available in a variety of grades, and good ones also are stable, take pencil, ink, and stickups, and endure some erasing. Both Mylar and tracing paper are transparent, so blueprints can be made from them.

Papers of different types and grades are used for printing maps made with CAD systems. When CAD is used, the paper quality can be relaxed somewhat because erasing, stickup lettering, and so on, will not be necessary. *If accurate measurements are to be extracted from the maps, however, then paper with good dimensional stability should be used.*

17-17 AUTOMATED MAPPING AND COMPUTER-AIDED DRAFTING SYSTEMS

Digital computers have had a profound impact in all areas of surveying, and mapping is certainly no exception. Automated mapping (AM), and computer-aided drafting (CAD) systems have now become commonplace in surveying and engineering offices throughout the world. Generic CAD systems developed for general drafting and engineering work are widely used for map drafting, but in addition, special AM systems have been designed specifically for surveying, mapping, and GIS work.

The hardware necessary for AM and CAD systems varies, but as a minimum it will include a computer with a hard drive, at least one disk drive, and a high-resolution monitor; an input device such as a digitizer or mouse; and a plotting device. Figure 17-9 shows a personal computer–based system interfaced with a drum plotter. The

Figure 17-9 IBM PC/AT interfaced with automated plotter. (Courtesy University of Wisconsin, Department of Civil and Environmental Engineering.)

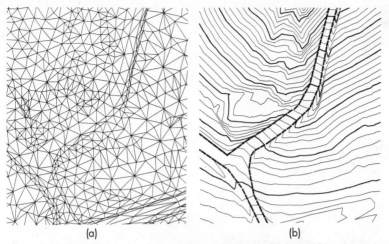

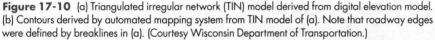

Figure 17-10 (a) Triangulated irregular network (TIN) model derived from digital elevation model. (b) Contours derived by automated mapping system from TIN model of (a). Note that roadway edges were defined by breaklines in (a). (Courtesy Wisconsin Department of Transportation.)

most important component of any CAD system is its software. This enables an operator to interact with the computer and activate the system's various functions.

CAD systems enable operators to design and draw manuscript maps in real time using the computer. A visual display of the manuscript can be examined on the monitor as it is being compiled, and any additions, deletions, or changes can be made as needed. Lines can be added, deleted, or their styles altered; placement of symbols and lettering modified; and lettering sizes and styles varied. Parts of the drawing may be "picked up" and moved to other areas to resemble a "cut and paste" operation that is handy for subdivision design or placement of frequently occurring symbols. A zoom feature allows more complicated or crowded parts of the manuscript to be magnified for a better view. In the end, the map can be checked for completeness and accuracy, and when the operator is satisfied that all requirements are met and the design is optimum, the automatic plotter can be actuated to draft the final product.

Required input to a computer for automated mapping includes a set of specific mapping instructions and a file of point locations and elevations. The instructions will include map scale, contour interval, line styles, lettering sizes and styles, symbols, and other items of information. Point locations are usually entered as a file of $X, Y,$ and Z coordinates, but angle and distance data can be entered and the coordinates computed. Special CAD systems for mapping with data collected by total station instruments and portable GPS units have been developed.

As explained in Section 16-7, most automated mapping systems draw contours after constructing a *triangulated irregular network* (TIN) model. These are networks of triangles which represent the individual facets of the terrain. The computer interpolates for contour crossings along the edges of the triangles and then draws the contours. A portion of a TIN model from an actual mapping project is illustrated in Figure 17-10(a), and the contours constructed from it are shown in Figure 17-10(b).

Contours compiled automatically should be carefully edited for correctness. In certain areas *breaklines* (see Section 16-7) may have to be added or changed to obtain the

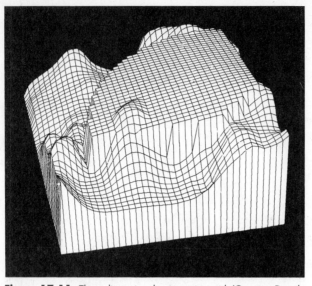

Figure 17-11 Three-dimensional perspective grid. (Courtesy Donohue and Associates.)

proper terrain representation. Incorrect interpolation often occurs along the outer edges of automatically contoured areas, so these areas require special checking. Also, because of this, it is good practice to carry the field survey somewhat beyond the area of interest and "trim" the edges of the map.

With terrain information stored in the computer in the form of TIN models, profiles and cross sections along selected lines can be derived automatically and plotted if desired. By including grade lines and design templates, earthwork computations can be made and stakeout information automatically derived for projects such as highways, railroads, and canals.

The *three-dimensional perspective grid* is an alternative form of terrain representation. It can also be produced by the computer from TIN models. The example shown in Figure 17-11 illustrates its important advantage—it gives a very vivid impression of relief.

Figure 17-12 shows a dual-screen CAD workstation. With this device a small-scale version of an entire map or diagram can be displayed on one screen and an enlarged portion of it on the other, where the operator performs the drafting or design. An operator controls the system by making entries on the keyboard, or by selecting instructions from a command board (*menu*) with a cursor.

Figure 17-13 is a portion of a topographic map for an engineering design project made using a CAD system, and Figure 17-14 is a subdivision plat, also designed and drawn using CAD. The condominium plat shown in Figure 22-6 is another example of a product designed and drafted with CAD.

There are numerous advantages derived from using CAD systems in map design and drafting. A major one is increased speed in completing projects. Others include reduction or elimination of errors, increased accuracy, and preparation of a consistently more uniform final product. With completed maps stored in digital form, copies can be quickly reproduced at any time and revisions easily made.

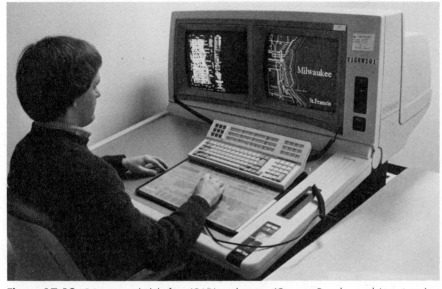

Figure 17-12 Computer-aided drafting (CAD) workstation. (Courtesy Donohue and Associates.)

Map data compiled using CAD systems can be stored in a data bank, with different numerical codes for each of the various kinds of features. They can be retrieved later for plotting in total, or in so-called layers, or parts, for special-purpose maps. A city engineer, for example, may only be interested in a map showing the roads and utilities, while the assessor may want only property boundaries. This concept of layered maps is fundamental to land and geographic information systems (see Chapter 29).

Another significant advantage of producing maps in digital form is that they can be transmitted electronically from one office to others at remote locations. As an example, the Wisconsin Department of Transportation produces digital maps photogrammetrically for roadway design at its central office in Madison. Using a telephone modem, these maps can then be transmitted instantaneously to any of nine district offices located throughout the state, where they are immediately available to engineers there for computer-aided design, or hard copies can be printed.

In spite of the many improvements made in automated mapping systems, there is still a possibility that mistakes can occur. For this reason it is good practice to have the field party chief, who is familiar with the area, review the completed maps.

The different AM and CAD systems all have varying individual capabilities. Books and brochures available from the manufacturers provide detailed descriptions.

17-18 IMPACTS OF MODERN LAND AND GEOGRAPHIC INFORMATION SYSTEMS ON MAPPING

Land Information Systems (LISs), Geographic Information Systems (GISs), and automated mapping and facilities management (AM/FM) systems all require enormous quantities of position-related land data. From this information, maps and other special-purpose graphic displays can be made and analyzed. A typical LIS or GIS, for example, may include attribute data such as political boundaries, land ownership, topography,

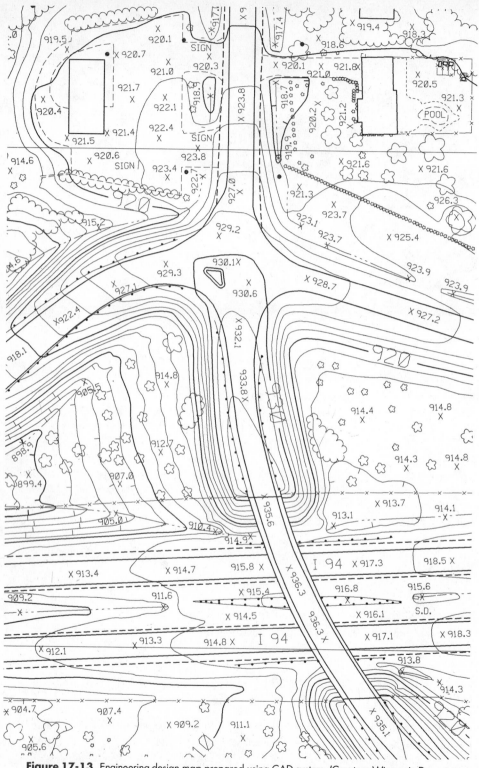

Figure 17-13 Engineering design map prepared using CAD system. (Courtesy Wisconsin Department of Transportation.)

RABBIT ESTATES
UNIT THREE
SECTION 29, T. 20 N., R. 4 E., E.M.
CITY OF BREWER, PENNSYLVANIA

THIS PLAT INCLUDES THE FOLLOWING:
1. LOT 3, P.C. SHORT PLAT 80-92.
2. A PARTIAL REPLAT OF LOT 2, P.C. SHORT PLAT 80.
3. A PARTIAL REPLAT OF LOTS 1-6 AND ALL OF LOT 8,
   BLOCK 23, PUYALLUP GARDENS SUPPLEMENT

NUMBER	RADIUS	ARC	DELTA	TANGENT	CHORD	CHORD-BEARING
1	800.00	283.96	020°20'14"	143.49	282.47	N11°40'07"E
2	25.00	39.27	090°00'00"	25.00	35.36	N46°30'00"E
3	25.00	39.27	090°00'00"	25.00	35.36	N43°30'00"W
4	440.00	56.63	007°22'26"	28.35	56.59	N05°11'13"E
5	25.00	34.22	078°25'48"	20.40	31.61	N48°05'20"E
6	60.00	259.20	247°31'14"	---	99.76	N36°27'23"W
7	500.00	159.55	018°17'00"	80.46	158.88	S10°38'30"W
8	25.00	39.27	090°00'00"	25.00	35.36	S46°30'00"W
9	25.00	39.27	090°00'00"	25.00	35.36	N43°30'00"W
10	760.00	284.64	021°27'32"	144.01	282.98	N12°13'46"E
11	25.00	39.27	090°00'00"	25.00	35.36	N46°30'00"E
12	25.00	39.27	090°00'00"	25.00	35.36	S43°30'00"E
13	630.00	287.96	026°11'19"	146.54	285.46	N14°35'40"E

Figure 17-14 Subdivision map compiled automatically by computer-driven plotter. (Courtesy Technical Advisors, Inc.)

land use, soil types, natural resources, transportation routes, utilities, and many others. From the stored information a user can display a map of each attribute category (or *layer*) on a screen, or several layers of data can be merged to produce combination maps. This merging, or *overlaying,* concept, discussed in Chapter 29, greatly facilitates data analysis and aids significantly in management and decision making. If printed maps of any selected layers or combinations are desired, they can be produced rapidly using automated drafting equipment.

The position-related land attribute data needed for LISs and GISs can be collected from a variety of sources and entered in the computer. These include field surveys (see Chapter 16), aerial photographs (see Chapter 28), and existing maps. Modern surveying instruments such as the total stations, GPS units, and digital photogrammetric plotters can produce huge quantities of digital terrain data in *X, Y,* and *Z* coordinate form rapidly and economically. New devices called *raster scanners* (see Section 29-7.2) are able to systematically scan existing maps and other printed documents line by line, and convert the information to numerical form. The processes of collecting and digitizing the data to support LISs and GISs are expected to place a heavy work load on surveyors for many years to come.

17-19 SOURCES OF ERROR IN MAPPING

Sources of error in mapping include:

1. Errors in the distances, angles, and coordinates used in plotting
2. Scaling errors in laying out lengths
3. Errors in plotting by protractor
4. Using a soft pencil, or one with a blunt point, for plotting
5. Variations in the dimensions of map sheets due to temperature and moisture

17-20 MISTAKES

Some common mistakes in mapping are:

1. Selecting an inappropriate scale or contour interval for the map
2. Failing to check grids by measuring diagonals, and not checking points plotted from coordinates by measuring distances between them
3. Using the wrong edge of an engineer's scale
4. Making the north arrow too large, complex, or black
5. Omitting the scale or necessary notes
6. Failing to balance the sheet by making a preliminary sketch
7. Drafting the map on a poor-quality medium
8. Failing to realize that errors are also magnified when maps are enlarged photographically, with CAD systems, or by any other means
9. Operating AM and CAD systems without sufficient prior training

PROBLEMS

17-1 Discuss the factors that must be considered when selecting the scale for a map.
17-2 What factors influence the choice of contour interval to use for a topographic map?
17-3 What are digital line graphs (DLGs) and digital elevation models (DEMs)?
17-4 Give the terms to which the acronyms CAD, LIS, and GIS apply.
*17-5 On a map sheet having a scale of 1:4800, what is the smallest distance that can be plotted with an engineer's scale? (Minimum scale graduations are $\frac{1}{60}$ in.)
17-6 Discuss important guidelines to follow regarding selection of map lettering sizes and their placement on the map.
17-7 What ratio scales are suitable to replace the following equivalent scales: 1 in. = 10 ft, 1 in. = 20 ft, 1 in. = 40 ft, 1 in. = 100 ft, 1 in. = 200 ft, and 1 in. = 500 ft?
*17-8 An area that varies in elevation from 1282 to 1405 ft is being mapped. What contour intervals will be drawn if a 10-ft interval is used? Which lines are drawn heavier?
17-9 Similar to Problem 17-8, except elevations vary from 661 to 702 ft and a 5-ft interval is used.
*17-10 For mapping average terrain, what contour interval would you recommend for map scales of 1 in. = 100 ft, 1:500, and 1:12,000?
17-11 List four convenient classifications of items that can be presented on general-purpose maps.

*Asterisks indicate problems that have answers given in Appendix G.

17-12 Draw the conventional symbols for national, state, and county lines.

17-13 Draw 1-ft contours for the data in Plate D-4 of Appendix D.

17-14 Sketch 5-ft contours for the notes in Plate D-7 of Appendix D.

17-15 Plot the topography covered in Plate D-8 of Appendix D. ~~~~~~~~~ 2' intervals

17-16 If elevations on a map must be interpolated to the nearest ±5 ft, what contour interval is necessary according to the National Map Accuracy Standards?

*17-17 If an area having an average slope of 8% is mapped using a scale of 1:1000 and contour interval of 2-ft, how far apart will contours be on the map?

17-18 Similar to Problem 17-17, except average slope is 5°, map scale is 500 ft/in., and contour interval is 10 ft.

17-19 List items that should be included within the title of a map.

17-20 Why are titles usually placed in the lower right-hand corner of most map sheets, and the meridian arrow on the top or right side?

17-21 What are the advantages of using stickups on maps?

17-22 List the essentials of a good meridian arrow on a survey map.

17-23 Discuss the advantages of computerized mapping systems.

17-24 Is there any difference between a map and an aerial photograph? Explain.

17-25 What is a coordinatograph? A raster scanner?

17-26 Discuss the dangers associated with enlarging maps photographically or by other means.

17-27 Discuss the impacts of modern Land Information Systems upon mapping.

*Asterisks indicate problems that have answers given in Appendix G.

BIBLIOGRAPHY

ASPRS. 1987. *Large Scale Mapping Guidelines.* Bethesda, Md.: American Society for Photogrammetry and Remote Sensing.

Bauer, K. W. 1987. "Mapping for Municipal Planning and Engineering—Back to Basics." *ASCE, Journal of Surveying Engineering* 113 (No. 1):41.

Brand, R. 1990. "Digitized Maps—Beware of Their Misleading Coordinate Values." *Point of Beginning* 16 (No. 2):52.

Coleman, D. J., and J. D. McLaughlin. 1988. "Shift to Digital Mapping: Points of Impact on Management." *ASCE, Journal of Surveying Engineering* 114 (No. 2):59.

Dix, W. 1981. "Mapmakers Shift to 'Digitization.' " *Bulletin, American Congress on Surveying and Mapping* (No. 72):43.

Harvey, P. 1980. *The History of Topographical Maps.* Rancho Cordova, Calif.: Landmark.

Hebrank, A. J. 1981. "Overlay Drafting for Surveyors." *Surveying and Mapping* 41 (No. 2):151.

Iler, W. H. 1981. "The Use of Computer Mapping in Planning and Public Works Management: The Case for Chicago." *The American Cartographer* (No. 8):115.

Jacober, R. P., Jr. 1980. "Map Content and Symbols," Chapter V of "Proposed Manual on Map Uses, Scales and Accuracies for Engineering and Associated Purposes." *ASCE, Journal of the Surveying and Mapping Division* 106 (No. SU1):41.

Jocober, R. P., Jr. 1981. "Standard for Symbology of Engineering Scale Maps." *ASCE, Journal of the Surveying and Mapping Division* 107 (No. SU1):21.

Kerski, J. J. 1993. "USGS Generates Digital Maps for Thematic Cartographic Projects," *Geo Info Systems,* 3 (No. 1):36.

Loxton, J. 1980. *Practical Map Production.* Rancho Cordova, Calif.: Landmark.

Robinson, A. H., et al. 1984. *Elements of Cartography,* 5th ed. New York: John Wiley and Sons.

Schaefer, J. E. 1984. "Computerized Municipal Utility System Mapping." *ASCE, Journal of Surveying Engineering* 110 (No. 1):8.

Thompson, M. M. 1976. *Maps for America.* Rancho Cordova, Calif.: Landmark.

Thorpe, J. A. 1990. "Digital Mapping for Utility Companies." *ASCE, Journal of Surveying Engineering* 116 (No. 1):13.

Wattles, G. 1981. *Surveying Drafting.* Rancho Cordova, Calif.: Landmark.

Zarzycki, J. M. 1984. "Standards for Digital Topographic Data—The Canadian Experience." *Surveying and Mapping* 44 (No. 1):29.

18

ASTRONOMICAL OBSERVATIONS

18-1 INTRODUCTION

Astronomical observations in surveying consist of measuring positions of the sun or certain stars. The principal purpose of this measurement in plane surveying is to determine the direction of the true meridian (astronomic north). True bearings and true azimuths can then be calculated using this meridian. These are needed to establish directions of new property lines so parcels can be adequately described; to retrace old property boundaries whose descriptions include bearings; to specify directions of tangents on route surveys; and for many other purposes. Other important but less frequently performed astronomical observations are made to determine latitudes and longitudes of points.

To expand on the definition of the true meridian given in Section 8-4, at any point it is a line tangent to, and in the plane of, the great circle which passes through the point and the earth's north and south geographic poles. This is illustrated in Figure 18-1, where P and P' are the poles located on the earth's axis of rotation. Arc PAP' is the great circle through A, and line AN the true meridian (tangent to the great circle at A in plane $POP'A$). With a true meridian established, the true azimuth α of any line, such as AB of Figure 18-1, can readily be obtained by determining horizontal angle NAB.

Figure 18-1 also illustrates latitude and longitude. These two terms give the unique location of any point on the earth. As illustrated in the figure, the latitude of any point A is ϕ_A, measured in the meridian plane containing A, between the equatorial plane and the ellipsoid normal at A. (The ellipsoid is defined in Section 19-1.) The longitude of point A is angle λ_A, measured in the equatorial plane between the Greenwich meridian (assumed in the figure to be arc $QPQ'P'$) and the meridian through A.

Astronomical observations are not necessarily required on every project where true bearings or true azimuths are needed. If a pair of intervisible control monuments from a previous survey exist in the area, and the true bearing or azimuth is known for that line, new directions can be referenced to it. Also, the new Global Positioning Systems (GPS), as described in Chapter 20, could be used to establish the positions of the two

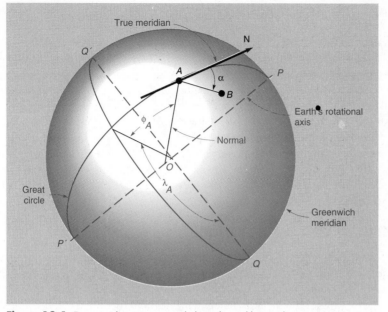

Figure 18-1 True meridian, true azimuth, latitude, and longitude.

Figure 18-2 TM20C theodolite and GP1 gyro attachment. (Courtesy Sokkia Corporation.)

end points of a project line. This would enable the line's direction to be calculated and all other project lines referenced to it.

North-seeking gyroscopes, or "gyros," are another alternative that can automatically and quickly ascertain the direction of true north without observing the sun or a star. This is achieved when the axis of the spinning gyro has aligned itself with the earth's axis of rotation. The theodolite's line of sight, to which the gyro is attached (Figure 18-2), can then be turned from north as determined by the gyro to measure a line's azimuth. The process takes only a few minutes and can yield accuracies of $\pm 20''$ or better.

When the gyro unit is not in use, it can be removed from the theodolite, as it is in Figure 18-2, so the theodolite can be used for its normal applications. Gyros can now also be obtained interfaced with total station instruments. With this combination, once the gyro has found true north, it is automatically transferred to the total station's horizontal circle so that true azimuths are displayed for any direction the telescope is pointed. Gyros are especially useful on projects where astronomical observations cannot be made—for example, in mine and tunnel surveys.

Existing control monuments are often not available for reference, and GPS equipment or north-seeking gyros are relatively expensive. Thus astronomical observations that can be made with standard surveying equipment are extremely important. Field and office procedures involved in this process are the subjects of this chapter.

18-2 SIMPLE METHODS OF DETERMINING THE MERIDIAN

To introduce the subject of meridian determination, two simple methods requiring no computation are described. Both are based on the knowledge that the sun appears to move in essentially a circular path through the sky on any given day, and achieves its highest altitude as it crosses the local meridian. Thus if the direction to the sun at its maximum altitude can be defined, the true meridian is located.

18-2.1 Shadow Method

The true meridian can be established by the shadow method using only a straight rod or pole and a piece of string. In Figure 18-3 points A, B, C, D, E, and F mark the end of the shadow of a plumbed staff or a telephone pole at intervals of perhaps 30 min throughout the period from 9 A.M. to 3 P.M. A smooth curve is sketched through the marks. With the staff or pole as a center and any appropriate radius, a circular arc is swung to obtain two intersections, x and y, with the shadow curve. A line from the staff through m, the midpoint of xy, approximates the meridian. If the ground is level, the pole plumb, and the shadow points are carefully marked, the angle between the line established and the true meridian can be obtained to an accuracy of within about $\pm 30'$.

Figure 18-3 Determination of azimuth by shadow method.

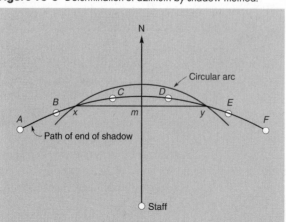

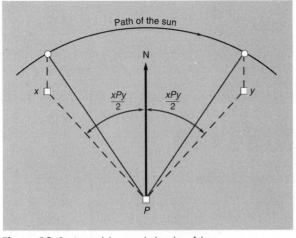

Figure 18-4 Azimuth by equal altitudes of the sun.

18-2.2 Equal Altitudes of the Sun

Determination of the meridian by equal altitudes of the sun requires a total station, theodolite, or transit, but the method is similar in principle to the shadow method. Assume that a meridian is to be passed through point P in Figure 18-4, over which the instrument is set up. At some time between 8 and 10 A.M., say about 9 A.M., with a dark glass over the eyepiece or objective lens, the sun's disk is bisected by both the horizontal and vertical cross hairs. The vertical (or zenith) angle is read, the telescope depressed, and point x set at least 500 ft from the instrument. Shortly before 3 P.M., with the vertical (or zenith) angle previously read placed on the circle, the sun is followed until the vertical and horizontal cross hairs again simultaneously bisect the sun. The telescope is depressed and a point y set at approximately the same distance from P as x. The bisector of angle xPy is the true meridian.

Perfect results cannot be attained with either of these simplified methods, because the sun's changing declination (see Section 18-5 for the definition of declination) gives it a path slightly skewed to the equator instead of parallel with it. Other disadvantages are the time and delay required, and the possibility of clouds obscuring the sights. Because of their drawbacks, these procedures are seldom used by surveyors, but they do illustrate uncomplicated methods of meridian determination.

18-3 OVERVIEW OF USUAL PROCEDURES FOR ASTRONOMICAL AZIMUTH DETERMINATION

The general field procedures employed by surveyors to define the direction of true north consist of the following steps: **(1)** a total station, theodolite, or transit is set up and leveled at one end of the line whose azimuth is to be determined, like point A of Figure 18-5; **(2)** the horizontal, and sometimes vertical, circles are read when pointing on a celestial body S; **(3)** the precise time of pointing is recorded; and **(4)** a horizontal angle is measured from the celestial body to a point on the other end of the line, like angle θ of Figure 18-5 from S to B. Office work involves **(1)** obtaining the precise location of the body in the heavens at the instant sighted from an *ephemeris* (almanac of celestial body positions); **(2)** computing the celestial body's azimuth (angle Z in

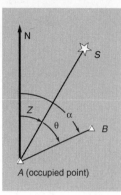

Figure 18-5 Azimuth determination from observation on a celestial body.

Figure 18-5) based on observed and ephemeris data; and **(3)** calculating the line's azimuth by applying the measured horizontal angle to the computed azimuth of the body ($\alpha = Z + \theta$ in Figure 18-5).

Any visible celestial body for which ephemeris data are available can be employed in the procedures outlined. However, the sun and, in the United States, Polaris (north star) are almost always selected. The sun permits observations to be made in lighted conditions during normal daytime working hours; Polaris is preferred for higher order accuracy.

Accuracies attainable in astronomical azimuths depend on many variables, including **(1)** the precision of the instrument used, **(2)** ability and experience of the observer, **(3)** weather conditions, **(4)** quality of the clock or chronometer used to measure the time of sighting, **(5)** celestial body sighted and its position when observed, and **(6)** accuracy of ephemeris and other data available. Polaris observations provide the most accurate results and, with several repetitions of measurements utilizing first-order instruments, accuracies to within $\pm 1''$ are possible. Sun observations yield a lower order of accuracy, but values accurate to within about $\pm 10''$ or better can be obtained if careful repeated measurements are made.

18-4 EPHEMERIDES

As noted previously, ephemerides are books containing tabular data on the positions of the sun and stars versus time. A variety of them are published annually and are available to surveyors. Some of the most useful ones are: The *Sokkia Celestial Observation Handbook and Ephemeris,* published by Elgin, Knowles, & Senne, Inc., P. O. Box 3371, Fayetteville, Ark. 72702;[1] *The Apparent Place of Polaris and Apparent Sidereal Time,* available from the U.S. Department of Commerce, National Geodetic Information Center, 1315 East-West Highway, Silver Spring, Md. 20910; *The Nautical Almanac,* published by the U.S. Naval Observatory and available from the U.S. Government Printing Office, Washington, D.C. 20402; and *Apparent Places of Fundamental Stars,* published by Astronomisches Rechen-Institut, Heidelberg, Germany.

---

[1]The Sokkia ephemeris is published annually and is available from the Lietz Company, 9111 Barton, Box 2934, Overland Park, Kans. 66021; telephone: (800)255-3913, Ext. 913.

In addition to published ephemerides, computer programs are also available which solve for the positions of celestial bodies. Their major advantages are that they provide accurate results without table lookup and can be used year after year. However, they must occasionally be updated.

The 1993 *Sokkia Celestial Observation Handbook and Ephemeris,* which contains data for the sun, Polaris, and eight other bright stars, is used for reference in examples given in this chapter. Table 18-1, which contains data for the month of June 1993, is an excerpt from that publication. It is recommended that students obtain a current-year copy of this ephemeris.

Values tabulated in ephemerides are given for universal time (Greenwich civil time), so before extracting data from them, standard or daylight times normally recorded for observations must be converted. This topic is discussed further in Section 18-7.

18-5 DEFINITIONS

In making and computing astronomical observations, the sun and stars are assumed to lie on the surface of a *celestial sphere* of infinite radius having the earth as its center. Because of the earth's rotation on its axis, all stars appear to move around centers that are on the extended earth's rotational axis, which is also the axis of the celestial sphere. Figure 18-6 is a sketch of the celestial sphere and illustrates some terms used in field astronomy. Here *O* represents the earth and *S* a heavenly body, as the sun or a star whose apparent direction of movement is indicated by an arrow. Students may find it helpful to sketch the various features on a basketball or globe. Definitions of terms pertinent to the study of field astronomy follow.

The *zenith* is located where a plumb line projected upward meets the celestial sphere. On a diagram it is usually designated by *Z*. Stated differently, it is the point on the celestial sphere vertically above the observer.

The *nadir* is the point on the celestial sphere vertically beneath the observer and exactly opposite the zenith. In Figure 18-6 it is at *N*.

The *north celestial pole* is point *P* where the earth's rotational axis, extended from the north geographic pole, intersects the celestial sphere.

The *south celestial pole* is point *P'* where the earth's rotational axis, extended from the south geographic pole, intersects the celestial sphere.

A *great circle* is any circle on the celestial sphere whose plane passes through the center of the sphere.

A *vertical circle* is any great circle of the celestial sphere passing through the zenith and nadir, and represents the line of intersection of a vertical plane with the celestial sphere. In Figure 18-6, *ZSS'N* is a vertical circle.

The *celestial equator* is the great circle on the celestial sphere whose plane is perpendicular to the axis of rotation of the earth. It corresponds to the earth's equator enlarged in diameter. Half of the celestial equator is represented by *Q'EQ* in Figure 18-6.

An *hour circle* is any great circle on the celestial sphere which passes through the north and south celestial poles. Therefore hour circles are perpendicular to the plane of the celestial equator. They correspond to meridians (longitudinal lines) and are used to measure hour angles. In Figure 18-6, *PSS"P'* is an hour circle.

The *horizon* is a great circle on the celestial sphere whose plane is perpendicular to the direction of the plumb line. In surveying, the plane of the horizon is determined by a level vial. Half of the horizon is represented by *H'EH* in Figure 18-6.

TABLE 18-1 APPARENT PLACES OF THE SUN AND POLARIS FOR 0 HOUR UT, JUNE 1993

DAY	GHA (SUN)	DECLINATION	EQ. OF TIME APPT-MEAN (M S)	SEMI-DIAM.	GHA (POLARIS)	DECLINATION	GREENWICH TRANSIT (H M S)
1 TU	180 34 17.2	22 01 29.3	02 17.14	15 47.9	213 31 13.5	89 13 57.00	9 44 19.
2 W	180 32 00.1	22 09 31.3	02 08.01	15 47.7	214 30 02.0	89 13 56.75	9 40 25.
3 TH	180 29 37.5	22 17 10.2	01 58.50	15 47.6	215 28 49.2	89 13 56.52	9 36 30.
4 F	180 27 09.4	22 24 25.7	01 48.63	15 47.5	216 27 36.1	89 13 56.31	9 32 36.
5 SA	180 24 36.2	22 31 17.7	01 38.41	15 47.3	217 26 23.4	89 13 56.12	9 28 41.
6 SU	180 21 57.9	22 37 46.2	01 27.86	15 47.2	218 25 11.7	89 13 55.96	9 24 46.
7 M	180 19 14.9	22 43 50.8	01 17.00	15 47.1	219 24 01.2	89 13 55.80	9 20 52.
8 TU	180 16 27.4	22 49 31.6	01 05.83	15 47.0	220 22 51.8	89 13 55.64	9 16 57.
9 W	180 13 35.6	22 54 48.2	00 54.37	15 46.9	221 21 43.3	89 13 55.47	9 13 02.
10 TH	180 10 39.7	22 59 40.8	00 42.65	15 46.7	222 20 35.0	89 13 55.30	9 09 07.
11 F	180 07 40.2	23 04 09.0	00 30.68	15 46.6	223 19 26.4	89 13 55.11	9 05 13.
12 SA	180 04 37.2	23 08 12.8	00 18.48	15 46.5	224 18 17.2	89 13 54.91	9 01 18.
13 SU	180 01 31.2	23 11 52.2	00 06.08	15 46.4	225 17 06.9	89 13 54.70	8 57 23.
14 M	179 58 22.4	23 15 07.0	-00 06.50	15 46.3	226 15 55.4	89 13 54.49	8 53 29.
15 TU	179 55 11.2	23 17 57.2	-00 19.25	15 46.2	227 14 42.4	89 13 54.28	8 49 34.
16 W	179 51 58.0	23 20 22.7	-00 32.13	15 46.1	228 13 28.1	89 13 54.09	8 45 40.
17 TH	179 48 43.1	23 22 23.4	-00 45.12	15 46.1	229 12 12.6	89 13 53.91	8 41 45.
18 F	179 45 27.0	23 23 59.4	-00 58.20	15 46.0	230 10 56.4	89 13 53.75	8 37 51.
19 SA	179 42 10.0	23 25 10.6	-01 11.34	15 45.9	231 09 40.0	89 13 53.61	8 33 57.
20 SU	179 38 52.5	23 25 57.0	-01 24.50	15 45.8	232 08 24.0	89 13 53.50	8 30 03.
21 M	179 35 35.0	23 26 18.6	-01 37.67	15 45.8	233 07 09.0	89 13 53.41	8 26 08.
22 TU	179 32 17.9	23 26 15.4	-01 50.81	15 45.7	234 05 55.3	89 13 53.32	8 22 14.
23 W	179 29 01.6	23 25 47.5	-02 03.89	15 45.7	235 04 42.7	89 13 53.23	8 18 19.
24 TH	179 25 46.6	23 24 54.8	-02 16.90	15 45.6	236 03 30.9	89 13 53.14	8 14 25.
25 F	179 22 33.2	23 23 37.3	-02 29.79	15 45.6	237 02 19.1	89 13 53.02	8 10 30.
26 SA	179 19 21.9	23 21 55.2	-02 42.54	15 45.5	238 01 06.5	89 13 52.89	8 06 36.
27 SU	179 16 13.1	23 19 48.4	-02 55.13	15 45.5	238 59 52.4	89 13 52.75	8 02 41.
28 M	179 13 07.0	23 17 17.0	-03 07.53	15 45.5	239 58 36.6	89 13 52.60	7 58 47.
29 TU	179 10 04.1	23 14 21.2	-03 19.73	15 45.5	240 57 19.0	89 13 52.47	7 54 53.
30 W	179 07 04.5	23 11 00.9	-03 31.70	15 45.5	241 56 00.2	89 13 52.36	7 50 59.

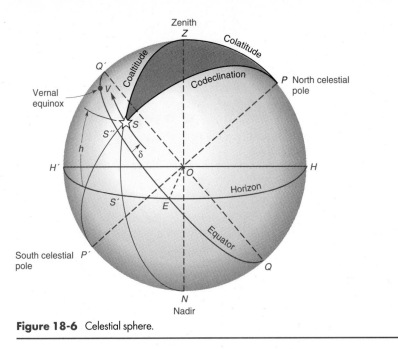

Figure 18-6 Celestial sphere.

A *celestial meridian,* interchangeably called *local meridian,* is that unique hour circle containing the observer's zenith. It is both an hour circle and a vertical circle. The intersection of the celestial meridian plane with the horizon plane is line $H'OH$ in Figure 18-6, which defines the direction of true north. Thus it is the astronomic meridian line used in plane surveying. Since east is 90° clockwise from true north, line OE in the horizon plane is a true east line. The celestial meridian is composed of two branches; the *upper branch* contains the zenith and is the semicircle $PZQ'H'P'$ in Figure 18-6, and the *lower branch* includes the nadir and is arc $PHQNP'$.

A *diurnal circle* is the complete path of travel of the sun or a star in its apparent daily orbit about the earth. Four terms describe specific positions of heavenly bodies in their diurnal circles: **(1)** *lower culmination*—the body's position when it is exactly on the lower branch of the celestial meridian; **(2)** *eastern elongation*—when the body is farthest east of the celestial meridian with its hour circle and vertical circle perpendicular; **(3)** *upper culmination*—when it is on the upper branch of the celestial meridian; and **(4)** *western elongation*—when the body is farthest west of the celestial meridian with its hour circle and vertical circle perpendicular.

An *hour angle* exists between a meridian of reference and the hour circle passing through a celestial body. It is measured by the angle at the pole between the meridian and hour circle, or by the arc of the equator intercepted by those circles. Hour angles are measured westward (in the direction of apparent travel of the sun or star) from the upper branch of the meridian of reference.

The *Greenwich hour angle* of a heavenly body at any instant of time is the angle, measured westward, from the upper branch of the meridian of Greenwich to the meridian over which the body is located at that moment.[2] In the ephemeris it is designated by

---

[2]The meridian of Greenwich, England, is internationally accepted as the reference meridian for specifying longitudes of points on earth and for giving positions of celestial bodies.

GHA. *Local hour angle* (LHA) is similar to GHA, except it is measured from the upper branch of the observer's celestial meridian.

A *meridian angle* is like a local hour angle, except it is measured either eastward or westward from the observer's meridian, and thus its value is always between 0° and 180°.

The *declination* of a heavenly body is the angular distance (measured along the hour circle) between the body and the equator; it is plus when the body is north of the equator, and minus when south of it. Declination is usually denoted by δ in formulas, and represented by arc $S''S$ in Figure 18-6.

The *polar distance* or *codeclination* of a body is equal to 90° minus the declination. In Figure 18-6 it is arc PS.

The *position* of a heavenly body with respect to the earth at any moment may be given by its Greenwich hour angle and declination.

The *altitude* of a heavenly body is its angular distance measured along a vertical circle above the horizon, $S'S$ in Figure 18-6. It is generally obtained by measuring a vertical angle with a total station, theodolite, or transit, and correcting for refraction, and parallax if the sun is observed (see Section 18-6). Altitude is usually denoted in formulas by h.

The *coaltitude* or *zenith distance* is arc ZS in Figure 18-6 and equals 90° minus the altitude.

The *astronomical* or *PZS triangle* (darkened in the figure) is the spherical triangle whose vertices are the pole P, zenith Z, and astronomical body S. Because of the body's movement through its diurnal circle, the three angles in this triangle are constantly changing.

The *azimuth* of a heavenly body is the angle measured in the horizon plane, clockwise from either the north or south point, to the vertical circle through the body. An azimuth from north is represented by arc HS' in Figure 18-6, and it equals the Z angle of the PZS triangle.

The *latitude* of an observer is the angular distance, measured along the meridian, from the equator to the observer's position. In Figure 18-6 it is arc $Q'Z$. It is also the angular distance between the polar axis and horizon, or arc HP. Depending on the observer's position, latitude is measured north or south of the equator. Formulas in this book denote it as ϕ. Colatitude is $(90° - \phi)$.

The *vernal equinox* is the intersection point of the celestial equator and the hour circle through the sun at the instant it reaches zero declination (about March 21 each year). For any calendar year it is a fixed point on the celestial sphere (the astronomer's zero-zero point of coordinates in the sky) and moves with the celestial sphere just as the stars do. In Figure 18-6 it is designated by V.

The *right ascension* of a heavenly body is the angular distance measured eastward from the hour circle through the vernal equinox to the hour circle of a celestial body. It is arc VS'' in Figure 18-6. Right ascension frequently replaces Greenwich hour angle as a means of specifying the position of a star with respect to the earth. In this system, however, the Greenwich hour angle of the vernal equinox must also be given.

18-6 REFRACTION AND PARALLAX

As illustrated in Figure 18-13, *refraction* is the angular increase in the apparent (observed) altitude of a heavenly body due to bending of light rays that pass obliquely through the earth's atmosphere. It varies from zero for an altitude of 90° to a maximum of about 35′ at the horizon, and for intermediate positions its value, in minutes, is roughly equal to the natural cotangent of the observed altitude. Its precise value depends on atmospheric pressure and temperature; these variables are therefore measured and recorded if altitude angles are observed. Refraction makes observations on heavenly bodies near the horizon less reliable than those taken at higher altitudes. Some ephemerides list refraction corrections for varying altitudes, temperatures, and pressures. The correction can also be computed using the following formula:

$$C_r = \frac{16.38\, b}{(460 + F)\tan V} \tag{18-1}$$

where C_r is the refraction correction, in minutes; b the barometric pressure, in inches of mercury; F the temperature, in degrees Fahrenheit; and V the observed altitude of the body. The correction is always subtracted from observed altitudes.

Parallax, also illustrated in Figure 18-13, results from observations being made from the earth's surface instead of its center. It causes a small angular decrease in apparent altitude; hence the correction is always added. Parallax is insignificant when a star is observed but must be included for sun altitudes. Some ephemerides also publish parallax correction tables for the sun. Alternatively, they may be computed using the following formula:

$$C_p = 8.79 \cos V \tag{18-2}$$

where C_p is the parallax correction for the sun in seconds, and V the sun's observed altitude. Sections 18-18 and 18-20 give example computations that illustrate how to make both refraction and parallax corrections.

18-7 TIME

Four kinds of time may be used in making and computing an astronomical observation.

1. *Sidereal time.* A sidereal day is the interval of time between two successive upper culminations of the vernal equinox over the same meridian. Sidereal time is star time. At any location for any instant, it is equal to the local hour angle of the vernal equinox.

2. *Apparent solar time.* An apparent solar day is the interval of time between two successive lower culminations of the sun. Apparent solar time is sun time, and the length of a day varies somewhat because the rate of travel of the sun is not constant. Since the earth revolves about the sun once a year, there is one fewer day of solar time in a year than sidereal time. Thus the length of a sidereal day is shorter than a solar day by approximately 3 min 56 sec. The relationship between sidereal and solar time is illustrated in Figure 18-7. (*Note:* The earth's orbit is actually elliptical, but for simplicity it is shown circular in the figure.)

3. *Mean solar, or civil, time.* This time is related to a fictitious sun, called the "mean" sun, which is assumed to move at a uniform rate. It is the basis for watch time and the 24-h day.

The *equation of time* is the difference between "apparent" solar and "mean" solar time. Its value changes continually as the apparent sun gets ahead of, and then falls

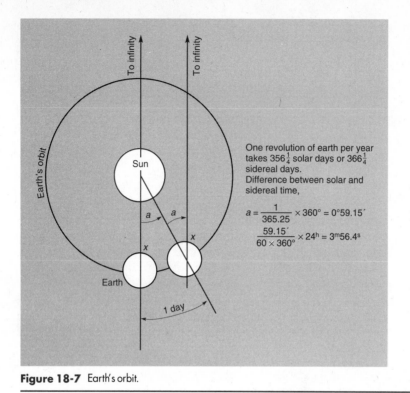

One revolution of earth per year takes $356\frac{1}{4}$ solar days or $366\frac{1}{4}$ sidereal days.
Difference between solar and sidereal time,

$$a = \frac{1}{365.25} \times 360° = 0°59.15'$$

$$\frac{59.15'}{60 \times 360°} \times 24^h = 3^m56.4^s$$

Figure 18-7 Earth's orbit.

behind, the mean sun. Values for each day of the year are given in an ephemeris (see Table 18-1). Local apparent time is obtained by adding the equation of time to local civil time. If the apparent sun is ahead of the mean sun the equation is plus, if behind, it is minus.

4. *Standard time.* This is the mean time at meridians 15° or 1 h apart, measured eastward and westward from Greenwich. Eastern standard time (EST) at the 75th meridian differs from universal time (UT), or Greenwich civil time (GCT), by 5 h (earlier, since the sun has not yet traveled from the meridian of Greenwich to the United States). Standard time was adopted in the United States in 1883, replacing some 100 local times used previously. *Daylight saving time* (DST) in any zone is equal to the standard time in the zone to the *east;* thus central daylight time is equivalent to eastern standard time.[3]

As previously noted, sun and star positions tabulated in ephemerides are given in UT. Observation times, on the other hand, may be recorded in the standard or daylight times of an observer's location and must therefore be converted to UT. Conversion is based on the longitude of the standard meridian for the time zone. Table 18-2 lists the different time zones in the United States, the longitudes of their standard meridians, and the number of hours to be added for converting standard and daylight time to UT.

---

[3]Daylight saving time officially begins at 2:00 A.M. on the first Sunday of April and ends at 2:00 A.M. on the last Sunday of October each year.

TABLE 18-2 LONGITUDES OF STANDARD MERIDIANS IN THE UNITED STATES AND TIME DIFFERENCES FROM GREENWICH

STANDARD TIME ZONE (AND ABBREVIATION)	LONGITUDE OF STANDARD MERIDIAN	CORRECTION IN HOURS, TO ADD TO OBTAIN UT	
		STANDARD TIME	DAYLIGHT TIME
Atlantic (AST)	60°	4	3
Eastern (EST)	75°	5	4
Central (CST)	90°	6	5
Mountain (MST)	105°	7	6
Pacific (PST)	120°	8	7
Yukon (YST)	135°	9	8
Alaska/Hawaii (AHST)	150°	10	9
Bering Sea (BST)	165°	11	10

In making civil time conversions based on longitude differences, the following relationships are helpful:

$$360° \text{ of longitude} = 24 \text{ h}$$
$$15° \text{ of longitude} = 1 \text{ h}$$
$$1° \text{ of longitude} = 4 \text{ min (of time)}$$

18-8 TIMING OBSERVATIONS

In the United States, the Institute of Standards and Technology (NIST), formerly the National Bureau of Standards, broadcasts time signals from station WWV in Fort Collins, Colo., on frequencies of 2.5, 5, 10, 15, and 20 MHz. These signals can be received with short-wave radios, including inexpensive so-called *time kubes* especially designed for this purpose. They can also be received over the telephone by dialing (303) 499-7111. To broaden coverage, signals are also transmitted from station WWVH in Hawaii on the same frequencies. These signals are broadcast as audible ticks with a computerized voice announcement of UT at each minute. In Canada EST is broadcast from station CHU on frequencies of 3.33, 7.335, and 14.67 MHz. This can be converted to UT by adding 5 h.

The time that is broadcast by WWV is actually *coordinated universal time* (UTC), whereas the time used for tabulating sun and star positions in ephemerides is a corrected version known as UT1. UTC is a uniform time at Greenwich which, unlike UT1, does not vary with changing rates of rotation and other irregular motions of the earth. *Leap seconds* are added to UTC as necessary to account for the gradual slowing of the earth's rotation rate and thus keep UTC within ∓ 0.7 sec of UT1 at all times. For precise astronomical work a difference correction (DUT) can be added to UTC to obtain UT1. The required DUT correction is given by means of double ticks broadcast by WWV and CHU during the first 15 sec following each minute tone. Each double tick represents a 0.1-sec correction. A plus correction is applied for double ticks that occur during the first 7 sec after the minute tone, while a negative correction is made for double ticks that are heard during sec 9 through 15. Thus if double ticks occurred for the first 5 sec after the minute tone, 0.5 sec would be added to broadcast UTC to obtain UT1. If double ticks were heard only on the ninth- and tenth-second ticks, 0.2 sec would be subtracted from UTC. Because the DUT correction is quite small, it can be ignored for most observations on Polaris or other stars with very high declinations. The correction should be considered, however, for more accurate observations on the sun, and stars of lower declination.

Digital watches, watches with sweep-second hands, and stop watches are all suitable for timing most astronomical observations in surveying. Pocket calculators equipped with time modules are especially convenient since they can serve not only as timing devices, but also for recording data and computing. Regardless of the timepiece used, it should be checked against WWV before starting observations, and either set to agree exactly with UTC, or the number of seconds it is fast or slow recorded. The time a check is made should also be recorded. When all observations are completed, the clock check should be repeated and any change recorded. Then intermediate observation times can be corrected in proportion to the elapsed time since the original check. With a stop watch, checks can be made before and after each individual observation.

EXAMPLE 18-1   A sweep-second hand watch was checked against WWV at 7:04 P.M. and recorded as 5.0 sec fast. The watch was checked again at 11:20 P.M. and found to be 9.5 sec fast. What is the corrected time for an observation recorded at 9:13:26 P.M.?

SOLUTION
(*Note:* The watch is fast; thus the correction is negative.)
Elapsed time between watch checks,

$$11:20 - 7:04 = 4:16 = 4.2667 \text{ h}$$

Elapsed time between first check and observation,

$$9:13:26 - 7:04:00 = 2:09:26 = 2.1572 \text{ h}$$

$$\text{Correction} = -5.0 + \frac{2.1572}{4.2667}[(-9.5) - (-5.0)] = -7.3 \text{ sec.}$$

Corrected time = 9:13:26 − 0:00:07.3 = 9:13:18.7.

18-9 STAR POSITIONS

If the pole could be seen as a definite point in the sky—that is, if a star were there—a meridian observation would require only a simple sighting. Since the pole is not so marked, observations must be made on stars—preferably those close to the pole—whose positions at given times are listed in an ephemeris.

If the polar distance of a star is less than the latitude of an observer's position, that star will never set; that is, if *PS* in Figure 18-6 is smaller than *PH*, the star will never move below the horizon plane and be invisible. Thus it can always be seen during night-time hours. Such stars are called *circumpolar stars*.

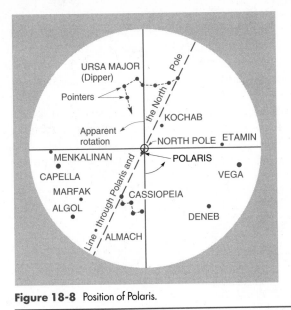

Figure 18-8 Position of Polaris.

The brightest star nearest the north pole is *Polaris,* which is part of the constellation Ursa Minor, also called the Little Dipper. Its polar distance changes slowly from day to day and from year to year, and in December 1993 it was approximately 0°45.5′. Polaris is a circumpolar star for all latitudes in the United States. In most of the southern hemisphere, stars of the constellation *Southern Cross* are circumpolar (their diurnal circles are close to the south celestial pole), and thus they are most frequently used for azimuth observations there.

Polaris is located in the sky by first finding the Big Dipper in Ursa Major. The two stars of the dipper farthest from the handle are the "Pointers." Polaris is the nearest bright star along the line through the Pointers; it is also on the line through the end star of the Big Dipper handle and the left star of Cassiopeia when viewed as a "W," as shown in Figure 18-8.

18-10 AZIMUTH FROM POLARIS OBSERVATIONS

Three different methods have been used in making Polaris observations for azimuth: (**1**) *Polaris at any hour angle,* (**2**) *Polaris at culmination,* and (**3**) *Polaris at elongation.* As indicated by their names, these procedures differ principally in the location of the star in its diurnal circle when observed. The method most frequently used by surveyors is observing Polaris at any hour angle, since it can be done at a convenient time beginning about dusk. Furthermore, repeated observations can be made with the instrument both direct and reversed, which increases precision.

The advantage of observing Polaris at culmination, just when crossing the meridian, is that its direction is true north then, and no subsequent calculations are required. However, computations must precede the observation to determine the exact time of culmination. A disadvantage of the method is that the star's apparent horizontal movement is maximum at culmination, thereby making the observations slightly more difficult and less accurate. Also, repeated observations cannot be made.

Advantages of observing Polaris at elongation are **(1)** the star's apparent motion is exactly vertical at those times, thus making observations easier and more accurate, and **(2)** the computations needed to determine its azimuth are somewhat simplified. Again, the time of elongation must be calculated prior to observing. Because the star's azimuth remains nearly constant for 15 to 20 min before and after elongation, repeated sightings can be made.

One serious disadvantage of both culmination and elongation methods is that observations are restricted to the fixed times of these events, which often occur during the daytime or at inconvenient hours late at night or early in the morning. Thus these methods are seldom used, and discussion in this book concentrates on Polaris at any hour angle.

18-11 COMPUTATIONS FOR AZIMUTH FROM OBSERVATIONS OF POLARIS AT ANY HOUR ANGLE

In this method only the horizontal circle reading and precise time have to be recorded when the star is sighted. A vertical circle reading for at least one pointing is recommended, however, to ensure that the correct star has been sighted. To make the observations, the instrument (total station, theodolite, or transit) is set up and leveled on one end of a line whose azimuth is to be determined. In the usual field procedure, the line's other end is first sighted, and then the horizontal angle measured to the star. To eliminate the effects of instrumental errors, equal numbers of direct and reversed observations are taken and the results averaged.

Computations after field work require the solution for angle Z in the astronomical (PZS) triangle (see Figure 18-6). The formula for Z, derived from laws of spherical trigonometry, is

$$Z = \tan^{-1}\left(\frac{\sin t}{\cos \phi \tan \delta - \sin \phi \cos t}\right) \qquad (18\text{-}3)$$

The geometry on which Eq. (18-3) is based is shown more clearly in Figure 18-9, where the PZS triangle is again darkened. The latitude ϕ of the observer's position is

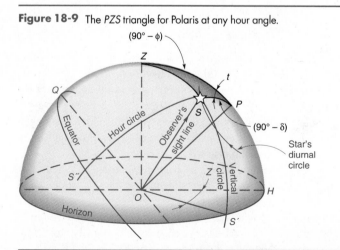

Figure 18-9 The PZS triangle for Polaris at any hour angle.

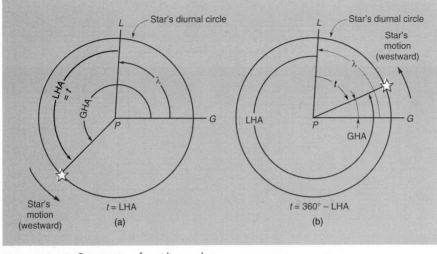

Figure 18-10 Computation of meridian angle t.

arc *HP;* thus arc *PZ* is $(90° - \phi)$, or *colatitude*. Declination δ of the star is arc *S″S,* so *SP* is $(90° - \delta)$, or *polar distance*. Angle *ZPS* in Figure 18-9 is *t,* the *meridian angle.* Descriptions of methods for obtaining the three variables in Eq. (18-3) follow.

The latitude of an observer's position can be scaled from a USGS quadrangle map and, with reasonable care, obtained to within ±2 sec. The "Topo Aid," a transparent template placed over a USGS quad sheet, can also be used to measure latitude and longitude. Alternatively, latitude could be obtained from astronomical observations, as explained in Section 18-20. Declination is extracted from an ephemeris for the instant of sighting.

Meridian angle *t* equals the local hour angle (LHA) of the star if it is west of north, or it is 360° − LHA if east of north. The local hour angle is between 0° and 180° when the star is west of north, and between 180° and 360° if the star is east of north. Diagrams such as those of Figure 18-10 are helpful in determining *t* angles. These diagrams show the north celestial pole *P* at the center of the star's diurnal circle *as viewed from the observer's position within the sphere.* The apparent motion of stars is counterclockwise. Angle λ between the meridian of Greenwich *(G)* and the local meridian *(L)* through the observer's position is the longitude of the station occupied. The star's Greenwich hour angle (GHA) for the observation time is taken from an ephemeris. Longitude is normally scaled from a USGS map, although it could be observed as described in Section 18-20. Sketching λ and GHA approximately to scale on a diagram such as those of Figure 18-10 immediately makes clear the star's position.

For the following set of field notes obtained on 3 June 1993, compute the azimuth of line *AB.*[4]

EXAMPLE 18-2

[4]A computer program written in BASIC, which computes azimuths from observations at any hour angle, is given in Section F-3 in Appendix F. Example 18-2 has been solved using the program to illustrate its application. A listing of the solution is also given in Appendix F.

FIELD DATA

Polaris Observation for Azimuth
Occupied Station A
Latitude A = 43°05′24″N, Longitude A = 89°26′00″W

OBSN NO.	INST D/R	STA SIGHTED	OBSN TIME (P.M. CDT)	HORIZONTAL CIRCLE
1*	D	*B*		0°00′00″
		Polaris	8:30:49	49°36′12′
2	R	*B*		180°00′01″
		Polaris	8:39:00	229°38′11″
3	D	*B*		0°00′00″
		Polaris	8:44:33	49°39′28″
4	R	*B*		180°00′00″
		Polaris	8:49:46	229°40′50″

---

*Zenith angle of 47°33′24″ observed for star verification.

When the observations occur over a short period of time, average values for observation time and horizontal angle can be used and one reduction made. This assumes the star moves linearly throughout the time between the first and last observations. For more accurate results, or if the elapsed time exceeds about 20 min, each observation should be reduced independently and the resulting azimuths averaged. In this example, the reductions are performed independently.

SOLUTION FOR FIRST OBSERVATION

1. UT of observation,

$$8:30:49 \quad \text{P.M. CDT, 3 June 1993}$$

$$+ 12 \quad \text{correction for P.M.}$$

$$+ \underline{5} \quad \text{correction to Greenwich}$$

$$25:30:49$$

$$1:30:49 = \text{UT of observation, 4 Jun 1993}$$

2. GHA of Polaris. From Table 18-1,

$$216°27′36″ = \text{GHA at 0h UT, 4 June 1993 (nearest second)}$$

$$217°26′23″ = \text{GHA at 0h UT, 5 June 1993}$$

Change in GHA for 24 h,

$$217°26′23″ + 360° - 216°27′36″ = 360°58′47″$$

(*Note:* This calculation accounts for the fact that the star traveled through its complete diurnal circle in 24 h, thus passing 360° in the process.)

Increase in GHA of Polaris for $1^h30^m49^s$,

$$1 + \frac{30}{60} + \frac{49}{3600} = 1.51361h$$

$$\frac{1.51361}{24} \times 360°58'47'' = 22°45'57''$$

GHA of Polaris at observation,

$$216°27'36'' + 22°45'57'' = 239°13'33''$$

3. Declination of Polaris. From Table 18-1,

$$89°13'56.31'' = \delta \text{ at 0h UT, 4 June 1993}$$

$$89°13'56.12'' = \delta \text{ at 0h UT, 5 June 1993}$$

(*Note:* Declination is taken to 0.01 sec because small changes in its value can have significant effects on accuracies of computed azimuths, especially when the LHA is near 90 or 270°.)

Change in δ for 24 h,

$$89°13'56.12'' - 89°13'56.31'' = -0.19''$$

Change for $1^h30^m49^s$,

$$\frac{1.51361}{24} \times (-0.19) = -0.01''$$

Declination of Polaris at observation,

$$89°13'56.31'' - 0.01'' = 89°13'56.30''$$

4. Local hour angle of Polaris,

$$239°13'33'' = \text{GHA}$$

$$\underline{89°26'00'' = \lambda}$$

$$149°47'33'' = \text{LHA} = t \text{ (Polaris is west of north)}$$

[*Note:* A sketch similar to Figure 18-10(a) applies.]

5. Azimuth of Polaris. By Eq. (18-3),

$$Z = \tan^{-1}\left[\frac{\sin 149°47'33''}{\cos 43°05'24'' \tan 89°13'56.30'' - \sin 43°05'24'' \cos 149°47'33''}\right]$$

$$= \tan^{-1}\left[\frac{0.503091}{(0.730282)(74.629107) - (0.683146)(-0.864233)}\right]$$

$$= \tan^{-1}(0.00913282)$$

$$= 0°\ 31'\ 24'' \text{ (west of north)}$$

Azimuth of Polaris,

$$360° - 0°31'24'' = 359°28'36''$$

6. Azimuth of line *AB*,

$$359°28'36'' = \text{azimuth of Polaris}$$

$$-49°36'12'' = \text{horizontal angle from } B \text{ to Polaris}$$

$$309°52'24'' = \text{azimuth of line } AB = \alpha$$

(*Note:* A sketch similar to Figure 18-5 will aid this calculation.)

The other three observations were calculated in the manner just demonstrated, and yielded azimuths for line *AB* of 309°52'23'', 309°52'26'', and 309°52'21'', respectively. The mean of the four values is 309°53'23.5''—a good set of observations.

18-12 VERIFICATION OF FIELD OBSERVATIONS

If possible it is advisable to verify the correctness of observations before leaving the field. Lap-top computers or hand-held programmable calculators are convenient for this work. Programs can be written or obtained to perform azimuth reductions, which can be done while still occupying the station. An azimuth should be computed for each pointing on the star, and all values compared. If obvious mistakes or large errors exist, these observations can be eliminated before averaging. If too many observations are rejected, some can be repeated.

When rejecting observations, equal numbers of direct and reversed measurements should always be retained in final averaging to eliminate the effects of instrumental errors. Performing these reductions on location reduces the possibility of having to return for additional observations.

If observation times and horizontal angles are averaged and a single azimuth computation is made, a plot of time versus angle as illustrated in Figure 18-15, will verify the correctness of individual observations.

18-13 LOCATING POLARIS IN THE TELESCOPE

If Polaris is observed at any hour angle, field operations can begin at dusk and be completed when some daylight still exists. This greatly assists in handling the instrument and recording notes. Polaris cannot be seen with the naked eye before dark, but it is visible through the telescope of a total station, theodolite, or transit. To bring the star into the field of view, it is very helpful to compute its altitude for the anticipated time of observation. This can be done using the following formula derived for the *PZS* triangle from the cosine law of spherical trigonometry:

$$h = \sin^{-1}(\sin \phi \sin \delta + \cos \phi \cos \delta \cos t) \qquad (18\text{-}4)$$

Equation (18-4) yields the star's true altitude, rather than the vertical angle containing refraction necessary to point at the star. However, since the field of view for most telescopes is approximately 1.5°, and refraction is less than a few minutes, true altitude is sufficient to enable locating Polaris.

An observation on Polaris is planned to begin at 8:00 P.M. CDT on 17 June 1993 at a station whose latitude and longitude are 42°35′N and 89°28′W, respectively. Determine the zenith angle at which the vertical circle should be set to find the star.

EXAMPLE 18-3

SOLUTION

(*Note:* The altitude angle need only be computed to the nearest minute. Thus values are rounded accordingly.)

1. UT of planned observation,

8:00 P.M. CDT, 17 June 1993

+ 12   correction for P.M.

+ 5   correction to Greenwich

25:00, 17 June 1993

1:00 UT, 18 June 1993

2. GHA of Polaris. From Table 18-1,

230°11′ = GHA at 0h UT, 18 June 1993 (nearest minute)

231°10′ = GHA at 0h UT, 19 June 1993

Increase in GHA for 24 h,

$$231°10′ + 360° - 230°11′ = 360°59′$$

Increase in GHA for 1 h,

$$\frac{1}{24} \times 360°59′ = 15°02′$$

GHA of Polaris at planned observation time,

$$230°11' + 15°02' = 245°13'$$

3. Meridian angle t of Polaris,

$$245°13' = \text{GHA of Polaris}$$
$$-\ \underline{89°28'} = \lambda$$
$$155°45' = \text{LHA} = t$$

[*Note:* A sketch similar to Figure 18-10(a) applies.]

4. Declination of Polaris. From Table 18-1,

$$89°14' = \delta \text{ at 0h UT, 18 June 1993 (nearest minute)}$$

(*Note:* The change in δ in 24 h is negligible for this calculation.)

5. True altitude of Polaris. By Eq. (18-4),

$$h = \sin^{-1}(\sin 42°35' \sin 89°14' + \cos 42°35' \cos 89°14' \cos 155°45')$$

$$= 41°53' \text{ (approximate vertical angle to Polaris)}$$

$$z = 90° - h = 48°07' \text{ (approximate zenith angle to Polaris)}$$

(*Note:* If the temperature and pressure were assumed to be 70°F and 30 in., respectively, the refraction correction would be about 1′, which is negligible for finding Polaris.)

An altitude calculation should be done before going into the field. Then when an instrument has been set up in preparation for observing, the telescope is focused on some faraway object. (The moon is excellent if visible; otherwise any distant point will suffice.) The computed zenith or vertical angle is set on the instrument, the telescope turned in the approximate direction of north (which can be determined by reading a compass and applying magnetic declination), and the sky slowly scanned until the tiny speck of light, Polaris, is found. Scanning need span only a few degrees in each direction from north as measured by compass, since Polaris is never more than about 2° in azimuth from the meridian for latitudes up to nearly 70°. Once the star has been found, the telescope may be plunged and a mark set on-line a few hundred feet from the instrument to provide a rough reference direction for resighting the star.

18-14 PRACTICAL SUGGESTIONS FOR POLARIS OBSERVATIONS

Although different individuals may prefer other techniques, with repeating instruments a recommended observing procedure consists in setting the horizontal circle to 0°00′00″. Then with the telescope direct, sight an azimuth mark with the lower motion. Open the upper motion, sight the star, and record the precise time of pointing and the horizontal and vertical circle readings. Plunge the scope and with the upper motion resight the mark. (The reading should theoretically be exactly 180°; any difference represents instrumental or personal errors.) The upper motion is again used to sight the star, whereupon similar data are recorded. Several repetitions of direct and reversed readings are made, the number being a function of the required accuracy. The notes of Example 18-2 should clarify the procedure.

The following additional suggestions make observations on Polaris easier to perform:

1. Precompute the star's altitude for the anticipated time of beginning observations.
2. Prepare the noteforms in advance of starting field work.
3. Set up the instrument in daylight if possible and begin observations before dark. To find Polaris, use its precomputed altitude for the planned starting time.
4. Be sure the azimuth mark is visible at night and recognizable and definite in the daytime. It should be at least 1000 ft away if possible to reduce pointing errors and simplify refocusing between mark and star, and vice versa.
5. Level the instrument very carefully. If it is just $10''$ out of level, an azimuth error of $6''$ can result at latitude 30°N, and $10''$ of error can occur at 45°N.
6. Have a good timepiece available and check it against WWV radio signals before and after field operations.
7. To enable pointing after dark, illuminate the cross hairs. Many newer instruments have a special internal battery-operated illumination system to light the cross wires, horizontal circle, micrometer scale, and vertical arc. If one of these systems is used, be sure to have extra batteries. If the instrument does not have internal illumination, a flashlight can be directed obliquely into the objective end of the telescope. The light must not be too bright or the star becomes invisible, and it cannot be too dim or the cross hairs fade.

18-15 AZIMUTH FROM SUN OBSERVATIONS

Two different methods may be used for determining the azimuth of a line from sun observations: **(1)** the *hour-angle method* and **(2)** the *altitude method.* The hour-angle method uses Eq. (18-3) and generally follows the same computational procedure as demonstrated for Polaris in Example 18-2. The altitude method requires the sun's vertical angle to be measured as a part of the field operations, in addition to observing the horizontal angle from the sun to an azimuth mark.

When the altitude method is employed to determine azimuth, the following equation, derived from the cosine law of spherical trigonometry, is used to compute the Z angle of the *PZS* triangle:

$$Z = \cos^{-1}\left(\frac{\sin \delta - \sin h \sin \phi}{\cos h \cos \phi}\right) \qquad (18\text{-}5)$$

An alternate form of Eq. (18-5) is

$$Z = \cos^{-1}\left(\frac{\sin \delta}{\cos \phi \cos h} - \tan \phi \tan h\right) \qquad (18\text{-}6)$$

where h is the true altitude of the sun and all other terms are as defined previously.

Again, latitude can be taken from a USGS quadrangle map or observed, and declination extracted from an ephemeris for the time of observation. In solving Eq. (18-5) or (18-6), either a positive or a negative value can be obtained for cos Z. For morning observations the azimuth of the sun is equal to Z; in the afternoon, $360° - Z$.

Advantages and disadvantages exist for both the hour-angle and the altitude methods. The main advantage of the altitude method is that timing of observations need not be extremely accurate because the only time-related variable used is declination, and it changes very slowly (never more than about one arc second per minute of time). A secondary advantage is that longitude is not needed.

The major disadvantage of the altitude method is that the sun's true altitude is required, which creates the following three problems:

1. It is more difficult to make the field observations because both the horizontal and the vertical cross hairs must be simultaneously aligned on the sun. In contrast, the hour-angle method requires aligning the vertical wire only, which is considerably easier, faster, and produces more accurate pointings.

2. A refraction correction which depends on temperature and pressure is necessary. These variables are measured at the observation station and can differ substantially along the ray path, thereby causing the refraction correction to be inaccurate. The problem becomes especially severe for low altitudes; hence observation times must be restricted.

3. Because the sun's altitude changes only slightly with time when it is close to the meridian, the solution for Z in the *PZS* triangle by Eq. (18-5) or (18-6) becomes very weak near noon. Thus observation times are further restricted.

For the reasons given, suitable times to observe the sun for azimuth using the altitude method, within latitudes of the conterminous United States, are between 8 and 10 A.M., or 2 and 4 P.M. standard time (9 to 11 A.M. and 3 to 5 P.M. daylight saving time). At high latitudes the method should not be used. There are no restrictions on observation time if the hour-angle method is used, but accurate time is required. All things considered, if a suitable timepiece is available, the hour-angle method is preferred as it will produce more accurate results in less time.

18-16 FIELD PROCEDURES FOR MAKING SUN OBSERVATIONS

Observations on the sun can be made directly by placing a dark glass filter (designed for solar viewing) over the telescope objective lens or eyepiece. They can also be made indirectly by focusing the sun's image on an unlined white card held behind the eyepiece. *Looking at the sun directly through the telescope without a dark glass can result in permanent eye injury.*

If a white card is used, the telescope is rotated in a horizontal plane until the shadow of the standards and the telescope are symmetrical. Then the telescope is turned up and down until the sun's image is visible and sharp at infinite distance focus. The eyepiece must also be adjusted to obtain a clear image of the cross hairs on the card.

Total station instruments require an objective lens filter to protect EDM components. Do not use the indirect white card procedure with these instruments.

The *Roelof solar prism* is a device designed especially for sighting the sun. It is easily mounted on the objective end of the telescope, and by means of prisms, produces four overlapping images of the sun in the pattern shown in Figure 18-11(a). While viewing the sun, which is possible due to a dark glass filter, an observer can accurately center the intersection of cross hairs in the small diamond-shaped area in the middle of the field of view. Because of symmetry, this is equivalent to sighting the sun's center.

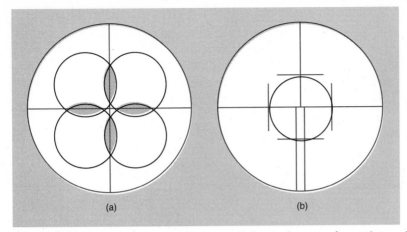

Figure 18-11 (a) Image of sun viewed through Roelof prism. (b) Image of sun with special interchangeable reticle.

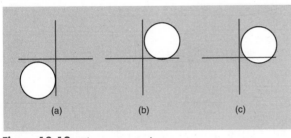

Figure 18-12 Observation on the sun.

To aid in pointing at the sun, the telescopes of many instruments can have their standard reticles replaced with special ones. As shown in Figure 18-11(b), one type has a circular ring or lines within which the sun's image just fits. Of course, a dark glass filter still must be used or, alternatively, the card method employed.

The Roelof prism and other solar reticles are designed for taking simultaneous horizontal and vertical readings. These accessories are not necessary for the hour-angle method where only horizontal readings are required.

The sun's diameter as viewed from earth is approximately 32' of arc. If a Roelof prism or other special reticle is not available, bisecting such a large object moving both horizontally and vertically is difficult. A better method is to sight the sun's edges (limbs). If the altitude method is being employed, the disk is brought tangent to both cross lines, first in one quadrant and then in the diagonally opposite one, as illustrated in Figure 18-12(a) and (b). Averaging the two readings practically eliminates consideration of the sun's semidiameter.

To avoid coordinating the movement of both tangent screws with the sun, it is simpler to follow the disk by keeping the vertical cross hair tangent to it and letting the sun come tangent to the horizontal wire. If the hour-angle method is used, the sun is placed in the field of view and allowed to move to the vertical cross hair. Time is recorded at the instant its limb touches the hair. Since the vertical angle is of no concern, an alignment like that shown in Figure 18-12(c) is satisfactory. Rather than taking point-

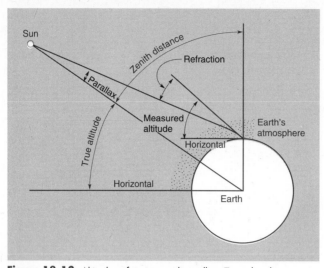

Figure 18-13 Altitude, refraction, and parallax. True altitude = measured altitude − refraction + parallax. (Courtesy Keuffel & Esser Company.)

ings on both edges and averaging, all pointings can be made on the trailing edge, with a horizontal semidiameter correction being computed from the following:

$$C_{SD} = \frac{\text{sun's semidiameter}}{\cos h} \qquad (18\text{-}7)$$

The sun's semidiameter for Eq. (18-7) is obtained from ephemeris tables (see Table 18-1), and h is either observed or computed from Eq. (18-4). For mid and high latitudes in the northern hemisphere, the correction is added to the measured clockwise horizontal angle from the line to the sun's trailing edge. Example 18-4 illustrates this correction.

If the sun's image is focused on a white card with an erecting telescope, its image will be upside down as viewed on the card. Thus if the sun's image were viewed on a card as in Figure 18-12(a), the sun would actually be above and to the right of the cross hairs. To avoid confusion as to how the sun was actually sighted, sketches such as those in the leftmost column of the field notes of Example 18-5 should be used.

The measured altitude of the sun is its angular distance above the horizon. If the altitude method is employed and a transit is used, the altitude angle will be measured. If a theodolite or a total station is employed, the zenith angle will be observed and the corresponding altitude computed as $h = 90° − z$. To obtain true altitude, this observed value must be corrected for index error if it exists, and for refraction and parallax as shown in Figure 18-13. It must also be corrected for semidiameter if its limb is sighted in only one quadrant of the cross hairs. Refraction and parallax corrections can be calculated using Eqs. (18-1) and (18-2) or obtained from tables. Example 18-5 illustrates use of the equations for making these corrections.

With either the hour-angle or the altitude method, errors of adjustment of the standards and cross hairs are eliminated by averaging equal numbers of sightings with the telescope both direct and reversed (face left and face right). Careful leveling is required. At least four observations, two direct and two reversed, should be made in a minimum of time. Precision is improved, however, if the number is increased.

In the usual procedure for observing the sun, the instrument is oriented with zero or a known value on the horizontal circle by sighting along a fixed line from the observer's position to a mark. The telescope (direct) is then turned to the sun, and after recording times and measured angles for four or more observations (2D and 2R), the mark is resighted (with the telescope reversed) to be certain the circle has not moved. The final reading should differ by exactly 180° from the original one, but slight variations will usually occur due to instrumental and personal errors. The field notes of Examples 18-4 and 18-5 illustrate this procedure.

18-17 COMPUTATIONS FOR AZIMUTH BY THE HOUR-ANGLE METHOD

The following example is presented to illustrate the method of computing azimuth from sun observations by the hour-angle method.

For the field data given below, calculate the azimuth of line Rover-Harris using the hour-angle method.

EXAMPLE 18-4

THEODOLITE AT STA. ROVER

Latitude = 42°45'10"N, Longitude = 73°56'30"W
3 June 1993

POINT SIGHTED	TELESCOPE D/R	WATCH TIME P.M. EDT	HORIZONTAL CIRCLE (CLOCKWISE)
△ Harris	D		0°00'00"
sun	D	$4^h46^m10^s$	212°05'40"
sun	D	$4^h46^m56^s$	212°13'54"
sun	R	$4^h49^m05^s$	32°37'24"
sun	R	$4^h49^m52^s$	32°45'42"
△ Harris	R		180°00'12"

SOLUTION

An azimuth is computed for each pointing. These are then compared, and acceptable values averaged. For illustration purposes, only calculations for the last observation are shown.

1. UT of observation,

4:49:52 P.M. EDT

+ 12   correction for P.M.

+ 4   correction to Greenwich

20:49:52 UT, 3 June 1993

2. GHA of sun. From Table 18-1,

$$180°29'38'' = \text{GHA at 0h UT, 3 June 1993 (nearest second)}$$
$$180°27'09'' = \text{GHA at 0h UT, 4 June 1993}$$

Change in GHA for 24h,

$$180°27'09'' + 360° - 180°29'38'' = 359°57'31''$$

Increase in GHA of sun for $20^h49^m52^s$ (20.83111 h),

$$\frac{20.83111}{24} \times 359°57'31'' = 312°25'51''$$

GHA of sun at observation,

$$180°29'38'' + 312°25'51'' = 492°55'29''$$

Since GHA is greater than 360°, subtract 360,

$$492°55'29'' - 360° = 132°55'29''$$

3. Local hour angle (LHA) of sun,

$$132°55'29'' = \text{GHA}$$
$$\underline{-73°56'30'' = \lambda}$$
$$58°58'59'' = \text{LHA} = t$$

4. Declination of sun. From Table 18-1,

$$22°17'10''\text{N} = \delta \text{ at 0h UT, 3 June 1993 (nearest second)}$$
$$22°24'26''\text{N} = \delta \text{ at 0h UT, 4 June 1993}$$

Change in δ for 24h,

$$22°24'26'' - 22°17'10'' = 0°07'16''$$

Change for $20^h49^m52^s$,

$$\frac{20.83111}{24} \times 0°07'16'' = 0°06'18''$$

δ of sun at observation,

$$22°17'10'' + 0°06'18'' = 22°23'28''$$

5. Azimuth of sun. By Eq. (18-3),

$$Z = \tan^{-1}\left[\frac{\sin 58°58'59''}{\cos 42°45'10'' \tan 22°23'28'' - \sin 42°45'10'' \cos 58°58'59''}\right]$$

$$= \tan^{-1}(-18.1265421)$$

$$= -86°50'32''$$

$$= 180° - 86°50'32'' = 93°09'28''$$

Since the observation was made in the afternoon, angle Z is counterclockwise from north,

$$AZ_{sun} = 360° - 93°09'28'' = 266°50'32''$$

6. Azimuth of line Rover-Harris (see Figure 18-14). Horizontal angle from Harris to sun's left edge,

$$32°45'42'' - 180°00'12'' + 360° = 212°45'30''$$

Semidiameter correction. By Eqs. (18-4) and (18-7),

$$h = \sin^{-1}(\sin 42°45'10'' \sin 22°23'28'' + \cos 42°45'10''$$

$$\cos 22°23'28'' \cos 58°58'59'')$$

$$= 37°28'35''$$

Figure 18-14 Azimuth of line Rover-Harris from sun observation.

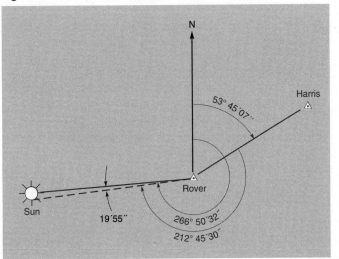

From Table 18-1,

$$SD = 0°15'48''$$

$$C_{SD} = \frac{0°15'48''}{\cos 37°28'35''} = 0°19'55''$$

Correction from left edge to sun's center is clockwise. Therefore horizontal angle from Harris to sun's center,

$$212°45'30'' + 0°19'55'' = 213°05'25''$$

Azimuth of line Rover-Harris,

$$266°50'32'' - 213°05'25'' = 53°45'07''$$

Calculations, done in order for the first three observations using the same procedure as above yield azimuths of 53°45'20'', 53°45'21'', and 53°45'04'', respectively. Comparison of the two azimuths obtained from direct pointings and the two from reverse indicates a good set of observations. The mean of the four azimuths is 53°45'13''.

This problem has also been solved using the computer program in Appendix F.

18-18 COMPUTATIONS FOR AZIMUTH BY THE ALTITUDE METHOD

The following example illustrates procedures for computing azimuth from sun observations by the altitude method. Mean values of time and vertical and horizontal angles are used in the solution.

EXAMPLE 18-5   For the following set of field notes, calculate the azimuth of line Rover-Harris using the altitude method.

THEODOLITE AT STA. ROVER

Latitude = 42°45'10''N, Longitude = 73°56'30''W
29 June 1993, Temperature 80°F, Pressure 28.7 in. Hg

POINT* SIGHTED	TELESCOPE D/R	WATCH TIME P.M. EDT	ZENITH ANGLE	HORIZONTAL ANGLE (CLOCKWISE)
△ Harris	D			0°00.0'
sun	D	4^{h}53^{m}10^s	51°26.3'	213°14.8'
sun	D	4^{h}54^{m}20^s	51°39.2'	213°27.2'
sun	R	4^{h}56^{m}37^s	307°23.7'	34°32.1'
sun	R	4^{h}57^{m}49^s	307°10.6'	34°44.7'
△ Harris	R			180°00.0'
Mean values		4^{h}55^{m}29^s	52°07.8'	213°59.7'

*Sketches give position of sun's image with respect to cross hairs as viewed on card. Erecting telescope used.

SOLUTION

1. Sun's true altitude,

Mean observed zenith angle	$= 52°07.8'$
Mean observed vertical angle $= 90° - 52°07.8'$	$= 37°52.2'$

Refraction correction, by Eq. (18-1),

$$C_r = \frac{16.38(28.7)}{(460 + 80)\tan 37°52'} \qquad\qquad = -01.1'$$

Parallax correction, by Eq. (18-2),

$$C_p = 8.79 \cos 37°55' = 6.9'' \qquad\qquad = \underline{+00.1'}$$

$$\text{True altitude} \qquad\qquad = 37°51.2'$$

2. Average UT of observation,

Mean EDT of observation times $=$	$4^h55^m29^s$
Correction for P.M.	$= 12$
Correction to Greenwich	$= \underline{\ 4}$
Average UT of observations	$= 20^h55^m29^s$

3. Sun's declination at average observation time. From Table 18-1,

$$\delta \text{ at 0h UT, 29 June 1993} = 23°14'21''\text{N}$$

$$\delta \text{ at 0h UT, 30 June 1993} = 23°11'01''\text{N}$$

Change in 24 h,

$$23°11'01'' - 23°14'21'' = -0°03'20''$$

Change in $20^h55^m29^s$,

$$20 + \frac{55}{60} + \frac{29}{3600} = 20.92 \text{ h}$$

$$\text{Change} = \frac{20.92}{24} \times (-03'20'') = -02'54''$$

$$\delta \text{ at observation time} = 23°14'21'' - 02'54'' = 23°11'27''$$

4. Sun's azimuth. By Eq. (18-5),

$$\cos Z = \frac{\sin 23°11'27'' - \sin 37°51'12'' \sin 42°45'10''}{\cos 37°51'12'' \cos 42°45'10''}$$

$$= -0.03926956$$

$$Z = 92°15'02''$$

Since the observation was made in the afternoon, angle Z is counterclockwise from north, and the minus sign indicates an angle greater than 90°. Thus the sun's azimuth at observation time is

$$AZ_{sun} = 360° - 92°15'02'' = 267°44'58''$$

5. Azimuth of line Rover-Harris,

Sun's azimuth	$= 267°44'58''$
Mean horizontal angle, Harris to sun	$= \underline{213°59'42''}$
Azimuth of line Rover-Harris	$= 53°45'16''$

When mean values for times and angles are used in the solution, as in the foregoing example, the correctness of individual readings can be verified by plotting horizontal and zenith (or vertical) angles as ordinates, versus time as abscissas. If a Roelof prism is used, all plotted zenith angles should lie on a straight line, and likewise all plotted horizontal angles should also define a straight line. If the indirect white card method is used, the line for direct readings of zenith (or vertical) angles should be parallel to the line for the reversed values, and the lines should be separated by the sun's diameter plus or minus any vertical index error. For horizontal angles, the direct and reversed lines should also be parallel, and separated by the sun's diameter divided by the cosine of the average vertical angle. Plotted observations that deviate from these conditions should be discarded and new ones taken. Figure 18-15 illustrates such a plot for the values of

Figure 18-15 Plot of observed horizontal and zenith angles to the sun versus observation times for verification of correctness.

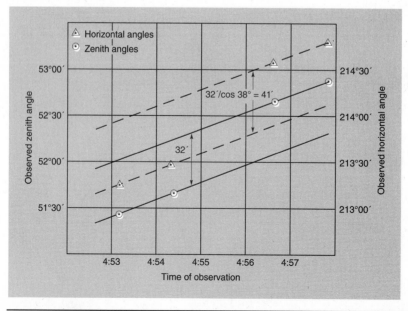

Example 18-5. These plots, done as the data are observed, will verify that the readings are acceptable before leaving the field.

The justification for making only one calculation of the azimuth based upon average values is that computing time is reduced. If a programmable calculator or a portable lap-top computer (equipped with appropriate software) can be taken into the field, a better way to verify the acceptability of readings is to process each individual observation for azimuth as the data are obtained. Then any observation that produces an azimuth which departs too far from the mean is discarded, and a new one taken in its place. In the end, the mean azimuth from all acceptable readings is taken, *but the mean must be computed from an equal number of direct and reversed readings.*

18-19 AZIMUTH BY SOLAR ATTACHMENT

A solar attachment which can be mounted on a transit gives a mechanical solution of the *PZS* triangle. By setting the latitude, declination, and hour angle on three scales and sighting the sun, the telescope is oriented in the north-south direction. Solar attachments were used quite extensively to establish meridians in surveys of the U.S. public lands, but now are seldom used. Unless the instrument is in very good adjustment, rather large azimuth errors can result in using a solar attachment.

18-20 DETERMINATION OF LATITUDE AND LONGITUDE

Latitude and longitude are important quantities because they provide the basis for specifying locations of points in geodetic surveying. These parameters were defined in Section 18-1, and other sections of this chapter have illustrated their importance in the solution for azimuth. Different methods are available for determining latitude and longitude. The Global Positioning System (see Chapter 20) currently provides the fastest method for accurately ascertaining these quantities. Prior to that technology, the only means for making direct measurements of latitude and longitude was by astronomical observations. Any one of several different procedures could be used, the choice depending on available equipment and desired accuracy. Textbooks that specialize in field astronomy describe them in detail. In this book only one simple method of determining each is presented to introduce students to these topics and provide a means of obtaining latitude and longitude for azimuth calculations if maps are unavailable for scaling them.

18-20.1 Latitude from Sun's Altitude at Culmination

The easiest method of observing for latitude is to measure the sun's altitude at culmination. An instrument can be set up on a station shortly before noon and carefully leveled. The sun is then followed as its altitude increases to a maximum, at which position it is on the meridian. The vertical circle reading and time are recorded. Time noted to the nearest few minutes is sufficient, since only declination, which changes slowly, is computed from it. Figure 18-16 illustrates the geometry, and from it the following equation for latitude is evident:

$$\phi = 90° - h + \delta \qquad (18\text{-}8)$$

where h is the sun's true altitude corrected for index error, refraction, parallax, and semidiameter if a limb is observed. Declination δ is taken from an ephemeris for the

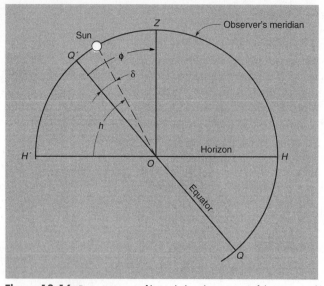

Figure 18-16 Determination of latitude by observation of the sun at culmination.

observation time. If declination is southerly, a negative sign must be affixed in computations. In low latitudes during summer, the sun's altitude at culmination will be too high for measurement with an ordinary theodolite unless a prismatic eyepiece is used.

EXAMPLE 18-6   On 1 June 1993 a theodolite was set up before noon at a point of unknown latitude and carefully leveled. By using a Roelof's prism, the zenith angle to the center of the sun at its maximum altitude was measured as 41°02′42″ at 1:03 P.M. EDT. Temperature and pressure were 75°F and 29.1 in. Hg, respectively. Compute the latitude of the point.

SOLUTION

1. True altitude,

Observed altitude = 90° − 41°02′42″   = 48°57′18″

Refraction correction, by Eq. (18-1),

$$C_r = \frac{16.38(29.1)}{(460 + 75)\tan 48°57′18″} = 0.776′ =   -47″$$

Parallax correction, by Eq. (18-2),

$$C_p = 8.79 \times \cos 48°57′18″   = +6″$$

Sun's true altitude   = 48° 56′37″

2. UT of observation,

$$\text{Observation time} \qquad = \quad 1{:}03$$

$$\text{Correction for P.M.} \qquad = \quad 12$$

$$\text{Correction to Greenwich} = \quad \underline{\ \ 4\ \ }$$

$$\text{Observation time} \qquad = \quad 17{:}03$$

3. Sun's declination. From Table 18-1,

$$\delta = 22°01'29''\text{N at 0h UT, 1 June 1993}$$

$$\delta = 22°09'31''\text{N at 0h UT, 2 June 1993}$$

Change in 24 h

$$22°09'31'' - 22°01'29'' = +08'02''$$

Change in $17^h03''$,

$$\frac{= 17 + 3/60}{24} \times 08'02'' = +05'42''$$

$$\delta \text{ at observation time} = 22°01'29'' + 05'42'' = 22°07'11''$$

4. Latitude calculation by Eq. (18-8),

$$\phi = 90° - 48°56'37'' + 22°07'11'' = 63°10'34''$$

Accuracies of latitudes determined by this method depend largely on instrument precision, but values to within about $\pm 30''$ are possible.

18-20.2 Longitude from Time of Sun's Culmination

A simple method of determining longitude is to measure the precise time at which the sun's center reaches culmination (crosses the true meridian). This time is then used to compute GHA. Since the local hour angle equals zero when the sun is on the meridian, longitude is equal to GHA. (Refer to Figure 18-10.)

The sun's center crossed the meridian at 12:57:19 CDT on 7 June 1993. Determine the station's longitude.

EXAMPLE 18-7

SOLUTION

1. UT of observation,

$$\text{CDT of observation} \qquad = \quad 12{:}57{:}19$$

$$\text{Correction to Greenwich} = \quad \underline{+\ 5}$$

$$\text{UT, 7 June 1993} \qquad = \quad 17{:}57{:}19$$

2. GHA of sun. From Table 18-1,

$$\text{GHA at 0h UT, 7 June 1993} = 180°19'15''$$

$$\text{GHA at 0h UT, 8 June 1993} = 180°16'27''$$

Change in GHA for 24 h,

$$180°16'27'' + 360° - 180°19'15'' = 359°57'12''$$

Increase in GHA for $17^h57^m19^s$ (17.95528 h),

$$\frac{17.95528}{24} \times 359°57'12'' = 269°17'39''$$

GHA of sun at observation,

$$180°19'15'' + 269°17'39'' = 449°36'54''$$

Since GHA is greater than 360°, subtract 360,

$$449°36'54'' - 360° = 89°36'54''$$

3. When the sun is due south, longitude = GHA,

$$\text{Longitude} = 89°36'54''$$

This method can produce longitudes accurate to within about $\pm 30''$ but timing must be very precise, since an error of $4''$ causes approximately a $1'$ error in longitude.

18-21 SOURCES OF ERROR IN ASTRONOMICAL OBSERVATIONS

Sources of error in astronomical observations are:

1. Instrument not perfectly leveled
2. Instrument not in good adjustment
3. Errors in pointing on the sun or stars
4. Time not correct, or not read exactly at the moment of observation
5. Parallax in readings taken at night
6. Inexact refraction corrections

18-22 MISTAKES

Some of the more common mistakes that occur in astronomical observations are:

1. Sighting on the wrong star
2. Use of a poor signal on the reference line

3. Index error not corrected for
4. Not averaging equal numbers of direct and reversed readings in the solution
5. Computational mistakes

18-1 Describe two simple methods of determining the true meridian that require no computations. Discuss the advantages and disadvantages of these methods.

18-2 A television tower 250 ft high and situated on level ground casts a 325-ft shadow. Compute the sun's altitude at that moment.

18-3 What is the approximate date of:
(a) The vernal equinox? (b) The winter solstice?

18-4 When is the sun's declination maximum? When is it positive?

*18-5 What is the declination of a star that passes through the zenith of an observer at latitude 39°47′N?

18-6 Do standard time zones follow 15° intervals exactly? Why?

18-7 What is the reason for having both apparent and mean solar time?

*18-8 A watch was checked against WWV at 8:13 P.M. and found to be 5.5 sec slow. At 11:45 P.M. it was checked again and recorded as 10.0 sec slow. What is the corrected time for an observation timed at 10:20:30 P.M. with this watch?

*18-9 Using Eq. (18-1), prepare a table of refraction corrections for altitude angles varying from 20° to 45° by increments of 5°. Assume temperature and pressure are 70°F and 29 in. Hg, respectively.

*18-10 Using Eq. (18-2), prepare a table of the sun's parallax corrections for altitude angles varying from 20° to 45° by increments of 5°.

18-11 Prepare a table of combined refraction and parallax corrections for altitude angles varying from 20° to 41° by increments of 2°. Assume temperature and pressure to be 60°F and 28.5 in. Hg, respectively.

18-12 At 12 noon universal time, what are the standard times at the following west longitudes:
(a) 74°20′ (b) 86°35′
*(c) 125°05′ (d) 169°45′

18-13 Express the following differences in longitude in terms of mean solar time:
(a) 85°10′ (b) 125°36′
(c) 130°20′30″ *(d) 16°42′12″

18-14 Explain what is meant by the equation of time. How is it used?

18-15 At what central daylight time on 10 June of this year will the apparent sun's center be on the meridian of 90°W longitude?

18-16 At what eastern standard time on 25 February of this year will the center of the apparent sun be on the meridian of 72°30′W longitude?

*18-17 For an observer located at 43°05′N latitude, what are the azimuth and altitude of the north celestial pole?

18-18 Explain why there is one more sidereal day than there are solar days in a year.

18-19 What are the standard times for the following locations and circumstances?
(a) EST in New York (longitude 73°57′30″W) when it is 9:10 A.M. PST in San Francisco (longitude 122°45′54″W)
*(b) PST in San Francisco (longitude 122°45′54″W) when it is 5:44 P.M. CST in Chicago (longitude 87°51′12″W)

18-20 The longitude of Cincinnati is 84°25′21″W. What is the central standard time there when local civil time is 2:52:18 P.M.?

18-21 List the advantages and disadvantages of making an azimuth observation on Polaris at:
(a) Culmination (b) Elongation (c) Any hour angle

18-22 Assume eight observations (4D and 4R) have been taken on Polaris for an azimuth determination. List the data that should be recorded and explain any checks available to verify correctness of the observations.

*Asterisks indicate problems that have answers given in Appendix G.

*18-23 What is the local hour angle of Polaris at a place of 103°12′W longitude when its Greenwich hour angle is 162°34′?

18-24 Compute the local hour angle of Polaris at 8 P.M. CST on 10 November of this year at latitude 35°15′N and longitude 89°48′W. What is the star's meridian angle t at that instant of time?

18-25 Compute the zenith angle that must be put on a theodolite's vertical circle in order to sight Polaris at 7:30 P.M. PST on 5 December of this year at a place of latitude 41°05′N and longitude 118°59′W.

18-26 Compute the true bearing of line AB for the following situations:

   *(a) Polaris was observed at upper culmination from station A with 0°00′00″ on the horizontal circle. After turning a clockwise angle to station B, the horizontal circle reading was 308°32′51″.

   (b) Polaris was observed at lower culmination from station A with 180°00′00″ on the horizontal circle. After turning a clockwise angle to station B, the horizontal circle reading was 217°21′34″.

   (c) Polaris was observed at any hour angle from station A with 0°00′00″ on the horizontal circle. After turning a clockwise angle to station B, the horizontal circle reading was 149°29′40″. Computations yielded a bearing of Polaris of N0°20′12″E for the instant sighted.

18-27 Determine the sun's declination for the following times this year:

   (a) 8:20 A.M. CST, 15 February   **(b)** 10:35 A.M. EDT, 13 June

   (c) 2:25 P.M. PDT, 20 August   **(d)** 4:10 P.M. MST, 25 December

*18-28 Direct (face left) and reversed (face right) observations on a fixed point yielded zenith angle readings of 60°52.0′ and 299°12.4′, respectively. With the same instrument, an azimuth observation made in the direct mode on the sun's center produced a zenith angle of 52°18.5′. If the temperature and pressure at the time of observation were 56°F and 29.6 in. Hg, respectively, what was the true altitude of the sun at the instant sighted?

18-29 Cite the advantages of observing Polaris for azimuth rather than the sun.

18-30 What data must be recorded in making observations on the sun for azimuth by: **(a)** The altitude method **(b)** The hour-angle method

18-31 List the sides of the PZS triangle used in computing solar observations for azimuth by the hour-angle method and give the angles opposite them.

18-32 In making sun observations for azimuth, cite the advantages of the hour-angle method over the altitude method.

18-33 Why are the hours of 8 to 10 A.M. and 2 to 4 P.M. standard time best to make solar observations for azimuth if the altitude method is used?

18-34 Compute the azimuth of line AB by the hour-angle method for the following data taken this year: station occupied A; date 22 May; average clock time and horizontal angle from the sun to B for four observations 3:00:30 P.M. (MDT) and 25°58′55″, respectively, latitude and longitude of station A 32°17′58″ and 106°48′05″, respectively.

18-35 Compute the azimuth of line CB by the hour-angle method for the following data taken this year: station occupied C; date 10 October; average clock time and horizontal angle from the sun to B for four observations 9:06:10 (EDT) and 101°35′20″, respectively; latitude and longitude of station C 39°40′50″ and 75°01′00″, respectively.

*18-36 The vertical angle to the sun's center, measured in the northern hemisphere as it crosses the observer's meridian is +42°31′ after all corrections have been made. If the sun's declination was −4°35′, what is the observer's latitude?

18-37 By using a Roelof prism, the zenith angle to the sun at its maximum altitude was measured in the direct mode as 53°59.6′ at 11:56:30 A.M. MST on 15 November of the current year. The temperature and pressure were recorded as 30°F and 29.0 in. Hg, respectively. A check on the instrument revealed an index error of +0°00.8′. Calculate the observer's latitude.

18-38 At what central standard time on 15 March of this year will the apparent sun cross an observer's meridian at longitude 90°41′00″?

*Asterisks indicate problems that have answers given in Appendix G.

*18-39 What is the latitude of an observer if the measured altitude of Polaris at upper culmination is 38°54′52″ when the polar distance to the star is 47′12″? (At the time of observation the pressure and temperature were 28.7 in. Hg and 58°F, respectively.)

18-40 Assume the center of the sun crossed the meridian of a place at 12:06:52 EST on 14 January of this year. What is the longitude of the place?

18-41 Similar to Problem 18-40, except the crossing occurred at 12:50:39 CDT on 26 May.

*Asterisks indicate problems that have answers given in Appendix G.

BIBLIOGRAPHY

Buckner, R. B. 1983. "Azimuth Determination by Solar Observation: New Perspectives on an Old Problem." *Surveying and Mapping* 43 (No. 3):307.

Buckner, R. B. 1984. *A Manual on Astronomic and Grid Azimuth.* Rancho Cordova, Calif.: Landmark.

Buckner, R. B. 1984. "From Astronomic to Grid Azimuth." *Point of Beginning* 10 (No. 2):24.

Knowles, D. W. 1984. "Advantages of the Hour Angle Method for Sun Observations." *Point of Beginning* 9 (No. 6):50.

Knowles, D. W. 1985. "Accurate Time for Celestial Observations Using an HP-41." *Point of Beginning* 11 (No. 2):26.

Mackie, J. B. 1985. *Elements of Astronomy for Surveyors,* 9th ed. London: Charles Griffin.

Mattson, D. F. 1991. "Double Sun Shots Without Latitude or Longitude." *Surveying and Land Information Systems* 51 (No. 3):171.

McDonnell, P. W., Jr. 1982. "Radio Time Signals for Astronomic Observations." *Point of Beginning* 7 (No. 2):44.

Rish, R. F. 1980. "Rate of Change Methods in Astronomical Surveying." *ASCE, Journal of the Surveying and Mapping Division* 106 (No. SU1):137.

Solaric, N., and D. Spoljaric. 1992. "Accuracy of the Automatic Astronomical Azimuth Determination by Polaris with Leica-Kern E2 Electronic Theodolite." *Surveying and Land Information Systems* 52 (No. 2):80.

Wijayratne, I. D. 1987. "A Note on Accurate Astronomic Azimuth Determination." *Surveying and Mapping* 47 (No. 2):143.

19
CONTROL SURVEYS

19-1 INTRODUCTION

Control surveys establish precise horizontal and vertical positions of reference monuments. These serve as the basis for originating or checking subordinate surveys for projects such as topographic and hydrographic mapping; property boundary delineation; and route and construction planning, design, and layout. They are also essential as a reference framework for giving locations of data entered in Land Information Systems (LISs) and Geographic Information Systems (GISs).

Traditionally there have been two general types of control surveys: *horizontal* and *vertical.* Horizontal surveys generally establish geodetic latitudes and longitudes of stations over large areas. From these values, plane rectangular coordinates, usually in a state plane or universal transverse Mercator (UTM) coordinate system (see Chapter 21), can be computed. On control surveys of smaller areas, plane rectangular coordinates may be determined directly without obtaining geodetic latitudes and longitudes.

To explain geodetic latitude and longitude, it is necessary to define the *ellipsoid,* which is a mathematical surface obtained by revolving an ellipse about the earth's polar axis. The ellipse dimensions are selected to give a good fit of the ellipsoid to the *geoid* over a large area. The geoid, for all practical purposes, is the earth's mean sea level surface, and it is everywhere perpendicular to the direction of gravity. Because of variations in the earth's mass distribution, the geoid has an irregular shape.

A two-dimensional view which illustrates conceptually the geoid and the ellipsoid is shown in Figure 19-1. As illustrated, the geoid contains nonuniform undulations (which are exaggerated in the figure for clarity) and is therefore not readily defined mathematically. The *Clarke Ellipsoid of 1866* fits the geoid in North America very well, and from 1879 until recently it was the ellipsoid used as a reference surface for specifying geodetic positions of points in the United States, Canada, and Mexico. Currently a new ellipsoid, known as the *Geodetic Reference System of 1980* (GRS80), is being employed (see Section 19-2.2). Sizes and shapes of ellipsoids can be specified by giving their

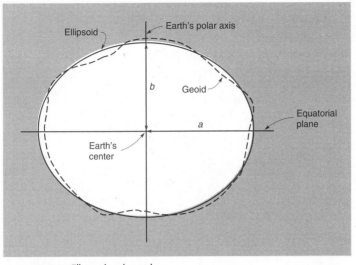

Figure 19-1 Ellipsoid and geoid.

equatorial semiaxes *a* and polar semiaxes *b,* as illustrated in Figure 19-1. These values for the two ellipsoids are:

ELLIPSOID	SEMIAXIS a (m)	SEMIAXIS b (m)
Clarke, 1866	6,378,206.4	6,356,583.8
GRS80	6,378,137.0	6,356,752.3

For either ellipsoid, the polar semiaxis is only about 21 km (13 mi) shorter than the equatorial semiaxis, so for some calculations involving moderate lengths (usually up to about 50 km) a true sphere of radius 6,371,000 m can be assumed.[1]

Figure 19-2 shows a three-dimensional view of the ellipsoid and illustrates geodetic latitude ϕ_A and geodetic longitude λ_A for any point *A* on its surface. Great circles on the circumference of the ellipsoid that pass through the north and south poles are called *meridians;* any plane containing a meridian and the polar axis is a *meridian plane.* Geodetic latitude ϕ_A is the angle, in the meridian plane containing *A,* between the equatorial plane and the *normal* to the ellipsoid at *A.* Geodetic longitude λ_A is the angle, in the equatorial plane, between the planes of the Greenwich meridian and the meridian through *A.* Precise latitudes and longitudes that accurately specify the relative positions of widely spaced points are determined by using geodetic field and computational techniques.

[1]This is the radius of a sphere having the same volume as the GRS80 reference ellipsoid. It is computed from $\sqrt[3]{a^2b}$.

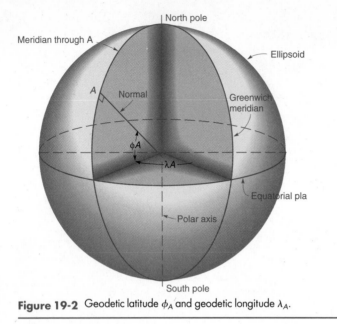

Figure 19-2 Geodetic latitude ϕ_A and geodetic longitude λ_A.

Field procedures used in horizontal-control surveys have traditionally been the ground methods of *triangulation, precise traversing, trilateration,* and combinations of these basic approaches. In addition, astronomical observations have been made to determine azimuths, latitudes, and longitudes. Rigorous photogrammetric techniques have also been used to densify control in limited areas. More recently satellite and inertial systems have been employed. The Global Positioning System (GPS) is currently gaining widespread use and is rapidly replacing all other methods for control surveying because of several advantages, an important one being its extremely high-accuracy capabilities.

Vertical-control surveys establish elevations for a network of monuments called *bench marks.* Depending on accuracy requirements, vertical-control surveys have traditionally been run by either *differential leveling* or *trigonometric leveling.* Satellite and inertial systems can also be used to establish vertical control, but the most accurate and widely applied method is precise differential leveling.

After presenting a general description of the National Geodetic Reference System, this chapter briefly describes traditional ground methods for conducting control surveys. These procedures are still employed on projects of limited extent. GPS and inertial surveying systems are described in Chapter 20.

19-2 REFERENCE DATUMS

Horizontal and vertical datums consist of a network of control monuments and bench marks whose horizontal positions and elevations have been determined by precise geodetic control surveys. These monuments serve as reference points for originating subordinate surveys of all types. The horizontal and vertical datums used in the immediate past and at present in the United States are described in the following subsections.

19-2.1 North American Horizontal Datum of 1927 (NAD27)

In 1927 a general least-squares adjustment was performed which incorporated all horizontal geodetic surveys that had been completed up to that date. It was referred to as the *North American Datum of 1927* (NAD27). The adjustment utilized the Clarke ellipsoid of 1866 and held fixed the latitude and longitude of station Meades Ranch in Kansas (the initial point), along with the azimuth to nearby station Waldo. The project yielded adjusted latitudes and longitudes for some 25,000 monuments existing at that time. Until the advent of the current datum (NAD83), positions of stations established after 1927 were adjusted to be consistent with these NAD27 monuments.

19-2.2 North American Horizontal Datum of 1983 (NAD83)

The National Geodetic Survey (NGS) began a new program in 1974 to perform another general adjustment of the North American horizontal datum. The project was originally scheduled for completion in 1983, hence its name *North American Datum of 1983* (NAD83), but it was not actually finished until 1986. The adjustment was a huge undertaking, incorporating approximately 270,000 stations and all geodetic surveying measurements on record—nearly 2 million of them. About 350 person-years of effort were required to accomplish the task.

The initial point in the new adjustment is not a single station such as Meades Ranch in Kansas. Rather, the earth's mass center, and numerous other points whose latitudes and longitudes had been precisely established from radio astronomy and satellite observations, were used. The GRS80 ellipsoid, centered at the earth's mass center, was employed since it fits the earth, in a global sense, more accurately than the Clarke ellipsoid of 1865. NAD83 positions are therefore consistent with satellite location systems (an important consideration with the rapidly expanding use of GPS), and with other latitudes and longitudes established worldwide on the GRS80 ellipsoid.

Within the United States, the adjusted latitudes and longitudes of all monuments in NAD83 differ from their NAD27 values. These differences result primarily because of the different ellipsoids and origins used, but part of the change is also due to the addition of the many post-NAD27 measurements in the NAD83 adjustment. The approximate magnitudes of these changes in the conterminous United States, expressed in meters, are illustrated in Figure 19-3. Here changes in latitude are shown in figure (a), while (b) shows longitude changes. From these diagrams it is seen that the changes are greatest along the west coast, where they reach or exceed 100 m. For Alaska and Hawaii (not shown in Figure 19-3), changes reach over 200 m and 400 m, respectively. In the Middle West, the changes are quite small. Of course, these differences have a significant impact on all existing control points and map products that were based on NAD27. Various mathematical models are being developed to transform NAD27 values to their NAD83 positions (see Section 21-11). A major benefit of NAD83 is that all control is now consistently uniform in accuracy and practically free from the distortions that had existed in the NAD27, owing to the piecemeal way additional points had been tied to the network.

19-2.3 National Geodetic Vertical Datum of 1929 (NGVD29)

Vertical datums for referencing bench-mark elevations are based on mean sea level. Until recently the vertical datum in use in the United States was the *National Geodetic Vertical Datum of 1929* (NGVD29), because that was the year that a general adjustment

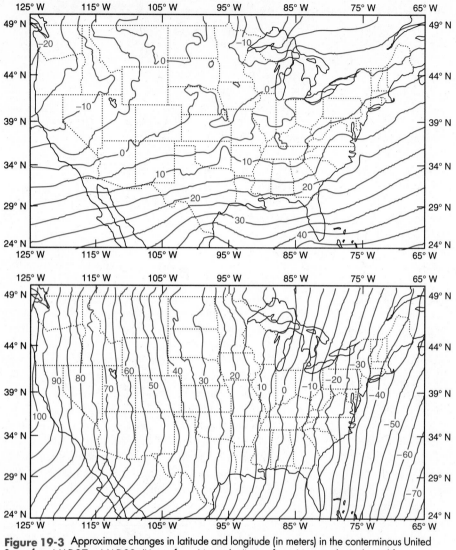

Figure 19-3 Approximate changes in latitude and longitude (in meters) in the conterminous United States from NAD27 to NAD83. (Upper figure) Latitude. (Lower figure) Longitude. (Adapted from National Geodetic Survey maps.)

of bench marks occurred. The NGVD29 was obtained from a best fit of mean sea-level observations taken at 26 gauging stations in the United States and Canada.

19-2.4 North American Vertical Datum of 1988 (NAVD88)

Since 1929 more than 625,000 km of additional control leveling has been run. Crustal movements and subsidence have changed the elevations of many bench marks. To incorporate the additional leveling and correct elevations of erroneous bench marks, a new general vertical adjustment was recently performed. It was originally scheduled for completion in 1988 and named the *North American Vertical Datum of 1988* (NAVD88), but it was not actually finished until 1991. This adjustment shifted the

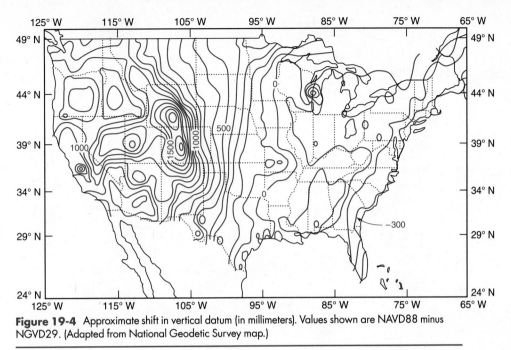

Figure 19-4 Approximate shift in vertical datum (in millimeters). Values shown are NAVD88 minus NGVD29. (Adapted from National Geodetic Survey map.)

position of mean sea level somewhat, and thus resulted in changes to the elevations of all bench marks in the vertical network. The magnitudes of these changes in the conterminous United States, expressed in millimeters, are given in Figure 19-4.

19-3 ACCURACY STANDARDS AND SPECIFICATIONS FOR CONTROL SURVEYS

The required accuracy for a control survey depends primarily on its purpose. Some major factors that affect accuracy are type and condition of equipment used, field procedures adopted, and experience and capabilities of available personnel. To guide surveyors, the Federal Geodetic Control Subcommittee (FGCS) has prepared and published a set of detailed standards of accuracy and specifications.[2] The rationale for them is twofold: **(1)** to provide a uniform set of standards specifying minimum acceptable accuracies of control surveys for various purposes, and **(2)** to establish specifications for instruments, field procedures, and misclosure checks to ensure that the intended level of accuracy is achieved.

Standards and specifications developed by the FGCS establish the following three distinct so-called *orders of accuracy* that govern traditional ground surveys, given in descending order: *first-order, second-order,* and *third-order*. For horizontal control surveys, second-order and third-order each have two separate accuracy categories, *class I* and *class II*. For vertical surveys, first-order and second-order each have class I and class II accuracy divisions. Thus a total of five classes are defined in the specifications for both horizontal and vertical control surveys.

[2]A booklet entitled *Standards and Specifications for Geodetic Control Networks* is available from the National Geodetic Information Center, NOAA, National Geodetic Survey, N/CG17, SSMC3 Station 09535, 1315 East-West Highway, Silver Spring, Md. 20910.

TABLE 19-1 HORIZONTAL CONTROL SURVEY ACCURACY STANDARDS

ORDER AND CLASS	RELATIVE ACCURACY REQUIRED BETWEEN POINTS
First order	1 part in 100,000
Second order	
Class I	1 part in 50,000
Class II	1 part in 20,000
Third order	
Class I	1 part in 10,000
Class II	1 part in 5000

TABLE 19-2 VERTICAL CONTROL SURVEY ACCURACY STANDARDS

ORDER AND CLASS	RELATIVE ACCURACY REQUIRED BETWEEN BENCH MARKS*
First order	
Class I	$0.5 \text{ mm} \times \sqrt{K}$
Class II	$0.7 \text{ mm} \times \sqrt{K}$
Second order	
Class I	$1.0 \text{ mm} \times \sqrt{K}$
Class II	$1.3 \text{ mm} \times \sqrt{K}$
Third order	$2.0 \text{ mm} \times \sqrt{K}$

*K is distance between bench marks, in kilometers.

Triangulation, trilateration, and traverse surveys are included in the FGCS horizontal control standards and specifications, and differential leveling is covered in the vertical control section. A separate set of guidelines has been prepared for GPS surveys (see Section 20-10).

Tables 19-1 and 19-2 give accuracy standards required by the FGCS for the various orders and classes of horizontal and vertical control surveys, respectively. Values in Table 19-1 are ratios of allowable relative positional errors of a pair of horizontal control points to the horizontal distance separating them. Thus two first-order stations located 100 km (60 mi) apart are expected to be correctly located with respect to each other to within ± 1 m. The FGCS is currently in the process of developing new standards, and surveyors should be alert for changes in the future.

Table 19-2 gives maximum relative elevation errors allowable between two bench marks, as determined by a weighted least-squares adjustment (see Appendix C). Thus elevations for two bench marks 25 km apart established by second-order class I standards should be correct to within $\pm 1.0\sqrt{25} = \pm 5$ mm. These standards are not the same as the maximum allowable loop misclosures for the five classes of leveling given in Section 7-5. Table 19-2 values specify relative accuracies of bench marks after adjustment, whereas loop misclosures enable assessment of results in differential leveling prior to adjustment.

The ultimate success of any engineering or mapping project depends on appropriate survey control. The higher the order of accuracy demanded, the more time and expense are required. It is therefore important to select the proper order of accuracy for a given project and carefully follow the specifications. Note that no matter how accurately a control survey is conducted, errors will exist in the computed positions of its stations, but a higher order of accuracy presumes smaller errors.

19-4 THE NATIONAL GEODETIC REFERENCE SYSTEM

To meet the various local needs of surveyors, engineers, and scientists, the federal government has established a National Geodetic Reference System (NGRS) consisting of more than 270,000 horizontal control monuments and approximately 600,000 bench marks throughout the United States. The National Geodetic Survey (NGS) began control surveying operations as the *Survey of the Coast* in 1807, changed to *Coast Survey* in 1836, to *Coast and Geodetic Survey* in 1878, and to a division of the *National Ocean Survey* (NOS) in 1970. It continues to assist with and to coordinate geodetic control surveying activities with other agencies and with all states to establish new control stations and upgrade and maintain existing ones. It also disseminates a variety of publications and software related to geodetic control surveying.

The NGRS is split into *horizontal* and *vertical* divisions. All control within each part is classified in a ranking scheme based on purpose and order of accuracy.

19-5 HIERARCHY OF THE NATIONAL HORIZONTAL CONTROL NETWORK

The hierarchy of control that has been established within the National Horizontal Control Network, from highest to lowest order, is as follows:

Primary control consists principally of east-west arcs of triangulation ideally spaced at about 100 km, crossed by north-south arcs having similar spacing. In addition to triangulation, traverse, and trilateration, satellite methods (see Chapter 20) have been used more recently to establish latitudes and longitudes of monuments. Primary control is established by first-order methods.

Secondary control densifies the network within areas surrounded by primary control, especially in high-value land areas. Secondary-control surveys are executed to second-order class I standards.

Supplemental control generally densifies monuments between the primary network in lightly developed areas. It is also located along coastlines and on extensive mapping and construction projects. Supplemental control surveys originate at stations of the primary and, occasionally, secondary networks, and are executed to second-order class II standards.

Local control provides reference points for local construction projects and small-scale topographic mapping. These surveys are referenced to higher order control monuments and, depending on accuracy requirements, may be third order class I or third order class II.

Figure 19-5 shows the locations of primary and secondary triangulation arcs and the traverse network that existed in the State of Wisconsin in 1986. Larger diagrams are available which show the triangulation and traverse networks for the entire United States.

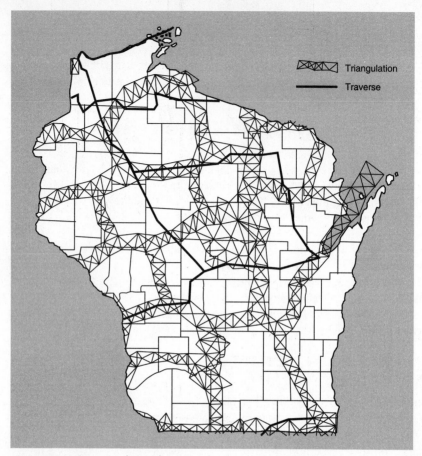

Figure 19-5 Primary and secondary triangulation arcs and traverse network in the State of Wisconsin in 1986. (Adapted from National Geodetic Survey map.)

19-6 HIERARCHY OF THE NATIONAL VERTICAL CONTROL NETWORK

The scheme of bench marks within the National Vertical Control Network may be classified as follows:

Basic framework is a uniformly distributed nationwide network of bench marks whose elevations are determined to the highest order of accuracy. It consists of nets *A* and *B*. In net *A* adjacent level lines are ideally spaced an average of about 100 to 300 km apart using first-order class I standards; in net *B* the average separation is ideally about 50 to 100 km, and first-order class II standards are specified. Bench marks are placed intermittently along the level lines at convenient locations.

Secondary network densifies the basic framework, especially in metropolitan areas and for large engineering projects. It is established to second-order class I standards.

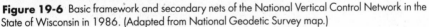

Figure 19-6 Basic framework and secondary nets of the National Vertical Control Network in the State of Wisconsin in 1986. (Adapted from National Geodetic Survey map.)

General area control consists of vertical control for local engineering, surveying, and mapping projects. It is established to second-order class II standards.

Local control provides vertical references for minor engineering projects and small-scale topographic mapping. Bench marks in this category satisfy third-order standards.

Figure 19-6 shows the basic framework and secondary net of the National Vertical Control Network that existed in Wisconsin in 1986.

19-7 GPS HIGH-ACCURACY REFERENCE NETWORKS (HARNS)

The internal consistency of first-order points adjusted in NAD83 was specified to be at least 1:100,000, but tests have verified that on average it is probably 1:200,000 or better. There are some areas, however, where relative accuracies fall below 1:100,000. As GPS technology improves, accuracies to 1:1,000,000 and better are consistently being obtained. There has been a growing concern among GPS users about how to fit their

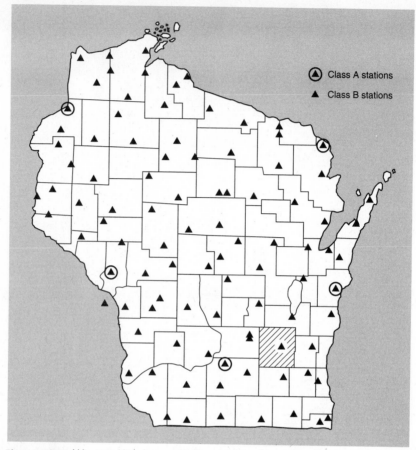

Class A stations
Class B stations

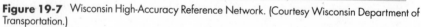

Figure 19-7 Wisconsin High-Accuracy Reference Network. (Courtesy Wisconsin Department of Transportation.)

measurements to a network of reference points whose inherent accuracy is less than that of this new technology.

Several states have sought to solve this problem by establishing their own GPS *High-Accuracy Reference Networks* (HARNs), also called *High Precision Geodetic Networks* (HPGNs). The NGS is assisting in this effort. As of January 1993, 14 states had already completed their networks, and all except eight of the remaining states were in the process. Wisconsin is one of the states whose network is completed, and the procedures used in establishing it are described here.

As discussed in Section 20-10, a new set of accuracy standards has been proposed by the FGCS for GPS surveys. It specifies orders A and B with relative accuracies of 1:10,000,000 and 1:1,000,000, respectively. The Wisconsin High-Accuracy Reference Network (WHARN) was created in two steps. The first consisted of adjusting extensive GPS observations that had been made at five stations uniformly spaced within the state. The adjustment of these five stations was tied to four so-called *fiducial* stations within the national network whose positions in NAD83 were known to have an extremely high precision. The observations and adjustment were performed to class A standards. In the second step, GPS observations made at an additional 93 stations in Wisconsin were

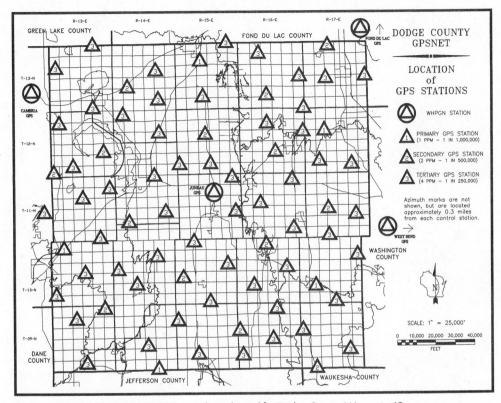

Figure 19-8 Densified NAD83 GPS control net planned for Dodge County, Wisconsin. (Courtesy Dodge County, Wisconsin.)

adjusted relative to the five class A stations. This set of observations and the adjustment conformed to class B standards. The WHARN stations (98 in total) are spaced uniformly throughout the state approximately 50 km (30 mi) apart, as shown in Figure 19-7. Coordinates on all stations derived from this adjustment have created an upgraded Wisconsin datum referred to as NAD83 (1991), since 1991 was the year the work was completed. Note that the WHARN provides a much better overall distribution of control stations than the triangulation network (shown in Figure 19-5) that existed in Wisconsin in 1986.

In states where GPS high-precision geodetic networks have been completed, some counties are now in the process of further densifying their networks of GPS stations. These stations are being tied to the state's HARN. An example is illustrated in Figure 19-8, which shows the planned locations of approximately 75 stations that will comprise the Dodge County, Wisconsin, GPS net (Dodge County is shown crosshatched in Figure 19-7). In the Dodge County net, station spacing ranges from approximately 5 to 10 km. An intervisible azimuth mark is being set near each station.

Stations with three different levels of accuracy will be established in the Dodge County GPS net: *level 1* at 1 part in 1,000,000, *level 2* at 1 part in 500,000, and *level 3*

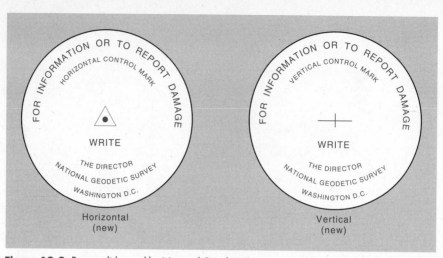

Figure 19-9 Bronze disks used by National Geodetic Survey to mark horizontal and vertical control stations.

at 1 part in 250,000. The different station types are shown within the monument symbols of Figure 19-8. At these precisions the relative positions of any two adjacent stations should be known within ±4 cm or less. This will practically eliminate any misclosure problems due to imprecise control for local surveyors who initiate surveys on one station and close on another.

Because of the relative ease with which accurate NAD83 coordinates can be determined using GPS, it can be expected that this trend of establishing more densely spaced control stations will continue into the future. In some states that were subdivided according to the U.S. Public Land Survey System (see Chapter 23), efforts are already under way to monument all section corners and determine their NAD83 coordinates using GPS.

19-8 CONTROL-POINT DESCRIPTIONS

To obtain maximum benefit from control surveys, horizontal stations and bench marks are placed in locations favorable to their subsequent use. The points should be permanently monumented and adequately described to ensure recovery by future potential users. Reference monuments placed by the NGS are marked by bronze disks about $3\frac{1}{2}$ in. in diameter set in concrete or bedrock. Figure 19-9 shows two of these disks.

The NGS makes complete descriptions of all its control stations available to local surveyors.[3] An example, excerpted from an actual NGS station description, is given in Figure 19-10. As illustrated in the example, these station descriptions give each station's general placement in relation to nearby towns, instructions on how to reach the station following named or numbered roads in the area, and the monument's precise location by means of distances and directions to several nearby objects. The station's specific description, namely, "a triangulation disk set in drill hole in rock outcrop," is given,

---

[3]Requests for control data in a given area should be made to the National Geodetic Information Center, NOAA, National Geodetic Survey, N/CG17, SSMC3 Station 09535, 1315 East-West Highway, Silver Spring, Md. 20910.

DEPARTMENT OF COMMERCE
NOAA - NOS - C&GS
NATIONAL GEODETIC SURVEY

PUBLICATION DATE: JANUARY 7, 1992
USGS QUAD SHEET: SPRINGFIELD CORNERS

ROCK
PID: OM0651

HORIZONTAL DATUM: NAD 83
VERTICAL DATUM: NGVD 29

STATE: WISCONSIN
COUNTY: DANE

LATITUDE:	43° 12' 01.93234" N
LONGITUDE:	089° 30' 05.36616" W

ORTHOMETRIC HEIGHT: {
323.1   METERS
1060.   FEET

GEOID HEIGHT: -34.36   METERS

HORIZONTAL NETWORK ORDER/CLASS: A

THE HORIZONTAL COORDINATES WERE ESTABLISHED BY GPS OBSERVATIONS
AND ADJUSTED BY THE NATIONAL GEODETIC SURVEY IN JUNE 1991
THE ORTHOMETRIC HEIGHT WAS DETERMINED BY GPS OBSERVATIONS.

PLANE COORDINATES

GRID	ZONE	NORTHING METERS	EASTING METERS	POINT SCALE FACTOR	CONVERGENCE DEG MIN SEC
SPC	WI S	133,489.006	640,513.767	0.99993865	+ 0 20 33.1
UTM	16	4,786,122.467	296,767.801	1.00010812	– 1 42 46.8

STATION MARK IS A . . . . TRIANGULATION STATION DISK
WITH SETTING . . . . SET IN DRILL HOLE IN ROCK OUTCROP
THE MARK IS STAMPED . . . ROCK 1981

STATION MARK HISTORY

YEAR MONUMENTED OR RECOVERED	CONDITION OF MARK	RECOVERED OR DESCRIBED BY
1981	STATION MONUMENTED	DANE COUNTY WISCONSIN
1987	GOOD	DANE COUNTY WISCONSIN
1990	GOOD	AERO SERVICE CORPORATION

STATION DESCRIPTION

DESCRIBED BY DANE COUNTY WISCONSIN 1987
THE STATION IS LOCATED ABOUT 5.6 KM (3.5 MI) SOUTH OF DANE, 4.8 KM
(3.0 MI) EAST OF SPRINGFIELD, AND 4.0 KM (2.5 MI) NORTHWEST OF WAUNAKEE
ON ROAD RIGHT-OF-WAY. OWNERSIP--TOWN OF SPRINGFIELD, LEONARD MEIER,
CHAIRMAN, 6535 FISHER ROAD, WAUNAKEE WI 53597, 608-831-8383.

TO REACH THE STATION FROM THE JUNCTION OF U.S. HIGHWAY 12 AND STATE
HIGHWAY 19, WHICH IS ABOUT 8.9 KM (5.5 MI) SOUTHWEST OF DANE, GO
SOUTHEAST ON U.S. HIGHWAY 12 AND STATE HIGHWAY 19 FOR 2.3 KM (1.45
MI) TO WHERE STATE HIGHWAY 19 TURNS LEFT.   TURN LEFT AND GO EAST
FOR 4.7 KM (2.9 MI) TO WHIPPERFURTH ROAD ON THE LEFT. TURN LEFT AND
GO NORTH FOR 1.4 KM (0.85 MI) TO THE STATION ON THE RIGHT.

THE STATION IS A STANDARD WI-025 DISK STAMPED---ROCK 1981---, SET IN
CONCRETE IN THE FLOOR OF AN OLD ROCK QUARRY. LOCATED 15.2 METERS
(50.0 FT) WEST OF A FIBERGLASS WITNESS POST, 7.6 METERS (25.0 FT)
WEST OF THE CENTER OF ROADWAY, 1.8 METERS (5.8 FT) EAST OF A
FIBERGLASS WITNESS POST, AND ABOUT 5 CM BELOW GROUND SURFACE.

DANE COUNTY, WISCONSIN 1987.

THIS STATION IS SUITABLE FOR GPS SURVEYS.

Figure 19-10 Description of a station in the National Geodetic Reference System. (Courtesy National
Geodetic Survey.)

along with a record of recovery history. Data supplied with horizontal control-point descriptions include the datum(s) used and the station's geodetic latitude and longitude. Also given are the state plane coordinates (in feet for NAD27, meters for NAD83), state plane convergence angle and scale factor, UTM coordinates (in meters and in NAD83 only), approximate elevation—also called *orthometric height* (in feet and meters), and geoid height (in meters). These latter terms are described in Chapter 21.

Some station descriptions give geodetic and plane azimuths to a nearby station or stations (from south for NAD27 and from north for NAD83). Geodetic and plane azimuths differ by an amount equal to the convergence angle, and therefore the appropriate azimuth must be selected for the particular surveying methods used.

Published bench-mark data include station locations and adjusted elevations in both meters and feet. Again the relevant datum, either NGVD29 or NAVD88, is identified.

Besides control within the national network set by the NGS, additional marks have also been placed in various parts of the United States by other federal agencies such as the USGS, Corps of Engineers, and Tennessee Valley Authority. Also, state, county, and municipal organizations may have added control. This work is frequently coordinated through the NGS, and descriptions of these stations are distributed by that agency.

As noted earlier, complete descriptions for all points in the National Geodetic Reference System can be obtained from the NGS. This includes both horizontal and vertical control points. The descriptions can be obtained in hard copy form, or in computer format on diskettes or compact disks. Only five compact disks are needed to store all NGRS data for the entire United State!

19-9 FIELD PROCEDURES FOR TRADITIONAL HORIZONTAL-CONTROL SURVEYS

As noted earlier, in spite of the increasing prominence of GPS, horizontal-control surveys over limited areas are still being accomplished by the traditional methods of triangulation, trilateration, precise traverse, or a combination of these techniques. These methods are described briefly in the sections that follow.

Traditional methods in horizontal control surveys require measurements of horizontal distances and horizontal angles. Basic theory, equipment, and procedures for making these measurements have been covered in earlier chapters. The following sections concentrate on procedures specific to control surveys, and on matters related to obtaining the higher orders of accuracy generally required for these types of surveys.

19-10 TRIANGULATION

Prior to the emergence of electronic distance-measuring equipment, triangulation was the preferred and principal method for horizontal-control surveys, especially if extensive areas were to be covered. Angles could be more easily measured compared with distances, particularly where long lines over rugged and forested terrain were involved, by erecting towers to elevate the operators and their instruments. Triangulation possesses a large number of inherent checks and closure conditions which help detect blunders and errors in field data and increase the possibility of meeting a high standard of accuracy.

As implied by its name, triangulation utilizes geometric figures composed of triangles. Horizontal angles and a limited number of sides called *base lines* are measured. By using the angles and base-line lengths, triangles are solved trigonometrically and positions of stations (vertices) calculated.

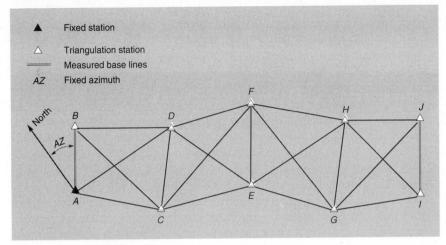

▲   Fixed station

△   Triangulation station

━━━   Measured base lines

AZ   Fixed azimuth

Figure 19-11   Chain of quadrilaterals.

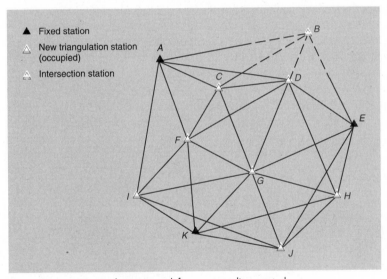

▲   Fixed station

△   New triangulation station
     (occupied)

△   Intersection station

Figure 19-12   Triangulation network for a metropolitan control survey.

Different geometric figures are employed for control extension by triangulation, but chains of quadrilaterals called *arcs* (Figure 19-11) have been most common. They are the simplest geometric figures that permit rigorous closure checks and adjustments of field observational errors, and they enable point positions to be calculated by two independent routes for computational checks. More complicated figures like that illustrated in Figure 19-12 have frequently been used to establish horizontal control by triangulation in metropolitan areas.

Arcs of triangulation originate from one or more stations of known or fixed position and require the azimuth of at least one line. If two or more stations are fixed, azimuth orientation of the network is automatically determined. Today fixed starting stations and initial azimuths are normally available from other previous higher order control surveys. Astronomical observations are also made at various intervals throughout extensive arcs to check and supplement angle and base-line measurements and help maintain true azimuth orientation. In Figure 19-11 the arc of triangulation originates at fixed station *A* and employs the known azimuth of line *AB*. All horizontal angles within the arc, and base lines *AB* and *IJ* have been measured. From this information the positions of stations *B* through *J* were calculated.

In executing triangulation surveys, *intersection stations* can be located as part of the project. In this process, angles are measured from as many occupied points as possible to tall prominent objects in the area such as church spires, smokestacks, or water towers. The intersection stations themselves are not occupied, but their positions are calculated, and thus they become available as local reference points. An example is station *B* in Figure 19-12.

One of the most important aspects of any control survey is the selection of station locations. In triangulation, factors to be considered are (**1**) strength of figure, (**2**) station intervisibility, (**3**) station accessibility, and (**4**) overall project efficiency. Careful attention must be given to each factor in planning the optimum triangulation network.

Strength of figure deals with the relative accuracies of computed station positions that result from using angles of various sizes in calculations. Triangulation computations are based on the trigonometric *law of sines*. The sine function changes rapidly for angles near 0 and 180°, so a small observational error in an angle close to these values produces a comparatively large difference in calculated positions. In general, if no angles smaller than 30° or larger than 150° exist in a network, it should have good strength. Because the locations of the triangulation stations fix angle sizes, they must be selected carefully. If local terrain or other conditions preclude establishing figures with strong angles, more frequent base-line measurements are needed.

Station intervisibility is vital in triangulation because lines of sight to all stations within each figure must be clear for measuring angles. Thus stations are normally placed on the highest points available in the area or, if necessary, towers are used to elevate observers and instruments.

As noted previously, the fundamental field measurements for triangulation are horizontal angles and base-line lengths. The angles can be measured using any of the theodolites or total station instruments described in Chapter 10. For most precise work, however, first-order directional theodolites like the Kern DKM-3 (Figure 19-13) or Wild T-3 can be used. Both permit readings to be made directly without estimation to the nearest 0.2″. At each station several "positions" are read, a position being a direct and reversed reading on each target visible from the station. To compensate for possible circle graduation errors, the circle is advanced by approximately 180°/*n* for each successive position (as described in Section 11-9), where *n* is the number of positions read at the station. Angles should be computed in the field from the directions, checked for acceptable misclosure, and any rejected ones repeated before leaving the station. The mean of all satisfactory values for each angle is used in the triangulation calculations.

Triangulation frequently involves triangles that are so large they can no longer be assumed plane. In these large spherical triangles, the sums of their angles exceed 180° by an amount called *spherical excess*. Its value is closely approximated by the following formula:

$$\epsilon'' = \frac{\rho ab \sin C}{2R_e^2}$$

(19-1)

Figure 19-13 Kern DKM-3 directional theodolite suitable for first-order control surveying. (Courtesy Leica, Inc.)

Where ϵ'' is spherical excess, in seconds; ρ the number of seconds per radian (206,265); a and b the lengths of two triangle sides, in meters; C the included angle; and R_e the earth's mean radius (6,371,000 m). For a triangle with sides of 10-mi average length, spherical excess is less than 0.6″, but for 50-mi-long sides it is about 14″. To check angle misclosures in large triangles, spherical excess is computed, added to 180°, and compared to the sum of the three measured angles.

Base lines are currently measured by electronic methods, although in the past precise Invar tapes were used. The mean of measurements made in both directions should be used. Slope distances must first be reduced to horizontal and then their ellipsoid lengths computed (see Section 19-14). If state plane coordinates of the triangulation stations are being computed, the ellipsoid lengths must be converted to grid lengths (see Chapter 21).

To compensate for the errors that occur in the measurements, triangulation networks must be adjusted. The most rigorous method utilizes least squares (see Section 2-26 and Appendix C). In that procedure all angle, distance, and azimuth observations are simultaneously included in the adjustment, and are given appropriate relative weights based on their precisions. The least-squares method not only yields the most probable adjusted station coordinates, but it also gives their precisions. Other less rigorous, or "approximate," methods for triangulation adjustment can be applied to standard figures such as quadrilaterals. They give satisfactory results and are described in some advanced surveying books.

19-11 THREE-POINT PROBLEM

This procedure, often interchangeably called *resection,* is a special case of simple triangulation. It locates a single point by measuring horizontal angles from it to three

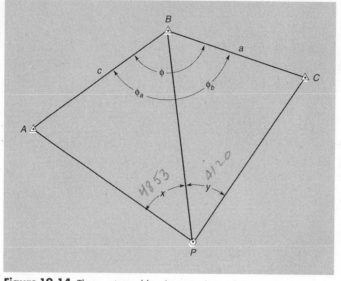

Figure 19-14 Three-point problem (resection).

visible stations whose positions are known. The situation is illustrated in Figure 19-14, where a theodolite occupies station P and angles x and y are measured. A summary of the method used to compute the coordinates of station P follows (refer to Figure 19-14):

1. From the coordinates of A, B, and C, calculate lengths a and c and angle ϕ at station B.

2. Summing angles x, y, and ϕ in figure $ABCP$, subtracting from 360°, and solving gives the sum of angles $A + C$,

$$A + C = 360° - (\phi + x + y) \qquad (19\text{-}2)$$

3. Calculate angles A and C using the following:

$$\cot A = \frac{c \sin y}{a \sin x \sin (A + C)} + \cot(A + C) \qquad (19\text{-}3)$$

$$\cot C = \frac{a \sin x}{c \sin y \sin (A + C)} + \cot (A + C) \qquad (19\text{-}4)$$

4. From angle A and azimuth AB, calculate azimuth AP. In triangle ABP solve for length AP using the law of sines, where $\phi_a = 180° - A - x$. Calculate the departure and latitude of AP, followed by the coordinates of P.

5. In the manner outlined in step 4, use triangle BCP to calculate the coordinates of P to obtain a check.

In Figure 19-14 angles x and y were measured as $48°53'12''$ and $41°20'35''$, respectively. Control points A, B, and C have coordinates $X_A = 5721.25$, $Y_A = 21,802.48$, $X_B = 12,963.71$, $Y_B = 27,002.38$, $X_C = 20,350.09$, and $Y_C = 24,861.22$ ft. Calculate the coordinates of P.

EXAMPLE 19-1

SOLUTION

1. By Eq. (13-16a),

$$a = \sqrt{(20,350.09 - 12,963.71)^2 + (24,861.22 - 27,002.38)^2} = 7690.46 \text{ ft}$$

$$c = \sqrt{(12,963.71 - 5721.25)^2 + (27,002.38 - 21,802.48)^2} = 8915.84 \text{ ft}$$

2. By Eq. (13-13),

$$AZ_{AB} = \tan^{-1}\left(\frac{12,963.71 - 5721.25}{27,002.38 - 21,802.48}\right) = 54°19'21.5''$$

$$AZ_{BC} = \tan^{-1}\left(\frac{20,350.09 - 12,963.71}{24,861.22 - 27,002.38}\right) = 106°09'56.8''$$

3. Calculate angle ϕ,

$$\phi = 180° - (106°09'56.8'' - 54°19'21.5'') = 128°09'24.7''$$

4. By Eq. (19-2),

$$A + C = 360° - 128°09'24.7'' - 48°53'12'' - 41°20'35''$$
$$= 141°36'48.3''$$

5. By Eq. (19-3),

$$\cot A = \frac{8915.84 \sin 41°20'35''}{7690.46 \sin 48°53'12'' \sin 141°36'48.3''} + \cot 141°36'48.3''$$

$$= 0.3746289$$

$$A = 69°27'45.4''$$

6. By Eq. (19-4),

$$\cot C = \frac{7690.46 \sin 48°53'12''}{8915.84 \sin 41°20'35'' \sin 141°36'48.3''} + \cot 141°36'48.3''$$

$$= 0.3220120$$

$$C = 72°09'02.9''$$

$$(A + C = 69°27'45.4'' + 72°09'02.9'' = 141°36'48.3'' \text{ check})$$

7. Calculate angle ϕ_a,

$$\phi_a = 180° - 69°27'45.4'' - 48°53'12'' = 61°39'02.6''$$

8. By the law of sines,

$$AP = \frac{\sin 61°39'02.6'' \ (8915.84)}{\sin 48°53'12''} = 10,414.72 \text{ ft}$$

9. By Eqs. (13-1) and (13-2),

$$\text{dep}_{AP} = 10,414.72 \sin 123°47'06.9'' = 8655.96 \text{ ft}$$

$$\text{lat}_{AP} = 10,414.72 \cos 123°47'06.9'' = -5791.43 \text{ ft}$$

10. By Eqs. (13-7),

$$X_p = 5721.25 + 8655.96 = 14,377.21 \text{ ft}$$

$$Y_P = 21,802.48 - 5791.43 = 16,011.05 \text{ ft}$$

11. As a check, triangle *BCP* was solved to obtain the same results.

The three-point problem just described is actually a special case of the more general *resection* problem and involved measuring two angles (*x* and *y*) to solve for two unknowns (the *X* and *Y* coordinates of station *P*). To provide redundancy and enable a least-squares solution, in addition to the angles, distances from *P* to one or more control stations could also have been measured. Other possible variations in resection that provide redundancy include measuring **(1)** one angle and two distances to two control stations; **(2)** two angles and one, two, or three distances to three control points; or **(3)** the use of more than three control stations. Then all measurements can be included in a least-squares solution to obtain the most probable coordinates of point *P*.

Resection has become a popular method for quickly orienting total station instruments. The procedure is convenient because these instruments can readily measure both angles and distances, and their on-board microprocessors can instantaneously provide the least-squares solution for the instrument's position (see Section 24-9).

19-12 PRECISE TRAVERSE

Precise traversing is common among local surveyors for horizontal-control extension, especially for small projects. Field work consists of two basic parts: reading horizontal angles at the traverse hubs and measuring distances between stations. Angles can be measured with either a repeating or a directional instrument, and distances measured with EDM equipment. Total station instruments are more convenient, however, because both angles and distances can be measured from a single setup. Precise traverses always begin and end on stations established by higher order surveys.

The FGCS has defined standards and specifications for five orders of accuracy for traverses. First-order and second-order classes have been used to supplement the

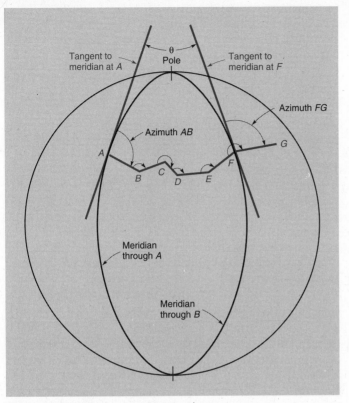

Figure 19-15 Meridian convergence on long east-west traverses.

National Horizontal Control Network, particularly where a greater density of control was needed than that afforded by triangulation. Second- and third-order traverses are run extensively to solidify control in metropolitan areas for engineering and construction projects, property surveys, aerial photogrammetric surveys, and numerous other projects.

Unlike triangulation, in which stations are normally widely separated and placed on the highest ridges and peaks in an area, traverse routes generally follow the cleared rights-of-way of highways and railroads, with stations located relatively close together. Besides easing field work, this provides a secondary benefit in accessibility of the stations. Traverses lack the automatic checks inherent in triangulation, and extreme observational caution must therefore be exercised to avoid blunders. Also, since traverses generally run along single lines, they are not as good as triangulation for establishing control over large areas.

For long traverses, checks on the measured horizontal angles can be obtained by making periodic astronomical azimuth observations. These should agree with the values computed from the direction of the starting line and the measured horizontal angles. However, if a traverse extends an appreciable east-west direction, as illustrated in Figure 19-15, meridian convergence will cause the two azimuths to disagree. In Figure 19-15, for example, azimuth *FG* obtained from direction *AB* and the measured horizontal angles should equal astronomic azimuth $FG + \theta$, where θ is the meridian convergence. A very good approximation for meridian convergence between two points on a traverse is

$$\theta'' = \frac{\rho d \tan \phi}{R_e} \tag{19-5}$$

where θ'' is meridian convergence, in seconds; d the east-west distance between the two points in meters; R_e the mean radius of the earth (6,371,000 m); ϕ the mean latitude of the two points; and ρ the number of seconds per radian (206,265).

Because of meridian convergence, forward and back azimuths of long east-west lines do not differ by exactly 180°, but rather by $180° \pm \theta$. (A sketch of the situation will clarify whether the sign should be plus or minus.) From Eq. (19-5) an east-west traverse of 1-mi length at latitude 30° produces a convergence angle of approximately 30″. At latitude 45°, convergence is approximately 51″/mi east-west. These calculations illustrate that the magnitude of convergence can be appreciable and must be considered when astronomic observations are made in connection with plane surveys that assume the Y axis parallel throughout the survey area.

Procedures for precise traverse computation vary, depending on whether a geodetic or a plane reference system is used. In either case it is necessary first to adjust angles and distances for observational errors. Closure conditions are enforced for (1) azimuths or angles, (2) departures, (3) and latitudes. The most rigorous process, the least-squares method (see Section 2-26 and Appendix C), should be used because it simultaneously satisfies all three conditions and gives residuals having the greatest likelihood according to the theory of probability. Other less precise procedures such as the compass (Bowditch) rule (see Section 13-7) are sometimes used.

In calculating state plane coordinates it is necessary to reduce observed horizontal distances to their grid lengths before making the calculations, as described in Section 21-9.

In the 1960s and 1970s the NGS conducted special "ultrahigh-accuracy transcontinental traverses" with results approaching 1 part per million. These surveys were originally begun to provide base lines for a worldwide satellite triangulation program, and were subsequently used to upgrade the horizontal control network. Theodolites and EDM equipment of highest precision were used and rigorous field procedures adopted to meet the extremely stringent specifications.

19-13 TRILATERATION

Trilateration, a method for horizontal control surveys based exclusively on measured horizontal distances, has gained acceptance because of EDM instrumentation. Both triangulation and traversing require time-consuming horizontal angle measurement. Hence trilateration surveys often can be executed faster and produce equally acceptable accuracies.

The geometric figures used in trilateration, although not as standardized, are similar to those employed in triangulation. Stations should be intervisible and therefore placed on the highest peaks, perhaps with towers to elevate instruments and observers.

Strength of figure in trilateration is less quantified than for triangulation. However, slender figures are weakest in the direction transverse to their long dimensions. Hence networks covering essentially square areas are better, since they give stronger overall uniform accuracy.

Because of intervisibility requirements and the desirability of having essentially square networks, trilateration is ideally suited to densify control in metropolitan areas and on large engineering projects. In special situations where topography or other conditions require elongated narrow figures, the network can be strengthened by reading some horizontal angles. Also, for long trilateration arcs, astronomic azimuth observations prevent the network from deforming in direction. A complete and accurate estimate of strength of figure for trilateration networks is secured by presurvey leastsquares analysis of trial configurations, but he procedure is beyond the scope of this book.

As in triangulation, surveys by trilateration can be extended from one or more monuments of known position. If only a single station is fixed, at least one azimuth must be known or observed.

Trilateration computations consist of reducing measured slope distances to horizontal lengths, then to the ellipsoid, and finally to grid lengths if the calculations are being done in state plane coordinate systems. Observational errors in trilateration networks must be adjusted, preferably by the least-squares method.

19-14 REDUCTION OF SLOPE DISTANCES TO THE ELLIPSOID

In geodetic control survey computations involving triangulation, traverse, or trilateration, measured slope distances must first be reduced to horizontal and then to the surface of the ellipsoid. Measured distances in control surveys are often long, and the short-line reduction techniques given in Section 5-9 do not provide satisfactory accuracy. This is especially true for steeply inclined long lines.

Figure 19-16 illustrates a slope distance L measured from A to B. Points A and B represent an EDM instrument and a reflector, respectively, O is the earth's center, and R_e its radius. Vertical angles α and β were measured at A and B, respectively. Arc AB_2, which is closely approximated by its chord, is the required horizontal distance. If the short-line reduction procedures given in Section 5-9 were used, horizontal distance AB_1 would result, which would be in error by B_1B_2. Arc $A'B'$ is the required ellipsoid distance.

From Figure 19-16 the following equations can be written to compute required horizontal distance AB_2:

$$\delta = \frac{\alpha - \beta}{2} \tag{19-6}$$

$$AB_1 = L \cos \delta \tag{19-7}$$

$$BB_1 = L \sin \delta \tag{19-8}$$

$$\psi = \frac{AB_1}{R_e + h_A} \times \frac{180°}{\pi} \text{ (approx.)} \tag{19-9}$$

$$B_1B_2 = BB_1 \tan\left(\frac{\psi}{2}\right) \tag{19-10}$$

$$AB_2 = AB_1 - B_1B_2 \tag{19-11}$$

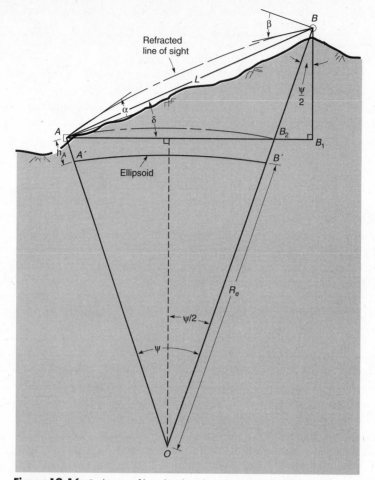

Figure 19-16 Reduction of long line lengths to horizontal and the ellipsoid.

Finally, ellipsoid length $A'B'$ can be computed from

$$A'B' = AB_2 \left(\frac{R_e}{R_e + h_A} \right) \tag{19-12}$$

where h_A is the *ellipsoid height* (the vertical distance from the ellipsoid to ground at point A). The ellipsoid height at any point is equal to the sum of the point's elevation, and geoid height at that location (see Section 21-9).

EXAMPLE 19-2   In Figure 19-16, slope distance L and vertical angles α and β were measured as 14,250.590 m, 4°32′18″, and −4°38′52″, respectively. If the ellipsoid height at A is 438.4 m, what is distance $A'B'$ reduced to the ellipsoid?

SOLUTION

Solving Eqs. (19-6) through (19-12) in sequence,

$$\delta = \frac{4°32'18'' - (-4°38'52'')}{2} = 4°35'35''$$

$$AB_1 = 14{,}250.590 \cos 4°35'35'' = 14{,}204.826 \text{ m}$$

$$BB_1 = 14{,}250.590 \sin 4°35'35'' = 1141.160 \text{ m}$$

$$\psi = \frac{14{,}204.826(180°)}{(6{,}371{,}000 + 438.4)\pi} = 0.127738° = 0°07'40''$$

$$B_1B_2 = 1141.160 \tan\left(\frac{0°07'40''}{2}\right) = 1.272 \text{ m}$$

$$AB_2 = 14{,}204.826 - 1.272 = 14{,}203.554 \text{ m}$$

$$A'B' = 14{,}203.554 \frac{6{,}371{,}000}{6{,}371{,}000 + 438.4} = 14{,}202.577 \text{ m}$$

(Note that if the short-line reduction procedure of Section 5-9 had been used, an error equal to B_1B_2, or 1.272 m, would have resulted.)

In the foregoing computations, the effects of refraction were eliminated by averaging reciprocal vertical angles α and β in Eq. (19-6). This procedure yields the best results. If the vertical angle were measured at only one end of the line, as angle α at A, then because refraction is approximately $\frac{1}{7}$ curvature, or $\frac{\psi}{7}$, a correction for its effect can be applied. In that case instead of using Eq. (19-6), angle δ is computed,

$$\delta = \alpha + \frac{\psi}{2} - \frac{\psi}{7} = \alpha + \frac{5\psi}{14} \qquad (19\text{-}13)$$

where

$$\psi = \frac{L \cos \alpha}{R_e + h_A} \times \frac{180°}{\pi}$$

Then Eqs. (19-7), (19-8), and (19-10) through (19-12) are solved as before.

Compute the ellipsoid length of the line in Example 19-2 using only the measured vertical angle at A.

EXAMPLE 19-3

SOLUTION

$$\psi = \frac{14{,}250.590 \cos 4°32'18''}{6{,}371{,}000 + 438.4} \times \frac{180°}{\pi} = 0.127748° = 07'40''$$

By Eq. (19-13),

$$\delta = 4°32'18'' + \frac{5}{14}(07'40'') = 4°35'02''$$

Then Eqs. (19-7), (19-8), and (19-10) through (19-12) are solved in order,

$$AB_1 = 14{,}250.590 \cos 4°35'02'' = 14{,}205.008 \text{ m}$$
$$BB_1 = 14{,}250.590 \sin 4°35'02'' = 1138.888 \text{ m}$$
$$B_1B_2 = 1138.888 \tan\left(\frac{07'40''}{2}\right) = 1.270 \text{ m}$$
$$AB_2 = 14{,}205.008 - 1.270 = 14{,}203.738 \text{ m}$$
$$A^1B^1 = 14{,}203.738\,\frac{6{,}371{,}000}{6{,}371{,}000 + 438.4} = 14{,}202.761 \text{ m}$$

Note that this answer differs by 0.184 m from the one obtained in Example 19-2. This can be expected, because refraction varies with atmospheric conditions, and the correction $\frac{\psi}{7}$ only approximates its effects.

If the slope is defined by differences in elevation ΔH, rather than a vertical angle, Eqs. (19-6) and (19-8) are replaced by

$$\delta = \sin^{-1}\left(\frac{\Delta H}{L}\right) \tag{19-14}$$

and

$$BB_1 = \Delta H \text{ (approx.)} \tag{19-15}$$

Equations (19-7) and (19-9) through (19-12) are then solved as before.

19-15 GEODETIC POSITION COMPUTATIONS

Two basic problems are included in geodetic position computations, the *direct* and the *inverse* problems. In the direct problem, given the latitude and longitude of station A and the ellipsoid length and azimuth of line AB, the latitude and longitude of station B are computed. In the inverse problem, the length of AB and its forward and back azimuths are calculated, given the latitudes and longitudes of stations A and B.

For long lines it is necessary to account for the ellipsoidal shape of the earth in these calculations to maintain suitable accuracy. Because modern surveying equipment is able to routinely reach out over increasingly longer distances, these procedures are becoming more important to the average surveyor. Many formulas are available for making direct and inverse calculations, some of which are simplified approximations that only apply for shorter lines. Development of these equations is beyond the scope of this text, but they can be found in the geodesy books cited in the bibliography at the end of this chapter.

To simplify position calculations for long lines, and yet maintain geodetic accuracy, state plane coordinate systems have been developed. These are described in Chapter 21.

19-16 VERTICAL CONTROL SURVEYS

Vertical-control surveys are generally run by either direct differential leveling or trigo-nometric leveling. The choice of method will depend primarily on the accuracy required, although the type of terrain over which the leveling will be done is also a factor. Direct differential leveling, described in Section 7-4, can produce the highest order of accuracy. The Global Positioning System (GPS) can be used for vertical-control surveys, but to get accurate elevations using this method, *geoid heights* in the area must be known (see Section 21-9).

Although trigonometric leveling produces a somewhat lower order of accuracy than direct differential leveling, the method is still suitable for many projects such as estab-lishing vertical control for topographic mapping or for lower order construction stake-out. It is particularly convenient in hilly or mountainous terrain where large differences in elevation are encountered. Field procedures for trigonometric leveling and methods for reducing the data are discussed in Section 6-5.4.

Differential leveling can produce varying levels of accuracy, depending on the pre-cautions taken. In this section only *precise differential leveling,* which produces the highest quality results, is considered.

As noted in Section 19-3 and Table 19-2, the FGCS has established accuracy stand-ards and specifications for various orders of differential leveling. To achieve higher orders, special care must be exercised to minimize errors, but the same basic principles apply.

For the most accurate work, precise tilting levels such as the Wild N3 of Figure 19-17 are used. This instrument has a split-image or coincidence bubble with a sensitiv-ity of 10″/2-mm division that is centered by means of a tilting screw. It has a parallel-plate glass micrometer which elevates or depresses the line of sight parallel to itself to set precisely on the nearest rod graduation. A micrometer shows the distance raised or lowered and permits rod readings to about $\frac{1}{100}$ of the smallest graduation, instead of $\frac{1}{10}$ division by estimation. The instrument reticle is shown in Figure 19-18. It has a single

Figure 19-17 Wild N3 precise tilting level. (Courtesy Leica, Inc.)

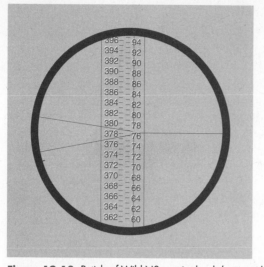

Figure 19-18 Reticle of Wild N3 precise level shown with dual metric-scale precise leveling rod. (Courtesy Leica, Inc.)

horizontal line on the right for making rod readings in the normal way, and two lines that form a wedge on the left to straddle a reading for utmost accuracy.

Special level rods are needed for precise work. They have scales graduated on Invar strips, which are only slightly affected by temperature variations. Precise level rods are equipped with rod bubbles to facilitate plumbing, and special braces aid in holding the rod steady. They usually have two separate graduated scales. One type of rod is divided in centimeters on an Invar strip on the rod's front side, with a scale in feet painted on the back for checking readings and minimizing blunders.

A second type of rod, shown in Figure 19-18, has two sets of centimeter graduations on the Invar strip, with the right one precisely offset from the left by a constant, thereby giving checks on readings.

Cloudy weather is preferable for precise leveling, but an umbrella can be used on sunny days to shade the instrument and prevent uneven heating which causes the bubble to run. (One design encases the vial in a Styrofoam shield.) Automatic levels are not as susceptible to errors caused by differential heating. Precise work should not be attempted on windy days. For best results, short equal backsight and foresight distances up to maximum of 50 m are recommended and their lengths balanced to within 2 m at each setup. Rodpersons pace or count rail lengths or highway slab joints to set sight distances, which are then checked for accuracy by three-wire stadia methods (see Section 7-8). Precise leveling demands good-quality turning points. Lines of sight should not pass closer than about $\frac{1}{2}$ m from the ground to minimize refraction. Readings at any setup must be completed in rapid succession; otherwise changes in atmospheric conditions might significantly alter refraction characteristics between them.

Three-wire leveling has been employed for much precise surveying in the United States. In this procedure, as described in Section 7-8, rod readings at the upper, middle, and lower cross hairs are taken and recorded for each backsight and foresight. The difference between the upper and middle readings is compared with that between the

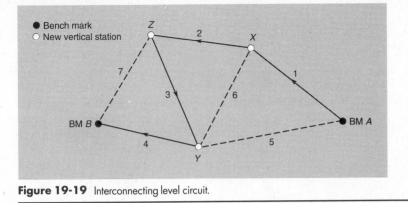

Figure 19-19 Interconnecting level circuit.

middle and lower values for a check, and the average of the three readings is used. A sample set of field notes for three-wire leveling is illustrated in Figure 7-8.

A second technique in precise leveling employs the parallel-plate micrometer attached to a precise leveling instrument, and a pair of precise rods like those described earlier. While this method has long been used in Europe, it was not adopted in the United States until the late 1960s.

It is generally advisable to design large-level networks so that several smaller circuits are interconnected to supply checks that isolate blunders or large errors. In Figure 19-19, for example, it is required to determine the elevations of points X, Y, and Z by commencing from BM A and closing on BM B. As a minimum, this could be done by running level lines 1 through 4, but if an unacceptable misclosure was obtained at BM B, it would be impossible to discover in which lines the blunder occurred. If additional lines 5, 6, and 7 are run, calculating differences in elevation by other routes through the network might isolate the blunder. Furthermore, by including supplemental measurements, precision of the resulting elevations at X, Y, and Z is increased.

In precise differential leveling, frequent calibration of the leveling instrument is necessary to determine its collimation error. A collimation error exists if, after leveling the instrument, its line of sight is inclined or depressed from horizontal. This causes errors in determining elevations when backsight and foresight distances are not equal. But they can be eliminated if the magnitude of the collimation error is known.

A method originated at the U.S. Coast and Geodetic Survey can be used to determine the collimation error. It requires a base line approximately 300 ft (90 m) long. Stakes are set at each end of the line and at two intermediate stations located approximately 20 ft (6 m) and 40 ft (12 m) from the two ends. Figure 19-20 shows an example set of field notes for determining the collimation error and includes a sketch illustrating the base-line layout. With the instrument at station 1, center cross-hair readings r_1 and R_1 are observed on stations A and B, respectively. If there were no collimation error, the true elevation difference ΔH from these observations would be $r_1 - R_1$. If a collimation error is present, however, each observation must be corrected by adding an amount proportional to the horizontal distance from the level to the rod. The horizontal distance is measured by the stadia interval (see Chapter 15). Introducing collimation corrections, the true elevation difference ΔH is

$$\Delta H = [r_1 + C(i_1)] - [R_1 + C(I_1)] \qquad \text{(a)}$$

		LEVEL CALIBRATION			
STA	A	INT	B	INT	
	4.516		5.576		
1	4.414	0.102	4.653	0.923	
	4.312	0.102	3.730	0.923	
	r_1	i_1	R_1	I_1	
	4.414	0.204	4.653	1.846	
	4.699		4.162		
2	3.825	0.874	4.022	0.140	
	2.950	0.875	3.881	0.141	
	R_2	I_2	r_2	i_2	
	3.825	1.749	4.022	0.281	
	(4.414 + 4.022) − (4.653 + 3.825)				
$C =$	(1.846 + 1.749) − (0.204 + 0.281)				
$C =$	$\dfrac{-0.042}{3.110}$		$= -0.0135$ ft/ft INT		

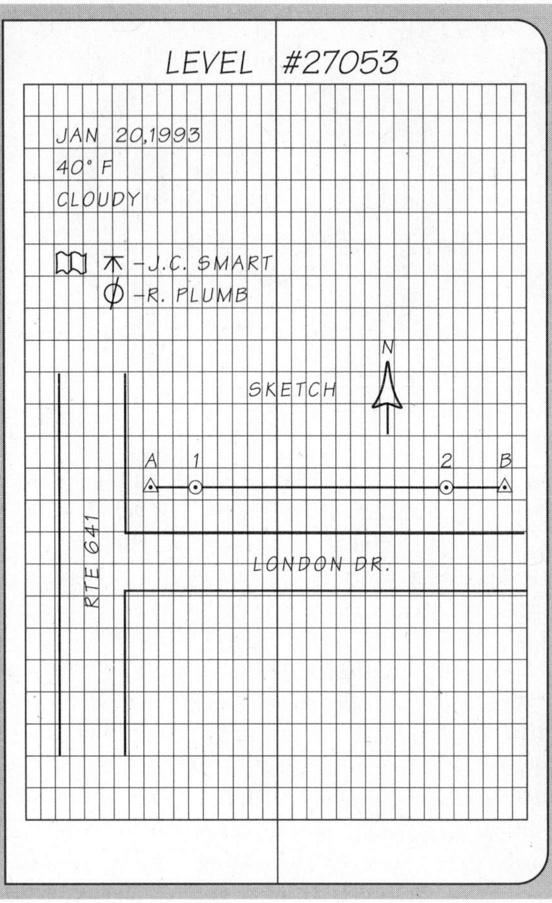

LEVEL #27053

JAN 20,1993
40° F
CLOUDY

📖 ⊼ −J.C. SMART
Φ −R. PLUMB

SKETCH
RTE 641
LONDON DR.

Figure 19-20 Field notes for determining collimation factor.

In Eq. (a), i_1 and I_1 are the stadia intervals (differences between top and bottom cross-hair values) for the rod readings on stations A and B, respectively, and C is the collimation factor (in feet per foot of stadia interval). A similar equation for the true elevation difference can be written for rod readings R_2 and r_2 taken on stations A and B, respectively, from station 2, or:

$$\Delta H = [R_2 + C(I_2)] - [r_2 + C(i_2)] \qquad (b)$$

Note that in Eqs. (a) and (b), uppercase R and I apply to the longer sights, and lowercase r and i are for the shorter sights. Equating the right sides of Eqs. (a) and (b) yields

$$C = \frac{(r_1 + r_2) - (R_1 + R_2)}{(I_1 + I_2) - (i_1 + i_2)} \qquad (19\text{-}16)$$

As previously noted, the units of the collimation factor calculated by Eq. (19-16) are feet per foot of stadia interval. Computation of the factor is illustrated in Figure 19-20 for the data of the field notes.

Because the collimation correction increases linearly with distance, it is unnecessary to apply it to each backsight and foresight. Rather the corrected elevation difference $\Delta H'$ for any loop or section leveled is computed from

$$\Delta H' = \Sigma\text{BS} - \Sigma\text{FS} + C(\Sigma\text{BS } I - \Sigma\text{FS } I) \qquad (19\text{-}17)$$

where ΣBS is the sum of the center cross-hair readings of backsights in the loop or section, ΣFS is the center cross-hair total of foresights, and $\Sigma\text{BS } I$ and $\Sigma\text{FS } I$ are the sums of the stadia intervals for the backsights and foresights, respectively.

EXAMPLE 19-4

The section from BM A to BM 3 is leveled using the instrument whose collimation factor of -0.0135 ft/ft of interval was determined in the field notes of Figure 19-20. The sum of the backsights is 125.59 ft, and the sum of the foresights is 88.33 ft. Backsight stadia intervals total 51.52 ft, while the sum of foresight intervals is 48.40 ft. Find the corrected elevation difference.

SOLUTION

By Eq. (19-17),

$$\Delta H' = (125.59 - 88.33) + (-0.0135)(51.52 - 48.40)$$

$$= 37.26 - 0.04 = 37.22 \text{ ft}$$

Regardless of precautions taken in field observations, errors accumulate in leveling and must be adjusted to provide perfect mathematical closure. For simple level loops, adjustment procedures presented in Section 7-6 can be followed; for interconnected level networks such as that of Figure 19-19, the method of least squares is preferable. An example least-squares adjustment of an interconnected network is given in Section C-7 of Appendix C.

PROBLEMS

19-1 Describe the terms *geoid* and *ellipsoid*.

19-2 Define the terms *geodetic latitude* and *geodetic longitude*.

*19-3 What is the difference between the equatorial circumference of the Clarke 1866 ellipsoid and that of the GRS80 ellipsoid?

19-4 What different field methods are used in horizontal control surveying? What factors influence the choice of field method to use?

19-5 Discuss the reference datums used for horizontal and vertical control surveys.

19-6 What are the major differences between NAD27 and NAD83? What important benefits result from NAD83?

*Asterisks indicate problems that have answers given in Appendix G.

19-7 What were the reasons for developing the new NAVD88 to replace NGVD29?

19-8 List the orders and classes of accuracy of horizontal control surveys and give their relative accuracy requirements.

19-9 Name the orders and classes of accuracy of vertical control surveys and give their relative accuracy requirements.

*19-10 To within what tolerance should the elevations of two bench marks 30 km apart be established if first-order class I standards were used to set them? What should it be if second-order class II standards were used?

19-11 Discuss the motivation for establishing the new GPS High-Accuracy Reference Networks.

19-12 Explain why it is important to permanently monument and adequately describe control stations. State the contents of a good control station description.

19-13 Why are quadrilaterals the most commonly used geometric figure in triangulation?

19-14 Discuss the factors that must be considered in triangulation reconnaissance.

19-15 Obtain a USGS quadrangle map of your area. On the map, lay out a quadrilateral having sides from 5 to 8 km long. Check all lines for intervisibility by plotting their profiles, and revise station positions if necessary to provide unobstructed lines of sight.

19-16 Explain the reason for establishing triangulation figures that contain angles between 30 and 150°.

19-17 Discuss the special precautions taken in observing angles on high-order triangulation.

19-18 The mean values of observed angles A, B, and C in a large triangle were $51°47'26.8''$, $60°57'08.7''$, and $67°15'25.8''$, respectively. Sides AB and AC were 65,578.52 and 72,961.38 ft long, respectively. Calculate the spherical excess, angular misclosure, and adjusted angles.

*19-19 Similar to Problem 19-18, except angles A, B, and C were $51°36'24.0''$, $71°52'26.3''$, and $56°31'25.0''$, respectively, and sides AB and BC were 78,571.55 and 95,275.68 m, respectively.

19-20 In Figure 19-14, assume that angles x and y were measured as $58°32'25''$ and $42°35'51''$, respectively. Control coordinates are $X_A = 1410.78$, $Y_A = 10,890.19$, $X_B = 6774.16$, $Y_B = 13,487.85$, $X_C = 10,165.85$, and $Y_C = 12,418.81$ ft. Compute the coordinates of point P.

*19-21 Similar to Problem 19-20, except angles x and y were $45°28'16''$ and $41°26'22''$, respectively, and the control coordinates are $X_A = 7631.25$, $Y_A = 29,831.67$, $X_B = 18,038.89$, $Y_B = 37,466.73$, $X_C = 28,235.75$, and $Y_C = 34,494.95$ ft.

19-22 Outline the advantages and disadvantages of traverse control surveys as compared to triangulation for horizontal networks.

19-23 Explain the differences between triangulation and trilateration. List their similarities.

*19-24 Compute the back azimuth of a line 21,341.52 ft long, at a mean latitude of $38°52'02''$, whose forward azimuth is $83°24'37.5''$ from north.

19-25 In Figure 19-15 the azimuth of AB is $102°36'20''$, and the angles to the right measured at B, C, D, E, and F are $132°01'04''$, $241°45'12''$, $141°15'01''$, $162°09'24''$, and $202°33'43''$, respectively. An astronomic observation yielded an azimuth of $82°24'03''$ for line FG. The mean latitude of the traverse is $42°16'$, and the total departure between points A and F was 24,986.26 ft. Compute the angular misclosure and the adjusted angles.

19-26 In Figure 19-16 slope distance L and vertical angles α and β were measured as 76,953.82 ft, $+4°18'42''$, and $-4°26'28''$, respectively. Ellipsoid height of point A is 1672.21 ft. What is length $A'B'$ on the ellipsoid?

*19-27 In Figure 19-16 slope distance L was measured as 19,875.28 m. The mean sea level elevations of points A and B were 659.23 and 1807.04 m, respectively, and the geoid height at both stations was -30.5 m. Calculate ellipsoid distance $A'B'$.

19-28 In Figure 19-16, slope distance L and vertical angle β at station B were measured as 26,432.83 m and $-6°17'18''$, respectively, to station A. If the elevation of station B is 1327.81 m and the ellipsoid heights at stations B and A are both -29.0 m, what is ellipsoid length $A'B'$?

*Asterisks indicate problems that have answers given in Appendix G.

19-29 List the special precautions taken on high-order differential leveling.

19-30 Describe two different types of level rods used to prevent reading blunders in precise leveling.

19-31 Discuss the advantages of the three-wire leveling procedures.

*19-32 Compute the collimation correction factor C for the following field data, taken in accordance with the example and sketch in the field notes of Figure 19-20. With the instrument at station 1, high, middle, and low cross-hair readings were 5.612, 5.501, and 5.490 ft on station A and 4.978, 3.728, and 2.576 ft on station B. With the instrument at station 2, high, middle, and low readings were 7.211, 6.053, and 4.894 ft on A and 4.561, 4.358, and 4.155 ft on B.

19-33 A leveling instrument having a collimation factor of $+0.0225$ ft/ft of interval was used to run a section of three-wire differential levels from BM A to BM B. The sums of backsights and foresights for the section were 16.385 and 65.290 ft, respectively. Backsight stadia intervals totaled 39.256 ft, while the sum of foresight intervals was 44.696 ft. What is the corrected elevation difference from BM A to BM B?

19-34 The relative error of the difference in elevation between two bench marks directly connected in a level circuit and located 56 mi apart is ± 0.009 m. What order and class of leveling does this represent?

*19-35 Similar to Problem 19-34, except the relative error is ± 0.004 ft for bench marks located 45 km apart.

*Asterisks indicate problems that have answers given in Appendix G.

BIBLIOGRAPHY

Bossler, J. D. 1982. "New Adjustment of the North American Datum." *ASCE, Journal of the Surveying and Mapping Division* 108 (No. SU2):47.

Burkholder, E. F. 1991. "Computation of Horizontal/Level Distances." *ASCE, Journal of Surveying Engineering* 117 (No. 3):104.

Burkholder, E. F. 1993. "Design of a Local Coordinate System for Surveying, Engineering and LIS/GIS," *Surveying and Land Information Systems* 53 (No. 1):29.

Cain, J. D. 1992. "Statewide High-Precision GPS Networks Support Local GIS Accuracy," *GIS World* 5 (No. 6):46.

Colcord, J. E. 1981. "The Surveying Engineer and NAD-83." *ASCE, Journal of the Surveying and Mapping Division* 107 (No. SU1):25.

El-Hassan, I. M. 1986. "An Analytical Solution of the Resection Problem." *ASCE, Journal of the Surveying Engineering Division* 112 (No. 1):30.

Ethridge, M. M. 1989. "The National Geodetic Reference System of the Future." *Point of Beginning* 14 (No. 2):46.

Ethridge, M. M. 1989. "Does the National Geodetic Reference System Need to Be Upgraded?" *Point of Beginning* 15 (No. 1):26.

Frodge, S. L. 1992. "Deformation Surveys." *Bulletin, American Congress on Surveying and Mapping* (No. 138):32.

Graff, D. R. 1988. "Coordinate Conversion from NAD27 to NAD83." *ASCE, Journal of Surveying Engineering* 114 (No. 3):125.

Holdohl, S. R. 1992. "A Dynamic Vertical Reference System." *Surveying and Land Information Systems* 52 (No. 2):92.

Kunnar, M. 1993. "World Geodetic System 1984: A Reference Frame for Global Mapping, Charting, and Geodetic Applications." *Surveying and Land Information Systems* 53 (No. 1):53.

Leick, A. 1993. "Accuracy Standards for modern Three-Dimensional Geodetic Networks." *Surveying and Land Information Systems* 53 (No. 2):111.

Leick, A. 1992. "Geodesy in Review." *Bulletin, American Congress on Surveying and Mapping* (No. 135):45.

Lippold, H. R., Jr. 1980. "Readjustment of the National Geodetic Vertical Datum." *Surveying and Mapping* 40 (No. 2):155.

National Geodetic Survey. 1982. "Bulletin to Users of Compensator-Type Leveling Instruments." *Bulletin, American Congress on Surveying and Mapping* (No. 79):35.

Slater, J. A. 1993. "A New GPS Geodetic Control Network for Paraguay." *Surveying and Land Information Systems* 53 (No. 3):179.

Smith, J. W. 1988. *Basic Geodesy.* Rancho Cordova, Calif.: Landmark.

Soler, T., and L. D. Hothem. 1988. "Coordinate Systems Used in Geodesy: Basic Definitions and Concepts." *ASCE, Journal of Surveying Engineering* 114 (No. 2):84.

Stoughton, H. W. 1992. "Some Notes on the U.S. Coast and Geodetic Survey's Collimation Error Test." *Surveying and Land Information Systems* 52 (No. 3):155.

Torge, W. 1980. *Geodesy: An Introduction.* Berlin: de Gruyter.

Vanicek, P., and E. Krakiwsky. 1986. *Geodesy: The Concepts.* Amsterdam: Elsevier Science Publishers.

Wade, E. B. 1986. "Impact of North American Datum of 1983." *ASCE, Journal of the Surveying Engineering Division* 112 (No. 1):49.

Whalen, C. T. 1982. "The New Adjustment of the North American Vertical Datum." *Bulletin, American Congress on Surveying and Mapping* (No. 78):39.

20

SATELLITE AND INERTIAL SURVEYING SYSTEMS

20-1 INTRODUCTION

In recent years two new and unique approaches to surveying have emerged—*satellite surveying systems* and *inertial surveying systems*. Both have resulted from research and development conducted by the military. Either can be operated day or night, rain or shine, and neither requires cleared lines of sight between stations. This represents a revolutionary departure from conventional surveying procedures, which rely on measured angles and distances for determining point positions.

Satellite surveying systems grew out of the space program and, more specifically, from U.S. Navy activities related to navigation for its Polaris submarine fleet. Development of the program began in 1958, and the first nonclassified use of artificial satellites for surveying purposes occurred in 1967. Early instruments were bulky and expensive, the observation sessions were lengthy, and the accuracy achieved was only moderate. Through research and development, however, the systems have been improved continually, and their size and cost substantially reduced. Field procedures have also been modified so that now very high accuracies can be achieved in real time. The current-generation satellite surveying system is called the *Global Positioning System* (GPS). It is being used in all areas of surveying and has been revolutionizing the field, making many traditional procedures obsolete.

Inertial surveying systems (ISSs), also sometimes called inertial positioning systems (IPSs), were originally developed as guidance systems for ships, aircraft, and missiles. Since 1975, when the first commercial ISS unit became available for nonmilitary use, these systems have been used in a variety of special types of surveys. The U.S. Bureau of Land Management, for example, has employed them extensively to establish new section corners throughout vast areas of Alaska. Drawbacks to these systems are that their initial cost is relatively high, the equipment bulky, and their accuracy less than that attainable with GPS. Thus in the future, use of ISSs will likely continue, but only on special types of projects. An application currently being researched is the simultaneous use of both ISS and GPS units in an aircraft to provide needed ground control for photogrammetric mapping from aerial photos.

Part I of this chapter discusses satellite surveying systems, while Part II covers inertial surveying systems.

PART I
SATELLITE SURVEYING SYSTEMS

20-2 FIRST-GENERATION SATELLITE SURVEYING SYSTEMS

The first generation of satellite systems employed for surveying operated on the *Doppler* principle and utilized the Navy's so-called TRANSIT navigation satellites. In this system, receivers located at ground stations measured changes in frequencies of radio signals transmitted from satellites operating in polar orbits at altitudes of about 1075 km. A Doppler receiver is shown in Figure 20-1.

The operating principle of satellite Doppler systems is illustrated in the simplified diagram of Figure 20-2. A precisely controlled radio frequency is continuously transmitted from the satellite as it passes in orbit above an observing station [see Figure 20-2(a)]. As the transmitter approaches a receiver, the received signal has a frequency higher than that transmitted, but it is decreasing. Then as the satellite moves away from the station, the frequency decreases below the transmitted level. This phenomenon,

Figure 20-1 Receiver of a satellite Doppler surveying system. (Courtesy Magnavox.)

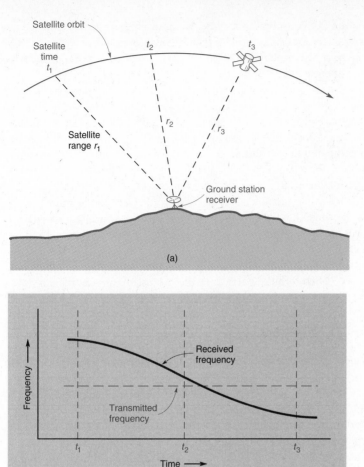

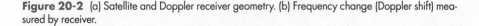

Figure 20-2 (a) Satellite and Doppler receiver geometry. (b) Frequency change (Doppler shift) measured by receiver.

illustrated in Figure 20-2(b), can be appreciated by anyone who has listened to the change in pitch of a train whistle as it approaches and then moves past. The magnitude of frequency change (*Doppler shift*), which is a function of the range (distance to the satellite), is measured by a receiver. With the transmitting frequency known, together with accurate satellite orbital position data and precise timing of observations, the position of a receiving station can be computed from the measured Doppler shifts.

Two basic methods were used in making satellite Doppler observations and reducing them to obtain station positions: *point positioning* and *translocation*. In point positioning, a receiver at a single station of unknown position collected data from multiple satellite passes. Based on the ranges determined from the measured data, the receiver's position was first reduced to its location in a coordinate system relative to the satellite's position, and then transformed into one of the conventional coordinate systems used by surveyors (see Section 20-8). In this procedure, by observing up to 30 or 40 passes, receiver positions with errors somewhat less than about 1 m could be achieved. The time required to observe this number of passes, however, was excessive, requiring from 2 to 8 days, depending on satellite visibility.

In the translocation method, receivers located at two or more stations tracked a satellite simultaneously. The position of at least one occupied station was known. In this procedure, the control station was first treated as an unknown, and its coordinates computed using the point-positioning technique just described. These computed coordinates were then compared to the station's known coordinates, with differences indicating errors in the system. Assuming that the conditions producing these errors were constant at all stations during the observing period, corrections for the errors were applied to the locations of the unknown stations that were determined simultaneously using point-positioning methods. Using translocation, positions accurate to within a few tenths of a meter could be obtained by observing 25 to 30 satellite passes, but again this required a lengthy observing session.

Because of the superiority of the GPS, satellite Doppler surveying programs have now been phased out. Therefore the balance of the discussion on satellite surveying systems in this text concentrates on GPS.

20-3 THE GLOBAL POSITIONING SYSTEM

Because of the success of the Doppler program, the U.S. Department of Defense began development of a new navigation and positioning system, called the *Global Positioning System* (GPS). In 1978 the first NAVSTAR (NAVigation Satellite Timing And Ranging) satellite to support the development and testing of the system was launched. This program, although designed for military applications, ushered in the current generation of satellite technology for surveying purposes.

The GPS, like the Doppler program, is based on observations of signals transmitted from satellites whose orbits are precisely known. These signals are picked up at ground stations by receivers like those shown in Figures 20-3, 20-4, and 1-5. As with Doppler systems, precise distances (ranges) from the satellites to the receivers are determined from the timing and signal information, enabling receiver positions to be computed. The transmitted signals, as well as methods of determining distances from them, are considerably different than those for the Doppler program, however.

The GPS is first of all a military operation, which is used only secondarily for civilian purposes. Therefore, of necessity, it is a "one-way" ranging system, that is, signals are transmitted only by the satellites, not by the receivers, which could give away strategic ground locations. Because it is one-way, the transmitters and receivers must both have accurate clocks in order to resolve distances from the broadcast signals.

GPS satellites are in near-circular orbits at altitudes of 20,200 km above the earth. Operating farther from the earth, they are less susceptible to gravitational anomalies than were the TRANSIT satellites, which enables more accurate orbital positions to be determined. Current plans call for a total of 24 NAVSTAR satellites when the GPS is fully operational. As of this writing, however, it is uncertain whether all 24 will be operational, or if up to three may be inactive but held in orbits and available as spares if needed. Figure 20-5 illustrates schematically the planned constellation of satellites. It will consist of six different orbital planes, with four satellites orbiting in each. The

Figure 20-3   Ashtech P-12 GPS receiver. (Courtesy Wisconsin Department of Transportation.)

orbits are inclined at an angle of 55° to the equatorial plane and spaced 60° apart. Each satellite will travel around the earth twice every sidereal day (which is 3 min and 56 sec shorter than a solar day), and will be visible to an observer for a maximum of approximately 10 h per day. The geometry and dynamics of this constellation ensure that at nearly every location on earth, at any given time, from four to six (and sometimes more) satellites will always be visible. As of May 1993, 23 satellites were already in orbit, three of which were prototypes scheduled to be replaced. Plans call for launching four more, the last scheduled for about March 1994.

As noted, GPS position determinations require precise satellite orbital information. These data are compiled by the GPS *Operational Control System* (OCS), which is monitoring the satellites from five tracking stations spaced uniformly in longitude throughout the world. The master control station is at Colorado Springs, with others on the islands of Hawaii, Ascension, Diego Garcia, and Kwajalein. From the monitored data, precise near-future predictions of all satellite orbits are determined, uploaded into the satellites as often as three times daily, and broadcast with their signals. Receipt of this so-called *broadcast ephemeris* enables the GPS receivers to make real-time positioning computations.

Figure 20-4 Magellan NAV 5000 PRO GPS hand-held receiver. (Courtesy Magellan Systems Corporation.)

For the most precise geodetic surveys, the broadcast ephemeris may not be accurate enough. With these types of surveys, however, there is usually no need to process positions in real time, thus the so-called *precise ephemeris* may be used for computations. The precise ephemeris gives extremely accurate post-orbital information, as determined by monitoring the satellites not only from the five OCS stations but from many others as well. One large tracking network which contributes data for generating the precise ephemeris is the *Cooperative International GPS Network* (CIGNET), which consists of 29 stations distributed around the world.

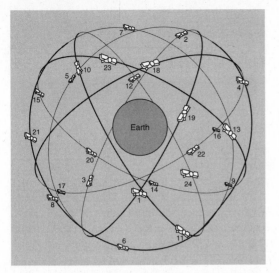

Figure 20-5 Planned constellation of GPS NAVSTAR satellites.

20-4 SIGNALS TRANSMITTED BY GPS SATELLITES

As the GPS satellites are orbiting, each broadcasts unique signals continuously on two *carrier frequencies*. The carriers, which are transmitted in the L band of microwave radio frequencies, are identified as the *L1* signal with a frequency of 1575.42 MHz, and the *L2* signal at a frequency of 1227.60 MHz. A variety of information is superimposed upon these carrier waves, thus making it possible for ground receivers to measure ranges (distances) to the satellites and to compute their positions from them.

In order for receivers to determine their positions independently in real time, it was necessary to devise a system for accurate measurement of signal travel time from satellite to receiver. This was accomplished by modulating the carriers with *pseudorandom noise* (PRN) codes. The PRN codes consist of unique sequences of binary values (zeros and ones) that appear to be random but, in fact, are generated according to a special mathematical algorithm. Two different PRN codes are transmitted by each satellite. The L1 signal is modulated with the so-called *precision code,* or *P code,* and also with the so-called *coarse/acquisition code,* or *C/A code.* The L2 signal is modulated only with the P code. Each satellite broadcasts its own unique codes, so that receivers are able to identify the origins of received signals. This is important when tracking several different satellites simultaneously.

The C/A code has a frequency of 1.023 MHz and a wavelength of about 300 m. It is accessible to all users. The P code, with a frequency of 10.23 MHz and a wavelength of about 30 m, is ten times more accurate for positioning than the C/A code. Note that all GPS frequencies are derivatives of the fundamental P code, whose frequency is generated by a high-precision oscillator within each satellite. Table 20-1 lists the GPS frequencies and gives their P code factors.

To meet military objectives, the Department of Defense has stated its intention to degrade the accuracy with which positions can be determined using GPS. This will be done in two ways; **(1)** by invoking so-called *selective availability* (SA), in which errors

TABLE 20-1 FREQUENCIES TRANSMITTED BY GPS

CODE NAME	FREQUENCY (MHz)	FACTOR OF P
C/A	1.023	Divide by 10
P	10.23	—
L1	1575.42	Multiply by 154
L2	1227.60	Multiply by 120

are deliberately introduced into satellite clocks; and **(2)** by implementing *antispoofing* (AS), in which the P code is encrypted with a secret code, making it receivable only to users authorized by the military. SA is currently being enforced, and although AS is not presently being implemented, it could be at any time. However, as will be discussed later in this chapter, accurate positions for surveying purposes can still be obtained when these denials are enforced by using so-called *differential positioning techniques.*

In addition to broadcasting the PRN codes, each satellite also transmits other unique information, including its orbital data or *ephemeris* for use in position computations, and the offset of its clock time with respect to precise GPS time. [GPS time is equivalent to coordinated universal time (see Section 18-8), except that leap seconds, which account for slowing in the earth's rotation rate, are not included.] Each satellite also transmits an almanac of all other operating satellites that includes crude descriptions of their orbits, so that the receivers can quickly find them.

20-5 BASIC PRINCIPLES OF GPS POSITION DETERMINATION

The practical objectives of GPS are twofold: navigation and precise positioning. Of the two, precise positioning is of greater concern to surveyors. GPS procedures for determining precise point positions consist fundamentally in measuring distances from points of unknown location to satellites whose positions are known at the instant the distances are measured. Conceptually, this is identical to performing conventional resection using distances measured by taping or EDM instruments from a station of unknown position to several control stations. The basic difference is that in GPS, the control stations are satellites. Also, the distances are determined in a somewhat different manner.

Two fundamental methods are employed by GPS receivers in measuring distances to satellites: *pseudoranging* and *carrier phase measurements.* These are described in the subsections that follow.

20-5.1 Positioning by Pseudoranging

The pseudoranging technique of measuring distances with GPS uses the PRN codes described in Section 20-4. The procedure can be explained in a rudimentary manner with reference to the diagram of Figure 20-6. Assume that the satellite and a receiver located on the ground station both generate an identical series of binary codes simultaneously. In the figure, these transmitted codes are illustrated schematically by sawtooth waves of varying lengths. It will take the satellite's signal some time to travel to the ground receiver, where it is picked up and compared to the signal being generated there. Because the frequency and functional relationship of the binary code being transmitted are known precisely, the ground receiver is able to make a precise determi-

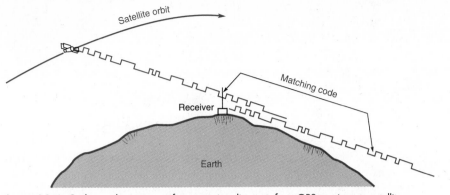

Satellite orbit

Matching code

Receiver

Earth

Figure 20-6 Code-matching concept for measuring distances from GPS receivers to satellites.

nation of the time that elapsed since it generated the exact portion of code that it just received from the satellite. Based on this signal travel time and the velocity of propagation of electromagnetic energy through the atmosphere (approximately 186,000 mi/sec), the distance to the satellite can be computed. This general distance-measuring procedure in GPS terminology is called *ranging*.

For accurate distance determination, the procedure just described depends on nearly exact synchronization of clocks within the satellite and receiver. As previously noted, extremely precise atomic clocks are used in the satellites, and furthermore any deviation of a satellite's clock from precise GPS time can be accounted for based on correction information transmitted with the signal. GPS receivers, on the other hand, employ less expensive, lower accuracy clocks. Thus exact clock synchronization cannot be accomplished, and hence there is a so-called *clock bias,* or time difference, between the two clocks. Therefore measured elapsed times are somewhat inaccurate, causing distance determinations to be also in error. Distances uncorrected for this timing error are called pseudoranges.

From the ephemeris information being transmitted simultaneously by the satellite, its position at the instant the matching portion of the signal was transmitted can be determined. With this information, together with pseudoranges measured from one ground station to three or more satellites, the receiver's position can be found. In GPS terminology this procedure is called *pseudoranging.*

A more detailed explanation of position determination by pseudoranging is illustrated with the following example. Assumed that for the four satellite positions shown in Figure 20-7, pseudoranges r_1, r_2, r_3, and r_4 have been measured. Assume also that the satellite locations, given by their three-dimensional rectangular coordinates X_1, Y_1, Z_1; X_2, Y_2, Z_2; X_3, Y_3, Z_3; and X_4, Y_4, Z_4, are known. (This three-dimensional rectangular coordinate system is described in Section 20-8). The receiving station A has three unknown coordinates X_A, Y_A, and Z_A. For any three of the satellites, for example, for 1, 2, and 3, the following equations can be written:

$$
\begin{aligned}
r_1 &= [(X_1 - X_A)^2 + (Y_1 - Y_A)^2 + (Z_1 - Z_A)^2]^{1/2} \\
r_2 &= [(X_2 - X_A)^2 + (Y_2 - Y_A)^2 + (Z_2 - Z_A)^2]^{1/2} \\
r_3 &= [(X_3 - X_A)^2 + (Y_3 - Y_A)^2 + (Z_3 - Z_A)^2]^{1/2}
\end{aligned}
\qquad (20\text{-}1)
$$

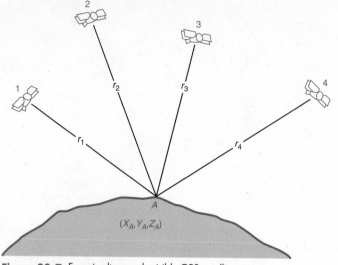

Figure 20-7 Four simultaneously visible GPS satellites.

Equations (20-1) can be solved simultaneously for the unknowns X_A, Y_A, and Z_A. This procedure has not accounted for the clock bias, however, and thus the computed coordinates for station A will contain substantial errors. Let the receiver's clock bias be Δt, and assume that is has remained constant while pseudoranges were determined to each of the four satellites. Because of clock bias, the error Δr in each range will be equal to the product $V(\Delta t)$, where V is the transmitted signal velocity. Four equations can be written, which incorporate the clock bias as follows:

$$
\begin{aligned}
r_1 + \Delta r &= [(X_1 - X_A)^2 + (Y_1 - Y_A)^2 + (Z_1 - Z_A)^2]^{1/2} \\
r_2 + \Delta r &= [(X_2 - X_A)^2 + (Y_2 - Y_A)^2 + (Z_2 - Z_A)^2]^{1/2} \\
r_3 + \Delta r &= [(X_3 - X_A)^2 + (Y_3 - Y_A)^2 + (Z_3 - Z_A)^2]^{1/2} \\
r_4 + \Delta r &= [(X_4 - X_A)^2 + (Y_4 - Y_A)^2 + (Z_4 - Z_A)^2]^{1/2}
\end{aligned}
\tag{20-2}
$$

In Eqs. (20-2) there are four unknowns, Δr, X_A, Y_A, and Z_A; they can be solved simultaneously to obtain the unknowns. The procedure eliminates the error due to clock bias.

In practice, a receiver will normally remain at a station long enough to make repeated measurements of pseudoranges to as many satellites as practical. For each measurement an equation like one of those of Eq. (20-2) can be added. The redundant system of equations is then solved by the method of least squares (see Section 2-26 and Appendix C) to improve the accuracy of determining X_A, Y_A, and Z_A.

Although the error due to receiver clock bias can be removed by observing four satellites as demonstrated, several other errors remain (see Section 20-7). As a result, accuracies attainable from C/A code measurements by pseudoranging using a single receiver are generally limited to on uncertainty of about ± 30 m without selective availability, and approximately ± 100 m if selective availability is activated. Accuracy can be improved somewhat by performing so-called *differential pseudoranging*. This is similar in concept to the Doppler method of "translocation." In the procedure, receivers operating simultaneously collect satellite signals at two (or more) ground stations, one of which is a control point whose position is known. The errors remaining at both receivers, after removal of the clock bias effects, are essentially the same. Thus after taking the differences in measured coordinates between the two stations, these errors cancel the resulting coordinate differences are added to the known coordinates of the control station, yielding coordinates for the unknown station.

Differential pseudoranging improves accuracy, but position errors on the order of ±3 m remain, so the procedure is still not acceptable for precise surveying work. Nevertheless, it yields accuracies suitable for navigation and certain other low-order work. The small hand-held Magellan 5000 receiver shown in Figure 20-4, when operated in differential ranging mode, yields position accuracies of approximately ±3 m.

20-5.2 Positioning from Carrier Phase Measurements

In this procedure the observed quantity is the phase change that occurs as a result of the carrier wave's travel from satellite to receiver. In principle this is similar to the method of phase shift measurement used by electronic distance measuring (EDM) instruments (see Chapter 5). One major difference, of course, is that the signal is not returned to a stationary transmitter for the "true" phase shift measurement, as it is in EDM work. Instead phase shift must be measured at the receiver. Although phase changes can be measured accurately (to within $\frac{1}{100}$ of a cycle), the satellite and receiver clocks must be perfectly synchronized to obtain the true phase change that results from the signal's travel. Since the clocks are not perfect, especially the receiver's, this problem must be overcome to obtain accuracy with the method.

Another problem that exists in the GPS carrier phase measurement technique is that only the indicated phase shift within the last cycle of the carrier wave being received (last wavelength) is measured, and the number of full wavelengths contained in the travel path of the signal from satellite to receiver is unknown. In GPS terminology this is called *cycle ambiguity*. Recall that the same problem existed in EDM work, and that the ambiguity was eliminated by broadcasting energy with longer wavelengths. In GPS surveying this is not possible because the satellites are moving constantly.

Ignoring errors in the system, a simplified equation, which expresses measurements of ranges from receivers to satellites using the phase change procedure, is:

$$R = (N + \phi) \lambda \qquad (20\text{-}3)$$

where R is the range from receiver to satellite and is given in a Cartesian coordinate system by the right-hand sides of Eqs. (20-1) and (20-2), N the number of full wavelengths of the carrier signal, ϕ the phase change within the last cycle of the signal, and λ the wavelength of the carrier.

Of course, as stated previously, measurements of ϕ are erroneous due to imperfect clock synchronization and other factors. The effect of a receiver's clock error can be eliminated, however, by *differencing* (taking the difference between) measurements made simultaneously to two different satellites. The situation is illustrated in Figure 20-8(a). Errors due to the satellite clock, although smaller, can be eliminated by differencing measurements made simultaneously from two stations to one satellite, a procedure that also accounts for the majority of orbital errors and atmospheric delays. This situation is shown in Figure 20-8(b). These procedures are called *single differencing*.

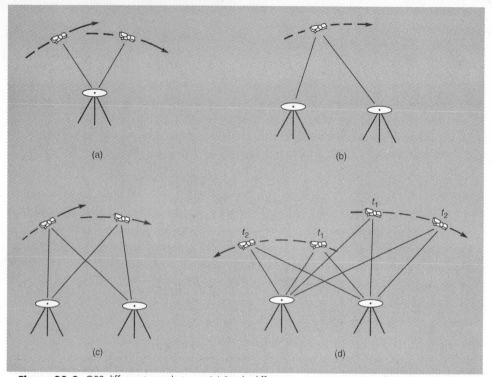

Figure 20-8 GPS differencing techniques. (a) Single differencing removes receiver clock errors by differencing readings made simultaneously with one receiver to two satellites. (b) Single differencing eliminates satellite clock errors by simultaneously measuring with two receivers to one satellite. (c) Double differencing removes clock and other system errors by simultaneously reading to two satellites with two receivers. (d) Triple differencing eliminates most system errors and resolves the ambiguity (number of full cycles) by differencing readings made at two different times to two satellites with two receivers.

Figure 20-8(c) illustrates a technique that combines the procedures of Figure 20-8(a) and (b) and involves simultaneously taking measurements from two stations to two satellites and differencing the results. This procedure, called *double differencing,* eliminates the effects of most of the errors in the system.

An extension of these differencing procedures, which eliminates system errors and also resolves the cycle ambiguity, is *triple differencing.* It consists in taking the difference between two double differences, and therefore involves making measurements at two different times to two satellites from two stations. Figure 20-8(d) illustrates this technique.

The differencing techniques just described are minimum conditions necessary to eliminate errors and resolve the cycle ambiguity. In practice four or more satellites are normally observed simultaneously with each receiver. This enables a large number of redundant equations to be formed, and many difference combinations to be computed using the method of least squares.

The Ashtech P-12 and Trimble 4000SE receivers shown in Figures 20-3 and 1-5, respectively, can determine positions by employing either the ranging techniques described in Section 20-5.1 or the carrier phase differencing techniques just described. Both instruments are capable of tracking several satellites simultaneously.

20-6 GPS FIELD PROCEDURES

In practice, field procedures employed on GPS surveys depend on the capability of receivers used and the type of survey. Some specific field procedures currently being used in surveying include the *static, rapid static, kinematic, pseudokinematic* and *real-time kinematic* methods. These are described in the subsections that follow. Each is based on carrier phase measurements and employs *relative positioning* techniques; that is, two (or more) receivers, occupying different stations, simultaneously make observations to several satellites. The vector (distance) between receivers is called the *base line,* and its ΔX, ΔY, and ΔZ coordinate difference components (in a three-dimensional rectangular system described in Section 20-8) are computed as a result of the observations.

20-6.1 Static GPS Surveying

For highest accuracy, for example geodetic control surveys, static GPS procedures are used. Two (or more) receivers are employed, and the process begins with one situated on an existing control station and the other on the first unknown point. Simultaneous measurements are made from both stations to four or more satellites for a time period of an hour or more depending on the base-line length. (Longer base lines require greater observing times.). The receiver on the control station is then moved up to the second unknown station, and the other remains on the first for another observing session. Upon completion of this session, the receiver from the first unknown station would move to the third, with the other remaining at the second. This leap frog technique continues until a receiver reaches another existing control station.

Most receivers have an internal memory for storing the observed data. After all measurements are completed, the data are transferred to a larger computer for *post-processing.* Figure 20-9 shows an Ashtech P-12 receiver that has been brought into the office and connected directly to a personal computer for downloading field data. Post-processing of static GPS surveying data normally uses the "broadcast" ephemeris, but if desired, the "precise" ephemeris can be used to obtain an even higher level of accuracy. Coordinate differences between stations are computed, and beginning from the first control station, they are added progressively to yield coordinates of all stations, including the closing control station. Any misclosure is adjusted back through the network. Relative accuracies with this method are generally in the range of about $\pm(5 \text{ mm} + 1 \text{ ppm})$.

20-6.2 Rapid Static GPS Surveying

This procedure is similar to static surveying, except that one receiver always remains on the first control station while the other is moved progressively from one unknown point to the next. An observing session is conducted for each point, but the sessions are shorter than for the static method. The procedure is suitable for measuring shorter base lines, and thus is applicable for lower order project control, control for mapping, and for boundary surveys. It yields accuracies on the order of about $\pm(10 \text{ mm} + 1 \text{ ppm})$ or better.

20-6.3 Kinematic GPS Surveys

This procedure also requires two (or more) receivers. To begin, the receivers must be *initialized.* This process includes determining the cycle ambiguity. (Cycle ambiguity was described in Section 20-5.2.) Initialization can be done in different ways. One

Figure 20-9 Downloading data from an Ashtech P-12 receiver directly into a personal computer. (Courtesy Wisconsin Department of Transportation.)

method requires two control stations. A short static observing session (from approximately 2 to 15 min) is conducted using both instruments simultaneously on the two stations. Because the base-line coordinate differences are known, differencing of the observations, as described in Section 20-5.2, will yield the unknown cycle ambiguity. These differencing computations are performed in a postprocessing operation by computer using the data from both receivers. If only one control station is available, a second one can be set using the static survey procedures described in Section 20-6.1. This then will allow initialization by the differencing procedure just described.

An alternative initialization procedure, called *antenna swap,* is also suitable if only one control station is available. Here receiver *A* is placed on the control point and receiver *B* on a nearby unknown point. For convenience, the unknown point can be within about 30 ft (10 m) of the control station. After a few minutes of data collection with both receivers, their positions are interchanged while keeping them running. In the interchange process, care must be exercised to make certain continuous tracking, or "lock" is maintained on at least four satellites. After a few more minutes of observing, the receivers are interchanged again, returning them to their starting positions. This enables the baseline coordinate differences and the cycle ambiguity to be determined, again by the differencing technique.

After initialization, one receiver, called the "base," remains at its control station, while the other, called the "rover," moves from point to point or along a line. The rover's antenna positions are determined at intervals as short as 1 sec. The accuracy of intermediate points is in the centimeter range. In kinematic surveys, both receivers must maintain lock on at least four satellites throughout the session. If lock is lost, the receivers must be reinitialized. Thus care must be taken to avoid obstructing the antenna by carrying it close to buildings, beneath trees or bridges, and so on.

At the end of the survey, the rover is returned to its initial control station, or another, for a check. Kinematic GPS is applicable for topographic surveys, hydrographic surveys, and construction layout. Newer applications include **(1)** providing supplemental control for photogrammetric mapping (see Section 28-17) by placing the rover in an aircraft and determining the camera's position precisely at the instant of each photo exposure, and **(2)** placing the rover on a bulldozer to provide real-time computer control of earthwork grading operations. Inertial Surveying Systems (see Part II of this chapter) are being researched as a means of regaining initialization, should loss of lock occur in kinematic GPS surveys.

20-6.4 Pseudokinematic GPS Surveys

This procedure, like the others, requires a minimum of two receivers. It eliminates the need to maintain constant lock on the satellites, which can be a problem. In pseudokinematic surveying, two relatively short observation sessions (around 5 min in duration) are conducted on each station, about an hour apart. Reduction procedures are similar to those of the static mode, and accuracies approach those of static surveying.

20-6.5 Real-Time Kinematic GPS Surveys

The kinematic and pseudokinematic GPS surveying procedures described in the preceding sections both require that data collected by the receivers be temporarily stored until the field work is completed. The data is then post-processed to obtain positions for the surveyed points. Real-time kinematic (RTK) surveying, however, as implied by its name, enables positions of points to be determined instantaneously as the roving receiver (or receivers) occupy each point.

Like the other kinematic methods, real-time kinematic surveying requires that two (or more) receivers be operated simultaneously. The unique aspect of this newest GPS surveying procedure is that radio modems are used. One receiver occupies a reference station and broadcasts raw GPS observations to the roving unit (or units). At the rover, the GPS measurements from both receivers are processed in real-time by the unit's onboard computer to produce an immediate determination of its location.

Tests of real-time kinematic surveying have shown that it yields position accuracies equivalent to those obtained by kinematic methods that employ post-processing. Because point positions with high accuracy are immediately known, real-time kinematic surveying is applicable for construction stakeout (see Section 24-2.3). It is also convenient for locating details for topographic mapping (see Section 16-9.6), and for many other surveying tasks.

20-7 GPS ERRORS

In addition to the satellite and receiver clock errors already discussed, several other sources of error exist in GPS surveying. These include

1. Satellite ephemeris errors
2. Errors due to atmospheric conditions

3. Receiver errors
4. Multipath errors
5. Errors in centering the antenna over a station and measuring its height

Satellite ephemeris errors are uncertainties that exist in the satellite orbits. These uncertainties result in erroneous values being computed for X_1, Y_1, Z_1; X_2, Y_2, Z_2; and so on, in Eqs. (20-1) and (20-2). Satellite positions computed from the broadcast ephemeris may contain errors as large as ± 20 m. In the pseudoranging technique, these satellite position errors are translated directly into the computed positions of ground stations. The error sizes can be reduced to less than 5 m by upgrading the orbital data, using information from tracking stations. One disadvantage of this is the delay in obtaining upgraded ephemeris data. Satellite position errors are of little consequence in the differential pseudoranging and carrier phase differencing procedures because they are effectively subtracted out of the solution.

Errors due to atmospheric conditions stem from the fact that satellite signals travel at different velocities through portions of the atmosphere having different transmission characteristics. These conditions change between day and night, as well as with the seasons. The angular direction of signals through the atmosphere also affects their travel. Delays in the propagation of signals through the *ionosphere* are inversely proportional to the squares of their frequencies. Therefore by making measurements using both L1 and L2 frequencies, and comparing the results, much of the error due to ionospheric conditions can be detected and eliminated from the solution. Instruments capable of measuring carrier phases of both L1 and L2 signals are called *dual-frequency* receivers. The Ashtech P-12 and the Trimble 4000SE receivers of Figures 20-3 and 1-5, respectively, both have that capability.

Receiver errors are caused by electrical noise and errors in matching the transmitted signals. Multipath errors occur when a transmitted signal that was initially not traveling toward the ground receiver is reflected from another object and then received. Travel times of these reflected signals will be greater than the direct straight-line travel times, and thus yield erroneous ranges. These particular errors are quite small, however.

To reduce errors due to antenna centering, bull's-eye bubbles of optical plummets must be checked regularly and adjusted if necessary, or a plumb bob must be used with care. The erroneous effect of a possible offset in the antenna's phase center can be minimized in relative positioning work by always orienting the antenna the same in azimuth. This can be done, for example, by always facing a reference mark on the antenna toward north using a compass. Antenna heights should be measured carefully both before and after the observing session, and both values entered on the data sheet.

20-8 REFERENCE COORDINATE SYSTEMS

In determining the positions of points on earth from satellite observations, at least three different reference coordinate systems are involved. First of all, satellite positions at the instant they are observed are specified in "space-related" three-dimensional rectangular coordinate systems defined by their orbits. A satellite's X_S, Y_S, Z_S orbital coordinate system is illustrated in Figure 20-10. The orbit is elliptical and has one of its two foci at the earth's center of gravity G. The *perigee* and *apogee* points are where the satellite is closest to and farthest away from G, respectively, in its orbit. The so-called *line of apsides* joins these two points, passes through the two foci, and is the reference X_S axis.

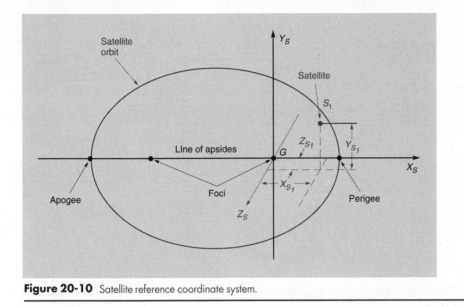

Figure 20-10 Satellite reference coordinate system.

The axis system's origin is at G; the Y_S axis is in the mean orbital plane; and Z_S is perpendicular to this plane. Values of Z_S coordinates represent departures of the satellite from its mean orbital plane and normally are very small. A satellite at position S_1 would have coordinates X_{S_1}, Y_{S_1}, and Z_{S_1}, as shown in Figure 20-10. For any instant of time these coordinates can be calculated from the satellite's orbital parameters.

In the reduction process it is next necessary to convert these X_S, Y_S, Z_S satellite coordinates into a *terrestrial,* or "earth-related," three-dimensional rectangular system X_e, Y_e, Z_e. For this system (see Figure 20-11), the Z_e axis coincides with the mean polar axis of the earth, X_e passes through the Greenwich meridian, and the origin is at the center of a selected reference ellipsoid, which also coincides with the earth's mass center. To make the conversion from satellite to terrestrial coordinates, four angular parameters are required which define the relationship between the satellite's orbital coordinate system and key reference planes and lines on the earth. As shown in Figure 20-11, these parameters are **(1)** the *inclination* angle i (angle between the orbital plane and the earth's equatorial plane), **(2)** the *argument of perigee* ω (angle measured in the orbital plane from the equator to the line of apsides), **(3)** *right ascension of the ascending node* Ω (angle measured in the plane of the earth's equator from the vernal equinox to the line of intersection between the orbital and equatorial planes), and **(4)** the *Greenwich hour angle of the vernal equinox* GHAγ (angle measured in the equatorial plane from the Greenwich meridian to the vernal equinox).

In a final step, the X_e, Y_e, and Z_e coordinates are converted into geodetic coordinates of latitude, longitude, and *geodetic height* (height above the ellipsoid). These coordinates are referenced to the World Geodetic Reference System of 1984, which uses the WGS84 ellipsoid. The origin of this ellipsoid is the earth's mass center, and for practical purposes it is the same as the GRS80 ellipsoid (see Section 19-1). From latitude and longitude values, state plane coordinates, which are more convenient for use by local surveyors, can be computed (see Chapter 21).

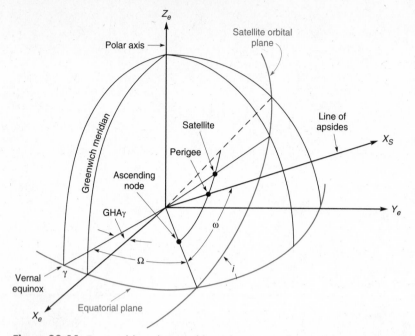

Figure 20-11 Terrestrial three-dimensional coordinate system for specifying satellite positions

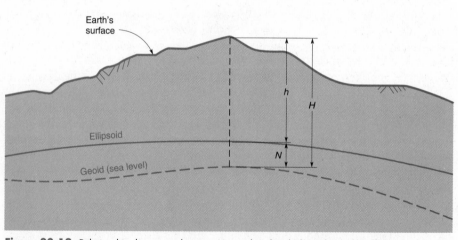

Figure 20-12 Relationships between elevation *H*, geodetic height *h*, and geoid height *N*.

It is important to note that geodetic heights obtained with GPS are measured with respect to the ellipsoid. As shown in Figure 20-12, these are not equivalent to *elevations* (also called *orthometric heights*) given with respect to the geoid. (Recall from Chapter 6 that the geoid is the mean sea level surface datum for elevations.) To convert geodetic heights to elevations, *geoid heights* (the vertical distance between ellipsoid and geoid) must be known (see Figure 20-12). Then elevations can be expressed as:

$$H = h - N \qquad (20\text{-}4)$$

where H is elevation above mean sea level, h the geodetic height (measured by GPS), and N the geoid height. *Figure 20-12 shows the correct relationships of the geoid and ellipsoid in the continental United States. Here the ellipsoid is above the geoid, and geoid heights (measured from the ellipsoid) are negative.* The geoid height at any point can be estimated from mathematical models developed from a network of points where geoid heights have been measured.[1]

EXAMPLE 20-1

Compute the elevation (orthometric height) for a station whose ellipsoid height is 59.1 m, if the geoid height in the area is -21.3 m.

[handwritten: $h = H + N$]
[handwritten: $59.1 =$]

SOLUTION
By Eq. (20-4):

$$H = 59.1 - (-21.3) = 80.4 \text{ m}$$

20-9 PLANNING GPS SURVEYS

Many factors must be taken into account in planning GPS surveys. One important consideration is the selection of station locations to be reasonably accessible by either land vehicles or aircraft that transport the GPS hardware. Stations can be somewhat removed from vehicle access points because GPS components are relatively small and portable. In addition, the receiver antenna is the only hardware component that must be accurately centered over the ground station. It is easily hand carried and can be separated from the other receiver hardware by a length of electric cable, as shown in Figure 20-3. Each station should be clearly marked and adequately described in advance so that valuable time is not lost by the GPS crew when it arrives.

In addition, because satellites are being observed, stations must be selected with an overhead view free of obstructions. As a minimum, it is recommended that visibility be clear in all directions from an altitude angle of 15° to the zenith. (Satellites are not normally observed below a 15° elevation angle.) In some cases, careful station placement will enable this visibility criterion to be met without difficulty; in other situations clearing around the stations may be necessary.

Preparing observational programs is another important activity in planning GPS surveys. This consists of determining which satellites will be visible from a given ground station during a proposed observing period. To aid in this activity, azimuth and elevation angles to each visible satellite can be predetermined by computers for times within the planned period. Required computer input, in addition to observing time, includes the station's latitude and longitude.

Table 20-2 gives azimuth and elevation angles (in degrees) to satellites that were visible from Madison, Wis., from 15:00 to 17:30 hours (3:00 to 5:30 P.M.) local time

---

[1]A diskette containing a high-resolution geoid height model known as *GEOID93* is available from the National Geodetic Information Center, NOAA, National Geodetic Survey, N/CG17, SSMC3 Station 09535, 1315 East West Highway, Silver Spring, Md. 20910, telephone (301) 713-3242.

TABLE 20-2 GPS MULTI-SITE MISSION PLANNING SATELLITE VISIBILITY

SV Number	3	16	17	21	23	24	26	28
Local Time	EL:AZ	EL:AZ	EL:AZ	EL:AZ	EL:AZ	EL:AZ	EL:AZ	EL:AZ
15:00:00	61:327	54: 50	31:291	—|—	—|—	19: 97	40:150	—|—
15:05:00	63:327	52: 51	33:293	—|—	—|—	18: 99	43:149	—|—
15:10:00	66:327	50: 51	34:294	—|—	—|—	16:100	45:147	—|—
15:15:00	68:326	48: 51	36:296	—|—	16:239	—|—	47:145	—|—
15:20:00	71:324	46: 52	38:297	—|—	18:241	—|—	49:143	—|—
15:25:00	74:322	43: 53	40:299	—|—	20:242	—|—	51:141	—|—
15:30:00	76:318	41: 53	42:300	—|—	21:243	—|—	53:138	—|—
15:35:00	78:312	39: 54	44:301	—|—	23:245	—|—	55:135	—|—
15:40:00	81:303	37: 55	46:303	—|—	25:246	—|—	56:132	—|—
15:45:00	82:288	35: 56	48:304	—|—	27:248	—|—	58:128	—|—
15:50:00	83:265	33: 57	50:306	—|—	29:250	—|—	59:124	—|—
15:55:00	83:240	31: 58	52:307	—|—	31:251	—|—	61:120	—|—
16:00:00	81:221	29: 59	54:309	—|—	32:253	—|—	62:115	—|—
16:05:00	79:209	27: 60	56:310	—|—	34:255	—|—	63:110	—|—
16:10:00	77:202	26: 61	58:311	—|—	36:257	—|—	63:104	—|—
16:15:00	74:197	24: 62	60:313	—|—	38:258	—|—	63: 99	—|—
16:20:00	72:194	22: 63	63:314	—|—	40:260	—|—	63: 93	—|—
16:25:00	69:192	20: 64	65:316	—|—	41:262	—|—	63: 88	—|—
16:30:00	66:190	18: 65	67:317	—|—	43:265	—|—	62: 83	—|—
16:35:00	63:189	17: 66	69:319	—|—	45:267	—|—	61: 78	—|—
16:40:00	61:188	15: 68	71:320	—|—	47:269	—|—	60: 74	—|—
16:45:00	58:188	—|—	74:322	—|—	49:271	—|—	59: 70	—|—
16:50:00	55:187	—|—	76:324	—|—	50:274	—|—	57: 66	—|—
16:55:00	52:187	—|—	78:327	15:264	52:276	—|—	56: 63	15:320
17:00:00	50:187	—|—	81:330	17:265	54:279	—|—	54: 61	16:318
17:05:00	47:187	—|—	83:334	18:267	56:282	—|—	52: 58	18:317
17:10:00	44:186	—|—	85:343	20:269	57:285	—|—	50: 56	19:315
17:15:00	41:186	—|—	88: 9	22:270	59:288	—|—	48: 55	20:313
17:20:00	39:186	—|—	88: 92	23:272	61:292	—|—	46: 53	21:312
17:25:00	36:186	—|—	86:124	25:273	62:295	—|—	44: 52	22:310
17:30:00	33:186	—|—	83:133	27:275	64:299	—|—	42: 51	23:308

Source: Tabular data generated by computer using Ashtech's Mission Planning Software and made available courtesy Wisconsin Department of Transportation.

on September 20, 1992. As indicated by the tabulated data, satellite vehicle (SV) numbers 3, 16, 17, 21, 23, 24, 26, and 28 were visible for part or all of that period. Note that five satellites were visible simultaneously from 15:15 through 16:40 that day, and six were available from 16:55 through 17:30.

Azimuths and elevations to visible satellites from a given location can be represented graphically using *polar plots* (also called *sky plots*) for rapid and easy interpretation. A polar plot, as illustrated in Figure 20-13, consists of a series of concentric circles. The circumference of the outer circle is graduated from 0 to 360° to represent satellite azimuth angles. The outer circle also corresponds to an elevation angle of 15° above

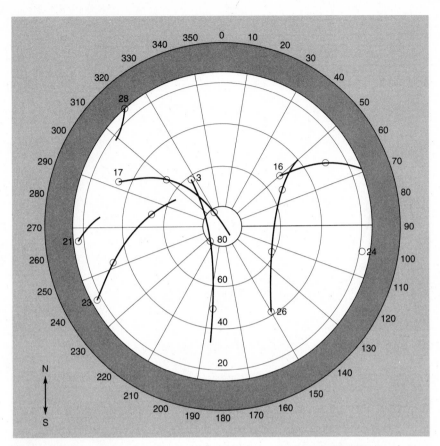

Figure 20-13 Polar plot of satellites visible from Madison, Wis. (latitude 43°04′N, longitude 89°23′W) for 3:00 to 5:30 P.M., September 20, 1992. (Plot produced using Ashtech's mission planning software and made available courtesy Wisconsin Department of Transportation.)

the horizon. Each successive concentric circle progressing toward the center represents an increment in elevation angle, with the radius point corresponding to the zenith.

The polar plot of Figure 20-13 was generated by a computer using Ashtech's mission planning software. To plot the figure, the satellite azimuths and elevations of Table 20-2 were plotted for uniform time increments. For each satellite, the SV number is plotted beside its first data point, and arcs connect successive plotted positions. Thus their travel paths in the sky are shown. Polar plots are valuable in GPS planning because they enable operators to quickly visualize not only the number of satellites visible during a planned observation period, but also their geometric distribution in the sky. This is important because highest accuracy in GPS work requires that observations be taken on groups of four or more widely spaced satellites that form a strong geometric intersection at the observing station. This condition is illustrated in Figure 20-14(a). Weaker geometry, as shown in Figure 20-14(b), should be avoided if possible as it will yield lower accuracy.

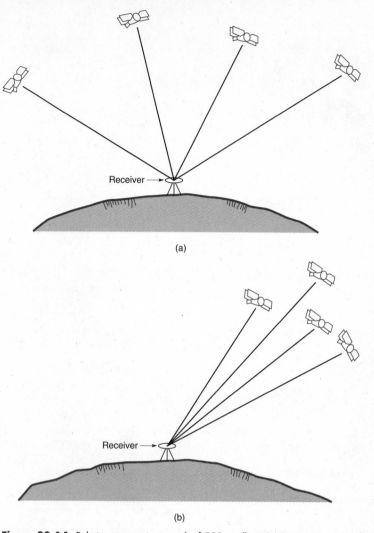

(a)

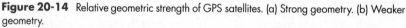

(b)

Figure 20-14 Relative geometric strength of GPS satellites. (a) Strong geometry. (b) Weaker geometry.

The relative geometric strength of a GPS position determination can be expressed quantitatively using a term called *position dilution of precision* (PDOP). Varying with time, PDOP is affected by both the number of visible satellites and their varying locations in the sky with respect to an observation station. For any given time it can be computed from a preliminary least-squares position solution based on the satellite locations in the constellation at the time. Figure 20-15 shows a plot of the number of visible satellites and PDOP for local times between 3:00 and 5:30 P.M. at Madison, Wis., on September 20, 1992. Note the strong negative correlation between the two curves; that is, when the number of visible satellites decreases, PDOP increases. PDOP values below about 5.0 are considered satisfactory for static GPS surveys—if a higher number exists, observations should be delayed until a more favorable satellite constellation is available.

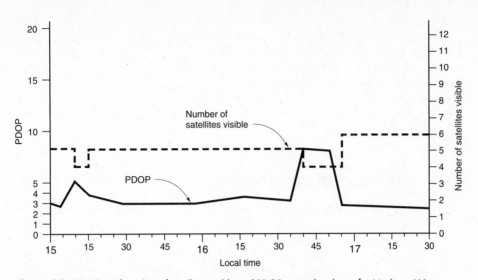

Figure 20-15 Plots of number of satellites visible and PDOP versus local time for Madison, Wis., on September 20, 1992. (Produced using Ashtech's Mission Planning software and made available courtesy Wisconsin Department of Transportation.)

Modern GPS receivers, with their built-in microprocessors, can derive polar plots and compute PDOP values in the field and display the results on their screens.

20-10 SPECIFICATIONS FOR GPS SURVEYS

The Federal Geodetic Control Subcommittee (FGCS) has published a preliminary document entitled "Geometric Geodetic Accuracy Standards and Specifications for Using GPS Relative Positioning Techniques."[2] The document specifies seven different orders of accuracy for GPS relative positioning and provides guidelines on instruments and field and office procedures to follow to achieve them. Table 20-3 lists these orders of accuracy.

TABLE 20-3 GPS RELATIVE POSITIONING ORDERS OF ACCURACY

ORDER	ALLOWABLE ERROR RATIO
AA	1:100,000,000
A	1:10,000,000
B	1:1,000,000
C-1	1:100,000
C-2-I	1:50,000
C-2-II	1:20,000
C-3	1:10,000

[2]Available from the National Geodetic Information Center, NOAA, National Geodetic Survey, N/CG17, SSMC3 Station 09535, 1315 East West Highway, Silver Spring, Md. 20910, telephone (301) 713-3242.

The document also makes recommendations concerning categories of surveys for which the different orders of accuracy are appropriate. Some of these recommendations are: order AA for global and regional geodynamics and deformation measurements; order A for "primary" networks of the National Geodetic Reference System (NGRS), and regional and local geodynamics; order B for "secondary" NGRS networks and high-precision engineering surveys; and the various classes of order C for mapping control surveys, property surveys, and engineering surveys. The allowable error ratios given in these standards imply the extremely high accuracies that are now possible with GPS.

20-11 FUTURE OUTLOOK FOR GPS

The future of GPS in the practice of surveying seems assured. The systems have already demonstrated reliability and a capability to yield extremely high accuracy. Although the most common application of GPS was originally for control work, the systems have now been used in virtually every type of survey, including property surveying, topographic mapping, and construction stakeout. GPS has been especially useful in solving some heretofore rather difficult problems, such as determining positions for hydrographic soundings, locating dredging rigs, and fixing positions of offshore oil wells. New applications are being tested on a regular basis, and research and development to improve the systems continue.

Even though the initial cost of GPS equipment is relatively high compared to other traditional surveying equipment, its cost-effectiveness has been adequately demonstrated. As technology improves, these systems will become smaller, more accurate, and less expensive. Their cost has already decreased significantly, and predictably they will soon be affordable to all surveyors.

Surveying by GPS provides many advantages over other traditional methods, including speed, accuracy, and operational capability by day or night and in any weather. Also intervisibility between stations is no longer required. For these reasons and others, GPS should be used increasingly in the future for all types of surveys except those with unavoidable obstructions to satellite visibility, such as in mining and tunneling surveys, and surveys in urban areas having high-rise structures.

PART II
INERTIAL SURVEYING SYSTEMS

20-12 OPERATING PRINCIPLE OF INERTIAL SURVEYING SYSTEMS

The operating principle of Inertial Surveying Systems (ISSs) consists fundamentally in making measurements of accelerations over time. This is done while the instrument is carried from point to point in a land vehicle or helicopter. An ISS mounted in a helicopter is shown in Figure 20-16. The acceleration and time measurements are taken independently in three mutually orthogonal planes which are oriented (1) north-south, (2) east-west, and (3) in the direction of gravity (up-down). From the acceleration and time data, components of the instrument's movement (distances and directions) in each of

Figure 20-16 Inertial surveying system mounted in a helicopter. (Courtesy Shell Canada Resources, Ltd.)

the three reference directions can be computed, and from this data the relative positions of points determined.

20-13 COMPONENTS OF ISSs

The major components of an ISS are *accelerometers, gyroscopes,* and a *computer.* In addition, the system contains units for *control and display, data storage,* and *power supply* (typically, a 24-V battery).

The operating principle of an accelerometer in an ISS can be illustrated with a simplified diagram like Figure 20-17. Here the accelerometer consists of a pendulum of mass M equipped with a feedback system. If the accelerometer is at rest or moving at a constant rate, the pendulum will hang in its rest position. When acceleration or deceleration occurs as a result of movement, the pendulum will have a tendency to swing away from its rest position. A detector senses this tendency to move and sends feedback signals to a forcing system that applies torque forces just sufficient to prevent the pendulum from swinging.

The magnitudes of force applied, together with the length of time over which they occur, are measured. Newton's fundamental law of motion states that force equals mass times acceleration. Therefore, since both mass and force are known, acceleration can be computed, that is, $a = F/M$, where a is acceleration, F is force, and M is pendulum mass. Finally, with acceleration and time available, the distance traveled is determined. The system's computer converts the combinations of force and time to movement components in the three reference directions and applies them to the unit's initial position, giving its location at any time.

Each ISS includes three accelerometers. To perform accurate surveying measurements, these must be carefully maintained along three mutually orthogonal axes: E

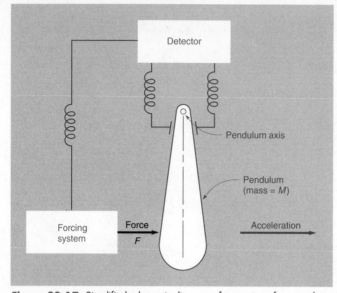

Figure 20-17 Simplified schematic diagram of operation of an accelerometer.

(east-west), N (north-south), and Z (gravity). Orientation of the accelerometer axes in the three required reference directions is accomplished by precision gyroscopes.

Each gyroscope is mounted on a gimbal,[3] enabling it to rotate freely as the vehicle turns in moving from point to point. A sensing mechanism operates with the computer to engage "torquing motors," which maintain the desired orientation of the accelerometer axes in spite of vehicle movements. In addition to rotations resulting from vehicle turns, the accelerometers and gyroscopes also rotate as a result of the earth's rotation on its axis, as well as rotations that occur in transporting the unit on the curved surface of the earth. The torquing motors also account for these effects in maintaining proper orientation of the accelerometers.

Gyroscope rotations necessary to counteract the earth's rotation on its axis, and rotations due to gyroscope transport are illustrated in Figure 20-18. Rotation of the earth on its axis occurs at a rate of approximately 15° per hour. In Figure 20-18 assume that at a given instant of time a gyro at A has its axis oriented toward O, the earth's gravitational center. At some time later, due to the earth's rotation, the gyro is at point B. Unless an axis correction is made, the gyro axis would be aligned in the direction of the arrow at B. However, the sensing mechanism will have activated the torquing motors and corrected the axis orientation at a uniform rate so that at B, and everywhere along the arc from A to B, it would point toward O. The figure shows that corrections required for maintaining the orientation of an accelerometer due to this cause vary with latitude, are maximum at the equator, and decrease to zero at the poles.

Rotations that result from transporting the unit are similar to those caused by the earth's rotation, and occur because the earth's surface is curved. Again, from Figure 20-18, if a gyro is carried from A to B, the same angular axis correction would be

[3]A gimbal is a mechanical device similar to a universal joint, which allows rotations to occur freely in any direction.

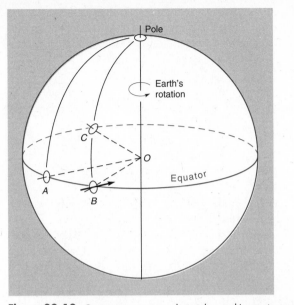

Figure 20-18 Gyroscope rotations due to the earth's rotation and gyroscope transport.

required as that resulting from earth's rotation through angle *AOB*. Similarly, for movement in a latitudinal direction, as in carrying the gyro from *B* to *C*, an axis correction of angle *BOC* is required. Thus it is evident that both latitudinal and longitudinal correction components would be needed simultaneously if the unit were moved obliquely from, say, *A* to *C*. The angular rate of correction depends on the speed at which the unit is transported and on its location.

Together the accelerometers, gyroscopes, sensing mechanisms, and torquing motors comprise the fundamental measuring system of an ISS, which is called the inertial measuring unit (IMU). It is mounted on the so-called *platform* (precision gimbal support) to isolate it from vehicle movements.

Each ISS has its own dedicated computer which controls and monitors the system's operation. The control and display unit includes devices that enable an operator to input data and communicate with the system. An operator can initiate the instrument's calibration and measurement phases with this unit. Also, the latitude, longitude, and elevation of the starting station can be entered when a traverse is being run. The data storage unit, consisting of a high-quality magnetic tape recorder, records all operational data, including instrument and survey information.

20-14 ISS MEASUREMENT THEORY

As stated previously, the basic operating principle of an ISS consists of measuring accelerations over time along three mutually perpendicular axes. Acceleration is the rate of change of velocity with respect to time. It normally varies as the instrument progresses from one point to another. As the ISS is moved, at precise incremental time intervals, it measures changing values of acceleration.

To help clarify the measuring procedure, a simple example is presented. Assume that an ISS moves from a stopped position at point *A* to another point *B*, and the direction of measurement is due east and along a level course. In moving from A to B, assume the unit first accelerates for 5 sec at a constant rate of 10 ft/sec/sec. Its E axis accelerometer would record this 10-ft/sec/sec acceleration at each time interval over this 5-sec time span. After the first second of time, the unit's velocity will be 10 ft/sec. After the next second it will be 20 ft/sec, and so on, until its velocity reaches 50 ft/sec after the fifth second. Assume then that the unit stops accelerating (maintains zero acceleration) for the next 10 sec. During this period its velocity does not change, but remains at 50 ft/sec. Next assume that the unit decelerates at a rate of 10 ft/sec/sec for 3 sec, and then decelerates at a rate of 5 ft/sec/sec for 4 sec, whereupon it stops at point B.

The total travel time in this example is 22 sec; accelerations, velocities, and distances run, after each 1-sec increment of time, are listed in Table 20-4. Note in the table that distances traveled are calculated as the product of the time increment and average velocity. Thus the distance covered during the first second is $1 \times [(10 + 0)/2] = 5$ ft. Similarly, the incremental distance moved during the next second is $1 \times [(10 + 20)/2] = 15$, and the cumulative distance covered during the first 2 sec is $5 + 15 = 20$ ft. By continuing these calculations through the 22 sec of travel time, the total distance

TABLE 20-4 ISS EXAMPLE ACCELERATION, VELOCITY, AND DISTANCE AFTER EACH SECOND

ELAPSED TIME (sec)	ACCELERATION (ft/sec/sec)	VELOCITY (ft/sec)	INCREMENTAL DISTANCE TRAVELED (ft)	CUMULATIVE DISTANCE TRAVELED (ft)
0	0	0	0	0 = A
1	10	10	5	5
2	10	20	15	20
3	10	30	25	45
4	10	40	35	80
5	10	50	45	125
6	0	50	50	175
7	0	50	50	225
8	0	50	50	275
9	0	50	50	325
10	0	50	50	375
11	0	50	50	425
12	0	50	50	475
13	0	50	50	525
14	0	50	50	575
15	0	50	50	625
16	− 10	40	45	670
17	− 10	30	35	705
18	− 10	20	25	730
19	− 5	15	17.5	747.5
20	− 5	10	12.5	760
21	− 5	5	7.5	767.5
22	− 5	0	2.5	770 = B

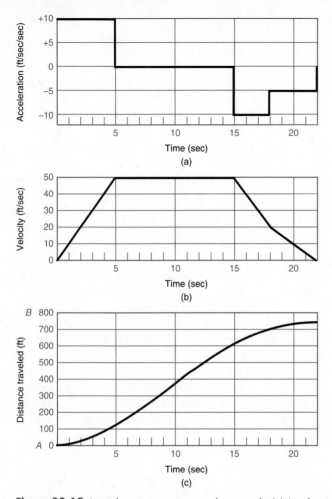

Figure 20-19 Inertial surveying system graphic example. (a) Acceleration. (b) Velocity. (c) Distance traveled.

between points *A* and *B,* 770 feet, is obtained. Figure 20-19 consists of three parts and illustrates the data of Table 20-4, graphically showing the relationships of acceleration, velocity, and distance traveled.

This example was simplified by assuming accelerometer measurements at 1-sec increments. (In reality, however, ISS devices measure at very short time intervals, that is, less than 0.02 sec.) Recall also that to retain simplicity, the example assumed travel on a level surface in a due easterly direction. Thus only the accelerometer aligned along the *E* axis would be recording accelerations, and the other two would show zero accelerations. The total distance from *A* to *B* in the example therefore consists of only an easterly distance component. In the general case of movements from point to point, each accelerometer would be recording accelerations, and three components of movement would be obtained. These are vector quantities. Thus the total horizontal distance would be the square root of the sum of the squares of the *E* and *N* components, and the elevation difference would be the *Z* component.

20-15 ISS FIELD PROCEDURES

As noted earlier, when ISS units are in operation, they are normally carried from point to point in either land vehicles or helicopters; thus surveys can proceed rapidly. Land vehicles most commonly used are vans or panel trucks, with four-wheel drives preferred since off-road travel is often necessary. If helicopters are employed, as is being done for much surveying in Alaska, widely spaced points can be readily located, even though they may be separated by inaccessible, rugged, and heavily vegetated terrain.

Before each day's operation begins, the ISS is calibrated and checked. The process is initialized by an operator but controlled by the computer. The procedure, which may take about an hour, ensures that the system's components are operating correctly and that the accelerometer axes are aligned properly with the directions of local north, local east, and gravity. During calibration the unit must be absolutely stationary.

To run a traverse, the ISS is first initialized at a control station. This is done by positioning the vehicle near the control station, then aligning an offset peep sight over the mark. The height as well as the length and direction of offset from the IMU to the station are measured and entered in the computer. The latitude and longitude (or coordinates) and the elevation of the control station are also input. These data provide the necessary starting position, and are also needed because latitude affects rotational corrections that must be applied to the accelerometers in compensating for the earth's rotation and system transport. Elevation influences gravitational pull (acceleration) and must be included for the vertical accelerometer.

After calibration and alignment are completed, the ISS is carried and temporarily parked beside each new traverse station whose position is required. Alignment over each point is achieved with the peep sight, and preliminary coordinates and elevations are measured and recorded. Finally, the traverse loop is closed by either returning to the initial station or taking the ISS to another control point.

As the unit is carried from place to place, various instrumental errors will tend to accumulate, causing the accelerometers to become misaligned and hence record erroneous accelerations. These errors are detected by making frequent so-called *zero-velocity updates* (ZUPTs). They consist of bringing the unit to a standstill and observing accelerations and hence velocities. Departures from zero velocity while stopped represent system errors. These are noted by the computer, whereupon it corrects the preliminary positions back to the previous ZUPT.

When a traverse loop is closed (by either returning to the starting station or visiting another control station), and all corrections have been made for ZUPTs, the measured coordinates and elevation of the closing station are compared with their respective control values. Misclosure errors will be revealed and processed together with all previous ZUPT information in a final adjustment to obtain coordinates and elevations of all new traverse stations.

Figure 20-20 shows an actual ISS traverse network. This network used five control points, four of which were exterior and one interior. Runs between control points were made in generally east-west and north-south directions to create an interlocking network. Some runs were repeated to increase accuracy.

Table 20-5 is a computer listing for an approximately 11-mi run (one way) between control stations "Brooklyn" and "Ian Eccentric" in Figure 20-20, and back. In the process, locations of intermediate stations 1235 and 1110 were determined. The listing

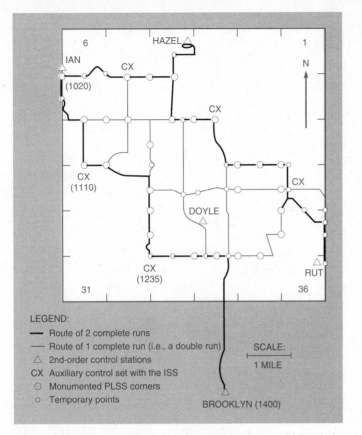

Figure 20-20 Inertial surveying system traverse network established to provide control for a regional land records project. (Courtesy Dane County Land Records Project.)

gives preliminary coordinates of the intermediate points for each run. When all courses had been run forward and back, a final simultaneous adjustment of all data was made. Note that even preliminary coordinates for the two new stations on the forward and back runs differ only by a few thousandths of a second in latitude and longitude (where 0.001″ is roughly equivalent to 0.1 ft).

20-16 PLANNING CONSIDERATIONS FOR ISS SURVEYS

In designing traverses for survey by ISS, points should be selected for convenient access by the transport vehicle. If the vehicle is a helicopter, considerable open space near the points is necessary; for land vehicles, points near roads are preferred. The points should be clearly marked, adequately referenced, and described so they can be quickly found by an ISS field crew. This is especially important for inertial surveys because errors accumulate with time, and thus delays in looking for points must be minimized. Movements from point to point should be as directional as possible, sharp turns should be avoided, and vehicle travel speed should be maintained as uniform as possible.

TABLE 20-5 ISS DATA FOR RUN FROM STATION "BROOKLYN" TO STATION "IAN ECCENTRIC" OF FIGURE 20-22, AND RETURN

I. FIRST RUN (BROOKLYN TO IAN ECCENTRIC)

1400 BROOKLYN

LAT 42°49′41.226″	TAPE ID 1400 LONG 89°24′42.490″	TIME 05H 00M 13S ELEV 0335.34 M

1235

LAT 42°52′15.195″	TAPE ID 1235 LONG 89°26′53.493″	TIME 05H 06M 50S ELEV 0273.17 M

1110

LAT 42°53′59.727″	TAPE ID 1110 LONG 89°28′40.015″	TIME 05H 10M 58S ELEV 0279.64M

1020 IAN ECC.

LAT 42°55′43.126″	TAPE ID 1020 LONG 89°29′39.920″	TIME 05H 16M 15S ELEV 0330.30 M

II. RETURN RUN (IAN ECCENTRIC TO BROOKLYN)

1020 IAN ECC.

LAT 42°55′43.124″	TAPE ID 1020 LONG 89°29′39.921″	TIME 05H 56M 27S ELEV 0330.27 M

1110

LAT 42°53′59.728″	TAPE ID 1110 LONG 89°28′40.011″	TIME 06H 00M 28S ELEV 0279.30M

1235

LAT 42°52′15.201″	TAPE ID 1235 LONG 89°26′53.478″	TIME 06H 04M 18S ELEV 0273.16 M

1400 BROOKLYN

LAT 42°49′41.226″	TAPE ID 1400 LONG 89°24′42.490″	TIME 06H 10M 19S ELEV 0335.30 M

Source: Courtesy U.S. Bureau of Land Management, Division of Cadastral Survey.

Traverses should be designed to begin on one control station and close on another where practical. Closing back on the initial traverse station reveals misclosures, but systematic errors that may be inherent in the measuring procedure cannot be detected. To improve accuracy, four control points are desirable. Also, traverses should be repeated and run forward and back between control points in a crisscrossing pattern. In

Figure 20-20, for example, four peripheral control point locations were selected, and a pattern of generally north-south and east-west courses was run.

Accuracy in ISS traversing is further enhanced by shortening the intervals between ZUPTs and limiting the total traverse time to not more than 2 h. Very high accuracies have been achieved when special precautions are taken. Positions and elevations correct to within 1 cm or less have been obtained where ZUPTs were made at about 1-min intervals on traverses run forward and back between control points spaced up to 4 km apart. For ZUPTs at approximately 3- to 5-min intervals (and exercising reasonable care), positions and elevations accurate to within 2 to 3 cm are the rule.

PROBLEMS

20-1 Explain the operating principle of satellite Doppler systems.

20-2 Name and briefly describe the two different basic methods for locating points using satellite Doppler observations.

20-3 Describe the planned fully operational constellation of GPS satellites, including a brief description of the satellite orbits.

20-4 Why is the fully operational GPS satellite constellation designed so that from four to six satellites will always be visible from any location on earth?

20-5 What is the purpose of the *GPS Operational Control System* and CIGNET?

20-6 Discuss the different frequencies used for carrier and code signals of GPS, and give their relationships to one another.

20-7 What are the purposes of *selective availability* (SA) and *antispoofing* (AS)?

20-8 In addition to transmitting PRN codes, what other information is broadcast by GPS satellites?

20-9 Describe the general procedure of pseudoranging for point position determination by GPS.

20-10 Same as Problem 20-9, except for differential pseudoranging.

20-11 Discuss the general concept of positioning by GPS carrier phase measurements.

20-12 What errors affect the accuracy of GPS surveys?

20-13 How can a receiver's clock error be eliminated in GPS positioning with carrier phase measurements?

20-14 How can a satellite's clock error be eliminated in GPS positioning with carrier phase measurements?

20-15 What is GPS double differencing? Triple differencing?

20-16 Explain the differences between **(a)** static, **(b)** rapid static, **(c)** kinematic, **(d)** pseudokinematic and **(e)** real-time kinematic GPS surveying procedures.

20-17 Describe the orbital coordinate system used to define satellite positions. Support your description with a sketch.

20-18 Using a sketch, define the angular parameters necessary to convert satellite positions from their orbital coordinate systems to a terrestrial or earth-based coordinate system. Why is this conversion necessary?

20-19 Define the terms "ellipsoid height," "geoid height," and "elevation." Give their relationships.

20-20 What is a GPS polar plot and how is it used?

20-21 What is position dilution of precision?

20-22 Discuss factors to consider when selecting the location of a GPS receiver station.

20-23 Explain conceptually how an accelerometer operates.

*20-24 Review Table 20-4 and Figure 20-19. Develop a similar table for 10 sec of constant acceleration at 5 ft/sec/sec, then decelerations at 10 ft/sec/sec for 3 sec, and 5 ft/sec/sec for 4 sec. (Assume all movements to be along the *E* axis only.)

*Asterisks indicate problems that have answers given in Appendix G.

20-25 Develop a table similar to Table 20-4, and draw diagrams similar to Figure 20-19 for a hypothetical ISS measurement from point *A* to point *B*. Assume an acceleration of 5 ft/sec/sec for 8 sec, zero acceleration for 10 sec, and then deceleration at 4 ft/sec/sec for 10 sec. (Assume all movements to be along the *N* axis only.)

20-26 What angular correction (to the nearest minute) will be required by torquing motors of an ISS to compensate for earth rotation over a period of:
  *(a) 1 h 30 min of operation?
  (b) 2 h 20 min of operation?

20-27 What angular value of gyroscope axis correction (to the nearest second) will be required due to transport of an ISS:
  *(a) Easterly a distance of 150 mi at 45° North latitude?
  (b) Westerly a distance of 200 mi at 32°30′ North latitude?

20-28 A control station has state plane coordinates of $X = 2,785,371.72$ ft and $Y = 89,538.02$ ft. If the peep sight for aligning an ISS over the station was offset by a distance of 5.75 ft on an azimuth of 257°32′ from the IMU, what are the coordinates of the IMU?

20-29 Locate a published article on an application of GPS in surveying. Give the name of the publication, the article's title and author, and write a summary of the article's content.

20-30 Similar to Problem 20-29, except for an application of ISS in surveying.

*Asterisks indicate problems that have answers given in Appendix G.

BIBLIOGRAPHY

Adkins, K. and J. Lyon. 1991. "Use of Aerial Photos to Identify Suitable GPS Survey Stations." *Photogrammetric Engineering and Remote Sensing* 42 (no. 7):933.

Alexander, L. 1992 "Differential GPS in Operation Desert Storm." *GPS World* 3 (No. 6):36.

Archinal, B., and I. Mueller. 1986. "GPS: An Overview." *Professional Surveyor* 6 (No. 3):7.

ASCE. 1988. "Special Issue: GPS 88." *ASCE, Journal of Surveying Engineering* 114 (No. 4):145.

ASCE. 1989. "Special Issue: GPS 88." *ASCE, Journal of Surveying Engineering* 115 (No. 1):1.

Blair, B. R. 1989. "Practical Applications of Global Positioning Systems." *ASCE, Journal of Surveying Engineering* 115 (No. 2):218.

Burkholder, E. F. 1993. "Using GPS Results in True 3-D Coordinate Systems." *ASCE, Journal of Surveying Engineering* 119 (No. 1): 1.

Cain, J. D. 1992. "Statewide, High-Precision GPS Networks Support Local GIS Accuracy." *GIS World* 5 (No. 6):46.

Collins, J. 1987. "GPS Surveying Techniques." *Point of Beginning* 11 (No. 4):22.

Collins, J. 1989. "Fundamentals of GPS Baseline and Height Determinations." *ASCE, Journal of Surveying Engineering* 115 (No. 2):223.

Ewing, B. D. 1990. "Pseudo-Kinematic GPS for the Surveyor." *GPS World* 1 (No. 5):50.

Fiedler, J. 1992. "Orthometric Heights from Global Positioning System." *ASCE, Journal of Surveying Engineering* 118 (No. 3):70.

Gerdan, G. P. 1992. "Efficient Surveying with the Global Positioning System." *Surveying and Land Information Systems* 52 (No. 1):34.

Goad, C. C. 1989. "On the Move with GPS." *Point of Beginning* 14 (No. 4):36.

Hatch, R. R., and E. V. Avery. 1989. "Strategic Planning Tool for GPS Surveys." *ASCE, Journal of Surveying Engineering* 115 (No. 2):207.

Haug, M. D., et al. 1980. "A Simplified Explanation of Doppler Positioning." *Surveying and Mapping* 40 (No. 1):47.

Hern J. 1989. *GPS: A Guide to the Next Utility.* Sunnyvale, Calif.: Trimble Navigation Ltd.

Heuverman, H. R., and W. J. Senus. 1983. "Navstar Global Positioning System." *ASCE, Journal of the Surveying Engineering Division* 109 (No. 2):73.

Hofmann-Wellenhof, B., et al. 1992. *GPS Theory and Practice.* New York: Springer-Verlag.

Klobuchar, J. A. 1991. "Ionospheric Effects on GPS." *GPS World* 2 (No. 4):48.

Langley, R. B. 1991. "The GPS Receiver: An Introduction." *GPS World* 2 (No. 1):50.

Langley, R. B. 1991. "The Orbits of GPS Satellites." *GPS World* 2 (No. 3):50.

Langley, R. B. 1993. "The GPS Observables." *GPS World* 4 (No. 4):52.

Leick, A. 1990. *GPS Satellite Surveying.* New York: John Wiley and Sons.

May, M. B. 1993. "Inertial Navigation and GPS." *GPS World* 4 (No. 6):56.

Milbert, D. G. 1992. "GPS and Geoid 90—The New Level Rod." *GPS World* 3 (No. 2):38.

Mooyman K., and C. A. Quirion. 1989. "High Production Kinematic GPS Surveying." *Surveying and Land Information Systems* 49 (No. 3):135.

Reilly, J. P. 1993. "P.O.B. 1993 GPS Equipment Survey." *Point of Beginning* 18 (No. 5):55.

Remondi, B. W. 1992 "GPS the Surveying Perspective." *Professional Surveyor* 12 (No. 4):13.

Roof, E. F. 1983. "Inertial Survey Systems." *ASCE, Journal of the Surveying Engineering Division* 109 (No. 2):116.

Schwarz, K. P. 1990. "Kinematic Positioning—Efficient New Tool for Surveying." *ASCE, Journal of Surveying Engineering* 116 (No. 4):181.

Schwarz, K. P. and M. G. Sideris. 1993. "Heights and GPS," *GPS World* 4 (No.2):50.

Treftz, W. H. 1981. "An Introduction to Inertial Positioning as Applied to Control and Land Surveying." *Surveying and Mapping* 41 (No. 1):59.

Van Wert, J. 1989. "Kinematic Stop and Go GPS." *Point of Beginning* 14 (No. 4):46.

Vicenty, T. 1989. "How to Obtain Elevations by GPS." *Point of Beginning* 15 (No. 1):62.

Wells, D., et al. 1986. *Guide to GPS Positioning.* Ottawa, Ont.: Canadian Institute of Surveying and Mapping.

Zilkoski, D. B., and L. D. Hothem. 1989. "GPS Satellite Surveys and Vertical Control." *ASCE, Journal of Surveying Engineering* 115 (No. 2):262.

21

STATE PLANE
COORDINATES

21-1 INTRODUCTION

Most surveys of small areas are based on the assumption that the earth's surface is a plane. As explained in Chapter 19, however, for large-area surveys it is necessary to consider earth curvature. This is done by computing the horizontal positions of widely spaced stations in terms of geodetic latitudes and longitudes. Unfortunately the calculations necessary to determine geodetic positions from survey measurements and get distances and azimuths from them are lengthy. Practicing surveyors often are not familiar with this procedure. Clearly, a system for specifying positions of geodetic stations using plane rectangular coordinates is desirable, since it allows computations to be made using relatively simple coordinate geometry formulas, such as those presented in Appendix B. The National Geodetic Survey met this need by developing a state plane coordinate system for each state.

A state plane coordinate system provides a common datum of reference for horizontal control of all surveys in a large area, just as mean sea level furnishes a single datum for vertical control. It eliminates having individual surveys based on different assumed coordinates, unrelated to those used in other adjacent work. State plane coordinates are available for all control points in the National Geodetic Reference System (NGRS), and for many other control points as well. They are widely used as the reference points for initiating surveys of all types, including those for highway construction projects, property boundary delineation, and photogrammetric mapping.

There are many examples illustrating the value of state plane coordinates. They make it possible for extensive surveys on highway projects to begin on one control station and close on another which is tied to the same coordinate system. On boundary surveys, if a parcel's corners are referenced to the state plane coordinate system, their locations are basically indestructible. The iron pipes, posts, or other monuments marking their positions may disappear, but their original locations can be restored from surveys initiated at other nearby monuments referenced to the state plane coordinate system. For this reason, some states require that state plane coordinates be included on all new subdivisions. State plane coordinates are highly recommended as the reference framework for entering maps and other data into Land and Geographic Information Systems, so that all data are referenced to a common system and thus can be accurately registered and overlaid for analysis purposes.

21-2 PROJECTIONS USED IN STATE PLANE COORDINATE SYSTEMS

As discussed in Chapter 19, the earth's curved or mean sea level surface closely approximates an *ellipsoid* (derived by mathematically revolving an ellipse about its semi-minor axis). To convert geodetic positions of a portion of the earth's surface to plane rectangular coordinates, points are projected mathematically from the ellipsoid to some imaginary *developable surface*—a surface that can conceptually be developed or "unrolled and laid out flat" without distortion of shape or size. A rectangular grid can be superimposed on the *developed* plane surface and the positions of points in the plane specified with respect to *X* and *Y* grid axes. A plane grid developed using this mathematical process is called a *map projection.*

Two basic projections are used in state plane coordinate systems: the *Lambert conformal conic projection* and the *transverse Mercator projection.* The Lambert projection utilizes an imaginary cone and the Mercator a fictitious cylinder as their developable surfaces. These are shown in Figure 21-1(a) and (b). The cone and cylinder are *secant* to the ellipsoid in the state plane coordinate systems; that is, they intersect the ellipsoid along two arcs *AB* and *CD* as shown. With this placement, the conical and cylindrical surfaces conform better to the ellipsoid over larger areas than they would if placed tangent.

Figure 21-1(c) and (d) illustrate plane surfaces developed from the cone and cylinder. Here, points are projected mathematically from the ellipsoid to the surface of the imaginary cone or cylinder based on their geodetic latitudes and longitudes. For discussion purposes this may be considered a radial projection from the earth's center. Figure 21-2 illustrates this process diagramatically and displays the relationship between the length of a line on the ellipsoid and its extent when projected onto the surface of either a cone or a cylinder. Note that distance $a'b'$ on the projection surface is greater than ab on the ellipsoid, and similarly $g'h'$ is longer than gh. From this observation it is clear that map projection scale is larger than true ellipsoid scale where the cone or cylinder is outside the ellipsoid. Conversely, distance $d'e'$ on the projection is shorter than de on the ellipsoid, and thus map scale is smaller than true ellipsoid scale when the projection surface is inside the ellipsoid. Points c and f occur at the intersection of projection and ellipsoid surfaces, and therefore map scale equals true ellipsoid scale along the lines of intersection. These relationships of map scale to true ellipsoid scale for various positions on the two projections are indicated in Figure 21-1(c) and (d). As will be discussed later, these length differences are accounted for by means of a so-called *scale factor.*

From the foregoing discussion it should be clear that points cannot be projected from the ellipsoid to developable surfaces without introducing distortions in the lengths of lines or the shapes of areas. These distortions are held to a minimum, however, by selected placement of the cone or cylinder secant to the ellipsoid, by choosing a *conformal projection* (one that preserves true angular relationships around points), and also by limiting the zone size or extent of coverage on the earth's surface for any one projection. If the width of zones is held to a maximum of 158 mi (254 km), and if two-thirds of this zone width is between the secant lines, distortions (differences in line

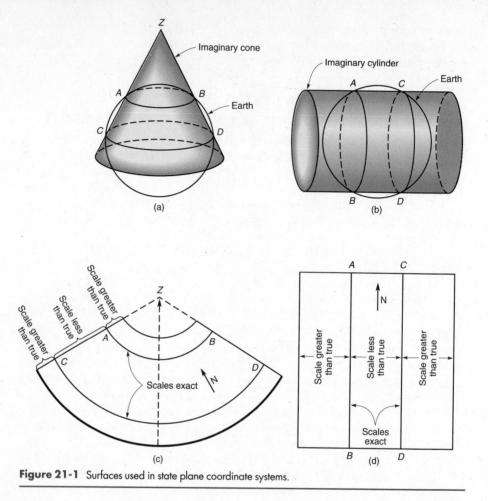

Figure 21-1 Surfaces used in state plane coordinate systems.

lengths on the two surfaces) are kept to 1 part in 10,000 or less. The NGS intended this accuracy in its development of the state plane coordinate systems.

For small states such as Connecticut and Delaware, one state plane coordinate zone is sufficient to cover the entire state. Larger states require several zones to encompass them; for example, Alaska has 10, California 6, and Texas 5. Where multiple zones are needed to cover a state, adjacent zones overlap each other. As explained in Section 21-10, this is important for surveys near zone edges, or when lengthy surveys extend from one zone to another. Figures 21-3 and 21-4 show the coverages of zones in Pennsylvania and Indiana, respectively. Both states have two zones, with Pennsylvania using the Lambert conformal conic projection and Indiana the transverse Mercator projection.

21-3 LAMBERT CONFORMAL CONIC PROJECTION

The Lambert conformal conic projection, as its name implies, is a projection onto the surface of an imaginary cone. The term *conformal*, as noted earlier, means that true angular relationships are retained around all points. Scale on a Lambert projection varies from north to south but not from east to west, as shown on Figure 21-1(c). Zone

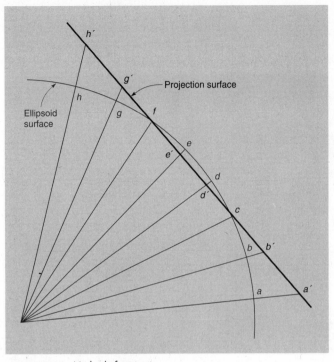

Figure 21-2 Method of projection.

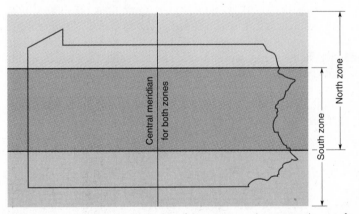

Figure 21-3 Coverages and overlap of zones in Pennsylvania's Lambert conformal conic state plane coordinate system.

widths in the projection are therefore limited north-south, but not east-west. The Lambert projection is thus ideal for mapping states that are narrow north-south, but which extend long distances in an east-west direction—for example, Kentucky, Pennsylvania, and especially Tennessee.

Figure 21-5 shows the portion of the developed cone of a Lambert projection covering an area of interest. In the Lambert projection, the cone intersects the ellipsoid along two parallels of latitude, called *standard parallels,* at one-sixth of the zone width

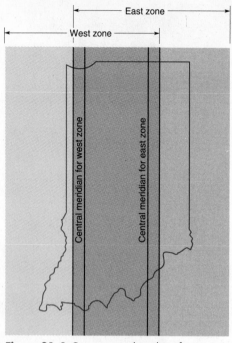

Figure 21-4 Coverages and overlap of zones in Indiana's transverse Mercator state plane coordinate system.

from the north and south zone limits. All meridians are straight lines converging at *Z*, the apex of the cone. An example is *ZM*, which is the so-called *central meridian*. All parallels of latitude are the arcs of concentric circles having centers at the apex. The projection is located in a zone in an east-west direction by assigning the central meridian a longitude value that is near the middle of the area to be covered. The direction of the central meridian on the projection establishes *grid north*. All lines parallel with the central meridian point in the direction of grid north. Except at the central meridian, therefore, directions of "true" and "grid" north do not coincide because true meridians converge. As shown in Figure 21-5, the *X* and *Y* coordinates of points are measured perpendicular and parallel to the central meridian, respectively, from a reference axis system which is offset to the west and south.

21-4 TRANSVERSE MERCATOR PROJECTION

The transverse Mercator projection is also a conformal projection, but is based on an imaginary secant cylinder as its developable surface. As illustrated in Figure 21-1(d), scale in the transverse Mercator projection varies from east to west and not from north to south. Thus this projection is used for states like Illinois and Indiana, which are narrow east-west and longer north-south.

In developing the tranverse Mercator projection, the axis of the imaginary cylinder is placed in the plane of the earth's equator. The cylinder cuts the spheroid along two small circles equidistant from the central meridian. On the developed plane surface (Figure 21-6) all parallels of latitude, and all meridians except for the central meridian are curves (shown in light broken lines). As with the Lambert projection, the central

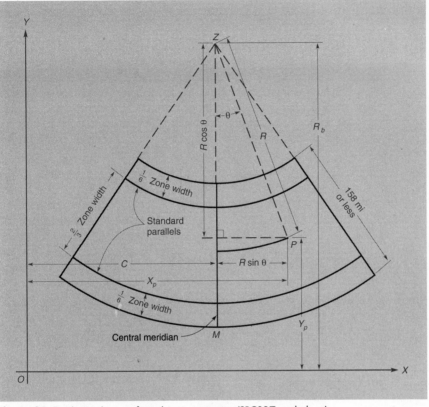

Figure 21-5 The Lambert conformal conic projection (SPCS27 symbology).

meridian establishes the direction of grid north, and the zone is centered over an area of interest by assigning the central meridian a longitude value that applies approximately at the center of the region to be mapped. Also, the X and Y coordinates of points are measured perpendicular to and parallel with the central meridian, respectively, from an axis system offset to the west and south.

21-5 STATE PLANE COORDINATES IN NAD27 AND NAD83

The first state plane coordinate system was developed by the NGS in 1933 for North Carolina. Systems for all other states followed shortly thereafter. As stated earlier, state plane coordinates of points are computed from geodetic latitudes and geodetic longitudes. Section 19-1 and Figure 19-2 define and illustrate these two terms and explain that geodetic latitudes and longitudes are defined with respect to a reference ellipsoid and associated datum. From 1927 until the inception of NAD83, the reference datum used in the United States was NAD27, based on the Clarke 1866 ellipsoid. All original state plane coordinates were therefore developed by the NGS in accordance with that ellipsoid and datum. This system is referred to as the *State Plane Coordinate System of 1927* (SPCS27).

As noted in Chapter 19, NAD83 employs a different set of defining parameters than NAD27, and it uses a reference surface of slightly different dimensions, the GRS80

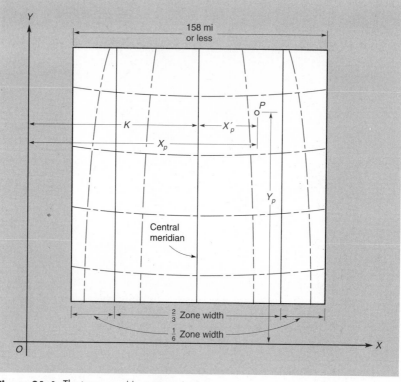

Figure 21-6 The transverse Mercator projection.

ellipsoid. Thus the latitudes and longitudes of points in NAD83 are somewhat different from their values in NAD27. (Figures 19-3 (a) and (b) illustrate these approximate latitude and longitude differences.) Because of these changes, the constants and variables which define the state plane coordinate systems also changed. Thus following completion of NAD83, it was necessary to develop a new state plane coordinate system. This has been completed by the NGS and is called the *State Plane Coordinate System of 1983* (SPCS83).

In the SPCS83 most states retained the same projections they had used in SPCS27, with basically the same central meridian positioning. There were changes, however; some major ones being **(1)** Montana reduced its number of zones from three to one; and **(2)** Nebraska and South Carolina reduced their number of zones from two to one. In SPCS83, 29 states employ the Lambert conformal conic projection, 18 use the transverse Mercator, and Alaska, Florida, and New York utilize both. In addition to using eight transverse Mercator zones for its mainland and a Lambert conformal conic for the Aleutian Islands, Alaska also employs an *oblique Mercator* projection for the southeast part of the state (see Section 21-13).

Although the same fundamental approach was used to develop both the SPCS27 and the SPCS83, as noted above the parameters which define the two systems are different. Accordingly, different symbols are used, and the equations for computing coordinates in SPCS83 have been refined. Thus slightly different procedures are used in making computations in the two systems. These procedures are discussed in separate sections that follow.

21-6 COMPUTING SPCS27 COORDINATES IN THE LAMBERT CONFORMAL CONIC SYSTEM

Procedures for computing X and Y coordinates of points from their geodetic latitudes and longitudes in the Lambert conformal conic system are illustrated in Figure 21-5. This process of converting from geodetic to plane coordinates is called the *direct problem*. The fundamentals described here apply to both SPCS27 and SPCS83, although the symbology used in Figure 21-5 and this section is applicable only to SPCS27. (A computation in the reverse manner can also be performed, i.e., calculating geodetic latitude and longitude from a given set of X, Y state plane coordinates. This reverse form of calculation, called the *inverse problem*, is discussed in Section 21-10.)

In Figure 21-5, line ZM is the central meridian of the projection, and point O the origin of rectangular coordinates. A constant C is adopted to offset the Y grid axis from the central meridian and make X coordinates of all points positive. The Y coordinate of the cone's apex is the constant R_b, also selected large enough to make all Y coordinates positive.

The coordinates X_p and Y_p of point P, whose geodetic latitude ϕ_p and geodetic longitude λ_p are known, are to be determined. Line ZP represents a portion of the meridian through point P with its length designated as R. Angle θ between the central meridian and the meridian ZP represents the amount of convergence between these two meridians. It is termed the *mapping angle*.

From Figure 21-5 the following equations of the direct problem can be solved for the X and Y coordinates of point P:

$$X_p = R \sin \theta + C$$

$$Y_p = R_b - R \cos \theta \tag{21-1}$$

In Eqs. (21-1), if θ is to the left of the central meridian, its algebraic sign is negative; if to the right, it is positive. Except where a line of reference azimuth exceeds about 5 mi in length, grid azimuth may be calculated with sufficient accuracy from geodetic azimuth using the following equation:

$$\text{grid azimuth} = \text{geodetic azimuth} - \theta \tag{21-2}$$

For lines longer than about 5 mi, a small second term is involved in making the conversions from grid to geodetic azimuth, and vice versa. The equation, as well as an example problem, is given in tables (published by the NGS and described later in this section).

To solve Eqs. (21-1), values for C, R_b, R, and θ must be known. The quantities C and R_b are constants for any zone. In SPCS27, C has been arbitrarily assigned a value of 2,000,000 ft for most states; R_b is computed and depends on the size and shape of the reference ellipsoid used and the zone's latitude. The magnitude of R_b decreases with increasing latitude. Values for R and θ also depend on the ellipsoid used and vary with changing locations of points in the zone; R changes with latitude, θ with longitude.

To aid in the solution of Eqs. (21-1), the NGS has computed and published individual SPCS27 booklets of projection tables for each state. These give the constants for each zone and tabulate R and θ values versus latitude and longitude, respectively. Thus given the geodetic latitude and longitude of any point, R and θ for that point can be precisely interpolated from the tables for use in Eqs. (21-1). The equations are solved using a calculator. This computational procedure is called the *tabular* method. Example problems provided with the tables illustrate the process.

For many years, use of these tables was the principal method for computing state plane coordinates. The emergence of digital computers, however, quickly demonstrated that several advantages would result from making a direct *computer* solution. These include **(1)** a significant time savings since table lookup and interpolation are avoided, and **(2)** assurance that once a computer program has been prepared and tested, if a point's geodetic latitude and longitude are correctly entered into the computer, the results obtained will be free from calculation mistakes. Accordingly, a computer program that makes SPCS27 coordinate conversions can be obtained from the NGS. Or, if preferred, the NGS has published a booklet (Special Publication 62-4) that contains the complete SPCS27 formulas and constants for every zone in the United States and its territories.[1] Sample problems are included to clarify the computational procedures. Using this booklet surveyors can develop their own computer programs.

It should be emphasized that computation of coordinates in the SPCS27 can only be done using points whose geodetic positions are given in NAD27.

21-7 COMPUTING SPCS27 COORDINATES IN THE TRANSVERSE MERCATOR SYSTEM

The NGS has also published SPCS27 transverse Mercator tables for those states which use that projection. All necessary constants and variables are tabulated, and instructions together with sample problems are given to illustrate the computational procedures.

With reference to Figure 21-6 and appropriate transverse Mercator projection tables, the following SPCS27 equations yield the solution of the *direct problem*, that is, obtaining the X_p and Y_p coordinates of any point P from its geodetic coordinates (Note that the symbols used in Figure 21-6 and this section are for SPCS27.):

$$X'_p = H \times \Delta\lambda'' \pm ab$$

$$X_p = X'_p + K$$

$$Y_p = Y_o + V\left(\frac{\Delta\lambda''}{100}\right)^2 \pm c \tag{21-3}$$

In these equations X'_p is the distance to point P either east or west of the central meridian. The difference in seconds between longitudes of the central meridian and point P is $\Delta\lambda''$, its algebraic sign being negative if P is west and positive if P is east of the central meridian. Constant K (500,000 ft for most states) offsets the Y axis from the central meridian, so all X coordinates are positive. Values of H, a, Y_o, and V are tabulated versus the latitude of point P in projection tables, and b and c are listed versus $\Delta\lambda''$. A negative sign for product ab decreases $H \times \Delta\lambda''$; a positive one increases it.

Except where a reference azimuth line exceeds 5 mi in length, the grid azimuth can be calculated with sufficient accuracy from the geodetic azimuth using the following equation:

[1]Individual booklets for each state for making SPCS27 tabular solutions, the computer program, and Special Publicaiton 62-4 can all be obtained from the National Geodetic Information Center, NOAA, National Geodetic Survey, N/CG174, SSMC3 Station 09535, 1315 East West Highway, Silver Spring, Md. 20910, telephone (301) 713-3242.

$$\text{grid azimuth} = \text{geodetic azimuth} - \Delta\alpha'' \qquad (21\text{-}4)$$

In Eq. (21-4), $\Delta\alpha'' = \Delta\lambda'' \sin \phi_P + g$, where g is listed versus $\Delta\lambda''$ in projection tables, and ϕ_P is the latitude of point P. As with the Lambert projection, a small second term is needed with Eq. (21-4) for lines over about 5 mi in length. The equation for this second term together with an example problem illustrating its use is given in the published NGS tables.

As an alternative to using the tables, the aforementioned computer program, available from the NGS, also solves the SPCS27 transverse Mercator equations. Again, if preferred, Special Publication 62-4 can be obtained and individual programs written for computer solution.

21-8 COMPUTING SPCS83 COORDINATES

Just as they had done for the SPCS27, the NGS has developed materials to allow two different approaches in computing SPCS83 coordinates: the use of **(1)** published tables, or **(2)** a computer program. Either approach enables solving both the *direct problem* (computing state plane coordinates from their geodetic positions) and the *inverse problem* (determining geodetic values from their plane coordinates) for both the Lambert conformal conic and the transverse Mercator projections.

Whereas coordinates in the SPCS27 were specified as X and Y and listed in feet, in SPCS83 they are called *eastings* and *northings,* and their values are listed in meters. Other changes include use of the symbol E_o for the offset of the central meridian from the Y axis; and assigning the symbol γ to the meridian convergence. These changes and others have been adopted to eliminate confusion over which system has been applied.

The SPCS83 "tabular solution" requires a special set of tables developed by the NGS for each zone, and is designed for use with a hand-held or a desk-top calculator. The method uses basically the same equations and procedures as explained for the SPCS27 in the two preceding sections of this chapter. The only differences are that, as noted earlier, some terms are identified with different symbols, and of course the tabulated constants and variables are different. Sample problems included with the tables illustrate the computation procedures.

The SPCS83 "computer solution" uses special polynomial equations developed and programmed by the NGS. The polynomial terms are computed from the zone constants and point positions. The equations are quite lengthy and thus not given here. They are listed in an NGS publication entitled *NOAA Manual NOS NGS 5,* "State Plane Coordinate System of 1983." This manual also lists all zone constants and gives a detailed explanation of all facets of the SPCS83. *It should be in the possession of all surveyors.*[2]

In the direct problem, in addition to getting eastings and northings, the computer solution also yields the *convergence angle* (equivalent to the mapping angle in SPCS27) and the scale factor. With the inverse problem the computer solution gives latitude, longitude, convergence angle, and scale factor.

The data in Table 21-1 give sample output from the NGS SPCS83 computer program, and indicate the simplicity with which solutions can be obtained for either the

---

[2]The SPCS83 tables, *NOAA Manual NOS NGS 5,* and the computer programs can be obtained from the National Geodetic Information Center, NOAA, National Geodetic Survey, N/CG174, SSMC3 Station 09535, 1315 East West Highway, Silver Spring, Md. 20910, telephone (301) 713-3242.

TABLE 21-1 SAMPLE OUTPUT FROM NGS SPCS83 COMPUTER PROGRAM
PART A: DIRECT PROBLEM (GEODETIC POSITIONS TO PLANE COORDINATES)

NATIONAL GEODETIC SURVEY
GP TO PC PROGRAM
1983 DATUM

STATION NAME	LATITUDE (NORTH)	LONGITUDE (WEST)	NORTHING (Y) METER	EASTING (X) METER	ZONE	CONVERGENCE D	M	S	SCALE FACTOR
ADA	45 22 21.47158	088 44 41.16664	23668.802	698327.343	WI N	0	54	19.75	1.00004087
			171809.767	698321.361	WI C	0	53	8.38	0.99997823
AUBURN 1943	32 36 22.83559	085 28 56.51479	233594.422	232942.745	AL E	0	11	20.85	0.99997338

PART B: INVERSE PROBLEM (PLANE COORDINATES TO GEODETIC POSITIONS)

NATIONAL GEODETIC SURVEY
PC TO GP PROGRAM
1983 DATUM

STATION NAME	NORTHING (Y) METER	EASTING (X) METER	LATITUDE (NORTH)	LONGITUDE (WEST)	ZONE	CONVERGENCE D	M	S	SCALE FACTOR
ADA	171809.767	698321.361	45 22 21.47159	088 44 41.16666	WI C	0	53	8.38	0.9999782
	23668.802	698327.343	45 22 21.47158	088 44 41.16665	WI N	0	54	19.75	1.0000409
AUBURN 1943	232942.745	233594.422	32 36 22.83558	085 28 56.51479	AL E	0	11	20.87	0.9999734

direct or the inverse problems. Table 21-1A shows output for the direct problem for two stations, *ADA* in Wisconsin's north zone (Lambert projection) and *AUBURN 1943* in Alabama's east zone (transverse Mercator projection). Input was done interactively by following screen prompts and consisted of station name, latitude, longitude, and zone number wherein the point is located. (Zone numbers are contained in *NOAA Manual NOS NGS 5.*) Output by column, from left to right, consists of station name, latitude, longitude, northing, easting, zone identity, convergence angle, and scale factor.

Table 21-1B gives sample output for the SPCS83 inverse problem for the same two points. Interactive input consisted of station name, northing, easting, and zone number. Output by column, from left to right, includes station name, northing, easting, latitude, longitude, zone identity, convergence angle, and scale factor. Note also that station ADA exists in the overlap area of Wisconsin's north and central zones. Thus the computer gave solutions for each zone to both the direct and the inverse problems. Note that except for roundoff error, the inverse solution reproduces the correct latitudes and longitudes of the two stations (i.e., the same values that were input for Part A).

21-9 COMPUTING STATE PLANE COORDINATES OF TRAVERSE STATIONS

Determining state plane coordinates of new traverse stations is a problem routinely solved by local surveyors. Normally it requires only that traverses (or triangulation or trilateration surveys) start and end on existing stations having known state plane coordinates, and from which known grid azimuth lines have been established. Generally these data are available for immediate use, but if not, they can be calculated as indicated in the preceding sections when latitude and longitude are known. State plane coordinates and the grid azimuth to a nearby azimuth mark in both SPCS27 and SPCS83 are published by the NGS for most stations in the national horizontal network. In most areas, many other stations set by local surveyors exist which also have state plane coordinates and reference grid azimuths.

It is important to note that if a survey begins with a given grid azimuth and ties into another, directions of all intermediate lines are automatically grid azimuths. This is true for either SPCS27 or SPCS83. Thus corrections for convergence of meridians are not necessary when the state plane coordinate system is used throughout the survey.

Assuming that starting stations meeting the above described conditions are available, then the only difference between making traverse computations in state plane coordinates and the procedures given for plane surveys in Chapter 13 is that distances must first be reduced to the state plane coordinate grid. In both SPCS27 and SPCS83 this is a two-step process of **(1)** reducing ground lengths to the ellipsoid and **(2)** applying scale factors to convert the ellipsoid lengths to grid lengths. A slightly different procedure is necessary for performing the first step, depending on whether the computations are in SPCS27 or SPCS83.

As noted earlier, SPCS27 is based on NAD27, which uses the Clarke 1866 ellipsoid. Because this ellipsoid conforms so closely to the geoid in North America (they depart from one another by only a few meters), the reduction of measured ground lengths to mean sea level practically reduces them also to the ellipsoid. Therefore Eq. (19-12), repeated here in slightly modified form, can be used in making calculations in SPCS27:

$$L_e = L_m \left[\frac{R_e}{R_e + H} \right] \tag{21-5}$$

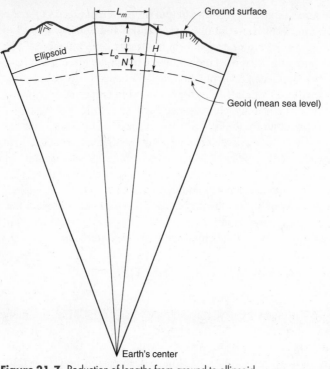

Figure 21-7 Reduction of lengths from ground to ellipsoid.

where L_e is the ellipsoid line length, L_m the measured line length, R_e the mean radius of the earth (approximately 20,902,000 ft or 6,371,000 m), and H the average elevation of the measured line above mean sea level. The ratio $R_e/(R_e + H)$ is commonly called the *elevation factor.*

SPCS83 is based on NAD83, which uses the GRS80 ellipsoid. This ellipsoid departs from the geoid by significant amounts in North America, the separation being from about 25 to 35 m within the conterminous 48 states. This variation is called *geoid height* (see Section 20-8), and because of its magnitude, except for low-order surveys, distances reduced to mean sea level do not approximate ellipsoid lengths satisfactorily. The situation is illustrated in Figure 21-7, where ground elevation H and geoid height N are shown. Geoid height is considered positive if the geoid is above the ellipsoid. It is shown in the figure as it exists in the conterminous United States; that is, *the geoid is below the ellipsoid and thus geoid heights are negative in all states (except Alaska).* The *ellipsoid height h* in Figure 21-7 is equal to $(H + N)$, where as noted, N is negative. (See Example Problem 20-1.)

To reduce a measured ground distance L_m to its ellipsoid length L_e in SPCS83 computations, the following equation is used:

$$L_e = L_m \left[\frac{R_e}{R_e + H + N} \right] \qquad (21\text{-}6)$$

In Eq. (21-6) all terms are as defined previously. The ratio $R_e/(R_e + H + N)$ is also called the *elevation factor.* Equation (21-6) in fact also applies for SPCS27, because in that case N is considered equal to zero.

Having calculated ellipsoid lengths, these are next multiplied by a *scale factor* to obtain grid lengths. The purpose of scaling was illustrated conceptually in Figure 21-2. Scale factors are obtained from tables or computed according to the particular area of the zone where the line falls. The range of reduction or increase in ellipsoid length resulting from this step varies from zero along the two secant lines of exact scale, to maximum and minimum values determined by the zone size. In the small state of Connecticut, for example, the correction is never more than about 1 part in 40,000, but it reaches 1 part in 10,000 and slightly less in a few large zones.

If a traverse is small enough, the scale factor may not change appreciably over its extent, and thus the same value may be applied to all lines. Also if the area has a fairly uniform elevation, the elevation factor may be treated as a constant. If conditions are such that both factors can be considered uniform, then a single so-called *combined factor* (product of elevation and scale factors) may be computed and applied to all lines. It may be ignored if it is near enough to 1.000000.

To illustrate the procedure of computing a traverse in the SPCS83, the following example is solved step by step.

The traverse illustrated in Figure 21-8 originates from station ROCKY and closes on station ROAD, both in Alabama's east zone. The traverse lies in an area of relatively flat terrain. The NAD83 coordinates, grid azimuths to nearby azimuth marks, elevations, geoid heights, and grid scale factors of the two stations are given in the figure. Measured horizonal distances and angles to the right for the traverse are given in Table 21-2, columns (2) and (3). Compute the SPCS83 coordinates for all stations of the traverse.

EXAMPLE 21-1

SOLUTION

1. Based on the fixed starting grid azimuth from ROCKY to $Az\ Mk_1$, and using the measured angles to the right, preliminary grid azimuths of all other lines are computed, including the closing line from ROAD to $Az\ Mk_2$. [These are listed in column (4) of Table 21-2.]
2. The computed azimuth of the line ROAD to $Az\ Mk_2$ is compared with its fixed control value. The difference ($+10''$) represents the traverse angular misclosure. This misclosure is divided by the number of angles (five) to get the correction per angle ($-02''$). [This calculation is shown at the bottom of Table 21-2.]
3. Corrected grid azimuths are computed based on the adjusted angles. [These azimuths are listed in column (3) of Table 21–3.]
4. The combined factor (product of elevation factor and scale factor) is computed. Because the traverse area is relatively flat, a single elevation factor based on the mean elevation of the two control points (243 m) is used for all lines. Also the mean geoid height of -28.3 m is applied. Thus by Eq. (21-6), the elevation factor EF is:

$$EF = \frac{6,371,000}{6,371,000 + 243 - 28.3} = 0.9999663$$

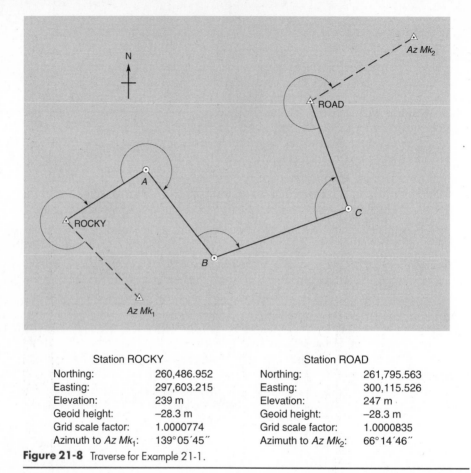

Station ROCKY		Station ROAD	
Northing:	260,486.952	Northing:	261,795.563
Easting:	297,603.215	Easting:	300,115.526
Elevation:	239 m	Elevation:	247 m
Geoid height:	−28.3 m	Geoid height:	−28.3 m
Grid scale factor:	1.0000774	Grid scale factor:	1.0000835
Azimuth to *Az Mk₁*:	139° 05′ 45″	Azimuth to *Az Mk₂*:	66° 14′ 46″

Figure 21-8 Traverse for Example 21-1.

Because of the limited size of this traverse, the mean value of the grid scale factor *SF* for the two stations may be used for all lines. Its value is

$$SF = \frac{1.0000774 + 1.0000835}{2} = 1.0000805$$

Finally, the combined factor *CF* is computed as

$$CF = 0.9999663 \times 1.0000805 = 1.0000468$$

5. Using the combined factor computed in step 4, the ellipsoid lengths of all lines are computed. [These calculations have been done and the results entered in column (2) of Table 21-3.]

6. Traverse computations are performed using the same steps as described in Chapter 13. The procedure, shown in Table 21-3, includes **(a)** calculation of departures and latitudes [tabulated in columns (4) and (5)]; **(b)** adjustment of the departures and latitudes [corrections are listed in columns (6) and (7), and adjusted departures and latitudes in columns (8) and (9)]; and **(c)** determination of the station

TABLE 21-2 MEASURED HORIZONTAL DISTANCES AND ANGLES TO THE RIGHT, AND ANGLE MISCLOSURE FOR EXAMPLE 21-1

STATION (1)	HORIZONTAL DISTANCE (m) (2)	ANGLE TO THE RIGHT (3)	PRELIMINARY AZIMUTH (4)
AZ Mk$_1$			
			319°05'45" (fixed)
ROCKY		283°50'02"	
	996.718		62°55'47"
A		256°17'18"	
	1267.663		139°13'05"
B		98°12'41"	
	1684.165		57°25'46"
C		103°30'34"	
	1131.833		340°56'20"
ROAD		265°18'36"	
			66°14'56"
Az Mk$_2$			

Angular misclosure = 66°14'56" − 66°14'46" = +10"
Correction per angle = −10"/5 = −2"

coordinates [tabulated in columns (10) and (11)]. Adjustment of departures and latitudes in this example has been done using the compass (Bowditch) rule, but any method could be used. In the adjustment, the differences in eastings and northings between control points were computed and checked against their fixed values (2512.311 m and 1308.611 m, respectively) to obtain the misclosures in departure (+0.144 m) and latitude (+0.132 m). An adjustment was then made to correct these computed differences to the required totals.

As explained previously, for traverses in flat or gently rolling terrain, one elevation factor will suffice for all lines, but if elevations vary considerably, a separate factor may be needed for each line. A 100-m elevation change will result in a 16 ppm distance change. This is equivalent to 1.6 cm in a length of 1000 m. Also, for smaller traverses a single scale factor (equal to that of the control station for polygon traverses or the mean of the beginning and ending stations for link traverses) may be used. On long traverses, however, it may be necessary to apply different scale factors for each line. Scale factor changes in an area can be readily investigated by solving the inverse problem with coordinates that vary in the direction of scale change, that is, north-south for the Lambert projection and east-west for the transverse Mercator.

Procedures for computing traverses in SPCS27 follow the same steps as outlined for Example 2-1, except that the elevation factor in step 4 is computed using Eq. (21-5).

21-10 SURVEYS EXTENDING FROM ONE ZONE TO ANOTHER

Surveys in border areas often cross into different zones or even abutting states. This presents no unusual problem, however, because adjacent zones overlap by appreciable distances, as shown in Figures 21-3 and 21-4.

TABLE 21-3 TRAVERSE COMPUTATIONS FOR EXAMPLE 21-1

STATION (1)	GRID LENGTH (m) (2)	GRID AZIMUTH (3)	UNBALANCED DEPARTURE (m) (4)	UNBALANCED LATITUDE (m) (5)	CORRECTIONS DEPARTURE (m) (6)	CORRECTIONS LATITUDE (m) (7)	BALANCED DEPARTURE (m) (8)	BALANCED LATITUDE (m) (9)	COORDINATES EASTING (m) (10)	COORDINATES NORTHING (m) (11)
ROCKY									297,603.215	260,486.952
	996.764	62°55'45"	887.563	453.619	−0.030	−0.027	887.536	453.592		
A									298,490.748	260,940.544
	1267.722	139°13'01"	828.072	−959.904	−0.038	−0.035	828.034	−959.939		
B									299,318.782	259,980.605
	1384.230	57°25'40"	1166.509	745.217	−0.042	−0.039	1166.470	745.178		
C									300,485.249	260,725.783
	1131.886	340°56'12"	−369.689	1069.811	−0.034	−0.031	−369.710	1069.780		
ROAD									300,115.526	261,795.563
	Σ=4780.602		Σ=2512.455 −2512.311	Σ=1308.743 −1308.611	Σ=−0.144	Σ=−0.132	Σ=2512.311	Σ=1308.611		
		Misclosures:	0.144	0.132						

Linear misclosure = $\sqrt{(0.144)^2 + (0.132)^2} = 0.195$ m

Relative precision = $\dfrac{0.195}{4780.6} = \dfrac{1}{24,500}$

The general procedure for extending surveys from one zone to another, whether in the SPCS27 or in the SPCS83, requires that the survey proceed from the first zone into the overlap area with the second. Then the geodetic latitudes and longitudes are computed for two intervisible points using their plane coordinates of the first zone. (Recall that this conversion is called the *inverse problem*.) Using the geodetic positions of the two points, their state plane coordinates in the new zone are then computed. (This is the *direct problem*.) Finally the grid azimuth of a line in the zone entered can then be obtained from the new coordinates of the two points.

Suppose, for example, that a survey being computed in SPCS27 originates in southern Wisconsin, which uses the Lambert conformal conic projection, and extends into northern Illinois, which uses the transverse Mercator grid. With Wisconsin south zone SPCS27 coordinates of two intervisible points in the overlap area of the two zones known, Eqs. (21-1) are rewritten for the first point as follows:

$$R \sin \theta = X_p - C \qquad (21\text{-}7)$$
$$R \cos \theta = R_b - Y_p \qquad (21\text{-}8)$$

Dividing Eq (21-7) by Eq. (21-8),

$$\tan \theta = \frac{X_p - C}{R_b - Y_p} \qquad (21\text{-}9)$$

All terms on the right side of Eq. (21-9) are known, C and R_b being Wisconsin south zone constants, and X_p and Y_p Wisconsin south zone coordinates available from the survey. The mapping angle θ can thus be calculated, after which R can be found by substituting θ back into either Eq. (21-7) or Eq. (21-8). Finally the geodetic latitude and longitude of the point, which are functions of R and θ, respectively, may be obtained. If the tables are being used, they are determined by interpolation. Alternatively, a computer program may be utilized. The second point is handled the same way.

With the geodetic latitudes and longitudes of the two points known, Eqs. (21-3) are solved using constants for the appropriate Illinois zone entered to obtain their X and Y coordinates in that zone. From these coordinates the grid azimuth of the line joining the intervisible points can be calculated by using Eq. (13-13), and the survey can continue into Illinois.

Extending surveys from zone to zone in SPCS83 follows the same procedure. Solving the direct and inverse problems that are necessary in this procedure is most conveniently handled by using the computer programs described previously.

21-11 IMPACTS OF NAD83 AND SPCS83

Each individual state must decide whether or not to officially adopt the new state plane coordinate system (SPCS83) based on NAD83. Many have already passed such legislation, and others are in the legislative process leading to adoption. A few states may choose to retain their SPCS27 state plane coordinate system (based on NAD27).

Many factors are being considered in making the decision to convert to NAD83 and SPCS83. One major factor in favor of adopting NAD83 is that coordinates determined from Global Positioning System (GPS) surveys are in this new datum. The expected future impacts of this rapidly expanding technology in the surveying field provide significant incentives for adoption of NAD83. Another important consideration is that in the future, products developed and distributed by federal agencies, such as maps, digital elevation models, digital line graphs, and other data useful for Land and Geographic Information Systems, will be referenced to NAD83. Also NAD83 is a more accurate datum than NAD27. Therefore its use will enable surveyors to make better ties on surveys between control points and to obtain more accurate coordinates for their intermediate points.

On the other hand, a factor that favors retaining NAD27 is that many thousands of points exist which were not in the National Geodetic Reference System (NGRS) and thus were not included in the NAD83 readjustment. These points have coordinates available in SPCS27 but not in SPCS83. They were set by state, county, and municipal agencies; utilities; and private surveyors; and have been used to describe property, or to serve as the basis of already existing Land and Geographic Information Systems. The cost of conversion is a factor to be considered.

States that do adopt the new datum can convert the coordinates of existing reference points from SPCS27 to SPCS83, but it will be a difficult and time-consuming task. Conversion can be done in different ways, depending on accuracy requirements and other factors. The most accurate procedure is to readjust the original survey data to control points whose positions were included in the NAD83 general adjustment and thus known in SPCS83. This process is only possible if the original survey was tied to one or more of these control points. If such points do not exist, they can be established using GPS, or traversing as described earlier in this chapter.

A second method, which yields an intermediate level of accuracy, computes *expected* changes for all points in a large area based on known changes for a selected network of points in the region. Changes of the selected points, which should be uniformly distributed throughout the area, are obtained by computing differences between their SPCS27 and SPCS83 coordinates. Mathematical functions (polynomials) are then used to predict changes of all other points in the area based on the pattern of changes for known points. This procedure gives reasonably good accuracy, suitable for many needs.

A third lower order conversion procedure uses either linear interpolation of changes in X and Y coordinates between points of known change, or an average change for a given area. It produces results suitable for certain purposes, for example, correcting coordinate grids of small-scale maps and nautical charts.

In the coming years many decisions will be made on matters related to NAD83 and the state plane coordinate systems. New legislation in many states will doubtlessly change requirements and procedures regarding their use. All surveyors should anticipate these developments and be alert for changes.

21-12 THE UNIVERSAL TRANSVERSE MERCATOR PROJECTION

The universal transverse Mercator (UTM) system is another important map projection. Originally developed by the military primarily for artillery use, it provides worldwide coverage from 80°S latitude through the equator to 84°N latitude. Each zone has a 6° longitudinal width; thus 60 zones are required to encircle the globe.

The UTM system is a modified transverse Mercator projection. It has recently taken on added importance for surveyors, since UTM coordinates in metric units are now being included along with state plane and geodetic coordinates for all published NAD83 station descriptions. UTM grids are also being included on all maps in the national mapping program, and UTM coordinates are being used more frequently for referencing positions of data entered into Land Information Systems (LISs) and Geographic Information Systems (GISs).

UTM zones are numbered easterly from 1 through 60, beginning at longitude 180°W. The United States is covered from zone 10 (west coast) through zone 20 (east coast). The central meridian for each zone is assigned an easting of 500,000 m. A northing of zero is applied to the equator for the northern portion of each zone, and 10 million m is assigned at the equator for the southern part to avoid negative Y coordinates. To specify the position of any point in the UTM system, the zone number must be given as well as its northing and easting.

In the UTM system, each zone overlaps adjacent ones by 0°30′. Because zone widths of 6° are considerably larger than those used in state plane systems, lower accuracies result, and 1 part in 2500 applies at the center and edges of zones.

Equations for calculating X and Y UTM coordinates are the same as those for the transverse Mercator projection. As with the state plane systems, tables giving formulas and constants for the system are available. Also like the state coordinate system, the datum of reference must be specified. Because UTM coordinates are available for all points within the NAD83, calculations between very widely spaced points can readily be made. This is convenient and entirely consistent with current capabilities for conducting surveys of global extent with new devices such as GPS.

21-13 OTHER MAP PROJECTIONS

The Lambert conformal conic and transverse Mercator map projections are designed to cover areas extensive in east-west and north-south directions, respectively. These systems do not, however, conveniently cover circular areas or long strips of the earth that are skewed to the meridians. Two other systems, the *horizon stereographic* and the *oblique Mercator* projections, satisfy these problems.[3]

The horizon stereographic projection can be divided into two classes: *tangent plane* and *secant plane*. In either class, as illustrated in Figure 21-9, the projection point P is on the ellipsoid where a line perpendicular to the map plane and passing through center point O intersects the ellipsoid. In the tangent plane system, ellipsoid points a and b are projected outward to a' and b', respectively, on the map plane. For the secant plane system, ellipsoid points c and d are projected inward to c' and d' on the map plane. (If they were outside of the secant points, projection would be outward.) Horizon stereographic projections are not employed in the United States, but are used in Canada and other parts of the world. If point P is the north or south pole, the projection is called *polar stereographic;* if it is on the equator, *equatorial stereographic.*

The oblique Mercator projection is designed for areas whose major extent runs obliquely to a meridian, such as northwest to southeast. The projection is conformal and is developed by projecting points from the ellipsoid to an imaginary cylinder that is oriented with its axis skewed to the equator. It is used as the state plane coordinate projection for the southeast portion of the state of Alaska.

[3]The oblique Mercator projection is also called the Hotine skew orthomorphic projection, named after the English geodesist Martin Hotine.

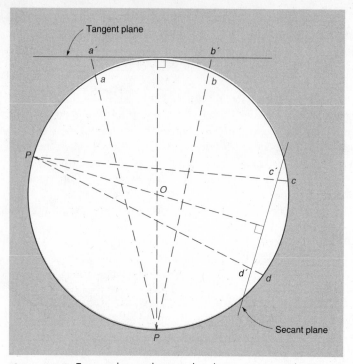

Figure 21-9 Tangent plane and secant plane horizon stereographic map projections.

PROBLEMS

21-1 Discuss the advantages of placing surveys on state plane coordinate systems.

21-2 Name the two basic projections used in state plane coordinate systems. What are their fundamental differences? Which one is preferred for states whose long dimensions are north-south? East-west?

21-3 Why was it necessary for the NGS to develop the new state plane coordinate systems (SPCS83) for each state following completion of NAD83?

21-4 Why do the states of New York, Florida, and Alaska use both the Lambert conformal conic and the transverse Mercator projections for their state plane coordinate systems?

21-5 What are the advantages of using a computer program to perform the state plane coordinate direct and inverse problems, as compared to doing the calculations manually using tables?

21-6 What corrections must be made to measured slope distances prior to computing state plane coordinates?

*21-7 Develop a table of SPCS27 elevation factors for ground elevations ranging from sea level to 6000 ft above sea level. Use increments of 500 ft.

21-8 Similar to Problem 21-7, except for ground elevations from sea level to 2000 m above sea level using 250-m increments.

*21-9 Develop a table of SPCS83 elevation factors for ground elevations ranging from 2000 to 4000 ft. Use increments of 200 ft and assume a geoid height of -30 m.

21-10 Similar to Problem 21-9, except for ground elevations from sea level to 1000 m by increments of 100 m. Assume a geoid height of -29 m.

21-11 Explain how surveys can be extended from one state plane coordinate zone to another, or from one state to another.

*Asterisks indicate problems that have answers given in Appendix G.

21-12 What accuracy in differences between ellipsoid and projection lengths was intended by the NGS in developing state plane coordinate systems? What maximum zone width is allowable to achieve the desired accuracy?

*21-13 The SPCS27 state plane coordinates of points *A* and *B* are as follows:

Point	X (ft)	Y (ft)
A	2,161,134.57	877,897.24
B	2,166,960.24	874,765.67

Calculate the grid length and grid azimuth of line *AB*.

21-14 Similar to Problem 21-13, except points *A* and *B* have the following SPCS83 state plane coordinates:

Point	E (m)	N (m)
A	235,920.486	217,482.527
B	240,001.842	213,041.826

21-15 In computing state plane coordinates for a project area whose mean elevation is 3840 ft above mean sea level, an average scale factor of 0.9999327 was used. The given distances between points in this project area were computed from SPCS27 state plane coordinates. What horizontal length would have to be measured to lay off these lines on the ground?
*(a) 895.00 ft
(b) 2312.05 ft

21-16 Similar to Problem 21-15, except that the mean project area elevation was 1235 m, the geoid height −28.5 m, the scale factor 0.9999502, and the computed lengths of lines from SPCS83 were:
(a) 765.128 m
*(b) 1642.950 m

*21-17 The horizontal ground lengths of a three-sided closed polygon traverse were measured as follows: *AB* = 2187.57, *BC* = 2991.96, and *CA* = 3923.51 ft. If the average elevation of the area is 2365 ft above sea level, calculate ellipsoid lengths of the lines suitable for use in computing SPCS27 coordinates.

*21-18 Assuming a scale factor for the traverse of Problem 21-17 to be 1.0000659, calculate grid lengths for the traverse lines.

*21-19 For the traverse of Problem 21-17, the grid azimuth of a line from *A* to a nearby azimuth mark was 150°31′16″ and the clockwise angle measured at *A* from the azimuth mark to *B*, 118°22′51″. The measured interior angles were *A* = 49°11′07″, *B* = 97°14′15″, and *C* = 33°35′02″. Balance the angles and compute grid azimuths for the traverse lines. (*Note:* Line *BC* bears southerly.)

*21-20 Using grid lengths of Problem 21-18 and grid azimuths from Problem 21-19, calculate departures and latitudes, linear misclosure, and relative precision for the traverse.

*21-21 If station *A* has SPCS27 state plane coordinates X = 2,306,294.81 and Y = 563,240.46, balance the departures and latitudes computed in Problem 21-20 using the compass (Bowditch) rule, and determine SPCS27 coordinates of stations *B* and *C*.

21-22 What is the combined factor for the traverse of Problems 21-17 and 21-18?

21-23 The horizontal ground lengths of a four-sided closed polygon traverse were measured as follows: *AB* = 479.549 m, *BC* = 830.616 m, *CD* = 685.983 m, and *DA* = 859.689 m. If the average elevation of the area is 1252 m above sea level, the geoid height is −32.0 m, and the scale factor for the traverse 0.9999574, calculate grid lengths of the lines for use in computing SPCS83 coordinates.

*Asterisks indicate problems that have answers given in Appendix G.

21-24 For the traverse of Problem 21-23, the grid bearing of line *BC* is N57°39′48″ W. Interior angles were measured as follows: $A = 120°26′27″$, $B = 73°48′55″$, $C = 101°27′00″$, and $D = 64°17′26″$. Balance the angles and compute grid bearings for the traverse lines. (*Note:* Line *CD* bears southerly.)

21-25 Using grid lengths from Problem 21-23 and grid bearings from Problem 21-24, calculate departures and latitudes, linear misclosure, and relative precision for the traverse. Balance the departures and latitudes by the compass (Bowditch) rule. If the SPCS83 state plane coordinates of point *B* are $E = 255,096.288$ m and $N = 280,654.342$ m, calculate SPCS83 coordinates for points *C, D,* and *A*.

21-26 Discuss the impact of NAD83 on the state plane coordinate systems.

21-27 What zone widths are used in the UTM system? What accuracy results in the differences between ellipsoid and grid lengths near zone edges for this system?

21-28 Imagine that you have been asked to give testimony before a legislative committee relating to a bill calling for adoption of SPCS83 in your state. Prepare a brief summary of the arguments you will present to the committee.

BIBLIOGRAPHY

Dracup, J. F. 1977. "The New Adjustment of the North American Datum: Plane Coordinate Systems." *Bulletin, American Congress on Surveying and Mapping* (No. 59):27.

Dracup, J. F. 1991. "Method for Converting Earlier Datum Plane Coordinates to North American Datum of 1983." *Surveying and Land Information Systems* 51 (No. 2):75.

Graff, D. R. 1988. "Coordinate Conversions from NAD27 to NAD83." *ASCE, Journal of Surveying Engineering* 114 (No. 3):125.

Hintz, R. J. and H. J. Onsrud. 1989. "Issues for the Bureau of Land Management to Consider in Converting from NAD27 to NAD83." *Point of Beginning* 15 (No. 2):90.

Hurlbert, D. D., and J. D. McDonald. 1976. "Horizontal Control in Kansas City by State Plane Coordinates." *ASCE, Journal of the Surveying and Mapping Division* 102 (No. SU1):9.

McDonnell, P. W., Jr. 1991. *Introduction to Map Projections.* Rancho Cordova, Calif.: Landmark.

Meade, B. K. 1973. "Coordinate Systems for Surveying and Mapping." *Surveying and Mapping* 33 (No. 3):337.

Meade, B. K. 1987. "Program for Computing UTM Coordinates for Latitudes North or South and Longitudes East or West." *Surveying and Mapping* 47 (No. 1):37.

Moffitt, F. H., and S. I. Jordan. 1987. "Conversion of State Plane Coordinates from NAD27 to NAD83 by Second Degree Conformal Transformation." *Bulletin, American Congress on Surveying and Mapping* (No. 106):13.

Pryor, W. J. 1973 "Plane Coordinates for Engineering and Cadastral Surveys." *Surveying and Mapping* 33 (No. 3):317.

Read, J. R. 1981. "A Coordinate System for North America Based upon the 6° UTM Zone." *Surveying and Mapping* 41 (No. 1):83.

Shrestha, R. L. 1989. "NAD83 Geodetic and State Plane Coordinates Computation." *Surveying and Land Information Systems* 49 (No. 2):87.

Shrestha, R. L., and S. E. Dicks. 1991. "Comparison of NAD27 to NAD83 State Plane Coordinate Transformation Models." *Surveying and Land Information Systems* 51 (No. 1):41.

Stem, J. E. 1984. "Status of Legislative Action to Enact the State Plane Coordinate System of 1983." *Bulletin, American Congress on Surveying and Mapping* (No. 88):33.

Stem, J. E. 1989. "State Plane Coordinate System of 1983." *NOAA Manual NOS NGS 5* Rockville, Md.: National Geodetic Information Center.

Stoughton, H. W., and R. M. Berry. 1985. "Simple Algorithms for Calculation of Scale Factors for Plane Coordinate Systems (1927 NAD and 1983 NAD)." *Surveying and Mapping* 45 (No. 3):247.

Vicenty, T. 1985. "Precise Determination of the Scale Factor from Lambert Conformal Conic Projection Coordinates." *Surveying and Mapping* 45 (No. 4):315.

Vicenty, T. 1986. "Use of Polynomial Coefficients in Conversions of Coordinates on the Lambert Conformal Conic Projection." *Surveying and Mapping* 46 (No. 1):15.

Vicenty, T. 1988. "Unraveling Some Mysteries About NAD83." *Point of Beginning* 13 (No. 5):46.

22

BOUNDARY SURVEYS

22-1 INTRODUCTION

The oldest types of surveys in recorded history are boundary surveys, which date back to about 1400 B.C., when plots of ground were subdivided in Egypt for taxation purposes. They still are one of the main areas of surveying practice.

From Biblical times,[1] when the death penalty was assessed for destroying corners, to the colonial days of George Washington,[2] who was licensed as a land surveyor by William and Mary College of Virginia, and through the years to the present, trees and other natural objects, or stakes driven into the ground, have been used to identify land parcel boundaries.

As property increased in value and owners disputed rights to land, the importance of more accurate surveys, permanent monuments, and written records became obvious. When Texas became a state in 1845, its public domain amounted to about 172,700,000 acres, which the U.S. government could have acquired by payment of the approximately $13,000,000 in debts accumulated by the Republic of Texas. However, Congress allowed the Texans to retain their land and pay their own debts—a good bargain at roughly 7.6 cents/acre! Recently land at Waikiki in Honolulu sold for more than $5,000,000/acre, and in Tokyo prices have reached over $30 million per acre.

The term *land tenure system* applies to the manner in which rights to land are held in any given country. Such a system, as a minimum, must provide (1) a means for transferring or changing the title and rights to the land, (2) permanently monumented or marked boundaries which enable parcels to be found on the ground, (3) officially retained records defining who possesses what rights to the land, and (4) an official legal description of each parcel. In the United States a two-tier land tenure system exists. At the federal level, records of surveys and rights to land are maintained by the U.S. Bureau of Land Management. At the state and local levels, official records concerning land tenure are held in county court houses.

---

[1]"Cursed be he that removeth his neighbor's landmark. And all the people shall say Amen." Deut. 27:17.
[2]"Mark well the land, it is our most valuable asset." George Washington.

Land titles in the United States are now transferred by written documents called *deeds* (*quit claim, agreement,* or *warranty*), which contain a description of the property. Property descriptions are prepared as the result of a *land survey.* The various methods of description include **(1)** metes and bounds, **(2)** block-and-lot number, **(3)** coordinate values for each corner, and **(4)** township, section, and smaller subdivisions of the *United States Public Land Survey System* (USPLSS). Often a property description will combine two or more of these methods. The first three methods are discussed briefly in this chapter, the USPLSS is covered in Chapter 23.

22-2 CATEGORIES OF LAND SURVEYS

Activities involved in the practice of land surveying can be classified into three categories: **(1)** *original surveys* to subdivide the remaining unsurveyed U.S. public lands, most of which are in Alaska, **(2)** *retracement surveys* to recover and monument or mark boundary lines that were previously surveyed, and **(3)** *subdivision surveys* to establish new smaller parcels of land within lands already surveyed. The last two categories are described in this chapter, the first is discussed in Chapter 23.

In establishing new property lines, and especially in retracing old ones, surveyors must exercise acute judgment based on education, practical experience, and knowledge of land laws. They must also be accurate and articulate in making measurements. This background must be bolstered by tenacity in searching the records of all adjacent property as well as studying descriptions of the land in question. In field work, surveyors must be untiring in their efforts to find points called for by the deed. Often it is necessary to obtain *parole evidence,* that is, testimony from people who have knowledge of accepted land lines and the location of corners, reference points, fences, and other information about the correct lines.

Modern-day land surveyors are confronted with a multitude of problems created over the past two centuries, under different technology and legal systems, that now require professional solutions. These include defective compass and chain surveys; incompatible descriptions and plats of common lines for adjacent tracts; lost or obliterated corners and reference marks; discordant testimony by local residents; questions of riparian rights; and a tremendous number of legal decisions on cases involving property boundaries.

The responsibility of a professional surveyor is to sift all evidence and try to obtain a meeting of minds among persons involved in any property-line dispute, although without legal authority to force a compromise or settlement. Fixing title boundaries must be done by agreement of adjacent owners or court action. Surveyors are often called upon to serve as expert witnesses in proceedings to establish boundaries, but to do so, they should be registered.

Because of the complicated technical judgment decisions that must be made, the increasing cost of land surveyors' professional liability insurance for "errors and omissions" has become a major part of operating expenses. Some clients demand that a surveyor have it for the protection of all parties.

22-3 HISTORICAL PERSPECTIVES

In the eastern part of the United States, individuals acquired the first land titles by gifts or purchase from the English Crown. Surveys and maps were completely lacking or inadequate, and descriptions could be given in only general terms. The remaining land in the 13 colonies was transferred to the states at the close of the Revolutionary War. Later this land was parceled out to individuals, generally in irregular tracts. Boundary lines were described by metes and bounds (see Section 22-4).

Many original transfers and subsequent ownerships and subdivisions were not recorded. Those that were usually had scanty or defective descriptions, since land was cheap and abundant. Trees, rocks, and natural landmarks defining the corners, as in the first example metes-and-bounds description (see Section 22-4), were soon disturbed. The intersection of two property lines might be described only as "the place where John killed a bear" or "the bend in a footpath from Jones's cabin to the river."

Numerous problems in land surveying stem from the confusion engendered by early property titles, descriptions, and compass surveys. The locations of thousands of corners have been established by compromise after resurveys, or by court interpretation of all available evidence pertinent to their original or intended positions. Other corners have been fixed by *squatters' rights, adverse possession,* and *riparian changes.* Many boundaries are still in doubt, particularly in areas having marginal land where the cost of a good retracement survey exceeds the property's value.

The fact that four corners of a field can be found, and the distances between them agree with the "calls" in a description, does not necessarily mean that they are in the proper place. Title or ownership is complete only when the land covered by a deed is positively identified and located on the ground.

Land law from the time of the Constitution has been held as a state's right, subject to interpretation by state court systems. Many millions of land parcels have been created in the United States over the past four centuries under different technology and legal systems. Some of the countless problems passed on to today's professional surveyors, equipped with immensely improved equipment, are discussed in this chapter and in Chapter 23.

Land surveying measurements and analysis follow basic plane surveying principles. But a land surveyor needs years of experience in a given state to become familiar with local conditions, basic reference points, and legal interpretations of complicated boundary problems. Methods used in one state for prorating differences between recorded and measured distances may not be acceptable in another. Rules on when and how fences determine property lines are not the same in all states or even in adjacent ones.

The term *practical location* is used by the legal profession to describe an agreement, either explicit or implied, in which two adjoining property owners settle a boundary dispute or mark out an ambiguous boundary. Fixed principles enter the process, and the boundary established becomes permanent.

Different interpretations are given locally to **(1)** the superiority or definiteness of one distance over another associated with it, **(2)** the position of boundaries shown by occupancy, **(3)** the value of corners in place in a tract and its subdivisions, and **(4)** many other factors. Registration of land surveyors is therefore required in all states to protect the public interest.

22-4 PROPERTY DESCRIPTION BY METES AND BOUNDS

As noted earlier, *metes and bounds* is one of the methods commonly used in preparing legal descriptions of property. Descriptions by metes (to measure, or assign by measure) and bounds (boundary lines or property limits) have a *point of beginning* (POB), such as a stake, fence post, road intersection, or some natural feature. Lengths and directions (bearings or azimuths) of successive lines from the point of beginning are given. Early distance units of chains, poles, and rods are now replaced by feet and decimals, and sometimes by metric units. Bearings or azimuths may be magnetic, true, or grid, the last two being preferable. Care must be exercised to indicate clearly which of these is the basis of directions so that no confusion arises. In the past *assumed* bearings or azimuths have sometimes been used, but many states no longer allow them because they are not readily reproducible. In some states, survey regulations call for exterior lines of new subdivisions to be based on the true meridian.

Metes-and-bounds property descriptions are written by surveyors and lawyers. In preparing them, extreme care must be exercised. A single mistake in transcribing a numerical value, or one incorrect or misplaced word or punctuation mark, may result in litigation for more than a generation, since the intentions of the *grantor* (person selling property) and the *grantee* (person buying property) may then be unclear. If numbers are both spelled out and given as figures, words control in the case of conflicts unless other proof is available. There is a greater likelihood of transposing than mis-spelling—and lawyers prefer words!

The importance of permanent monuments is evident. In fact, some states require pipes, iron pins, and/or concrete markers set deep enough to reach below the frost line at all property corners before surveys will be accepted for recording. Actually, almost anything can be called for as a monument. A map attached to the description clarifies it, and scaling provides a rough check on the angles and distances.

To increase precision in property surveys, large cities and some states have estab-lished a network of control monuments to supplement stations of the National Geodetic Reference System. Property corners can be tied to these control points, and boundary lines relocated with assurance.

Description of land by metes and bounds in a deed should always contain the follow-ing information in addition to the recital:

1. *Point of beginning* (POB). This point must be identifiable, permanent, well ref-erenced, and one of the property corners. Coordinates, preferably state plane, should be given if known or computable. Note that a POB is no more important than others and a called-for monument in place at the next corner establishes its position, even though bearing and distance calls to it may not agree.

2. *Definite corners.* Such corners are clearly defined points with coordinates if possible.

3. *Lengths and directions of the property sides.* All lengths in feet and decimals (or metric units), and directions by angles, true bearings, or azimuths must be stated to permit computation of any misclosure error. Omitting the length or direction of a closing line to the POB and substituting a phrase "and thence to the point of beginning" is not acceptable. The survey date is required and particularly important if directions are referred to magnetic north.

4. *Names of adjoining property owners.* These are helpful to show the intent of a deed in case an error in the description leaves a gap or creates an overlap. However, called-for monuments in place will control title over calls for adjoiners unless precluded by *senior rights.*

5. *Areas.* The included area is normally given as an aid in the valuation and identi-fication of a piece of property. Areas of rural land are given in acres or hectares and those of city lots in square feet or square meters. Because of differences in measure-

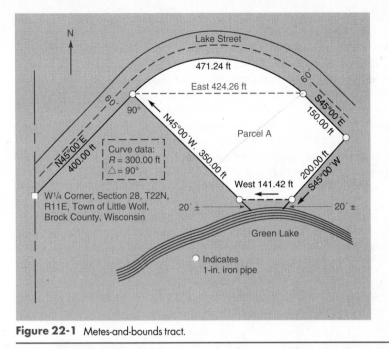

Figure 22-1 Metes-and-bounds tract.

ments, and depending on the adjustment method used for a traverse (compass, transit, least squares, and so on), one surveyor's calculated area, angles, and distances may differ slightly from another's.

The expression "more or less," which may follow a computed area, allows for only minor errors. It avoids nuisance suits for insignificant variations.

A partial metes-and-bounds description for the tract shown in Figure 22-1 is given as an example.

That part of the SW$\frac{1}{4}$ of the NW$\frac{1}{4}$ of Section 28, T 22 N, R 11 E, Town of Little Wolf, Brock County, Wisconsin, described as follows: Commencing at a stone monument at the W$\frac{1}{4}$ corner of said Section 28; thence N45°00′E, 400.00 feet along the Southerly R/W line of Lake Street to a 1″ iron pipe at the point of beginning of this description, said point also being the point of curvature of a curve to the right having a central angle of 90°00′ and radius of 300.00 feet; thence Easterly, 471.24 feet along the arc of the curve, the long chord of which bears East, 424.26 feet, to a 1″ iron pipe at the point of tangency thereof, said arc also being the aforesaid Southerly R/W line of Lake Street; thence continuing along the Southerly R/W line of Lake Street, S45°00′E, 150.00 feet to a 1″ iron pipe; thence S45°00′W 200.00 feet to a 1″ iron pipe located N45°00″E, 20 feet, more or less, from the water's edge of Green Lake, and is the beginning of the meander line along the lake; thence West 141.42 feet along the said meander line to a 1″ iron pipe at the end of the meander line; said pipe being located N45°00′W, 20 feet, more or less from the said water's edge; thence N45°00′W, 350.00 feet to a 1″ iron pipe at the point of beginning . . . including all lands lying between the meander line herein described and the Northerly shore of Green Lake, which lie between true extensions of the Southeasterly and Southwesterly boundary lines of the parcel herein described, said parcel containing 2.54 acres, more or less. Bearings are based on astronomic north.

Two examples of *old* metes-and-bounds descriptions from the eastern United States will be given. The first, part of an early deed registered in Maine, is:

> Beginning at an apple tree at about 5 minutes walk from Trefethens Landing thence easterly to an apple tree, thence southerly to a rock, thence westerly to an apple tree, thence northerly to the point of beginning.

With numerous apple trees and an abundance of rocks in the area, the dilemma of a surveyor trying to retrace the boundaries many years later is obvious.

The second, a more typical *old* description of a city lot showing lack of comparable precision in angles and distances, follows:

> Beginning at a point on the west side of Beech Street marked by a brass plug set in a concrete monument located one hundred twelve and five tenths (112.5) feet southerly from a city monument No. 27 at the intersection of Beech Street and West Avenue; thence along the west line of Beech Street S15°14′30″E fifty (50) feet to a brass plug in a concrete monument; thence at right angles to Beech Street S74°45′30″W one hundred fifty (150) feet to an iron pin; thence at right angles N15°14′30″W parallel to Beech Street fifty (50) feet to an iron pin; thence at right angles N74°45′30″E one hundred fifty (150) feet to place of beginning; bounded on the north by Norton, on the east by Beech Street, on the south by Stearns, and on the west by Weston.

22-5 PROPERTY DESCRIPTION BY BLOCK-AND-LOT SYSTEM

As cities grow, adjoining parcels of land are subdivided into blocks and lots according to a specific plan. Each new subdivision is assigned a name and annexed by the city. The individual lots within the subdivided area are identified by *block and lot number, tract and lot number,* or *subdivision name and lot number.* Examples are:

> Lot 34 of Tract 12314 as per map recorded in book 232, pages 23 and 24 of maps, in the office of the county recorder of Los Angeles County.

> Lot 9 except the North 12 feet thereof, and the East 26 feet of Lot 10, Broderick's Addition to Minneapolis. [Parts of two lots are included in the parcel described.]

> That portion of Lot 306 of Tract 4178 in the City of Los Angeles, as per Map recorded in Book 75, pages 30 to 32 inclusive of maps in the office of the County Recorder of said County, lying Southeasterly of a line extending Southwesterly at right angles from the Northeasterly line of said Lot, from a point in said Northeasterly line distant Southeasterly 23.75 feet from the most Northerly corner of said Lot.

Map books in the city or county recorder's office give the location and dimensions of all the blocks and lots. It is now standard practice to require subdividers to file a map with the proper office, showing the type and location of monuments, size of lots, and other pertinent information such as the dedication of streets. It is evident that if the boundary lines of a tract are in doubt, the individual lot lines must also be questioned.

The block-and-lot system is a short and unique method of describing property for tax purposes as well as for transfer of title. Identification by street and house number is satisfactory only for tax-assessment purposes.

Figure 22-2 is an example of a small hand-drafted block-and-lot subdivision. Figure 17–14 shows a portion of a subdivision map produced using a CAD system.

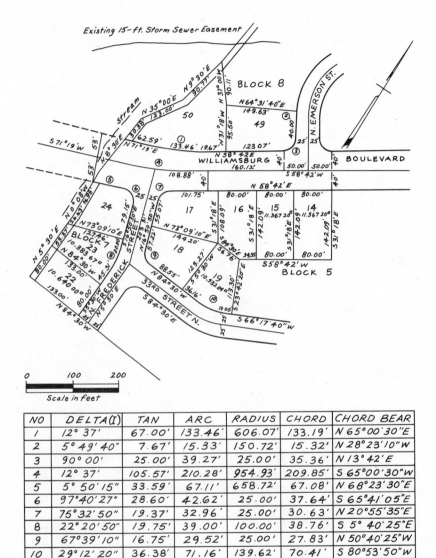

NO	DELTA(I)	TAN	ARC	RADIUS	CHORD	CHORD BEAR
1	12° 37'	67.00'	133.46'	606.07'	133.19'	N 65°00'30"E
2	5° 49'40"	7.67'	15.33'	150.72'	15.32'	N 28°23'10"W
3	90° 00'	25.00'	39.27'	25.00'	35.36'	N 13°42'E
4	12° 37'	105.57'	210.28'	954.93'	209.85'	S 65°00'30"W
5	5° 50'15"	33.59'	67.11'	658.72'	67.08'	N 68°23'30"E
6	97°40'27"	28.60'	42.62'	25.00'	37.64'	S 65°41'05"E
7	75°32'50"	19.37'	32.96'	25.00'	30.63'	N 20°55'35"E
8	22°20'50"	19.75'	39.00'	100.00'	38.76'	S 5° 40'25"E
9	67°39'10"	16.75'	29.52'	25.00'	27.83'	N 50°40'25"W
10	29°12'20"	36.38'	71.16'	139.62'	70.41'	S 80°53'50"W

Figure 22-2 Small subdivision plat.

22-6 PROPERTY DESCRIPTION BY COORDINATES

State plane coordinate systems provide a common reference system for surveys in large regions, even entire states (see Chapter 21). Several advantages result from using them on property surveys. One of the most significant is that they greatly facilitate the relocation of lost and obliterated corners. Every monument which has known state plane coordinates becomes a "witness" to other corner markers whose positions are given in the same system. State plane coordinates also enable the evaluation of adjoiners with less field work. These and other advantages have led to their legal acceptance in property descriptions.

A coordinate description of corners may be used alone, but is usually prepared in conjunction with an alternative method. Wider use of state plane coordinates will be made as more reference points become available to local surveyors, and because they serve as an excellent common reference system for correlating data entered into modern multipurpose Land Information Systems (see Chapter 29).

A description by coordinates of a parcel in California follows.

> A parcel of tide and submerged land, in the State-owned bed of Seven Mile Slough, Sacramento County, California, in projected Section 10, T 3 N, R 3 E, Mt. Diablo Meridian, more particularly described as follows:
>
> BEGINNING at a point on the southerly bank of said Seven Mile Slough which bears S62°37'E, 860 feet from a California State Lands Commission brass cap set in concrete stamped "JACK 1969," said point having coordinates of $X = 2,106,973.68$ and $Y = 164,301.93$ as shown on Record of Survey of Owl Island, filed October 6, 1969, in Book 27 of Surveys, Page 9, Sacramento County Records, thence to a point having coordinates of $X = 2,107,196.04$ and $Y = 164,285.08$; thence to a point having coordinates of $X = 2,107,205.56$, $Y = 164,410.72$; thence to a point having coordinates of $X = 2,106,983.20$, $Y = 164,427.57$; thence to the point of beginning.
>
> Coordinates, bearings, and distances in the above description are based on the California Coordinate System, Zone II.

When the preceding description was prepared, the writer could not have anticipated that an ambiguity would later arise concerning the datum of reference. From the dates given in the description, of course it can be concluded that the coordinates are in SPCS27. In preparing coordinate descriptions nowadays, however, the coordinates should be identified as being either in SPCS27 or SPCS83, as appropriate, to avoid any confusion.

Earthquakes in Alaska, California, and Hawaii, and the subsidence due to withdrawal of oil and groundwater in many states, have caused ground shifts that move corner monuments and thereby change their coordinates. The monuments, rather than the coordinates, then have greater weight in ownership rights.

22-7 RETRACEMENT SURVEYS

Retracement surveys are run for the purpose of relocating or reestablishing previously surveyed boundary lines. They are perhaps the most challenging of all types of surveys. The fundamental precept governing retracement surveys is that the monuments as originally placed and agreed to by the grantee and grantor constitute the correct boundary location. The objective of resurveys therefore is to restore boundary markers to their original locations, and this should guide all of the surveyor's actions.

In making a retracement survey, written evidence of title for the parcel involved should first be obtained. This will normally be in the form of a deed, but could also be obtained from an abstract or title policy. Even if the deed is available, it is good practice to trace it back to its creation in order to ensure that no transcription errors have been made, and to check for possible modifiers (such as the term "surface measure"). Deeds of all adjoining properties should also always be obtained and matched to **(1)** determine if any gaps or overlaps exist, and **(2)** understand any junior-senior rights that might possibly apply. The possibility that the written title could be supplanted by an *unwritten conveyance* must also be investigated. In the absence of any alterations of the written title by unwritten means, an evaluation of all evidence related to the written conveyance should be made in order to properly establish the property boundaries.

Various types of evidence are considered and used when retracing boundary surveys. When conflicts exist between the different types, the order of importance, or weight generally assigned in evaluating that evidence, is as follows:

1. *Senior rights.* When parcels of land are conveyed in sequence, the one created first receives all that was specified in the written documents, and in case of any overlap of descriptions, the second receives the remainder. In case of overlaps in other subsequent conveyances, the elder receives the benefits.
2. *Intent of the parties.* The intent of the grantee and grantor at the time of conveyancing must be considered in resurveys. Usually the best evidence of intent is contained in the written documents themselves.
3. *Call for a survey.* If the written documents describe a survey, an attempt should be made to locate the stakes or monuments placed as a result of that survey.
4. *Monuments.* If the written documents describe original monuments that were set to mark the boundaries, these must be searched out. When there are conflicts in monuments, natural ones such as trees or streams receive more weight than artificial ones such as stakes and iron pipes.
5. *Measurements.* Courts have consistently ruled that measurements called for in a description merely describe the positions of the corners. Consequently they generally receive the least weight in interpreting a conveyance. When measurements are evaluated as evidence, the order of importance that generally applies to them is **(1)** distance, **(2)** direction, **(3)** area, and **(4)** coordinates. (This order may change in the future because of the new instruments now being used on land surveys. Usually when total stations or GPS units are employed, for example, distance and direction measurements are not recorded but rather coordinates obtained directly.)

It should be noted that the foregoing are "general rules" for evaluating conflicting evidence, and it is possible, in certain circumstances, for a supposedly inferior element to control a superior one.

Good judgment is especially important when old lines are being restored and the original corners are lost. Then all possible evidence that applies to the original location must be found, and a decision made as to what evidence is good and what can or should be discarded. All found evidence should be recorded in the resurvey field notes, and reasons for using or rejecting any of it noted. Then if the survey is questioned in the courts, all actions taken can be justified. Familiarity with state and local laws, and past court decisions affecting surveys in an area, are valuable when making evaluations of evidence.

A corner that has been preserved is the best evidence of the original location of a line in question. The original notes are used as guides in locating these monuments, but a marker's actual position is the governing factor. Thus if a monument is found that is and has been accepted for years as the location of a particular corner, it should also be accepted by the surveyor.

A monument should not be assumed lost until every possible source of information has been exhausted and no trace of it can be found. Even then the surveyor should hesitate before disturbing settled possessions. It may be possible, for example, that where one or more corners are lost, all concerned parties have acquiesced in lines or corners based upon some other corner or landmark. These acquiesced corners may not be the original ones, but it would be unreasonable to discredit them when the people concerned do not question them. In a legal controversy, the law as well as common sense will normally declare that a boundary line, long acquiesced in, is better evidence of where the real line should be than one established by survey long after the original monuments have disappeared.

In retracement surveys, a determination must be made as to whether directions cited in the documents are true or magnetic. If they are magnetic, the declination at the time of the original survey must be determined so that true bearings or azimuths can be determined and lines retraced following them. A good assumption to adopt in retracing lines is that the boundaries are where the description says they are. If they aren't, some indications of their actual locations will probably be found when the distances and directions in the description are followed.

Testimony of persons who remember boundary locations is always valuable, but not always reliable. Therefore when such testimony is taken, a careful search must be made to find some corroborative evidence. Old fences or decayed wood at the location where a stake was purportedly originally set are examples of extremely valuable corroborative evidence. When two possibilities exist for establishing a boundary, evidence for rejecting one is often as important as evidence for accepting the other.

In retracement surveys, measurements made with a total station or EDM instrument between found monuments may not agree with distances on record. This situation provides a real test for surveyors. Perhaps the original chain or tape had an uncorrected systematic error, or a mistake was made. Alternatively, the marks may have been disturbed, or are the wrong ones. If differences exist between distances of record and measured values for found monuments, it may be helpful to get a *calibrated tape* length, which relates actual measurements to the original recorded distances. In relocating monuments, this calibrated length can then be used to lay off the actual distances noted by the original surveyor to search for monuments and evidence.

Measurements may indicate that bearings and angles between adjacent sides also do not fit those called for in the writings. These discrepancies could be due to a faulty original compass survey, incorrect corner marks, or other causes.

If, while conducting a resurvey, discrepancies are found between adjacent descriptions which cannot be resolved, they should be called to the attention of the owners so that steps can be taken to harmonize the land actually owned with that contained in the titles. Once the location of the original boundary is decided upon, it should be carefully marked and witnessed so that it can be easily found in the future.

22-8 SUBDIVISION SURVEYS

Subdivision surveys consist in establishing new smaller parcels of land within larger previously surveyed tracts. In these types of surveys, one or only a few new parcels may be created, in which case they may be described using the metes-and-bounds system. Conversely, in areas where new housing is planned, a block-and-lot subdivision survey can be conducted, thus creating many small lots simultaneously. Laws governing

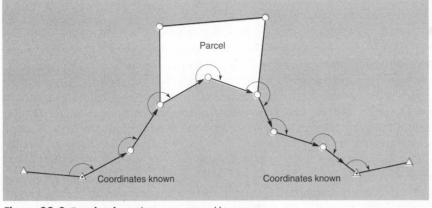

Figure 22-3 Transfer of coordinates to a parcel by traverse.

subdivision surveys vary from state to state, and surveyors must follow them carefully in performing these types of surveys.

Whether a single new lot is created or a block-and-lot survey performed, generally a closed traverse is first run around the larger tract, with all corners being occupied if possible. Fences, trees, shrubbery, hedges, common (party) walls, and other obstacles may necessitate running the traverse either inside or outside the property. If corners cannot be occupied, stub (side shot) measurements can be made to them and their coordinates computed, from which side lengths and bearings can be computed (see Section 13-13). All measurements should be made with a precision suited to the specifications and land values.

Before traversing around the larger tract, it is generally first necessary to establish a reference direction for one of the tract's lines. Also, if state plane coordinates are to be used in the new parcel's description, a set of starting coordinates must be determined for one of the parcel's corners. A reference direction can be established by making an astronomic observation. Alternatively, as illustrated in Figure 22-3, both a reference direction and coordinates can be transferred to the parcel by including one of its lines in a separate traverse initiated at existing nearby control monuments. A check is secured on the traverse by closing back on the starting station, or on a second nearby monument, as indicated in the figure.

After the traverse around the larger exterior parcel has been completed and adjusted, the new parcels can be surveyed. Where a single new parcel is being created, its corners are established according to the owner's specifications within requirements of the statutes. From the survey, the parcel description is prepared, certified, and recorded.

Block-and-lot subdivisions must not only conform to state statutes, but in addition many municipalities have laws covering these types of surveys. Regulations may specify minimum lot size, allowable misclosures for surveys, types of corner marks to be used, minimum width of streets and the procedure for dedicating them, rules for registry of plats, procedures for review, and other matters. Often several jurisdictions and agencies, each with its own laws and regulations, may have authority over the subdivision of land. In these cases, if standards conflict, the most stringent usually applies. The mismatched street and highway layouts of today could have largely been eliminated by suitable subdivision regulations and by a thorough review of them in past years.

A small block-and-lot subdivision is shown in Figure 22-2. (Some lot areas have been deleted so that their calculations can become end-of-chapter problems.) Computers with appropriate software such as coordinate geometry (COGO) and CAD greatly reduce the labor of computing subdivisions. They are especially valuable for large plats and designs with curved streets. Automatic plotters using stored computer files make drafting the final plat map accurate, simple, and fast.

Critical subdivision design and layout considerations include creating good building sites, an efficient street and utility layout, and assured drainage. Furthermore, subdivision rules and regulations must be followed and the developer's desires met as nearly as possible. A subdivision project involves a survey of the *exterior* boundaries of the tract to be divided, followed by a topographic survey, design, and layout of the *interior* of the tract. Following is a brief outline of the steps to be accomplished in these procedures.

1. Exterior survey
   (a) Obtain accurate descriptions of the tract of land to be subdivided, and of all adjoiners, from the Register of Deeds office. Note any discrepancies between the tract and its adjoiners.
   (b) Search for monuments marking corners of the tract, those of its adjoiners where necessary, and, where appropriate, for U.S. Public Land Survey monuments to which the survey may be referred or tied. Resolve any discrepancies with adjoiners.
   (c) Make a closed survey of the tract and adequately reference it to existing monuments.
   (d) Compute departures, latitudes, and misclosures to see whether the survey meets requirements. Balance the survey if the misclosure is within allowable limits.
   (e) Resolve, if possible, any encroachments on the property, or differences between occupation lines and title lines, so there will be no problems later with the final subdivision.

2. Interior survey, design, and layout
   (a) Perform a topographic survey of the area within the tract. From the survey data, prepare a topographic map for evaluating drainage, and for use in preliminary street layout and lot design.
   (b) Develop a preliminary plat showing the streets and the blocks and lots into which the tract is to be subdivided. Compute departures and latitudes on every block and lot to ensure their perfect mathematical closures.
   (c) Obtain the necessary approvals of the preliminary plat.
   (d) Set block and lot corners of the tract according to the approved preliminary plat. Block corners should be located first, and lot corners set by measuring along lines between block ends. (In some cases final stakeout may be delayed until utilities are placed so corners are not destroyed in this process.)
   (e) Prepare a final plat map that will conform to state and city platting regulations, and include whatever certificates may be required.
   (f) Have the certificates executed and witnessed, and the final plat approved and recorded.

22-9 PARTITIONING LAND

A common problem in property surveys is partitioning land into two or more pieces for sale or distribution to family members, heirs, and so on. Prior to partitioning, a boundary survey is run, departures and latitudes computed, the traverse balanced, and the total enclosed area calculated. Computational procedures in partitioning vary depending on conditions. Some parcel shapes and requirements permit formula solutions, often by using analytical geometry. Others require trial-and-error methods.

Figure 22-4 illustrates cutting off 12 acres by means of a line *EF* parallel to base *AD.* EXAMPLE 22-1
The distances in this old farm deed were given and balanced only to the nearest 0.1 ft, and thus, without exceeding the number of significant figures in the data, a "more or less" area of 12 acres = 522,700 ft^2 is required.

SOLUTION

Let $x = BC = 790.7$ ft. Then, required area = $522,700 = xy + \dfrac{y^2 \tan 8°17'}{2}$.

$$= 790.7y + \frac{y^2(0.1455872)}{2}.$$

Then $0.0727936y^2 + 790.7y - 522,700 = 0$.
Solve the quadratic equation

$$y = 625.1 \text{ ft}$$

$$z = y \tan 8°17' = 91.0 \text{ ft}$$

$$\text{base } EF = 790.7 + 91.0 = 881.7 \text{ ft}$$

$$CF = \frac{625.1}{\cos 8°17'} = 631.7 \text{ ft}$$

$$\text{area } EBCF = 625.1\left(\frac{790.7 + 881.7}{2}\right) = 522,700 \text{ ft}^2 \text{ (check)}$$

(*Note:* with lengths of sides measured to only tenths of a foot, a computed area to more than four significant figures is not justified.)

Another possible approach in the solution of Example 22-1 is to extend *AB* and *DC* to meet at a point, then use the resulting triangle in which simple proportions are readily visualized.

Cutoff lines to separate a certain area from the parcel may have (1) a specific starting point (distance from one corner of the tract polygon and run to the midpoint, or any other location on the opposite side); or (2) a required direction (parallel with, perpendicular to, or on a designated bearing angle from a selected line). These cases can often be handled by trial-and-error solutions involving an initial assumption such as the cutoff line direction or the starting point. Certain problems are amenable to solution using coordinate formulas for the intersection of two lines (see Appendix B).

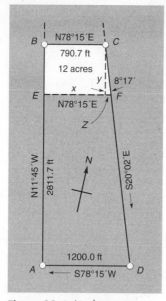

Figure 22-4 Land partitioning.

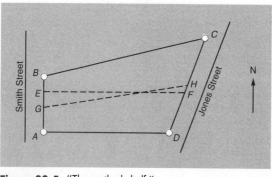

Figure 22-5 "The southerly half."

Figure 22-5 shows that a statement "the southerly half" of tract *ABCD* can have a number of meanings—the *most* southerly half, half the frontage, or half the actual acreage. An important consideration is the final shape of each lot. Connecting midpoints *G* and *H* leaves the southerly "half" smaller than the northerly "half," but provides equal frontage for both parts on the two streets. Course *EF,* parallel with *AD,* produces one trapezoidal lot, but a poorly shaped northerly parcel with meager frontage on Smith Street. The intent of the deed should therefore be clearly stated.

22-10 REGISTRATION OF TITLE

To remedy difficulties arising from inaccurate descriptions and disputed boundary claims, some states provide for registration of property titles under rigid rules. The usual requirements include marking each corner with standardized monuments referenced to established points, and recording a plat drawn to scale and containing specified items. Titles are then guaranteed by the court under certain conditions.

A number of states have followed Massachusetts' example and maintain separate *land courts* dealing exclusively with land titles. As the practice spreads, the accuracy of property surveys will be increased and transfer of property simplified.

Title insurance companies search, assemble, and interpret official records, laws, and court decisions affecting ownership of land, and then will insure purchasers against loss regarding defects, liens, encumbrances, restrictions, assessments, and easements. Defense in lawsuits is provided by the company against threats to a clear title from claims shown in public records and not exempted in the policy. The location of corners and lines is not guaranteed; hence it is necessary to establish on the ground the exact boundaries called for by the deed and title policy. Close cooperation between surveyors and title insurance companies is necessary to prevent later problems for their clients.

Many technical and legal problems are considered before title insurance is granted. In some states, title companies refuse to issue a policy covering a lot if fences in place are not on the property line, and exclude from the contract "all items that would be disclosed by a property survey."

To guide surveyors in their conduct of land title surveys, the American Land Title Association (ALTA) and the American Congress on Surveying and Mapping (ACSM) recently adopted a new set of standards. Named the "ALTA/ACSM Land Title Survey Standards," they set forth concise guidelines on what must be included in property surveys for title insurance purposes, and they also include minimum standards of accuracy for four different classes of surveys. The classes are based on land use, and include *urban, suburban, rural,* and *mountain* or *marshland.* For these four classes linear misclosures cannot exceed 1:15,000; 1:10,000; 1:7,500; and 1:5,000, respectively. Other accuracy criteria specified relate to the required precisions of instruments used and acceptable field procedures. Benefits derived from these guidelines are clarification of the exact requirements of land title surveys so that uniformly high-quality results are obtained in the future.

22-11 ADVERSE POSSESSION

Adverse rights can be obtained against all except the public by occupying a parcel of land for a period of years specified by state law, and performing certain acts. To claim land or rights to it by *adverse possession,* its occupation or use must be **(1)** actual, **(2)** exclusive, **(3)** open and notorious, **(4)** hostile, and **(5)** continuous. It may also be necessary for the property to be held under *color of title* (a claim to a parcel of real property based on some written instrument, though a defective one). In some states all taxes must be paid. The time required to establish a claim of adverse possession varies from a minimum of 7 years in Florida to a maximum of 60 years for urban property in New York. The customary period is 20 years. In Wisconsin it is 10 years under color of title, 20 years without.

The occupation and use of land belonging to a neighbor but outside his or her apparent boundary line as defined by a fence may lead to a claim of adverse possession. Continuous use of a street, driveway, or footpath by an individual or the general public for a specified number of years results in the establishment of a right-of-way privilege which cannot be withheld by the original owner.

An *easement* is a right, by grant or agreement, which allows a person or persons to use the land of another for a specific purpose. It always implies an interest in the land on which it is imposed. Black's *Law Dictionary* lists and defines 18 types of easements; hence the exact purpose of an easement should be clearly stated.

The discussion of property surveys has necessarily been condensed in this text, but it provides helpful information to readers while deterring inexperienced people from attempting to run boundary lines. For more extensive coverage, references are listed in the Bibliography.

22-12 CONDOMINIUM SURVEYS

The term *condominium* refers to a type of property ownership in which an individual owns a single unit within a larger apartment building or complex of apartments or buildings. The term also applies to the apartment or complex itself. Every owner receives a deed describing their property, is able to mortgage or sell the unit independent of the owners of other units, and is taxed separately. Thus legal descriptions based on surveys are required.

Although the limits of ownership of condominium units vary, often they are the surfaces of the floor tops, interior walls, and ceiling bottoms. Condominium ownership usually also includes joint interest in *common elements* such as the land on which the building or buildings rest, water lines, electrical wiring, and stairways if the building is multistory. Other common elements of joint interest may include lobbies, swimming pools, tennis courts, and recreation halls. The owner of each unit may also hold interest in *limited common elements* such as a garage, parking area, or patio, with their use restricted to certain units only.

Condominium surveys differ from ordinary property surveys in a number of ways, but they also bear many similarities. Statutes govern the procedures and requirements in each state. In the usual procedures, a condominium is created by filing with the Register of Deeds an instrument usually called the *condominium declaration*. This document must be accompanied by a plat and plans. The plat is prepared by a registered land surveyor and describes the land parcel on which the building or buildings will be placed. It also shows the building(s) on the lot in relation to the parcel's boundaries, and gives the floor plan dimensions and area for each unit. An example condominium survey plat for a single-story building is shown in Figure 22-6.

For multistory occupancy, the plans must include elevations of the floors and ceilings. These must be tied to mean sea level, or a local datum, so leveling from a nearby bench mark is required. If a reference elevation is established at the base of the building, elevations can be transferred vertically to floors and ceilings above using a tape. For single-story occupancy, as illustrated in Figure 22-6, elevations of floors and ceilings are not required.

The dimensions of individual units must represent the structure as built. Because there are often differences between the architect's plan and the finished building, actual measurements must be made after construction.

22-13 PARCEL-BASED LAND INFORMATION SYSTEMS

As noted in Chapter 29, surveyors are playing a major role in the development and implementation of modern Land Information Systems (LISs), and this activity will continue in the future. These computerized data banks contain information, or *attributes,* about the land, including its ownership, easements, zoning, flood plain extent if any, land use, soil types, existence of mineral and water resources, and much more. The

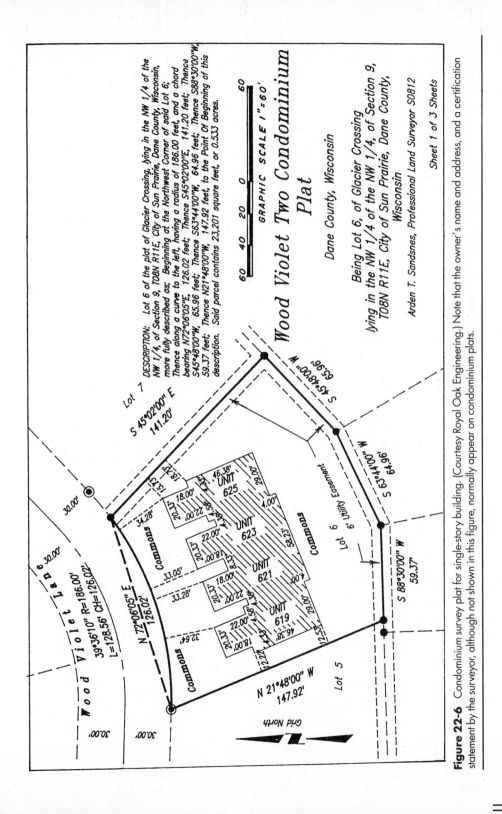

Figure 22-6 Condominium survey plat for single-story building. (Courtesy Royal Oak Engineering.) Note that the owner's name and address, and a certification statement by the surveyor, although not shown in this figure, normally appear on condominium plats.

information is available for rapid retrieval and is invaluable to surveyors, government officials, lawyers, developers, planners, environmentalists, and others.

Boundary surveys are fundamental to LISs since knowledge about the land is meaningless unless its position on earth is specified. In most modern LISs being developed, positional data are established by associating attributes about the land with individual lots or tracts of ownership. These are referred to as *parcel-based* LISs.

The most fundamental information associated with each individual parcel in an LIS is its legal description. This gives the unique location on the earth of the parcel, and thus provides the positional information needed to support the system. As described in previous sections, legal descriptions are written instruments, based on measurements and prepared to exacting standards and specifications. Thus the land surveyor's role in modern LISs is an important one. The subject of LISs is discussed in greater depth in Chapter 29.

22-14 SOURCES OF ERROR

Some sources of error in boundary surveys follow:

1. Errors in measured distances and directions
2. Corners not defined by unique monuments
3. Judgment errors in evaluating evidence

22-15 MISTAKES

Some typical mistakes in connection with boundary surveys are:

1. Use of the wrong corner marks
2. Failure to check deeds of adjacent property as well as the description of the parcel in question
3. Ambiguous deed descriptions
4. Omission of the length or direction of the closing line
5. Failure to close on known control
6. Magnetic bearings not properly corrected to the date of the new survey

---

PROBLEMS

22-1 Define the following terms:
    (a) Land tenure system   (b) Parole evidence
    (c) Retracement surveys   (d) Cutoff line

22-2 From the county courthouse or other local records, determine the types of property descriptions used in your area.

22-3 What are the essential elements of metes-and-bounds property descriptions?

22-4 Write a metes-and-bounds description for the house and lot where you live. Assume data if necessary. Sketch a map of the property.

22-5 In your city's residential section, what setbacks are required from property lines in the front, rear, and on the sides?

22-6 In a description by metes and bounds, what purpose may be served by the phrase "more or less" following the acreage?

*22-7 From the metes-and-bounds description of the lot in the town of Little Wolf, described in Section 22-4, compute the lot's misclosure.

---

*Asterisks indicate problems that have answers given in Appendix G.

22-8 What major advantage does the coordinate method of property description have over other methods?

22-9 Determine how the assignment of block, tract, subdivision, and lot numbers is made, and by whom, in your area.

22-10 List in their order of importance the following types of evidence when conducting retracement surveys: **(a)** measurements, **(b)** call for a survey, **(c)** intent of the parties, **(d)** monuments, and **(e)** senior rights.

22-11 In performing retracement surveys, list in their order of importance, the four different types of measurements called for in a description.

22-12 List all types of pertinent information or data that should appear on a completed property survey plat.

22-13 Telephone poles are set on two corners of a rectangular lot. Explain how to survey the lot, locate boundaries, and find the area.

22-14 Two disputing neighbors employ a surveyor to check their boundary line. Discuss the surveyor's authority if **(a)** the line established is agreeable to both clients, and **(b)** the line is not accepted by one or both of them.

22-15 What time period is required to claim adverse rights to property in your state?

22-16 Compute the area of lot 16 in Figure 22-2.

*22-17 Compute the misclosure of lot 19 in Figure 22-2. On the basis of your findings, would this plat be acceptable for recording? Explain.

22-18 Similar to Problem 22-17, except for lot 49 in Figure 22-2.

22-19 Similar to Problem 22-17, except for lot 50 in Figure 22-2.

22-20 Calculate the misclosure and area of lot 17 of Figure 22-2.

*22-21 Calculate the misclosure and area of lot 18 of Figure 22-2.

*22-22 For the accompanying figure; using a line perpendicular to AB through x, cut off two-fifths of the area to include corner B, and determine lengths xy and By.

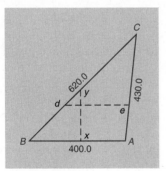

Problem 22-22

22-23 For the figure of Problem 22-22, calculate the length of line de, parallel to BA, which will divide the tract into two parts, two-fifths of the area being in the parcel containing corner C. Give lengths Bd and Ae.

22-24 Prepare a metes-and-bound description for the parcel shown. (See figure on page 496.) Assume all corners are marked with 2-in.-diameter iron pipes and a 12-ft meander line setback.

22-25 Draw a plat map of the parcel in Problem 22-24 at a convenient scale. Label all monuments and the lengths and directions of each boundary line on the drawing. Include a title, scale, North arrow, and legend.

22-26 Prepare a metes-and-bounds description for the property shown. (See figure on page 496.) Assume all corners are marked with 1-in.-diameter steel rods.

22-27 Create a 1.5-acre tract on the westerly side of the parcel in Problem 22-26 with a line parallel to the westerly property line. Give the lengths and bearings of all lines for both new parcels.

*Asterisks indicate problems that have answers given in Appendix G.

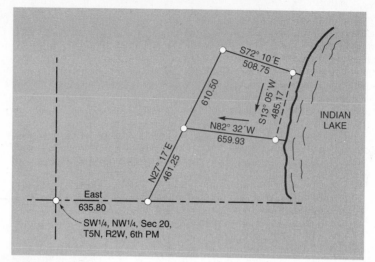

Problem 22-24

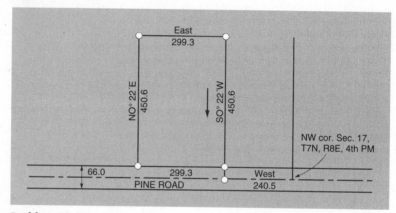

Problem 22-26

22-28 Discuss the ownership limits of a condominium unit.

22-29 What are parcel-based Land Information Systems?

*22-30 Compute the misclosure of, and area within, the external boundary of the condominium plat of Figure 22-6.

*Asterisks indicate problems that have answers given in Appendix G.

BIBLIOGRAPHY ═══ ASCE. 1993. ''Proposed Model Standard of Practice for Property Boundary Surveys.'' *Journal of Surveying Engineering* 119 (No. 3):111.

ASCE. 1991. *Right of Way Surveying.* New York: American Society of Civil Engineers.

Bales, R. F. 1993. ''The New 1992 ALTA/ACSM Land Title Survey Standards.'' *Point of Beginning* 18 (No. 3):48.

Black, Henry C. 1990. *Black's Law Dictionary,* 6th ed. St. Paul, Minn.: West Co.

Brown, C. M., W. G. Robillard, and D. A. Wilson. 1981. *Evidence and Procedures for Boundary Location,* 2nd ed. New York: John Wiley and Sons.

Brown, C. M., W. G. Robillard, and D. A. Wilson. 1986. *Boundary Control and Legal Principles,* 3rd ed. New York: John Wiley and Sons.

Buckner, R. B. 1975. "Reasons and Methods for Accurate Direction in Land Surveys." *Surveying and Mapping* 35 (No.4):305.

Conradi, K. 1992. "Boundary Surveys Are Possible with GIS." *Professional Surveyor* 12 (No. 2):24.

Duffy, M. A. 1991. "GPS and the Cadastral Survey." *GPS World* 2 (No. 5):38.

Elgin, R. L., and D. R. Knowles. 1984. "Arkansas Riparian Rights." *Surveying and Mapping* 44 (No. 1):39.

Epstein, E. and T. Duchesneau. 1990. "Use and Value of a Geodetic Reference System." *Journal of the Urban and Regional Information Systems Association,* 2 (No. 1):11.

Fant, J. E., A. R. Freeman, and C. Madson. 1981. "Metes and Bounds Description." *Surveying and Mapping* 41 (No.2):222.

Grimes, J. 1976. *Clark on Surveying and Boundaries,* 3rd ed. Rancho Cordova, Calif.: Landmark.

Jeffress, G. A. 1992. "GIS and Land Records." *Bulletin, American Congress on Surveying and Mapping* (No. 140):36.

Kellie, A. C. 1985. "The Surveyor and Written Boundary Agreements." *ASCE, Journal of the Surveying Engineering Division* 111 (No. 2):155.

Kellie, A. C. 1988. "Monuments in the Retracement." *Surveying and Land Information Systems* 48 (No. 4):247.

Kratz, K. E. 1981. "The Land Surveyor's Quasi-judicial Rights and Duties Regarding Boundary Location." *Surveying and Mapping* 41 (No. 2):213.

Lampert, L. L. 1980. "The Retracement." *Surveying and Mapping* 40 (No. 3):315.

McCall, C. E. 1980. "Subdivision of Land Today." *Surveying and Mapping* 40 (No. 4):415.

McEntyre, J. G. 1978. *Land Survey Systems.* New York: John Wiley and Sons.

Onsrud, H. J. 1985. "Choosing and Evaluation of Property Line Survey Methods." *Surveying and Mapping* 45 (No. 2):139.

Onsrud, H. J. 1992. "Mortgage Loan Surveys." *Point of Beginning* 17 (No. 4):54.

Roskay, T. 1992. "Automated Legal Descriptions." *Professional Surveyor* 12 (No. 1):33.

Tyler, D. A. 1987. "Position Tolerance in Land Surveying." *ASCE, Journal of Surveying Engineering* 113 (No. 3):152.

U.S. Department of the Interior, Bureau of Land Management. 1973. *Manual of Surveying Instructions 1973.* Washington, D.C.: U.S. Government Printing Office.

Von Meyer, N. 1991. "Cadastral Evidence Data in a Land Information System." *Journal of the Urban and Regional Information Systems Association* 3 (No. 1):14.

Wattles, G. H. 1976. *Writing Legal Descriptions.* New York: Parker.

SURVEYS OF THE PUBLIC LANDS

23-1 INTRODUCTION

The term *public lands* is applied broadly to the areas that have been subject to administration, survey, and transfer of title under the public-lands laws of the United States since 1785. These lands include those turned over to the federal government by the colonial states and the larger areas acquired by purchase from (or treaty with) the native Indians or foreign powers that had previously exercised sovereignty.

Thirty states, including Alaska, constitute the public-land survey states which have been, or will be, subdivided into rectangular tracts (see Figure 23-1). The area of these states represents approximately 72 percent of the United States.

Title to the vacant lands, and therefore direction over the surveys within their own boundaries, was retained by the colonial states, the other New England and Atlantic coast states (except Florida), and later by the states of West Virginia, Kentucky, Tennessee, Texas, and Hawaii. In these areas the U.S. public-land laws have not been applicable.

The beds of navigable bodies of water are not public domain and not subject to survey and disposal by the United States. Sovereignty is in the individual states.

The survey and disposition of the public lands were governed originally by two factors:

1. A recognition of the value of grid-system subdivision based on experience in the colonies and another large-scale systematic boundary survey—the 1656 Down Survey in Ireland.
2. The need of the colonies for revenue from the sale of public land. Monetary returns from their disposal were disappointing, but the planners' farsighted vision of a grid system of subdivision deserves commendation.

Although nearly a billion acres of public land has either been sold or granted since 1785, approximately one-third of the area of the country is still federally owned. The

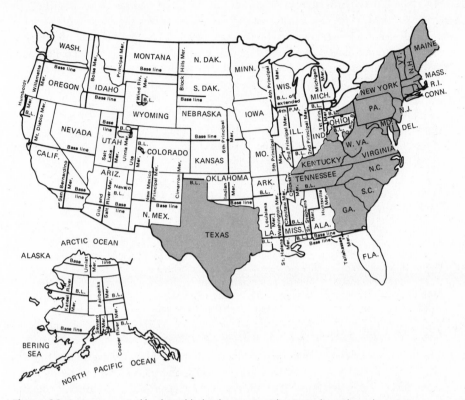

Figure 23-1 Areas covered by the public-lands surveys, with principal meridians shown. Areas excluded are shaded. (Hawaii, although not shown on this map, would also be shaded. Texas has a rectangular system similar to the U.S. public land system.)

Bureau of Land Management (BLM), within the Department of the Interior, was created in 1946 as a merger of the Grazing Service and the General Land Office, and is responsible for surveying and managing a significant portion of these lands.

23-2 INSTRUCTIONS FOR SURVEYS OF THE PUBLIC LANDS

The U.S. system of public-lands surveys was inaugurated in 1784, and the territory immediately northwest of the Ohio River in what is now eastern Ohio served as a test area. Sets of instructions for the surveys were issued in 1785 and 1796. Manuals of instructions were later issued in 1855, 1881, 1890, 1894, 1902, 1930, 1947, and 1973.

In 1796 General Rufus Putnam was appointed as the first surveyor general, and the numbering of sections changed to the system now in use (see the excerpt from the 1973 *Manual of Surveying Instructions* below, and Figure 23-7). Supplementary rules were promulgated by each surveyor general "according to the dictates of his own judgment" until 1836, when the General Land Office was reorganized. Copies of changes and instructions for local use were not always preserved and sent to Washington. As a result, no office in the United States has a complete set of instructions under which the original surveys were supposed to have been made.

Most later public-lands surveys have been run by the procedures to be described, or variations of them. The task of present-day surveyors consists primarily of retracing the original lines and further subdividing sections. To do so, they must be thoroughly familiar with the rules, laws, equipment, and field conditions that governed their predecessors in a given area.

Basically, the rules of survey stated in the 1973 *Manual of Surveying Instructions* are as follows:[1]

> The public lands shall be divided by north and south lines run according to the true meridian, and by others crossing them at right angles, so as to form townships six miles square. . . .
>
> The corners of the townships must be marked with progressive numbers from the beginning; each distance of a mile between such corners must be also distinctly marked with marks different from those of the corners.
>
> The township shall be subdivided into sections, containing as nearly as may be, six hundred and forty acres each, by running parallel lines through the same from east to west and from south to north at the distance of one mile from each other (originally at the end of every two miles but amended in 1800), and marking corners at the distance of each half mile. The sections shall be numbered, respectively, beginning with the number one in the northeast section, and proceeding west and east alternately through the township with progressive numbers until the thirty-six be completed.

Additional rules of survey covering field books, subdivision of sections, adjustment for excess and deficiency, and other matters are given in the manuals and special instructions. Private surveyors, on a contract basis, were paid $2/mi of line run until 1796 and $3/mi thereafter. Sometimes the amount was adjusted in accordance with the importance of a line, the terrain, location, and other factors. From this meager fee surveyors had to pay and feed a party of at least four while on the job and in transit to and from distant points. They had to brush out and blaze (mark trees by scarring the bark) the line, set corners and other marks, and provide satisfactory notes and one or more copies of completed plats. The contract system was completely discarded in 1910. Public-land surveyors are now appointed.

Since meridians converge, it is evident that the requirements that "lines shall conform to the true meridians and townships shall be 6 mi square" are mathematically impossible. An elaborate system of subdivision was therefore worked out as a practical solution.

Two principles furnished the legal background for stabilizing land lines:

1. Boundaries of public lands established by duly appointed surveyors are unchangeable.
2. Original township and section corners established by surveyors must stand as the true corners they were intended to represent, whether or not in the place shown by the field notes.

Expressed differently, the original surveyors had an official plan with detailed instructions for its layout, and presumably set corners to the best of their ability. After

---

[1]Since new surveys (particularly in Alaska) as well as retracements are still being carried out, the 1973 manual employs more than one tense, and this chapter does also.

title passed from the United States, their established corners (monuments), regardless of errors, became the lawful ones. Therefore if monuments have disappeared, *the purpose of resurveys is to determine where they were, not where they should have been.* Correcting mistakes or errors now would disrupt too many accepted property lines and result in an unmanageable number of lawsuits.

In general, the procedure in surveying the public lands provides for the following subdivisions:

1. Division into quadrangles (tracts) approximately 24 mi on a side (after about 1840)
2. Division of quadrangles into townships (16), approximately 6 mi on a side
3. Division of townships into sections (36), approximately 1 mi square
4. Subdivision of sections (usually by local surveyors)

It will be helpful to keep in mind that the purpose of the grid system was to obtain sections 1 mi on a side. To this end, surveys proceeded from south to north and east to west, and all discrepancies were thrown into the sections bordering the north and west township boundaries to get as many *regular sections* as possible.

Although the general method of subdivision outlined above was normally followed, detailed procedures were altered in surveys made at different times in various areas of the country. As examples, instructions for New Mexico said only township lines were to be run where the land was deemed unfit for cultivation, and in Wisconsin the first four correction lines north of the base line were 60 mi apart rather than 24.

Another current example relates to the surveys in Alaska, where the area's sheer vastness requires changes. When Alaska gained statehood in 1958, only 2 percent of its 375 million acres has been surveyed. Priorities have been set for conducting the remaining surveys, and current plans call for subdividing some 155 million acres to be transferred to that state and native Alaskans. To accelerate the project, the 18,651 townships in Alaska were first established on protraction diagrams, and latitude and longitude determined for each corner. In executing the surveys, markers are being set at 2–mi intervals in most areas, and satellite and inertial surveying systems are being utilized extensively. But even with this modern technology and relaxed procedures, at the current rates of progress, it is estimated the job will take 40 years to complete!

23-3 INITIAL POINT

Thomas Jefferson recognized the importance of surveys and served as chairman of a committee to develop a plan for locating and selling the western lands. His report to the Continental Congress in 1784, adopted as an ordinance on May 20, 1785, called for survey lines to be run and marked before land sales. Many of today's property disputes would have been eliminated if all property lines were resurveyed and monuments checked and/or set before sales became final!

Subdivision of the public lands became necessary in many areas as settlers moved in and mining or other land claims were filed. The early hope that surveys would precede settlement was not fulfilled.

As settlers pressed westward, in each area where a substantial amount of surveying was needed, an initial point was established and located by astronomical observations. The manual of 1902 was the first to specify an indestructible monument, preferably a

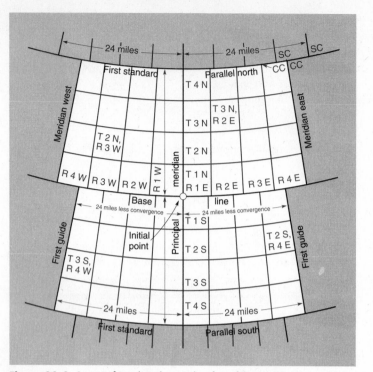

Figure 23-2 Survey of quadrangles. (Only a few of the standard corners and closing corners are identified.)

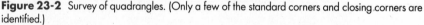

copper bolt, firmly set in a rock ledge if possible and witnessed by rock bearings. In all, thirty-seven initial points have been set, five of them in Alaska. A principal meridian and base line were passed through each initial point, such as the one in the center of Figure 23-2.

23-4 PRINCIPAL MERIDIAN

From each initial point, a true north-south line called a *principal meridian* (Prin. Mer. or PM) was run north and/or south to the limits of the area to be covered. Generally, a solar attachment—a device for solving mechanically the mathematics of the astronomical triangle—was used (see Section 18-19). Monuments were set for section and quarter-section corners every 40 ch and at the intersections with all meanderable bodies of water (streams 3 ch or more in width, and lakes covering 25 acres or more).

The line was supposed to be within 3′ of the cardinal direction. Two independent sets of linear measurements were required to check within 20 lk (13.2 ft)/80 ch, which corresponds to a precision ratio of only $\frac{1}{400}$. The allowable difference between sets of measurements is now limited to 7 lk/80 ch.

Areas within a principal meridian system vary greatly as depicted in Figure 23-1.

23-5 BASE LINE

From the initial point, a *base line* was extended east and/or west as a true parallel of latitude to the limits of the area to be covered. As required on the principal meridian, monuments were set for section and quarter-section corners every 40 ch and at the

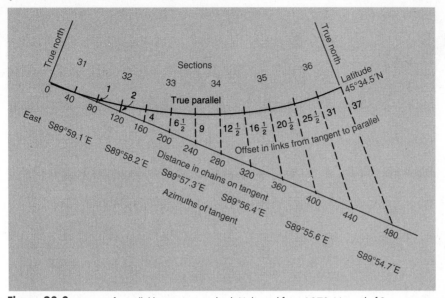

Figure 23-3 Layout of parallel by tangent method. (Adapted from 1973 *Manual of Surveying Instructions.*)

intersections with all meanderable bodies of water. Permissible closures were the same as those for the principal meridian.

Base lines are actually circular curves on the earth's surface, and were run with chords of 40 ch by the **(1)** solar method, **(2)** tangent method, or **(3)** secant method. These are briefly described as follows:

1. Solar method. An observation is made with a solar attachment to determine the direction of true north. A right angle is then turned off and a line extended 40 ch, where the process is repeated. The series of lines so established, with a slight change in direction every half-mile, closely approaches a true parallel. Obviously, if the sun is obscured, the method cannot be used.

2. Tangent method. This method of laying out a true parallel is illustrated in Figure 23-3. A 90° angle is turned to the east or to the west, as may be required from a true meridian, and corners are set every 40 ch. At the same time, proper offsets, which increase with increasing latitudes, are taken from *Standard Field Tables* issued by the BLM and measured north from the tangent to the parallel. In the example shown, the offsets in links are 1, 2, 4, $6\frac{1}{2}$, 9, $12\frac{1}{2}$, $16\frac{1}{2}$, $20\frac{1}{2}$, $25\frac{1}{2}$, 31, and 37. The error resulting from taking right-angle offsets instead of offsets along the converging lines is negligible. The main objection to the tangent method is that the parallel departs considerably from the tangent, so both the tangent and the parallel must be brushed out.

3. Secant method. This method of laying out a true parallel is shown in Figure 23-4. It actually is a modification of the tangent method in which a line parallel to the tangent at the 3-mi (center) point is passed through the 1- and 5-mi points to produce minimum offsets.

Field work includes establishing a point on the true meridian, south of the beginning corner, at a distance taken from the *Standard Field Tables* for the latitude of a desired parallel. The proper bearing angle from the same table is turned to the east or west

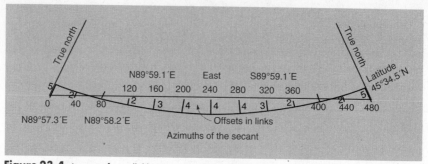

Figure 23-4 Layout of parallel by secant method. (Adapted from 1973 *Manual of Surveying Instructions.*)

from the true meridian to define the secant, which is then projected 6 mi. Offsets, which also increase with increasing latitudes, are measured north or south from the secant to the parallel. Advantages of the secant method are that the offsets are small and can be measured perpendicular to the secant without appreciable error, and the amount of clearing is reduced.

23-6 STANDARD PARALLELS (CORRECTION LINES)

After the principal meridian and the base line have been run, *standard parallels* (Stan. Par. or SP), also called *correction lines,* are run as true parallels of latitude 24 mi apart in the same manner as was the base line. All 40-ch corners are marked. Standard parallels are shown in Figure 23-2. In some early surveys, standard parallels were placed at intervals of 30, 36, or 60 mi.

Standard parallels are numbered consecutively north and south of the base line; examples are first standard parallel north and third standard parallel south.

23-7 GUIDE MERIDIANS

Guide meridians (GM) are run due north from the base line and the standard parallels at intervals of 24 mi east and west of the principal meridian, in the same manner as was the principal meridian, and with the same limits of error. Before work is started, the chain or tape must be checked by measuring 1 mi on the base line or standard parallel. All 40-ch corners are marked.

Because meridians converge, a *closing corner* (CC) is set at the intersection of each guide meridian and standard parallel or base line (see Figure 23-2). The distance from the closing corner to the *standard corner* (SC), which was set when the parallel was run, is measured and recorded in the notes as a check. Any error in the 24-mi-long guide meridian is put in the northernmost half-mile.

Guide meridians are numbered consecutively east and west of the principal meridian; examples are first guide meridian west and fourth guide meridian east. Correction lines and guide meridians, established according to instructions, created so-called *quadrangles* (or tracts) whose nominal dimensions are 24 mi on a side. These are shown in Figure 23-2.

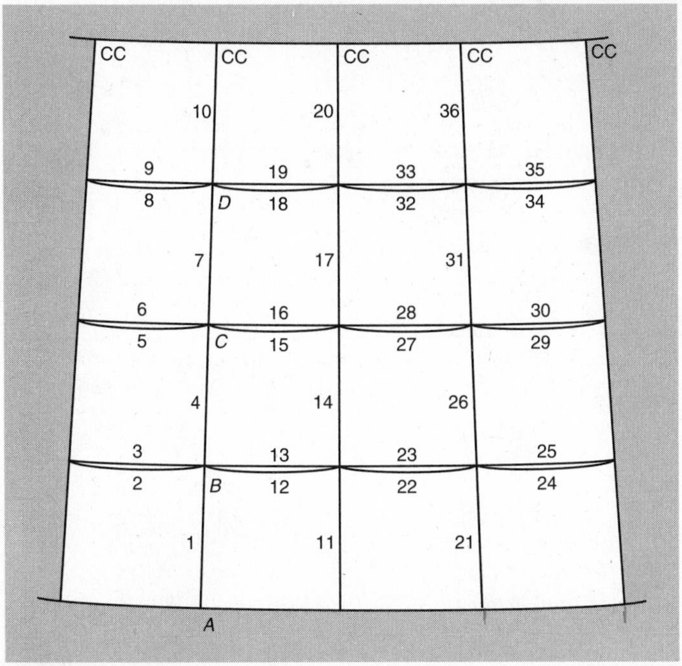

Figure 23-5 Order of running lines for the subdivision of a quadrangle into townships.

23-8 TOWNSHIP EXTERIORS, MERIDIONAL (RANGE) LINES, AND LATITUDINAL (TOWNSHIP) LINES

Division of a quadrangle, or tract, into townships is accomplished by running *range* (R) and *township* (T or Tp) lines.

Range lines are true meridians through the standard township corners previously established at intervals of 6 mi on the base line and standard parallels. They are extended north to intersect the next standard parallel or base line, and closing corners set (see Figures 23-2 and 23-5). Township lines are east–west lines that connect township corners previously established at intervals of 6 mi on the principal meridian, guide meridians, and range lines.

Formulas for convergence of meridians (derived in various texts on geodesy), are as follows:

$$\theta = 52.13d \tan \phi \qquad (23\text{-}1)$$

and

$$c = \tfrac{4}{3}Ld \tan \phi \quad \text{(slight approximation)} \qquad (23\text{-}2)$$

where θ is the angle of convergence, in seconds; d the distance between meridians, in miles, on a parallel; ϕ the mean latitude; c the linear convergence, in feet; and L the length of meridians, in miles.

EXAMPLE 23-1   Compute the angular convergence at 40°25′ North latitude between two adjacent guide meridians.

SOLUTION

By Eq. (23-1) (guide meridians are 24 mi apart),

$$\theta = 52.13(24) \tan 40°25′ = 1065″ = 17′45″$$

EXAMPLE 23-2   Determine the distance between the standard corner and its closing corner for a range line 12 mi east of the principal meridian, extended 24 mi north at a mean latitude of 43°10′.

SOLUTION

By Eq. (23-2),

$$c = \tfrac{4}{3}(24)(12) \tan 43°10′ = 360.18 \text{ ft.}$$

23-9 DESIGNATION OF TOWNSHIPS

A township is identified by a unique description based on the principal meridian governing it.

North and south rows of townships are called *ranges* and numbered in consecutive order east and west of the principal meridian as indicated in Figure 23-2.

East and west rows of townships are named *tiers* and numbered in order north and south of the base line. By common practice, the term *tier* is usually replaced by the township in designating the rows.

An individual township is identified by its number north or south of the base line, followed by the number east or west of the principal meridian. An example is Township 7 South, Range 19 East, of the Sixth Principal Meridian. Abbreviated, this becomes T 7 S, R 19 E, 6th PM.

23-10 SUBDIVISION OF A QUADRANGLE INTO TOWNSHIPS

The method to be used in subdividing a quadrangle into townships is fixed by regulations in the *Manual of Surveying Instructions*. Under the old regulations, township boundaries were required to be within 21′ of the cardinal direction. Later this was reduced to 14′ to keep interior lines within 21′ of the cardinal direction.

The detailed procedure for subdividing a quadrangle into townships can best be described as a series of steps designed to ultimately produce the maximum number of regular sections with minimum unproductive travel by the field party. The order of running the lines is shown by consecutive numbers in Figure 23-5. Some details are described in the following steps:

1. Begin at the southeast corner of the southwest township, point *A,* after checking the chain or tape against a 1-mi measurement on the standard parallel.

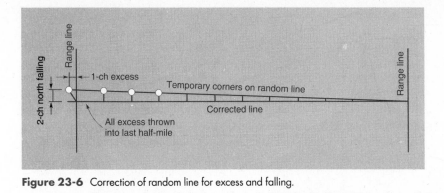

Figure 23-6 Correction of random line for excess and falling.

2. Run north on the true meridian for 6 mi (line 1), setting alternate section and quarter-section corners every 40 ch. Set township corner *B*.
3. From *B*, run a random line (line 2) due west to intersect the principal meridian. Set temporary corners every 40 ch.
4. If the random line has an excess or deficiency of 3 ch or less (allowing for convergence), and a falling north or south of 3 ch or less, the line is accepted. It is then corrected back (line 3), and all corners are set in their proper positions. Any excess or deficiency is thrown into the most westerly half-mile. The method of correcting a random line having an excess of 1 ch and a north falling of 2 ch is shown in Figure 23-6.
5. If the random line misses the corner by more than the permissible 3 ch, all four sides of the township must be retraced.
6. The same procedure is followed until the southeast corner *D* of the most northerly township is reached. From *D*, range line 10 is continued as a true meridian to intersect the standard parallel or base line, where a closing corner is set. All excess or deficiency in the 24 mi is thrown into the most northerly half-mile.
7. The second and third ranges of townships are run the same way, beginning at the south line of the quadrangle.
8. While the third range is being run, random lines are also projected to the east and corrected back, and any excess or deficiency is thrown into the most westerly half-mile. (All points may have to be moved diagonally to the corrected line, instead of just the last point, as in Figure 23-6.)

23-11 SUBDIVISION OF A TOWNSHIP INTO SECTIONS

Sections are now numbered from 1 to 36, beginning in the northeast corner of a township and ending in the southeast corner, as shown in Figure 23-7. The method used to subdivide a township can be described most readily as a series of steps to produce the maximum number of regular sections 1 mi on a side. Lines were run in the following order:

1. Set up at the southeast corner of the township, point *A,* and observe the meridian. Retrace the range line northward and the township line westward for 1 mi to compare the meridian, needle readings, and taped distances with those recorded.

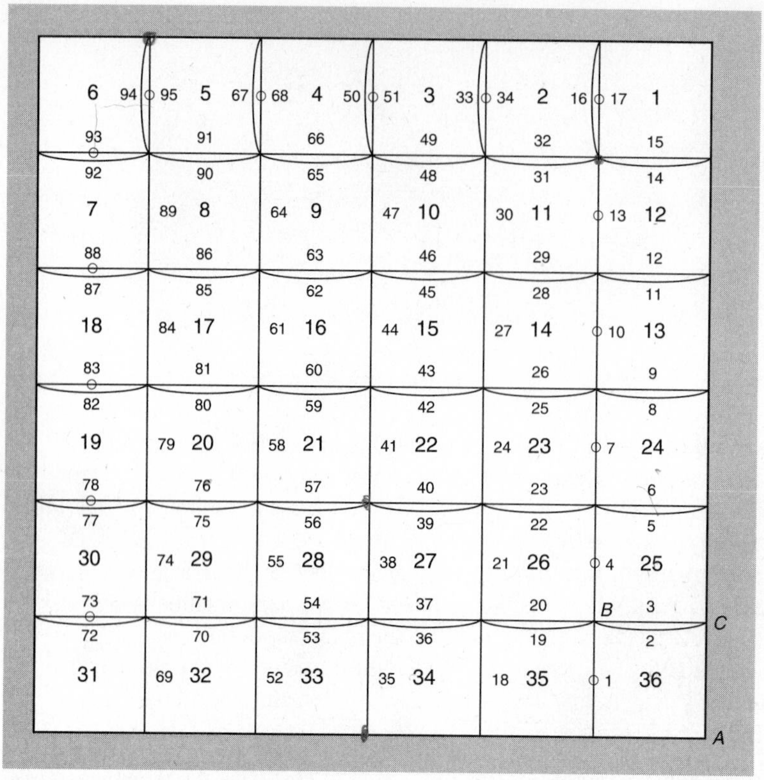

Figure 23-7 Order of running lines for the subdivision of a township into sections.

2. From the southwest corner of section 36, run north *parallel* with the east boundary of the township. Set quarter-section and section corners on line 1 (Figure 23-7).

3. From the section corner just set, run a random line *parallel* with the south boundary of the township eastward to the range line. Set a temporary quarter-corner at 40 ch.

4. If the 80-ch distance on the random line is within 50 lk, falling or distance, the line is accepted. The correct line is calculated and a quarter-corner located at the *midpoint* of line *BC* connecting the previously established corner *C* and the new section corner *B*.

5. If the random line misses the corner by more than the permissible 50 lk, the township lines must be rechecked and the error source determined.

6. The east range of sections is run in a similar manner until the southwest corner of section 1 is reached. From this point a random line is run northward to connect with the north township line section corner. A quarter-corner is set 40 ch from the south section corner (on line 17, corrected back by later manuals). All discrepancies in the 6 mi are thrown into the last half-mile.

7. Successive ranges of sections across the township are run until the first four have been completed. All north-south lines are parallel with the township east side. All east-west lines are run randomly parallel with the south boundary line and then corrected back (so parallelism may be destroyed).

8. When the fifth range is being run, random lines are projected to the west as well as to the east. Quarter-corners in the west range are set 40 ch from the east side of the section, with all excess or deficiency resulting from the errors and convergence being thrown into the most westerly half-mile.

9. If the north side of the township is a standard parallel, instead of running a random line to the north, lines parallel with the east township boundary are projected to the correction line and closing corners set. The distance to the nearest corner is measured and recorded.

10. True bearings of interior north-south section lines for any latitude can be obtained by applying corrections from tables for the convergence at a given distance from the east boundary.

By throwing the effect of meridian convergence into the westernmost half-mile of the township and all errors to the north and west, 25 regular sections nominally 1 mi^2 are obtained. Also, the south half of sections 1, 2, 3, 4, and 5; the east half of sections 7, 18, 19, 30, and 31; and the southeast quarter of section 6 are regular size.

23-12 SUBDIVISION OF SECTIONS

A section was the basic unit of the General Land Office system, but land was often patented in parcels smaller than a section. Subdivision of sections was performed by local surveyors and others as the owner took up the land. The BLM provides guidelines on the proper and intended way a section should be subdivided.

To divide a section into quarter-sections (nominally 160 acres), straight lines are run between opposite quarter-section corners previously established or reestablished. This rule holds whether or not the quarter-section corners are equidistant from the adjacent section corners.

To divide a quarter-section into quarter-quarter-sections (nominally 40 acres), straight lines are run between opposite quarter-quarter-section corners established at the midpoints of the four sides of the quarter section. The same procedure is followed to obtain smaller subdivisions.

If the quarter-sections are on the north or west side of a township, the quarter-quarter-section corners are placed 20 ch from the east or south quarter-section corners—or by single proportional measurement (see Section 23-19) on line if the total length on the ground is not equal to that on record.

23-13 FRACTIONAL SECTIONS

In sections made fractional by rivers, lakes, or other bodies of water, lots are formed bordering on the body of water and numbered consecutively through the section (see section 8 in Figure 23-8). Boundaries of lots usually follow the quarter-section and quarter-quarter-section lines, but extreme lengths or narrow widths are avoided, as are areas of fewer than 5 acres or more than 45 acres.

Quarter-sections along the north and west boundaries of a township, made irregular by discrepancies of measurements and convergence of the range lines, are usually numbered as lots (see Figure 23-8). One such quarter-section in a Wisconsin township contains 640 acres!

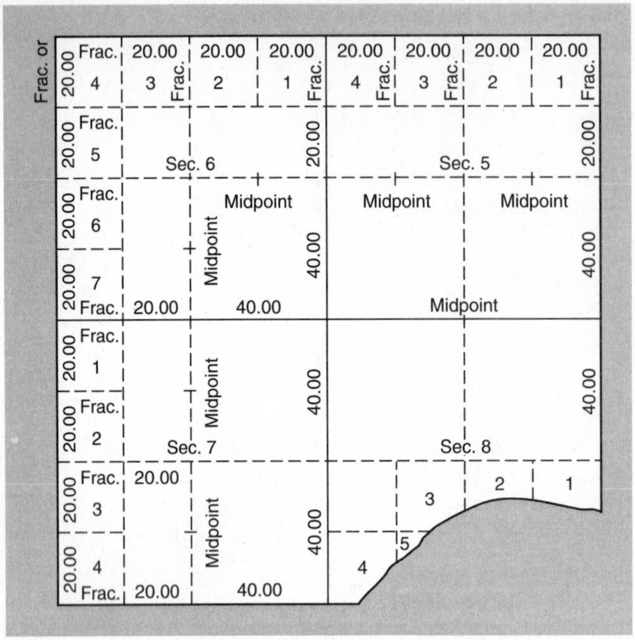

Figure 23-8 Subdivision of regular and fractional sections.

Lot lines are not actually run in the field. Like quarter-section lines, they are merely indicated on the plats by *protraction* (subdivisions of parcels on paper only). Lot areas are computed from plats. Many quarter-section lines are not indicated on plats.

23-14 NOTES

Specimen field notes for each of the several kinds of lines to be run are shown in various instruction manuals. Actual recording had to closely follow the model sets.

The original notes, or copies of them, are maintained in a land office in each state for the benefit of all interested persons.

23-15 OUTLINE OF SUBDIVISION STEPS

Pertinent points in the subdivision of quadrangles into townships, and townships into sections, are summarized in Table 23-1.

23-16 MARKING CORNERS

Various materials were approved and used for monuments in the original surveys. These included pits and mounds, stones, posts, charcoal, and bottles. A zinc-coated, alloyed iron pipe with $2\frac{1}{2}$ in. outside diameter, 30 in. long, is now standard. The bottom end of the pipe is split for several inches to form flanges, and a brass cap is securely fastened to its top. Substitutions for this standard monument are permitted when authorized. Specially manufactured markers—for example, a cast-aluminum type with a breakaway cap and attached small underground magnetic device to aid in recovery with a metal detector—are now becoming commonplace. In rock outcrop, a $3\frac{1}{4}$-in.-diameter brass tablet with $3\frac{1}{2}$-in. stem is specified.

TABLE 23-1 SUBDIVISION STEPS

ITEM	SUBDIVISION OF A QUADRANGLE	SUBDIVISION OF A TOWNSHIP
Starting point	SE corner of SW township	SW corner of SE section (36)
Meridional lines		
Name	Range line	Section line
Direction	True north	North, parallel with east range line
Length	6 mi = 480 ch	1 mi = 80 ch
Corners set	Quarter-section and section corners at 40 and 80 ch alternately	Quarter-section corner at 40 ch; section corner at 80 ch
Latitudinal lines		
Name	Township line	Section line
Direction of random	True east-west parallel	East, parallel with south side of section
Length	6 mi less convergence	1 mi
Permissible error	3 ch, length or falling	50 lk, length or falling
Distribution of error		
Falling	Corners moved proportionately from random to true line	Corners moved proportionately from random to true line
Distance	All error thrown into west quarter-section	Error divided equally between quarter-sections

[Work repeated until north side of area is reached. Subdivision of last area on the north of the range of townships and sections follows.]

Case I. When Line on the North Is a Standard Parallel

ITEM	SUBDIVISION OF A QUADRANGLE	SUBDIVISION OF A TOWNSHIP
Direction of line	True north	North, parallel with east range line
Distribution of error in length	Placed in north quarter-section	Placed in north quarter-section
Corner placed at end	Closing corner	Closing corner
Permissible errors	Specified in *Manual of Surveying Instructions*	Specified in *Manual of Surveying Instructions*

Case II. When Line on the North Is Not a Standard Parallel

ITEM	SUBDIVISION OF A QUADRANGLE	SUBDIVISION OF A TOWNSHIP
Direction of line	No case	Random north and correct back to section corner already established
Distribution of error in length		Same as case I

[Other ranges of townships and sections continued until all but two are laid out.]

Location of last two ranges	On east side of tract	On west side of township
Next-to-last range subdivided	As before	As before
Last range		
Direction of random	True east	Westerly, parallel with south side of section
Nominal length	6 mi less convergence	1 mi less convergence
Correction for temporary corners	Corners moved proportionately from random to true line	Corners moved proportionately from random to true line
Distribution of error of closure	Corners moved westerly (or easterly) to place error in west quarter-section	Corners placed on the true line so total error falls in west quarter-section

Stones and posts were marked with one to six notches on one or two faces. The arrangements identify a monument as a particular section or township corner. Each notch represents 1 mi of distance to a township line or corner. Quarter-section monuments were marked with the fraction "$\frac{1}{4}$" on a single face.

In prairie country, where large stones and trees were scarce, a system of pits and mounds was used to mark corners. Different groupings of pits and mounds, 12 in. deep and 18 in. square, designated corners of several classes. Unless perpetuated by some other type of mark, these corners were lost in the first plowing.

23-17 WITNESS CORNERS

Whenever possible, monuments were witnessed by two or three adjacent objects such as trees and rock outcrops. *Bearing trees* were blazed on the side facing the corner and marked with scribing tools.

When a regular corner fell in a creek, pond, swamp, or other place where it was impracticable to place a mark, *witness corners* (WC) were set on all lines leading to the corner. Letters WC were added to all other marks normally placed on the corner, and the WC then witnessed also.

23-18 MEANDER CORNERS

A *meander corner* (MC) was established on survey lines intersecting the bank of a stream having a width greater than 3 ch, or a lake, bayou, or other body of water of 25 acres or more. The distance to the nearest section corner or quarter-section corner was measured and recorded in the notes. A monument was set and marked MC on the side facing the water, and the usual witness noted. If practicable, the line was carried across the stream or other body of water by triangulation to another corner set in line on the farther bank.

A traverse joined successive meander corners along the banks of streams or lakes and followed as closely as practicable the sinuosities of the bank. The traverse was checked out by calculating the position of the new meander corner and comparing it with its known position on a surveyed line.

Meander lines follow the mean high-water mark and are used only for plotting and protraction of the area. They are *not* boundaries defining the limits of property adjacent to the water.

23-19 LOST AND OBLITERATED CORNERS

A common problem in resurveys of the public lands is the replacement of lost or obliterated corners. This difficult task requires a combination of experience, hard work, and ample time to reestablish the location of a wooden stake or post, set perhaps 150 years ago, and with all witness trees long since cut or burned by apathetic owners.

An *obliterated corner* is one for which there are no remaining traces of the monument or its accessories, but whose location has been perpetuated or can be recovered beyond reasonable doubt. The corner may be restored from the acts or testimony of interested landowners, surveyors, qualified local authorities, witnesses, or from written evidence. Satisfactory evidence has value in the following order:

1. Evidence of the corner itself
2. Bearing trees or other witness marks

3. Fences, walls, or other evidence showing occupation of the property to the lines or corners
4. Testimony of living persons

A *lost corner* is one whose position cannot be determined, beyond reasonable doubt, either from traces of the original marks or from acceptable evidence or testimony that bears on the original position. It can be restored only by rerunning lines from one or more independent corners (existing corners that were established at the same time and with the same care as the lost corner). Proportionate measurements distribute the excess or deficiency between a recently measured distance d separating the nearest found monuments that straddle the lost point, and the record distance D given in the original survey notes between these monuments. Then the distance x from one of the found monuments required to set the lost point is calculated by proportion as $x = X(d/D)$, where X is the record distance from that monument.

Single-proportionate measurement follows the procedure just described, and is used to relocate lost corners that have a specific alignment in one direction only. These include standard corners on base lines and standard parallels, intermediate section corners on township boundaries, all quarter-section corners, and meander corners established originally on lines carried across a meanderable body of water.

Figure 23-9 illustrates a lost quarter-section corner a on the line between sections 2 and 3. Section corners b and c have been found. The record distances for lines ba and ac are 40.00 and 39.57 ch, respectively. The measured distance between found corners b and c was 5246.25 ft. Describe the process for restoring lost corner a.   EXAMPLE 23-3

SOLUTION

1. Since point a is a quarter-section corner, it is replaced on line bc by single proportionate measurement.
2. Distance ba that must be laid off from section corner b to restore lost corner a is

$$ba = \left(\frac{40}{79.57}\right) \times 5246.25 = 2637.30 \text{ ft.}$$

Double-proportionate measurements are used to establish lost corners located originally by specific alignment in two directions, such as interior section corners and

---

Figure 23-9 Example of single-proportionate measurement.

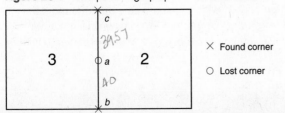

× Found corner

○ Lost corner

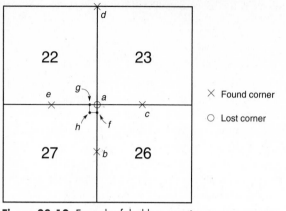

Figure 23-10 Example of double-proportionate measurement.

corners common to four townships. The general procedure for single-proportionate measurement is used, but in two directions. It will establish two points: one on the north-south line and another on the east-west line. The lost corner is then located where lines from the two points, perpendicular to their respective north-south and east-west lines, intersect.

EXAMPLE 23-4 Figure 23-10 illustrates lost corner *a* which is common to sections 22, 23, 26, and 27. Corners *b*, *c*, *d*, and *e* have been found. Record and measured distances are as follows:

RECORD		MEASURED	
LINE	DISTANCE (CH)	LINE	DISTANCE (FT)
ba	40.16	*bd*	7925.49
ca	40.00	*ec*	5293.24
da	79.20		
ea	39.72		

Describe the process of restoring lost corner *a*.

SOLUTION

1. Corner *a* is an interior section corner, which is constrained in alignment in two directions. Thus it must be restored by double-proportionate measurement.
2. First establish temporary point *f* by laying off distance *bf* along line *bd*, where *bf* is computed as

$$bf = \left(\frac{40.16}{119.36}\right) \times 7925.49 = 2666.62 \text{ ft}$$

3. Then locate temporary point *g* by laying off distance *eg* along line *ec*, where *eg* is computed as

$$eg = \left(\frac{39.72}{79.72}\right) \times 5293.24 = 2637.32 \text{ ft}$$

4. Establish point *h,* the restored lost corner, where an east-west line through *f* intersects a north-south line through *g.*

When the original surveys were run, *topographic calls* (distances along each line from the starting corner to natural features such as streams, swamps, and ridges) were recorded. Using the recorded distances to any of these features found today, and applying single- or double-proportionate measurements to them, may help locate an obliterated corner or produce a more reliable reestablished lost corner.

23-20 ACCURACY OF PUBLIC-LANDS SURVEYS

The accuracy required in the early surveys was a very low order. Frequently it fell below what the notes indicated. A small percentage of the surveys were made by drawing on the imagination in the comparative comfort of a tent; no monuments were set, and the notes serve only to confuse the situation for present-day surveyors and land owners. A few surveyors threw in an extra chain length at intervals to assure a full measure.

Many surveys in one California county were fraudulent. In another state, a meridian 108 mi long was run without including the chain handles in its 66-ft length. When discovered, it was rerun without additional payment.

The poor results obtained in various areas were due primarily to the following:

1. Lack of trained personnel; some contracts were given to persons with no technical training.
2. Very poor equipment by today's standards.
3. Work done by contract at low prices.
4. Surveys made in piecemeal fashion as the Indian titles and other claims were extinguished.
5. Marauding Indians, swarms of insects, dangerous animals and reptiles.
6. Lack of appreciation for the need to do accurate work.
7. Erratic or missing field inspection.
8. Magnitude of the problem.

In general, considering the handicaps listed, the work was reasonably well done in most cases.

23-21 DESCRIPTIONS BY TOWNSHIP, SECTION, AND SMALLER SUBDIVISION

Description by the U.S. public land sectional system offers a means of defining boundaries uniquely, clearly, and concisely. Several examples of acceptable descriptions are listed.

Sec. 6, T 7 S, R 19 E, 6th PM.

Frac. Sec. 34, T 2 N, R 5 W, Ute Prin. Mer.

The $SE\frac{1}{4}$, $NE\frac{1}{4}$, Sec. 14, Tp. 3 S, Range 22 W, SBM [San Bernardino Meridian].

$E\frac{1}{2}$ of $NE\frac{1}{4}$ of Sec. 20, T 15 N, R 10 E, Indian Prin. Mer.

E 80 acres of $NE\frac{1}{4}$ of Sec. 20, T 15 N, R 10 E, Indian Prin. Mer.

Note that the last two descriptions do not necessarily describe the same land. A California case in point occurred when the owner of a Southwest quarter-section, nominally 160 acres but actually 162.3 acres, deeded the westerly portion as "the West 80 acres" and the easterly portion as "the East $\frac{1}{2}$."

Sectional land that is privately owned may be partitioned in any legal manner at the option of the owner. The metes-and-bounds form is preferable for irregular parcels. In fact, metes and bounds are required to establish the boundaries of mineral claims and various grants and reservations.

Differences between the physical and legal (or record) ground locations and areas may result because of departures from accepted procedures in description writing, loose and ambiguous statements, or dependence on the accuracy of early surveys.

23-22 BLM LAND INFORMATION SYSTEM

As noted in Section 23-1, approximately one-third of the United States land area remains in the public domain. The Bureau of Land Management (BLM) is responsible for managing a significant portion of this vast acreage. Rapid access to accurate information related to this public land is in demand now more than ever before. As an example of its importance, consider that millions of dollars worth of oil and mineral royalties can be gained or lost by relatively small changes in property boundaries.

To assist in the monumental task of managing this enormous quantity of diverse land, the BLM is implementing a modern multipurpose Land Information System (LIS), two portions of which are called the *Automated Land and Mineral Record System* (ALMRS), and the *Automated Resource DATA* (ARD). The ALMRS program catalogs data related to land parcel descriptions, including land status, mining claims, oil and gas leases, and other records, while ARD includes various natural resource and wildlife information.

To support ALMRS and ARD, the BLM is in the process of creating a new *Geographic Coordinate Data Base* (GCDB). This database will consist of a digital layer of information on the *Public Land Survey System* (PLSS) and provide the positional components necessary for correlating all other information in the LIS. Included in the GCDB will be geographic coordinates of all PLSS corners and estimates of their reliability, identifications of the surveyors who set the corners, the types of corners set and dates placed, any records of resurveys, a full record of ownership of each parcel, ownership of abutting parcels, and much more information.

The objectives of a Bureau-wide LIS are to (**1**) streamline responses to land and mineral inquiries, (**2**) facilitate processing of applications and permits, (**3**) simplify access to all records, (**4**) improve the physical security of records and protect proprietary and confidential information, (**5**) ensure accurate and consistent data, (**6**) augment planning, monitoring, and evaluation of programs, and (**7**) reduce costs of records and land management. To meet these objectives the LIS is designed to be compatible with other automated Land and Geographic Information Systems, permitting shared data and minimizing duplication of effort.

When fully operational, the system will enable the preparation of graphic illustrations of complex land record combinations, and significantly aid in the management of all aspects of the public lands. Much of the data within the system will be accessible to personnel of the BLM, other federal agencies, state and local governments, business and industry, special interest groups, and private citizens.

23-23 SOURCES OF ERROR

Some of the many sources of error in retracing the public-lands surveys follow:

1. Discrepancy between length measured with an early surveyor's chain and one obtained with a tape or EDM instrument
2. Change in magnetic declination, local attraction, or both
3. Lack of agreement between field notes and actual measurements
4. Change in water courses
5. Nonpermanent objects used for corner marks
6. Loss of witness corners

23-24 MISTAKES

Some typical mistakes in the retracement of boundaries in public-land surveys are:

1. Failing to follow the general rules and special instructions of procedure governing the original survey
2. Neglecting to calibrate the tape or EDM instrument used against record distances for marks in place
3. Treating corners as lost when they are actually obliterated
4. Resetting corners without exhausting every means of relocating the original corners
5. Failing to recognize that restored corners *must* be placed in their original locations regardless of deviations by the original surveyor from the general rules and special instructions

---

PROBLEMS

23-1 What primary differences exist between original and present government instructions and practices concerning spacing of section corners and running section lines?

23-2 Why are the boundaries of public lands established by duly appointed surveyors unchangeable, even though incorrectly set in the original surveys?

23-3 List the advantages and disadvantages of running base lines and standard parallels by each of the three methods described in Section 23-5.

23-4 What is the convergence in feet of meridians for the following conditions:
  *(a) 6 mi apart, extended 12 mi, at mean latitude 35°30′N.
  (b) 24 mi apart, extended 24 mi, at mean latitude 45°15′N.

23-5 What is the angular convergence, in seconds, for the two meridians defining a township exterior at a mean latitude of:
  *(a) 30° 46′N.
  (b) 46°10′N.

*Asterisks indicate problems that have answers given in Appendix G.

23-6 What is the nominal distance in miles between the following:
   (a) Third Guide Meridian East, and the East Range Line of R25E.
   *__(b)__ SE corner of Sec. 21, T5N, R3E, Indian PM, and the NW corner of Sec. 12, T6N, R5E, Indian PM.
   (c) NE corner of Sec. 6, T10N, R14E, 4th PM, and the SW corner of Sec. 34, T11N, R17E, 4th PM.
23-7 In a sketch, show the locations of closing corners on township lines.
23-8 What steps in public-lands subdivision are left to the local surveyor?

Sketch and label pertinent lines and legal distances, and compute nominal areas of the parcels described in Problems 23-9 through 23-11.

23-9 $S\frac{1}{2}$, $NW\frac{1}{4}$, Sec. 13, T2N, R3W, Boise PM.
*__23-10__ $SE\frac{1}{4}$, $SW\frac{1}{4}$, Sec. 26, T3N, R1W, 4th PM.
23-11 $NE\frac{1}{4}$, $SW\frac{1}{4}$, $SW\frac{1}{4}$, Sec. 29, T2N, R2W, Indian PM.
23-12 What are the nominal dimensions and acreages of the following parcels:
   *__(a)__ $NE\frac{1}{4}$, $SW\frac{1}{4}$, Sec. 10.
   (b) $S\frac{1}{2}$, $SW\frac{1}{4}$, Sec. 1.
   (c) $SE\frac{1}{4}$, $SE\frac{1}{4}$, $SE\frac{1}{4}$, Sec. 14.
*__23-13__ How many rods of fence are required to enclose the following:
   (a) A parcel including the $SW\frac{1}{4}$, $SW\frac{1}{4}$, Sec. 16, and the $SE\frac{1}{4}$, $SE\frac{1}{4}$, Sec. 17, T12N, R5W?
   *__(b)__ A parcel consisting of Secs. 14, 15, 16, 21, 22, and 28 of T4N, R2W?
23-14 What lines of the U.S. public-land system were run as random lines?
23-15 In subdividing a township, which section line is run first? Which last?
23-16 Corners of the $NE\frac{1}{4}$ of the $SW\frac{1}{4}$ of Sec. 22 are to be monumented. If all section and quarter-section corners originally set are in place, explain the procedure to follow, and sketch all lines to be run and corners set.
*__23-17__ The southern boundary of a township lies on a standard parallel at latitude 43°24′N. What is the theoretical length of its northern boundary?
23-18 The quarter-section corner between Secs. 15 and 16 is found to be 40.30 ch from the corner common to Secs. 9, 10, 15, and 16. Where should the quarter-quarter-section corner be set along this line in subdividing Sec. 15?
23-19 As shown in the figure, in a normal township the exterior dimensions of Sec. 6 on the west, north, east, and south sides are 78, 78, 79, and 79 ch, respectively. Explain with a sketch how to divide the section into quarter-sections. (See figure below.)
23-20 The problem figure shows original record distances. Corners A, B, C, and D are found, but corner E is lost. Measured distances are $AB = 10,662.45$ ft and $CD = 10,525.60$ ft. Explain how to establish corner E. (See figure below.)

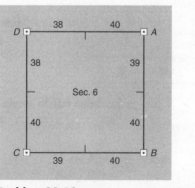

Problem 23-19

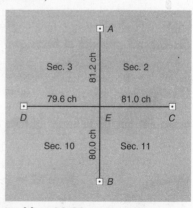

Problem 23-20

*Asterisks indicate problems that have answers given in Appendix G.

To restore the corners in Problems 23-21 through 23-24, which method is used, single proportion or double proportion?

23-21 Township corners on guide meridians; section corners on range lines.
23-22 Section corners on section lines; township corners on township lines.
*23-23 Quarter-section corners on range lines.
23-24 Quarter-quarter-section corners on section lines.
23-25 Why are meander lines not accepted as the boundaries defining ownership of lands adjacent to a stream or lake?
23-26 What is a witness corner? A meander corner?
23-27 Define the terms "obliterated corner" and "lost corner."
23-28 Explain the value of using proportionate measurements from topographic calls in relocating obliterated or lost corners.
23-29 Are deeds and agreements involving land effective if not in writing?
23-30 Why are the areas of many public-lands sections smaller than the nominal size?
23-31 What are the objectives of the automated Land Information Systems that are being developed by the BLM?

*Asterisks indicate problems that have answers given in Appendix G.

BIBLIOGRAPHY

Arndorfer, R. W., and J. L. Pinkerton. 1991. "Cadastral Survey Technology in Alaska from Statehood to Present." *Surveying and Land Information Systems* 51 (No. 4):233.

Bowman, L. J. 1978. "Subdivision of Sections of Rectangular Surveys of the United States." *Surveying and Mapping* 33 (No. 4):307.

Brown, C. M., W. G. Robillard, and D. A. Wilson. 1981. *Evidence and Procedures for Boundary Location,* 2nd ed. New York: John Wiley and Sons.

Brown, C. M., W. G. Robillard, and D. A. Wilson. 1986. *Boundary Control and Legal Principles,* 3rd ed. New York: John Wiley and Sons.

Crossfield, J. K. 1984. "Evolution of the United States Public Land System." *Surveying and Mapping* 44 (No. 3):259.

Glenn, W. W., and J. E. Glenn. 1991. "Early Public Land Surveyors in the Northwest." *Surveying and Land Information Systems* 51 (No. 4):204.

Livermore, M. 1991. "Following in the Footsteps of a Fraudulent Survey." *Point of Beginning* 16 (No. 4):64.

McCall, C. E. 1980. "Subdivision of Land Today." *Surveying and Mapping* 40 (No. 4):415.

McEntyre, J. G. 1978. *Land Survey Systems.* New York: John Wiley and Sons.

McEntyre, J. G. 1981. "Subdivision of a Section." *Survey and Mapping* 41 (No. 4):385.

Minnick, R. 1986. *Plotters and Patterns of American Land Surveying.* Rancho Cordova, Calif.: Landmark.

O'Callaghan, J. A. 1985. "Significance of the U.S. Public Land Survey System." *Bulletin, American Congress on Surveying and Mapping* (No. 95):13.

Robillard, W. and L. Bouman. 1992. *Clark on Surveying and Boundaries,* 6th ed. Charlottesville, Va.: The Michie Co.

Shartle, S. M. 1979. "Re: Subdivision of Sections of Rectangular Surveys of the United States." *Surveying and Mapping* 38 (No. 4):307.

U.S. Department of the Interior, Bureau of Land Management. 1973. *Manual of Surveying Instructions.* Washington, D.C.: U.S. Government Printing Office.

U.S. Department of Interior, Bureau of Land Management. 1974. *Restoration of Lost or Obliterated Corners and Subdivision of Sections.* Washington, D.C.: U.S. Government Printing Office.

Whittlesey, C. 1977. "Origin of the American System of Land Surveys." *Surveying and Mapping* 37 (No. 2):129.

William, M. G. 1981. "Sir William Petty, Thomas Jefferson, and the Down Survey: A Fresh Perspective on the U.S. Public Lands System." *Surveying and Mapping* 41 (No. 1):77.

24
CONSTRUCTION SURVEYS

24-1 INTRODUCTION

Construction is one of the largest industries in the United States, so surveying, as the basis for it, is extremely important. It is estimated that 60% of all hours spent in surveying are on location-type work, giving line and grade. Nevertheless, insufficient attention is frequently given to this type of survey.

An accurate topographic survey and site map are the first requirements in designing streets, sewer and water lines, and structures. Surveyors then lay out and position these facilities according to the design plan. A final "as-built" map, incorporating any modifications made to the design plans, is prepared during and after construction, and filed. Such maps are extremely important, especially where underground utilities are involved, to assure that the utilities can be located quickly if trouble develops, and that they will not be disturbed by later improvements.

Construction surveying involves establishing both *line* and *grade* by means of *stakes* and *string lines* placed on the construction site. These guide the contractor so that the facility is constructed according to plan. Placement of the stakes involves making the fundamental measurements of horizontal distances, horizontal and vertical angles, and differences in elevation using the basic equipment and methods described in earlier chapters of this text. This chapter describes some additional specialized procedures, and discusses a few common types of construction surveys with which every surveyor, engineer, and architect should be familiar. Chapters 25, 26, and 27 also relate to construction surveys, particularly those for transportation routes.

Construction surveying is best learned on the job by adapting fundamental principles to the undertaking at hand. Since each project may involve individual problems, textbook coverage tends to be limited to introductory material.

24-2 EQUIPMENT FOR CONSTRUCTION SURVEYS

The layout of stakes for line and grade to guide construction operations has traditionally been accomplished using the surveyor's standard equipment—levels, theodolites, tapes, and EDM instruments. Although these devices are still used widely, recent advances in modern technology have produced some innovative new equipment and procedures that have improved, simplified, and greatly increased the speed with which construction stakes can be set. New devices include visible laser-beam instruments, total stations that can be operated in a "tracking" mode, and, most recently, portable GPS units. These are described briefly in the subsections that follow.

24-2.1 Laser Instruments

The fundamental purpose of laser instruments is to create a visible line of known orientation, or a plane of known elevation, from which measurements for line and grade can be made. Two general types of lasers are described here:

1. *Single-beam lasers* project visible "string lines" or "plumb lines" utilized in linear and vertical alignment applications such as tunneling, sewer pipe placement, and building construction.
2. *Rotating-beam lasers* are merely single-beam lasers with spinning optics that rotate the beam in azimuth, thereby creating planes of reference. They expedite the placement of grade stakes over large areas such as airports, parking lots, and subdivisions, and they are also useful for topographic mapping.

The instrument shown in Figure 24-1 is a single-beam type laser. It combines a standard theodolite with a laser, creating an instrument useful for a variety of construction layout applications. The laser beam is projected collinear with the theodolite's line of sight, a feature that facilitates aligning it in prescribed directions. The instrument can be used to project string lines for distances up to about 1000 m. Also it can be turned about the theodolite's vertical axis to generate a horizontal plane, or about the horizontal axis to define a vertical plane.

Figure 24-1 Laser theodolite. (Courtesy TOPCON Instrument Corporation of America.)

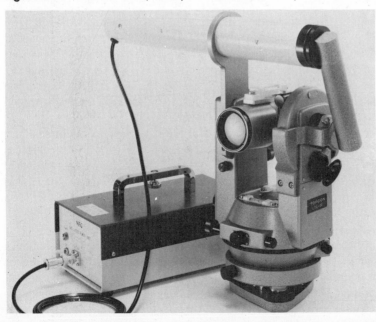

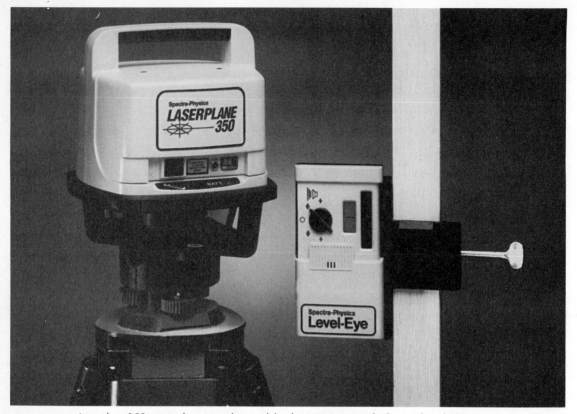

Figure 24-2 Laserplane 350 rotating-beam type laser with level-eye receiver attached to grade rod. (Courtesy Spectra-Physics.)

Figure 24-2 shows a rotating-beam type laser. It projects a beam up to 350 m while rotating at 600 rpm. The laser signal can be picked up by one or more receivers attached to grade rods or staffs (shown on the right in Figure 24-2). The instrument is self-leveling and quickly set up. If somehow bumped out of level, the laser beam shuts off and does not come back on until it relevels. It can be operated with the laser plane oriented horizontally for setting footings, floors, and so on, or the beam can be turned 90° and used vertically for plumbing walls or columns.

Because laser beams are not readily visible to the naked eye in bright sunlight, special detectors attached to a hand-held rod are often used. To lay out horizontal planes with either of these devices, the HI must be established. Then the height on a graduated rod that a reference mark or detector must be set is the difference between the HI and the plane's required elevation.

24-2.2 Total Station Instruments

As described in Chapter 10, total station instruments measure simultaneously horizontal and vertical angles as well as slope distances. Their built-in microcomputers reduce the measured slope distances to their horizontal and vertical components and display the results in real-time. These features make total stations very convenient for construction stakeout.

Smaller, less expensive total stations are now manufactured specifically for construction surveying. These instruments generally have shorter ranges and lower angle reso-

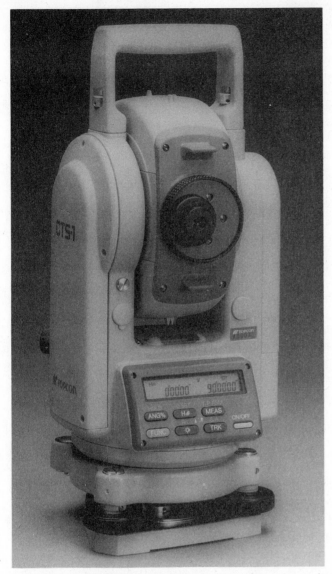

Figure 24-3 TOPCON CTS-1 total station instrument for construction layout. (Courtesy TOPCON Instrument Corporation of America.)

lutions than those made for control or land survey work. The TOPCON CTS-1 shown in Figure 24-3, for example, has a range of 500 m with a single prism, and its angle resolution is $+20''$. A more detailed discussion of the use of total stations for construction stakeout is given in Section 24-9.

24-2.3 GPS Equipment

Great strides have been made in the development of GPS surveying since the technology was first introduced in the early 1980s. GPS receivers have become smaller, more easily operated, and less expensive. Furthermore, field observing techniques and data reduction procedures have been improved so that point locations having extremely high accuracies can now be obtained in real-time. The technique called *real-time kinematic* GPS, described in Section 20-6.5 for example, yields subcentimeter accuracies in horizontal positions instantaneously (in real-time).

In employing real-time kinematic GPS procedures for construction staking, a minimum of two receivers are needed. Each is equipped with a radio modem. One receiver occupies a nearby control station (a point having known coordinates), and the other, called the "rover," is moved from one point to be set to another. (The points to be staked must have had their required coordinates computed before stakeout begins.) The receiver at the control station broadcasts its raw GPS signals to the rover. At the rover an on-board computer processes the signals from both receivers in real-time using relative positioning techniques. This yields a highly accurate determination of the rover's location immediately. If its measured coordinates do not agree with the required values for the point being staked, the rover is adjusted in position until agreement is reached, where the stake is set. GPS is particularly useful in staking widely spaced points, especially in areas where terrain or vegetation makes it difficult to conduct traditional ground surveys. GPS of course requires overhead clearance so that satellites will be visible.

Although excellent horizontal stakeout accuracy can be achieved using GPS, elevations are less reliable. GPS receivers determine ellipsoid heights to subcentimeter accuracy, but to get an elevation related to mean sea level, the geoid height must be subtracted (see Section 21-9). Unfortunately geoid heights are not precisely known for most areas, and therefore GPS elevations may contain inaccuracies of from several centimeters to a half-meter or more. The severity of the problem can be reduced by placing GPS receivers at one or more bench marks in an area to obtain an approximation of the geoid. Then elevations accurate to within a few centimeters can be determined, which is suitable for a great deal of stakeout work. Slopestaking is an example. Researchers are actively working on a more precise geoid definition.

24-3 HORIZONTAL AND VERTICAL CONTROL

The importance of a good framework of horizontal and vertical control in a project area cannot be overemphasized. It provides the basis for positioning structures, utilities, roads, and so on, in both the design and construction stages. Too often attempts have been made to skimp on establishing proper marks and monumenting them for preservation.

The surveyor in charge should receive copies of the plans well in advance of construction to become familiar with the job and have time to "tie out" or "transfer" any established control points that might be destroyed during building operations. The methods of Figures 12-4 and 12-5 are especially applicable and should be used with intersection angles as close to 90° as possible.

On most projects, additional horizontal and vertical control is required to supplement any already available in the job area. The points must be:

1. Convenient for use by the contractor's personnel, that is, located sufficiently close to the item being built so that workers using relatively simple equipment such as carpenters' levels and string lines can accurately transfer alignment and grade.
2. Far enough from the actual construction to ensure working room for the contractor and to avoid possible destruction of stakes.

3. Clearly marked and understood by the contractor in the absence of a surveyor.
4. Supplemented by guard stakes to deter removal, and referenced to facilitate restoring them. Contracts usually require the owner to pay the cost of setting initial control points and the contractor to replace damaged or removed ones.
5. Suitable for securing the accuracy agreed on for construction layout (which may be to only the nearest foot for a manhole, 0.01 ft for an anchor bolt, or 0.001 ft for a critical feature).

Construction stakes can be set in their required horizontal positions by making measurements of horizontal angles or horizontal distances from established control points. Any of the procedures discussed in Section 16-8 and illustrated in Figure 16-5 can be used. Radial stakeout by angle and distance from one control point is often most expedient, but the choice will depend on the project's nature and extent. Frequent checks should be made on points set, which can be done with measurements from other control stations, or checks should be made between staked points to verify the correctness of their relative positions.

Grade stakes and reference elevations are most often set using a leveling instrument whose HI has been established by differential leveling. For convenience, enough bench marks are generally placed on construction sites so that at least one is readily accessible at any location in the area. Then the HI of the level can be established with just a single backsight to the bench mark. After grade stakes are set, a closing foresight is taken back to the bench mark for a check. This practice can be dangerous, however, because the instrument operator will have a tendency to expect the closing foresight to agree with the initial backsight, and therefore could read it carelessly. As a result, a serious mistake in the initial backsight could go undetected, resulting in a faulty setting of grade stakes. Therefore, even though it requires more time, it is recommended that level circuits for setting grade stakes always begin on one bench mark and close on another.

24-4 STAKING OUT A PIPELINE

Pipelines are used to carry water for human consumption, storm water, sewage, oil, natural gas, and other fluids. Pipes which carry storm runoff are called *storm sewers;* those which transport sewage, *sanitary sewers.* Flow in these two types of sewers is usually by gravity, and therefore their alignments and grades must be carefully set. Flow in pipes carrying city water, oil, and natural gas is generally under pressure, so usually they need not be aligned to as high an order of accuracy.

In pipeline construction, trenches are usually opened along the required alignment to the prescribed depth (slightly below if pipe bedding is required), the pipe is installed according to plan, and the trench backfilled. Pipeline grades are fixed by a variety of existing conditions, topography being a critical one. A profile like that of Figure 7-10 is usually used to analyze the topography and assist in designing the grade line for each pipe segment. To minimize construction difficulties and costs, excavation depths are minimized, but at the same time a certain minimum cover over the pipeline must be maintained to protect it from damage by heavy loading from above and to prevent freezing in cold climates. Minimum grades also become an important design factor for pipes under gravity flow. Accordingly, a grade of at least 0.5% is recommended for storm sewers, but slightly higher grades are needed for sanitary sewers. In designing pipe grade lines, other existing underground elements often must be avoided, and due

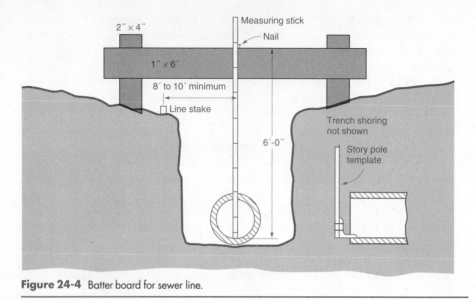

Figure 24-4 Batter board for sewer line.

regard must also be given to the grades of connecting lines and the vertical clearances need to construct manholes, catch basins, and outfalls.

Prior to staking a pipeline, the surveyor and contractor should discuss details of the project. An understanding must be reached concerning the planned trench width, where the installation equipment will be placed, and how and where the excavated material will be stockpiled. Then a reference offset line can be appropriately established that will **(1)** meet the contractor's needs, **(2)** be safe from destruction, and **(3)** not interfere with operations.

The alignment and grade for the pipeline are taken from the plans. An offset reference line parallel to the required centerline is established, usually at 25- or 50-ft stations when the ground is reasonably uniform. Marks should be closer together on horizontal and vertical curves than on straight segments. For pipes of large diameter, stakes may be placed for each pipe length—say, 6 or 8 ft. On hard surfaces where stakes cannot be driven, points are marked by paint, spikes, or scratch marks.

Precise alignment and grade for pipe placement are guided by either laser beams or *batter boards*. Figure 24-4 shows one arrangement of a batter board for a sewer line. It is constructed using 1 × 4-in., 1 × 6-in., or 2 × 4-in. boards nailed to 2 × 4-in. posts which have been pointed and driven into the ground on either side of the trench. The top of the batter board is generally placed a full number of feet above the invert (flow line or lower inside surface) of the pipe. Nails are driven into the board tops so a string stretched tightly between them will define the pipe centerline. A graduated pole or special rod is used to measure the required distance from the string to the pipe invert. Thus the string gives both line and grade. It can be kept taut by hanging a weight on each end after wrapping it around the nails.

In Figure 24-4, instead of a fixed batter board, a two-by-four carrying a level vial can be placed on top of the offset-line stake whose elevation is known. Measurement is made from the underside of a leveled two-by-four with a tape or graduated pole to establish the flow line.

On some jobs that have a deep wide cut, a level or laser instrument is set up in the trench to give line and grade. If the pipe is large enough, the laser device can be placed inside it.

24-5 STAKING PIPELINE GRADES

Staking pipeline grades is essentially the reverse of running profiles, although in both operations the centerline must first be marked and stationed in horizontal location. The actual profiling and staking are on an offset line.

Information conveyed to the contractor on stakes for laying pipelines usually consists of two parts: (1) giving the depth of cut (or fill), normally only to the nearest 0.1 ft, to enable a rough trench to be excavated; and (2) providing precise grade information, generally to the nearest 0.01 ft, to guide in the actual placement of the pipe invert at its planned elevation. Cut (or fill) values for the first part are vertical distances from ground elevation at the offset stakes to the pipe invert. After the pipe's grade line has been computed and the offset line run, cuts (or fills) can be determined by a leveling process, illustrated in Figure 24-5 and the corresponding field notes given in Plate D-9 of Appendix D. The process is summarized as follows:

1. List the stations staked on the pipeline (column 1 of the field notes).
2. Compute the flow line or invert elevation at each station (column 6).
3. Set up the level and get an HI by reading a plus sight on a BM; for example, HI = 2.11 + 100.65 = 102.76 (see Figure 24-5).
4. Obtain the elevation at each station from a rod reading on the ground at every stake (column 4)—for example, 4.07 at station 1 + 00 (see Figure 24-5)—and subtract it from the HI (column 5); for example, 102.76 − 4.07 = 98.69 at station 1 + 00.

Figure 24-5 Leveling process to determine cut or fill and set batter boards for laying pipelines.

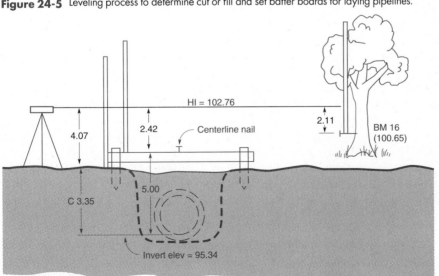

5. Subtract the pipe elevation from the ground elevation to get cut $(+)$ or fill $(-)$ (column 7); for example, $98.69 - 95.34 = C\ 3.35$ (see Figure 24-5).

6. Mark the cut or fill (using a permanent marking felt pen or keel) on an offset stake facing the centerline; the station number is written on the other side.

In another variation, which produces the same results, *grade rod* (difference between HI and pipe invert) is computed, and *ground rod* (reading with rod held at stake) is subtracted from it to get cut or fill. For station $1 + 00$, grade rod $= 102.76 - 95.34 = 7.42$, and $7.42 - 4.07 = C\ 3.35$.

After the trench has been excavated based on cuts and fills marked on the stakes, batter boards are set. Marks needed to place them can be made with a pencil or felt pen on the offset stakes during the same leveling operation used to obtain cut and fill information. Figure 24-5 also illustrates the process. Suppose that at station $1 + 00$, the batter board will be set so its top is exactly 5.00 ft above the pipe invert. The rod reading necessary to set the batter board is obtained by subtracting the pipe invert elevation plus 5.00 ft from the HI; thus $102.76 - (95.34 + 5.00) = 2.42$ ft (see Figure 24-5). The rod is held at the stake and adjusted in vertical position by commands from the level operator until a rod reading of 2.42 ft is obtained; then a mark is made at the rod's base on the stake. (To facilitate this process, a rod target or a colored rubber band can be placed on the rod at the required reading.) The board is then fastened to the stake with its top at the mark using nails or C clamps, and a carpenter's level is used to align it horizontally across the trench. A nail marking the pipe centerline is set by measuring the stake's offset distance along the board.

24-6 STAKING OUT A BUILDING

The first task in staking out a building is to locate it properly on the correct lot by making measurements from the property lines. Most cities have an ordinance establishing setback lines from the street and between houses to improve appearance and provide fire protection.

Stakes may be set initially at the exact building corners as a visual check on the positioning of the structure, but obviously such points are lost immediately when excavation is begun on the footings. A set of batter boards and reference stakes, placed as shown in Figure 24-6, is therefore erected near each corner, but out of the way of construction. The boards are nailed a full number of feet above the footing base or at first-floor elevation. (The procedure of setting boards at a desired elevation was described in the preceding section.) Nails are driven into the batter-board tops so that strings stretched tightly between them define the outside wall or form line of the building. The layout is checked by measuring diagonals and comparing them with each other (for symmetric layouts) or to their computed values.

Figure 24-7 illustrates the placement on a lot and staking of a slightly more complicated building. The following are recommended steps in the procedure:

1. Set hubs *A* and *B* 5.00 ft inside the east lot line, with hub *A* 20.00 ft from the south lot line and hub *B* 70.00 ft from *A*. Mark the points precisely with nails.

2. Set a theodolite or total station over hub *A*, backsight on hub *B*, and turn a 90° angle to set batter-board nails 1 and 2 and stakes *C* and *D*.

3. Set the instrument over hub *B*, backsight on hub *A*, and turn a 90° angle. Set batter-board nails 3 and 4 and stakes *E* and *F*.

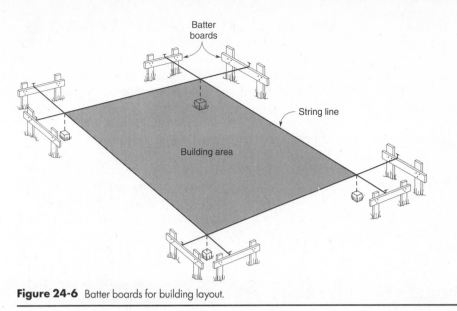

Figure 24-6 Batter boards for building layout.

Figure 24-7 Location of building on lot and batter-board placement.

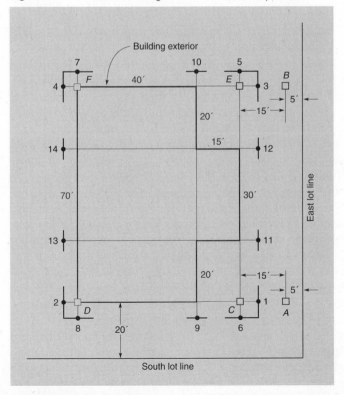

4. Measure diagonals *CF* and *DE,* and adjust if the error is small or restake if large.
5. Set the instrument over *C,* backsight on *E,* and set batter-board nail 5. Plunge the instrument and set nail 6.
6. Set the instrument over *D,* backsight on *F,* and set nail 7. Plunge and set nail 8.
7. Set batter-board nails 9, 10, 11, 12, 13, and 14 by measurements from established points.
8. Stretch the string lines to create the building's outline, and check all diagonals.
(*Note:* When each batter board is set, it must be placed with its top at the proper elevation.)

As an alternative to this building stakeout procedure, radial methods (described in Section 12-10) can be used. This can substantially reduce the number of instrument setups and stakeout time required. In the radial method, coordinates of all building corners are computed in the same coordinate system as the lot corners. Then the theodolite or total station is set on any convenient control point and oriented in azimuth by sighting another intervisible control point. Angles and distances, computed from coordinates, are then laid off to mark each building corner. The layout is checked by measuring diagonals. (An example illustrating radial stakeout of a circular curve is given in Section 25-9.) After constructing the batter boards and setting the cross pieces at the desired elevations, the alignment nails on the batter boards can be set by pulling taught string lines across established corners. In Figure 24-7, for example, with corners *D* and *F* marked, a line stretched across these two points enables placing nails 7 and 8 on the boards. With the strings in place after setting batter-board nails, diagonals between corners should again be checked.

One method of handling difficult jobs is to stake all (or many) points from a single instrument setup using precalculated angles and distances. Figure 24-8 shows an unusual building shape which was laid out rapidly using only two setups by choosing an initial one to reach half the corners, and having the same calculated angles and distances (Figure 24-9) for both ends of the structure. In using this method, it is essential that enough building dimensions be checked by measuring between marked corners to ensure that no large errors or mistakes were made.

To control elevations on a construction site, a bench mark (two or more on large projects) should be set outside the construction limits but within easy sight distance. Rotating lasers (see Section 24-2.1) can be used to control elevations for the tops of footings, floors, an so on.

Permanent foresights are helpful in establishing the principal lines of the structure. Targets or marks on nearby existing buildings can be used if movement due to thermal effects or settlement is considered negligible. On formed concrete structures, such as retaining walls, offset lines are necessary because the outside wall face is obstructed.

Positions of such things as interior footings, anchor bolts for columns, and special piping or equipment can first be marked by 2 × 2-in. stakes with tacks. Survey disks, scratches on bolts or concrete surfaces, and steel pins are also used. Batter boards set inside the building dimensions for column footings have to be removed as later construction develops.

On multistory buildings, care is required to ensure vertical alignment in the construction of walls, columns, elevator shafts, structural steel, etc. One method of checking plumbness of constructed members is to carefully aim a theodolite's line of sight on a reference mark at the base of the member. The line of sight is then plunged to its

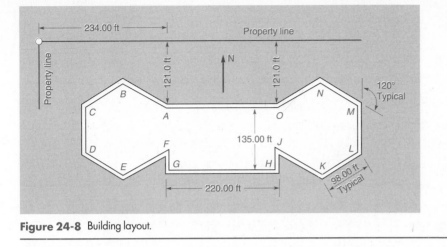

Figure 24-8 Building layout.

top. For an instrument that has been carefully leveled and that is in proper adjustment, the line of sight will define a vertical plane as it is plunged. It should not be assumed that the instrument is in good adjustment, however, and the line should be plunged in both the face left and face right positions. It is necessary to check plumbness in two perpendicular directions when using this procedure. To guide construction of vertical members in real-time, two instruments can be set up with their lines of sight oriented perpendicular to each other, and verticality monitored as construction progresses. Alternatively, lasers (see Section 24-3.1) can be used to guide and monitor vertical construction.

Staking out a building can be a time-consuming and lengthy process if the surveyor does not give sufficient forethought to the basic control points required and the best method to establish them. The number of instrument setups should be minimized to conserve time, and calculations made in the office if possible, rather than in the field while a survey party waits.

24-7 STAKING OUT A HIGHWAY

Alignments for highways, railroads, and other transportation routes are designed after careful study of existing maps, air photos, and preliminary surveys covering the area. From alternative routes, the one best meeting overall objectives while minimizing costs and environmental impacts is selected. Before construction can begin, the surveyor must transfer that alignment (either the centerline or an offset reference line) to the ground.

Normally staking will commence at the initial point where the first straight segment (*tangent*) is run, placing stakes at 100-ft intervals (*full stations*). Stationing (this subject is described in Section 7-9.1) continues until the planned alignment changes direction at the first *point of intersection* (PI). The deflection angle is measured there and the second tangent stationed forward to the next PI, where the deflection angle there is measured. The process continues to the terminal point. Staking continuously from the initial point to the terminus may result in large amounts of accumulated error on long projects. Therefore work should be checked by making frequent ties to intermediate

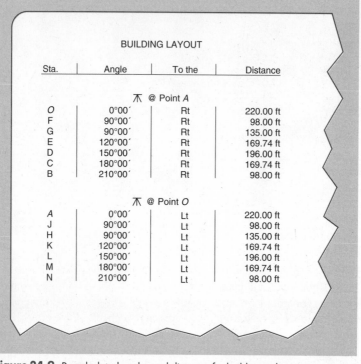

BUILDING LAYOUT

Sta.	Angle	To the	Distance
⊼ @ Point A			
O	0°00′	Rt	220.00 ft
F	90°00′	Rt	98.00 ft
G	90°00′	Rt	135.00 ft
E	120°00′	Rt	169.74 ft
D	150°00′	Rt	196.00 ft
C	180°00′	Rt	169.74 ft
B	210°00′	Rt	98.00 ft
⊼ @ Point O			
A	0°00′	Lt	220.00 ft
J	90°00′	Lt	98.00 ft
H	90°00′	Lt	135.00 ft
K	120°00′	Lt	169.74 ft
L	150°00′	Lt	196.00 ft
M	180°00′	Lt	169.74 ft
N	210°00′	Lt	98.00 ft

Figure 24-9 Precalculated angles and distances for building stakeout.

horizontal control points, and adjustments should be made as necessary. Alternatively, on smaller projects the alignments can be run from both ends to a point near the middle.

After tangents are established, horizontal curves (usually circular arcs) are inserted at all PIs according to plan. The subject of horizontal alignments, including methods for computing and laying out horizontal curves, is discussed in detail in Chapter 25. Vertical alignments are described in Chapter 26.

After the centerline or reference line (including curves), has been established, the PIs, intermediate *points on tangent* (POTs) on long tangents, and points where horizontal curves begin (*PCs*) and end (*PTs*) are referenced using procedures described in Section 12-5. Points used in referencing must be located safely outside the construction limits. Referencing is important because the centerline points will be destroyed during various phases of construction and will need to be replaced several times. Bench marks are also established at regular spacing (usually not more than about 1000 ft apart) along the route. These are placed on the right of way, far enough from the centerline to be safe from destruction, but convenient for access.

After the centerline or reference line has been established, stakes marking the right-of-way should be set. This is normally done by carefully measuring perpendicular offsets from the established reference line. The right-of-way is staked at every change in its width, at all changes in alignment, including each PC and PT, and at sufficient other intermediate points along the tangents so that it is clearly delineated.

When the reference line and right-of-way have been staked, the limits of actual construction are marked so that the contractor can clear to them. Next some contractors

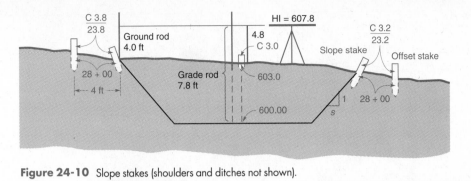

Figure 24-10 Slope stakes (shoulders and ditches not shown).

want points set on the right-of-way with subgrade elevations, showing cut or fill to a given elevation, for use in performing rough grading and preliminary excavation of excess material.

To guide a contractor in making final excavations and embankments, *slope stakes* are driven at the *slope intercepts* (intersections of the original ground and each side slope), or offset a short distance, perhaps 4 ft (see Figure 24-10). The cut or fill at each location is marked on the slope stake. Note that actually there is *no* cut or fill at a slope stake—the value given is the vertical distance from the ground elevation at the slope stake to grade.

Slope stakes can be set at slope intercept locations predetermined in the office from cross-sectional data. (Methods for determining slope intercepts from cross-sections are described in Chapter 27.) If predetermined slope intercepts are used, the ground elevation at each stake must still be checked in the field to verify its agreement with the cross section. If a significant discrepancy in elevation exists, the stake's position must be adjusted by a trial-and-error method to be described. The amount of cut or fill marked on the stake is computed from the actual difference in elevation between the ground elevation at the slope stake and grade elevation.

If slope intercepts have not been precalculated from cross-section data, slope stakes are located by a trial-and-error method based on mental calculations involving HI, grade rod, ground rod, half-roadway width, and side slopes. One or two trials are generally sufficient to fix the stake position within an allowable error of 0.3 to 0.5 ft for rough grading. The infinite number of ground variations prohibits use of a standard formula in slope staking. An experienced surveyor employs only mental arithmetic, without scratch paper or hand calculator. Whether using the method to be described or any other, systematic procedures must be followed to avoid confusion and mistakes.

Grade stakes are set at points that have the same ground and grade elevation. Three transition sections normally occur in passing from cut to fill (or vice versa), and a grade stake is set at each one (see Figures 24-11 and 27-1). A line connecting grade stakes, perhaps scratched out on the ground, defines the change from cut to fill, as line *ABC* in Figure 27-1.

Example 24-1 lists the sequential steps to be taken in slope staking, *assuming, for simplicity, academic conditions* of a level roadway. In practice, travel lanes and shoulders of modern highways have lateral slopes for drainage, then a steeper slope to a ditch in cut, and another slope up the hillside to the slope intercept. Transition sections may have half-roadway widths in cuts different from those in fills to accommodate ditches,

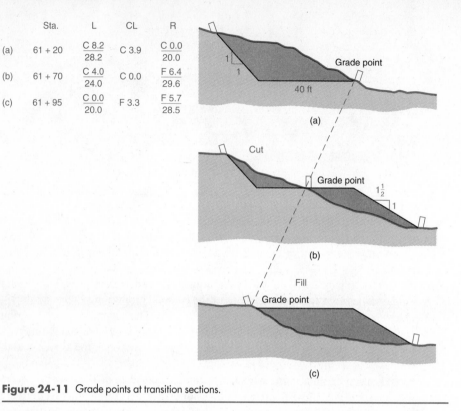

	Sta.	L	CL	R
(a)	61 + 20	$\dfrac{C\,8.2}{28.2}$	C 3.9	$\dfrac{C\,0.0}{20.0}$
(b)	61 + 70	$\dfrac{C\,4.0}{24.0}$	C 0.0	$\dfrac{F\,6.4}{29.6}$
(c)	61 + 95	$\dfrac{C\,0.0}{20.0}$	F 3.3	$\dfrac{F\,5.7}{28.5}$

Figure 24-11 Grade points at transition sections.

and flatter side slopes for fills that tend to be less stable than cuts. But the same basic steps still apply and can be extended by students after learning the fundamental approach.

EXAMPLE 24-1   List the field procedures, including calculations, necessary to set slope stakes for a 40-ft-wide level roadbed with side slopes of 1:1 in cut and fill (see Figures 24-10 and 24-11).

SOLUTION

1. Compute the cut at the centerline stake from profile and grade elevations (603.0 − 600.0 = C 3.0 in Figure 24-10). Check in field by grade rod minus ground rod = 7.8 − 4.8 = C 3.0 ft. Mark the stake C 3.0/0.0. (On some jobs the center stake is omitted and stakes are set only at the slope intercepts.)
2. Estimate the difference in elevation between the left-side slope-stake point (20+ ft out) and the center stake. Apply the difference—say, +0.5 ft—to the center cut and get an estimated cut of 3.5 ft.
3. Mentally calculate the distance out to the slope stake, 20 + 1(3.5) = 23.5 ft, where 1 is the side slope.

4. Hold the zero end of a cloth tape at the center stake while the rodperson goes out at right angles (turned by prism or extended-arm method) with the other end and holds the rod at 23.5 ft.

5. *Forget all previous calculations to avoid confusion of too many numbers and remember only the grade-rod value.*

6. Read the rod with the level and get the cut from grade rod minus ground rod, perhaps $7.8 - 4.0 = C\ 3.8$ ft.

7. Compute the required distance out for this cut, $20 + 1(3.8) = 23.8$ ft.

8. Check the tape to see what is actually being held and find it is 23.5 ft.

9. The distance is within a few tenths of a foot and close enough. Move out to 23.8 ft if the ground is level and drive the stake. Move farther out if the ground slopes up, since a greater cut would result, and thus the slope stake must be beyond the computed distance, or not so far if the ground has begun to slope down, which gives a smaller cut.

10. If the distance has been missed badly, make a better estimate of the cut, compute a new distance out, and take a reading to repeat the procedure.

11. In going out on the other side, the rodperson lines up the center and left-hand slope stake to get the right-angle direction.

12. To locate grade stakes at the road edge, one person carries the zero end of the tape along the centerline while the rodperson walks parallel, holding the 20-ft mark until the required ground-rod reading is found by trial. Note that the grade rod changes during the movement but can be computed at 5- or 10-ft intervals. The notekeeper should have the grade rod listed in the field book for quick reference at full stations and other points where slope stakes are to be set.

13. Grade points on the centerline are located using a starting estimate determined by comparing cut and fill at back and forward stations.

Practice varies for different organizations, but often the slope stake is set 4 ft beyond the slope intercept. It is marked with the required cut or fill, distance out from the centerline to the slope-stake point, side-slope ratio, and base half-width noted on the side facing the centerline. Stationing is given on the back side. A reference stake having the same information on it may also be placed 6 ft or more farther out of the way of clearing and grading. On transition sections, grade-stake points are marked.

Total station instruments, with their ability to automatically reduce measured slope distances to horizontal and vertical components, speed slope staking significantly, especially in rugged terrain where slope intercept elevations differ greatly from centerline grade. GPS receivers operating in the real-time kinematic mode (see Section 24-3.3) can also be advantageously used in these types of terrain if overhead clearances exist so that satellites are visible.

Slope staking should be done with utmost care, for once cut and fill embankments are started, it is difficult and expensive to reshape them if a mistake is discovered.

After rough grading has shaped cuts and embankments to near final elevation, finished grade is set more accurately from *blue tops* (stakes whose tops are driven to grade elevation and then marked with a blue keel or spray paint). These are not normally offset, but rather driven directly on centerline. The procedure for setting blue tops at required grade elevation is described in Section 26-7.

Highway and railroad grades can often be rounded off to multiples of 0.05 or 0.10% without appreciably increasing earthwork costs or sacrificing good drainage. Streets need a minimum 0.50% grade for drainage from intersection to intersection, or from midblock both ways to the corners. They are also crowned to provide for lateral flow to gutters. Drainage profiles, prepared to verify or construct drainage cross sections, can be used to locate drainage structures and easements accurately. An experienced engineer, when asked a question regarding the three most important items in highway work, replied "drainage, drainage, and drainage." This requirement must be satisfied by good surveying and design.

To ensure unobstructed drainage after construction, culverts must be placed in most fill sections so that water can continue to flow in its normal pattern from one side of the embankment to the other. In staking culverts, their locations, skew angles if any, lengths, and invert elevations are taken from the plans. Required pipe alignments and grades are marked using stakes, offset from each end of the pipe's extended centerline. The invert elevation (or an even number of feet above or below it) is noted on the stake. This field procedure, like setting slope stakes, requires marking a point on the stake where a rod reading equals the difference between the required grade and the current HI of a leveling instrument.

After the subgrade has been completed, if the highway is being surfaced with rigid concrete pavement, *paving pins* will be necessary to guide this operation. They are usually about $\frac{1}{2}$-in.-diameter steel rods, driven to mark an offset line parallel to one edge of the required pavement. This line is usually staked at 50-ft increments, but closer spacing may be used on sharp curves. The finished grade (or one parallel to it but offset vertically above) is marked on the pins using tape, or affixing a special stringline holding device. Again in this operation, elevations are set by marking a point on the stake where a rod reading equals the difference between the finished grade and a current HI. (The need for frequent bench marks at convenient locations on a highway project is obvious.)

Utility relocation surveys may be necessary in connection with highway construction; for example, manhole or valve-box covers have to be set at correct grade before earthwork begins so they will conform to finished grade. Here differential elevation resulting from transverse surface slope must be considered. Utilities are located by centerline station and offset distance.

Location staking for railroads, rapid-transit systems, and canals follows the same general methods outlined above for highways.

24-8 OTHER CONSTRUCTION SURVEYS

For the planning and construction of causeways, bridges, and offshore oil platforms, it is often necessary to perform hydrographic surveys (see Section 16-11). These types of projects require special procedures to solve the problem of establishing horizontal positions and depths where it is impossible to hold a rod or reflector. Modern surveying equipment and procedures and sonar mapping devices are used to plot dredging cross sections for underwater trenching and pipe laying. Today more pipelines are crossing wider rivers than ever before. Mammoth pipeline projects now in progress to transport crude oil, natural gas, and water have introduced numerous new problems and solutions. Permafrost, extremely low temperature, and the need to provide animal crossings are examples of special problems that have been encountered in Alaska pipeline construction.

Large earthwork projects such as dams and levees require widespread permanent control for quick setups and frequent replacement of slope stakes, all of which may disappear under fill in one day. Fixed signals for elevation and alignment painted or mounted on canyon walls or hillsides can mark important reference lines. Failures of some large structures, such as the Teton Dam, demonstrate the need for monitoring them periodically so that any necessary remedial work can be done.

Underground surveys in tunnels and mines necessitate transferring lines and elevations from the ground above, often down shafts. Directions of lines in mine tunnels can be most conveniently established using north-seeking gyros (see Section 18-1). In another, and still practiced method, two heavy plumb bobs hung on wires (and damped in oil or water) from opposite sides of the surface opening can be aligned by theodolite there and in the tunnel. (A vertical collimator will also provide two points on line below ground.) A theodolite or laser is wiggled in on the short line defined by the two plumb-bob wires, a station mark set in the tunnel ceiling above the instrument, and the line extended. Later setups are made beneath *spads* (surveying nails with hooks) anchored in the ceiling. Elevations are brought down by taping or other means. Bench marks and instrument stations are set on the ceiling, out of the way of equipment.

Surveys are run at intervals on all large jobs to check progress for periodic payments to the contractor. And finally, an *as-built* survey is made to determine compliance with plans, note changes, and to make terminal contract payment.

Airplane and ship construction requires special equipment and methods as part of a unique branch of surveying, *optical tooling.* The precise location and erection of offshore oil drilling platforms many miles from a coast utilizes new surveying technology, principally GPS satellite positioning systems (see Chapter 20).

24-9 CONSTRUCTION SURVEYS USING TOTAL STATION INSTRUMENTS

In Section 24-2.2 it was noted that total station instruments are very convenient for construction stakeout. This section provides details on their use in these types of surveys. The procedures described here apply to most total station instruments, although some may require interfaced data collectors to perform the operations described.

Before using a total station for stakeout, it is necessary to orient the instrument. Depending on the type of project, *horizontal* or both *horizontal and vertical* orientation may be needed. For example, if just the lot corners of a subdivision are being staked, then only horizontal orientation (establishing the instrument's position and direction of pointing) is needed. If grade stakes are to be set, then the instrument must also be oriented vertically (its HI determined).

With total station instruments, three methods are commonly used for horizontal orientation: **(1)** *azimuth,* **(2)** *coordinates,* and **(3)** *resection.* The first two apply where an existing control point is occupied, and the latter is used when the instrument is set up at a noncontrol point. In azimuth orientation, the coordinates of the occupied control station and the known azimuth to a backsight station are entered into the instrument. If the occupied station's coordinates have been downloaded into the instrument prior to going into the field, it is only necessary to input its point number. The backsight station is then sighted, and when completed, the azimuth of the line is transferred to the total station by a keyboard stroke, whereupon it appears in the display.

The coordinate method of orientation uses the same approach, except that the coordinates of both the occupied and the backsight station are entered. Again these data could have been downloaded previously so that it would only be necessary to key in the numbers identifying the two stations. The instrument computes the backsight line's azimuth from the coordinates, displays it, and prompts the operator to sight the backsight station. Upon completion of the backsight, the azimuth is transferred to the instrument with a keystroke, and it appears on the display.

In the resection procedure, a station whose position is unknown is occupied and the instrument's position determined by sighting two or more control stations. This is very convenient on projects where a certain point of high elevation in an open area gives good visibility to all (or most) points to be staked. As noted, two or more control points must be sighted. Measurements of angles or of angles and distances are made to the control stations. The microprocessor then computes the instrument's position by the method of least squares. Orientation could be determined, for example, using the three-point resection method (illustrated in Figure 16-5(f) and demonstrated with an example in Section 19-11).

Project conditions will normally dictate which orientation procedure to use. The resection method should definitely be kept in mind, however, because it often provides the most efficient approach to staking. Regardless of the procedure selected, after orientation is completed, a check should be made by sighting another control point and comparing the observed azimuth and distance against their known values. If there is a discrepancy, the orientation procedure should be repeated. It is also a good idea to recheck orientation at regular intervals after stakeout has commenced, especially on large projects. In fact, if possible a reflector should be left on a control point just for that purpose.

Vertical orientation of a total station can be achieved using one of two procedures. The simplest case occurs if the elevation of the occupied station is known, as then it is only necessary to measure *carefully* and add the *hi* (height of instrument above the point) to the elevation of the point. If the occupied station's elevation is unknown, then another station of known elevation must be sighted. The situation is illustrated in Figure 24-12, where the instrument is located at station A of unknown elevation, and station B whose elevation is known is sighted. From slope distance S and zenith angle z the instrument computes V. Then its HI is

$$\text{HI} = \text{elev}_B + h_p - V \qquad (24\text{-}1)$$

where h_p is the prism height above station B.

As with horizontal orientation, it is good practice to check the instrument's vertical orientation by sighting a second vertical control point.

Once orientation is completed, project stakeout can begin. In general, staking is either a two- or three-dimensional problem. Staking lots of a subdivision or layout of construction alignments is generally two-dimensional. Slope staking, blue-top setting, pipeline layout, and batter-board placement require both horizontal position and elevation and are therefore three-dimensional.

For two-dimensional stakeout, after the file of coordinates for all control stations and points to be staked is downloaded and the instrument is oriented horizontally, the identifying number of a point to be staked is entered into the instrument through the keyboard. The microprocessor immediately calculates the horizontal distance and azi-

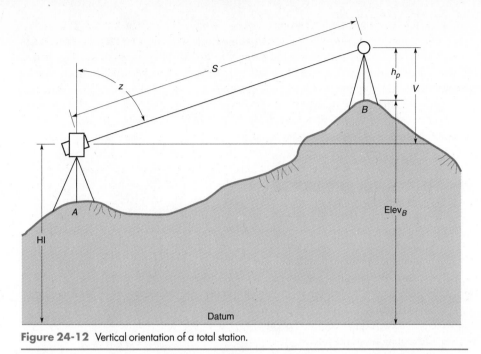

Figure 24-12 Vertical orientation of a total station.

muth required to stake the point. The difference between the instrument's current direction of pointing and that required is displayed. The operator turns the telescope until the difference becomes zero to achieve the required direction. The reflector is then directed onto this alignment and a horizontal distance reading taken, whereupon the difference ΔD between it and that required is displayed. The reflector is then directed inward or outward, as necessary, until the distance difference is zero and the stake placed there. A two-way radio is invaluable for communicating with the reflector person in this operation. This procedure for stakeout is discussed further in Section 25-12 and an example problem presented.

Special tracking systems have been developed to aid the reflector person in getting on line. The Lumiguide system of Nikon's Top Gun total station, for example, utilizes prisms to emit a solid beam of red light to the right of the line of sight, and a flashing pattern to the left. An operator at the prism sees one or the other and immediately knows what direction to move to get on line.

For three-dimensional staking, the total station must be oriented vertically as well as horizontally. The initial part of three-dimensional stakeout is exactly like that described for the two-dimensional procedure; that is, the horizontal position of the stake is set first. Then simultaneously with the measurement of the stake's horizontal position, its vertical component, and thus its elevation is determined. The difference ΔZ between the required elevation and the stake's elevation is displayed with a plus or minus sign, the former indicating fill, the latter cut. This information is communicated to the reflector person for marking the stake, or incrementally driving it further down until the required grade is reached.

With the high order of accuracy possible using total station instruments, stakes quite distant from the instrument can be laid out, and thus many points set from a single setup. Often, in fact, an entire project can be staked from one location. This is made possible in many cases because of the flexibility that resection orientation provides in instrument placement. It should be remembered that if long distances are involved in three-dimensional staking, earth curvature and refraction can become factors that must be accounted for (see Section 6-4). Also with total stations, each point is set independently of the others, and thus no inherent checks are available. Checks should therefore be made by either repeating the measurements, checking placements from different control stations, or measuring between staked stations to ascertain their relative accuracies.

24-10 SOURCES OF ERROR

Important sources of error in construction surveys are:

1. Inadequate number and/or location of control points on the job site
2. Errors in establishing control
3. Measurement errors in layout
4. Failure to double-center in laying out angles or extending lines, and failure to check vertical members by plunging the instrument
5. Careless referencing of key points
6. Movement of stakes and marks
7. Failure to use tacks for proper line where justified

24-11 MISTAKES

Typical mistakes often made in construction surveys are:

1. Lack of foresight as to where construction will destroy points
2. Notation for cut (or fill) and stationing on stake not checked
3. Wrong datum for cuts, whether cut is to finished grade or subgrade
4. Arithmetic mistakes, generally due to lack of checking
5. Use of incorrect elevations, grades, and stations
6. Failure to check the diagonals of a building
7. Carrying out computed values beyond field accuracy possible (one good hundredth is better than all the bad thousandths)
8. Reading the rod on top of stakes instead of on the ground beside them in profiling and slope making

PROBLEMS

24-1 Give examples of construction projects where visible laser-beam instruments are useful for stakeout.
24-2 Discuss how line and grade can be set with a total station instrument operated in the tracking mode.
24-3 For what types of stakeout projects, or conditions, is real-time kinematic GPS surveying most advantageous?
24-4 What problem limits the accuracy with which elevations can be staked using GPS? How can this problem be overcome?
24-5 Should stakes for constructing a sewer on a curve be closer together or farther apart than for a straight section? Explain.

24-6 State two conflicting requirements that enter the decision on how far offset stakes should be set beyond the construction line.

24-7 Why is it good practice in staking grades by differential leveling to initiate the leveling circuit on one bench mark and close on another?

*24-8 A sewer pipe is to be laid from station 10 + 00 to station 13 + 20 on a −0.80% grade, starting with invert elevation 852.30 ft at 10 + 00. Calculate invert elevations at each 50-ft station along the line.

24-9 A sewer pipe must be laid from a starting invert elevation of 1360.75 ft at station 9 + 25 to an ending invert elevation of 1354.10 ft at station 14 + 10. Determine the uniform grade needed, and calculate invert elevations at each 50-ft station.

*24-10 Grade stakes for a pipeline running between stations 0 + 00 and 6 + 46 are to be set at each full station. Elevations of the pipe invert must be 1057.14 ft at station 0 + 00 and 1050.94 ft at 6 + 46, with a uniform grade between. After staking an offset centerline, an instrument is set up nearby, and a backsight of 4.60 taken on BM *A* (elevation 1062.14 ft). The following foresights are taken with the rod held on ground at each stake: (0 + 00, 5.51); (1 + 00, 5.67); (2 + 00, 5.03); (3 + 00, 7.16); (4 + 00, 7.92); (5 + 00, 8.80); (6 + 00, 9.10); and (6 + 46, 9.25). Prepare a set of suitable field notes for this project (see Plate D-9) and compute the cut required at each stake. Close the level circuit back to the bench mark.

*24-11 If batter boards are to be set exactly 8.00 ft above the pipe invert at each station on the project of Problem 24-10, calculate the necessary rod readings for placing the batter boards. Assume the instrument has the same HI as in Problem 24-10.

24-12 How are streets and street grades arranged for drainage in a city with flat terrain?

24-13 List the most important considerations in designing a grade line to fit a ground profile.

24-14 By means of a sketch, show how and where batter boards should be located: **(a)** For an L-shaped building **(b)** For an E-shaped structure

*24-15 A building in the shape of an L must be staked. Corners *ABCDEF* all have right angles. Proceeding clockwise around the building, the required outside dimensions are *AB* = 100.00 ft, *BC* = 40.00 ft, *CD* = 45.00 ft, *DE* = 60.00 ft, *EF* = 55.00 ft, and *FA* = 100.00 ft. After staking the batter boards for this building and stretching string lines taut, check measurements of the diagonals should be made. What should be the values of *AC, AD, AE, FB, FC, FD,* and *BD*?

24-16 Compute the floor area of the building in Problem 24-15.

*24-17 Compute the floor area of the building in Figure 24-8.

*24-18 The design floor elevation for a building to be constructed is 1086.40. An instrument is set up nearby, leveled, and a backsight of 4.62 taken on BM *A* whose elevation is 1087.09 ft. If batter boards are placed exactly 1.00 ft above floor elevation, what rod readings are necessary on the batter-board tops to set them properly?

24-19 Compute the diagonals necessary to check the stakeout of the building in Figure 24-7.

24-20 Can the corner of a building be plumbed using a theodolite? Explain.

24-21 Which building line of a structure is shown on a record plat, the outside or the center-line of a wall?

24-22 Discuss the suitability of a 0.00% grade for a street or highway.

24-23 A rapid-transit horizontal alignment survey is to be run by deflection angles. Which is the most important theodolite adjustment to check prior to performing this work?

24-24 Discuss the importance of tying in and referencing critical centerline points on high-way construction surveys.

24-25 Explain why slope stakes are placed at an offset distance from slope intercepts. What offset distance is recommended?

24-26 What information is normally lettered on slope stakes?

24-27 Describe a field procedure for setting slope stakes.

24-28 Discuss the procedure and advantages of using total station instruments for slope staking.

*24-29 A highway centerline subgrade elevation is 985.20 ft at station 12 + 00 and 993.70 ft at 17 + 00 with a smooth grade in between. To set blue tops for this portion of the

*Asterisks indicate problems that have answers given in Appendix G.

centerline, a level is set up in the area and a backsight of 6.19 ft taken on a bench mark whose elevation is 990.05 ft above sea level. From that HI, what rod readings will be necessary to set the blue tops for the full stations from 12 + 00 through 17 + 00?

24-30 Similar to Problem 24-29, except the elevations at stations 12 + 00 and 17 + 00 are 1713.35 and 1707.10 ft, respectively, the BM elevation is 1710.84 ft, and the backsight is 5.28 ft.

24-31 Describe the procedure for staking out a highway culvert.

24-32 Explain the resection procedure of orienting a total station instrument for construction stakeout.

BIBLIOGRAPHY

Barry, B. A. 1988. *Construction Measurements,* 2nd ed. New York: John Wiley and Sons.

Braunstein, P. 1986. "An Introduction to Industrial Surveying." *Point of Beginning* 10 (No. 1):38.

Brinker, R. C., and R. Minnick. 1987. *The Surveying Handbook.* New York: Van Nostrand Reinhold, chap. 26.

Culver, G. 1988. "Staking—Pioneers of Construction." *Professional Surveyor* 8 (No. 6):26.

Evans, M. 1984. "Subway Surveying." *Point of Beginning* 9 (No. 6):6.

Greulich, G. 1978. "Underground Utilities." *Bulletin, American Congress on Surveying and Mapping* (No. 60):7.

Harrelson, N. 1987. "Computer-Aided Surveying in Urban Building Construction." *ASCE, Journal of Surveying Engineering* 113 (No. 1):50.

Hole, S. W. 1979. "Pipelines—Positioning and Survey Problems." *ASCE, Journal of the Surveying and Mapping Division* 105 (No. SU1):15.

Hopper, A. G. 1982. "Laser Technology and the Inverted Plumb Bob." *Professional Surveyor* 2 (No. 2):107.

Kavanagh, B. 1992. *Surveying with Construction Applications.* Englewood Cliffs, N.J.: Prentice-Hall.

Keene, D. F. 1974. "Precise Dam Surveys—Los Angeles County Flood Control District." *ASCE, Journal of the Surveying and Mapping Division* 100 (No. SU2):99.

Kellie, A. C. et al. 1984. "Trilateration as Basis for Radial Staking." *ASCE, Journal of the Surveying Engineering Division* 110 (No. 1):49.

Posch, M. 1980. "Survey Work in Constructing the Arlberg Road Tunnel." *Bulletin, American Congress on Surveying and Mapping* (No. 69):19.

Roth, A., and I. Gonzalez. 1989. "Rotating Beam Laser Leveling Systems—Applications for Surveyors." *Point of Beginning* 14 (No. 1):78.

Thompson, B. J. 1974. "Planning Economical Tunnel Surveys." *ASCE, Journal of the Surveying and Mapping Division* 100 (No. SU2):95.

Vanden Berg, J., and A. Lindberg. 1983. *Measuring Practice on the Building Site.* Gaule, Sweden: National Swedish Institute for Building Research, Bulletin M83:16.

Warne, T. R. 1990. "Contractor Staking: Saving Public Works Dollars." *ASCE, Journal of Surveying Engineering* 116 (No. 2):82.

25

HORIZONTAL CURVES

25-1 INTRODUCTION

Straight (tangent) sections of most types of transportation routes, such as highways, railroads, and pipelines, are connected by curves in both the horizontal and vertical planes. An exception is a transmission line, in which a series of straight lines is used with abrupt angular changes at tower locations if needed.

Curves used in horizontal planes to connect two straight tangent sections are called *horizontal curves.* Two types are used: circular arcs and spirals. Both are readily laid out in the field with standard surveying equipment. A *simple curve* [Figure 25-1(a)] is a circular arc connecting two tangents. It is the type most often used. A *compound curve* [Figure 25-1(b)] is composed of two or more circular arcs of different radii tangent to each other, with their centers on the same side of the alignment. The combination of a short length of tangent (less than 100 ft) connecting two circular arcs that have centers on the same side [Figure 25-1(c)] is called a *broken-back curve.* A *reverse curve* [Figure 25-1(d)] consists of two circular arcs tangent to each other, with their centers on opposite sides of the alignment. Compound, broken-back, and reverse curves are unsuitable for modern high-speed highway, rapid-transit, and railroad traffic, and should be avoided if possible. They are sometimes necessary, however, in mountainous terrain to avoid excessive grades or very deep cuts and fills. Compound curves

Figure 25-1 Circular curves.

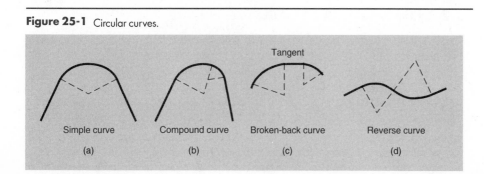

| Simple curve | Compound curve | Broken-back curve | Reverse curve |
| (a) | (b) | (c) | (d) |

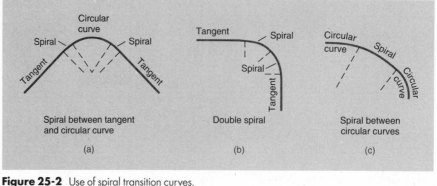

Figure 25-2 Use of spiral transition curves.

are often used on lower speed exit and entrance ramps of interstate highways and expressways.

Easement curves are desirable, especially for railroads and rapid-transit systems, to lessen the sudden change in curvature at the junction of a tangent and a circular curve. A *spiral* makes an excellent easement curve because its radius decreases uniformly from infinity at the tangent to that of the curve it meets. Spirals are used to connect a tangent with a circular curve, a tangent with a tangent (double spiral), and a circular curve with a circular curve. Figure 25-2 illustrates these arrangements.

The effect of centrifugal force on a vehicle passing around a curve can be balanced by *superelevation,* which raises the outer rail of a track or outer edge of a highway pavement. Correct transition into superelevation on a spiral increases uniformly with the distance from the beginning of the spiral, and is in inverse proportion to the radius at any point. Properly superelevated spirals ensure smooth and safe riding with less wear on equipment. As noted, spirals are used for railroads and rapid-transit systems. This is because trains are constrained to follow the tracks, and thus a smooth, safe, and comfortable ride can only be assured with properly constructed alignments that include easement curves. On highways, spirals are less frequently used because drivers are able to overcome abrupt directional changes at circular curves by steering a spiraled path as they enter and exit the curves.

Although this chapter concentrates on circular curves, methods of computing and laying out spirals are briefly introduced in Section 25-16.

25-2 DEGREE OF CIRCULAR CURVE

In European practice and for the majority of American highway work, circular curves are designated by their radius—for example, 1500-m curve or 1000-ft curve. American railroads and some highway departments prefer to identify curves by their *degree,* using either arc or chord definition.

In most highway work, *degree of curve* is the central angle subtended by a circular *arc* of 100 ft. This is called the *arc definition* [Figure 25-3(a)]. In railroad practice (and in early highway construction) the degree of curve has been the angle at the center of a circular arc subtended by a *chord* of 100 ft. This is the *chord definition* [Figure 25-3(b)]. The formulas relating radius R and degree D are shown next to the illustrations.

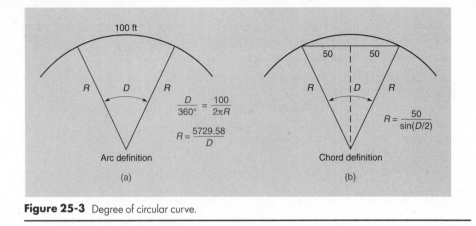

Figure 25-3 Degree of circular curve.

Using the equations given in Figure 25-3, radii of arc- and chord-definition curves for values of D from 1 to 10° have been computed and are given in columns (2) and (5) of Table 25-1. Although radius differences between the two definitions appear to be small in this range, they are significant.

Arc-definition curves have the advantage that computations are somewhat simplified as compared to the chord definition, and, as will be shown later, the formula for curve length is exact, which simplifies preparing right-of-way descriptions. A disadvantage with the arc definition is that most measurements between full stations are shorter than a full 100-ft tape length, but this is of no consequence if a total station instrument is used for stakeout. With the chord definition, full stations are separated by chords of exactly 100.00 ft regardless of the value of D.

For a given value of D, arc and chord definitions give practically the same result when applied to the flat curves common on modern highways, railroads, and rapid-transit systems.

TABLE 25-1 FUNCTIONS OF CIRCULAR CURVES (LENGTHS IN FEET)

DEGREE OF CURVE D	ARC DEFINITION			CHORD DEFINITION		
	RADIUS R	TRUE CHORD, FULL STA	TRUE CHORD, HALF-STA	RADIUS R	ARC LENGTH, FULL STA	TRUE CHORD, HALF-STA
(1)	(2)	(3)	(4)	(5)	(6)	(7)
1	5729.58	100.00	50.00	5729.65	100.00	50.00
2	2864.79	99.99	50.00	2864.93	100.01	50.00
3	1909.86	99.99	50.00	1910.08	100.01	50.00
4	1432.39	99.98	50.00	1432.69	100.02	50.01
5	1145.92	99.97	50.00	1146.28	100.03	50.01
6	954.93	99.95	50.00	955.37	100.05	50.02
7	818.51	99.94	50.00	819.02	100.06	50.02
8	716.20	99.92	49.99	716.78	100.08	50.03
9	636.62	99.90	49.99	637.27	100.10	50.04
10	572.96	99.88	49.98	573.69	100.13	50.05

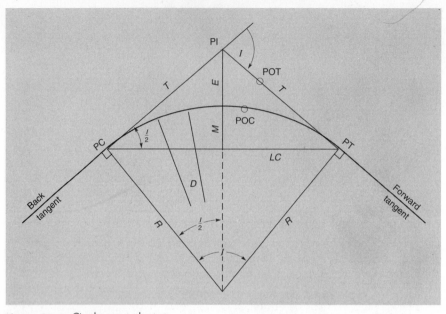

Figure 25-4 Circular curve elements.

25-3 DEFINITIONS AND DERIVATION OF CIRCULAR CURVE FORMULAS

Circular curve elements are shown in Figure 25-4. The *point of intersection* PI of the two tangents is also called the *vertex V*. The *back tangent* precedes the PI, the *forward tangent* follows it. The beginning of the curve, or *point of curvature* PC, and the end of the curve, or *point of tangency* PT, are also sometimes called BC and EC, respectively. Other expressions for these points are *tangent to curve* TC and *curve to tangent* CT. The curve radius is *R*. Note that the radii at the PC and PT are perpendicular to the back tangent and forward tangent, respectively.

The distance from PC to PI and from PI to PT is the *tangent distance T*. A line connecting the PC and PT is the *long chord* LC. The *length of the curve L* is the distance from PC to PT measured along the curve for the arc definition, or by 100-ft chords for the chord definition.

The *external distance E* is the length from vertex to curve midpoint on a radial line. The *middle ordinate M* is the (radial) distance from the midpoint of the long chord to the curve's midpoint. Any *point on curve* is POC; any *point on tangent*, POT. The degree of any curve is D_a (*arc definition*) or D_c (*chord definition*). The change in direction of two tangents is the intersection angle *I*, which is also equal to the central angle subtended by the curve.

By definition, and from inspection of Figure 25-4, relations for the *arc definition* follow:

$$L = 100 \frac{I^\circ}{D^\circ} \quad \text{(ft)} \qquad (25\text{-}1)$$

$$L = \frac{I^\circ}{D^\circ} \quad \text{(sta)} \qquad (25\text{-}1a)$$

$$L = RI \quad (I \text{ in radians}) \tag{25-2}$$

$$R = \frac{5729.58}{D} \text{ (ft)} \tag{25-3}$$

$$T = R \tan \frac{I}{2} \tag{25-4}$$

$$LC = 2R \sin \frac{I}{2} \tag{25-5}$$

$$\frac{R}{R+E} = \cos \frac{I}{2} \quad \text{and} \quad E = R \left[\frac{1}{\cos (I/2)} - 1 \right] \tag{25-6}$$

$$\frac{R-M}{R} = \cos \frac{I}{2} \quad \text{and} \quad M = R \left(1 - \cos \frac{I}{2} \right) \tag{25-7}$$

Other convenient formulas that can be derived:

$$E = T \tan \frac{I}{4} \tag{25-8}$$

$$M = E \cos \frac{I}{2} \tag{25-9}$$

If a hand-held calculator is used, it is expedient to compute R, T, E, and M in the sequence of Eqs. (25-3), (25-4), (25-8), and (25-9), because the previously computed value is in the calculator and available for each succeeding calculation.

The formulas for T, L, LC, E, and M also apply to a chord-definition curve. However, L calculated by Eq. (25-1) implies the total length as if measured along the 100-ft chords of an inscribed polygon. Also, the following formula is used for relating R and D for a chord-definition curve:

$$R = \frac{50}{\sin (D/2)} \tag{25-10}$$

25-4 CIRCULAR CURVE STATIONING

Normally an initial route survey consists of establishing PIs, laying out tangents, and establishing continuous stationing along them from the start of the project through each PI to the project end. After this, I angles are measured and curves computed and staked. The stationing of points on any curve is based on that of its PI. To compute the PC station, tangent distance T is subtracted from the PI station, and to calculate the PT station, curve length L is added to the PC station.

Assume that $I = 8°24'$, the station of the PI is 64 + 27.46, and terrain conditions require the maximum degree of curve permitted by the specifications, which is, say, 2°00' (arc definition). Calculate the PC and PT stationing, and the external and middle ordinate distances for this curve.

EXAMPLE 25-1

SOLUTION

By Eq. (25-1), $L = 100 \times \dfrac{8.40°}{2°} = 420.00$ ft.

By Eq. (25-3), $R = \dfrac{5729.58}{2} = 2864.79$ ft.

By Eq. (25-4), $T = 2864.79 \times 0.73435 = 210.38$ ft.
Calculate stationing:

$$
\begin{array}{rcl}
\text{PI station} & = & 64 + 27.46 \\
-T & = & \underline{2 + 10.38} \\
\text{PC station} & = & 62 + 17.08 \\
+L & = & \underline{4 + 20.00} \\
\text{PT station} & = & 66 + 37.08
\end{array}
$$

Also, by Eq. (25-8), $E = 210.38 \times 0.036668 = 7.71$ ft
and by Eq. (25-9), $M = 7.71 \times 0.99731 = 7.69$ ft.

Calculation for the stations of the PC and PT should be arranged as shown. Note that the station of the PT *cannot* be obtained by adding the tangent distance to the station of the PI, although the location of the PT on the ground is determined by measuring the tangent distance from the PI. Points representing the PC and PT must be carefully marked and placed exactly on the tangent lines at the correct distance from the PI, so other computed values will fit their fixed positions on the ground.

Route surveys are often staked originally as a series of tangents having continuous stationing. Then an adjustment has to be made at each PT after curves are inserted which shorten the route. Thus for final stationing at the PT, there is a "station equation," which relates the stationing *back along the curve* to that *forward along the tangent*. For Example 25-1 it would be $66 + 37.08$ back $= 66 + 37.84$ ahead, where $66 + 37.08 = \text{PI} - T + L$, and $66 + 37.84 = \text{PI} + T$. The difference between the ahead and back stations represents the amount the route was shortened. If the curves are run in and stationed at the time of staking the original alignment, continuous stationing along the route results, and station equations at PTs are avoided.

The curve used in a particular situation is selected to fit ground conditions and specification limitations on maximum D or minimum R. Normally the value of I and the station of the PI are available from field measurements on the preliminary line. Then a value of D or R, suitable for the highway or railroad, is chosen. Sometimes the distance E or M required to miss a stream or steep slope outside or inside the PI is measured, and D or R computed. Tangent distance governs infrequently. (One exception is to make a railroad, bus, or subway station fall on a tangent rather than a superelevated curve.) The length of curve practically never governs.

25-5 GENERAL PROCEDURE OF CIRCULAR CURVE LAYOUT BY DEFLECTION ANGLES WITH THEODOLITE AND TAPE

Except for unusual cases, the radii of curves on route surveys are too large to permit swinging an arc from the curve center. Circular curves are therefore laid out by

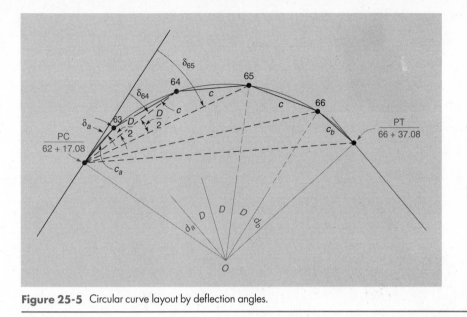

Figure 25-5 Circular curve layout by deflection angles.

(1) deflection angles, (2) tangent offsets, (3) chord offsets, (4) middle ordinates, and (5) coordinates. Stakeout by deflection angles has been the standard method.

Layout of a curve by deflection angles can be done using either a theodolite (or transit) and tape, or a total station. Methods vary somewhat depending on the instrument used. The procedure using a theodolite and tape herein called the *incremental chord method* is illustrated in Figure 25-5 and will be discussed first. Layout procedures when using a total station instrument, herein called the *total chord method,* are described in Section 25-10.

In Figure 25-5 assume that a theodolite (or transit) is set up over the PC (station 62 + 17.08 in Example 25-1). Each full station is to be marked along the curve, since cross sections are normally taken, construction stakes set, and computations of earthwork made at these points. (Half-stations or any other critical points can also be established, of course.) The first station to be set in this example is 63 + 00. To mark that point from the PC, a backsight is taken on the PI with zero set on the instrument's horizontal circle. Deflection angle δ_a to station 63 + 00 is then turned; two tapepersons measure chord c_a from the PC and set 63 + 00 at the end of the chord on the instrument's line of sight. With station 63 + 00 set, the tapepersons measure the chord length c from it and stake station 64 + 00, where the line of sight of the instrument, now set to δ_{64}, intersects the end of that chord. This process is repeated until the entire curve is laid out.

25-6 COMPUTING DEFLECTION ANGLES AND CHORDS

From the preceding discussion it is clear that deflection angles and chords are important values which must be calculated if a curve is to be run by the deflection-angle method. To stake the first station, which is normally an odd distance from the PC (shorter than a full-station increment), *subdeflection* angle δ_a and *subchord* c_a are needed. These are

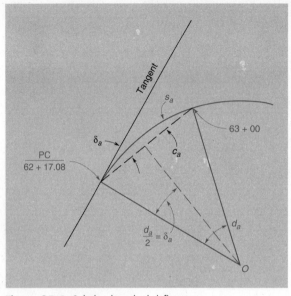

Figure 25-6 Subchords and subdeflections.

detailed in Figure 25-6. In this figure, central angle d_a subtended by arc s_a from the PC to 63 + 00 is calculated by proportion according to the definition of D as

$$\frac{d_a}{s_a} = \frac{D}{100} \quad \text{from which} \quad d_a = \frac{s_a D}{100} \tag{25-11}$$

where s_a is the difference in stationing between the two points. A fundamental theorem of geometry helpful in circular curve computation and stakeout is that *the angle at a point between a tangent and any chord is equal to half the central angle subtended by the chord.* Thus subdeflection angle δ_a needed to stake station 63 + 00 is $d_a/2$, or

$$\delta_a = \frac{s_a D}{200} \quad \text{(degrees)} \tag{25-12}$$

Since subdeflection angles are usually small and expressed in minutes, Eq. (25-12) becomes

$$\delta_a = 0.3 s_a D \quad \text{(minutes)} \tag{25-13}$$

The length of the subchord c_a can be represented in terms of δ_a and the curve radius as

$$\sin \delta_a = \frac{c_a}{2R} \quad \text{from which} \quad c_a = 2R \sin \delta_a \tag{25-14}$$

Since the arc between full stations subtends the central angle D, from the earlier stated geometric theorem, deflection angles to each full station beyond 63 + 00 are found by adding $D/2$ to the previous deflection angle. Full chord c, which corresponds to 100 ft of curve length, is calculated using Eq. (25-14), except that $D/2$ is substituted for δ_a. Equations (25-13) and (25-14) are also used to compute the last subdeflection angle δ_b and subchord c_b, but the difference in stationing s_b between the last full station and the PT replaces arc length s_a.

Compute subdeflection angles and subchords δ_a, c_a, δ_b, and c_b and also calculate chord c of Example 25-1.

EXAMPLE 25-2

SOLUTION
By Eq. (25-13),

$$\delta_a = 0.3(82.92)2° = 49.75' = 0°49'45''$$
$$\delta_b = 0.3(37.08)2° = 22.25' = 0°22'15''$$

By Eq. (25-14),

$$c_a = 2(2864.79) \sin 0°49'45'' = 82.92 \text{ ft}$$
$$c_b = 2(2864.79) \sin 0°22'15'' = 37.08 \text{ ft}$$
$$c = 2(2864.79) \sin 1°00'00'' = 99.99 \text{ ft}$$

For curves up to about 2°00' (arc definition), the lengths of arcs and their corresponding chords are nearly the same. On sharper curves chords are shorter than corresponding arc lengths. This is verified by the data in columns (3) and (4) of Table 25-1, which give true chord lengths for full- and half-station increments for varying values of D (arc definition).

An alternate method of computing deflection angles is to multiply the deflection angle per foot of arc or chord by the length of the arc or chord being laid out. If 1 ft is substituted for s_a in Eq. (25-13), then the deflection (in minutes) per foot of stationing is simply $0.3D$. In Example 25-2 the deflection angle per foot of arc is $0.3 \times 2.00 = 0.60'/\text{ft}$. Then the deflection angle δ_a from the PC to station $63 + 00$ is $0.60 \times 82.92 = 49.75' = 49'45''$.

Computations for deflection angles and chords on chord-definition curves use the same formulas, but R is calculated by Eq. (25-10). Note that for a given degree of curve, R is longer for chord definition than for arc definition. Also arc lengths for full stations are longer than their nominal 100.00-ft value, and true subchords are longer than their "nominal values" (differences in stationing). A check of columns (6) and (7) in Table 25-1 verifies these facts.

25-7 NOTES FOR CIRCULAR CURVE LAYOUT BY DEFLECTION-ANGLE AND INCREMENTAL CHORD METHOD

Based on principles discussed, the deflection-angle and incremental chord data for stakeout of the complete curve of Examples 25-1 and 25-2 with theodolite and tape have been computed and listed in Table 25-2. Normally, as has been done in this case, the data are prepared for stakeout from PC, although field conditions may not allow the curve to be completely run from there. This problem is discussed in Section 25-9.

TABLE 25-2 DEFLECTION-ANGLE AND INCREMENTAL CHORD DATA FOR EXAMPLE CURVE

STATION	INCREMENTAL CHORD	DEFLECTION INCREMENT	DEFLECTION ANGLE
66 + 37.08 (PT)	37.08	0°22'15"	4°12'00"
66 + 00	99.99	1°00'00"	3°49'45"
65 + 00	99.99	1°00'00"	2°49'45"
64 + 00	99.99	1°00'00"	1°49'45"
63 + 00	82.92	0°49'45"	0°49'45"
62 + 17.08 (PC)			

Values of deflection angles are normally carried out to several decimal places for checking purposes, and to avoid accumulating small errors when D is a noninteger number, such as, perhaps, $3°17.24'$. (In early railroad work with $1'$ transits, D was generally rounded off to a multiple of $2'$.) Note in Table 25-2 that the deflection angle to the PT is $4°12'$, exactly half the I angle of $8°24'$. This comparison affords an important check on the calculations of all deflection angles.

Field notes for the curve of this example are recorded in Plate D-10 in Appendix D as they would appear in a field book. Notes run up the page to simplify sketching while looking in a forward direction. Digital computers can conveniently perform all necessary computations and list the notes for curve stakeout by deflection angles.

In many cases it is desirable to *back in* a curve by setting up over the PT instead of the PC. One setup is thereby eliminated, and the long sights are taken on the first measurements. In precise work it is better to run in the curve from both ends to the center, where small errors can be adjusted more readily. On long or very sharp circular curves, or if obstacles block sights from the PC or PT, setups on the curve are necessary (see Section 25-9).

25-8 DETAILED PROCEDURES FOR CIRCULAR CURVE LAYOUT BY DEFLECTION ANGLES WITH A THEODOLITE AND TAPE

Regardless of the method used to stake intermediate curve points, the first steps in curve layout are: **(1)** establishing the PC and PT, normally by measuring tangent distance T from the PI along both the back and forward tangents, and **(2)** measuring the total deflection angle at the PC from PI to PT. This latter step should be performed whenever possible, since the measured angle must check $I/2$; if it does not, an error exists in either measurement or computation, and time should not be wasted running an impossible curve.

It is also good practice to stake the midpoint of a curve before beginning to set intermediate points, especially on long curves. The midpoint can be set by bisecting the angle $180° - I$ at the PI and laying off the external distance from there. A check of the deflection angle from the PC to the midpoint should yield $I/4$. When staking intermediate points along the curve has reached the midpoint, a chord check measurement should be made to it.

The remaining steps in staking intermediate curve points by the deflection-angle and incremental chord method with a theodolite (or transit) and tape are presented with reference to the curve of Examples 25-1 and 25-2. With the instrument set up and leveled over the PC, it is oriented by backsighting on the PI, or on a point along the back tangent, with 0°00′ on the circle. The subdeflection angle of 49′45″ is then turned. Meanwhile the 17-ft mark of the tape is held on the PC. The 100-ft end of the (cut) tape is swung until the line of sight hits a point 0.08 ft back from the 100-ft mark. This is station 63 + 00. To stake station 64 + 00, the rear tapeperson next holds the zero mark on station 63, and the forward tapeperson sets station 64 at distance 99.99 ft by direction from the instrument operator, who has placed an angle of 1°49′45″ on the circle. An experienced forward tapeperson will walk along the extended first full chord, know or estimate the chord offset, and from an outside-the-chord position will be holding the tape end and a stake within a foot or so of the correct location when the instrument operator has the deflection angle ready.

After placing the final full station (66 + 00 in this example), to determine any misclosure in staking a curve, the *closing* PT should be staked using the final deflection angle and subchord. This will rarely agree with the PT established by measuring distance *T* along the forward tangent from PI because of accumulated errors. Any misclosure ("falling") should be measured; then the field precision can be expressed as a numerical ratio like that used in traverse checks. The measured falling distance is the numerator, and $L + 2T$ the denominator. If the misclosure of this example was 0.25 ft, the precision would be 0.25/(420.00 + 2 × 210.38) = 1/3300.

25-9 SETUPS ON CURVE

Obstacles that prevent visibility along curve chords and extremely long sight distances sometimes make it necessary to set up on the curve after it has been partially staked. The simplest procedure to follow is one that permits use of the same notes computed for running the curve from PC. In this method the following rule applies: *The instrument is moved forward to the last staked point, backsighted on any previously set curve station with the telescope inverted and the circle set to the deflection angle from the PC for the station sighted. The telescope is then plunged to the normal position, and deflection angles previously computed from the PC for the various stations are used.*

The example of the preceding sections is used to illustrate this rule. If a setup is required at station 65, place 0°00′ on the instrument and sight to PC with the telescope inverted. Plunge, set the angle to read the deflection angle 3°49′45″, and stake station 66.

To prove this geometric rule, assume the instrument were set on station 65 + 00 and the PC backsighted with 0°00′00″ on the circle. If the telescope were plunged and turned 2°49′45″ in azimuth, the line of sight would then be tangent to the curve. To stake the next station (66 + 00), an additional angle of *D*/2 or 1°00′00″ would need to be turned. The sum of 2°49′45″ and 1°00′00″ is of course the deflection angle to station 66 + 00 from the PC. The rule also applies for backsights to any previously set stations, not just the PC. Thus with the instrument at 65 + 00, station 63 + 00 could be sighted with 0°49′45″ on the circle and then turned to 3°49′45″ to set station 66 + 00. Further study of the geometry illustrated in Figure 25-5 should clarify this procedure.

25-10 CIRCULAR CURVE LAYOUT BY DEFLECTION ANGLES WITH TOTAL STATION INSTRUMENTS

The total station instruments described in Chapter 10 are extremely convenient for curve layout by deflection angles. With these instruments the field party size is reduced from three to two persons. Deflection angles are calculated and laid off as in the preceding example, but the chords are all measured electronically as radial distances (total chords) from the PC or other station where the instrument is placed. If stakeout is planned from the PC, total chords from there are the dashed lines of Figure 25-5. They are calculated by Eq. (25-14), except that the deflection angle for each station is substituted for δ_a to obtain the corresponding chord. The total chords necessary to stake the curve of the preceding example using a total station instrument set up at PC are 82.92 ft for 63 + 00, 182.89 ft for 64 + 00, 282.80 ft for 65 + 00, 382.63 ft for 66 + 00, and 419.62 ft for PT, which is the long chord LC given by Eq. (25-5).

To stake curves using a total station, the instrument is placed in its tracking mode. The deflection angle to each station is turned and the required chord to that station entered in the instrument. The instrument operator directs the person with the reflector to the proper alignment. The reflector is then moved forward or back as necessary, until the proper total chord distance is achieved, where the stake is set.

If intermediate setups are required on the curve using this method, the instrument is oriented as described in Section 25-9. New radial chords to be measured from the intermediate station would then have to be calculated.

Although curves can be staked rapidly with total stations using this method, an associated danger is that each stake is set individually and does not depend on previous stations. Thus a check at the end of the curve is not achieved, as in the incremental chord method with theodolite and tape described in Section 25-8, and mistakes in angles or distances at intermediate stations could go undetected. Large blunders can usually be discovered by visual inspection of the curve stakes, but quickly taping the chords between adjacent stations gives a better check.

25-11 CIRCULAR CURVE LAYOUT BY OFFSETS

For short curves, when a theodolite or transit is not available, or for checking purposes, one of four offset-type methods can be used for circular curves: *tangent offsets* (TO), *chord offsets* (CO), *middle ordinates* (MO), and *ordinates from the long chord*. Figure 25-7 shows the relationship of CO, TO, and MO. Visually and by formula comparison, the chord offset for full station is

$$CO = 2c \sin \frac{D}{2} = \frac{c^2}{R} = 2TO \approx 8MO \tag{25-15}$$

Since $\sin 1° = 0.0175$ (approx.), $CO = c(0.0175)D = 0.01\frac{3}{4}D$ (approx.), where D is in degrees and decimals.

The middle ordinate m for any subchord is $R(1 - \cos \delta)$, with δ being the deflection angle for the chord. A useful equation in the layout or checking of curves in place is

$$D \text{ (degrees)} = m \text{ (inches) for a 62-ft chord (approx.)} \tag{25-16}$$

The geometry of the tangent offset method is shown in Figure 25-8. The figure illustrates that the curve is most conveniently laid out in both directions from PC and PT to a common point near the middle of the curve. This procedure avoids long measurements and provides a check where small adjustments can be made more easily if necessary. To lay out a curve using this method, tangent distances are measured to established temporary points A, B, and C in Figure 25-8. From these points, right-angle

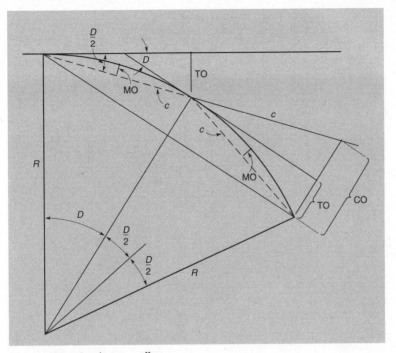

Figure 25-7 Circular curve offsets.

measurements (tangent offsets) are made to set the curve stakes. Tangent distances (TD) and tangent offsets (TO) are calculated using chords and angles in the following formulas:

$$TD = c \cos \delta \tag{25-17}$$

$$TO = c \sin \delta \tag{25-18}$$

where δ angles are calculated using either Eq. (25-12) or Eq. (25-13), and chords c are determined from Eq. (25-14).

Compute and tabulate the data necessary to stake, by tangent offsets, the half-stations of a circular curve having $I = 11°00'$, $D_c = 5°00'$ (chord definition), and PC = 77 + 80.00.

EXAMPLE 25-3

SOLUTION

By Eq. (25-1), the curve length is $L = 100\left(\dfrac{11}{5}\right) = 220$ ft.

Therefore the PT station is $(77 + 80) + (2 + 20) = 80 + 00$. Intermediate stations to be staked are 78 + 00, 78 + 50, 79 + 00, and 79 + 50, as shown in Figure 25-8.

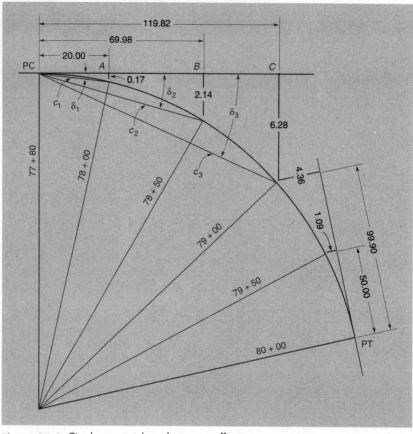

Figure 25-8 Circular curve stakeout by tangent offsets.

By Eq. (25-13), δ angles from the PC are

$$\delta_1 = 0.3(20)5 = 0°30'$$
$$\delta_2 = 0.3(70)5 = 105' = 1°45'$$
$$\delta_3 = 0.3(120)5 = 180' = 3°00'$$

By Eq. (25-10), the radius is

$$R = \frac{50}{\sin 2°30'} = 1146.28 \text{ ft}$$

By Eq. (25-14), chords from the PC are

$$c_1 = 2(1146.28) \sin 0°30' = 20.00 \text{ ft}$$
$$c_2 = 2(1146.28) \sin 1°45' = 70.01 \text{ ft}$$
$$c_3 = 2(1146.28) \sin 3°00' = 119.98 \text{ ft}$$

TABLE 25-3 TANGENT OFFSET DATA FOR EXAMPLE 25-3

STATION	DEFLECTION ANGLE δ	CHORD c	TANGENT DISTANCE $c \cos \delta$	TANGENT OFFSET $c \sin \delta$
80 + 00 (PT)				
79 + 50	1°15'	50.01	50.00	1.09
79 + 00	2°30'	100.00	99.90	4.36
79 + 00	3°00'	119.98	119.82	6.28
78 + 50	1°45'	70.01	69.98	2.14
78 + 00	0°30'	20.00	20.00	0.17
77 + 80 (PC)				

Now, using Eqs. (25-17) and (25-18), tangent distances and tangent offsets are calculated. Chords, δ angles, tangent distances, and tangent offsets to stake points from the PT are computed in the same manner. All data for the problem are listed in Table 25-3. The tangent distances tabulated are lengths from the PC or PT that must be measured to establish points *A, B, C,* and so on, and the tangent offsets are distances from these points needed to locate the curve stakes. Accurate curve layout by tangent offsets generally requires more instrument setups and a greater amount of time than stakeout by deflection angles.

25-12 CIRCULAR CURVE LAYOUT BY COORDINATES WITH TOTAL STATION INSTRUMENTS

The coordinate method can be used to advantage in many situations if a total station instrument is available. In this procedure the coordinates of each curve station to be staked are calculated. The instrument is placed at a nearby control station which gives a good vantage point of the entire area where the curve will be laid out. Azimuths and distances to each station are computed by inversing, using the coordinates of the occupied station and those of each curve station. The instrument is oriented by backsighting another visible control station, and the curve is staked by "radial" methods, measuring the computed azimuth and distance to each of the curve points.

Figure 25-9 illustrates a situation where a curve is being staked by the coordinate method. The total station instrument is placed at control station *B* because all curve points are visible from there. After a backsight on control station *A*, radial methods of measuring distance and direction are used to stake all curve points. The computations necessary for staking a curve by the coordinate method are illustrated with the following example.

Two tangents intersect at a PI station of 45 + 45.50 whose coordinates are $X = 5723.18$ ft and $Y = 3728.94$ ft. The intersection angle is 24°32' left, and the azimuth of the back tangent is 326°40'20". A curve of 5°20' (chord definition) will be used to join the tangents. Compute the data needed to stake the curve at half-station increments by coordinates using a total station. For staking, the instrument will be set at station *B*,

EXAMPLE 25-4

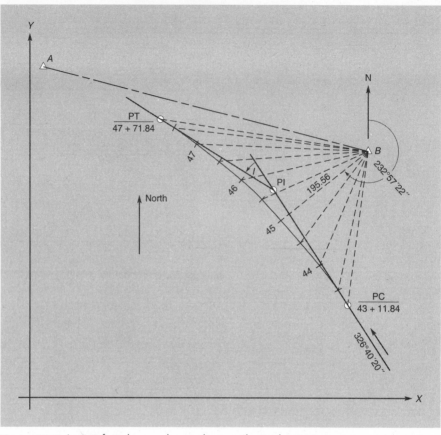

Figure 25-9 Layout of circular curve by coordinates with a total station instrument.

whose coordinates are $X = 5891.03$ ft and $Y = 3798.95$ ft, and a backsight taken on station A, whose coordinates are $X = 5308.29$ ft and $Y = 3945.10$ ft.

SOLUTION

By Eq. (25-1a), $L = \dfrac{24°32'}{5°20'} = 4.600$ sta.

By Eq. (25-10), $R = \dfrac{50}{\sin(2°40')} = 1074.68$ ft.

By Eq. (25-4), $T = 1074.68 \tan (12°16') = 233.66$ ft.

Curve stationing:

$$
\begin{array}{rl}
\text{PI} = & 45 + 45.50 \\
-T = & 2 + 33.66 \\
\hline
\text{PC} = & 43 + 11.84 \\
+L = & 4 + 60.00 \\
\hline
\text{PT} = & 47 + 71.84
\end{array}
$$

A tabular solution for curve point coordinates is given in Table 25-4. The differences in stationing from one curve point to the next are listed in column (2). Total deflection

TABLE 25-4 COMPUTATIONS FOR STAKING CURVE OF EXAMPLE 25-4 BY COORDINATES

STATION (1)	STATION DIFFERENCE (2)	TOTAL DEFLECTION (3)	TOTAL CHORD (4)	CHORD AZIMUTH (5)	ΔX (6)	ΔY (7)	X (8)	Y (9)
						COORDINATES		
43 + 11.84							5851.56	3533.71
43 + 50	38.16	1°01'03"	38.17	325°39'17"	−21.53	31.52	5830.03	3565.23
44 + 00	88.16	2°21'03"	88.17	324°19'17"	−51.42	71.62	5800.14	3605.33
44 + 50	138.16	3°41'03"	138.11	322°59'17"	−83.14	110.28	5768.42	3643.99
45 + 00	188.16	5°01'03"	187.98	321°39'17"	−116.62	147.43	5734.94	3681.14
45 + 50	238.16	6°21'03"	237.76	320°19'17"	−151.81	182.99	5699.75	3716.70
46 + 00	288.16	7°41'03"	287.40	318°59'17"	−188.60	216.86	5662.96	3750.57
46 + 50	338.16	9°01'03"	336.88	317°39'17"	−226.92	248.99	5624.64	3782.70
47 + 00	388.16	10°21'03"	386.19	316°19'17"	−266.71	279.30	5584.85	3813.01
47 + 50	438.16	11°41'03"	435.28	314°59'17"	−307.85	307.73	5543.71	3841.44
47 + 71.84	460.00	12°16'00"	456.66	314°24'20"	−326.24	319.53	5525.32	3853.24

angles are calculated using Eq. (25-13) and tabulated in column (3). Total chords, computed from Eq. (25-14) using these total deflection angles, are tabulated in column (4). From the azimuth of the back tangent and the deflection angles, the azimuth of each total chord is calculated and tabulated in column (5).

The coordinates of the PC are computed from the coordinates of the PI by applying the departure and latitude from the PI to the PC. The departure and latitude are computed by Eqs. (13-1) and (13-2) using the tangent distance and the back azimuth of the back tangent. Thus

$$X_{PC} = 5723.18 + 233.66 \sin(326°40'20'' - 180°) = 5851.56 \text{ ft}$$
$$Y_{PC} = 3728.94 + 233.66 \cos(326°40'20'' - 180°) = 3533.71 \text{ ft}$$

Using their lengths and azimuths, the departure ΔX and latitude ΔY of each total chord are calculated. These are added to the coordinates of the PC to obtain the coordinates of the curve points. Values of ΔX and ΔY are tabulated in columns (6) and (7), and the X and Y coordinates are listed in columns (8) and (9) of Table 25-4.

A check on the PT coordinates of Table 25-4 can be obtained by computing them independently using the azimuth and tangent length of the forward tangent. The azimuth of the forward tangent is calculated by subtracting the I angle from the back tangent's azimuth,

$$Az = 326°40'20'' - 24°32' = 302°08'20''$$

Then the X and Y coordinates of the PT are

$$X_{PT} = 5723.18 + 233.66 \sin(302°08'20'') = 5525.32 \text{ ft (check)}$$

$$Y_{PT} = 3728.94 + 233.66 \cos(302°08'20'') = 3853.24 \text{ ft (check)}$$

Computations for the lengths and azimuths of the radial lines needed to stake curve points from station B are listed in Table 25-5. Column (1) gives the curve stations, and columns (2) and (3) list the differences ΔX and ΔY between each curve point's X and Y

TABLE 25-5 COMPUTATIONS FOR RADIAL LENGTHS AND AZI-
MUTHS FOR STAKING CURVE OF EXAMPLE 25-4 BY COORDINATES

STATION (1)	ΔX (2)	ΔY (3)	L (4)	Az (5)
43 + 11.84 (PC)	− 39.47	− 265.24	268.16	188°27′50″
43 + 50	− 61.00	− 233.72	241.55	194°37′40″
44 + 00	− 90.89	− 193.62	213.89	205°08′47″
44 + 50	− 122.61	− 154.96	197.60	218°21′08″
45 + 00	− 156.09	− 117.81	195.56	232°57′22″
45 + 50	− 191.28	− 82.25	208.22	246°43′57″
46 + 00	− 228.07	− 48.38	233.14	258°01′25″
46 + 50	− 266.39	− 16.25	266.89	266°30′33″
47 + 00	− 306.18	14.06	306.50	272°37′45″
47 + 50	− 347.32	42.49	349.91	276°58′29″
47 + 71.84 (PT)	− 365.71	54.29	369.72	278°26′38″

coordinates and those of station B. Radial lengths computed from Eq. (13-16b) and azimuths computed by Eq. (13-13) are tabulated in columns (4) and (5).

For orienting the instrument it is necessary to calculate the azimuth of line BA. This, also by Eq. (13-13), is

$$Az_{BA} = \tan^{-1}\left(\frac{1{,}875{,}308.29 - 1{,}875{,}891.03}{103{,}945.10 - 103{,}798.95}\right) = 284°04′46″$$

After a backsight on station A, a value of 284°04′46″ is indexed on the total station's horizontal circle. Then each curve point is staked by measuring its radial distance and azimuth taken from Table 25-5. The radial lines are shown dashed in Figure 25-9. Note that to stake station 45 + 00, for example, a distance of 195.56 ft is measured on an azimuth of 232°57′22″, as shown in the figure.

While the calculations needed to stake a curve by coordinates may seem rather extensive, they are routinely handled by computers. After calculating the curve coordinate data, with most total stations and data collectors it is only necessary to download a file of point identifiers and coordinates for the job. Then in the field, the numbers of the station occupied and the point to be staked are entered via the instrument's keyboard. The microprocessor immediately displays the needed distance and azimuth for staking the point.

During stakeout the instrument is operated in the tracking mode. Note that each curve point is staked independently of the others, and thus there is no misclosure at the end to verify the accuracy of the work. Thus it must be done very carefully.

25-13 SPECIAL CIRCULAR CURVE PROBLEMS

Many special problems arise in the design and computation of circular curves. Three of the more common ones are discussed here, and each can be solved using the coordinate geometry formulas given in Appendix B.

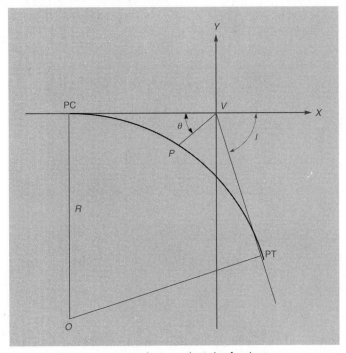

Figure 25-10 Passing a circular curve through a fixed point.

25-13.1 Passing a Circular Curve Through a Fixed Point

One problem that often occurs in practice is to determine the radius of a curve connecting two established tangents and going through a fixed point such as an underpass, overpass, or existing bridge. The problem can be solved by establishing an X-Y coordinate system, as shown in Figure 25-10, where the origin occurs at V (the PI) and X coincides with the back tangent. Coordinates of the radius point in this system are $X_O = -R \tan (I/2)$ and $Y_O = -R$. From measurements of distance PV and angle θ, the coordinates X_p and Y_p of point P can be determined. Then the following equation for a circle, obtained by substitution into Eq. (B-6) of Appendix B, can be written:

$$R^2 = \left(X_p + R \tan \frac{I}{2} \right)^2 + (Y_p + R)^2 \qquad (25\text{-}19)$$

With X_p, Y_p, and $I/2$ known, a solution for R can be found. The equation is quadratic, but can be solved using Eq. (B-8) of Appendix B.

25-13.2 Intersection of a Circular Curve and a Straight Line

Another frequently encountered problem involving curve computation is the determination of the intersection point of a circular curve and a straight line. An example is presented in Figure 25-11. The problem is solved by writing an equation for the straight line, Eq. (B-5), and one for the circle, Eq. (B-6), in which the unknown coordinates X_p and Y_p of the intersection point are included. The two equations are then solved

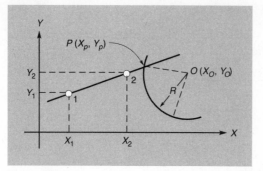

Figure 25-11 Intersection of circular curve and straight line.

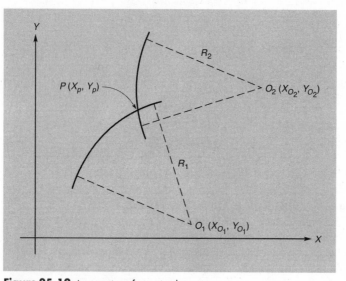

Figure 25-12 Intersection of two circular curves.

simultaneously for X_p and Y_p. In typical cases, coordinates X_1, Y_1, X_2, Y_2, X_O, and Y_O are known, as well as R. An example similar to that illustrated in Figure 25-11 is solved in Appendix *B*.

25-13.3 Intersection of Two Circular Curves

Figure 25-12 illustrates another common problem: computing the intersection point of two circular curves. This can also be handled by coordinate geometry and requires writing two circle equations [see Eq. (B-6) of Appendix B] in which unknown coordinates X_p and Y_p are included. The equations are solved to determine the unknowns. Coordinates X_{O1}, Y_{O1}, X_{O2}, and Y_{O2} are typically determined through survey, and R_1 and R_2 are selected based on design or topographic constraints. An example similar to that illustrated in Figure 25-12 is given in Appendix B.

The problems described in this section and the preceding one arise most often in the design of subdivisions and interchanges, and in calculating right-of-way points along highways and railroads.

25-14 COMPOUND AND REVERSE CURVES

Compound and reverse curves are combinations of two or more circular curves. They should be used only for low-speed traffic routes, and in terrain where simple curves cannot be fitted to the ground without excessive construction costs.

Special formulas have been derived to facilitate computations for such curves and are demonstrated in texts on route surveying. A compound curve can be staked with instrument setups at the beginning PC and ending PT, or perhaps with one setup at the *point of compound curvature* (PCC). Reverse curves are handled in similar fashion.

25-15 SIGHT DISTANCE ON HORIZONTAL CURVES

Highway safety requires certain minimum sight distances in zones where passing is permitted, and in nonpassing areas to assure a reasonable stopping distance if there is an object on the roadway. Specifications and tables list suitable values based on vehicular speeds, the perception and reaction times of an average individual, the braking distance for a given coefficient of friction during deceleration, and type and condition of the pavement.

A minimum stopping sight distance of 450 to 550 ft is desirable for a speed of 55 mi/hr. An approximate formula for determining the available sight distance can be derived from Figure 25-13, in which the clear sight distance past an obstruction is the length of the long chord *AS*, denoted by *C;* and the required clearance is the middle ordinate *PM,* denoted by *m*. Then in triangles *SPG* and *SOH,*

$$m : SP = \frac{SP}{2} : R \quad \text{and} \quad m = \frac{(SP)^2}{2R}$$

Usually *m* is small compared with *R*, and *SP* may be assumed equal to *C*/2. Then

$$C = \sqrt{8mR} \tag{25-20}$$

Figure 25-13 Sight distance.

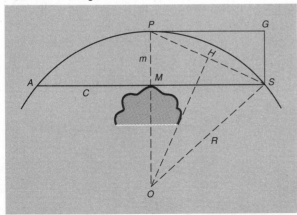

If distance *m* from the centerline of a highway to the obstruction is known or can be measured, the available sight distance *C* is calculated from the formula. Actually cars travel on either the inside or the outside lane, so sight distance *AS* is not exactly the true stopping distance, but the computed length is on the safe side and satisfactory for practical use. If the calculation reveals a sight distance restriction, the obstruction might possibly be removed, or a safe speed limit posted in the area.

25-16 SPIRALS

As noted in Section 25-1, spirals are used to provide gradual transitions in horizontal curvature. Their most common use is to connect straight sections of alignment with circular curves, thereby lessening the sudden change in direction that would otherwise occur at the point of tangency. Since spirals introduce curvature gradually, they afford the logical location for introducing superelevation to offset the centrifugal force experienced by vehicles entering curves.

25-16.1 Spiral Geometry

Figure 25-14 illustrates the geometry of spirals connecting tangents with a circular curve of radius *R* and degree of curvature *D*. The *entrance spiral* at the left begins on the back tangent at the TS (*tangent to spiral*) and ends at the SC (*spiral to curve*). The circular curve runs from the SC to the beginning of the *exit spiral* at the CS (*curve to spiral*), and the exit spiral terminates on the forward tangent at the ST (*spiral to tangent*).

The entrance and exit spirals are geometrically identical. Their length L_s is the arc distance from the TS to the SC, or CS to ST. The designer selects this length to provide sufficient distance for introducing the curve's superelevation.

If a tangent to the entrance spiral (and curve) at the SC is projected to the back tangent, it locates the *spiral point of intersection* SPI. The angle at the SPI between the two tangents is the *spiral angle* Δ_s. From the basic property of a spiral, stated in Section 25-1 (*its radius changes uniformly from infinity at TS to the radius of the circular curve at the SC*), it follows that the spiral's degree of curve changes uniformly from 0^0 at the TS to *D* at the SC. Since the change is uniform, the average degree of curve over the spiral's length is *D*/2. Thus from the definition of degree of curve, the spiral angle Δ_s is

$$\Delta_s = L_s\frac{D}{2} \tag{25-21}$$

where L_s is in stations and Δ_s and *D* are in degrees.

Assume in Figure 25-15 that *M* is at the midpoint of the spiral, so its distance from the TS is $L_s/2$. If this reasoning is continued, the degree of curvature at *M* is *D*/2, the average degree of curvature from the TS to *M* is (*D*/2)/2 = *D*/4, and the spiral angle Δ_M is

$$\Delta_M = \frac{L_s}{2}\frac{D}{4} = \frac{L_s D}{8} \tag{a}$$

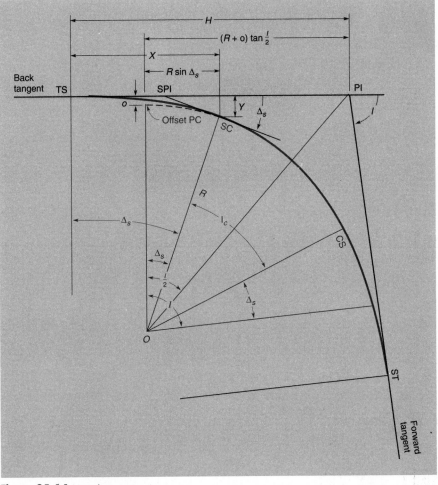

Figure 25-14 Spiral geometry.

Solving for D in Eq. (25-21) and substituting into Eq. (a) gives

$$\Delta_M = \frac{L_s}{8}\frac{2\Delta_s}{L_s} = \frac{\Delta_s}{4} \qquad \text{(b)}$$

According to Eq. (b), at $L_s/2$, the spiral angle is $\Delta_s/4$. This exemplifies another basic property of a spiral: *spiral angles at any point are proportional to the square of the distance from TS to the point*, or

$$\Delta_p = \left(\frac{L_p}{L_s}\right)^2 \Delta_s \qquad \text{(25-22)}$$

where Δ_p is the spiral angle at any point P whose distance from the TS is L_p.

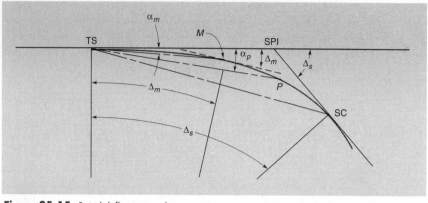

Figure 25-15 Spiral deflection angles.

25-16.2 Spiral Calculation and Layout

Spirals can be laid out by deflection angles and chords in a manner similar to that used for circular curves. In Figure 25-15 α_p is the deflection angle from the TS to any point P. From calculus it can be shown that for the gradual spirals used on transportation routes, *deflection angles are very nearly one-third their corresponding spiral angles.* Thus

$$\alpha_p = \left(\frac{L_p}{L_s}\right)^2 \frac{\Delta_s}{3} \qquad (25\text{-}23)$$

In Eq. (25-23) L_p is the distance from the TS to P, which is simply the difference in stationing from the TS to the point.

In Figure 25-14 the position of the SC is given by X and Y coordinates. In this coordinate system, the origin is at the TS, and the X axis coincides with the back tangent. It can be shown from calculus that

$$X = L_s(100 - 0.0030462\Delta_s^2) \quad \text{(ft)} \qquad (25\text{-}24)$$

$$Y = L_s(0.58178\Delta_s - 0.000012659\Delta_s^3) \quad \text{(ft)} \qquad (25\text{-}25)$$

where L_s is in stations and Δ_s is in degrees.

When a spiral is inserted ahead of a circular curve, as illustrated in Figure 25-14, the circular curve is shifted inward by an amount o, known as the *throw*. This can be visualized by constructing the circular curve back from the SC to the *offset* PC (the point where a tangent to the curve is parallel to the back tangent). The perpendicular distance from the offset PC to the back tangent is the throw, which from Figure 25-14 is

$$o = Y - R(1 - \cos \Delta_s) \qquad (25\text{-}26)$$

To lay out a spiral in the field, distance H of Figure 25-14 from PI to TS is measured back along the tangent to locate the TS. Then the SC can be established by laying off X and Y. From the geometry of Figure 25-14 distance H is

$$H = X - R \sin \Delta_s + (R + o) \tan \frac{I}{2} \qquad (25\text{-}27)$$

The following example will illustrate the computations required in laying out a spiral by the deflection-angle method.

A spiral of 300-ft length is used for transition into a 3°00′ circular curve (arc definition). Angle I at the PI station of $20 + 00$ is $60°00′$. Compute and tabulate the deflection angles and chords necessary to stake out this spiral at half-stations.

EXAMPLE 25-5

SOLUTION

By Eq. (25-3), $R = \dfrac{5729.58}{3.00} = 1909.86$ ft.

By Eq. (25-21), $\Delta_s = \dfrac{3(3.00)}{2} = 4.5° = 4°30′$.

By Eqs. (25-24) and (25-25),

$$X = 3[100 - 0.0030462(4.5)^2] = 299.81 \text{ ft}$$
$$Y = 3[0.58178(4.5) - 0.000012659(4.5)^3] = 7.86 \text{ ft}$$

By Eq. (25-26), $o = 7.86 - 1909.86(1 - \cos 4°30′) = 1.97$ ft
By Eq. (25-27),

$$H = 299.81 - 1909.86 \sin 4°30′ + (1909.86 + 1.97) \tan 30° = 1253.75 \text{ ft}$$

Calculate stationing:

$$
\begin{aligned}
\text{PI station} &= 20 + 00.00 \\
-H &= 12 + 53.75 \\
\hline
\text{TS station} &= 7 + 46.25 \\
+L_s &= 3 + 00.00 \\
\hline
\text{SC station} &= 10 + 46.25
\end{aligned}
$$

The deflection angles calculated using Eq. (25-23) are listed in column (3) of Table 25-6. The values for L_p used in the equation are given in column (2). If a total station is used for stakeout with a setup at the TS, the total chords of column (2) are used with the deflection angles of column (3). If a theodolite and tape are used, the deflection angles, and incremental chords listed in column (4) are used. The chords in columns (2) and (4) are simply station differences between points on the spiral, and are nearly exact for the relatively flat curvature of highway and railroad spirals.

After progressing through the spiral, the SC is finally staked. The falling between its position and the SC set by coordinates should be measured, the precision calculated, and a decision made to accept or repeat the work.

To continue staking the alignment, the instrument is moved forward to the SC. With $2\Delta_s/3$ set on the horizontal circle, a backsight is taken on the TS and the line of sight plunged. Turning the instrument to $0°00′$ orients its tangent to the circular curve, ready for laying off deflection angles. The circular curve is computed and laid out by methods given in earlier sections of this chapter, except that its intersection angle I_c is given by

$$I_c = I - 2\Delta_s \tag{25-28}$$

TABLE 25-6 DATA FOR STAKING SPIRAL OF EXAMPLE 25-5

STATION (1)	TOTAL CHORD DISTANCE FROM TS (ft) (2)	DEFLECTION ANGLE (3)	INCREMENTAL CHORD (ft) (4)
10 + 46.25 (SC)	300.00	1°30.0′	46.25
10 + 00	253.75	1°04.4′	50.00
9 + 50	203.75	0°41.5′	50.00
9 + 00	153.75	0°23.6′	50.00
8 + 50	103.75	0°10.8′	50.00
8 + 00	53.75	0°02.9′	50.00
7 + 50	3.75	0°00.0′	3.75
7 + 46.25 (TS)			

When staking reaches the circular curve's end, the CS is set. The exit spiral is calculated by the same methods described for the entrance spiral and laid out by staking back from the ST.

Spirals can be computed and staked by several different methods. In this brief treatment, only one commonly applied procedure has been discussed. Students interested in further study of spirals can consult the text on route surveying listed in the Bibliography at the end of this chapter.

25-17 SOURCES OF ERROR

Some sources of error in curve layout are:

1. Errors in setting up, leveling, and reading the instrument
2. Total station, theodolite, or transit out of adjustment
3. Bull's-eye bubble out of adjustment on prism pole used for stakeout with a total station

25-18 MISTAKES

Typical mistakes that occur in laying out a curve in the field are:

1. Failure to take equal numbers of direct and reversed measurements of the deflection angle at the PI before computing or laying out the curve
2. Adding the tangent distance to the PI station to get the PT station
3. Using 100.00-ft chords to lay out arc-definition curves having D greater than 2°
4. Taping subchords of nominal length for chord-definition curves having D greater than 5° (a *nominal* 50-ft subchord for a 6° curve requires a measurement of 50.02 ft)
5. Failure to check curve points staked using a total station and either total chord or the coordinate method
6. Mistake in backsight when staking by the coordinate method

7. Failure to stake the PT independently by measuring the tangent distance forward from the PI

25-1 For the following circular curves having a radius R, what is their degree of curve by **(1)** arc definition and **(2)** chord definition?
   *(a) 750 ft   (b) 1000 ft   (c) 3500 ft

25-2 Describe and sketch compound, broken back, and reverse curves. Why are they objectionable for transportation route alignments?

Compute T, L, E, M, R or D, and stations of the PC and PT for the circular curves in Problems 25-3 through 25-6. Distances are in feet. Use the chord definition for railroad curves and the arc definition for highways.

25-3 Railroad curve with $D_c = 2°00'$, $I = 14°00'$, and PI station $= 56 + 45.00$
*25-4 Highway curve with $D_a = 3°40'$, $I = 24°20'$, and PI station $= 50 + 62.48$
25-5 Highway curve (arc definition) with $R = 1500.00$, $I = 18°30'$, and PI station $= 65 + 17.50$
*25-6 Railroad curve (chord definition) with $R = 2500.00$, $I = 18°15'$, and PI station $= 123 + 24.80$.

Tabulate R or D, T, L, E, M, PC, PT, deflection angles, and incremental chords to lay out the circular curves at full stations in Problems 25-7 through 25-14. Use the arc definition for highways and the chord definition for railroads.

*25-7 Highway curve with $D_a = 3°10'$, $I = 32°30'$, and PI station $= 29 + 64.20$
25-8 Railroad curve with $D_c = 2°10'$, $I = 18°18'$, and PI station $= 58 + 65.42$
25-9 Highway curve (arc definition) with $R = 2600$ ft, $I = 16°00'$, and PI station $= 132 + 81.61$
25-10 Railroad curve (chord definition) with $R = 2400$ ft, $I = 14°50'$, and PI station $= 106 + 32.72$
25-11 Highway curve (arc definition) with $L = 850$ ft, $R = 2800$ ft, and PI station $= 86 + 35.00$
*25-12 Railroad curve with $L = 1000$ ft, $D_c = 1°50'$, and PI station $= 39 + 52.90$
25-13 Highway curve (arc definition) with $T = 305.00$ ft, $R = 1200$ ft, and PI station $= 87 + 20.75$
*25-14 Railroad curve with $T = 235.00$ ft, $D_c = 2°35'$, and PI station $= 106 + 95.12$

In Problems 25-15 through 25-18 tabulate the curve data, deflection angles, and total chords needed to lay out the following circular curves at full-station increments using a total station instrument set up at the PC.

25-15 The curve of Problem 25-7
25-16 The curve of Problem 25-8
*25-17 The curve of Problem 25-9
25-18 The curve of Problem 25-10
25-19 A streetcar line on the center of an 80-ft street makes a 70°24' turn into another street of equal width. The corner curb line has $R = 12$ ft. What is the largest R that can be given a circular curve for the track centerline if the law requires it to be at least 15 ft from the curb?

Tabulate all data required to lay out by deflection angles and incremental chords, at full-station increments, the circular curves of Problems 25-20 and 25-21.

25-20 The R for a highway curve (arc definition) will be rounded off to the nearest larger multiple of 100 ft. Field conditions require M to be approximately 24 ft to avoid an embankment. The PI $= 84 + 71.80$ and $I = 22°00'$.

*Asterisks indicate problems that have answers given in Appendix G.

25-21 The D_a for a highway curve will be rounded off to the nearest multiple of 20′. Field measurements show that T should be approximately 200 ft to avoid an overpass. The PI $= 63 + 56.40$ and $I = 13°20′$.

*25-22 A highway survey PI falls in a pond, so a cutoff line $AB = 275.21$ ft is run between the tangents. In the triangle formed by points A, B, and PI, the angle at $A = 16°26′$ and at $B = 22°12′$. The station of A is $54 + 92.30$. Calculate and tabulate curve notes to run, by deflection angles and incremental chords, a $4°15′$ (arc definition) circular curve at half-station increments to connect the tangents.

25-23 In the figure, a single circular highway curve (arc definition) will join tangents XV and VY and also be tangent to BC. Calculate R, L, and the stations of the PC and PT.

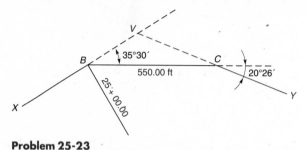

Problem 25-23

*25-24 Compute R_x to fit requirements of the figure and make the tangent distances of the two curves equal.

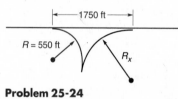

Problem 25-24

25-25 After a backsight on the PC with $0°00′$ set on the instrument, what is the deflection angle to the following circular curve points?
(a) Setup at curve midpoint, deflection to the PT
(b) Instrument at curve midpoint, deflection to the $\frac{3}{4}$ point
(c) Setup at $\frac{1}{4}$ point of curve, deflection to $\frac{3}{4}$ point

25-26 In surveying a construction alignment, why should the I angle be measured by repetition?

25-27 Tabulate the curve data, deflection angles, and incremental chords needed to layout the curve of Problem 25-7 at half-station intervals.

25-28 Similar to Problem 25-27, except for the curve in Problem 25-8.

25-29 Assume the curve of Problem 25-7 will be staked at half-station increments from the PC using a total station instrument. Compute and tabulate the curve data, deflection angles, and total chords needed.

*25-30 Similar to Problem 25-29, except for the curve of Problem 25-8.

25-31 A highway curve (arc definition) to the right, having $D_a = 3°20′$ and $I = 18°40′$, will be laid out by coordinates with a total station setup at the PI. The PI station is $38 + 55.20$, and its coordinates are $X = 75,428.86$ ft and $Y = 36,007.43$ ft. The azimuth (from north) of the back tangent proceeding toward the PI is $48°17′12″$. To orient the total station, a backsight will be made on a POT on the back tangent at station $32 + 00$. Compute azimuths and radial chords necessary to stake the curve at full stations.

*Asterisks indicate problems that have answers given in Appendix G.

For Problems 25-32 and 25-33, compute and tabulate the data necessary to stake out a circular railroad curve (chord definition) by tangent offsets. Full and 50-ft stations are to be set and approximately one-half the curve laid out from each tangent.

25-32 Station of PI $= 54 + 25.50$, $D_c = 5°15'$, and $L = 460.00$ ft

25-33 $L = 300.00$ ft, $D_c = 4°30'$, and station of PC $= 38 + 52.00$

25-34 A running track must consist of two semicircles and two tangents, and be exactly 5000 ft along its centerline. The two curves are to constitute one-half its total length. Calculate L, R, and D_a for the curves.

*25-35 In the figure, the coordinates (in feet) of points A and O are $X_A = 600.00$, $Y_A = 960.00$, and $X_o = 1510.00$, $Y_o = 1800.00$. If the azimuth of line AB is $40°15'$ and the circular curve radius 430.00 ft, calculate the coordinates of intersection point P.

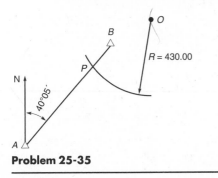

Problem 25-35

25-36 The coordinates of circle center O_1 are $X_{O1} = 960.00$ and $Y_{O1} = 960.00$ ft, for circle center O_2, $X_{O2} = 1510.00$ and $Y_{O2} = 470.00$ ft, and $R_1 = 380$ and $R_2 = 425$ ft. Compute the coordinates of intersection point P shown in the figure.

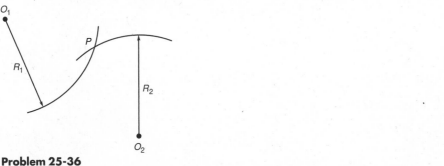

Problem 25-36

What sight distance is available if there is an obstruction on a radial line through the PI inside the curves in Problems 25-37 and 25-38?

25-37 For Problem 25-7, obstacle 30 ft from curve

*25-38 For Problem 25-12, obstacle 25 ft from curve

25-39 If the misclosure for the curve of Problem 25-7 computed as described in Section 25-8 is 0.18 ft, what is the field layout precision?

25-40 Assume that a 200-ft entry spiral will be used with the curve of Problem 25-7. Compute and tabulate curve notes to stake out the alignment from the TS to ST at full stations using a theodolite and tape and the deflection-angle incremental chord method.

*Asterisks indicate problems that have answers given in Appendix G.

*25-41 Same as Problem 25-40, except use a 300 ft spiral for the curve of Problem 25-8.

25-42 Same as Problem 25-40, except for the curve of Problem 25-9, and staking at half-stations using a total station instrument set up at the TS.

25-43 Compute the area bounded by the two arcs and tangent in Problem 25-24.

25-44 Write a computer program for calculating circular curve notes by the deflection-angle incremental chord method.

25-45 Write a computer program to calculate circular curve notes for layout with a total station by deflection angles and total chords from a set up at the PC.

*Asterisks indicate problems that have answers given in Appendix G.

BIBLIOGRAPHY

Brinker, R. C., and R. Minnick. 1987. *The Surveying Handbook*. New York: Van Nostrand Reinhold, chap. 24.

Chen, C., and L. Hwang. 1992. "Solving Circular Curves Using Newton–Raphson's Method." *ASCE, Journal of Surveying Engineering* 118 (No. 1):24.

Labio, S. 1983. "Some Aspects of Transition Spirals of Interest to Land Surveyors." *Surveying and Mapping* 43 (No. 4):365.

Meyer, C. F., and D. W. Gibson. 1980. *Route Surveying and Design,* 5th ed. New York: Harper & Row.

Moffitt, F. H., and H. Bouchard. 1992. *Surveying,* 9th ed. New York: Harper-Collins, chap. 13.

Pearson, F. 1983. "Non-Centerline Layout." *ASCE, Journal of Surveying Engineering* 109 (No. 1):24.

26

VERTICAL CURVES

26-1 INTRODUCTION

Curves are needed to provide smooth transitions between grade lines of tangent sections for highways and railroads. Because these curves exist in vertical planes, they are called *vertical curves.* An example is illustrated in Figure 26-1, which shows the profile view of a proposed section of highway to be constructed from *A* to *B*. A grade line consisting of three tangent sections has been designed to fit the ground profile. Two vertical curves are needed: curve *a* to join tangents 1 and 2, and curve *b* to connect tangents 2 and 3. The function of each curve is to provide a gradual change in grade from the initial (*back*) tangent to the grade of the second (*forward*) tangent. Because parabolas provide a *constant rate of change of grade,* they are ideal and almost always applied for vertical alignments used by vehicular traffic.

Two basic types of vertical curves exist, *crest* and *sag.* These are illustrated in Figure 26-1. Curve *a* is a crest type, which by definition undergoes a negative change in grade; that is, the curve turns downward. Curve *b* is a sag type, in which the change in grade is positive and the curve turns upward.

There are several factors that must be taken into account when designing a grade line of tangents and curves on any highway or railroad project. They include (**1**) providing a good fit with the existing ground profile, thereby minimizing the depths

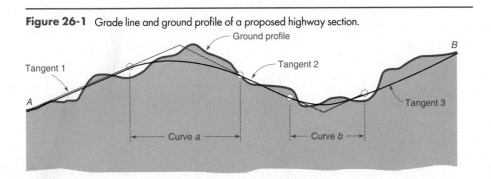

Figure 26-1 Grade line and ground profile of a proposed highway section.

of cuts and fills, **(2)** balancing the volume of cut material against fill, **(3)** maintaining adequate drainage, **(4)** not exceeding maximum specified grades, and **(5)** meeting fixed elevations such as intersections with other roads. In addition, the curves must be designed to **(1)** fit the grade lines they connect, **(2)** have lengths sufficient to meet specifications covering maximum rate of change of grade (which affects the comfort of vehicle occupants), and **(3)** provide sufficient sight distance for safe vehicle operation (see Section 26-11).

Elevations at selected points (usually full or half-stations) along vertical parabolic curves can be computed by either the *tangent-offset method* or the *chord-gradient method*. In this brief treatment, only the tangent-offset method will be demonstrated because it is simple, straightforward, conveniently performed with calculators and computers, and self-checking.

After the elevations of curve points have been computed, they are staked in the field to guide construction operations so the route can be built according to plan.

26-2 GENERAL EQUATION OF A VERTICAL PARABOLIC CURVE

The general mathematical expression of a parabola, with respect to an XY rectangular coordinate system, is given by

$$Y_p = a + bX_p + cX_p^2 \qquad (26\text{-}1)$$

where Y_p is the ordinate at any point p on the parabola located a distance X_p from the origin of the curve, and a, b, and c are constants. Figure 26-2 shows a parabola in an

Figure 26-2 Terms for a parabola.

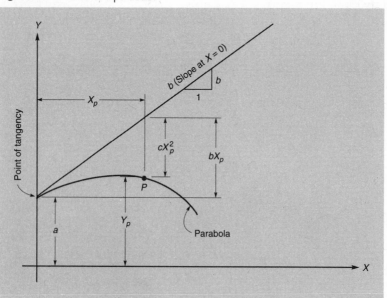

XY rectangular coordinate system and illustrates the physical significance of the terms in Eq. (26-1). Note from the figure that a is the ordinate at the beginning of the curve where $X = 0$, b the slope of a tangent to the curve at $X = 0$, bX_p the change in ordinate along the tangent over distance X_p, and cX_p^2 the parabola's departure from the tangent (*tangent offset*) in distance X_p. When the terms a, bX_p, and cX_p^2 are combined as in Eq. (26-1) and shown in Figure 26-2, they produce Y_p. For the crest curve of Figure 26-2, b has positive algebraic sign and c is negative.

26-3 EQUATION OF AN EQUAL-TANGENT VERTICAL PARA-BOLIC CURVE IN SURVEYING TERMINOLOGY

Figure 26-3 shows a parabola that joins two intersecting tangents of a grade line. The parabola is essentially identical to that in Figure 26-2, except that the terms used are those commonly employed by surveyors and engineers. In the figure, BVC denotes the *beginning of vertical curve,* sometimes called the VPC (vertical point of curvature); V is the *vertex,* often called the VPI (vertical point of intersection); and EVC denotes the *end of vertical curve,* interchangeably called the VPT (vertical point of tangency). The percent grade of the back tangent is g_1, that of the forward tangent is g_2. The curve length L is the horizontal distance (in stations) from the BVC to the EVC. The curve of Figure 26-3 is called *equal tangent* because the horizontal distances from the BVC to V and from V to the EVC are equal, each being $L/2$. Proof of this is given in Section 26-5.

On the XY axis system in Figure 26-3, X values are horizontal distances, in stations, measured from the BVC, and Y values are elevations, in feet (or meters), measured from

Figure 26-3 Vertical parabolic curve relationships.

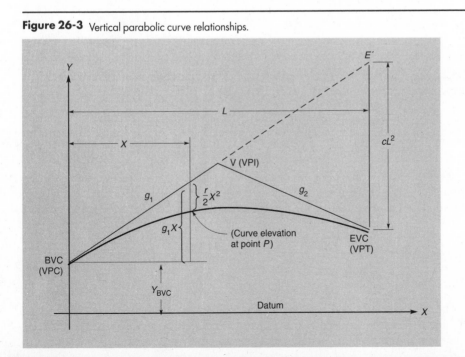

datum, usually sea level. By substituting this surveying terminology into Eq. (26-1), the parabola can be expressed as

$$Y = Y_{BVC} + g_1 X + c X^2 \qquad (26\text{-}2)$$

The correspondence of terms in Eq. (26-2) to those of Eq. (26-1) is $a = Y_{BVC}$ (elevation of BVC) and $bX_p = g_1 X$ (change in elevation along the back tangent with increasing X). To express the constant c of Eq. (26-2) in surveying terminology, consider the tangent offset from E' on the extended back tangent to the EVC in Figure 26-3. Its value (which is negative for the crest curve shown) is cL^2, where L (horizontal distance from BVC to E') is substituted for X. From the figure, cL^2 can be expressed in terms of horizontal lengths (in stations) and percent grades as follows:

$$cL^2 = g_1 \left(\frac{L}{2}\right) + g_2 \left(\frac{L}{2}\right) - g_1 L \qquad (a)$$

Solving Eq. (a) for constant c gives

$$c = \frac{g_2 - g_1}{2L} \qquad (b)$$

Substituting Eq. (b) into Eq. (26-2) results in the following equation for an equal-tangent vertical curve in surveying terminology:

$$Y = Y_{BVC} + g_1 X + \left(\frac{g_2 - g_1}{2L}\right) X^2 \qquad (26\text{-}3)$$

The *rate of change of grade* r of an equal-tangent parabolic curve equals the total grade change from BVC to EVC divided by length L, in stations, over which the change occurs, or

$$r = \frac{g_2 - g_1}{L} \qquad (26\text{-}4)$$

As mentioned earlier, the value of r (which is negative for a crest curve and positive for a sag type) is an important design parameter because it controls rider comfort. To incorporate it in the equation for parabolic curves, Eq. (26-4) is substituted into Eq. (26-3),

$$Y = Y_{BVC} + g_1 X + \left(\frac{r}{2}\right) X^2 \qquad (26\text{-}5)$$

Figure 26-3 illustrates how the terms of Eq. (26-5) combine to give the curve elevation at point P. Because the last term of the equation is the curve's offset from the back tangent, this formula is commonly called the *tangent offset equation*.

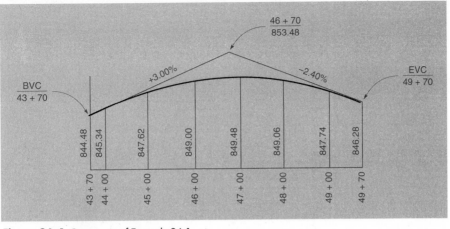

Figure 26-4 Crest curve of Example 26-1.

26-4 VERTICAL-CURVE COMPUTATION USING THE TANGENT OFFSET EQUATION

In designing grade lines, the locations and individual grades of the tangents are normally selected first. This produces a series of intersection points V, each defined by its station and elevation. A curve is then chosen to join each pair of intersecting tangents. The parameter selected in vertical-curve design is length L. Having chosen it, the station of the BVC is obtained by subtracting $L/2$ from the vertex station. The EVC station is then determined by adding L to the BVC station.

Computations for vertical parabolic curves are normally done in tabular form. The following example for a crest curve illustrates these procedures.

A grade g_1 of $+3.00\%$ intersects grade g_2 of -2.40% at a vertex whose station and elevation are $46 + 70$ and 853.48, respectively. An equal-tangent parabolic curve 600 ft long has been selected to join the two tangents. Compute and tabulate the curve for stakeout at full stations. (Figure 26-4 shows the curve.)

EXAMPLE 26-1

SOLUTION

By Eq. (26-4), $r = \dfrac{-2.40 - 3.00}{6} = -0.90\%/\text{station}$.

Stationing:

$$
\begin{aligned}
V &= 46 + 70 \\
-L/2 &= \underline{3 + 00} \\
\text{BVC} &= 43 + 70 \\
+L &= \underline{6 + 00} \\
\text{EVC} &= 49 + 70
\end{aligned}
$$

$$\text{elev}_{\text{BVC}} = 853.48 - 3.00(3) = 844.48 \text{ ft}$$

TABLE 26-1 NOTES FOR CURVE OF EXAMPLE 26-1

STATION	X (sta)	g_1X	$\dfrac{rX^2}{2}$	CURVE ELEVATION	FIRST DIFFERENCE	SECOND DIFFERENCE
49 + 70 (EVC)	6.0	18.00	− 16.20	846.28		
49 + 00	5.3	15.90	− 12.64	847.74	− 1.32	
48 + 00	4.3	12.90	− 8.32	849.06	− 0.42	− 0.90
47 + 00	3.3	9.90	− 4.90	849.48	0.48	− 0.90
46 + 00	2.3	6.90	− 2.38	849.00	1.38	− 0.90
45 + 00	1.3	3.90	− 0.76	847.62	2.28	− 0.90
44 + 00	0.3	0.90	− 0.04	845.34		− 0.90
43 + 70 (BVC)	0.0	0.00	− 0.00	844.48		

$$\text{Check: EVC} = \text{ELEV}_v - g_2\left(\frac{L}{2}\right) = 853.48 - 2.40(3) = 846.28$$

The remaining calculations utilize Eq. (26-5) and are listed in Table 26-1.

A check on curve elevations is obtained by computing the first and second differences between the elevations of full stations, as shown in the right-hand columns of the table. Unless disturbed by rounding off, all second differences (rate of change) should be equal. For full-station increments the second differences equal r; for half-station increments they equal $r/4$.

It is sometimes desirable to calculate the elevation of the curve's center point. This can be done using $X = L/2$ in Eq. (26-5). For Example 26-1 it is

$$Y_{center} = 844.48 + 3.00(3) - \left(\frac{0.90}{2}\right)(3^2) = 849.43 \text{ ft}$$

This can be checked by employing the property of a parabolic curve that *the curve center falls halfway between the vertex and the midpoint of the long chord* (line from BVC to EVC). The elevation of the midpoint of the long chord (LC) is simply the average of the elevations of the BVC and EVC. For Example 26-1 it is

$$Y_{midpoint\ LC} = \frac{844.48 + 846.28}{2} = 845.38 \text{ ft}$$

By the property just stated, the elevation of the curve center for Example 26-1 is

$$Y_{center} = \frac{845.38 + 853.48}{2} = 849.43 \text{ ft (check)}$$

26-5 EQUAL-TANGENT PROPERTY OF A PARABOLA

The curve defined by Eqs. (26-3) and (26-5) has been called an *equal-tangent parabolic curve*, which means the vertex occurs a distance $X = L/2$ from the BVC. Proof of this property is readily made with reference to Figure 26-5, which illustrates a sag curve. In the figure, assume the horizontal distance from BVC to V is an unknown value X; thus

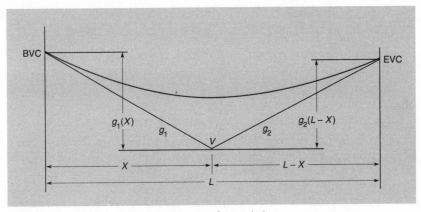

Figure 26-5 Proof of equal-tangent property of a parabola.

the remaining distance from V to EVC is $L - X$. Two expressions can be written for the elevation of the EVC. The first, using Eq. (26-3) with $X = L$, yields

$$Y_{EVC} = Y_{BVC} + g_1 L + \left(\frac{g_2 - g_1}{2L} \right) L^2 \qquad \text{(c)}$$

The second, using changes in elevation that occur along the tangents, gives

$$Y_{EVC} = Y_{BVC} + g_1 X + g_2 (L - X) \qquad \text{(d)}$$

Equating Eqs. (c) and (d) and solving, $X = L/2$. Thus distances BVC to V and V to EVC are equal—hence the term *equal-tangent parabolic curve*.

26-6 CURVE COMPUTATIONS BY PROPORTION

The following basic property of a parabola can be used to simplify vertical curve calculations: *offsets from a tangent to a parabola are proportional to the squares of the distances from the point of tangency.* This property is conveniently utilized by calculating the offset E at the midpoint of the curve, and then computing offsets at any other distance X from the BVC by proportion according to the following formula:

$$\text{offset}_X = E \left(\frac{X}{\frac{1}{2}L} \right)^2 \qquad \text{(26-6)}$$

The value of E in Eq. (26-6) is simply the difference in elevation from the curve's midpoint to the VPI. Computation of the curve's midpoint elevation was discussed in Section 26-4, and for Example 26-1 it was 849.43 ft. To illustrate the use of Eq. (26-6), the offset for station $47 + 00$, where $X = 3.3$ stations, will be computed:

$$\text{offset}_{47+00} = (849.43 - 853.48) \left(\frac{3.3}{3} \right)^2 = -4.90 \text{ ft}$$

This checks the value listed in Table 26-1 for station 47 + 00. This simplified procedure is convenient for computing vertical curves with a hand-held calculator.

26-7 STAKING A VERTICAL PARABOLIC CURVE

Prior to initiating construction on a route project, the planned centerline, or an offset one, will normally be staked at full or half-stations, as well as other critical horizontal alignment points such as PCs and PTs. Then, as described in Section 24-7, slope stakes will be set out perpendicular to the centerline at or near the slope intercepts to guide rough grading. Excavation and embankment construction then proceed and continue until the grade is near plan elevation. The centerline stations are then staked again, this time using sharpened 2-in.-square wooden hubs, usually about 10 in. long. These are known as "blue tops," so called because when their tops are driven to grade elevation, they are colored blue. Contractors request blue tops when excavated areas are still slightly high and embankments somewhat low. After blue tops are set to mark the precise grade, final grading is completed.

To set blue tops at grade, a circuit of differential levels is run from a nearby bench mark to establish the HI of a leveling instrument in the project area. The difference between the HI and any station's grade is the required rod reading on that stake. Assume, for example, an HI of 856.20 exists, and station 45 + 00 of Example 26-1 is to be set. The required rod reading is 856.20 − 847.62 = 8.58 ft. With the stake initially driven firmly into the ground, and the rod held on its top, suppose a reading of 8.25 is obtained. The stake then must be driven down an additional 8.58 − 8.25 = 0.33 ft further, and the notetaker so indicates. After the stake is driven a distance estimated to be somewhat less than 0.33 ft, the rod reading is checked. This is repeated until the required reading of 8.58 is achieved, whereupon the stake is colored blue using keel or spray.

This process is continued until all stakes are set. The required rod reading at station 46 + 00, for an HI of 856.20, is 7.20 ft. If a rock is encountered and the stake cannot be driven to grade, a vertical offset of, say, 1.00 ft above grade can be marked and noted on the stake.

When the level is too far away from the station being set, a turning point is established and the instrument brought forward to establish a new HI. Whenever possible, level circuit checks should be made by closing on bench marks, and the circuit *must* be closed when completed to verify that no mistakes were made.

26-8 COMPUTATIONS FOR AN UNEQUAL-TANGENT VERTICAL CURVE

An unequal-tangent vertical curve is simply a compounded pair of equal-tangent curves, where the EVC of the first is the BVC of the second. This point is called CVC, point of *compound vertical curvature*. In Figure 26-6 a −2.00% grade intersects a +1.60% grade at station 87 + 00 and elevation 743.24. A 400-ft vertical curve is to be extended back from the vertex, and a 600-ft curve run in forward to closely fit ground conditions.

To perform calculations for this type of curve, connect the midpoints of the tangents for the two curves, stations 85 + 00 and 90 + 00, to obtain line *AB*. Compute elevations for *A* and *B* and, using them, calculate the grade of *AB* by dividing the difference

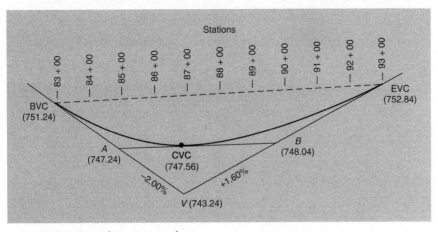

Figure 26-6 Unequal-tangent vertical curve.

in elevation between B and A by the distance in stations separating these two points. From grade AB, determine the CVC elevation.

Now compute two vertical curves, one from BVC to CVC and another from CVC to EVC, by the methods of Section 26-4. Since both curves are tangent to the same line AB at point CVC, they will be tangent to each other and form a smooth curve.

For the configuration of Figure 26-6, compute and tabulate the notes necessary to stake the unequal-tangent vertical curve at full stations.

EXAMPLE 26-2

SOLUTION

1. Calculate elevations of BVC, EVC, A, B, and CVC, and grade AB:

$$\text{elev}_{BVC} = 743.24 + 4(2.00) = 751.24 \text{ ft}$$
$$\text{elev}_A = 743.24 + 2(2.00) = 747.24 \text{ ft}$$
$$\text{elev}_{EVC} = 743.24 + 6(1.60) = 752.84 \text{ ft}$$
$$\text{elev}_B = 743.24 + 3(1.60) = 748.04 \text{ ft}$$
$$\text{grade}_{AB} = \left(\frac{748.04 - 747.24}{5}\right) = +0.16\%$$
$$\text{elev}_{CVC} = 747.24 + 2(0.16) = 747.56 \text{ ft}$$

These elevations are shown in Figure 26-6.

2. In computing the first curve, the grade of AB will be g_2 in the formulas, and for the second curve it will be g_1. The rates of change of grade for the two curves are, by Eq. (26-4),

$$r_1 = \frac{0.16 - (-2.00)}{4} = +0.54\%/\text{station}$$

$$r_2 = \frac{1.60 - 0.16}{6} = +0.24\%/\text{station}$$

3. Equation (26-5) is now solved in tabular form, and the results are listed in Table 26-2.

Vertical-curve computations by themselves are quite simple, hardly a challenge to an electronic computer. But when vertical curves are combined with horizontal curves, spirals, and superelevation in complex highway interchanges, programming saves time.

26-9 HIGH OR LOW POINT ON A VERTICAL CURVE

To investigate drainage conditions, clearance beneath overhead structures, cover over pipes, or sight distance, it may be necessary to determine the elevation and location of the low (or high) point on a vertical curve. At the low or high point, a tangent to the curve will be horizontal and its slope equal to zero. Based on this fact, by taking the derivative of Eq. (26-3) and setting it equal to zero, the following formula is readily derived:

$$X = \frac{g_1 L}{g_1 - g_2} \tag{26-7}$$

where X is the distance in stations from the BVC to the high or low point of the curve, g_1 the tangent grade through the BVC, g_2 the tangent grade through the EVC, and L the curve length, in stations.

If g_2 is substituted for g_1 in the numerator of Eq. (26-7), distance X is measured back from the EVC. By substituting Eq. (26-4) into Eq. (26-7), the following alternate formula for locating the high or low point results:

$$X = \frac{-g_1}{r} \tag{26-8}$$

TABLE 26-2 NOTES FOR CURVE OF EXAMPLE 26-2

STATION	X (sta)	$g_1 X$	$rX^2/2$	CURVE ELEVATION	FIRST DIFFERENCE	SECOND DIFFERENCE
93 + 00 (EVC)	6	0.96	4.32	752.84 (check)		
					1.48	
92 + 00	5	0.80	3.00	751.36		0.24
					1.24	
91 + 00	4	0.64	1.92	750.12		0.24
					1.00	
90 + 00	3	0.48	1.08	749.12		0.24
					0.76	
89 + 00	2	0.32	0.48	748.36		0.24
					0.52	
88 + 00	1	0.16	0.12	747.84		0.24
					0.28	
87 + 00 (CVC)	4	−8.00	4.32	747.56 (check)		
					−0.11	
86 + 00	3	−6.00	2.43	747.67		0.54
					−0.65	
85 + 00	2	−4.00	1.08	748.32		0.54
					−1.19	
84 + 00	1	−2.00	0.27	749.51		0.54
					−1.73	
83 + 00 (BVC)	0	0.00	0.00	751.24		

Compute the station and elevation of the curve's high point in Example 26-1.   EXAMPLE 26-3

SOLUTION

By Eq. (26-8), $X = \dfrac{-3.00}{-0.90} = 3.3333$ stations

Then the station of the high point is

$$\text{station}_{\text{high}} = 43 + 70 + (3 + 33.33) = 47 + 03.33$$

By Eq. (26-3), the elevation at this point is

$$844.48 + 3.00(3.3333) + \frac{-2.40 - 3.00}{2(6)}(3.3333)^2 = 849.48$$

Note that in using Eq. (26-8) and all other equations of this chapter, correct algebraic signs must be applied to grades g_1 and g_2.

26-10 DESIGNING A CURVE TO PASS THROUGH A FIXED POINT

The problem of designing a parabolic curve to pass through a point of fixed station and elevation is frequently encountered in practice. It occurs, for example, where a new grade line must meet existing railroad or highway crossings, or when a minimum vertical distance must be maintained between the grade line and underground utilities or drainage structures.

Given the station and elevation of the VPI, and grades g_1 and g_2 of the back and forward tangents, respectively, the problem consists of calculating the curve length required to meet the fixed condition. It is solved by substituting known quantities into Eq. (26-3) and reducing the equation to a quadratic containing only L as an unknown. Two values will satisfy the quadratic equation, but the correct one will be obvious.

In Figure 26-7, grades $g_1 = -4.00\%$ and $g_2 = +3.80\%$ meet at VPI station $52 + 00$ EXAMPLE 26-4
and elevation 1261.50. Design a parabolic curve to meet a railroad crossing which exists at station $53 + 50$ and elevation 1271.20.

SOLUTION

In referring to Figure 26-7 and substituting known quantities into Eq. (26-3), the following equation is obtained:

$$1271.20 = \left[1261.50 + 4.00\left(\frac{L}{2}\right)\right] + \left[-4.00\left(\frac{L}{2} + 1.5\right)\right]$$
$$+ \left[\frac{3.80 + 4.00}{2L}\left(\frac{L}{2} + 1.5\right)^2\right]$$

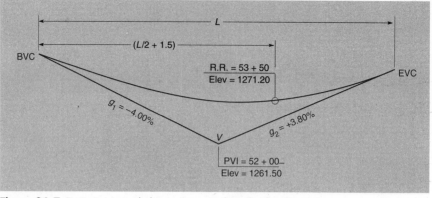

Figure 26-7 Designing a parabolic curve to pass through a fixed point.

In this expression the value of X for the railroad crossing is $L/2 + 1.5$, and the terms within successive brackets are Y_{BVC}, g_1X, and $(r/2)X^2$, respectively. Reducing the equation to quadratic form gives

$$0.975L^2 - 9.75L + 8.775 = 0$$

Solving for L gives 9.1152 stations. To check the solution, $L = 9.1152$ stations and $X = (9.1152/2 + 1.5)$ stations are used in Eq. (26-3) to calculate the elevation at station $53 + 50$. A value of 1271.20 checks the computations.

26-11 SIGHT DISTANCE

The formula for sight distance S, with a vehicle on a vertical curve and S shorter than length L, is

$$S = \sqrt{\frac{2L(\sqrt{h_1} + \sqrt{h_2})^2}{g_1 - g_2}} \qquad (26\text{-}9)$$

where S is the sight distance, in stations; L the length of curve, in stations; h_1 the height of the driver's eye, in feet; and h_2 the height of an object sighted on the roadway ahead, in feet. For design, the American Association of State Highway and Transportation Officials recommends 3.50 ft for h_1. The value used for h_2 can vary from 0.5 ft representing the size of an object that would damage a vehicle, to a larger value, say 4.25 ft, representing the top of an oncoming or parked car.

Then for a crest curve having a length of 600 ft, and grades of $+2.00\%$ and -1.60%, if $h_1 = 3.50$ ft and h_2 is taken as 4.25 ft, Eq. (26-9) gives

$$S = \sqrt{\frac{2 \times 6 (\sqrt{3.5} + \sqrt{4.25})^2}{2.00 - (-1.60)}} = 7.18 \text{ stations} = 718 \text{ ft}$$

Since this distance is greater than the length of the curve and thus not in agreement with the assumption used in deriving the formula, a different expression must be employed.

If the vehicle is off the curve and on the tangent to it, and S is greater than L, the applicable sight distance formula is

$$S = \frac{L}{2} + \frac{(\sqrt{h_1} + \sqrt{h_2})^2}{g_1 - g_2} \qquad (26\text{-}10)$$

Then in the preceding example, with $h_1 = 3.50$ ft and $h_2 = 4.25$ ft,

$$S = \frac{6}{2} + \frac{(\sqrt{3.50} + \sqrt{4.25})^2}{2.00 - (-1.60)} = 7.29 \text{ stations}$$

For a combined horizontal and vertical curve, the sight distance is the smaller of the two values computed independently for each curve. (See Section 25-15 for a discussion of horizontal sight distance.)

26-12 SOURCES OF ERROR

Some sources of error in staking out parabolic curves are:

1. Making errors in measuring horizontal distances when staking the centerline
2. Not holding the level rod plumb when setting blue tops
3. Using a leveling instrument which is out of adjustment

26-13 MISTAKES

Some typical mistakes made in computations for vertical curves include the following:

1. Arithmetic mistakes
2. Failure to properly account for the algebraic signs of g_1 and g_2
3. Subtracting offsets from tangents for a sag curve, or adding them for a crest curve
4. Failure to make the second-difference check

26-1 Why is a parabola rather than a circular arc used for vertical curves on grade lines used for vehicular travel?   **PROBLEMS**

26-2 What is meant by the "rate of grade change" on vertical curves, and why is it important?

Tabulate station elevations for an equal-tangent parabolic curve for the data given in Problems 26-3 through 26-8. Check by second differences.

*26-3 A $+2.75\%$ grade meets a -2.25% grade at station $50 + 50$ and elevation 975.50 ft, 1000-ft curve, stakeout at full stations.

 26-4 A -3.5% grade meets a $+1.50\%$ grade at station $63 + 00$ and elevation 754.80 ft, 1200-ft curve, stakeout at full stations.

*26-5 A 550-ft curve, grades of $g_1 = -2.90\%$ and $g_2 = +1.25\%$, VPI at station $89 + 30$, and elevation 130.00 ft, stakeout at half-stations.

 26-6 An 800-ft curve, grades of $g_1 = -5.00\%$ and $g_2 = -1.00\%$, VPI at station $50 + 00$, and elevation 1020.00 ft, stakeout at half-stations.

---

*Asterisks indicate problems that have answers given in Appendix G.

26-7 A 1400-ft curve, $g_1 = +5.25\%$, $g_2 = -1.50\%$, VPI station $= 12 + 75$, VPI elevation $= 4531.20$ ft, stakeout at full stations.

26-8 A 300 ft curve, $g_1 = -1.75\%$, $g_2 = +1.05\%$, VPI station $= 102 + 00$, VPI elevation $= 145.00$ ft, stakeout at quarter-stations.

Field conditions require a highway curve to pass through a fixed point. Compute a suitable equal-tangent vertical curve and full-station elevations for Problems 26-9 through 26-11.

26-9 Grades of $g_1 = -3.50\%$ and $g_2 = +0.50\%$, VPI elevation 870.00 ft at station 55 + 00. Fixed elevation 872.75 ft at station 55 + 00.

26-10 Grades of $g_1 = -2.50\%$ and $g_2 = +1.50\%$, VPI elevation 2430.00 ft at station 315 + 00. Fixed elevation 2436.50 ft at station 314 + 00.

*26-11 Grades of $g_1 = +5.00\%$ and $g_2 = +1.50\%$, VPI station 36 + 00 and elevation 620.00 ft. Fixed elevation 618.70 ft at station 37 + 00.

26-12 A -1.10% grade meets a $+0.60\%$ grade at station 46 + 00 and elevation 900.00 ft. The $+0.60\%$ grade then joins a 2.40% grade at station 49 + 00. Compute and tabulate the notes for an equal-tangent vertical curve, at half-stations, that passes through the midpoint of the 0.60% grade.

Compute and tabulate full-station elevations for an unequal-tangent vertical curve to fit the requirements in Problems 26-13 through 26-16.

*26-13 A $+4.00\%$ grade meets a -2.50% grade at station 70 + 00 and elevation 1260.00 ft. Length of first curve 600 ft, second curve 400 ft.

26-14 Grade $g_1 = +1.25\%$, $g_2 = 4.75\%$, VPI at station 52 + 00 and elevation 2310.00 ft, $L_1 = 800$ ft and $L_2 = 400$ ft.

26-15 Grades of $+3.75\%$ and -2.20% meet at the VPI at station 42 + 00 and elevation 862.42 ft. Lengths of curves are 400 and 600 ft.

26-16 A -0.80% grade meets a $+3.30\%$ grade at station 95 + 00 and elevation 646.80 ft. Length of first curve is 200 ft, of second curve, 600 ft.

26-17 Write a computer program for vertical-curve computations.

26-18 When is it advantageous to use an unequal-tangent vertical curve instead of an equal-tangent one?

*26-19 A manhole is 10 ft from the centerline of a 40-ft-wide street that has a 6-in. parabolic crown. The street center at the station of the manhole is at elevation 712.58 ft. What is the elevation of the manhole cover?

26-20 A 60-ft-wide street has an average parabolic crown from the center to each edge of $\frac{1}{4}$ in./ft. How much does the surface drop from the street center to a point 6 ft from the edge?

*26-21 Determine the station and elevation at the high point of the curve in Problem 26-3.

26-22 Calculate the station and elevation at the low point of the curve in Problem 26-4.

26-23 Compute the station and elevation at the low point of the curve of Problem 26-5.

26-24 What are the station and elevation of the high point of the curve of Problem 26-7?

26-25 Compute the sight distance available in Problem 26-3. (Assume $h_1 = 3.50$ ft and $h_2 = 4.25$ ft.)

26-26 Similar to Problem 26-25, except $h_2 = 0.50$ ft.

26-27 Similar to Problem 26-25, except for the data of Problem 26-7.

26-28 In determining sight distances on vertical curves, how does the designer determine whether the cars or objects are on the curve or tangent?

What is the minimum length of a vertical curve to provide a required sight distance for the conditions given in Problems 26-29 through 26-31? (Assume $h_1 = 3.50$ ft and $h_2 = 4.25$ ft.)

26-29 Grades of $+3.50\%$ and -2.25%, sight distance 700 ft.

*26-30 A crest curve with grades of $+4.60\%$ and -3.05%, sight distance 850 ft.

26-31 Sight distance of 1200 ft, grades of $+0.75\%$ and -1.10%.

*Asterisks indicate problems that have answers given in Appendix G.

26-32 A backsight of 6.92 ft is taken on a bench mark whose elevation is 1276.35 ft. What rod reading is needed at that HI to set a blue top at grade elevation of 1280.50 ft?

*26-33 A backsight of 7.52 ft is taken on a bench mark whose elevation is 856.82. A foresight of 3.18 ft and a backsight of 6.04 ft are then taken in turn on TP$_1$ to establish an HI. What rod reading will be necessary to set a blue top at a grade elevation of 859.67?

*Asterisks indicate problems that have answers given in Appendix G.

BIBLIOGRAPHY

American Association of State Highway and Transportation Officials. 1990. *A Policy on Geometric Design of Highways and Streets.* Washington, D.C.: AASHTO.

Brinker, R. C., and R. Minnick. 1987. *The Surveying Handbook.* New York: Van Nostrand Reinhold, chap. 24.

Easa, S. M. 1991. "Unified Design of Vertical Parabolic Curves." *Surveying and Land Information Systems* 51 (No. 2):105.

Meyer, C. F., and D. W. Gibson. 1980. *Route Surveying and Design,* 5th ed. New York: Harper & Row.

Moffitt, F. H., and H. Bouchard. 1992. *Surveying,* 9th ed. New York: Harper-Collins, chap. 13

27
VOLUMES

27-1 INTRODUCTION

Surveyors are often called on to determine volumes of various types of material. Quantities of earthwork and concrete are needed, for example, on many types of construction projects. Volume computations are also required to determine the capacities of bins, tanks, reservoirs, and buildings, and to check stockpiles of coal, gravel, and other materials. The determination of quantities of water discharged by streams and rivers, per unit of time, is also important.

The most common unit of volume is a cube having edges of unit length. Cubic feet, cubic yards, and cubic meters are used in surveying calculations, although the cubic yard is most common for earthwork: $1 \text{ yd}^3 = 27 \text{ ft}^3$; $1 \text{ m}^3 = 35.315 \text{ ft}^3$. The *acre-foot* (the volume equivalent to an acre of area 1 ft deep) is commonly used for large quantities of water, while cubic feet per second is the usual unit for water flow measurement.

27-2 METHODS OF VOLUME MEASUREMENT

Direct measurement of volumes is rarely made in surveying, since it is difficult to actually apply a unit of measure to the material involved. Instead, indirect measurements are obtained by measuring lines and areas that have a relationship to the volume desired.

Three principal systems are used: **(1)** the cross-section method, **(2)** the unit-area, or borrow-pit, method, and **(3)** the contour-area method.

27-3 THE CROSS-SECTION METHOD

The cross-section method is employed almost exclusively for computing volumes on linear construction projects such as highways, railroads, and canals. In this procedure, after the centerline has been staked, ground profiles called cross sections are taken (at right angles to the centerline), usually at intervals of 50 or 100 ft. Cross-sectioning consists of measuring ground elevations and their corresponding distances left and right perpendicular to the centerline. Readings must be taken at the centerline, at high and low points, and at locations where slope changes occur to determine the ground profile accurately. This can be cone in the field using a level, level rod, and tape. Plate D-8 in Appendix D illustrates a set of field notes for cross-sectioning.

Much of the field work formerly involved in running preliminary centerlines, getting cross-section data, and making slope-stake and other measurements on long route surveys is now being done more efficiently by photogrammetry. Research has shown that earthwork quantities computed from photogrammetric cross-sectioning agree to within about 1 or 2% of those obtained from good-quality field cross sections. It is not intended to discuss photogrammetric methods in this chapter; rather, basic field and office procedures for determining and calculating volumes will be presented briefly. Chapter 28 discusses the subject of photogrammetry.

In Section 16-7 the subject of terrain representation by means of digital elevation models (DEMs) was introduced, and concepts for deriving triangulated irregular networks (TINs) from DEMs were presented. It was noted that once a TIN model is created for a region, profiles and cross sections anywhere within the area can be readily derived using a computer. This can be of significant advantage where the general location of a proposed road or railroad has been decided upon, but the final alignment is not yet fixed. In those situations, controlling points and breaklines can be surveyed in the region where the facility is expected to be located, and a DEM for the area generated. Either ground or photogrammetic methods can be employed for deriving the terrain data. From the DEM information, a TIN model can be created, and then cross sections for the analysis of any number of alternate alignments are obtained automatically by computer.

After cross sections have been taken and plotted, *design templates* (outlines of base widths and side slopes of the planned excavation or embankment) are superimposed on each plot to define the excavation or embankment to be constructed at each station. Areas of these sections, called *end areas,* are obtained by computation or planimetering. Nowadays, using computers, end areas are calculated directly from field cross-section data and design information. From the end areas, volumes are determined by the *average-end-area,* or *prismoidal,* formula, discussed later in this chapter.

Figure 27-1 portrays a section of planned highway construction and illustrates some of the points just discussed. Centerline stakes are shown in place, and mark locations where cross sections are taken, in this instance at full stations. End areas, based on the planned grade line, size of roadway, and selected embankment and excavation slopes, are superimposed at each station and shown shaded. Areas of these shaded sections are determined, whereupon volumes are computed using formulas given in Section 27-5 or 27-8. Note in the figure that embankment, or *fill,* is planned from stations 10 + 00 through 11 + 21, a transition from fill to excavation or *cut* occurs from station 11 + 21 to 11 + 64, and cut is required from stations 11 + 64 through 13 + 00.

27-4 TYPES OF CROSS SECTIONS

The types of cross sections commonly used on route surveys are shown in Figure 27-2. In flat terrain the *level section* (a) is suitable. The *three-level section* (b) is generally used where ordinary ground conditions prevail. Rough topography may require a *five-level section* (c), or more practically an *irregular section* (d). A *transition section* (e) and a *side-hill section* (f) occur in passing from cut to fill and on side-hill locations. In Figure 27-1, transition sections occur at stations 11 + 21 and 11 + 64, while a side-hill section exists at 11 + 40.

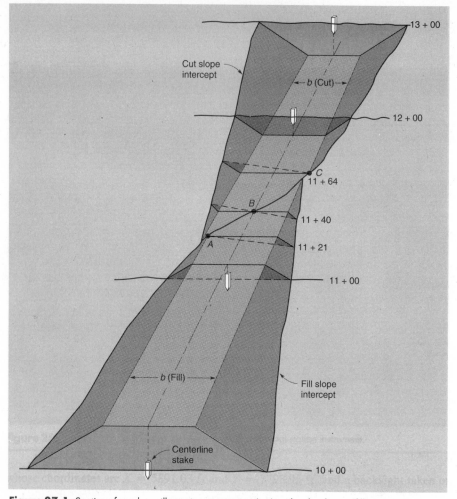

Figure 27-1 Section of roadway illustrating excavation (cut) and embankment (fill).

The width of base *b,* the finished roadway, is fixed by project requirements. As shown in Figure 27-1, it is usually wider in cuts than on fills to provide for drainage ditches. The side slope *s* [the horizontal dimension required for a unit vertical rise and illustrated in Figure 27-2(a)] depends on the type of soil encountered. Side slopes in fill usually are flatter than those in cuts where the soil remains in its natural state.

Cut slopes of 1:1 (1 horizontal to 1 vertical) and fill slopes of $1\frac{1}{2}$:1 might be satisfactory for ordinary loam soils, but $1\frac{1}{2}$:1 in excavation and 2:1 in embankment are common. Even flatter proportions may be required—one cut in the Panama Canal area was 13:1—depending on type of soil, rainfall, and other factors. Formulas for areas of sections are readily derived and listed with some of the sketches in Figure 27-2.

27-5 AVERAGE-END-AREA FORMULA

Figure 27-3 illustrates the concept of computing volumes by the average-end-area method. In the figure, A_1 and A_2 are end areas at two stations separated by a horizontal

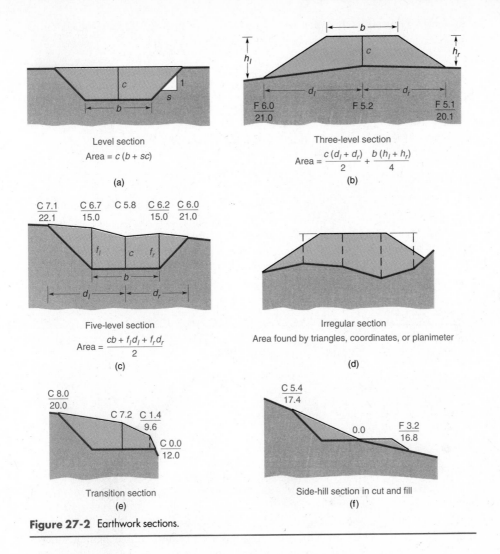

Level section
Area = $c(b + sc)$

(a)

Three-level section

$$\text{Area} = \frac{c(d_l + d_r)}{2} + \frac{b(h_l + h_r)}{4}$$

(b)

Five-level section

$$\text{Area} = \frac{cb + f_l d_l + f_r d_r}{2}$$

(c)

Irregular section
Area found by triangles, coordinates, or planimeter

(d)

Transition section

(e)

Side-hill section in cut and fill

(f)

Figure 27-2 Earthwork sections.

Figure 27-3 Volume by average-end-area method.

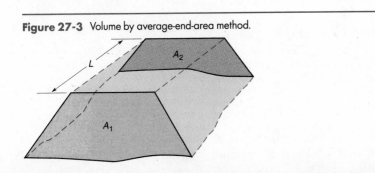

distance L. The volume between two stations is equal to the average of the end areas multiplied by the horizontal distance L between them. Thus

$$V_e = \frac{A_1 + A_2}{2} \times \frac{L}{27} \quad \text{yd}^3 \tag{27-1}$$

where V_e, the average-end-area volume, is in cubic yards, L in feet, and A_1 and A_2 in square feet. If L is 100 ft, as for full stations, Eq. (27-1) becomes

$$V_e = 1.85(A_1 + A_2) \tag{27-2}$$

Equations (27-1) and (27-2) are approximate and give answers that generally are slightly larger than the true prismoidal volumes (see Section 27-8). They are used in practice because of their simplicity, and contractors are satisfied because pay quantities are generally slightly greater than true values. Increased accuracy is obtained by decreasing the distance L between sections. When the ground is irregular, cross sections must be taken closer together.

EXAMPLE 27-1

Compute the volume of excavation between station 24 + 00, with an end area of 711 ft², and station 25 + 00, with an end area of 515 ft².

SOLUTION
By Eq. (27-2), $V = 1.85(A_1 + A_2) = 1.85(711 + 515) = 2268$ yd³.

27-6 DETERMINING END AREAS

End areas can be determined either graphically or by computation. In graphic methods, the cross section and template are plotted to scale on grid paper; then the number of small squares within the section can be counted and converted to area, or the area within the section can be measured using a planimeter (see Section 14-8.4). Computational procedures consist of either dividing the section into simple figures such as triangles and trapezoids and computing and summing these areas, or using the coordinate formula (see Section 14-5). These computational methods are discussed in the sections that follow. Most such calculations are now done by computer—usually by the coordinate method, which is general and readily programmed.

27-6.1 End Areas by Simple Figures

To illustrate the procedures of calculating end areas by simple figures such as triangles or trapezoids, assume the following excerpt of field notes applies to the cross section and end area shown in Figure 27-4:

$$HI = 879.29$$

		867.3	870.9	874.7	876.9	869.0	872.8
24 + 00	Lt	12.0	8.4	4.6	2.4	10.3	6.5
		50	36	20	₵	12	50

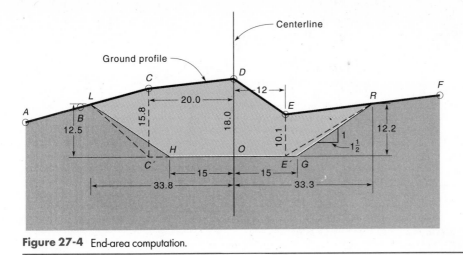

Figure 27-4 End-area computation.

In this excerpt, the top numbers are elevations obtained by subtracting rod readings (middle numbers) from the leveling instrument's HI. Bottom numbers are distances from centerline, beginning from the left. Assume the design calls for a level roadbed of 30-ft width, cut slopes of $1\frac{1}{2}{:}1$, and a subgrade elevation at station $24 + 00$ of 858.9. A corresponding design template is superimposed over the plotted cross section in Figure 27-4. Subtracting the subgrade elevation from cross-section elevations at C, D, and E yields the ordinates of cut required at those locations. Elevations and distances out from centerline to the *slope intercepts* at L and R must be either scaled from the plot or computed. Assuming they have been scaled (methods for computing them are given in Section 27-7), the following tabulation of distances from centerline and required cut ordinates at each point to subgrade elevation was made:

STATION	H	L	C	D	E	R	G
24 + 00	0	C 12.5	C 15.8	C 18.0	C 10.1	C 12.2	0
	15	33.8	20	0	12	33.3	15

Numbers above the lines (preceded by the letter C) are cut ordinates in feet; those below the lines are distances out from the centerline. Fills are denoted by the letter F. Using C instead of plus for cut, and F instead of minus for fill, eliminates confusion.

From the cut ordinates and distances from centerline shown, the area of the cross section in Figure 27-4 is computed by summing the individual areas of triangles and trapezoids. A list of the calculations is given in Table 27-1.

27-6.2 End Areas by Coordinates

The coordinate method for computing end areas can be used for any type of section and has many engineering applications. The procedure was described in Section 14-5 as a way to determine the area contained within a closed polygon traverse.

To demonstrate the method in end-area calculation, the example of Figure 27-4 will be solved. Coordinates of each point of the section are calculated, in an axis system

TABLE 27-1 END AREA BY SIMPLE FIGURES

FIGURE	COMPUTATION	AREA
ODCC'	-(18.0 + 15.8)20	338
C'CL	-(15.8)13.8	109
HLC'	$-\frac{1}{2}(5)12.5$	-31
ODEE'	-(18.0 + 10.1)12	169
EE'R	-(10.1)21.3	108
E'RG	-(3)12.2	18
		Area = 711 ft²

TABLE 27-2 END AREA BY COORDINATES

POINT	X	Y	PLUS +	MINUS −
O	0	0		
H	−15	0	0	0
L	−33.8	12.5	0	+188
C	−20	15.8	−250	+534
D	0	18.0	0	+360
E	12	10.1	216	0
R	33.3	12.2	336	−146
G	15	0	183	0
O	0	0	0	0
			+485	+936
			+936	
			2)1421	
			711 ft² (nearest ft²)	

having point *O* as its origin, using the earlier listed data on cuts and distances from centerline. In computing coordinates, distances to the right of centerline and cut values are considered plus; distances left and fill values are minus. Beginning with point *O* and proceeding clockwise around the figure, the coordinates of each point are listed in sequence. Point *O* is repeated at the end (see Table 27-2). Then Eq. (14-7) is applied, with products of diagonals downward to the right (solid arrows) considered minus, and diagonal products down to the left (dashed arrows) plus. Algebraic signs of the coordinates must be considered. Thus a positive product (dashed arrow) having a negative coordinate will actually be minus. The total area is obtained by dividing the absolute value of the algebraic summation of all products by 2. The calculations are illustrated in Table 27-2.

It is necessary to make separate computations for cut and fill end areas when they occur in the same section (as at station 11 + 40 of Figure 27-1), since they must always be tabulated independently for pay purposes. Payment is normally made only for excavation (its unit price includes making and shaping the fills) except on projects consisting primarily of embankment such as levees, earth dams, some military fortifications, and highways built up by continuous fill in flat areas.

27-7 COMPUTING SLOPE INTERCEPTS

The elevations and distances out from the centerline of slope intercepts can be calculated using cross-section data and the cut or fill slope values. In Figure 27-4, for example, intercept R, occurs between ground profile point E (distance 12 ft right and elevation 869.0) and point F (distance 50 ft right and elevation 872.8). The cut slope is $1\frac{1}{2}{:}1$, or 0.67 ft/ft. A more detailed diagram illustrating the geometry for calculating slope intercept R is given in Figure 27-5.

The slope along ground line EF is $(872.8 - 869.0)/38 = 0.10$ ft/ft, where 38 ft is the horizontal distance between the points. The elevation of G' (point vertically above G) is $869.0 + 0.10(3) = 869.3$; thus ordinate GG' is $(869.3 - 858.9) = 10.4$ ft. Lines EF and GR converge at a rate equal to the difference in their slopes (because they are both sloping upward) at $0.67 - 0.10 = 0.57$ ft/ft. Dividing ordinate GG' by this convergence yields horizontal distance GR, or $10.4/0.57 = 18.3$ ft. Adding 18.3 to distance OG yields $18.3 + 15 = 33.3$ ft, which is the distance from centerline to slope intercept R. Finally, to obtain the elevation of R, the increase in elevation from E to R is added to the elevation of E, or $0.10(21.3) + 869.0 = 871.1$. The cut ordinate at R equals $871.1 - 858.9 = 12.2$ ft. Recall that 33.3 and 12.2 were the X and Y coordinates, respectively, used in the end-area calculations of Section 27-6.2 and Table 27-2.

The elevation and distance from centerline of the slope intercept L of Figure 27-4 are calculated in a similar manner, except the rate of convergence of lines CB and HL is the sum of their slopes because CB slopes downward and HL upward. Calculations of slope intercepts are somewhat laborious, but routine when programmed for solution by computer. If a computer is not used for computing end areas and volumes, the usual

Figure 27-5 Computation of slope intercept R of Figure 27-4.

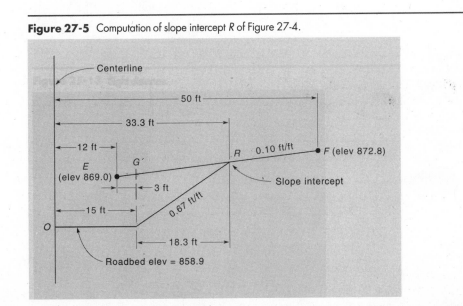

procedure is to plot the cross sections and templates, determine the end area by planimeter, and scale the slope intercepts from the plot. Slope intercepts are essential since the placement of slope stakes that guide construction operations is based on them.

27-8 PRISMOIDAL FORMULA

The prismoidal formula applies to volumes of all geometric solids that can be considered prismoids. A prismoid, illustrated in Figure 27-6, is a solid having ends that are parallel but not congruent, and trapezoidal sides that are also not congruent. Most earthwork solids obtained from cross section data fit this classification. From a practical standpoint, however, the differences in volumes computed by the average-end-area method and the prismoidal formula are usually so small as to be negligible. Where extreme accuracy is needed, such as in expensive rock cuts, the prismoidal method can be used.

One arrangement of the prismoidal formula is

$$V_p = \frac{L(A_1 + 4A_m + A_2)}{6 \times 27} \qquad (27\text{-}3)$$

where V_p is the prismoidal volume in cubic yards, A_1 and A_2 are areas of successive cross sections taken in the field, A_m is the area of a "computed" section midway between A_1 and A_2, and L is the horizontal distance between A_1 and A_2.

To use this formula it is necessary to know area A_m of the section halfway between stations. This is found by the usual computation *after averaging the heights and widths of the end sections*. Obviously, the middle area is *not* the average of the end areas, since there would then be no difference between the results of the end-area formula and the prismoidal formula.

The prismoidal formula generally gives a volume smaller than that found by the average-end-area formula. For example, the volume of a pyramid by the prismoidal formula is $Ah/3$ (the exact value), whereas by the average-end-area method it is $Ah/2$. An exception occurs when the center height is great but the width narrow at one station, and the center height small but the width large at the adjacent station. Figure 27-6 illustrates this condition.

The difference between the volumes obtained by the average-end-area formula and the prismoidal formula is called the *prismoidal correction C_p*. Various books on route

Figure 27-6 Sections for which the prismoidal correction is added to the end-area volume.

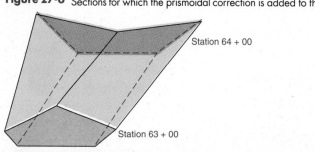

Station 64 + 00

Station 63 + 00

surveying give formulas and tables for computing prismoidal corrections, which can be applied to average-end-area volumes to get prismoidal volumes. A prismoidal correction formula, accurate for three-level sections and close enough for most others, is

$$C_p = \frac{L}{12 \times 27}(c_1 - c_2)(w_1 - w_2) \qquad (27\text{-}4)$$

where C_p is the volume of the prismoidal correction in cubic yards, c_1 and c_2 are center heights in cut (or in fill), and w_1 and w_2 are widths of sections (from slope intercept to slope intercept) at adjacent sections.

If the product of $(c_1 - c_2)(w_1 - w_2)$ is minus, as in Figure 27-6, the prismoidal correction is added rather than subtracted from the end-area volume.

27-9 VOLUME COMPUTATIONS

Volume calculations for route construction projects are usually done by computer and arranged in tabular form. To illustrate this procedure, assume that end areas listed in columns (2) and (3) of Table 27-3 apply to the section of roadway illustrated in Figure 27-1. By using Eq. (27-1), cut and fill volumes are computed and tabulated in columns (4) and (5).

The volume computations illustrated in Table 27-3 include the transition sections of Figure 27-1. This is normally not done when preliminary earthwork volumes are being estimated (during design and prior to construction) because the exact locations of the transition sections and their configurations are usually unknown until slope staking occurs. Thus for calculating preliminary earthwork quantities, an end area of zero would be used at the station of the centerline grade point (station 11 + 40 of Figure 27-1), and transition sections (stations 11 + 21 and 11 + 64 of Figure 27-1) would not

TABLE 27-3 TABULAR FORM OF VOLUME COMPUTATION

STATION (1)	END AREA (ft)^2 CUT (2)	FILL (3)	VOLUME (yd)^3 CUT (4)	FILL (5)	FILL VOLUME +25% (yd)^3 (6)	CUMULATIVE VOLUME (yd)^3 (7)
10 + 00		992				0
				2614	3268	
11 + 00		421				−3268
				190	238	
11 + 21	0	68				−3506
			12	29	37	
11 + 40	34	31				−3531
			79	14	17	
11 + 64	144	0				−3469
			553			
12 + 00	686					−2916
			2967			
13 + 00	918					+51

appear in the computations. After slope staking (procedures for slope staking are described in Section 24-7) the locations and end areas of transition sections are known, and they should be included in final volume computations, especially if they significantly affect the quantities for which payment is made.

In highway and railroad construction, excavation or cut material is used to build embankments or fill sections. Unless there are other controlling factors, a well-designed grade line should nearly balance total cut volume against total fill volume. To accomplish a balance, either fill volumes must be expanded or cut volumes shrunk.[1] This is necessary because, except for rock cuts, embankments are compacted to a density greater than that of material excavated from its natural state, and to balance earthwork this must be considered. (Rock cut expands to occupy a greater fill volume; thus either the cut must be expanded or the fill shrunk to obtain a balance.) The rate of expansion depends on the type of material and can never be estimated exactly. However, samples and records of past projects in the immediate area are helpful in assigning reasonable factors. Column (6) of Table 27-3 lists expanded fills for the example of Figure 27-1, where a 25% factor was applied.

To investigate whether or not an earthwork balance is achieved, *cumulative volumes* are computed. This involves adding cut and expanded fill volumes algebraically from project beginning to end, with cuts considered positive and fills negative. Cumulative volumes are listed in column (7) of Table 27-3. In this example there is a cut volume excess of 51 yd^3 between stations $10 + 00$ and $13 + 00$ or, in other words, there is a surplus of that much excavation.

To analyze the movement of earthwork quantities on large projects, *mass diagrams* are constructed. These are plots of cumulative volumes for each station as the ordinate, versus the stations on the abscissa. Horizontal (balance) lines on the mass diagram then determine the limit of economic haul and the direction of movement of material. Mass diagrams are described more thoroughly in books on route surveying.

If there is insufficient material from cuts to make the required fills, the difference must be *borrowed* [obtained from borrow pits or other sources such as by "day-lighting" curves (flattening cut slopes to improve visibility)]. If there is excess cut, it is *wasted* or perhaps used to extend and flatten the fills.

For projects with more than a few cross sections, computer programs are available and are generally used for earthwork computations, but surveyors and engineers must still understand the basic methods.

27-10 UNIT-AREA, OR BORROW-PIT, METHOD

On many projects, except long linear route construction, the quantity of earth, gravel, rock, or other material excavated or filled can often best be determined by borrow-pit leveling. The quantities computed form the basis for payment to the contractor or materials supplier. The number of cubic yards of coal or other loose materials in stockpiles can be found in the same way.

As an example, assume the area shown in Figure 27-7 is to be graded to an elevation of 358.0 for a building site. Notes for the field work are shown in Plate D-4 of Appendix

---

[1]Expansion of fill volumes is generally preferred, since payment is usually based on actual volumes of material excavated.

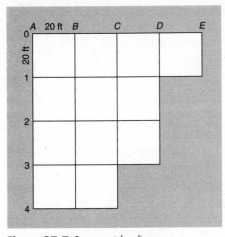

Figure 27-7 Borrow-pit leveling.

D. The area to be covered in this example is staked in squares of 20 ft, although 10, 50, 100, or more feet could be used, with the choice depending on project size and accuracy desired. A total station, theodolite (or transit) and tape, or only a tape may be used for the layout. A bench mark of known or assumed elevation is established outside the area in a place not likely to be disturbed.

After the area is laid out in squares, to determine elevations, a level is set up at any convenient location, a plus sight taken on the bench mark, and minus sights read on corners of the squares. If the terrain is not too rough, it may be possible to select a point near the area center and take sights on all corners from the same setup, as in the example of Plate D-4. For rough terrain it may be most convenient to determine the corner elevations by radial surveying from one well-chosen setup using a total station instrument.

Corners of the squares are designated by letters and numbers, such as A-1, C-4, and D-2. For site grading to a specified elevation, say 358.0 ft, the amount of cut or fill at each corner is obtained by subtracting 358.0 from its elevation. For each square, then, the average height of the four corners of the prism of cut or fill is determined and multiplied by the base area, 20×20 ft $= 400$ ft^2, to get the volume. The total volume is found by adding the individual values for each block and dividing by 27 to obtain the result in cubic yards.

To simplify calculations, the cut at each corner multiplied by the number of times it enters the volume computation can be shown in a separate column. The column sum is divided by 4 and multiplied by the base area of one block to get the volume. In equation form, this procedure is given as

$$V = \Sigma(h_{i,j}\, n)\, \frac{A}{4 \times 27} \quad (\text{yd}^3) \tag{27-5}$$

where $h_{i,j}$ is the corner height in row i and column j, and n the number of squares to which that height is common. The corner at C-4, for example, is common to only one square, *D*-2 is common to two, *D*-1 is common to three, and *C*-1 is common to four. $\Sigma(h_{i,j}\, n)$ is the sum of the products of the height and the number of common squares, and *A* is the area of one square. An example illustrating the use of Eq. (27-5) is given in the field notes of Plate D-4.

27-11 CONTOUR-AREA METHOD

Volumes based on contours can be obtained from contour maps by planimetering the area enclosed by each contour and multiplying the average of the areas of adjacent contours by their spacing, the contour interval, using Eq. (27-1). Use of the prismoidal formula is seldom, if ever, justified in this type of computation.

Instead of determining areas enclosed within contours by planimeter, they can be obtained using the coordinate formula [Eq. (14-7) or (14-8)]. In this procedure a tablet digitizer like the one shown in Figure 29-8 is first used to measure the coordinates along each contour at enough points to define its configuration satisfactorily.

The contour-area method is suitable for determining volumes over large areas, for example, computing the amounts and locations of cut and fill in the grading for a proposed airport runway to be constructed at a given elevation. Another useful application of the contour-area method is in determining the volume of water that will be impounded in the reservoir created by a proposed dam.

EXAMPLE 27-2   Compute the volume of water impounded by the proposed dam illustrated in Figure 27-8. Map scale is 500 ft/in. and the proposed spillway elevation 940 ft.

SOLUTION

The crosshatched portion of Figure 27-8 represents the area that will be inundated with water when the reservoir is full. The solution is presented in Table 27-4. Column (2) gives the planimetered areas for each contour in square inches, and in column (3) these areas have been converted to acres based on map scale, that is, 1 in.^{2} = $[(500)^2]/43,560$ = 5.739 acres. Column (4) gives the volumes between adjacent contours, computed by Eq. (27-1). The sum of column (4), 1285.9 acre-ft, is the volume of the reservoir.

27-12 MEASURING VOLUMES OF WATER DISCHARGE

The volumes of water discharge in streams and rivers is a matter of vital concern and must be monitored regularly. In the usual procedure, the stream's cross section is broken into a series of uniformly spaced vertical sections, as illustrated in Figure 27-9. The U.S. Geological Survey recommends use of from 25 to 30 sections, with not more than 5% of the total flow occurring in any one section. Depths and current velocities are measured at each ordinate using a *current meter*. (There are various types available.) The discharge volume for each section is the product of its area and average current velocity. The sum of all section discharges is the total volume of water passing through the stream at the cross-section location. Units of section areas and current velocities are ft^2 and ft/sec, respectively; thus the discharge unit is ft^3/sec.

Current velocities can be measured at every 0.1 of the depth at each ordinate and the average taken. Alternatively, a good average results from the mean of the 0.2- and 0.8-

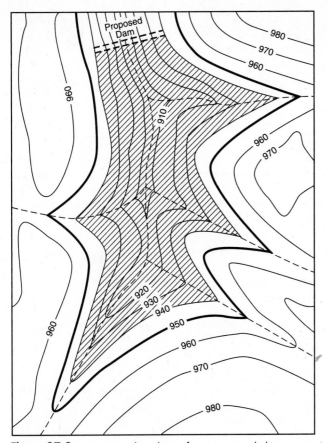

Figure 27-8 Determining the volume of water impounded in a reservoir by the contour-area method.

TABLE 27-4 VOLUME COMPUTATION BY
CONTOUR-AREA METHOD

| CONTOUR | AREA | | VOLUME |
| | (in.²) | (acres) | (acre-ft) |
(1)	(2)	(3)	(4)
910	1.683	9.659	—
920	5.208	29.889	197.7
930	11.256	64.598	472.4
940	19.210	110.246	615.8
			$\Sigma = 1285.9$

depth velocities, or a single measurement at the 0.6-depth point. For depths up to $2\frac{1}{2}$ ft, the U.S. Geological Survey uses the 0.6 method; for deeper sections the 0.2 and 0.8 procedure is employed.

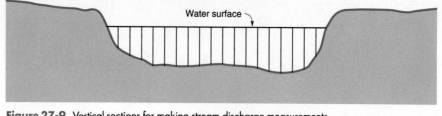

Figure 27-9 Vertical sections for making stream discharge measurements.

The cross section should be taken at right angles to the stream, and in a straight reach with solid bottom and uniform flow. In shallow streams, measurements can be made by wading, in which case the current meter is held upstream free from eddies caused by the wader's legs. In deeper streams and rivers, measurements are taken from boats, bridges, or overhead cable cars. In these situations, the current meter, with a heavy weight attached to its bottom, is suspended by a cable and thus doubles as a leadline for measuring depths.

27-13 SOURCES OF ERROR

Some common errors in determining areas of sections and volumes of earthwork are:

1. Making errors in measuring cross sections
2. Failing to use the prismoidal formula where it is justified
3. Carrying out areas of cross sections beyond the nearest square foot, or beyond the limit justified by the field data
4. Carrying out volumes beyond the nearest cubic yard

27-14 MISTAKES

Some typical mistakes made in earthwork calculations are:

1. Confusing algebraic signs in end-area computations using the coordinate method
2. Using Eq. (27-2) for full-station volume computation when partial stations are involved
3. Using end-area volumes for pyramidal or wedge-shaped solids
4. Mixing cut and fill quantities

PROBLEMS

27-1 Why is a roadway in cut normally wider than the same roadway in fill?

*27-2 Prepare a table of end areas versus depths of fill from 0 to 20 ft by increments of 2 ft for level sections, a 30-ft-wide level roadbed, and side slopes of 3:1.

27-3 Similar to Problem 27-2, except use side slopes of 4:1.

Draw the cross sections and compute V_e for the data given in Problems 27-4 through 27-7.

27-4 Two level sections of 100-ft stations with center heights of 6.7 and 8.3 ft in cut, base width 40 ft, side slopes 2:1.

*Asterisks indicate problems that have answers given in Appendix G.

*27-5 Two level sections 75 ft apart with center heights 5.0 and 7.6 ft in fill, base width 30 ft, side slopes 3:1.

27-6 The end area at station 36 + 00 is 265 ft². Notes giving distance from centerline and cut ordinates for station 36 + 40 are C 4.8/19.2; C 5.9; C 6.8/22.2. Base is 24 ft.

*27-7 An irrigation ditch with b = 16 ft and side slopes of 2:1. Notes giving distances from centerline and cut ordinates for stations 52 + 00 and 53 + 00 are C 2.4/12.8; C 3.0; C 3.7/15.4; and C 3.1/14.2; C 3.8; C 4.1/16.2.

27-8 Why must cut and fill volumes be totaled separately?

27-9 For the data tabulated, calculate the volume of excavation in cubic yards between stations 10 + 00 and 15 + 00.

STATION	CUT END AREA (ft²)
10 + 00	275
11 + 00	392
12 + 00	486
13 + 00	420
14 + 00	244
15 + 00	180

*27-10 For the data listed, tabulate cut, fill, and cumulative volumes in cubic yards between stations 10 + 00 and 20 + 00. Use an expansion factor of 1.25 for fills.

STATION	END AREA (ft²)	
	CUT	FILL
10 + 00	0	
11 + 00	158	
12 + 00	403	
13 + 00	560	
14 + 00	426	
14 + 60	0	0
15 + 00		124
16 + 00		283
17 + 00		350
18 + 00		356
19 + 00		183
20 + 00		126

27-11 Compute the section areas in Problem 27-4 by the coordinate method.

27-12 Calculate the section area of station 36 + 40 in Problem 27-6 by the coordinate method.

*27-13 Determine the section areas in Problem 27-7 by the coordinate method.

27-14 Compute C_p and V_p for Problem 27-4. Is C_p significant?

*27-15 Calculate C_p and V_p for Problem 27-7. Would C_p be significant in rock cut?

27-16 From the following excerpt of field notes, plot the cross section on graph paper and superimpose on it a design template for a 28-ft-wide level roadbed with fill slopes of 3:1 and a subgrade elevation at centerline of 969.60 ft. Determine the end area graphically by counting squares.

*Asterisks indicate problems that have answers given in Appendix G.

HI = 968.11

20 + 00	Lt	5.2/50	4.8/22	6.6/0	5.9/12	7.0/30	8.1/50

27-17 For the data of Problem 27-16, determine the end area by planimeter.

*27-18 For the data of Problem 27-16, calculate slope intercepts and determine the end area by the coordinate method. Check by computing areas of triangles and trapezoids.

27-19 From the following excerpt of field notes, plot the cross section on graph paper and superimpose on it a design template for a 38-ft-wide level roadbed with cut slopes of 4:1 and a subgrade elevation of 1239.75 ft. Determine the end area graphically by counting squares. Check by planimeter.

HI = 1254.60

46 + 00	Lt	8.0/60	7.9/27	5.5/10	4.9/0	6.6/24	7.5/60

27-20 For the data of Problem 27-19, calculate slope intercepts and determine the end area by the coordinate method. Check by computing areas of triangles and trapezoids.

*27-21 Complete the following notes and compute V_e and V_p. The roadbed is level, the base is 30 ft, and the side slopes are 4:1.

Station 88 + 00	C 6.4	C 3.6	C 5.7
Station 87 + 00	C 3.1	C 4.9	C 4.3

27-22 Similar to Problem 27-21, except the base is 36 ft and the side slopes are 3:1.

27-23 Calculate V_e and V_p for the following notes. Base in fill is 24 ft, base in cut is 34 ft, side slopes are 2:1.

12 + 90	C 3.4	C 2.0	0.0	F 3.0
	23.8	0	6.0	18.0

12 + 30	C 2.2	0.0	F 5.0
	21.4	0	22.0

*27-24 Calculate V_e, C_p, and V_p for the following notes. Base in cut is 40 ft, side slopes are 4:1.

46 + 00	C 4.2	C 2.0	C 3.6
	36.8	0	34.4

45 + 00	C 2.4	C 3.0	0.0
	29.6	0	20.0

*Asterisks indicate problems that have answers given in Appendix G.

For Problems 27-25 and 27-26 compute the reservoir capacity (in acre-ft) between highest and lowest contours for planimetered areas on a topographic map.

27-25

Elevation (ft)	860	870	880	890	900	910
Area (ft^2)	1730	2860	3493	4950	6540	10,605

*27-26

Elevation (ft)	1015	1020	1025	1030	1035	1040
Area (ft^2)	1815	3957	6000	9164	13,336	21,677

27-27 A calibrated polar planimeter gives an average reading of 1.476 revolutions of the roller over a 4-in.-diameter circle; 20-ft contours planimetered on a reservoir site map to a scale of 1 in. = 500 ft give the values tabulated. Calculate the reservoir volume in acre-ft.

Contour	720	740	760	780	800
Planimeter reading	0.180	1.116	1.806	4.509	8.094

27-28 State two situations where prismoidal corrections are most significant.

27-29 Write a computer program to calculate slope intercepts and end areas by the coordinate method, given cross-section notes and roadbed design information. Use the program to calculate the slope intercepts for the data of Problem 27-16.

*27-30 Distances (ft) from the left bank, corresponding depths (ft), and velocities (ft/sec), respectively, are given for a stream discharge measurement. What is the volume in ft^3/sec? 0, 1.0, 0; 10, 2.3, 1.30; 20, 3.0, 1.54; 30, 2.7, 1.90; 40, 2.4, 1.95; 50, 3.0, 1.60; 60, 3.1, 1.70; 70, 3.0, 1.70; 80, 2.8, 1.54; 90, 3.3, 1.24; 100, 2.0, 0.58; 108, 2.2, 0.28; 116, 1.5, 0.

*Asterisks indicate problems that have answers given in Appendix G.

BIBLIOGRAPHY

Brinker, R. C., and R. Minnick. 1987. *The Surveying Handbook.* New York: Van Nostrand Reinhold, chap. 19.

Chambers, D. W. 1989. "Estimating Pit Excavation Volume Using Unequal Intervals." *ASCE, Journal of Surveying Engineering* 115 (No. 4):390.

Chen, C., and H. Lin. 1990. "Estimating Pit Excavation Volume Using Cubic Spline Volume Formula." *ASCE, Journal of Surveying Engineering* 117 (No. 2):51.

Chen, C., and H. Lin. 1992. "Estimating Excavation Volumes Using New Formulas." *Surveying and Land Information Systems* 52 (No. 2):104.

Easa, S. M. 1988. "Estimating Pit Excavation Volume Using Nonlinear Ground Profile." *ASCE, Journal of Surveying Engineering* 114 (No. 2):71.

Hebrank, A. J. 1984. "An Application of Volume Determination by Vertical Prismoids." *Surveying and Mapping* 44 (No. 4):323.

Meyer, C. F., and D. W. Gibson. 1980. *Route Surveying and Design,* 5th ed. New York: Harper & Row.

Siyam, Y. M. 1987. "Precision in Cross-Sectional Area Calculations on Earthwork Determination." *ASCE, Journal of Surveying Engineering* 113 (No. 3):139.

28
PHOTOGRAMMETRY

28-1 INTRODUCTION

Photogrammetry may be defined as the science, art, and technology of obtaining reliable information from photographs. It encompasses two major areas of specialization: *metrical* and *interpretative*. The first area is of principal interest to surveyors, since it is applied to determine distances, elevations, areas, volumes, and cross sections, and to compile topographic maps from measurements made on photographs. *Aerial* photographs (exposed from aircraft) are normally used, although for certain special work, *terrestrial* photos (taken from earth-based cameras) are employed.

Interpretative photogrammetry involves recognizing objects from their photographic images and judging their significance. Critical factors considered in identifying objects are the shapes, sizes, patterns, shadows, tones, and textures of their images. This area of photogrammetry was traditionally called *photographic interpretation* because initially it relied on aerial photos. More recently, other sensing and imaging devices such as multispectral scanners, thermal scanners, radiometers, and side-looking airborne radar have been developed which aid in interpretation. These instruments sense energy in wavelengths beyond those which the human eye can see or ordinary photographic film can record. They are often carried in aircraft as remote as satellites; hence a new term, *remote sensing*, is now generally applied to the interpretative area of photogrammetry.

In this chapter metrical photogrammetry using aerial photographs will be emphasized because it is the phase most frequently applied in surveying work. Remote sensing is rapidly gaining importance, however, and the subject is discussed further in Section 28-19.

28-2 USES OF PHOTOGRAMMETRY

Photography dates back to 1839, and the first attempt to use photogrammetry in preparing a topographic map occurred a year later. Photogrammetry is now the chief method of topographic mapping. The U.S. Geological Survey, for example, employs the procedure almost exclusively in compiling its quadrangle maps. Cameras, films, plotting instruments, and techniques have been improved continually so that photogrammetrically prepared maps today meet very high accuracy standards. Other advantages of this method of mapping are **(1)** speed of coverage of an area, **(2)** relatively low cost, **(3)** ease of obtaining topographic details, especially in inaccessible areas, and **(4)** reduced likelihood of omitting data due to the tremendous amount of detail shown in photographs.

Photogrammetry presently has many applications in surveying and engineering. It is used, for example, in land surveying to compute coordinates of section corners, boundary corners, or points of evidence that help locate these corners. Large-scale maps are made by photogrammetric procedures for many uses, one being subdivision design. Photogrammetry is used to map shorelines in hydrographic surveying, to determine precise ground coordinates of points in control surveying, and to develop maps and cross sections for route and engineering surveys. Photogrammetry is playing an important role in developing the necessary data for modern Land and Geographic Information Systems.

Photogrammetry is also being successfully applied in many nonengineering fields, for example, geology, archeology, forestry, agriculture, conservation, planning, military intelligence, traffic management, and accident investigation. It is beyond the scope of this chapter to describe all the varied applications of photogrammetry. Use of the science has increased dramatically in recent years, and its future growth for solving measurement and mapping problems is assured.

28-3 AERIAL CAMERAS

Aerial mapping cameras are perhaps the most important photogrammetric instruments, since they expose the photographs on which the science depends. To understand photogrammetry, especially the geometrical foundation of its equations, it is essential to have a fundamental understanding of cameras and how they operate.

Aerial cameras must be capable of exposing a large number of photographs in rapid succession while moving in an aircraft at high speed; so a short cycling time, fast lens, efficient shutter, and large-capacity magazine are required.

Single-lens frame cameras are the type most often used in metrical photogrammetry. These cameras expose the entire frame or format simultaneously through a lens held at a fixed distance from the focal plane. Generally they have a format size of 9×9 in. and lenses with focal lengths of 6 in., although $3\frac{1}{2}$-, $8\frac{1}{4}$-, and 12-in. lengths are also used. A single-lens frame camera is shown in Figure 28-1.

The principal components of a single-lens frame camera are shown in the diagram of Figure 28-2. These include the *lens* (the most important part) which gathers incoming light rays and brings them to focus on the focal plane; the *shutter* to control the interval of time that light passes through the lens; a *diaphragm* to regulate the size of lens opening; a *filter* to reduce the effect of haze and distribute light uniformly over the format; a *camera cone* to support the lens-shutter-diaphragm assembly with respect to the focal plane and prevent stray light from striking the film; a *focal plane,* the surface on which the film lies when exposed; *fiducial marks* (not shown in Figure 28-2), four or eight in number to define the photographic principal point (described later); a *camera body* to house the drive mechanism that cocks and trips the shutter, flattens the film, and advances it between exposures; and a *magazine,* which holds the supply of exposed and unexposed film.

An aerial camera shutter can be operated manually by an operator or by an *intervalometer,* which automatically trips the shutter at specified intervals. A level vial

Figure 28-1 Aerial camera. (Courtesy Carl Zeiss.)

Figure 28-2 Principal components of a single-lens frame aerial camera.

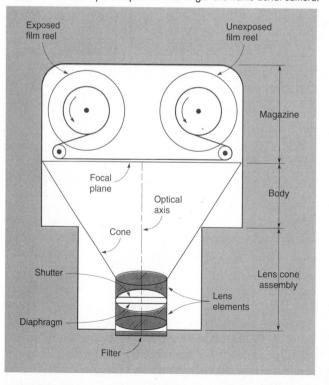

Figure 28-3 Vertical aerial photograph. (Courtesy Ayres & Associates, Inc.)

attached to the camera enables an operator to keep the optical axis of the camera lens (which is perpendicular to the focal plane) nearly vertical in spite of any slight tip and tilt of the aircraft. Polyester roll film is normally used with magazine capacities of 200 ft or more.

Images of the fiducial marks are printed on the photographs, and lines joining opposite pairs intersect at or very near the *principal point,* defined as the point where a perpendicular from the emergent nodal point of the camera lens strikes the focal plane. Fiducial marks may be located in the corners, as shown in Figure 28-3, on the sides, or preferably in both places, as shown in Figure 28-4.

Aerial mapping cameras are laboratory calibrated to get precise values for the focal length and lens distortions. Flatness of the focal plane, relative position of the principal point with respect to the fiducial marks, and fiducial mark locations are also specified. These calibration data are necessary for precise photogrammetric work.

28-4 TYPES OF AERIAL PHOTOGRAPHS

Aerial photographs exposed with single-lens frame cameras are classified as *vertical* (taken with the camera axis aimed vertically downward, or as nearly vertical as

Figure 28-4 Low oblique aerial photograph. (Courtesy Wisconsin Department of Transportation.)

possible) and *oblique* (made with the camera axis intentionally inclined at an angle between the horizontal and vertical). Oblique photographs are further classified as *high* if the horizon shows on the picture, and *low* if it does not. Figures 28-3 and 28-4 show examples of vertical and low oblique photographs, respectively. As illustrated by these examples, aerial photos clearly depict all natural and cultural features within the region covered such as roads, railroads, buildings, rivers, bridges, trees, and cultivated lands.

Vertical photographs are the principal mode of obtaining imagery for photogrammetric work. Obliques are seldom used for mapping or metrical applications, but are advantageous in interpretative work and for reconnaissance.

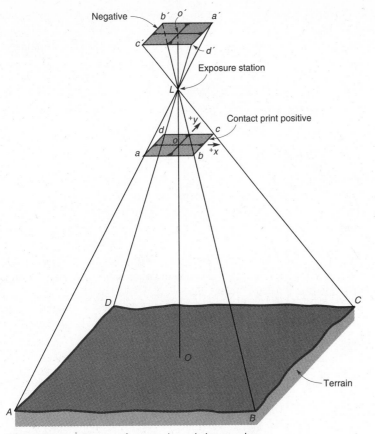

Figure 28-5 Geometry of a vertical aerial photograph.

28-5 VERTICAL AERIAL PHOTOGRAPHS

A *truly vertical photograph* results if the axis of the camera is exactly vertical when exposure is made. Despite all precautions, small tilts, generally less than 1° and rarely greater than 3°, are invariably present, and the resulting photos are called *near-vertical* or *tilted* photographs. Photogrammetric principles and practices have been developed to handle tilted photos, and accuracy is not sacrificed in compiling maps from them.

Although vertical photographs look like maps to laypersons, they are not true orthographic projections of the earth's surface. Rather, they are perspective views, and the principles of perspective geometry must be applied to prepare maps from them. Figure 28-5 illustrates the geometry of a vertical photograph taken at exposure station L. The photograph, considered a *contact print positive,* is a 180° exact reversal of the negative. The positive shown in Figure 28-5 is used to develop photogrammetric equations in subsequent sections.

Distance oL (Figure 28-5) is the camera focal length. The x and y reference axis system for measuring photographic coordinates of images is defined by straight lines joining opposite-side fiducial marks shown on the positive of Figure 28-5. The x axis, arbitrarily designated as the line most nearly parallel with the direction of flight, is positive in the direction of flight. Positive y is 90° counterclockwise from positive x.

Vertical photographs for topographic mapping are taken in strips which normally run lengthwise over the area to be covered. The strips or *flight lines* generally have a *sidelap* (overlap of adjacent flight lines) of about 30%. *Endlap* (overlap of adjacent photographs in the same flight line) is usually about 60 ± 5%. Figures 28-18(a) and (b) illustrates endlap and sidelap. If endlap is greater than 50%, all ground points will appear in at least two photographs and some will show in three. Images common to three photographs permit *aerotriangulation* to extend or densify control through a strip of photographs using only minimal existing control.

28-6 SCALE OF A VERTICAL PHOTOGRAPH

Scale is ordinarily interpreted as the ratio of a distance on a map to that same length on the ground. It is uniform throughout because a map is an orthographic projection. The scale of a vertical photograph is the ratio of photo distance to ground distance. Since a photograph is a perspective view, scale varies from point to point with variations in terrain elevation.

In Figure 28-6, L is the exposure station of a vertical photograph taken at an altitude H above datum. The camera focal length is f, and o is the photographic principal point. Points A, B, C, and D, which lie at elevations above datum of h_A, h_B, h_C, and h_D, respectively, are imaged on the photograph at a, b, c, and d. The scale at any point can be expressed in terms of its elevation, the camera focal length, and the flying height

Figure 28-6 Scale of a vertical photograph.

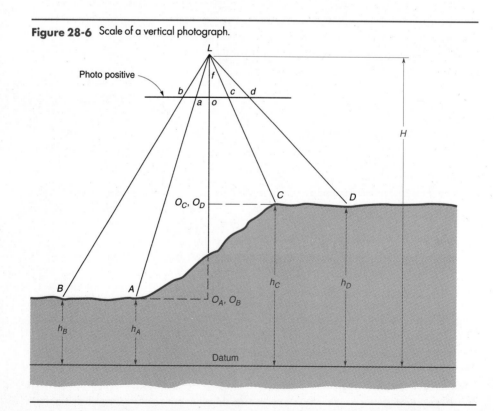

above datum. From Figure 28-6, for similar triangles *Lab* and *LAB*, the following expression can be written:

$$\frac{ab}{AB} = \frac{La}{LA} \qquad\qquad (a)$$

Also from similar triangles *Loa* and *LO$_A$A*, a similar expression results:

$$\frac{La}{LA} = \frac{f}{H - h_A} \qquad\qquad (b)$$

Equating (a) and (b), recognizing that *ab/AB* equals photo scale at *A* and *B*, and considering *AB* to be infinitesimally short, the equation for the scale at *A* is

$$S_A = \frac{f}{H - h_A} \qquad\qquad (c)$$

Scales at *B*, *C*, and *D* may be expressed similarly as $S_B = f/(H - h_B)$, $S_C = f/(H - h_C)$, and $S_D = f/(H - h_D)$.

It is apparent from these relationships that photo scale increases at higher elevations and decreases at lower ones. This concept is seen graphically in Figure 28-6. Ground lengths *AB* and *CD* are equal, but photo distances *ab* and *cd* are not, *cd* being longer and at larger scale than *ab* due to the higher elevation of *CD*. In general, by dropping subscripts, the scale *S* at any point whose elevation above datum is *h* may be expressed as

$$S = \frac{f}{H - h} \qquad\qquad (28\text{-}1)$$

where *S* is the scale at any point on a vertical photo, *f* the camera focal length, *H* the flying height above datum, and *h* the elevation of the point.

Use of an average photographic scale is frequently desirable, but must be accepted with caution as an approximation. For any vertical photographs taken of terrain whose average elevation above datum is h_{avg}, the average scale S_{avg} is

$$S_{avg} = \frac{f}{H - h_{avg}} \qquad\qquad (28\text{-}2)$$

The vertical photograph of Figure 28-6 was exposed with a 6-in. focal length camera at a flying height of 10,000 ft. above mean sea level **(a)** What is the photo scale at point *a* if the elevation of point *A* on the ground is 2500 ft above mean sea level? **(b)** For this photo, if the average terrain is 4000 ft above mean sea level, what is the average photo scale?

EXAMPLE 28-1

SOLUTION

(a) From Eq. (28-1),

$$S_A = \frac{f}{H - h_A} = \frac{6 \text{ in.}}{10{,}000 - 2500} = \frac{1 \text{ in.}}{1250 \text{ ft}} = 1{:}15{,}000$$

(b) From Eq. (28-2),

$$S_{avg} = \frac{f}{H - h_{avg}} = \frac{6 \text{ in.}}{10{,}000 - 4000} = \frac{1 \text{ in.}}{1000 \text{ ft}} = 1{:}12{,}000$$

The scale of a photograph can be determined if a map is available of the same area. This method does not require the focal length and flying height to be known. Rather, it is necessary only to measure the photographic distance between two well-defined points also identifiable on the map. Photo scale is then calculated from the equation

$$\text{photo scale} = \frac{\text{photo distance}}{\text{map distance}} \times \text{map scale} \tag{28-3}$$

In using Eq. (28-3), the distances must be in the same units, and the answer is the scale at average elevation of the two points used.

EXAMPLE 28-2   On a vertical photograph, the length of an airport runway measures 4.24 in. On a map plotted to a scale of 1:9600 it extends 7.92 in. What is the photo scale at the runway elevation?

SOLUTION

From Eq. (28-3),

$$S = \frac{4.24}{7.92} \times \frac{1}{9600} = \frac{1}{17{,}900} \quad \text{or} \quad 1 \text{ in.} = 1490 \text{ ft}$$

The scale of a photograph can also be computed readily if lines whose lengths are common knowledge appear in the photograph. Section lines, a football or baseball field, and so on, can be measured on the photograph and an approximate scale at that elevation ascertained as the ratio of measured photo distance to known ground length. With an approximate photographic scale known, rough determinations of the lengths of lines appearing in the photo can be made.

EXAMPLE 28-3   On a certain vertical aerial photo, a section line (assumed to be 5280 ft long) appears. Its photographic length is 3.32 in. On this same photo, a rectangular parcel of land measures 1.74 by 0.83 in. Calculate the approximate ground dimensions of the parcel and its acreage.

SOLUTION

1. Approximate photo scale $= \dfrac{3.32}{5280} = \dfrac{1 \text{ in.}}{1590 \text{ ft}}$ or 1 in. = 1590 ft

2. Parcel dimensions and area:

$$\text{length} = 1590 \times 1.74 = 2770 \text{ ft}$$
$$\text{width} = 1590 \times 0.83 = 1320 \text{ ft}$$
$$\text{area} = \dfrac{2770 \times 1320}{43,560} = 84 \text{ acres}$$

28-7 GROUND COORDINATES FROM A SINGLE VERTICAL PHOTOGRAPH

Ground coordinates of points whose images appear in a vertical photograph can be determined with respect to an arbitrary ground-axis system. The arbitrary X and Y ground axes are in the same vertical planes as photographic x and y, respectively, and the system's origin is in the datum plane vertically beneath the exposure station. Ground coordinates of points determined in this manner are used to calculate horizontal distances, horizontal angles, and areas.

Figure 28-7 illustrates a vertical photograph taken at flying height H above datum. Images a and b of ground points A and B appear on the photograph. The measured

Figure 28-7 Ground coordinates from a vertical photograph.

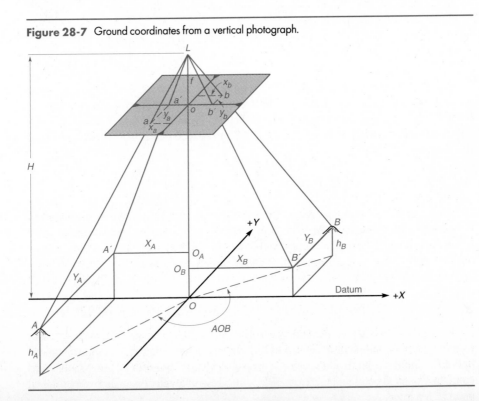

photographic coordinates are x_a, y_a, x_b, and y_b; the ground coordinates are X_A, Y_A, X_B, and Y_B. From similar triangles LO_AA' and Loa',

$$\frac{oa'}{O_AA'} = \frac{f}{H - h_A} = \frac{x_a}{X_A}$$

Then

$$X_A = \frac{(H - h_A)x_a}{f} \tag{28-4}$$

Also from similar triangles $LA'A$ and $La'a$,

$$\frac{a'a}{A'A} = \frac{f}{H - h_A} = \frac{y_a}{Y_A}$$

and

$$Y_A = \frac{(H - h_A)y_a}{f} \tag{28-5}$$

Similarly,

$$X_B = \frac{(H - h_B)x_b}{f} \tag{28-6}$$

$$Y_B = \frac{(H - h_B)y_b}{f} \tag{28-7}$$

Note that Eqs. (28-4) through (28-7) require point elevations h_A and h_B for their solution. These are normally either taken from existing contour maps, or they can be obtained by differential or trigonometric leveling. From the X and Y coordinates of points A and B, the horizontal length of line AB can be calculated using the Pythagorean theorem,

$$AB = \sqrt{(X_B - X_A)^2 + (Y_B - Y_A)^2}$$

Areas can be determined from X and Y coordinates by the method discussed in Chapter 14. The advantage of calculating lengths and areas by the coordinate formulas, rather than by average scale as in Example 28-3, is that greater accuracy results because differences in elevation, which cause the photo scale to vary, are rigorously accounted for.

28-8 RELIEF DISPLACEMENT ON A VERTICAL PHOTOGRAPH

Relief displacement on a vertical photograph is the shift or movement of an image from its theoretical datum location caused by the object's relief—that is, its elevation above or below datum. Relief displacement on a vertical photograph occurs along radial lines from the principal point and increases in magnitude with greater distance from principal point to the image.

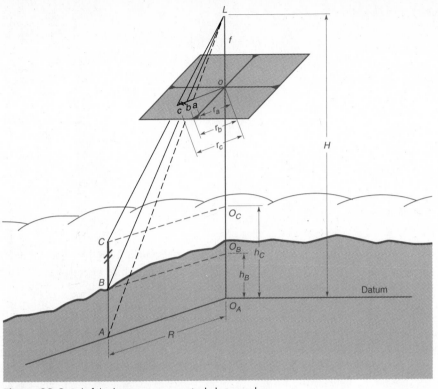

Figure 28-8 Relief displacement on a vertical photograph.

The concept of relief displacement in a vertical photograph taken from a flying height H above datum is illustrated in Figure 28-8. The camera focal length is f, and the principal point o. Points B and C are the base and top, respectively, of a pole with images at b and c on the photograph. A is an imaginary point on the datum plane vertically beneath B with corresponding imaginary position a on the photograph. Distance ab on the photograph is the image displacement due to h_B, the elevation of B above datum, and bc is the image displacement due to the height of the pole.

From similar triangles of Figure 28-8, an expression for relief displacement is formulated. First, from triangles LO_AA and Loa,

$$\frac{r_a}{R} = \frac{f}{H}$$

and rearranging,

$$r_aH = fR \tag{d}$$

Also from similar triangles LO_BB and Lob,

$$\frac{r_b}{R} = \frac{f}{H - h_B} \quad \text{or} \quad r_b(H - h_B) = fR \tag{e}$$

Equating (d) and (e),

$$r_a H = r_b(H - h_B)$$

and rearranging,

$$r_b - r_a = \frac{r_b h_B}{H}$$

If $d_b = r_b - r_a$ is the relief displacement of image b, then $d_b = r_b h_b/H$ can be written in general terms as

$$d = \frac{rh}{H} \qquad (28\text{-}8)$$

where d is the relief displacement, r the photo radial distance from the principal point to the image of the displaced point, h the height above datum of the displaced point, and H the flying height above that same datum.

Equation (28-8) can be used to locate the datum photographic positions of images on a vertical photograph. True horizontal angles may then be scaled directly from the datum images, and if the photo scale at datum is known, true horizontal lengths of the lines can be measured directly. The datum position is located by scaling the calculated relief displacement d of a point along a radial line to the principal point (inward for a point whose elevation is above datum).

Equation (28-8) can also be applied in computing heights of vertical objects such as buildings, church steeples, radio towers, trees, and power poles. To determine heights using the equation, images of both the top and bottom of an object must be visible.

EXAMPLE 28-4　In Figure 28-8 radial distance r_b to the image of the base of the pole is 75.23 mm, and radial distance r_c to the image of its top is 76.45 mm. The flying height H is 4000 ft above mean sea level, and the elevation of B is 450 ft. What is the height of the pole?

SOLUTION
The relief displacement is $r_c - r_b = 76.45 - 75.23 = 1.22$ mm. Selecting a datum at the pole's base and applying Eq. (28-8),

$$d = \frac{rh}{H} \quad \text{so} \quad 1.22 = \frac{76.45h}{4000 - 450}$$

Then

$$h = \frac{3550(1.22)}{76.45} = 56.6 \text{ ft}$$

The relief displacement equation is particularly valuable to photo interpreters, who are usually interested in relative heights rather than absolute elevations.

Figure 28-3 vividly illustrates relief displacements. This vertical photo taken over the national capital shows the relief displacement of the Washington Monument in the upper right-hand portion of the format. This displacement, as well as that of other buildings throughout the photograph, occurs radially outward from the principal point.

28-9 FLYING HEIGHT OF A VERTICAL PHOTOGRAPH

From previous sections it is apparent that the flying height above datum is an important parameter in solving basic photogrammetry equations. For rough computations, flying heights can be taken from altimeter readings if available. An approximate H can also be obtained by using Eq. (28-1) if a line of known length appears on a photograph.

The length of a section line is measured on a vertical photograph as 4.15 in. Find the approximate flying height above the terrain if $f = 6$ in.   EXAMPLE 28-5

SOLUTION
Assuming the datum at the section line elevation, Eq. (28-1) reduces to

$$\text{scale} = \frac{f}{H} \quad \text{and} \quad \frac{4.15}{5280} = \frac{6}{H}$$

from which

$$H = \frac{5280 \times 6}{4.15} = 7630 \text{ ft above the terrain}$$

If the images of two ground control points A and B appear on a vertical photograph, the flying height can be determined more precisely from the Pythagorean theorem,

$$L^2 = (X_B - X_A)^2 + (Y_B - Y_A)^2$$

Equations (28-4) through (28-7) are substituted in this expression,

$$L^2 = \left[\frac{(H - h_B)x_b - (H - h_A)x_a}{f}\right]^2 + \left[\frac{(H - h_B)y_b - (H - h_A)y_a}{f}\right]^2 \quad (28-9)$$

where L is the horizontal length of ground line AB, H the flying height above datum, h_A and h_B are the elevations of the control points above datum, and x and y the measured photo coordinates of the control points.

In Eq. (28-9) all variables except H are known. Hence a direct solution can be found for the unknown flying height. The equation is quadratic, so there are two solutions, but the incorrect one will be obvious and can be discarded.

28-10 STEREOSCOPIC PARALLAX

Parallax is defined as the apparent displacement of the position of an object with respect to a frame of reference due to a shift in the point of observation. For example, a person looking through the view finder of an aerial camera in an aircraft as it moves forward sees images of objects moving across the field of view. This apparent motion (parallax) is due to the changing location of the observer. By using the camera format as a frame of reference, it can be seen that parallax exists for all images appearing on successive photographs due to forward motion between exposures. Points closer to the camera (of higher elevation) will appear to move faster and have greater parallaxes than lower ones. For 60% endlap, the parallax of images on successive photographs should average approximately 40% of the focal plane width.

Parallax of a point is a function of its relief, and consequently measuring it provides a means of calculating elevations. It is also possible to compute X and Y ground coordinates from parallax.

Movement of an image across the focal plane between successive exposures takes place in a line parallel with the direction of flight. Thus to measure parallax, that direction must first be established. For a pair of overlapping photos, this is done by locating positions of the principal points and *corresponding principal points* (that is, principal points transferred to their places in the overlap area of the other photo). A line on each print ruled through these points defines the direction of flight. It also serves as the photographic x axis for parallax measurement. The y axis for making parallax measurements is drawn perpendicular to the flight line passing through each photo's principal point. The x coordinate of a point is scaled on each photograph with respect to the axes so constructed, and the parallax of the point is then calculated from the expression

$$p = x - x_1 \tag{28-10}$$

Photographic coordinates x and x_1 are measured on the left-hand and right-hand prints, respectively, with due regard given for algebraic signs.

Figure 28-9 illustrates an overlapping pair of vertical photographs exposed at equal flight heights H above datum. The distance between exposure stations L and L_1 is called B, the *air base*. The small inset figure shows the two exposure stations L and L_1 in superposition to make the similarity of triangles $La_1'a'$ and $LA'L_1$ more easily recognized. When these two similar triangles are equated, there results

$$\frac{p}{f} = \frac{B}{H - h}$$

from which

$$H - h = \frac{Bf}{p} \tag{28-11}$$

Also from similar triangles LOA' and Loa',

$$X = \frac{x}{f}(H - h) \tag{f}$$

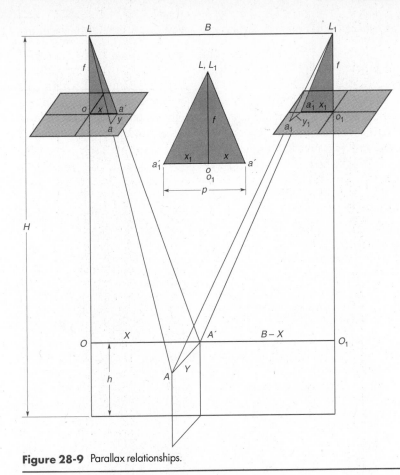

Figure 28-9 Parallax relationships.

Substituting Eq. (28-11) into (f) gives

$$X = \frac{B}{p}x \qquad (28\text{-}12)$$

and from triangles LAA' and Laa', with substitution of Eq. (28-11),

$$Y = \frac{B}{p}y \qquad (28\text{-}13)$$

In Equations (28-12) and (28-13) X and Y are ground coordinates of a point with respect to an origin vertically beneath the exposure station of the left photograph, with positive X coinciding with the direction of flight. Positive Y is 90° counterclockwise to positive X. The parallax of the point is p, x and y are the photographic coordinates of a point on the left-hand print, H is the flying height above datum, h the point's elevation above the same datum, and f the camera lens focal length.

Equations (28-11) through (28-13), commonly called the *parallax equations,* are useful in calculating horizontal lengths of lines and elevations of points. They also provide the fundamental basis for the design and operation of stereoscopic plotting instruments.

EXAMPLE 28-6

The length of line AB and elevations of points A and B, whose images appear on two overlapping vertical photographs, are needed. The flying height above mean sea level was 4050 ft, and the air base, 2410 ft. The camera had a 6-in. focal length. Measured photographic coordinates (in inches) on the left-hand print are $x_a = 2.10$, $x_b = 3.50$, $y_a = 2.00$, and $y_b = 1.05$; on the right-hand print, $x_{1a} = -2.25$ and $x_{1b} = -1.17$.

SOLUTION

From Eq. (28-10),

$$p_a = x_a - x_{1a} = 2.10 - (-2.25) = 4.35 \text{ in.}$$

$$p_b = x_b - x_{1b} = 3.50 - (-1.17) = 4.67 \text{ in.}$$

By Eqs. (28-12) and (28-13),

$$X_A = \frac{B}{p_a} x_a = \frac{2410 \times 2.10}{4.35} = 1160 \text{ ft}$$

$$X_B = \frac{2410 \times 3.50}{4.67} = 1810 \text{ ft}$$

$$Y_A = \frac{B}{p_a} y_a = \frac{2410 \times 2.00}{4.35} = 1110 \text{ ft}$$

$$Y_B = \frac{2410 \times (-1.05)}{4.67} = -542 \text{ ft}$$

From the Pythagorean theorem, length AB is

$$AB = \sqrt{(1810 - 1160)^2 + (-542 - 1110)^2} = 1780 \text{ ft}$$

By Eq. (28-11), the elevations of A and B are

$$h_A = H - \frac{Bf}{p_a} = 4050 - \frac{2410 \times 6}{4.35} = 726 \text{ ft}$$

$$h_B = 4050 - \frac{2410 \times 6}{4.67} = 954 \text{ ft}$$

28-11 STEREOSCOPIC VIEWING

The term *stereoscopic viewing* means seeing an object in three dimensions, a process requiring a person to have normal *binocular* (two-eyed) vision. In Figure 28-10 two *eyes* L and R are separated by a distance b called the *eye base*. When the eyes are

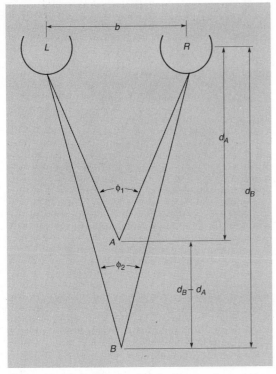

Figure 28-10 Parallactic angles in stereoscopic viewing.

focused on point *A*, their optical axes converge to form angle ϕ_1, and when sighting on *B*, ϕ_2 is produced. Angles ϕ_1 and ϕ_2 are called *parallactic angles*, and the brain associates distances d_A and d_B with them. The depth $d_B - d_A$ of the object is perceived from the brain's unconscious comparison of these parallactic angles.

If two photographs of the same subject are taken from two different perspectives or camera stations, the left print viewed with the left eye and simultaneously the right print seen with the right eye, a mental impression of a three-dimensional model results. In normal stereoscopic viewing (not using photos) the *eye base* gives a true impression of parallactic angles. While looking at aerial photographs stereoscopically, the exposure station spacing simulates an eye base so the viewer actually sees parallactic angles comparable with having one eye at each of the two exposure stations. This creates a condition called *vertical exaggeration*, which causes the vertical scale of the three-dimensional model to appear greater than its horizontal scale; that is, objects are perceived to be too tall. The condition is of concern to photo interpreters who often estimate heights of objects and slopes of surfaces. The amount of vertical exaggeration varies with percent endlap and the camera's format dimensions and focal length. A factor of about 4 results if endlap is 60% and a camera of 9-in. format and 6-in. focal length is used.

The *stereoscope* shown in Figure 28-11 permits viewing photographs stereoscopically by enabling the left and right eyes to focus comfortably on the left and right prints, respectively, assuming proper orientation has been made of the overlapping pair

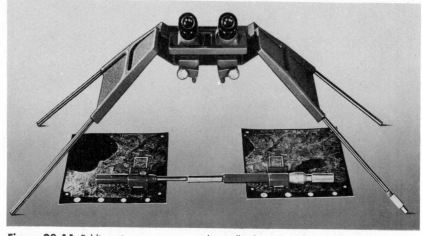

Figure 28-11 Folding-mirror stereoscope with parallax bar. (Courtesy Leica, Inc.)

of photographs under the instrument. Correct orientation requires the two photographs to be laid out in the same order they were taken, with the stereoscope so set that the line joining its lens centers is parallel with the direction of flight. The print spacing is varied, carefully maintaining this parallelism, until a clear three-dimensional view (*stereoscopic model*) is obtained.

28-12 STEREOSCOPIC MEASUREMENT OF PARALLAX

The parallax of a point can be measured while viewing stereoscopically with the advantage of speed and, because binocular vision is used, greater accuracy. As the viewer looks through a stereoscope, imagine that two small identical marks etched on pieces of clear glass, called *half-marks*, are placed over each photograph. The viewer simultaneously sees one mark with the left eye and the other with the right eye. Then the positions of the marks are shifted until they seem to fuse together as one mark that appears to lie at a certain elevation. The height of the mark will vary or "float" as the spacing of the half-marks is varied; hence it is called the *floating mark*. Figure 28-12 demonstrates this principle and also illustrates that the floating mark can be set exactly on particular points such as *A*, *B*, and *C* by placing the half-marks (small black dots) at *a* and *a′*, *b* and *b′*, and *c* and *c′*, respectively.

Based on the floating mark principle, the parallax of points is measured stereoscopically with a parallax bar, as shown beneath the stereoscope in Figure 28-11. It is simply a bar to which two half-marks are fastened. The right mark can be moved with respect to the left one by turning a micrometer screw to register the displacement on a dial. When the floating mark appears to rest on a point, a micrometer reading is taken and added to the parallax bar *setup constant* to obtain the parallax.

When a parallax bar is used, two overlapping photographs are oriented properly for viewing under a mirror stereoscope and fastened securely with respect to each other using drafting tape. The parallax bar constant for the setup is determined by measuring the photo coordinates for a discrete point and applying Eq. (28-10) to obtain its parallax.

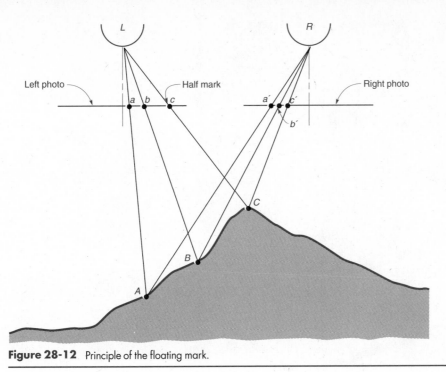

Figure 28-12 Principle of the floating mark.

The floating mark is placed on the same point, the micrometer read, and the constant for the setup found by

$$C = p - r \qquad (28\text{-}14)$$

where C is the parallax bar setup constant, p the parallax of a point determined by Eq. (28-10), and r the micrometer reading obtained with the floating mark set on that same point.

Once the constant has been determined, the parallax of any other point can be computed by adding its micrometer reading to the constant. Thus a single measurement gives the parallax of a point. Each time another pair of photos is oriented for parallax measurements, a new parallax bar setup constant must be determined. A major advantage of the stereoscopic method is that parallaxes of nondiscrete points can be determined. Thus elevations of hill tops, points in fields, and so on, can be calculated using Eq. (28-11), even though their x coordinates cannot be measured for use in Eq. (28-10).

28-13 STEREOSCOPIC PLOTTERS

The primary use of stereoscopic plotters is to compile topographic maps from overlapping aerial photographs, but they are also used extensively to take cross sections, and to record digital elevation models. Two classifications of stereoscopic plotters are *optical projection* instruments and *mechanical projection* instruments. Each type incorporates the following components: **(1)** a projection system (to create a stereomodel), **(2)** a viewing system (which enables an operator to view the stereomodel),

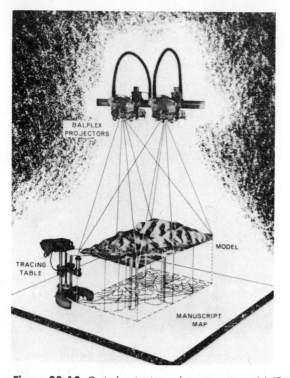

Figure 28-13 Optical projection and stereoscopic model. (Courtesy Bausch and Lomb, Inc.)

and (3) a measuring/tracing system (for measuring or mapping the stereomodel). Optical projection stereoplotters are now nearly obsolete, but nevertheless preferred for introducing beginning students to the subject because of their relatively simple concept and design. Mechanical projection plotters are still widely used in practice, but these are now gradually being replaced by *analytical plotters* (see Section 28-14) and "softcopy" systems (see Section 28-16).

The concept of an optical projection-type stereoplotter is depicted in Figure 28-13. With this instrument, *diapositives* (positives developed on film or glass plates) of a pair of overlapping photographs are projected so that light rays carrying images common to both intersect to form a stereomodel. Projectors used in optical projection stereoscopic plotters resemble ordinary slide projectors. However, they are much more precise and can be adjusted in angular orientation and position to recreate the aerial camera's spatial location and attitude when the overlapping photos were exposed. This produces a "true" model of the terrain in the overlap area. The scale of the stereomodel is, of course, greatly reduced, and is the ratio of the *model base* (distance between projector lenses) to the *air base* (actual distance between the two exposure stations).

Plotter viewing systems must provide a stereoscopic view and hence be designed so that the left and right eyes see only projected images of the corresponding left and right diapositives. One method of accomplishing this for optical projection instruments is to place a blue filter in one projector and a red filter in the other. A pair of spectacles with corresponding blue and red lenses is worn by the operator. This system of viewing stereoscopically is called the *anaglyphic* method. Another system, called the *stereo image alternator* (SIA), operates with rapidly rotating shutters, located on the projectors, and a viewing eyepiece. The shutters are synchronized so that the left and right eyes can see only the images from the corresponding left and right projectors. A third method, similar to the anaglyphic system and called the *polarized platen viewing* (PPV) system, uses filters of opposite polarity.

Various measuring and tracing systems have been devised for plotters. In optical projection instruments (see Figure 28-13), light rays carrying corresponding images are intercepted on the *tracing table platen* (a small white circular disk). A light emitted through a pinhole in the platen, which can be seen by both the left and right eyes and fuses, provides the floating mark. It can be made to appear to rest exactly on a point of the model by raising or lowering the tracing table. A tracing pencil located directly below the floating mark permits positioning the point planimetrically. A counter linked to the tracing table responds to up-and-down motion, so the elevation for any setting of the floating mark is read directly.

When diapositives are placed in the projectors and the lights turned on, corresponding rays will not intersect to form a clear model because of tilt in the photographs and unequal flying heights. The projectors can be moved linearly along the X, Y, and Z axes and can also be rotated about each of them until they duplicate the relative tilts and flying heights that existed when the photographs were taken. This is called *relative orientation*, and when accomplished, corresponding rays will intersect to form a perfect three-dimensional model.

The model is brought to required scale by making the rays of at least two, but preferably three, ground control points intersect at their positions plotted on a manuscript map prepared at the desired scale. It is leveled by adjusting the projectors so the counter reads the correct elevations of each of a minimum of three, but preferably four, corner ground control points when the floating mark is set on them. *Absolute orientation* is a term applied to the processes of scaling and leveling the model.

When orientation is completed, a map can be made from the model or cross sections compiled. In mapping, planimetric details are located first by bringing the floating mark into contact with objects in the model and tracing them. A pencil directly beneath the reference mark records their locations on the manuscript map below. Contours are traced by setting the elevation counter successively at each contour elevation and moving the reference mark over the model while keeping it in contact with the terrain. Again, the pencil draws the contours. When a manuscript is completed, it is examined for omissions and mistakes and field-checked. The final map is drafted or scribed by tracing it.

Mechanical projection stereoplotters use two precisely made metal space rods to simulate light rays. The diapositives are viewed through binoculars via an optical train of lenses and prisms. The floating mark, composed of a pair of half-marks superimposed in the optical train, is moved up or down by turning a hand screw or foot disk, and impelled in the X and Y directions, either manually or by means of hand wheels. Instruments equipped with a coordinatograph or pantograph allow planimetry to be traced directly on the manuscript, and X and Y coordinates read. Figure 28-14 shows the Kern PG2 mechanical projection stereoplotter with attached pantograph on the right on which the map is compiled.

Advantages of mechanical projection instruments over the optical types are that they (**1**) can be operated in a lighted room by an operator who sits rather than stands, (**2**) are more versatile, (**3**) have greater stability, and (**4**) produce more accurate results.

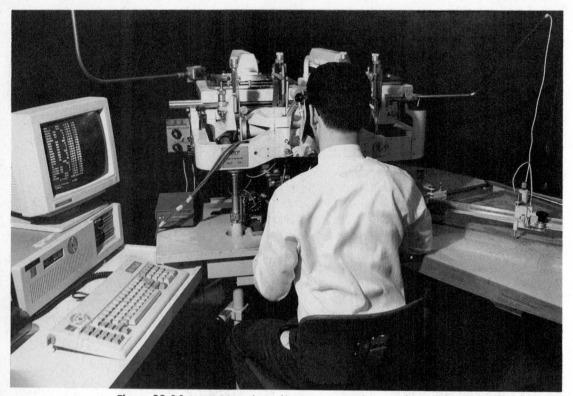

Figure 28-14 Kern PG2 mechanical projection stereoplotter interfaced with computer for digital mapping, and equipped with pantograph for direct manuscript plotting. (Courtesy Department of Civil and Environmental Engineering, University of Wisconsin-Madison.)

In recent years computer graphics systems have been incorporated with mechanical projection stereoplotters to facilitate mapping and recording measurements greatly. With these systems, after an operator sets the floating mark on a point in the stereo-model, a footpedal or key is depressed and the object's X, Y, and Z coordinates are automatically recorded. Cross sections, profiles, and digital elevation models (DEMs) are digitized in this way rapidly and conveniently. Planimetric maps can also be compiled using these systems. In this mode the operator indicates the type of object being mapped by keying in a certain identifying number code. A tree, for example, might be identified with the number 5, while 7 could indicate a building corner. To create lines such as fences, roads, and the like, connectivity between points is also keyed in. Then when the floating mark is set and the footpedal depressed, this information is recorded along with the object's ground coordinates. As the mapping progresses, a current view of work accomplished appears on the computer's monitor. When the map is completed, a hardcopy can be obtained by transferring the digital data file to an automatic plotter. Maps compiled in digital form have many advantages. They can be plotted at any scale, are easily edited and updated, and can be integrated with other digital data for use in computer-aided design (CAD) or Geographic Information System (GIS) projects. The computer and monitor shown to the left of the PG2 stereoplotter of Figure 28-14 is being used to support an on-line computer graphics system.

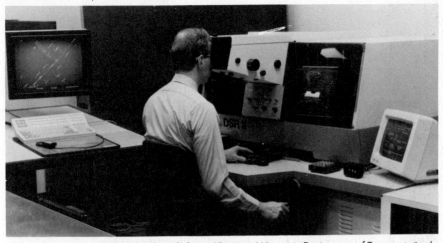

Figure 28-15 Kern DSR11 analytical plotter. (Courtesy Wisconsin Department of Transportation.)

28-14 ANALYTICAL PHOTOGRAMMETRY

With the advent of high-speed electronic computers, much work done with stereoscopic plotters can now be performed economically by means of analytical photogrammetry. This science involves precise measurements of photographic coordinates of images, and construction of a *mathematical model* which can be solved by numerical methods. The procedures of analytical photogrammetry are highly accurate and particularly well adapted to establishing vertical and horizontal control through *aerotriangulation*. Accuracies of $\frac{1}{15,000}$ of flying height are readily obtained for computed X and Y ground coordinates, and $\frac{1}{10,000} H$ for Z coordinates. Thus for photos exposed from 6000 ft, X and Y would be calculated correctly to within approximately 0.4 ft, and Z to within about 0.6 ft.

Analytical plotters combine a precise stereoscopic system for measuring photo coordinates, and a digital computer. When an operator places the floating mark on ground points, their x and y photo coordinates are fed directly to the computer, which calculates their X, Y, and Z positions in real time. The data are thus in digital form and readily available for a variety of computations. They can also be incorporated directly into CAD systems and Land and Geographic Information Systems.

Figure 28-15 shows an analytical plotter with a monitor screen on the left to give the operator a visual record of work performed, and to permit review and editing of the digitized data. With their many advantages, including speed of operation, accuracy, and versatility, analytical plotters are gradually replacing mechanical stereoplotters.

28-15 ORTHOPHOTOS

As implied by their name, orthophotos, are orthographic representations of the terrain in picture form. They are derived from aerial photos in a process called *differential rectification*, which removes scale variations and image displacements due to relief and tilt. Thus the imaged features are shown in their true planimetric positions.

Instruments used for differential rectification vary considerably in design. Earlier versions were basically modified stereoscopic plotters with either optical or mechanical

Figure 28-16 Zeiss/Intergraph PS1 PhotoScan. (Courtesy Carl Zeiss.)

projection. Optical projection instruments derived an orthophoto by systematically *scanning* a stereomodel and photographing it in a series of adjacent narrow strips. *Rectification* (removal of tilt) was accomplished by leveling the model to ground control prior to scanning, and scale variations due to terrain relief were removed by varying the projection distance during scanning. As the instrument automatically traversed back and forth across the model, exposure was made through a narrow slit onto an ortho-negative below. An operator, viewing the model in three dimensions, continually mon-itored the scans and adjusted the projection distance to keep the exposure slit in contact with the model. Because the model itself had uniform scale throughout, the resulting *orthonegative* (from which the orthophoto was made) was also of uniform scale. Ortho-photo systems based on modified mechanical projection stereoplotters functioned in a similar fashion.

Contemporary instruments are computer operated and produce orthophotos by *dig-ital image processing*. With these systems, the entire contents of a photo is scanned using a *scanning microdensitometer* to create a grid (raster) of tiny *picture elements* (pixels) arranged systematically in rows and columns. Each pixel is assigned a digital value that represents its density (degree of darkness or lightness). Figure 28-16 shows the PhotoScan, developed jointly by Carl Zeiss and Intergraph Corporation for digitiz-ing photographs. The digitized file of scanned data is input to a computer, which modifies pixel locations to account for scale variations and image displacements, then prints them electronically to produce an orthophoto.

Orthophotos combine the advantages of both aerial photos and line maps. Like photos, they show features by their actual images rather than as lines and symbols, thus making them more easily interpreted and understood. Like maps, orthophotos show the features in their true planimetric positions. Therefore true distances, angles, and areas can be scaled directly from them. *Orthophotomaps* (maps produced from orthophotos) are used for a variety of applications, including planning and engineering design. They have been particularly valuable in cadastral and tax mapping, because the identification of property boundaries is greatly aided through visual interpretation of fence lines, roads, and other evidence.

Orthophotos can generally be prepared more rapidly and economically than line or symbol planimetric maps. With their many significant advantages, orthophotos have superseded conventional maps for many uses.

28-16 DIGITAL OR "SOFTCOPY" PHOTOGRAMMETRY

Continued photogrammetric research and development have resulted in a new computerized approach to mapping known as *digital photogrammetry*, also often called *softcopy photogrammetry*. Like contemporary instruments for orthophoto production, softcopy photogrammetry is based on digital image processing and utilizes rasters of digitized image densities. In softcopy photogrammetry, the digitized image files of an overlapping stereopair of photos are input to a computer or photogrammetric workstation for processing. The overlapping photos can be displayed on the computer's screen and viewed stereoscopically.

One type of softcopy photogrammetric work station, the digital video plotter (DVP), is shown in Figure 28-17. As seen in the figure, it is simply a standard personal computer with a stereoscopic viewing system attached. With this device, the left and right photos are displayed simultaneously on the left and right portions, respectively, of the computer's monitor. An operator looking through the stereoscope can view the stereomodel in three dimensions. Using the cursor of an interfaced digitizer or the keyboard's arrow keys, the left and right images can be shifted on the screen with respect to each other. The floating marks, made of two reference pixels, one on each image, can also be moved about the screen. While viewing stereoscopically, an operator shifts the imagery and moves the floating mark until it appears to rest exactly on the point of interest. This identifies to the computer the image pixels which correspond to the point in both the left and right photos. The computer converts the pixel row and column locations to x and y photo coordinates and then calculates the point's ground coordinates by employing standard analytical photogrammetry equations.

Other softcopy photogrammetric systems are also available. Some rely on polarizing lenses for stereoviewing, while others use a system of electronically synchronized shutters. All are controlled by sophisticated software. Some advanced systems have become almost totally automated. They employ a process called *digital image correlation*. In this process the computer automatically matches points in the left photo of the stereopair to their corresponding or *conjugate* points in the right photo. This is done by comparing the patterns of image densities in the point's immediate area on both photos. Thus an operator's process of making measurements while stereoviewing is eliminated.

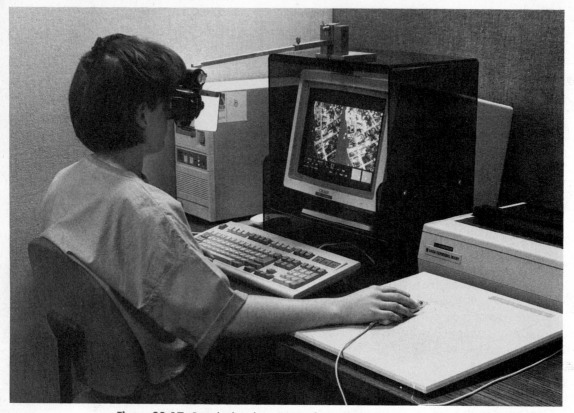

Figure 28-17 Digital video plotter (DVP) softcopy photogrammetric workstation. (Courtesy Dr. P. A. Gagnon, Laval University, Quebec, Canada.)

Softcopy photogrammetry has several advantages. A major one is that expensive single-purpose stereoplotters are replaced by standard computers, which can also perform many other tasks. Digital maps produced by softcopy systems are created in a computer environment, and are therefore directly available for introduction to CAD systems and Geographic Information Systems. (Other advantages of digital maps were stated earlier in Section 28-13.) Softcopy photogrammetry systems not only enable the production of maps, but they can also be used to derive digital elevation models, they can be employed for a variety of image interpretation problems, and they can support the production of mosaics and orthophotos.

28-17 GROUND CONTROL FOR PHOTOGRAMMETRY

As pointed out in preceding sections, almost all phases of photogrammetry depend on ground control (points of known positions and elevations with identifiable images on the photograph). Ground control can be *basic control*—traverse, triangulation, or trilateration monuments and bench marks already in existence and marked prior to photography to make them visible on the photos; or it can be *photo control*—natural points having images recognizable on the photographs, and positions that are subsequently determined by ground surveys originating from basic control. Instruments and procedures used in the ground surveys were described in earlier chapters. Ordinarily, photo control points are selected after photography to ensure their satisfactory location and positive identification. Premarking points with artificial targets is sometimes necessary in areas that lack natural objects to provide definite images.

As pointed out in Section 28-13, scaling and leveling stereomodels for mapping with stereoplotters requires a practical minimum of three horizontal control points and four vertical points in each model. For large mapping jobs, therefore, the cost of establishing the required ground control is substantial. In these situations, analytical aerotriangulation (see Section 28-14) is used to establish many of the needed control points from only a sparse network of ground surveyed points. This reduces costs significantly.

Research is now active in exploring the use of GPS for establishing the needed ground control for photogrammetric mapping. The kinematic GPS surveying procedure is being employed (see Section 20-6), which requires two GPS receivers. One unit is stationed at a ground control point, the other in the aircraft carrying the camera. During the flight, aircraft positions are continuously determined at time intervals of a few seconds using the GPS units, and precise timing of each photo exposure is also recorded. From this information, the precise location of each exposure station, in the ground coordinate system, can be calculated. Test projects have already shown that highly accurate results can be obtained by employing this procedure, supplemented with only a few ground control points. It is plausible that in the near future the method will be perfected so that no photo control points at all will be needed on the ground.

28-18 FLIGHT PLANNING

Certain factors, depending generally on the purpose of the photography, must be specified to guide a flight crew in executing its mission of taking aerial photographs. Some of them are **(1)** boundaries of the area to be covered, **(2)** required scale of the photography, **(3)** camera focal length and format size, **(4)** endlap, and **(5)** sidelap. Once these elements have been fixed, it is possible to compute the entire flight plan and prepare a flight map on which the required flight lines have been delineated. The pilot flies specified flight lines by choosing and correlating headings on existing natural features shown on the flight map.

Purpose of the photography is the paramount consideration in flight planning. In taking aerial photos for topographic mapping using a stereoplotter, for example, endlap should optimally be 60% and sidelap between 25 and 35%. The required scale and contour interval of the final map must be evaluated to settle flying height. Enlargement capability from photo scale to map is restricted for stereoplotters, and for many of them the optimum ratio is $5\times$. With such an instrument, if the required map scale is 200 ft/in., the photo scale becomes fixed at 1000 ft/in. If the camera focal length is 6 in., by Eq. (28-2) the flying height is established at $6 \times 1000 = 6000$ ft above average terrain.

The *C factor* (the ratio of flight height above ground to contour interval which is practical for any specific stereoplotter) is the criterion used to select the flying height in relation to the required contour interval. The *C* factor has been determined for various plotters to be in the range of from approximately 800 to about 2000. Thus if a plotter has a *C* factor of, say, 1200, and a map is to be compiled with a 5-ft contour interval, a flight height of not more than $1200 \times 5 = 6000$ ft above the terrain should be sustained.

Information ordinarily calculated in flight planning includes **(1)** flying height above mean sea level, **(2)** distance between exposures, **(3)** number of photographs per flight line, **(4)** distance between flight lines, **(5)** number of flight lines, and **(6)** total number of photographs. A flight plan is prepared based on these items.

EXAMPLE 28-7

A flight plan for an area 10 mi wide and 15 mi long is required. The average terrain in the area is 1500 ft above mean sea level. The camera has a 6-in. focal length with 9 × 9-in. format. Endlap is to be 60%, sidelap 25%. The required scale of the photography is 1:12,000 (1000 ft/in.).

SOLUTION

1. Flying height above datum from Eq. (28-2):

$$\text{scale} = \frac{f}{H - H_{\text{avg}}} \quad \text{so} \quad \frac{1}{1000} = \frac{6}{H - 1500} \quad \text{and} \quad H = 7500 \text{ ft}$$

2. Distance between exposures: Endlap is 60%, so the linear advance per photograph is 40% of the total coverage of 9 in. × 1000 ft/in. = 9000 ft. Thus the distance between exposures is 0.40 × 9000 = 3600 ft.

3. Total number of photographs per flight line:

$$\text{length of each flight line} = 15 \text{ mi} \times 5280 \text{ ft/mi} = 79{,}200 \text{ ft}$$

$$\text{number of photos per flight line} = \frac{79{,}200 \text{ ft}}{3600 \text{ ft/photo}} + 1 = 23$$

Adding two photos on each end to ensure complete coverage, the total is 23 + 2 + 2 = 27 photos per flight line.

4. Distance between flight lines: Sidelap is 25%, so the lateral advance per flight line is 75% of the total photographic coverage,

$$\text{distance between flight lines} = 0.75 \times 9000 \text{ ft} = 6750 \text{ ft}$$

5. Number of flight lines:

$$\text{width of area} = 10 \text{ mi} \times 5280 \text{ ft/mi} = 52{,}800 \text{ ft}$$

$$\text{number of spaces between flight lines} = \frac{52{,}800 \text{ ft}}{6750 \text{ ft/line}} = 7.8 \text{ (say 8)}$$

$$\text{total flight lines} = 8 + 1 = 9$$

$$\text{planned spacing between flight lines} = \frac{52{,}800}{8} = 6600 \text{ ft}$$

Note: The first and last flight lines should either coincide with or be near the edges of the area, thus providing a safety factor to ensure complete coverage.

6. Total number of photos required:

$$\text{total photos} = 27 \text{ per flight line} \times 9 \text{ flight lines} = 243 \text{ photos}$$

Figures 28-18(a) and (b) illustrate endlap and sidelap, (c) shows the flight map.

Figure 28-18 Endlap, sidelap, and flight map.

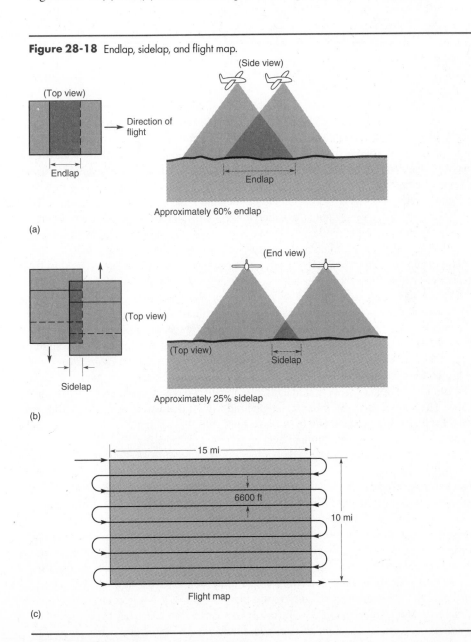

(a)

(b)

(c)

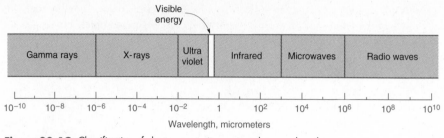

Figure 28-19 Classification of electromagnetic spectrum by wavelength.

28-19 REMOTE SENSING

In general, remote sensing can be defined as any methodology employed to study the characteristics of objects using data collected from a remote observation point. More specifically, and in the context of surveying and photogrammetry, it is the extraction of information about the earth and our environment from imagery obtained by various sensors carried in aircraft and satellites. Satellite imagery is unique because it affords a practical means of monitoring our entire planet on a regular basis.

Remote sensing imaging systems operate much the same as the human eye, but they can sense or "see" over a much broader range than humans. Cameras that expose various types of film are among the best types of remote sensing imaging systems. Nonphotographic systems such as *multispectral scanners* (MSS), *radiometers, side-looking airborne radar*, and *passive microwave* are also commonly employed. Their manner of operation and uses of the imagery are described briefly below.

The sun and other sources emit a wide range of electromagnetic energy called the *electromagnetic spectrum.* X-rays, visible light rays, and radio waves are some familiar examples of energy variations within the electromagnetic spectrum. Energy is classified according to its wavelength (see Figure 28-19). Visible light (that energy to which our eyes are sensitive) has wavelengths from about 0.4 to 0.7 μm and thus comprises only a very small portion of the spectrum.

Within the wavelengths of visible light, the human eye is able to distinguish different colors. The primary colors—blue, green, and red—consist of wavelengths in the ranges of 0.4 to 0.5, 0.5 to 0.6, and 0.6 to 0.7 μm, respectively. All other hues are combinations of the primary colors. To the human eye, an object appears a certain color because it reflects energy of wavelengths producing that color. If an object reflects all energy in the visible range, it will appear white, and if it absorbs all energy, it will be black. If an object absorbs all green and red energy but reflects blue, that object will appear as blue.

Just as the retina of the human eye can detect variations in wavelengths, photographic films or *emulsions* can also be made with wavelength sensitivity variations. Normal color emulsions are sensitive to blue, green, and red energy; others respond to energy in the near-infrared range. These are called *infrared* (IR) emulsions. They make it possible to photograph energy that is invisible to the human eye. An early application of IR film was in camouflage detection, where it was found that dead foliation or green netting reflected infrared energy differently than normal vegetation, even though both appeared green to the human eye. Infrared film is now widely used for a variety of applications, such as detection of crop stress, and identification and mapping of tree species.

Nonphotographic imaging systems used in remote sensing are able to detect energy variations over a broad range of the electromagnetic spectrum. MSS systems, for example, are carried in satellites and can operate within wavelengths from about 0.3 to 14 μm. In a manner similar to the way humans detect colors, MSS units isolate incoming energy into discrete spectral categories, or *bands*, then convert them into electric signals that can be represented by digits. The scanned scene below the satellite's path consists of contiguous rows and columns of pixels. The digits associated with each pixel represent intensities of the various bands of energy within them. This digital format is ideal for computer processing and analysis, and enables prints to be made by electronic processing. The bands of a scene can be analyzed separately, which is extremely useful in identifying and interpreting objects sensed, or all bands can be combined into a composite.

Figure 28-20 shows an image obtained from an MSS system carried in a Landsat satellite. It was taken at an altitude of 560 miles and shows a large portion of northern Wisconsin and the upper Michigan peninsula. The western portion of Lake Superior is in the upper left. Imagery of this type is useful for a variety of applications. As examples, geologic formations over vast areas can be studied; the number of lakes in the area, their relative positions, shapes, and acreages can be determined readily; acreages of croplands and forests, with a breakdown of coniferous and deciduous tree types, can be obtained; and small-scale maps showing these different land-use classifications can be prepared. All are amenable to computer processing.

Remote sensing technology has been applied to locate forest fires, detect diseased crops and trees, study wildlife, monitor floods and flood damage, analyze population growth and distribution, determine locations and extents of oil spills, monitor water quality and detect the presence of pollutants, and accomplish numerous other tasks over large areas for the benefit of humankind.

Resolution of the earlier Landsat MSS imaging systems (used to take Figure 28-20) was 80 m. This has been improved to 30 m for the current U.S. Landsat thematic mapper (TM), and more recently reduced to 10 m with the French SPOT satellite imaging system. This enhanced resolution permits detection and analysis of much smaller objects, and leads to a host of new applications. Figure 28-21 shows an image taken by Landsat TM near Madison, Wisconsin. It dramatically illustrates the improved resolution as compared to Figure 28-20. Note, for example, the clarity of the roads and the detail with which urban areas and agricultural crops are shown.

In the future, surveyors will be called on to map land-use activities and extract a variety of other positional types of information from satellite images. This kind of data is needed to develop modern Land and Geographic Information Systems. Remote sensing will play a significant future role in providing data to assess the impacts of human activities on our air, water, and land resources. It can provide information to assist in making sound decisions and formulating policies related to resource management and development and land-use activities.

28-20 SOURCES OF ERROR IN PHOTOGRAMMETRY

Some sources of error in photogrammetric work are:

1. Measuring instruments not standard length or not calibrated
2. Inaccurate location of principal and corresponding principal points

Figure 28-20 Multispectral scanner image taken over northern Wisconsin and upper Michigan from Landsat satellite.

Figure 28-21 Landsat thematic mapper image taken near Madison, Wisconsin. (Note the improved resolution, compared to the Landsat MSS image of Figure 28-20.)

3. Failure to use camera calibration data
4. Assumption of vertical exposures when photographs are actually tilted
5. Presumption of equal flying heights when they were unequal
6. Disregard of differential shrinkage or expansion of photographic prints
7. Incorrect orientation of photographs under a stereoscope or in a stereoscopic plotter
8. Faulty setting of the floating mark on a point

28-21 MISTAKES

Some mistakes that occur in photogrammetry are:

1. Incorrect reading of measuring scales
2. Mistake of units—for example, inches instead of millimeters
3. Confusion in identifying corresponding points on different photographs
4. Neglect of relief displacement
5. Failure to provide proper control or use of erroneous control coordinates
6. Attachment of an incorrect sign (plus or minus) to a measured photographic coordinate
7. Blunder in computations
8. Misidentification of control-point images

PROBLEMS

28-1 List the principal components of a single-lens frame aerial camera and describe their functions.

28-2 Describe the differences between vertical, low oblique, and high oblique aerial photos.

28-3 Discuss the advantages of preparing maps photogrammetrically.

28-4 The distance between two points on a vertical photograph is ab and the corresponding ground distance AB. For the following data, compute the average photographic scale along the line ab.
 (a) $ab = 2.49$ in.; $AB = 1986$ ft
 (b) $ab = 7.68$ in.; $AB = 3105$ ft
 *(c) $ab = 107.36$ mm; $AB = 1284$ m

28-5 On a vertical photograph of flat terrain, section corners appear a distance d apart. If the camera focal length is f, compute flying height above average ground for the following data:
 (a) $d = 5.45$ in.; $f = 8\frac{1}{4}$ in.
 *(b) $d = 80.47$ mm; $f = 152.4$ mm

28-6 On a vertical photograph of flat terrain, the scaled distance between two points is ab. Find the average photographic scale along ab if the measured length between the same line is AB on a map plotted at a scale of S for the following data.
 (a) $ab = 1.01$ in.; $AB = 3.67$ in.; $S = 1{:}10{,}000$
 (b) $ab = 116.04$ mm; $AB = 2.40$ in.; $S = 1{:}24{,}000$

28-7 What are the average scales of vertical photographs for the following data, given flying height above sea level H, camera focal length f, and average ground elevation h?
 (a) $H = 7250$ ft; $f = 152.4$ mm; $h = 1250$ ft
 (b) $H = 5800$ ft; $f = 8\frac{1}{4}$ in.; $h = 925$ ft
 *(c) $H = 3200$ m; $f = 88.90$ mm; $h = 615$ m

28-8 The length of a football field from goal post to goal post scales 49.15 mm on a vertical photograph. Find the approximate dimensions of a large rectangular building that also appears on this photo and whose sides measure 1.62 in. by 1.35 in.

*28-9 Compute the area in acres of a triangular parcel of land whose sides measure 47.94 mm, 75.65 mm, and 64.40 mm on a vertical photograph taken from 6075 ft above average ground with a 152.4-mm-focal-length camera.

28-10 Calculate the flight height above average terrain that is required to obtain vertical photographs at an average scale of S if the camera focal length is f for the following data:
 (a) $S = 1{:}10{,}000$; $f = 152.4$ mm
 (b) $S = 1{:}5000$; $f = 3\frac{1}{2}$ in.

28-11 Determine the horizontal distance between two points A and B whose elevations above datum are $h_A = 1650$ ft and $h_B = 1475$ ft, and whose images a and b on a vertical

*Asterisks indicate problems that have answers given in Appendix G.

photograph have photo coordinates $x_a = 3.24$ in., $y_a = 2.55$ in., $x_b = -1.80$ in., and $y_b = -2.93$ in. The camera focal length was 152.4 mm and the flying height above datum 7550 ft.

*28-12 Similar to Problem 28-11, except that the camera focal length was $3\frac{1}{2}$ in., the flying height above datum 4050 ft, and elevations h_A and h_B 925 ft and 1085 ft, respectively. Photo coordinates of images a and b were $x_a = 108.81$ mm, $y_a = -73.73$ mm, $x_b = -87.05$ mm, and $y_b = 52.14$ mm.

28-13 On the photograph of Problem 28-11, the image c of a third point C appears. Its elevation $h_c = 1390$ ft, and its photo coordinates are x_c 3.26 in. and $y_c = -2.59$ in. Compute the horizontal angles in triangle ABC.

*28-14 On the photograph of Problem 28-12, the image d of a third point D appears. Its elevation is $h_D = 1030$ ft, and its photo coordinates are $x_d = 79.75$ mm and $y_d = 74.01$ mm. Calculate the area, in acres, of triangle ABD.

28-15 Determine the height of radio towers which appear on a vertical photograph for the following conditions of flying height above the tower base H, distance on the photograph from principal point to tower base r_b, and distance from principal point to tower top r_t.
    *(a) $H = 4500$ ft; $r_b = 2.85$ in.; $r_t = 3.10$ in.
    (b) $H = 6500$ ft; $r_b = 90.67$ mm; $r_t = 97.52$ mm

*28-16 On a vertical photograph, images a and b of ground points A and B have photographic coordinates $x_a = 3.00$ in., $y_a = 2.08$ in., $x_b = -1.79$ in., and $y_b = -2.32$ in. The horizontal distance between A and B is 5350 ft, and the elevations of A and B above datum are 652 ft and 785 ft, respectively. Calculate the flying height above datum for a camera having a focal length of 152.4 mm.

28-17 Similar to Problem 28-16, except $x_a = -49.90$ mm, $y_a = 66.19$ mm, $x_b = 24.99$ mm, $y_b = -56.33$ mm, line length $AB = 4695$ ft, and elevations of points A and B are 925 and 900 ft, respectively.

*28-18 An air base of 3190 ft exists for a pair of overlapping vertical photographs taken at a flying height of 5540 ft above MSL with a camera having a focal length of 152.4 mm. Photo coordinates of points A and B on the left photograph are $x_a = 41.60$ mm, $y_a = 42.50$ mm, $x_b = 24.69$ mm, and $y_b = -59.15$ mm. The x photo coordinates on the right photograph are $x_a = -59.58$ mm and $x_b = -69.19$ mm. Calculate horizontal length AB.

28-19 Similar to Problem 28-18, except the air base is 7250 ft, the flying height above mean sea level is 14,920 ft, the x and y photo coordinates on the left photo are $x_a = 39.08$ mm, $y_a = 49.54$ mm, $x_b = 25.73$ mm, and $y_b = -45.09$ mm, and the x photo coordinates on the right photo are $x_a = -51.07$ mm and $x_b = -62.77$ mm.

28-20 Calculate the elevations of points A and B in Problem 28-18.

*28-21 Compute the elevations of points A and B in Problem 28-19.

28-22 A pair of overlapping vertical photographs is oriented and secured under a mirror stereoscope, and a parallax-bar reading of $r_a = 14.32$ mm is obtained on point a. The x photo coordinate of point a is 2.42 in. on the left photograph and -1.16 in. on the right photograph. The camera focal length is 152.4 mm, the flying height 8,540 ft above datum, and the air base 3150 ft. Determine the difference in elevation between points B and C if their parallax-bar readings are $r_b = 16.70$ mm and $r_c = 14.87$ mm.

28-23 Compare an orthophoto with a conventional line and symbol map.

28-24 Discuss the advantages of orthophotos as compared to maps.

Aerial photography is to be taken of a tract of land that is X mi square. Flying height will be H ft above average terrain, and the camera has focal length f. If the focal plane opening is 9×9 in., and minimum sidelap is 30%, how many flight lines will be needed to cover the tract for the data given in Problems 28-25 and 28-26?

*28-25 $X = 18$; $H = 6000$; $f = 6$ in.
28-26 $X = 40$; $H = 7000$; $f = 3\frac{1}{2}$ in.

---

*Asterisks indicate problems that have answers given in Appendix G.

Aerial photography was taken at a flying height H ft above average terrain. If the camera focal plane dimensions are 9×9 in., the focal length is f, and the spacing between adjacent flight lines is X ft, what is the percent sidelap for the data given in Problems 28-27 and 28-28?

28-27 $H = 4950; f = 152.4$ mm; $X = 5170$
28-28 $H = 6300; f = 3\frac{1}{2}$ in.; $X = 12,200$

Photographs at a scale of S are required to cover an area X mi square. The camera has a focal length f and focal plane dimensions of 9×9 in. If endlap is 60% and sidelap 30%, how many pictures will be required to cover the area for the data given in Problems 28-29 and 28-30?

28-29 $S = 1:6000; X = 10; f = 152.4$ mm
*28-30 $S = 1:9600; X = 22; f = 3\frac{1}{2}$ in.
28-31 What is "softcopy" photogrammetry, and what are its advantages?
28-32 How can GPS reduce or eliminate ground control surveys in photogrammetry?
28-33 To what wavelengths of electromagnetic energy is the human eye sensitive? What wavelengths produce the colors blue, green, and red?
28-34 Discuss the uses and advantages of satellite imagery.

*Asterisks indicate problems that have answers given in Appendix G.

BIBLIOGRAPHY

ASPRS 1980. *Manual of Photogrammetry,* 4th ed. Bethesda, Md.: American Society for Photogrammetry and Remote Sensing.

Bobbe, T. 1992. "Real-Time Differential GPS for Aerial Surveying and Remote Sensing." *GPS World* 3 (No. 7):18.

Chamard, R. R. 1983. "Photogrammetric Mapping for Highways: Western U.S." *ASCE, Journal of the Surveying Engineering Division* 109 (No. 1):1.

Colvocoresses, A. 1984. "Landsat: Looking at Earth from Space." *Professional Surveyor* 4 (No. 4):14.

Cowden, R. W., and R. F. Brinkman. 1993. "Digital Orthophotos: A New Alternative for Creating Base Maps." *Geo Info Systems* 3 (No. 5):53.

Cracknell, A., and L. Hayes. 1991. *Introduction to Remote Sensing.* London: Taylor and Francis.

Derenyi, E. E., and A. Maarek. 1974. "Photogrammetric Control Extension for Route Design." *ASCE, Journal of the Surveying and Mapping Division* 100 (No. SU1):49.

Gruen, A. 1989. "Digital Photogrammetric Processing Systems." *Photogrammetric Engineering and Remote Sensing* 40 (No. 5):581.

Karara, H. 1989. *Non-Topographic Photogrammetry,* 2nd ed. Bethesda, Md.: American Society for Photogrammetry and Remote Sensing.

Keel, G., et al. 1989. "A Test of Airborne Kinematic GPS Positioning for Aerial Photography." *Photogrammetric Engineering and Remote Sensing* 40 (No. 12):1727.

Lapine, L. A. 1990. "Practical Photogrammetric Control by Kinematic GPS." *GPS World* 1 (No. 3):44.

Light, D. 1992. "The New Camera Calibration System at the U.S. Geological Survey." *Photogrammetric Engineering and Remote Sensing* 43 (No. 2):185.

Lillesand, T. M., and R. W. Kiefer. 1987. *Remote Sensing and Image Interpretation,* 2nd ed. New York: John Wiley and Sons.

Miller, S., U. Helava, and K. Devenecia. 1992. "Softcopy Photogrammetric Workstations." *Photogrammetric Engineering and Remote Sensing* 43 (No. 1):77.

Moffitt, F. H., and E. Mikhail. 1980. *Photogrammetry,* 3rd ed. New York: Harper & Row.

Morain, S., et al. 1992. "U.S. National Report—Status of Photogrammetry, Remote Sensing and Geographic Information Systems in the United States." *Photogrammetric Engineering and Remote Sensing* 43 (No. 8):1073.

Nolette, C., P.-A. Gagnon, and J.-P. Agnard. 1992. "The DVP: Design, Operation and Performance." *Photogrammetric Engineering and Remote Sensing* 43 (No. 1):65.

Novak, K. 1992. "Rectification of Digital Imagery." *Photogrammetric Engineering and Remote Sensing* 43 (No. 3):339.

van der Vegt, J. W. 1989. "Differential GPS: Efficient Tool in Photogrammetry." *ASCE, Journal of Surveying Engineering* 115 (No. 3):285.

Wolf, P. R. 1983. *Elements of Photogrammetry,* 2nd ed. New York: McGraw-Hill.

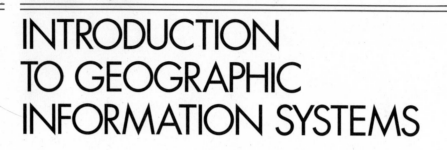

INTRODUCTION TO GEOGRAPHIC INFORMATION SYSTEMS

29-1 INTRODUCTION

A *Geographic Information System* (GIS) can, in general, be defined as a system of hardware, software, data, and organizational structure for collecting, storing, manipulating, and spatially analyzing "geo-referenced" data, and displaying information resulting from those processes. A more detailed definition (Hanigan, 1988) describes a GIS as

> "any information management system which can:
>
> 1. Collect, store, and retrieve information based on its spatial location;
> 2. Identify locations within a targeted environment which meet specific criteria;
> 3. Explore relationships among data sets within that environment;
> 4. Analyze the related data spatially as an aid to making decisions about that environment;
> 5. Facilitate selecting and passing data to application-specific analytical models capable of assessing the impact of alternatives on the chosen environment; and
> 6. Display the selected environment both graphically and numerically either before or after analysis."

The thread that is common to both definitions just given is that in a GIS, decisions are made based on *spatial analyses* performed on data sets that are referenced in a common geographical system. The geographic referencing system used could be a state plane or UTM coordinate system, latitude and longitude, or other suitable local coordinate system. In any GIS, the accuracy of the spatial analyses and, hence, the validity of decisions reached as a result of those analyses are directly dependent on the quality of the spatially related data used. It is therefore important to realize at the outset of this chapter that the surveyor's role in developing accurately positioned data sets is a critical one in GIS activity.

The term *Geographic Information System* is a relatively new one, which first appeared in published literature in the mid-1960s. But although the term is new, many of its concepts have long been in existence. The *map overlay* concept, for example, which is one of the important tools used in GIS spatial analysis, was used by French cartographer Louis-Alexandre Berthier more than 200 years ago. He prepared and overlaid a series of maps to analyze troop movements during the American Revolution. Another early example illustrating the use and value of the overlay concept was demonstrated in 1854 by Dr. John Snow. He overlaid a map of London showing where cholera deaths had occurred with another, giving locations of wells in that city to demonstrate the relationship between those two data sets. These early examples illustrate fundamentals that still comprise the basis of our modern GISs, that is, making decisions based on the simultaneous analysis of data of differing types, all located spatially in a common geographic reference system. The full capabilities and benefits of our modern GISs could not occur, however, until the advent of the computer.

A generalized concept of how data of different types or "layers" are collected and overlaid in a GIS is illustrated in Figure 29-1. In that figure, maps A through G represent some of the different layers of spatially related information that can be digitally recorded and incorporated into a GIS, and include parcels of different land ownership A, zoning B, floodplains C, wetlands D, land cover E, and soil types F. Map G is the geodetic reference framework, comprised of the network of survey control points in the area. Note that these control points are found in each of the other layers, thereby providing the means for spatially locating all data in a common reference system. Thus composite maps that merge two or more different data sets can be accurately created. In Figure 29-1, for example, bottom map H is the composite of all layers.

A GIS merges conventional database management software with software for manipulating spatial data. This combination enables the simultaneous storage, retrieval, overlay, and display of many different spatially related data sets in the manner illustrated by Figure 29-1. These capabilities, coupled with sophisticated GIS software to analyze and query the data sets that result from these different overlay and display combinations, provide answers to questions that never before were possible to obtain. As a result, GISs have become extremely important in planning, design, impact assessment, predictive modeling, and many other applications.

GISs have been applied in virtually every imaginable field of activity, from engineering to agriculture, and from the medical science of epidemiology to wildlife management. Flood forecasting on a large regional basis, such as statewide, is one particular example which illustrates some of the benefits that can be derived from using GIS. Critical location-related data entered into the GIS to support statewide flood forecasting would include the state's topography; soil; land cover; number, sizes, and locations of drainage basins; existing stream network with stream-gaging records; locations and sizes of existing bridges, culverts, and other drainage structures; data on existing dams and the water impoundment capacities of their associated reservoirs; and records of past rainfall intensity and duration. Given these and other data sets, together with a model to estimate runoff, a computer can be used to perform an analysis and predict locations of potential floods and their severity. In addition, experiments can be conducted in which certain input can be varied. Examples may include **(1)** input of an extremely intense rainfall for a lengthy duration in a given area to assess the magnitude of the resulting flood, and **(2)** the addition of dams and other flood-detention structures of varying sizes at specific locations, to analyze their impact on mitigating the disaster.

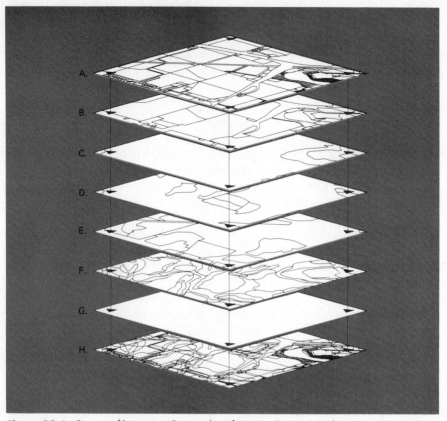

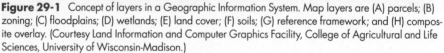

Figure 29-1 Concept of layers in a Geographic Information System. Map layers are (A) parcels; (B) zoning; (C) floodplains; (D) wetlands; (E) land cover; (F) soils; (G) reference framework; and (H) composite overlay. (Courtesy Land Information and Computer Graphics Facility, College of Agricultural and Life Sciences, University of Wisconsin-Madison.)

Many other similar examples could be given. Obviously, GISs are very powerful tools, and their use will increase significantly in the future.

The successful implementation of GISs relies on people with backgrounds and skills in many different disciplines, but none are more important than surveyors. Virtually every aspect of surveying, and thus all material presented in the preceding chapters of this book, bear upon GIS development, management, and use. Of special importance, however, are topographic surveying (Chapter 16), mapping (Chapter 17), control surveys (Chapter 19), satellite and inertial surveying systems (Chapter 20), reference coordinate systems (Chapter 21 and Appendix B), boundary or cadastral surveys (Chapter 22), the U.S. Public Land Survey System (Chapter 23), and photogrammetry and remote sensing (Chapter 28). In addition to surveying specialists, personnel in the fields of computer science, geography, soil science, forestry, landscape architecture, and many others play important roles in GIS development.

With GIS drawing upon so many areas in surveying, it should not be surprising that a chapter on this subject would be included in a modern textbook on surveying.

29-2 LAND INFORMATION SYSTEMS

The terms *Geographic Information System* (GIS) and *Land Information System* (LIS) are sometimes used interchangeably. They do have many similarities, but the distinguishing characteristic between the two is that an LIS has its focus directed primarily toward land records data. Information stored within an LIS for a given locality would include a spatial database of parcel information derived from property descriptions in the U.S. Public Land system; other types of legal descriptions such as metes and bounds or block and lot that apply to parcels in the area; and other cadastral data. It might include the actual deeds and other records linked to the spatial data. Information on improvements and parcel values would also be included.

An LIS and GIS can share data sources such as control networks, parcel ownership information, and municipal boundaries. However, a GIS will usually incorporate data over a broader range and might include layers such as topography, soil types, land cover, hydrography, depth to ground water, and so on. Because of its narrower focus, there is a tendency to consider an LIS as a subset of a GIS.

LISs are used to obtain answers to questions about who has ownership or interests in the land in a certain area, the particular nature of those interests, and the specific land affected by them. They can also provide information about what resources and improvements exist in a given area, and give their values. Answers to these questions are essential in making property assessments for taxation, transferring title to property, mortgaging, making investment decisions, resolving boundary disputes, and developing roads, utilities, and other services on the land that require land appraisal and property acquisitions. The data are also critical in policy development and land-use planning.

29-3 GIS DATA SOURCES AND CLASSIFICATIONS

As noted earlier, the capabilities and benefits of any GIS are directly related to the content and integrity of its database. Data that are entered into a GIS come from many sources and may be of varying quality. To support a specific GIS, some new information may be gathered expressly for its database. More than likely, however, much of the data will be obtained from existing sources such as maps, engineering plans, aerial photos, satellite images, and other documents and files that were developed for other purposes. Building the database is one of the most expensive and challenging aspects of developing a GIS. In fact, it has been estimated that this aspect may represent from about 60 to 80% of the total cost of implementing a GIS.

Two basic data classifications are used in GISs, (**1**) *spatial* and (**2**) *nonspatial*. These are described in sections that follow.

29-4 SPATIAL DATA

Spatial data, sometimes interchangeably called *graphic data,* consist in general of natural and cultural features that can be shown with lines or symbols on maps, or seen as images on photographs. In a GIS these data must be represented, and spatially located, in digital form, using a combination of fundamental elements called "simple spatial objects." The formats used in this representation are either *vector* or *raster.* The "relative spatial relationships" of the simple spatial objects are given by their *topology.*

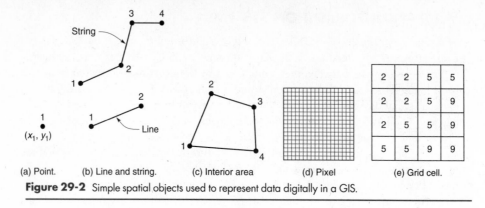

Figure 29-2 Simple spatial objects used to represent data digitally in a GIS.

These topics—simple spatial objects, formats, and topology—are described in the following subsections.

29-4.1 Simple Spatial Objects

The simple spatial objects most commonly used in spatially locating data are illustrated in Figure 29-2 and described as follows:

1. *Points* define single geometric locations. They are used to locate features such as houses, wells, mines, or bridges [see Figure 29-2(a)]. The spatial locations of points are given by their coordinates, commonly in state plane or UTM systems (see Chapter 21).
2. *Lines* and *strings* are obtained by connecting points. A line connects two points, and a string is a sequence of two or more connected lines. Lines and strings are used to represent and locate roads, streams, fences, property lines, and so on [see Figure 29-2(b)].
3. *Interior areas* consist of the continuous space within three or more connected lines or strings that form a closed loop [see Figure 29-2(c)]. Interior areas are used to represent and locate, for example, the limits of governmental jurisdictions, parcels of land ownership, different types of land cover, or large buildings.
4. *Pixels* are usually tiny squares that represent the smallest elements into which a digital image is divided [see Figure 29-2(d)]. The term *pixel* is derived phonetically from the words *pic*ture and *el*ement. Continuous arrays of pixels, arranged in rows and columns, are used to enter data from aerial photos, orthophotos, satellite images, and so forth. The distributions of colors or tones throughout the image are specified by assigning a numerical value to each pixel. Pixel size can be varied, and is usually specified by the number of *dots per inch* (dpi). As an example, 100 dpi would correspond to squares having dimensions of $\frac{1}{100}$ in. on each side. Thus 100 dpi yields 10,000 pixels per square inch.
5. *Grid cells* are single elements, usually square, within a continuous geographic variable. Similar to pixels, their sizes can be varied, with smaller cells yielding improved resolution. Grid cells may be used to represent slopes, soil types, land cover, water table depths, land values, population density, and so on. The distribution of a given data type within an area is indicated by assigning a numerical

value to each cell, for example, showing soil types in an area using the number 2 to represent sand, 5 for loam, and 9 for clay, as illustrated in Figure 29-2(e).

29-4.2 Vector and Raster Formats

The simple spatial objects described in Section 29-4.1 give rise to two different formats for storing and manipulating spatial data in a GIS—*vector* and *raster*. When data are depicted in the vector format, a combination of points, lines, strings, and interior areas is used. The raster format uses pixels and grid cells.

In the vector format, points are used to specify locations of objects such as survey control monuments, utility poles, or manholes; lines and strings depict linear features such as roads, transmission lines, or boundaries; and interior areas show regions having common attributes, for example, governmental entities or areas of uniform land cover. An example illustrating the vector format is given with Figure 29-3 and Table 29-1. Figure 29-3 shows two adjacent land parcels, one designated parcel I, owned by Smith, and the other identified as parcel II, owned by Brown. As shown, the configuration consists of points, lines, and areas.

Vector representation of the data can be achieved by creating a set of tables, which list these points, lines, and areas (Table 29-1). Data within the tables are linked using *identifiers,* and related spatially through the coordinates of points. As illustrated in Table 29-1(a), all points in the area are identified by a reference number. Similarly each line is described by its endpoints, as shown in Table 29-1(b), and the endpoint coordinates locate the various lines spatially. Areas in Figure 29-3 are defined by the lines which enclose them, as shown in Table 29-1(c). As before, coordinates of line endpoints locate the areas and enable determining their locations and magnitudes.

Figure 29-3 Vector representation of a simple graphic record.

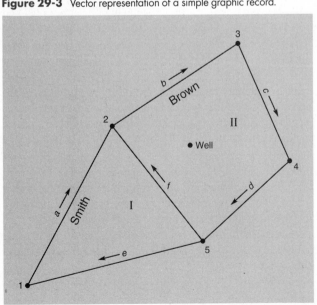

TABLE 29-1 VECTOR REPRESENTATION OF FIGURE 29-3

(a)		(b)		(c)	
POINT IDENTIFIER	COORDINATES	LINE IDENTIFIER	POINTS	AREA IDENTIFIER	LINES
1	$(X, Y)_1$	a	1, 2	I	a, f, e
2	$(X, Y)_2$	b	2, 3	II	b, c, d, f
3	$(X, Y)_3$	c	3, 4		
4	$(X, Y)_4$	d	4, 5		
5	$(X, Y)_5$	e	5, 1		
Well	$(X, Y)_{\text{well}}$	f	5, 2		

Other data types can also be represented in vector formats. Consider, for example, the simple case illustrated with the land cover map shown in Figure 29-4(a). In that figure, areas of different land cover (forest, marsh, and so on) are shown with standard topographic symbols (see Figure 17-6). A vector representation of this region is shown in Figure 29-4(b). Here lines and strings locate boundaries of regions having a common land cover. The stream consists of the string connecting points 1 through 10. By means of tables similar to Table 29-1 the data of this figure can also be entered into a GIS using the vector format. Having considered these simple vector representations, imagine the magnitude and complexity of entering data in vector format to cover a much larger area such as that shown in the map of Figure 16-2.

As an alternative to the vector approach, data can also be depicted in the raster format using grid cells (or pixels if the data are derived from images). Each equal-sized cell (or pixel) is uniquely located by its row and column numbers, and is coded with a numerical value or *code* that corresponds to the properties of the specific area it covers. In the raster format, a point would be indicated with a single grid cell, a line would be depicted as a sequence (linear array) of adjacent grid cells having the same code, and an area having common properties would be shown as a group of identically coded contiguous cells. It should be appreciated, therefore, that in general the raster method yields a coarser level of accuracy or definition of points, lines, and areas than the vector method.

In the raster format, the size of the individual cells defines the *resolution,* or precision, with which data are represented. The smaller the elements, the higher the resolution. Examples illustrating raster representation, and the degradation of resolution with increasing grid size, are given in Figure 29-4(c) and (d). In these figures the land cover data from Figure 29-4(a) have been entered as two different raster data sets. Each cell has been assigned a value representing one of the land cover classes, that is, F for forest, G for grassland, M for marsh, and S for stream. Figure 29-4(c) depicts the area with a relatively large-resolution grid and, as shown, it yields a coarse representation of the original points, lines, and areas. With a finer-resolution grid, such as in Figure 29-4(d), the points, lines, and areas are rendered with more precision. It is important to note, however, that as grid resolution increases, so does the volume of data (number of grid cells) required to enter the data.

Despite the coarser resolution present in a raster depiction of spatial features, this format is still often used in GISs. One reason is that many data are available in raster format. Examples include aerial photos, orthophotos, and images from the Landsat and SPOT satellites. Another reason for the popularity of the raster format is the ease with

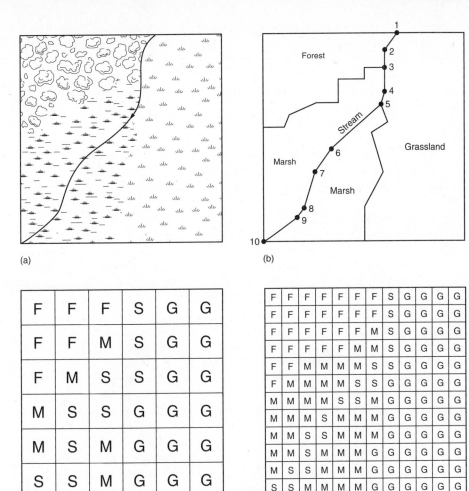

Figure 29-4 Land cover maps of a region. (a) Region using standard topographic symbols. (b) Vector representation of same region. (c) Raster representation of region using a coarse-resolution grid cell. (d) Raster representation using a finer-resolution grid cell.

which it enables collection, storage, and manipulation of data using computers. Furthermore, various refinements of raster images are readily made using available image processing software programs. Finally, for many data sets such as wetlands and soil types, boundary locations are rather vague and the use of the raster format does not adversely affect the data's inherent accuracy.

29-4.3 Topology

Topology is a branch of mathematics that describes how spatial objects are related to each other. The unique sizes, dimensions, and shapes of the individual objects are not addressed by topology. Rather, it is only their *relative relationships* that are specified.

In discussing topology, it is necessary to first define *nodes, chains,* and *polygons.* These are some additional simple spatial objects that are commonly used for specifying the topological relationships of information entered into GIS databases. Nodes define the beginnings and endings of chains, or identify the junctions of intersecting chains. Chains are similar to lines (or strings), and are used to define the limits of certain areas or delineate specific boundaries. Polygons are closed loops similar to areas, and are defined by a series of connected chains. In topology, sometimes single nodes exist within polygons for labeling purposes.

In GISs the most important topological relationships are:

1. *Connectivity.* Which chains are connected at which nodes.
2. *Direction.* Defining a "from node" and a "to node" of a chain.
3. *Adjacency.* Which polygons are on the left and which are on the right side of a chain.
4. *Nestedness.* What simple spatial objects are within a polygon. They could be nodes, chains, or other smaller polygons.

The topological relationships just described are illustrated and described by example with reference to Figure 29-3. In the figure, for example, through connectivity, it is established that nodes 2 and 3 are connected to form the chain labeled *b.* Connectivity would also indicate that at node 2, chains *a, b,* and *f* are connected. Topological relationships are normally listed in tables and stored within the database of a GIS. Table 29-2(a) summarizes all of the connectivity relationships of Figure 29-3.

Directions of chains are also indicated topologically in Figure 29-3. For example, chain *b* proceeds from node 2 to node 3. Directions can be very important in a GIS for establishing such things as the flow of a river or the direction traffic moves on one-way streets. In a GIS, often a consistent direction convention is followed, that is, proceeding clockwise around polygons. Table 29-2(b) summarizes the directions of all chains within Figure 29-3.

The topology of Figure 29-3 would also describe, through adjacency, that Smith and Brown share a common boundary, which is chain *f* from node 5 to node 2, and that Smith is on the left side of the chain and Brown is on the right. Obviously the chain's direction must be stated before left or right positions can be declared. Table 29-2(c) lists the adjacency relationships of Figure 29-3. Note that a zero has been used to designate regions outside of the polygons and beyond the area of interest.

TABLE 29-2 TOPOLOGICAL RELATIONSHIPS OF ELEMENTS IN THE GRAPHIC RECORD OF FIGURE 29-3

(a) CONNECTIVITY		(b) DIRECTION			(c) ADJACENCY			(d) NESTEDNESS	
NODES	CHAIN	CHAIN	FROM NODE	TO NODE	CHAIN	LEFT POLYGON	RIGHT POLYGON	POLYGON	NESTED NODE
1–2	*a*	*a*	1	2	*a*	0	I	II	Well
2–3	*b*	*b*	2	3	*b*	0	II		
3–4	*c*	*c*	3	4	*c*	0	II		
4–5	*d*	*d*	4	5	*d*	0	II		
5–1	*e*	*e*	5	1	*e*	0	I		
2–5	*f*	*f*	5	2	*f*	I	II		

Nestedness establishes that the well is contained within Brown's polygon. Table 29-2(d) lists that topological information.

The relationships expressed through the identifiers for points, lines, and areas of Table 29-1, and the topology in Table 29-2, conceptually yield a "map." With these types of information available to the computer, the analysis and query processes of a GIS are made possible.

29-5 NONSPATIAL DATA

Nonspatial data, also often called *attribute data,* describe geographic regions or define characteristics of spatial features within geographic regions. Nonspatial data are usually alphanumeric and provide information such as color, texture, quantity, quality, and value of features. Smith and Brown as the property owners of parcels I and II of Figure 29-3 and the land cover classifications of forest, marsh, grassland, and stream in Figure 29-4 are examples. Other examples could include the addresses of the owners of land parcels, their types of zoning, dates purchased, and assessed values; or data regarding a particular highway, including its route number, pavement type, number of lanes, lane widths, and year of last resurfacing. Nonspatial data are often derived from sources such as documents, files, and tables.

In general, spatial data will have related nonspatial attributes, and thus some form of linkage must be established between these two different types of information. Usually this is achieved with a *common identifier* that is stored with both the graphic and the nongraphic data. Identifiers such as a unique parcel identification number, a grid cell label, or the specific mile point along a particular highway may be used.

29-6 DATA FORMAT CONVERSIONS

In manipulating information within a GIS database, it is often necessary to either integrate vector and raster data, or convert from one form to the other. Integration of the two types of data, that is, using both types simultaneously, has become possible due to recent hardware and software improvements. It is usually accomplished by displaying vector data overlaid on a raster image background, as illustrated in Figure 29-5. In that figure, vector data representing the dwellings (points) that exist within the different subdivisions (areas) are overlaid on a satellite image of the same area. This graphic was developed as part of a population growth and distribution study being conducted for a municipality. The combination of vector and raster data is useful to provide a frame of reference and to assist GIS operators in interpreting displayed data.

Sometimes it is necessary or desirable to convert raster data to vector format, or vice versa. Procedures for accomplishing these conversions are described in subsections that follow.

29-6.1 Vector-to-Raster Conversion

Vector-to-raster conversion is also known as *coding* and can be accomplished in several ways, three of which are illustrated in Figure 29-6. Figure 29-6(a) is an overlay of the vector representation of Figure 29-4(b) with a coarse raster of grid cells. In one conversion method, called *predominant type coding,* each grid cell is assigned the value corresponding to the predominant characteristic of the area it covers. For example, the cell located in row 3, column 1 of Figure 29-6(a) overlaps two polygons, one of forest

Figure 29-5 Vector data overlaid on a raster image background. (Courtesy ERDAS.)

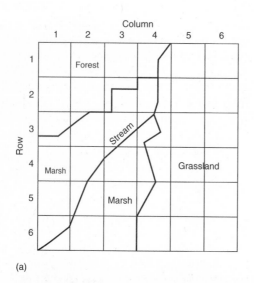

(a)

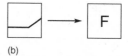

(b)

Predominant type conversion
of grid cell (3, 1)

(c)

Precedence method conversion
of grid cell (4, 3)

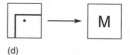

(d)

Center-point method conversion
of grid cell (2, 3)

Figure 29-6 Methods for converting data from vector to raster format. (a) Vector representation of Figure 29-4(b) overlaid on a coarse raster format. (b), (c), (d) Vector-to-raster conversion by predominant type, precedence method, and center-point method, respectively.

(type F) and one of marsh (type M). As shown in Figure 29-6(b), since the largest portion of this cell lies in forest, the cell is assigned the value F—the predominant type.

In another coding method, called *precedence coding,* each category in the vector data is ranked according to its importance or "precedence" with respect to the other categories. In other words, each cell is assigned the value of the highest ranked category present in the corresponding area of the vector data. A common example involves water. While a stream channel may cover only a small portion of a cell area, it is arguably the most important feature in that area. Also, it is important to avoid breaking up the stream. Thus for the cell in row 4, column 3 of Figure 29-6(a), which is illustrated in Figure 29-6(c), water would be given the highest precedence, and the cell coded S even though most of the cell is covered by marsh.

Center-point coding is a third technique for converting from vector to raster data. Here a cell is simply assigned the category value at the vector location corresponding to its center point. An example is shown in Figure 29-6(d), which represents the cell in row 2, column 3 of Figure 29-6(a). Here, since marsh exists at the cell's center, the entire cell is designated as category M. Note that the grid cell of row 3, column 4 would be classified by predominant type as grassland, by precedence as stream, and by center point as marsh. This illustrates how different conversion processes can yield different classifications for the same data.

The precisions of these vector-to-raster conversions depend on the size of the grid used. Obviously, using a raster of large cells would result in a relatively inaccurate representation of the original vector data. On the other hand, a fine-resolution grid can very closely represent the vector data, but would require a large amount of computer memory. Thus the choice of grid resolution becomes a tradeoff between computing efficiency and spatial precision.

29-6.2 Raster-to-Vector Conversion

Raster-to-vector conversions are more vaguely defined than vector to raster. The procedure involves extracting lines from raster data which represent linear features such as roads, streams, or boundaries of common data types. Whereas the approach is basically a simple one, and consists in identifying the pixels or cells through which vector lines pass, the resulting jagged or "staircase"-type outlines are not indicative of the true lines. One raster-to-vector converson example is illustrated in Figure 29-7(a), which shows the cells identifying the stream line of Figure 29-4(b). Having selected these cells, a problem that must be resolved is how to fit a line to these jagged forms. One solution consists in simply connecting adjacent cell centers [see Figure 29-7(b)] with line segments. Note, however, that the resulting line [see Figure 29-7(c)] does not agree very well with the original stream line of Figure 29-4(a). This example illustrates that some type of line smoothing is usually necessary to properly represent the gently curved boundaries that normally occur in nature. However, fitting smooth lines to the jagged cell boundaries is a complicated mathematical problem that does not necessarily have a unique solution. The decision ultimately becomes a choice between accuracy of representation and cost of computation.

No matter which conversion is performed, errors are introduced during the process, and some information from the original data is lost. Use of smaller grid cells improves the results. Nevertheless, as illustrated by the example of Figure 29-7, if a data set is converted from vector to raster and then back to vector (or vice versa), the final data

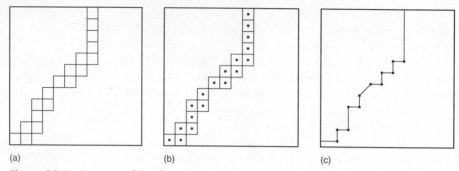

(a)   (b)   (c)

Figure 29-7 Conversion of data from raster to vector format. (a) Cells of Figure 29-4(b) that identify the stream line. (b) Cell centers. (c) Vector representation of stream line recreated by connecting adjacent cell centers.

set will not likely match the original. Thus it is important for GIS operators to be aware of how their data have been manipulated, and what can be expected if conversion is performed.

29-7 CREATING GIS DATABASES

Several important factors must be considered prior to developing the database for a GIS. These include the types of data that need to be obtained, optimum formats for these data, the reference coordinate system that will be used for spatially relating all data, and the necessary accuracy of each data type. Provisions for updating the database must also be considered. Having made these decisions, the next step is to locate data sources. Depending on the situation, it may be possible to utilize existing data, in which case a significant cost savings can result. In some circumstances it is necessary to collect new data to meet the needs of the GIS. Different methods are available for generating the computer files needed to support a GIS. These are discussed in the subsections that follow.

29-7.1 Digitizing Existing Graphic Materials

If sources such as maps, orthophotos, plans, diagrams, or other graphic documents already exist that will meet the needs of the GIS database, these can be conveniently and economically converted to digital files using a *tablet digitizer*. Many GIS software packages provide programs to support the procedure directly. A tablet digitizer, as shown in Figure 29-8, contains an electronic grid and an attached cursor. Movement of the cursor across the grid creates an electronic signal unique to the cursor's position. This signal is relayed to the computer which records the point's coordinates. Data identifiers or attribute codes can be associated with each point through the computer's keyboard, or by pressing numerical buttons on the cursor.

The digitizing process begins by securing the source document to the tablet digitizer. If the document is a map, the next step is to *register* its reference coordinate system to that of the digitizer. This is accomplished by digitizing a series of reference (tic) marks on the map for which geographic coordinates such as state plane, UTM, or latitude and longitude are precisely known. With both reference and digitizer coordinates known for these tic marks, a *coordinate transformation* (see Section B-7 of Appendix B) can

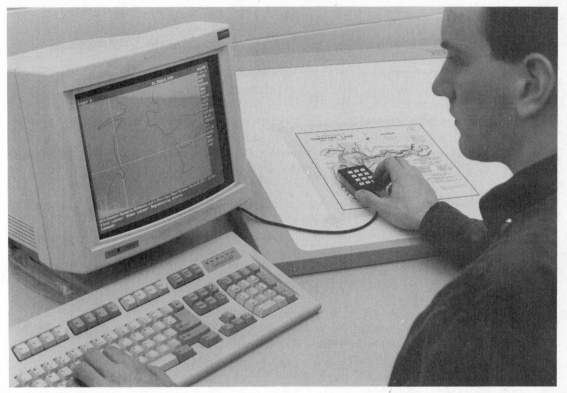

Figure 29-8 Tablet digitizer interfaced with personal computer. (Courtesy Department of Civil and Environmental Engineering, University of Wisconsin—Madison.)

be computed. This determines the parameters of scale change, rotation, and translation necessary to convert digitized coordinates into the reference geographic coordinate system. After this, any map features can be digitized, whereupon their coordinates are automatically converted to the selected system of reference and entered directly into the database.

Both planimetric features and contours can be digitized from the map. Planimetric features are recorded by digitizing the individual points, lines, or areas that identify them. As described in Section 29-4.2, this process creates data in a vector format. Elevation data can be recorded as a digital elevation model (DEM) by digitizing critical points along contours. From these data, *triangulated irregular networks* (TIN models) can be derived using the computer (see Section 16-7). From the TIN models, point elevations, profiles, cross sections, slopes, aspects (slope directions), and contours having any specified contour interval can be derived automatically using the computer.

29-7.2 Digitizing from Aerial Photos with Stereoplotters

Information from aerial photographs can also be entered directly into a GIS database in digital form using a digitized stereoplotter (see Section 28-13). In this procedure, both planimetric features and elevations can be recorded, and very high accuracy can be achieved. The data are registered to the selected reference coordinate system and vertical datum by *orienting* the stereoplotter to ground control points prior to beginning the digitizing. Then to record planimetric features, an operator views the stereomodel, points to objects of interest, enters any necessary feature identifiers or codes, and depresses a key or foot pedal to transfer the information to a file in an interfaced computer. To digitize elevations, a DEM is read directly from the three-dimensional stereomodel and stored in a computer file.

Information on the source and accuracy of each file created by digitizing, whether done by tablet digitizer or stereoplotter, should be included with the data so GIS users are aware of its quality. Resultant accuracies will depend mainly on the quality of the source materials, which of course is directly related to the manner in which the data were originally collected. Accuracies will also be affected by errors in control points used for registration or orientation, differential shrinkage and expansion of the materials upon which the graphic products are printed, pointing errors, and imperfections in the digitizing equipment.

Data sets generated by digitizing will usually need to be checked carefully to ensure that all desired features have been included. Also the data must be corrected or "cleaned" before being used in a GIS. In this process, unwanted points and line portions must be removed; and "unclosed" polygons, which result from imprecise pointing when returning to a polygon's starting node, must be closed. Finally, thin polygons or "slivers" created by lines being inadvertently digitized twice, but not precisely in the same location, must be eliminated. This editing process can be performed by the operator, or with a program that can find and remove certain features that fall within a set of user-defined tolerances. Digitizing and editing requires a relatively large investment in time and cost.

29-7.3 Scanning

Scanners are instruments which automatically convert graphic documents into digital format. Their principal advantages are that the tedious work of manual digitizing is eliminated, and the process of converting graphic documents to digital form is greatly accelerated. Scanners accomplish their objective by measuring the amount of light reflected from a document and encoding the information in pixels. This is possible because different areas of a document will reflect light in proportion to their tones, from a maximum for white through the various shades of gray to a minimum for black. The scanner of Figure 29-9, for example, uses a linear array of light sensors to capture the varying intensities of reflected light, line by line, as the document is fed through the system. This creates a raster data set. Its pixel size can be varied and made as small as $\frac{1}{500}$ in.^{2} (500 dpi). Large complicated documents can be scanned in a matter of a few minutes. The data are stored directly on the hard drive of an interfaced computer, and can be viewed on a screen, edited, and manipulated. Editing is an important and necessary step in the process, because the scanner will record everything, including blemishes, stains, and creases.

Documents such as subdivision plats, topographic maps, engineering drawings and plans, aerial photos, and orthophotos can be digitized using scanners. Then, if necessary, the raster data can be converted to vector form using techniques described in Section 29-6.2.

Accuracy of the raster file obtained from scanning depends somewhat on the instrument's precision, but pixel size or resolution is generally the major factor. A smaller pixel size will normally yield superior precision. However there are certain tradeoffs that must be considered. Whereas a large pixel size will result in a coarse representation

Figure 29-9 Intergraph ANAtech Eagle 4050 scanner. (Courtesy Wisconsin Department of Transportation.)

of the original, it will require less scanning time and computer storage. Conversely, a fine resolution, which generates a precise depiction of the original, requires more scanning time and computer storage. An additional problem is that at very fine resolution, the scanner will record too much "noise," that is, impurities such as specks of dirt.

29-7.4 Keyboard Entry and Coordinate Geometry

Data can be entered into a GIS file directly, using the keyboard on a computer. Often data input by this method are nonspatial, such as map annotations or numerical or tabular data. To facilitate keying in data, an intermediate file having a simple format is sometimes created. This file is then converted into a GIS-compatible format using special software.

Data for use in a GIS can also be generated directly from field survey data using coordinate geometry (COGO) software. This procedure is particularly convenient when a total station instrument with interfaced data collector is available. The field data can be downloaded from the data collector into a computer, processed with COGO, and entered directly into the GIS database. Data generated in this manner are generally very accurate.

29-7.5 Existing Digital Data Sets

Sometimes data are already available in digital format, so conversion is not necessary. Organizations located in the vicinity of the area covered by the new GIS may be able to provide such digital data. In addition, the federal government has produced digital data files of a wide variety of types. Examples are the *digital line graphs* (DLGs) and *digital elevation models* (DEMs) available from the U.S. Geological Survey (see Section 17-5). Often existing digital data must undergo conversion of file structures and formats to be usable with specific GIS software. Because of differences in the way data are represented by different software, it is possible that information can be lost, or that spurious data can creep in during this process.

29-8 GIS ANALYTICAL FUNCTIONS

Most GISs are equipped with a set of basic analytical functions that enable data to be manipulated, analyzed, and queried. These functions, coupled with appropriate databases, provide GISs with their powerful capabilities for supplying information that so significantly aids in planning, management, and decision making.

The specific functions available within the software of any particular GIS system will vary. They enable data to be stored, retrieved, viewed, analyzed spatially and computationally, and displayed. Some of the more common and useful spatial analysis and computational functions are briefly described in the subsections that follow.

29-8.1 Buffering

This spatial analysis function involves the creation of new polygons that are geographically related to nodes, chains, or existing polygons. *Point buffering,* also known as *radius searching,* is illustrated in Figure 29-10(a). It involves the creation of a circular buffer zone of radius *R* around a specific node. Information about the new zone can then be gathered and analyses made of the new data. A simple example illustrates its value. Assume that a polluted well has just been discovered. With appropriate databases available, all dwellings within a specified radius of the well can be located, the names, addresses, and telephone numbers of all individuals living within the point buffer zone tabulated, and the people quickly alerted to the possibility of their water also being polluted.

Line buffering, illustrated in Figure 29-10(b), creates new polygons along established lines such as streams and roads. To illustrate the use of line buffering, assume that to preserve the natural stream bank and prevent erosion, a zoning commission has set the construction setback distance from a certain stream at *D.* Line buffering can quickly identify the areas within this zone. *Polygon buffering,* illustrated in Figure 29-10(c), creates new polygons around an existing polygon. An example of its use could

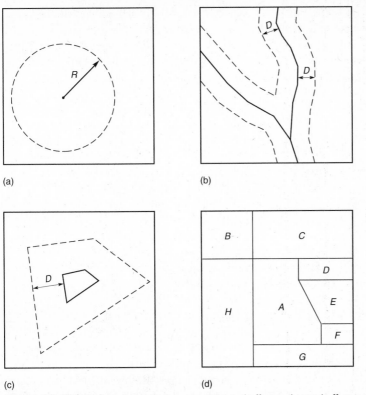

Figure 29-10 GIS spatial analysis functions. (a) Point buffering. (b) Line buffering. (c) Polygon buffering. (d) Adjacency analysis.

occur in identifying those land owners whose property lies within a certain distance D of the proposed site of a new industrial facility. Many other examples could be given, which illustrate the value of buffering for rapidly extracting information to support management and decision making.

29-8.2 Adjacency and Connectivity

If the topological relationships discussed in Section 29-4.3 have been entered into a database, certain analyses regarding relative positioning of features can be performed. *Adjacency* and *connectivity* are two such important spatial analysis functions that often assist significantly in management and decision making. An example of adjacency is illustrated in Figure 29-10(d) and relates to a zoning change requested by the owner of parcel A. Before taking action on the request, the jurisdiction's zoning administrators are required to notify all owners of adjacent properties B through H. If the GIS database includes the parcel descriptions with topology and other appropriate attributes, an adjacency analysis will identify the abutting properties and provide the names and addresses of the owners.

Connectivity involves analyses of the intersections or connections of linear features. The need to repair a city water main serves as an example to illustrate its value. Suppose that the decision has been made that these repairs will take place between the hours of 1:00 and 4:00 P.M. on a certain date. If infrastructure data are stored within the city's

GIS database, all customers connected to this line whose water service will be interrupted by the repairs can be identified and their names and addresses tabulated. The GIS can even print a letter and address labels to facilitate a mailing announcing details of the planned interruption to all affected customers. Many similar examples could be given to illustrate benefits that can result from adjacency and connectivity.

29-8.3 Overlay

This is one of the most widely used spatial analysis functions of a GIS. As indicated in Figure 29-1, GIS graphic data are usually divided into layers, with each containing data in a single category of closely related features. Nonspatial data or "attributes" are often associated with each category. The individual layers are spatially registered to each other through a common reference network or coordinate system. Any number of layers can be entered into a GIS database, and could include parcels, municipal boundaries, public land survey system, zoning, soils, road networks, topography, land cover, hydrology, and many others.

Having these various data sets available in spatially related layers makes the overlay function possible. Its employment in a GIS can be compared to using a collection of Mylar overlays in traditional mapping. However, much greater efficiency and flexibility are possible when operating in the computer environment of a GIS, and not only can graphic data be overlaid, but attribute information can be combined as well.

Many examples could be given to illustrate the applications and benefits of the GIS overlay process. Consider one case where the land in a particular area suitable for development must be identified. To perform an in-depth analysis of this situation, the evaluation would normally have to consider numerous variables within the area, including the topography (slope and aspect of the terrain), soil type, land cover, land ownership, and others. Certain combinations of these variables could make land unsuitable for development. Figure 29-11 illustrates the simple case of land suitability analysis involving only two variables, slope and soil type. Figure 29-11(a) shows polygons within which the average slope is either 5 or 10%. Figure 29-11(b) classifies the soils in the area as E (erodible) or S (stable). The composite of the two data sets, which results from a *polygon-on-polygon* overlay, is shown in Figure 29-11(c). It identifies

Figure 29-11 Example of GIS overlay function used to evaluate land suitability. (a) Polygons of differing slopes. (b) Varying soil types in an area. (c) Overlay of (a) and (b), identifying polygon I as an area combining lower slopes with stable soils that would be suitable for development.

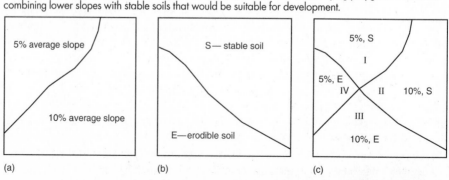

polygon I, which combines 5% slopes and low-erodibility class S soils. Since this combination does not present potential erosion problems, the area within polygon I is suitable for development, while areas II, III, and IV are not.

Another GIS overlay function is that of *point in polygon.* Here the question involves which point features are located in certain polygons where layers are combined. For example, to predict possible well contamination, a GIS operator may want to know which wells are located in an area of highly permeable soil. A similar overlay process, *line in polygon,* identifies specific linear features within polygons of interest. An example of its application would be the identification of all bituminous roads, paved more than 15 years ago, in townships whose roadway maintenance budgets are less than $250,000. Obviously such information would be valuable to support decisions concerning the allocation of state resources for local roadway maintenance.

The GIS functions just described can be used individually, as has been illustrated with the examples, or they can be employed in combination. The following example illustrates the simultaneous application of line buffering, adjacency, and overlay. The situation involves giving timely notice to affected persons of an impending flood that is predicted to crest at a stage of Z ft above a particular river's normal elevation. Here it is first necessary to identify the lands which lie at or below the expected flood stage. This can be done using line buffering, where the thread of the river is the line of reference. The width of the buffer zone, however, is variable and is determined by combining elevation data in the TIN model with the buffering process. The adjacency and overlay functions are then used to determine which land owners are next to or within the flood zone. Then the names, addresses, and telephone numbers of property owners and dwellers within and adjacent to the affected area can be tabulated. These people can then be notified of the impending situation, and emergency preparations such as construction of temporary levees can be performed, or, if necessary, the area can be evacuated. Should evacuation be required, the GIS may even be used to identify the best and safest escape routes. Data sets necessary for such analyses would include the topography in the area, including the river's location, normal stage, and floodplain cross sections; census data; property ownership; and the transportation network.

29-8.4 Other GIS Functions

In addition to the spatial analysis functions described in preceding subsections, many other functions are available with most GIS software. Some of these include the capability of computing (**1**) the number of times a particular type of point occurs in a certain polygon; (**2**) the distances between selected points, or from a selected line to a point; (**3**) areas within polygons; (**4**) locations of polygon centroids; and (**5**) volumes within polygons where depth or other conditions are specified. A variety of different mapping functions may be performed using GISs. These may include (**1**) performing map scale changes; (**2**) changing the reference coordinate system from, say, the state plane to the UTM system; (**3**) rotating the reference grid; and (**4**) changing the contour interval used to represent elevations.

Most GISs are also capable of performing several different digital terrain analysis functions. Some of these include (**1**) creating TIN models or other DEMs from randomly spaced X, Y, Z terrain data; (**2**) calculating profiles along designated reference lines, and determining cross sections at specified points along the reference line; (**3**) generating perspective views where the viewpoint can be varied; (**4**) analyzing visibility to determine what can or cannot be seen from a given vantage point; **(5)** computing slopes and aspects; and **(6)** making sun intensity analyses.

Output from GISs can be provided in graphic form as charts, diagrams, and maps; in numerical form as statistical tabulations; or in other files that result from computations and manipulations of the geographic data. These materials can be supplied in either printed (hardcopy) form, or on diskettes or tapes (softcopy).

29-9 GIS APPLICATIONS

As stated earlier, and as indicated by the examples in preceding sections of this chapter, the areas of GIS applications are widespread. Further evidence of the diversity of GIS applications can be seen by reviewing the bibliography at the end of this chapter. The technology is being used worldwide, at all levels of government, in business and industry, by public utilities, and in private engineering and surveying offices.

Some of the more common areas of application occur in **(1)** land-use planning; **(2)** natural resource mapping and management; **(3)** environmental impact assessment; **(4)** census, population distribution, and related demographic analyses; **(5)** route selection for highways, rapid-transit systems, pipelines, transmission lines, and so on; **(6)** displaying geographic distributions of events such as automobile accidents, fires, crimes, or facility failures; **(7)** routing buses or trucks in a fleet; **(8)** tax mapping and mapping for surveying and engineering purposes; **(9)** subdivision design; **(10)** infrastructure and utility mapping and management; and **(11)** urban and regional planning.

As the use of GIS technology expands, there will be a growing need for trained individuals who understand the fundamentals of these systems. Users should be aware of the manner in which information is recorded, stored, managed, retrieved, analyzed, and displayed using a GIS. System users should also have a fundamental understanding of each of the GIS functions, including their bases for operation, their limits, and their capabilities. *Perhaps of most importance, users must realize that information obtained from a GIS can be no better than the quality of the data from which it was derived.*

From the surveyor's perspective, it is important to underscore again the fact that the fundamental basis of GISs is a database of spatially related, digitally mapped data. Since accurate position determination and mapping are the surveyor's forte, in the future surveyors will continue to play key roles in designing, developing, implementing, and managing these systems. Their input will be particularly essential in establishing the necessary basic control frameworks, conducting ground and aerial surveys to locate geographic features and their attributes, compiling maps, and assembling the digital data files needed for these systems.

PROBLEMS

29-1 List the fundamental components of a GIS.

29-2 Describe the differences between a GIS and an LIS.

29-3 List the fields within surveying and mapping that are fundamental to the development and implementation of GISs.

29-4 Discuss the importance of the reference coordinate system layer in a GIS.

29-5 Name and describe the different simple spatial objects used for representing graphic data in digital form. Which objects are used in vector format representations?

29-6 Discuss the differences between vector and raster formats for entering data into a GIS. Cite the advantages and disadvantages of each.

29-7 How many pixels are required to convert the following documents to raster form for the conditions given:

   *(a) A 30-in. square map scanned at 100 dpi

   (b) A 9-in. square aerial photo scanned at 500 dpi

   (c) An orthophoto of 20 × 24 in. dimensions scanned at 200 dpi

29-8 Explain how data can be converted from:

   (a) Vector to raster format

   (b) Raster to vector format

29-9 For what types of data is the raster format best suited?

29-10 Discuss the compromising relationships between grid cell size and resolution in raster data representation.

29-11 Discuss the advantages and disadvantages of using tablet digitizers for converting maps and other graphic data to digital form.

29-12 What are the advantages of using scanners for converting graphic data to digital form?

29-13 Define the term *topology* and discuss its importance in GISs.

29-14 Develop identifier and topology tables similar to those of Tables 29-1 and 29-2 in the text for the vector representation of (see figures below):

   (a) Problem 29-14(a)   (b) Problem 29-14(b)

29-15 Compile a list of linear features for which the topological relationship of direction would be important.

29-16 Prepare a raster (grid cell) representation of the sample map of:

   (a) Problem 29-16(a), using a cell size of 0.20 in. square (see figure on page 666).

   (b) Problem 29-16(b), using a cell size of 0.10 in. square (see figure on page 666).

29-17 What is *connectivity* in GIS terminology? Cite an example GIS application illustrating its value.

29-18 Explain the concept of *adjacency* in GIS spatial analysis. Give an example illustrating its beneficial application.

29-19 If data are being represented in vector format, what simple spatial objects would be associated with each of the following topological properties:

   (a) Connectivity   (b) Direction

   (c) Adjacency   (d) Nestedness

Problem 29-14

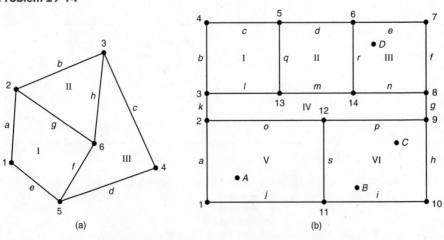

(a)   (b)

---

*Asterisks indicate problems that have answers given in Appendix G.

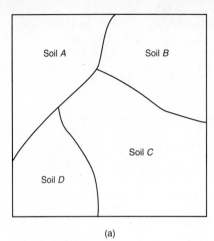

(a)

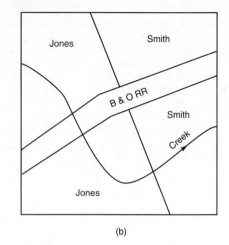

(b)

Problem 29-16

C	C	C	C	B	B	A	A
C	C	C	C	B	B	A	A
C	C	C	B	B	A	A	A
C	B	B	B	A	A	A	A
C	B	B	B	A	A	A	B
B	B	B	A	A	A	A	B
B	A	A	A	A	A	B	B
A	A	A	A	A	A	B	C

1	1	1	2	2	2	2	4
1	1	2	2	2	2	4	2
1	3	3	3	3	4	3	2
1	3	3	3	3	4	3	2
1	1	3	3	3	4	3	2
1	1	3	3	4	3	2	2
1	1	3	4	3	3	2	2
1	3	3	4	3	3	2	2

Annual rainfall legend:
A—25 in.
B—30 in.
C—35 in.

Land cover legend:
1—Agricultural
2—Woodland.
3—Floodplain
4—River

(a)   (b)

Problem 29-20

29-20 Using the technique discussed in Section 29-6.2, prepare a vector representation of the raster data given in (see figures above):
    (a) Problem 29-20(a)   **(b)** Problem 29-20(b)
29-21 What is the actual ground dimension of a pixel for the following conditions:
    *(a) A 1:12,000 scale orthophoto scanned at 200 dpi
    (b) An aerial photo having a scale of 1 in. = 500 ft scanned at 500 dpi
    (c) A 1:2400 map scanned at 100 dpi

*Asterisks indicate problems that have answers given in Appendix G.

29-22 Describe the GIS functions of *point buffering, line buffering,* and *polygon buffering.*

29-23 Figure 29-11 shows an example demonstrating one application of the value of the GIS overlay function. Cite two other examples where this function would be beneficial.

29-24 Identify possible sources in your state for locating the following data sets or layers of information for a GIS:

    (a) Land ownership parcels   **(b)** Zoning

    (c) Floodplains and wetlands   **(d)** Soil types

    (e) Land cover

29-25 Compile a list of data layers and attributes that would likely be included in an LIS.

29-26 Compile a list of data layers and attributes that would likely be included in a GIS for:

    (a) Selecting the optimum corridor for constructing a new rapid-transit system to connect two major cities

    (b) Selecting the optimum location for a new airport in a large metropolitan area

    (c) Routing a nationwide fleet of trucks

    (d) Selecting the fastest routes for reaching locations of fires from various fire stations in a large city

29-27 In Section 29-8.3, a flood warning example is given to illustrate the value of simultaneously applying more than one GIS analytical function. Describe another example.

29-28 Consult the literature on GISs and, based on your research, describe an example that gives an application of a GIS in:

    (a) Natural resource management   **(b)** Agriculture

    (c) Engineering   **(d)** Forestry

    (e) Demography   **(f)** Land records

    (g) Surveying and measurement

BIBLIOGRAPHY

American Congress on Surveying and Mapping. 1992. "Implementing the Spatial Data Transfer Standard." *Cartography and Geographic Information Systems,* 19 (No. 5):277.

Antennucci, J., et al. 1991. *Geographic Information Systems.* New York: Van Nostrand Reinhold.

Aronoff, S. 1989. *Geographic Information Systems: A Management Perspective.* Ottawa, Canada: WDL Publications.

Blair, B. R. 1992. "Time-Critical GIS: The Key to Emergency Response." *Geo Info Systems* 2 (No. 5):24.

Bossler, J. D., et al. 1991. "GPS and GIS Map the Nation's Highways." *Geo Info Systems* 1 (No. 3):26.

Carter, J. 1992. "Perspective on Sharing Data in Geographic Information Systems." *Photogrammetric Engineering and Remote Sensing* 43 (No. 11):1557.

Clapp, J. L. 1991. "Geographic Information Systems: State Perspective." *ASCE, Journal of Surveying Engineering* 117 (No. 3):117.

Dangermond, J. 1992. "Where Is GIS Technology Going?" *Professional Surveyor* 12 (No. 2):30.

Dueker, K. J. 1987. "Geographic Information Systems and Computer Aided Mapping." *Journal of the American Planning Association* 53 (No. 3):383.

Donnelly, J. 1991. "Geographic Information Systems in Map Making." *Bulletin, American Congress in Surveying and Mapping* (No. 134):30.

Doyle, D. 1991. "Cadastral Surveys in a GIS/LIS." *Bulletin, American Congress on Surveying and Mapping* (No. 134):38.

Epstein, E., and T. Duchesneau. 1990. "Use and Value of a Geodetic Reference System." *Journal of the Urban and Regional Information Systems Association* 2 (No. 1):11.

Falconer, A. 1992. "Geographic Information Technology Fulfills Need for Timely Data." *GIS World* 5 (No. 6):37.

Foust, B., and H. Botts, 1991. "Tourism Gets a Boost from Wisconsin's Recreational Resource GIS." *Geo Info Systems* 1 (No. 8):43.

Frank, A., et al. 1991. "A Perspective on GIS Technology in the Nineties." *Photogrammetric Engineering and Remote Sensing* 42 (No. 11):1431.

Fule, P., and S. Bradshaw. 1992. "Wildfire Management Planning for New Mexico's Native American Lands." *Geo Info Systems* 2 (No. 6):34.

Giovachino, D. 1993. "How to Determine the Accuracy of Your Graphic Digitizer." *Geo Info Systems* 3 (No. 3):50.

Hanigan, F. 1988. "GIS by Any Other Name Is Still . . ." *The GIS Forum* 1:6.

Hendricksen, C., and L. Hall. 1992. "GIS Measures Water Use in the Arid West." *Geo Info Systems* 2 (No. 7):63.

Hodgson, M. E., and R. Palm. 1992. "Attitude Toward Disaster: A GIS Design for Analyzing Human Response to Earthquake Hazards." *Geo Info Systems* 2 (No. 7):40.

Holmes, D. D. 1991. "Automated Data Capture for Geographic Information Systems." *Surveying and Land Information Systems* 51 (No. 2):87.

Huxhold, W. 1991. *An Introduction to Urban Geographic Information Systems.* New York: Oxford University Press.

Jaffe, B. A., et al. 1992. "Update from the Ashes: GIS Helps Oakland Organize to Rebuild." *Geo Info Systems* 2 (No. 7):68.

Kautz, R. S. 1992. "Satellite Imagery and GIS Help Protect Wildlife Habitat in Florida." *Geo Info Systems* 2 (No. 1):37.

Kilborn, K., et al. 1992. "Connecting Groundwater Models and GIS." *Geo Info Systems* 2 (No. 2):26.

Korte, G. B. 1992. "GIS Technology Trends." *Point of Beginning* 18 (No. 1):42.

Lee, Y. C., and G. Y. Zhana. 1989. "Developments in Geographic Information Systems Technology." *ASCE, Journal of Surveying Engineering* 115 (No. 3):304.

Light, D. 1993. "The National Aerial Photography Program as a Geographic Information System Resource." *Photogrammetric Engineering and Remote Sensing* 59 (No. 1):61.

Lindquist, R. C. 1991. "Illinois Cleans Up: Using GIS for Landfill Siting." *Geo Info Systems* 1 (No. 2):30.

Loukes, D. K., and J. McLaughlin. 1991. "GIS and Transportation: Canadian Perspective." *ASCE, Journal of Surveying Engineering* 117 (No. 3):123.

Maguire, D., et al. 1991. *Geographical Information Systems: Principles and Applications.* New York: John Wiley and Sons.

McLaughlin, J. D., and S. E. Nichols. 1987. "Parcel-Based Land Information Systems." *Surveying and Mapping* 47 (No. 1):155.

McRae, S. D., and P. M. Durgin. 1989. "An Introduction to GIS." *Point of Beginning* 15 (No. 1):10.

Mercready, R., et al. 1992. "GIS Provides Clues to Elusive Natural Gas Reservoirs." *Geo Info Systems* 2 (No. 2):42.

Mettel, C. 1992. "GIS and Satellite Images Provide Precise Calculations of Probable Maximum Flood." *Geo Info Systems* 2 (No. 6):44.

Moellering, H. 1992. "Spatial Data Transfer Standards." *Bulletin, American Congress on Surveying and Mapping* (No. 137):30.

Morain, S., et al. 1992. "U.S. National Report—Status of Photogrammetry, Remote Sensing and Geographic Information Systems in the United States." *Photogrammetric Engineering and Remote Sensing,* 43 (No. 8):1073.

National Institute of Standards and Technology. 1992. *Federal Information Processing Standard Publication 173: Spatial Data Transfer Standard.* U.S. Department of Commerce.

Nichols, S., and J. McLaughlin. 1993. "The Surveyor and the Information Highway." *Professional Surveyor* 13 (No. 1):4.

Onsrud, H., and D. Cook. 1990. *Geographic and Land Information Systems for Practicing Surveyors: A Compendium.* Bethesda, Md.: American Congress on Surveying and Mapping.

Padgett, D. A. 1992. "Assessing the Safety of Transportation Routes for Hazardous Materials." *Geo Info Systems* 2 (No. 2):46.

Pollock, R. J., and J. D. McLaughlin. 1990. "Data Base Management System Technology and Geographic Information Systems." *ASCE, Journal of Surveying Engineering* 117 (No. 1):9.

Ripple, W. 1986. *Geographic Information Systems for Resource Management.* Bethesda, Md.: American Congress on Surveying and Mapping.

Schulman, R. D. 1991. "Portable GIS: From the Sands of Desert Storm to the Forests of California." *Geo Info Systems* 1 (No. 8):24.

Slonecker, E. T., and M. J. Hewitt. 1991. "Evaluating Locational Point Accuracy in a GIS Environment." *Geo Info Systems* 1 (No. 6):36.

Smyrnew, J. M. 1990. "Trends in Geographic Information System Technology." *ASCE, Journal of Surveying Engineering* 116 (No. 2):105.

Sorenson, M. 1991. "Land Surveying: The Cornerstone of Project Planning and Design with GIS." *Bulletin, American Congress on Surveying and Mapping* (No. 131):41.

Ventura, S. 1991. *Land Information Systems in Local Government—Steps Toward Land Records Modernization in Wisconsin.* Madison, Wis.: Wisconsin State Cartographer's Office.

Vonderohe, A., et al. 1991. *Introduction to Local Land Information Systems for Wisconsin's Future.* Madison, Wis.: Wisconsin State Cartographer's Office.

von Meyer, N. 1991. "Cadastral Evidence Data in a Land Information System." *Journal of the Urban and Regional Information Systems Association* 3 (No. 1):14.

von Meyer, N., and R. Scruggs. 1991. "Cadastral Survey Data Modelling for the Bureau of Land Management Land Information Systems." *Surveying and Land Information Systems* 51 (No. 1):49.

Voss, D. A. 1991. "Integrating Pavement Management and GIS." *Geo Info Systems* 1 (No. 9):18.

APPENDIX A
INSTRUMENT TESTING AND ADJUSTING

A-1 INTRODUCTION

Surveying instruments are designed and constructed to give correct measurements, most often directly in horizontal and vertical planes. A good instrument, properly used and cared for, may stay in adjustment for months, or longer, and last a lifetime. Nevertheless, temperature changes, jarring, and improper handling can cause instruments to go out of adjustment. Therefore they should be tested periodically and adjusted when necessary to maintain accuracy. A level, for example, should be checked each day it is used on important work.

For many surveying measurements, proper field procedures, such as double centering and keeping backsights and foresights equal, permit accurate work to be done even though an instrument is out of adjustment. Frequently, however, a few minutes spent in making simple adjustments reduces the time and effort required to operate equipment and perform surveys efficiently.

Surveyors and engineers can conduct tests to determine whether any surveying instrument is out of adjustment; and they should be capable of making routine adjustments of dumpy levels, tilting dumpy levels, and transits. Sections of this appendix describe standard techniques. Except for bull's-eye level vials, optical plummets, and some instrument plate bubbles, adjustment of automatic levels, theodolites, EDM instruments, and total stations is generally best left to qualified specialists. Dismantling these instruments by inexperienced persons can result in serious damage. If tests described in the following sections reveal that adjustments are required and the operator or owner lacks the knowledge and skills necessary to perform these adjustments, the equipment should be carefully packed and shipped to the manufacturer or a laboratory where specialists are available. Maintenance agreements with surveying equipment manufacturers and sales organizations are now commonplace for modern surveying instruments. Under these arrangements, the instruments are returned on a scheduled (usually annual) basis for checking, cleaning, lubricating, and adjusting as necessary.

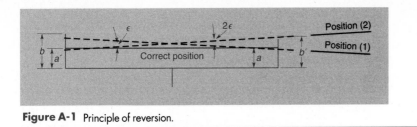

Figure A-1 Principle of reversion.

A-2 METHODS OF TESTING INSTRUMENTS

Most tests used to check instruments fall into one of the following two categories:

1. *Comparison with known values.* This method consists in making direct comparisons of measuring devices against a precisely calibrated standard. It is used in testing and calibrating tapes, EDM instruments, and level rods. A variation of this procedure was discussed in Section 5-8.2 in connection with testing and calibrating an EDM instrument. In that method a base line is first measured in total, then in two parts, and by comparing the sum of the parts to the total, the system constant is determined.

2. *Principle of reversion.* This procedure consists of reversing an instrument in position to double the error and make it more apparent. It is used extensively in testing levels, transits, theodolites, and total stations. To illustrate, assume that in Figure A-1 the angular error between the correct and unadjusted line ε is caused by the difference in lengths a and b, as shown in position (1). (An example would be the heights of the two ends of a level vial on a telescope.) After the telescope is turned 180° in azimuth, the unadjusted line occupies position (2) because a and b have changed places. Since the angle between positions (1) and (2) is 2ε, it follows that reversion doubles the error. Specific methods of making reversion tests with various surveying instruments are described in this appendix.

A-3 REQUIREMENTS FOR TESTING AND ADJUSTING LEVELS, TRANSITS, THEODOLITES, AND TOTAL STATION INSTRUMENTS

Before making adjustments, *careful* tests should be made to ensure that any apparent lack of adjustment is actually caused by the instrument's condition, not by test deficiencies. To properly check and adjust levels, transits, theodolites, and total station instruments in the field, the following rules should be followed:

1. Choose terrain that permits solid setups in a nearly level area enabling sights of at least 200 ft to be made in opposite directions. Three permanent points set approximately 200 ft apart in a straight line, on nearly level ground, and preferably at the same elevation, expedite adjustments. Organizations having a number of instruments in use, and surveyors working in an area over a long period of time, find it worthwhile to set permanent marks for this purpose.

2. Perform adjustments when good atmospheric conditions prevail, preferably on cloudy days free of heat waves. No sight line should pass through alternate sun and shadow, or be directed into the sun.

3. Place the instrument in shade, or shield it from direct rays of the sun.

4. Make sure the tripod shoes are tight and the instrument is firmly screwed onto the tripod. Spread the tripod legs well apart and position them so that the tripod plate is nearly level. Press the shoes firmly into the ground. With older conventional tripods, loosen the three tripod hinge screws to relieve stresses and then tighten again.

Standard methods and a prescribed order must be followed in adjusting surveying instruments. Correct positioning of parts is attained by loosening or tightening the proper adjusting nuts and screws with special tools and pins. Time is wasted if each adjustment is perfected on the first trial, since some adjustments affect others. The complete series of tests may have to be repeated several times if an instrument is badly off. A final check of all adjustments should be made to ensure that none has been disturbed. The simplest adjustment of all, removal of *parallax* by carefully focusing the objective lens and eyepiece, must be kept in mind at all times.

Tools and adjusting pins that fit the capstans and screws should be used, and the capstans and screws handled with care to avoid damaging the soft metal. Adjustment screws have been properly set when an instrument is shipped from the factory. Tightening them too much (or not enough) nullifies otherwise correct adjustment procedures and may leave the instrument in worse condition than it was before testing.

A-4 ADJUSTMENT OF A DUMPY LEVEL

A level in adjustment establishes a horizontal plane of sight when the telescope is revolved about a vertical axis. The principal lines of the dumpy level, as illustrated in Figure A-2, are **(1)** axis of sight, **(2)** axis of the level bubble, **(3)** axis of the level bar, and **(4)** vertical axis.

For perfect adjustment it is necessary that the axis of sight, the axis of the level bubble, and the axis of the level bar be parallel to each other, and perpendicular to the vertical axis. There are two adjustable parts: the cross hairs and the level vial. The adjustments should be made in the order given.

A-4.1 Adjustment of Level Vial

Purpose. To make the axis of the level bubble perpendicular to the vertical axis.

Figure A-2 Dumpy level.

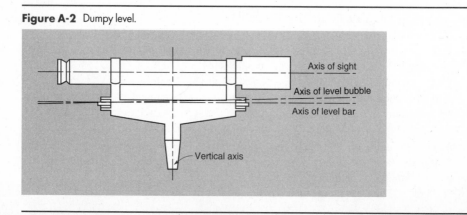

Test. Set up the level, center the bubble, and revolve the telescope 180° about the vertical axis. The distance the bubble moves off the central position is double the error.

Correction. Turn the capstan nuts at one end of the level vial to move the bubble *halfway back* to the centered position. Level the instrument using the leveling screws. Repeat the test until the bubble remains centered during a complete revolution of the telescope.

A-4.2 Preliminary Adjustment of Horizontal Cross Hair

Purpose. To make the horizontal cross hair truly horizontal when the instrument is leveled.

Test. Sight a sharply defined point with one end of the horizontal cross hair. Turn the telescope slowly on its vertical axis so that the cross hair moves across the point. If the cross hair does not remain on the point for its full length, it is out of adjustment.

Correction. Loosen the four capstan screws holding the reticle. Rotate the reticle in the telescope tube until the horizontal hair remains on the point as the telescope is turned. The screws should then be carefully tightened in their final position.

A-4.3 Line of Sight Adjustment

Purpose. To make the axis of sight perpendicular to the vertical axis and thus parallel to the axis of the level bubble. This adjustment is also called the *two-peg method* and the *direct adjustment.*

Test. Level the instrument over a point C halfway between two stakes A and B that are about 200 ft apart (see Figure A-3). Determine the difference in rod readings a_1 and b_1 on A and B, respectively. Since the distance to the two points is equal, the true difference in elevation is obtained even though the axis of sight is not exactly horizontal.

Then set the instrument at D on line with the stakes and close to one of them—A in this case—and level. With the eyepiece only a few inches from the rod, a reading a_2 on

Figure A-3 Peg adjustment.

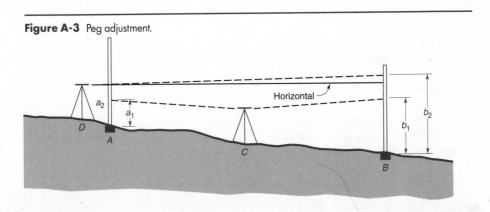

A is taken by sighting through the objective lens end of the telescope. Usually a pencil is centered in the small field of view to make the reading. A rod reading b_2 is also taken on *B*.

If the axis of sight is parallel to the axis of the level bubble (that is, horizontal), the rod reading b_2 should equal the rod reading at *A* plus the difference in elevation between *A* and *B*, or $a_2 + (b_1 - a_1)$. The difference, if any, between the computed and actual readings is the error to be corrected by adjustment.

Correction. Loosen the top (or bottom) capstan screw holding the reticle, and tighten the bottom (or top) screw to move the horizontal hair up or down and give the required reading on the rod at *B*. Several trials may be necessary to get an exact setting. (*Caution:* One screw should be loosened before the other is tightened on older instruments to avoid breaking the cross hair!)

An alternative method of testing the adjustment is by reciprocal leveling (see Section 7-7). A setup is made close to *A* and readings are taken on *A* and *B*. The level is moved to a position near *B* and similar sights are taken. The difference in elevation is computed and the reticle shifted to give a reading on the distant rod equal to the reading on the near point plus the difference in elevation of the points.

If the true difference in elevation of *A* and *B* is known, only one setup is required (near either point) for the adjustment.

A-5 ADJUSTMENT OF A TILTING DUMPY LEVEL

As described in Section 6-10, the vertical axis of a tilting dumpy level is oriented approximately vertical using a bull's-eye bubble, and the axis of sight is made horizontal for each individual backsight and foresight by carefully centering the main sensitive bubble. For a tilting dumpy level in good adjustment, the axis of its bull's-eye bubble should be approximately perpendicular to the vertical axis, and the line of sight must be parallel to the axis of the sensitive level bubble. The bull's-eye bubble can be tested and adjusted following procedures given in Section A-9. The remaining adjustments of tilting levels should be done in the order given below.

A-5.1 Adjustment of the Horizontal Cross Hair

Purpose, Test, and Correction. Same as the second adjustment of the dumpy level (see Section A-4.2).

A-5.2 Adjustment of the Sensitive Level Bubble

Purpose. To make the axis of the sensitive bubble parallel to the axis of sight.

Test. Perform the peg test as described in Section A-4.3. The telescope is brought to the rod reading required for the horizontal sight using the tilting knob. Maladjustment is indicated if the bubble is not centered in this position.

Correction. With the axis of sight horizontal, the adjusting screws on the sensitive bubble are raised or lowered as necessary to center the bubble, or to make the two ends of the bubble match for coincidence bubbles.

A-6 TESTING AUTOMATIC LEVELS

The automatic compensator of an automatic level, described in Section 6-11, can become maladjusted such that the line of sight it defines is not horizontal. This is the so-called *collimation error.* It would not cause errors in differential leveling so long as backsight and foresight distances are balanced. It will cause errors, however, where backsights and foresights are not balanced, which sometimes occurs in differential leveling, and cannot be avoiding in construction staking and profile leveling.

As noted in Section A-1, except for the bull's-eye bubble, adjustment of other parts of an automatic level is generally best left to qualified specialists. The possibility of an inclined line of sight can be easily tested, however, using the two-peg test described in Section A-4.3. Then if the test reveals a significant error, a correction can be computed and applied to eliminate the error on precise work (see Section 19-16), or the instrument can be adjusted by someone having the necessary knowledge and skills, or it can be returned to the manufacturer for adjustment.

A-7 ADJUSTMENT OF A TRANSIT[1]

Transits are designed to measure the horizontal and vertical projections of angles, and sometimes to serve as levels. Thus specific reference lines and axes must be precisely positioned with respect to each other. The principal axes of a transit are the same as those of the theodolite shown in Figure 11-16, except that transits have a telescope bubble and a theodolite does not. The reference axes of a transit are:

1. Axis of the plate bubble. (The line tangent to the upper circular arc of the plate level vial at its midpoint.)
2. Vertical axis, also called the *standing axis.* (Same as for the level.)
3. Horizontal axis. (The line through the center of rotation of the telescope axle and its bearings in the standards.)
4. Axis of sight. (Same as for the level.)
5. Axis of telescope bubble. (Same as for the level.)

For correct adjustment, **(1)** the axis of the plate bubble must be perpendicular to the vertical axis, **(2)** the horizontal axis parallel to the plate bubble axis, **(3)** the axis of sight perpendicular to the horizontal axis, and **(4)** the axis of the telescope bubble parallel to the axis of sight. To maintain these relationships, plate-level vials, cross hairs, standards, telescope-level vial, and vertical-circle vernier are adjustable.

A-7.1 Adjustment of Plate-Level Vials

Purpose. To make the axis of each plate-level bubble perpendicular to the vertical axis.

---

[1]The tests described in this section can also be used to check theodolites and total stations; however, as stated earlier, their adjustment is generally best left to specialists. As described in Section 10–13, some total stations can automatically apply numerical corrections to measured horizontal and vertical angles in real time, thus compensating for their systematic instrumental errors. First, however, a simple calibration procedure (outlined in the owner's manual) must be followed, which enables the instruments to compute and store the correction parameters.

Test. Set up the instrument, bring one plate-level vial over two opposite leveling screws, and center it. Revolve the instrument 180° about the vertical axis to place that level vial, turned end for end, over the same leveling screws. The distance the bubble moves from its central position is *double* the error.

Correction. Turn the capstan screws at one end of the level vial to move the bubble *halfway* back to the centered position. Center the bubbles with the leveling screws. Repeat the test until the bubble remains centered during a complete revolution of the instrument. Adjust the other bubble in the same manner.

   (This adjustment can also be performed for theodolites and total station instruments equipped with a plate bubble that has exposed adjusting screws.)

A-7.2 Preliminary Adjustment of Vertical Cross Hair

Purpose. To place the vertical cross hair in a plane perpendicular to the horizontal axis of the instrument.

Test. Sight on a well-defined point with one end of the vertical cross hair. Turn the telescope on its horizontal axis so the cross hair moves along the point. If it departs, the cross hair is not perpendicular to the horizontal axis.

Correction. Loosen all four capstan screws holding the reticle, and turn the reticle slightly until the vertical hair remains on the fixed point during rotation of the telescope. Tighten the capstan screws and recheck the adjustment.

A-7.3 Adjustment of Vertical Cross Hair

Purpose. To make the axis of sight perpendicular to the horizontal axis.

Test. This test employs the double-centering procedure for prolonging a straight line. Level the transit and backsight carefully on a well-defined distant point *A,* as in Figure A-4, clamping the plates. Plunge the telescope and set a foresight point *B* at approximately the same elevation as *A,* and at least 400 ft away if possible. With the telescope still in the inverted position, unclamp either plate, turn the instrument on the vertical

***

Figure A-4 Double reversion.

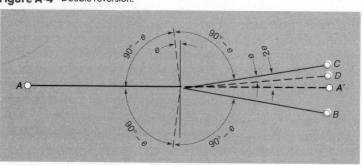

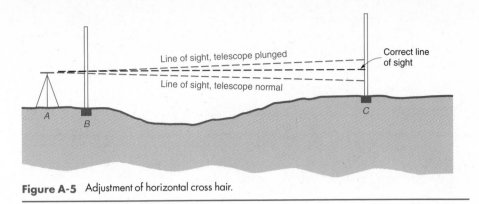

Figure A-5 Adjustment of horizontal cross hair.

axis, backsight on the first point A again, and clamp the plate. Plunge the telescope back to its normal position and set a point C beside the first foresight point B. The distance between B and C is *four times* the error of adjustment because of the double reversion.

(This test applies also for theodolites and total station instruments.)

Correction. Loosen one of the side capstan screws which hold the reticle to the telescope tube and tighten the opposite screw to move the vertical hair *one-fourth* of the distance CB to point D. Repeat the test until the telescope sights the same point A' after reversing from backsight A.

A-7.4 Adjustment of Horizontal Cross Hair

Purpose. To bring the horizontal cross hair into the optical axis of the telescope. This is necessary if the transit is to be used for leveling or measuring vertical angles.

Test. Set up and level the transit over point A, as in Figure A-5. Line in two stakes B and C at approximately the same elevation. Stake B should be at minimum focusing distance from A, perhaps 5 or 10 ft, and stake C at least 300 ft away.

Take readings of the horizontal cross hair on a rod held first on B, then on C. Plunge the telescope, turn the instrument about its vertical axis, and sight B again, setting the horizontal hair on the first rod reading. Then with vertical and horizontal axes clamped, read the rod held on far stake C. Any discrepancy between the two readings on the rod at C is approximately *double* the error.

Correction. By means of the top and bottom capstan screws holding the reticle, move the horizontal hair until it intercepts the rod *halfway* between the two readings on C. Repeat the test and adjustment until the horizontal-hair reading does not change for normal and plunged sights on the far point.

A-7.5 Adjustment of Standards

Purpose. To make the telescope's horizontal axis perpendicular to the transit's vertical axis.

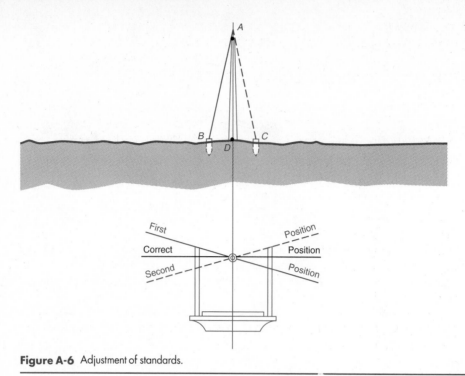

Figure A-6 Adjustment of standards.

Test. With the transit carefully leveled parallel to the horizontal axis, sight a well-defined high point A, as in Figure A-6, at a vertical angle of at least 30°, and clamp the plates. Depress the telescope and mark a point B near the ground. Plunge the telescope, unclamp either plate, turn the instrument about the vertical axis, sight point A again, and clamp the plate. Now depress the telescope and set another point, C, near B. Any discrepancy between B and C is the result of unequal standard heights and represents approximately *double* the error.

Correction. Set a point D approximately *halfway* between B and C and sight on it. With the plates clamped, elevate the telescope and bring the line of sight on point A by raising or lowering the movable block in one standard. To raise the horizontal axis, first loosen the friction screws holding the trunnion cap on the standard, then tighten the capstan screw below the block. To lower the axis, reverse the procedure. The friction screws holding the trunnion cap must be set carefully to prevent the telescope from being too loose or binding. Repeat the test and adjustment until the high and low points remain in the line of sight with the telescope in both normal and inverted positions.

A-7.6 Adjustment of Telescope-Level Vial

Purpose. To make the axis of the telescope level bubble perpendicular to the vertical axis and parallel to the axis of sight.

Test. Same as for the dumpy level (see Section A-4.3).

Correction. If rod reading b_2, as in Figure A-3, indicates an adjustment is necessary, the correct rod reading required to produce a horizontal axis of sight is set on the rod by means of the vertical-circle slow-motion screw. The telescope bubble is then centered by turning the capstan screws at one end of the level vial.

A-7.7 Adjustment of Vertical-Circle Vernier

Purpose. To make the vernier of the vertical circle read zero when the telescope-level bubble is centered, thereby producing a zero index error.

Test. Level the instrument with both plate-level bubbles. Using the vertical-circle slow-motion screw, center the telescope bubble. Any nonzero angle on the vertical-circle vernier is an index error.

Correction. Loosen the capstan screws holding the vernier plate and move it so the vertical circle and vernier zero marks coincide. Tighten the capstan screws. Avoid leaving a gap between the vernier and vertical circle, since such a space introduces errors in reading angles.

A-8   ADJUSTMENT OF AN OPTICAL PLUMMET

The line of sight of an optical plummet should coincide with the instrument's vertical axis. Two different situations exist: **(1)** the optical plummet is part of the instrument and rotates with it when turned in azimuth, or **(2)** the optical plummet is part of the tribrach, which is fastened to the tripod and does not turn in azimuth. To adjust an optical plummet for the first case, set the instrument over a fine point and aim the line of sight exactly at it by turning the leveling screws. (The instrument need not be level.) Rotate the instrument 180° in azimuth. If the optical plummet reticle moves off the point, bring it *halfway* back by means of the adjusting screws provided. Center the reticle on the point again with the leveling screws, and repeat the test.

For the second case, carefully level the instrument and suspend a plumb bob beneath it. Place a fine mark on the ground at the location of its point. Remove the plumb bob and sight through the optical plummet. If the mark is not on the plummet's cross hairs, bring the cross hairs to the ground point using the adjusting screws.

A-9   ADJUSTMENT OF BULL'S-EYE BUBBLES

If a bull's-eye bubble on a level, theodolite, or total station does not remain centered when the instrument is rotated in azimuth, the bubble is out of adjustment. It should be corrected, although precise adjustment is unnecessary because it does not control fine leveling of the reference axes. To adjust the bubble, carefully center it using the leveling screws and turn the instrument 180° in azimuth. *Half* of the bubble run is corrected by manipulating the vial-adjusting screws, the bubble centered by operating the leveling screws, and the test repeated.

Bull's-eye bubbles used on prism poles and level rods must be in good adjustment for accurate work. To adjust them, carefully orient the rod or pole vertically by aligning it parallel to a long plumb line, and fasten it in that position using shims and C-clamps. Then center the bubble in the vial using the adjusting screws.

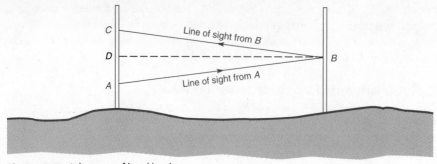

Figure A-7 Adjustment of hand level.

A-10 ADJUSTMENT OF A HAND LEVEL

The only adjustable part of a hand level is the horizontal cross hair. It should be in a position that makes the axis of sight horizontal when the bubble is centered. To test a hand level, hold the instrument against a solid support such as a post at elevation A of Figure A-7. With the bubble centered, sight to another post or building corner and mark point B. Distance AB should not be greater than about 100 ft. Then support the level at B, center the bubble, and note whether the line of sight strikes point A. If not, mark another point C. To adjust the level, *bisect* distance AC and set point D. With the level at B and the bubble centered, move the cross hair to D by means of the adjusting screws.

APPENDIX B
COORDINATE GEOMETRY IN SURVEYING CALCULATIONS

B-1 INTRODUCTION

Except for extensive geodetic control surveys, almost all other surveys are referenced to plane rectangular coordinate systems. State plane coordinates (see Chapter 21) are most frequently employed, although local arbitrary systems can be used. Advantages of referencing points in a rectangular coordinate system are (1) the relative positions of points are uniquely defined, (2) they can be conveniently plotted, (3) if lost in the field, they can readily be recovered from other available points referenced to the same system, and (4) computations are greatly facilitated. The last-named advantage is the subject of this appendix.

Computations involving coordinates are performed in a variety of surveying problems. Two situations were introduced in Chapter 13, where it was shown that the length and direction (azimuth or bearing) of a line can be calculated from the coordinates of its end points. Area computation using coordinates was discussed in Chapter 14. Additional problems that are conveniently solved using coordinates are determining the point of intersection of (1) two straight lines, (2) a line and a circle, and (3) two circles. These problems are often encountered in route surveys where it is necessary to compute intersections of tangents and circular curves in horizontal alignments, and in boundary and subdivision work where parcels of land are often defined by straight lines and circular arcs.

Solutions to these problems can be obtained by writing equations for the lines and circles involved; include in them unknown coordinates of the intersection points, then solve them simultaneously for the unknowns. The required equations, together with example problems, are presented in the following sections.

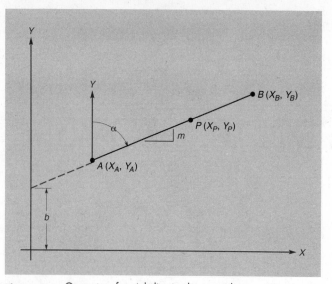

Figure B-1 Geometry of straight line in plane coordinate system.

B-2 COORDINATE FORMS OF EQUATIONS FOR LINES

In Figure B-1 straight line AB is referenced in a plane rectangular coordinate system. Coordinates of end points A and B are X_A, Y_A, X_B, and Y_B. Length AB and the azimuth α of this line in terms of these coordinates are

$$AB = \sqrt{(X_B - X_A)^2 + (Y_B - Y_A)^2} \tag{B-1}$$

$$\alpha = \tan^{-1}\left(\frac{X_B - X_A}{Y_B - Y_A}\right) \tag{B-2}$$

The general mathematical expression for a straight line is

$$Y_P = mX_P + b \tag{B-3}$$

where Y_P is the Y coordinate of any point P on the line whose X coordinate is X_P, m the slope of the line, and b the Y intercept of the line. Slope m can be expressed as

$$m = \frac{Y_B - Y_A}{X_B - X_A} = \cot \alpha \tag{B-4}$$

For any straight line, the slope is constant; that is, in Figure B-1 the slope between A and B is the same as that between A and P. Thus the following equation can be written intuitively from Eq. (B-4):

$$\frac{Y_B - Y_A}{X_B - X_A} = \frac{Y_P - Y_A}{X_P - X_A} = \cot \alpha \tag{B-5}$$

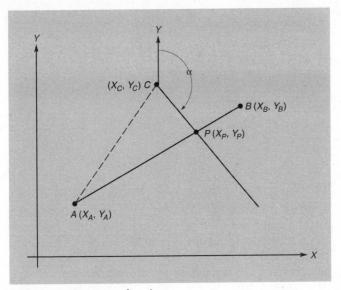

Figure B-2 Intersection of two lines.

B-3 INTERSECTION OF TWO LINES

Equation (B-5) is very useful in computing the point of intersection of two lines. Data generally known for this type of problem are coordinates of end points of lines, or fixed azimuths of lines, determined from survey or design data.

EXAMPLE B-1

In Figure B-2, assuming the following information is known for two lines, compute coordinates X_P and Y_P of the intersection point.

$$X_A = 1425.07 \qquad X_B = 7484.80 \qquad X_C = 4497.96 \qquad \alpha = 141°30'$$

$$Y_A = 1971.28 \qquad Y_B = 5209.64 \qquad Y_C = 6062.00$$

SOLUTION
By Eq. (B-5), an expression for line AB is

$$\frac{5209.64 - 1971.28}{7484.80 - 1425.07} = \frac{Y_P - 1971.28}{X_P - 1425.07} \qquad \text{(a)}$$

Also by Eq. (B-5), an expression for the line at C is

$$\frac{Y_P - 6062.00}{X_P - 4497.96} = \cot 141°30' \qquad \text{(b)}$$

Reducing expressions (a) and (b),

$$0.53441X_P - Y_P = -1209.71 \qquad \text{(c)}$$

$$1.25717X_P + Y_P = 11{,}716.43 \qquad \text{(d)}$$

Solving Eqs. (c) and (d) simultaneously yields

$$X_P = 5864.50 \text{ ft}$$

$$Y_P = 4343.76 \text{ ft}$$

An alternate solution for this problem can be obtained by solving oblique triangle *APC* in Figure B-2. In this procedure, azimuth *AB* can be computed using the coordinates of *A* and *B* in Eq. (B-2). Also from the coordinates of *A* and *C*, the length and azimuth of line *AC* can be calculated using Eqs. (B-1) and (B-2), respectively. Then since azimuth *CP* is given, the directions of all sides of triangle *CPA* are known, so all three triangles can be computed. Side *AP* in the triangle can thus be computed using the *law of sines,* and then the latitude and departure from *A* to *P* calculated using Eqs. (13-1) and (13-2). Adding these components to the coordinates of point *A* yields the coordinates of point *P.* An independent check on this solution can be obtained by also solving triangle *BPC* in the same manner.

B-4 COORDINATE FORM OF EQUATION FOR A CIRCLE

The general mathematical expression for a circle in rectangular coordinates is

$$R^2 = (X_P - X_O)^2 + (Y_P - Y_O)^2 \qquad \text{(B-6)}$$

In Eq. (B-6), and with reference to Figure B-3, *R* is the circle's radius, X_O and Y_O are the coordinates of radius point *O,* and X_P and Y_P the coordinates of any point *P* on the circle. For most problems the radius and coordinates of the radius point are known, *R* having been selected on the basis of design specifications or geometric constraints, and X_O and Y_O having been computed as a result of survey measurements, or scaled from a design map.

B-5 INTERSECTION OF A LINE AND CIRCLE

Figure B-3 shows line *AB* intersecting a circle at point *P.* The coordinates of points *A, B,* and *O* are known, as is the radius. To solve for the intersection point, an equation in the form of Eq. (B-5) can be written for the line, and one like Eq. (B-6) for the circle. These equations, when solved simultaneously, produce a quadratic expression for one of the unknowns, as

$$aY_P^2 + bY_P + c = 0 \qquad \text{(B-7)}$$

The solution for Y_P is then obtained from

$$Y_P = \frac{-b \pm \sqrt{b^2 - 4ac}}{2a} \qquad \text{(B-8)}$$

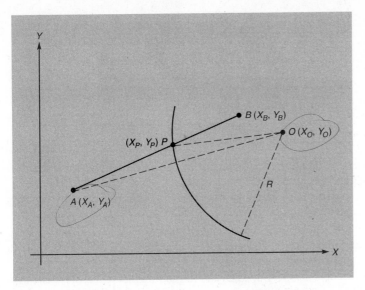

Figure B-3 Intersection of line and circle.

After calculating Y_P, its value can be substituted in one of the original equations to obtain X_P.

In Figure B-3, assume the coordinates of the circle center are $X_O = 500.00$ and $Y_O = 200.00$; for points A and B, $X_A = 100.00$, $Y_A = 130.00$, $X_B = 300.00$, and $Y_B = 200.00$; and $R = 150$ ft. Determine the coordinates of intersection point P.

EXAMPLE B-2

SOLUTION

From Eq. (B-5),

$$\frac{X_P - 100.00}{Y_P - 130.00} = \frac{300.00 - 100.00}{200.00 - 130.00} \qquad (e)$$

and by Eq. (B-6),

$$(X_P - 500.00)^2 + (Y_P - 200.00)^2 = (150.00)^2 \qquad (f)$$

Reducing Eq. (e),

$$X_P = 2.8571Y_P - 271.43 \qquad (g)$$

Substituting Eq. (g) into (f) and reducing,

$$Y_P^2 - 524.73Y_P + 66{,}856 = 0 \qquad (h)$$

Solving Eq. (h) with Eq. (B-8),

$$Y_P = \frac{524.73 \pm \sqrt{(524.73)^2 - 4(66,856)}}{2} = 217.87$$

Then substituting $Y_P = 217.87$ into Eq. (g),

$$X_P = 2.8571(217.87) - 271.43 = 351.05$$

In solving quadratic Eq. (h), the decision to add or subtract the value under the radical can be made on the basis of experience or use of a carefully constructed scaled diagram, which also provides a check on the computations. One answer will be unreasonable and is discarded.

As with Example B-1, an alternative solution to Example B-2 can also be obtained by solving an oblique triangle. In this procedure, the length and azimuth of line AO of Figure B-3 are computed using the coordinates of A and O in Eqs. (B-1) and (B-2). Also azimuth AB is calculated using Eq. (B-2) and the coordinates of A and B. Angle A in triangle APO of Figure B-3 is then computed from azimuths AP and AO. Side OP is the known circle radius, thus two sides and one angle of the triangle are known. Angle P can be calculated using angle A and sides OP and AO in the *law of sines*. Azimuth OP can be determined by applying angle P to azimuth AP. Then the departure and latitude of OP can be calculated, and these components applied to the coordinates of O to obtain the coordinates of point P. Again an independent check can be obtained by also solving triangle BPO.

B-6 INTERSECTION OF TWO CIRCLES

Occasionally, surveyors are required to compute the point of intersection of two circles having known radii and coordinates of their centers. The situation is illustrated in Figure B-4. This problem can be solved by writing equations in the form of Eq. (B-6), including unknown coordinates X_P and Y_P in both equations, then solving simultaneously for the unknowns. The unknowns will both appear as second-degree terms in the equations, however; thus the solution is somewhat difficult to obtain.

In an alternative approach, the length and azimuth of O_1O_2 of Figure B-4 can be obtained from Eqs. (B-1) and (B-2), after which angles β_1 and β_2 are calculated using the law of cosines. With β_1 and β_2 known, azimuths O_1P and O_2P are computed, and the problem is reduced to solving for the coordinates of point P given the length and direction from either known point O_1 or O_2.

EXAMPLE B-3   In Figure B-4, assume the following data are available and X_P and Y_P at point P are required:

$$X_{O_1} = 2851.28 \qquad Y_{O_1} = 299.40 \qquad R_1 = 2000 \text{ ft}$$

$$X_{O_2} = 3898.72 \qquad Y_{O_2} = 2870.15 \qquad R_2 = 1500 \text{ ft}$$

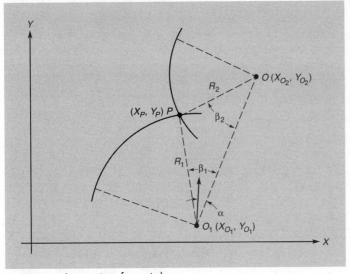

Figure B-4 Intersection of two circles.

SOLUTION
By Eq. (B-1),

$$O_1 O_2 = \sqrt{(3898.72) - 2851.28)^2 + (2870.15 - 299.40)^2} = 2775.95 \text{ ft}$$

By Eq. (B-2),

$$\alpha = \tan^{-1}\left(\frac{3898.72 - 2851.28}{2870.15 - 299.40}\right) = 22°10'05.5''$$

From the law of cosines,

$$\beta_1 = \cos^{-1}\left[\frac{(2000)^2 + (2775.95)^2 - (1500)^2}{2(2000)(2775.95)}\right] = 31°36'53.4''$$

$$\beta_2 = \cos^{-1}\left[\frac{(1500)^2 + (2775.95)^2 - (2000)^2}{2(1500)(2775.95)}\right] = 44°20'31.8''$$

$$\text{azimuth } O_1 P = 22°10'05.5'' - 31°36'53.4'' = -9°26'47.9''$$

$$= 360° - 9°26'47.9'' = 350°33'12.1''$$

$$\text{azimuth } O_2 P = 22°10'05.5'' + 180° + 44°20'31.8'' = 246°30'37.3''$$

With the coordinates of O_1, length R_1, and the direction of $O_1 P$ known, the coordinates of point P can be obtained directly as:

$$X_P = 2851.28 + 2000 \times \sin 350°33'12.1'' = 2523.02 \text{ ft}$$

$$Y_P = 299.40 + 2000 \times \cos 350°33'12.1'' = 2272.28 \text{ ft}$$

These coordinates can be checked by a similar computation from point O_2 using length R_2 and the direction of O_2P.

B-7 TWO-DIMENSIONAL CONFORMAL COORDINATE TRANSFORMATION

It is sometimes necessary to convert coordinates of points from one survey axis system to another. This happens, for example, if a survey is performed in some local-assumed or arbitrary coordinate system, and later it is desired to convert it to state plane coordinates. The process of making these conversions is called *coordinate transformation*. If only planimetric coordinates (i.e., Xs and Ys) are involved, and true shape is retained, it is called *two-dimensional conformal coordinate transformation*.

The geometry of a two-dimensional coordinate transformation is illustrated in Figure B-5. In the figure, $X–Y$ represents a local-assumed coordinate system, and $E–N$ a state plane coordinate system. Coordinates of points A through D are known in the $X–Y$ system, and those of A and B are also known in the $E–N$ system. Points such as A and B, whose positions are known in both systems, are termed *control points*. At least two

Figure B-5 Geometry of two-dimensional coordinate transformation.

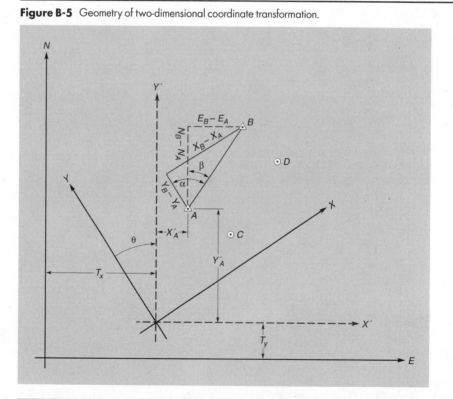

control points are required in order to determine *E–N* coordinates of other points such as *C* and *D*.

In general, three steps are involved in coordinate transformation: **(1)** *rotation,* **(2)** *scaling,* and **(3)** *translation.* As shown in Figure B-5, rotation consists in determining coordinates of points in the rotated *X'–Y'* axis system (shown dashed). The *X'–Y'* axes are parallel with *E–N,* but the origin of this system coincides with the origin of *X–Y.* In the figure, the rotation angle θ, between the *X–Y* and *X'–Y'* axis systems, is

$$\theta = \alpha - \beta \tag{B-9}$$

In Eq. (B-9), α and β are calculated from the two sets of coordinates of control points *A* and *B* using Eq. (B–2) as follows:

$$\alpha = \tan^{-1}\left[\frac{X_B - X_A}{Y_B - Y_A}\right]$$

$$\beta = \tan^{-1}\left[\frac{E_B - E_A}{N_B - N_A}\right]$$

In many cases a scale factor must be incorporated in coordinate transformations. This would occur, for example, in transforming from a local arbitrary coordinate system into a state plane coordinate grid. The scale factor relating any two coordinate systems can be computed according to the ratio of the length of a line between two control points obtained from *E–N* coordinates to that determined using *X–Y* coordinates. Thus

$$s = \frac{\sqrt{(E_B - E_A)^2 - (N_B - N_A)^2}}{\sqrt{(X_B - X_A)^2 - (Y_B - Y_A)^2}} \tag{B-10}$$

(*Note:* If the scale factor is unity, the two surveys are of equal scale, and it can be ignored in the coordinate transformation.)

With θ and *s* known, scaled *X'* and *Y'* coordinates of any point, for example, *A*, can be calculated from

$$X'_A = sX_A\cos\theta - sY_A\sin\theta$$

$$Y'_A = sX_A\sin\theta + sY_A\cos\theta \tag{B-11}$$

Individual parts of the rotation formulas [right-hand sides of Eqs. (B-11)] are detailed in Figure B-6.

Translation consists of shifting the origin of the *X'–Y'* axes to that in the *E–N* system. This is achieved by adding translation factors T_x and T_y (see Figure B-5) to *X'* and *Y'* coordinates to obtain *E* and *N* coordinates. Thus for point *A,*

$$E_A = X'_A + T_x$$

$$N_A = Y'_A + T_y \tag{B-12}$$

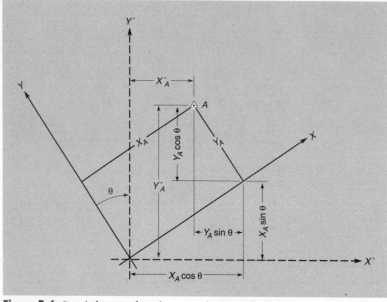

Figure B-6 Detail of rotation formulas in two-dimensional coordinate transformation.

Rearranging Eqs. (B-12), and using coordinates of one of the control points (such as A), numerical values for T_x and T_y can be obtained as

$$T_x = E_A - X'_A$$

$$T_y = N_A - Y'_A \tag{B-13}$$

The other control point (i.e., point B) should also be used in Eqs. (B–13) to calculate T_x and T_y and thus to obtain a computational check.

Substituting Eqs. (B-11) into Eqs. (B-12) and dropping subscripts, the following equations are obtained for calculating E and N coordinates of noncontrol points (such as C and D) from their X and Y values:

$$E = sX \cos \theta - sY \sin \theta + T_x$$

$$N = sX \sin \theta + sY \cos \theta + T_y \tag{B-14}$$

In summary, the procedure for performing two-dimensional conformal coordinate transformations consists of **(1)** calculating rotation angle θ using two control points and Eqs. (B-2) and (B-9), **(2)** solving Eqs. (B-10), (B-11), and (B-13) using control points to obtain scale factor s and translation factors T_x and T_y, and **(3)** using θ and T_x and T_y in Eqs. (B-14) to transform all noncontrol points. If more than two control points are available, an improved solution can be obtained using least squares (see Section 2-26 and Appendix C). Coordinate transformation calculations require a significant amount of time if done by hand, but are easily performed when programmed for computer solution.

In Figure B-5, the following E–N and X–Y coordinates are known for points A through D. Compute E and N coordinates for points C and D.

EXAMPLE B-4

POINT	STATE PLANE COORDINATES (ft)		ARBITRARY COORDINATES (ft)	
	E	N	X	Y
A	194,683.50	99,760.22	2848.28	2319.94
B	196,412.80	102,367.61	5720.05	3561.68
C			3541.72	897.03
D			6160.31	1941.26

SOLUTION

1. Determine α, β, and θ from Eqs. (B-2) and (B-9),

$$\alpha = \tan^{-1}\left[\frac{5720.05 - 2848.28}{3561.68 - 2319.94}\right] = 66°36'59.7''$$

$$\beta = \tan^{-1}\left[\frac{196{,}412.80 - 194{,}683.50}{102{,}367.61 - 99{,}760.22}\right] = 33°33'12.7''$$

$$\theta = 66°36'59.7'' - 33°33'12.7'' = 33°03'47''$$

2. Compute the scale factor from Eq. (B-10),

$$s = \frac{\sqrt{(196{,}412.80 - 194{,}683.50)^2 - (102{,}367.61 - 99{,}760.22)^2}}{\sqrt{(5720.05 - 2848.28)^2 - (3561.68 - 2319.94)^2}}$$

$$= 3128.73/3128.73 = 1.00000$$

(Since the scale factor is unity, it can be ignored.)

3. Determine T_x and T_y from Eqs. (B-11) and (B-13) using point A,

$$X'_A = 2848.28 \cos 33°03'47'' - 2319.94 \sin 33°03'47'' = 1121.39$$

$$Y'_A = 2848.28 \sin 33°03'47'' + 2319.94 \cos 33°03'47'' = 3498.18$$

$$T_x = 194{,}683.50 - 1121.39 = 193{,}562.11$$

$$T_y = 99{,}760.22 - 3498.18 = 96{,}262.04$$

4. Check T_x and T_y using point B,

$$X'_B = 5720.05 \cos 33°03'47'' - 3561.68 \sin 33°03'47'' = 2850.69$$

$$Y'_B = 5720.05 \sin 33°03'47'' + 3561.68 \cos 33°03'47'' = 6105.58$$

$$T_x = 196{,}412.80 - 2850.69 = 193{,}562.11 \text{ (Check!)}$$

$$T_y = 102{,}367.61 - 6105.58 = 96{,}262.03 \text{ (Check!)}$$

5. Solve Eqs. (B-14) for E and N coordinates of points C and D,

$$E_C = 3541.72 \cos 33°03'47'' - 897.03 \sin 33°03'47'' + 193{,}562.11$$

$$= 196{,}040.93$$

$$N_C = 3541.72 \sin 33°03'47'' + 897.03 \cos 33°03'47'' + 96{,}262.04$$

$$= 98{,}946.04$$

$$E_D = 6160.31 \cos 33°03'47'' - 1941.26 \sin 33°03'47'' + 193{,}562.11$$

$$= 197{,}665.81$$

$$N_D = 6160.31 \sin 33°03'47'' + 1941.26 \cos 33°03'47'' + 96{,}262.04$$

$$= 101{,}249.78$$

APPENDIX C

PROPAGATION OF RANDOM ERRORS AND LEAST-SQUARES ADJUSTMENT

PART I
ERROR PROPAGATION

C-1 PROPAGATION OF RANDOM ERRORS

In Section 2-22 the subject of error propagation was introduced, and simple equations were presented for determining the propagated error in a *sum, product,* and *mean* that were computed from quantities containing errors. These equations all stem from the *general law of error propagation.*

Let $a, b, c, \ldots, n$ be measured values containing random errors $E_a, E_b, E_c, \ldots, E_n$, respectively. Also let Z be a quantity derived by computation using these measured quantities in a function f, such that

$$Z = f(a, b, c, \cdots, n) \tag{C-1}$$

Then, according to the general law of error propagation, the error in the computed quantity Z is

$$E_Z = \pm \sqrt{\left(\frac{\partial f}{\partial a}\right)^2 (E_a)^2 + \left(\frac{\partial f}{\partial b}\right)^2 (E_b)^2 + \left(\frac{\partial f}{\partial c}\right)^2 (E_c)^2 + \cdots + \left(\frac{\partial f}{\partial n}\right)^2 (E_n)^2} \tag{C-2}$$

In Eq. (C-2) the terms $\partial f/\partial a$, $\partial f/\partial b$, $\partial f/\partial c$, . . . are partial derivatives of the function f taken with respect to the variables a, b, c, ... Consider the special case where $Z = a \times b$. Substituting $\partial Z/\partial a = b$ and $\partial Z/\partial b = a$ into Eq. (C-2), the propagated error E_Z in the computed value of Z is

$$E_Z = \pm \sqrt{b^2(E_a)^2 + a^2(E_b)^2} \tag{C-3}$$

Note that Eq. (C-3) is equivalent to Eq. (2-11) given in Chapter 2 for computing the propagated error in a computed area. In fact, each of the equations given in Section 2-22 is derived from this general law of propagation of error.

Equation (C-2) is general, and is applicable for computing the propagated error in any function of independently measured quantities containing random errors.

EXAMPLE C-1   The radius and height of a cylinder were measured as $R = 12.08$ ft and $H = 49.80$ ft, respectively. The standard errors in D and H were ± 0.02 ft and ± 0.10 ft, respectively. Compute the cylinder's volume and the standard error in the computed volume.

SOLUTION

The equation for the volume of a cylinder is $V = \pi R^2 H$. Thus the cylinder's volume in this example is

$$V = \pi(12.08)^2(49.80) = 22{,}830 \text{ ft}^3$$

To compute the error in the volume, Eq. (C-2) is applied. Partial derivatives of the volume function with respect to the measured variables R and H are $\partial V/\partial R = 2\pi RH$ and $\partial V/\partial H = \pi R^2$. Substituting into Eq. (C-2) and solving,

$$E_V = \pm \sqrt{[2\pi(12.08)(49.80)]^2(0.02)^2 + [\pi(12.08)^2]^2(0.10)^2} = \pm 88 \text{ ft}^3$$

EXAMPLE C-2   The slope distance and zenith angle measured from point A to point B were $S = 783.25$ ft and $Z = 83°17.3'$, respectively. Standard errors in S and Z were ± 0.10 ft and ± 0.50 min; respectively. Calculate the elevation difference h between A and B and its standard error.

SOLUTION

By Eq. (6-6), the elevation difference is

$$h = S \cos Z = 783.25 \cos 83°17.3' = 91.54 \text{ ft}$$

By Eq. (C-2), the standard error in h is

$$E_h = \pm \sqrt{(\cos 83°17.3')^2(0.10)^2 + [(-783.25)(\sin 83°17.3')]^2\left(\frac{0.50'}{3437'/\text{rad}}\right)^2}$$

$$= \pm 0.11 \text{ ft}$$

where

$$\frac{\partial h}{\partial S} = \cos Z \quad \text{and} \quad \frac{\partial h}{\partial Z} = -S \sin Z$$

PART II
LEAST-SQUARES ADJUSTMENT

C-2 INTRODUCTION

The subject of errors in measurements was introduced in Chapter 2, and the two types of errors, *systematic* and *random*, were defined. It was noted that systematic errors follow physical laws, and that if conditions producing them are measured, corrections can be computed to eliminate them. Random errors will still exist in all observed values, however. As explained in Chapter 2, experience has shown that random errors in surveying follow the mathematical laws of probability, and in any group of measurements they are expected to conform to a "normal distribution," as illustrated in Figure 2-5.

In surveying, after eliminating mistakes and making corrections for systematic errors, the presence of remaining random errors will be evident in the form of misclosures. Examples include sums of interior angles in closed polygons that do not total $(n-2)180°$, misclosures in closed leveling circuits, and traverse misclosures in departures and latitudes. To account for these misclosures, adjustments are applied to produce mathematically perfect geometric conditions. Although various techniques are used, the most rigorous adjustments are made by the method of *least squares*, which is based on the laws of probability.

Because of the lengthy calculations involved, least-squares adjustments were not commonly made prior to the availability of computers. Now the calculations are handled routinely, and making adjustments by this method is rapidly becoming indispensable in modern surveying. Least squares is currently being used to adjust all kinds of surveying measurements, including differences in elevation, horizontal distances, and horizontal and vertical angles. It has become essential in the adjustment of GPS observations, and is also widely used in adjusting photogrammetric data. Least-squares adjustment has taken on added importance with the most recent surveying accuracy standards prepared by the Federal Geodetic Control Subcommittee (FGCS). These standards include the use of statistical quantities that result from least-squares adjustment. Therefore evaluation of survey results for acceptance according to these standards cannot be done without first performing a least-squares adjustment. Since many surveying projects today require that FGCS standards be met, modern surveyors must be familiar with the method of least-squares adjustment.

In the sections that follow, the fundamental condition enforced in least squares is described, and elementary examples of least-squares adjustments are presented. Then systematic procedures for forming and solving least-squares equations using matrix methods are given and demonstrated with examples.

C-3 FUNDAMENTAL CONDITION OF LEAST SQUARES

It was shown through the discussion in Section 2-17, and the normal distribution curves illustrated in Figures 2-5 through 2-7, that small errors (residuals) have a higher probability of occurrence than large ones in a group of normally distributed observations. Also discussed was the fact that in such a set of measurements there is a specific probability that an error (residual) of a certain size will exist within a group of errors. In other words, there is a direct relationship between probabilities and residual sizes in a normally distributed set of observations. The method of least-squares adjustment is derived from the equation for the normal distribution curve. It produces that unique set of residuals for a group of measurements that have the highest probability of occurrence.

For a group of equally weighted observations, the condition enforced by least squares is that *the sum of the squares of the residuals is a minimum*. Suppose a group of m measurements of equal weight were taken having residuals $v_1, v_2, v_3, \ldots, v_m$. Then, in equation form, the fundamental condition of least squares is

$$\sum_{i=1}^{m} v_i^2 = v_1^2 + v_2^2 + v_3^2 + \cdots + v_m^2 = \text{minimum} \tag{C-4}$$

(The derivation of this equation is beyond the scope of this text.)

For any group of measured values, weights may be assigned to individual observations according to a priori estimates of their relative worths, or they may be obtained from the standard deviations of the measurements if available. An equation expressing the relation between standard deviations and weights, given in Section 2-25 and repeated here, is

$$p_i = \frac{1}{\sigma_i^2} \tag{C-5}$$

In Eq. (C-5), p_i is the weight of the ith observed quantity, and σ_i^2 is the square of the standard deviation or *variance* of that observation. *This equation states that weights are inversely proportional to variances.*

If measured values are to be weighted in least-squares adjustment, then the fundamental condition to be enforced is that *the sum of the weights times their corresponding squared residuals is minimized* or, in equation form,

$$\sum_{i=1}^{m} p_i v_i^2 = p_1 v_1^2 + p_2 v_2^2 + p_3 v_3^2 + \cdots + p_m v_m^2 = \text{minimum} \tag{C-6}$$

Some basic assumptions underlying least-squares theory are that (1) mistakes and systematic errors have been eliminated, so only random errors remain, (2) the number of observations being adjusted is large, and (3) as stated earlier, the frequency distribution of the errors is normal. Although these basic assumptions are not always met, least-squares adjustment still provides the most rigorous error treatment available.

C-4 LEAST-SQUARES ADJUSTMENT BY THE OBSERVATION-EQUATION METHOD

Two basic methods are employed in least-squares adjustments: **(1)** the *observation-equation* method and **(2)** the *condition-equation* approach. The former is most common, and is the one discussed in this appendix. In this method "observation equations" are written relating measured values to their residual errors and the unknown parameters. One observation equation is written for each measurement. For a unique solution, the number of equations must equal the number of unknowns. If redundant observations are made, the method of least squares can be applied. In that case, an expression for each residual error is obtained from every observation equation. The residuals are squared and added to obtain the function expressed in either Eq. (C-4) or Eq. (C-6).

To minimize the function in accordance with either Eq. (C-4) or Eq. (C-6), partial derivatives of the expression are taken with respect to each unknown variable and set equal to zero. This yields a set of so-called *normal equations*, which are equal in number to the number of unknowns. The normal equations are solved to obtain most probable values for the unknowns. The following elementary examples illustrate the procedures.

Using least squares, compute the most probable value for the equally weighted distance measurements of Example 2-1.

EXAMPLE C-3

SOLUTION

1. From rearranged Eq. (2-3), write the following observation equations,

$$\overline{M} = 538.57 + v_1$$
$$\overline{M} = 538.39 + v_2$$
$$\overline{M} = 538.37 + v_3$$
$$\overline{M} = 538.39 + v_4$$
$$\overline{M} = 538.48 + v_5$$
$$\overline{M} = 538.49 + v_6$$
$$\overline{M} = 538.33 + v_7$$
$$\overline{M} = 538.46 + v_8$$
$$\overline{M} = 538.47 + v_9$$
$$\overline{M} = 538.55 + v_{10}$$

2. Solve for the residual in each observation equation and form the function Σv^2,

$$\Sigma v^2 = (\overline{M} - 538.57)^2 + (\overline{M} - 538.39)^2 + (\overline{M} - 538.37)^2$$
$$+ (\overline{M} - 538.39)^2 + (\overline{M} - 538.48)^2 + (\overline{M} - 538.49)^2$$
$$+ (\overline{M} - 538.33)^2 + (\overline{M} - 538.46)^2 + (\overline{M} - 538.47)^2$$
$$+ (\overline{M} - 538.55)^2$$

3. Take the derivative of the function Σv^2 with respect to $\overline{M}$, set it equal to zero (this minimizes the function), and solve for $\overline{M}$,

$$\partial \Sigma v^2 / \partial \overline{M} = 0 = 2(\overline{M} - 538.57) + 2(\overline{M} - 538.39) + 2(\overline{M} - 538.37)$$
$$+ 2(\overline{M} - 538.39) + 2(\overline{M} - 538.48) + 2(\overline{M} - 538.49)$$
$$+ 2(\overline{M} - 538.33) + 2(\overline{M} - 538.46) + 2(\overline{M} - 538.47)$$
$$+ 2(\overline{M} - 538.55)$$

4. Reduce and solve for $\overline{M}$,

$$10\overline{M} = 5384.50$$
$$\overline{M} = \frac{5384.50}{10} = 538.45$$

Note that this answer agrees with the one given for Example 2–1 in Chapter 2. Note also that this procedure verifies the statement given earlier in Section 2–15, that *the most probable value for an unknown quantity, measured repeatedly using the same equipment and procedures, is simply the mean of the measurements.*

EXAMPLE C-4   In Figure 11–3 the three horizontal angles measured around the horizon are $x = 42°17'$, $y = 60°48'$, and $z = 256°52'$. Adjust these angles by least squares so that their sum equals the required geometric total of $360°$.

SOLUTION

1. Form the observation equations

$$x = 42°17' + v_1 \tag{a}$$
$$y = 60°48' + v_2 \tag{b}$$
$$z = 256°52' + v_3 \tag{c}$$

2. Write an expression that enforces the condition that the sum of the three adjusted angles total $360°$,

$$x + y + z = 360° \tag{d}$$

3. Substitute Eqs. (a), (b), and (c) into Eq. (d) and solve for v_3,

$$(42°17' + v_1) + (60°48' + v_2) + (256°52' + v_3) = 360°$$
$$v_3 = 03' - v_1 - v_2 \tag{e}$$

(Because of the $360°$ condition, if v_1 and v_2 are fixed, v_3 is also fixed. Thus there are only two independent residuals in the solution.)

4. Form the function Σv^2 which involves all three residuals but includes only the two independent variables v_1 and v_2,

$$\Sigma v^2 = v_1^2 + v_2^2 + (03' - v_1 - v_2)^2 \tag{f}$$

5. Take partial derivatives of the function [Eq. (f)] with respect to the variables v_1 and v_2 and set them equal to zero,

$$\partial\Sigma v^2/\partial v_1 = 0 = 2v_1 + 2(03' - v_1 - v_2)(-1); \qquad 4v_1 + 2v_2 = 06' \quad (g)$$
$$\partial\Sigma v^2/\partial v_2 = 0 = 2v_2 + 2(03' - v_1 - v_2)(-1); \qquad 2v_1 + 4v_2 = 06' \quad (h)$$

6. Solve Eqs. (g) and (h) simultaneously,

$$v_1 = 01' \quad \text{and} \quad v_2 = 01'$$

7. Substitute v_1 and v_2 into Eq. (e) to compute v_3,

$$v_3 = 03' - 01' - 01' = 01'$$

8. Finally substitute the residuals into Eqs. (a) through (c) to get the adjusted angles,

$$
\begin{aligned}
x &= 42°17' + 01' = 42°18' \\
y &= 60°48' + 01' = 60°49' \\
z &= 256°52' + 01' = \underline{256°53'} \\
& \Sigma = 360°00' \quad \text{(Check)}
\end{aligned}
$$

Note that this result verifies another basic procedure frequently applied in surveying, that *for equally weighted angles measured around the horizon, corrections of equal size are applied to each angle.* The same result occurs when the three equally weighted angles in a plane triangle are adjusted by least squares, that is, each receives an equal-size correction.

Examples C-3 and C-4 are indeed simple, hardly the type for which least squares is best suited. They do supply the bases for some commonly applied simple adjustments, however, and they also illustrate procedures involved in making least-squares adjustments without complicating the mathematics. The following example illustrates least-squares adjustment of distance measurements which are functionally related.

Adjust the three equally weighted distance measurements taken (in feet) between points A, B, and C of Figure C-1. **EXAMPLE C-5**

Figure C-1 Equally weighted distance measurements of Example C-5.

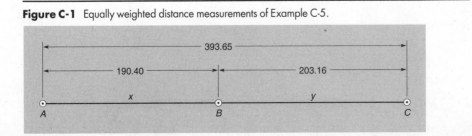

SOLUTION

1. Let the unknown distances AB and BC be x and y, respectively. These two unknowns are related through the measurements as follows:

$$x + y = 393.65$$
$$x = 190.40$$
$$y = 203.16 \tag{i}$$

2. Values for x and y could be obtained from any two of these equations so that the remaining equation is redundant. Notice, however, that values obtained for x and y will differ, depending on which two equations are solved. It is therefore apparent that the measurements contain errors. Equations (i) may be rewritten as observation equations by including residual errors as follows:

$$x + y = 393.65 + v_1$$
$$x = 190.40 + v_2$$
$$y = 203.16 + v_3 \tag{j}$$

3. To obtain the least-squares solution, the observation equations (j) are rearranged to obtain expressions for the residuals. These are squared and added to form the function given in Eq. (C-4) as follows:

$$\sum_{i=1}^{m} v_i^2 = (x + y - 393.65)^2 + (x - 190.40)^2 + (y - 203.16)^2 \tag{k}$$

4. Function (k) is minimized, enforcing the condition of least squares, by taking partial derivatives with respect to the unknowns x and y and setting them equal to zero. This yields the following two normal equations:

$$\frac{\partial \sum v^2}{\partial x} = 0 = 2(x + y - 393.65) + 2(x - 190.40)$$

$$\frac{\partial \sum v^2}{\partial y} = 0 = 2(x + y - 393.65) + 2(y - 203.16)$$

5. Reducing the normal equations and solving yields $x = 190.43$ ft and $y = 203.19$ ft. The residuals can now be calculated by substituting x and y into the original observation equations (j),

$$v_1 = 190.43 + 203.19 - 393.65 = -0.03 \text{ ft}$$
$$v_2 = 190.43 - 190.40 = +0.03 \text{ ft}$$
$$v_3 = 203.19 - 203.16 = +0.03 \text{ ft}$$

C-5 MATRIX METHODS IN LEAST-SQUARES ADJUSTMENT

It has been noted that least-squares computations are quite lengthy and therefore generally performed on a computer. The solution follows a systematic procedure that is conveniently adapted to matrix methods. In general, any group of observation equations may be represented in matrix form as

$$_m\mathbf{A}_n\ _n\mathbf{X}_l = _m\mathbf{L}_l + _m\mathbf{V}_l \qquad (C-7)$$

where $\mathbf{A}$ is the matrix of coefficients of the unknowns, $\mathbf{X}$ the matrix of unknowns, $\mathbf{L}$ the matrix of observations, and $\mathbf{V}$ the matrix of residuals. The detailed structures of these matrices are

$$\mathbf{A} = \begin{bmatrix} a_{11} & a_{12} & \cdots & a_{1n} \\ a_{21} & a_{22} & \cdots & a_{2n} \\ \vdots & & & \\ \vdots & & & \\ a_{m1} & a_{m2} & \cdots & a_{mn} \end{bmatrix} \quad \mathbf{X} = \begin{bmatrix} x_1 \\ x_2 \\ x_3 \\ \vdots \\ \\ x_n \end{bmatrix} \quad \mathbf{L} = \begin{bmatrix} l_1 \\ l_2 \\ l_3 \\ \vdots \\ \\ l_m \end{bmatrix} \quad \mathbf{V} = \begin{bmatrix} v_1 \\ v_2 \\ v_3 \\ \vdots \\ \\ v_m \end{bmatrix}$$

The normal equations that result from a set of equally weighted observation equations [Eqs. (C-7)] are given in matrix form by

$$\mathbf{A}^T\mathbf{A}\mathbf{X} = \mathbf{A}^T\mathbf{L} \qquad (C-8)$$

In Eq. (C-8) $\mathbf{A}^T\mathbf{A}$ is the matrix of normal equation coefficients for the unknowns. Premultiplying both sides of Eq. (C-8) by $(\mathbf{A}^T\mathbf{A})^{-1}$ and reducing, there results

$$\mathbf{X} = (\mathbf{A}^T\mathbf{A})^{-1}\,\mathbf{A}^T\mathbf{L} \qquad (C-9)$$

Equation (C-9) is the least-squares solution for equally weighted observations. The matrix $\mathbf{X}$ consists of most probable values for unknowns $x_1, x_2, x_3, \ldots, x_n$.

For a system of weighted observations, the following equation provides the $\mathbf{X}$ matrix:

$$\mathbf{X} = (\mathbf{A}^T\mathbf{P}\mathbf{A})^{-1}\mathbf{A}^T\mathbf{P}\mathbf{L} \qquad (C-10)$$

In Eq. (C-10) the matrices are identical to those of the equally weighted case, except that $\mathbf{P}$ is a diagonal matrix of weights defined as follows:

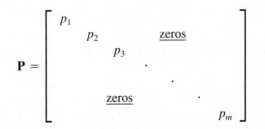

$$\mathbf{P} = \begin{bmatrix} p_1 & & & & & & \\ & p_2 & & & \text{zeros} & & \\ & & p_3 & & & & \\ & & & \cdot & & & \\ & & & & \cdot & & \\ & \text{zeros} & & & & \cdot & \\ & & & & & & p_m \end{bmatrix}$$

In the $\mathbf{P}$ matrix all off-diagonal elements are zeros. This is appropriate when the individual observations are independent and uncorrelated (not related to each other), which is usually the case in surveying.

If the observations in an adjustment are all of equal weight, Eq. (C-10) can still be used, but the $\mathbf{P}$ matrix becomes an identity matrix. It therefore reduces exactly to Eq.

(C-9). Thus Eq. (C-10) is general and can be used for both unweighted and weighted adjustments. It is readily programmed for computer solution.

EXAMPLE C-6 | Solve Example C-5 using matrix methods.

SOLUTION

1. The observation equations of Example C-5 can be expressed in matrix form as follows:

$$_3\mathbf{A}_2 \, _2\mathbf{X}_1 = \, _3\mathbf{L}_1 + \, _3\mathbf{V}_1$$

where

$$\mathbf{A} = \begin{bmatrix} 1 & 1 \\ 1 & 0 \\ 0 & 1 \end{bmatrix} \quad \mathbf{X} = \begin{bmatrix} x \\ y \end{bmatrix} \quad \mathbf{L} = \begin{bmatrix} 393.65 \\ 190.40 \\ 203.16 \end{bmatrix} \quad \mathbf{V} = \begin{bmatrix} v_1 \\ v_2 \\ v_3 \end{bmatrix}$$

2. Solving matrix equation (C-9),

$$\mathbf{A}^T\mathbf{A} = \begin{bmatrix} 1 & 1 & 0 \\ 1 & 0 & 1 \end{bmatrix} \begin{bmatrix} 1 & 1 \\ 1 & 0 \\ 0 & 1 \end{bmatrix} = \begin{bmatrix} 2 & 1 \\ 1 & 2 \end{bmatrix}$$

$$(\mathbf{A}^T\mathbf{A})^{-1} = \frac{1}{3}\begin{bmatrix} 2 & -1 \\ -1 & 2 \end{bmatrix} \qquad \mathbf{A}^T\mathbf{L} = \begin{bmatrix} 584.05 \\ 596.81 \end{bmatrix}$$

$$\mathbf{X} = (\mathbf{A}^T\mathbf{A})^{-1}\mathbf{A}^T\mathbf{L} = \frac{1}{3}\begin{bmatrix} 2 & -1 \\ -1 & 2 \end{bmatrix} \times \begin{bmatrix} 584.05 \\ 599.81 \end{bmatrix} = \begin{bmatrix} 190.43 \\ 203.19 \end{bmatrix}$$

Note that this solution yields $x = 190.43$ and $y = 203.19$, which are exactly the same values obtained through the algebraic approach of Example C-5.

C-6 MATRIX EQUATIONS FOR PRECISIONS OF ADJUSTED QUANTITIES

The matrix equation for calculating residuals after adjustment, whether the adjustment is weighted or not, is

$$\mathbf{V} = \mathbf{A}\mathbf{X} - \mathbf{L} \qquad\qquad (C\text{-}11)$$

The standard deviation of unit weight for an unweighted adjustment is

$$\sigma_0 = \sqrt{\frac{\mathbf{V}^T\mathbf{V}}{r}} \qquad\qquad (C\text{-}12)$$

The standard deviation of unit weight for a weighted adjustment is

$$\sigma_0 = \sqrt{\frac{\mathbf{V}^T\mathbf{PV}}{r}} \qquad \text{(C-13)}$$

In Eqs. (C-12) and (C-13) r is the number of degrees of freedom in an adjustment, which equals the number of observations minus the number of unknowns, or $r = m - n$.

Standard deviations of the individual adjusted quantities are as follows:

$$\sigma_{x_i} = \sigma_0\sqrt{Q_{x_i x_i}} \qquad \text{(C-14)}$$

In Eq. (C-14) σ_{x_i} is the standard deviation of the ith adjusted quantity, that is the quantity in the ith row of the $\mathbf{X}$ matrix; σ_0 is the standard deviation of unit weight as calculated by Eq. (C-12) or (C-13); and $Q_{x_i x_i}$ is the diagonal element in the ith row and the ith column of the matrix $(\mathbf{A}^T\mathbf{A})^{-1}$ in the unweighted case or the matrix $(\mathbf{A}^T\mathbf{PA})^{-1}$ in the weighted case. The $(\mathbf{A}^T\mathbf{A})^{-1}$ and $(\mathbf{A}^T\mathbf{PA})^{-1}$ matrices are the so-called *covariance matrices*, and usually symbolized by $\mathbf{Q}$.

Calculate the standard deviation of unit weight and the standard deviations of the adjusted quantities x and y for the unweighted problem of Example C-6.

EXAMPLE C-7

SOLUTION

1. By Eq. (C-11) the residuals are

$$\mathbf{V} = \begin{bmatrix} 1 & 1 \\ 1 & 0 \\ 0 & 1 \end{bmatrix} \begin{bmatrix} 190.43 \\ 203.19 \end{bmatrix} - \begin{bmatrix} 393.65 \\ 190.40 \\ 203.26 \end{bmatrix} = \begin{bmatrix} -0.03 \\ 0.03 \\ 0.03 \end{bmatrix}$$

2. By Eq. (C–12) the standard deviation of unit weight is

$$\mathbf{V}^T\mathbf{V} = [-0.03 \quad 0.03 \quad 0.03] \begin{bmatrix} -0.03 \\ 0.03 \\ 0.03 \end{bmatrix} = 0.0027$$

$$\sigma_0 = \sqrt{\frac{0.0027}{3 - 2}} = \pm 0.052$$

3. Using Eq. (C-14), the standard deviations of the adjusted values for x and y are

$$\sigma_x = \pm 0.052 \sqrt{\frac{2}{3}} = \pm 0.042 \text{ ft}$$

$$\sigma_y = \pm 0.052 \sqrt{\frac{2}{3}} = \pm 0.042 \text{ ft}$$

In part 3, the numbers $\frac{2}{3}$ under the radicals are the elements in row 1, column 1 and row 2, column 2, respectively, of the $(\mathbf{A}^T\mathbf{A})^{-1}$ matrix of Example C-6. The interpretation of the standard deviations computed under part 3 is that a 68% probability exists that the adjusted values for x and y are within ± 0.042 ft of their true values. Note that for this simple example, the three residuals calculated in part 1 were equal, and the standard deviations of x and y were equal in part 3. This is due to the symmetric nature of this particular problem (illustrated in Figure C-1), but it is not generally the case with more complex problems.

C-7 LEAST-SQUARES ADJUSTMENT OF LEVELING CIRCUITS

When control leveling is being done to establish new bench marks, for example, temporary ones for a construction project, it is common practice to create a network like that illustrated in Figure C-2. This enables each new bench mark to be established using redundant observations. In Figure C-2, for example, A and B are two new project bench marks being established near a construction site. Each could be set by running a single loop, such as from BM 1 to A and back to establish A, and from BM 3 to B and back to set B. To build redundancy into the survey, and to increase the precisions of the new bench marks, additional lines from other nearby bench marks can be run. Thus in Figure C-2, five loops are run rather than the minimum of two needed to establish A and B. These redundant observations enable a least-squares adjustment of the data to be performed.

In adjusting level networks, the observed difference in elevation for each course is treated as one observation containing a single random error. Observation equations are written which relate these measured elevation differences and their residual errors to the unknown elevations of the bench marks involved. These can then be processed through the matrix equations to obtain adjusted values for the bench marks and their standard deviations. The procedure is illustrated with the following example.

Figure C-2 Level net for least-squares Example C-8.

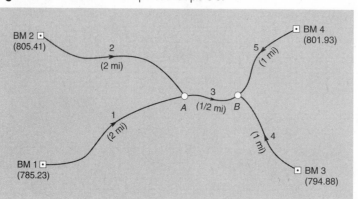

Adjust the level net of Figure C-2 by weighted least squares, and compute precisions of the adjusted bench marks. In the figure, the bench mark elevations (in feet above mean sea level) and course lengths (in miles) are shown in parentheses. Observed elevation differences for courses 1 through 5 (given in order) are $+10.97$ ft, -9.17 ft, $+3.58$ ft, $+4.91$ ft, and -2.20 ft. Arrows on the courses in the figure indicate the direction of leveling. Thus for course 1, having a length of 2 miles leveling proceeded from BM 1 to A and the observed elevation difference was 10.97 ft.

EXAMPLE C-8

SOLUTION

1. Observation equations are written relating each line's measured elevation difference to its residual error and the most probable values for unknown elevations A and B as follows:

$$A = BM\ 1 + 10.97 + v_1$$
$$A = BM\ 2 - 9.17 + v_2$$
$$B = A + 3.58 + v_3$$
$$B = BM\ 3 + 4.91 + v_4$$
$$B = BM\ 4 - 2.20 + v_5 \qquad \text{(C-15)}$$

2. Substituting the elevations of BM 1, BM 2, BM 3, and BM 4 into Eqs. (C-15) and rearranging gives

$$A = 796.20 + v_1$$
$$A = 796.24 + v_2$$
$$-A + B = 3.58 + v_3$$
$$B = 799.79 + v_4$$
$$B = 799.73 + v_5 \qquad \text{(C-16)}$$

3. The $\mathbf{A}, \mathbf{X}, \mathbf{L},$ and $\mathbf{V}$ matrices for this adjustment are

$$\mathbf{A} = \begin{bmatrix} 1 & 0 \\ 1 & 0 \\ -1 & 1 \\ 0 & 1 \\ 0 & 1 \end{bmatrix} \quad \mathbf{X} = \begin{bmatrix} A \\ B \end{bmatrix} \quad \mathbf{L} = \begin{bmatrix} 796.20 \\ 796.24 \\ 3.58 \\ 799.79 \\ 799.73 \end{bmatrix} \quad \mathbf{V} = \begin{bmatrix} v_1 \\ v_2 \\ v_3 \\ v_4 \\ v_5 \end{bmatrix}$$

4. Weights in differential leveling are inversely proportional to course lengths. Thus after inverting the lengths and multiplying by 2, the course weights are 1, 1, 4, 2, and 2, respectively. Thus the weight matrix is

$$\mathbf{P} = \begin{bmatrix} 1 & 0 & 0 & 0 & 0 \\ 0 & 1 & 0 & 0 & 0 \\ 0 & 0 & 4 & 0 & 0 \\ 0 & 0 & 0 & 2 & 0 \\ 0 & 0 & 0 & 0 & 2 \end{bmatrix}$$

5. The weighted matrix solution for the most probable values according to Eq. (C-10) is

$$\mathbf{A}^T\mathbf{P} = \begin{bmatrix} 1 & 1 & -1 & 0 & 0 \\ 0 & 0 & 1 & 1 & 1 \end{bmatrix} \begin{bmatrix} 1 & 0 & 0 & 0 & 0 \\ 0 & 1 & 0 & 0 & 0 \\ 0 & 0 & 4 & 0 & 0 \\ 0 & 0 & 0 & 2 & 0 \\ 0 & 0 & 0 & 0 & 2 \end{bmatrix} = \begin{bmatrix} 1 & 1 & -4 & 0 & 0 \\ 0 & 0 & 4 & 2 & 2 \end{bmatrix}$$

$$\mathbf{A}^T\mathbf{PA} = \begin{bmatrix} 1 & 1 & -4 & 0 & 0 \\ 0 & 0 & 4 & 2 & 2 \end{bmatrix} \begin{bmatrix} 1 & 0 \\ 1 & 0 \\ -1 & 1 \\ 0 & 1 \\ 0 & 1 \end{bmatrix} = \begin{bmatrix} 6 & -4 \\ -4 & 8 \end{bmatrix}$$

$$(\mathbf{A}^T\mathbf{PA})^{-1} = \begin{bmatrix} 0.250 & 0.125 \\ 0.125 & 0.188 \end{bmatrix}$$

$$\mathbf{A}^T\mathbf{PL} = \begin{bmatrix} 1 & 1 & -4 & 0 & 0 \\ 0 & 0 & 4 & 2 & 2 \end{bmatrix} \begin{bmatrix} 796.20 \\ 796.24 \\ 3.58 \\ 799.79 \\ 799.73 \end{bmatrix} = \begin{bmatrix} 1578.12 \\ 3213.36 \end{bmatrix}$$

$$\mathbf{X} = (\mathbf{A}^T\mathbf{PA})^{-1}\mathbf{A}^T\mathbf{PL} = \begin{bmatrix} 6 & -4 \\ -4 & 8 \end{bmatrix} \begin{bmatrix} 1578.12 \\ 3213.36 \end{bmatrix} = \begin{bmatrix} 796.20 \\ 799.77 \end{bmatrix}$$

Thus the adjusted bench-mark elevations are $A = 796.20$ ft and $B = 799.77$ ft.

6. The residuals by Eq. (C-11) are

$$\mathbf{V} = \mathbf{AX} - \mathbf{L} = \begin{bmatrix} 1 & 0 \\ 1 & 0 \\ -1 & 1 \\ 0 & 1 \\ 0 & 1 \end{bmatrix} \begin{bmatrix} 796.20 \\ 799.77 \end{bmatrix} - \begin{bmatrix} 796.20 \\ 796.24 \\ 3.58 \\ 799.79 \\ 799.73 \end{bmatrix} = \begin{bmatrix} 0.00 \\ -0.04 \\ -0.01 \\ -0.02 \\ 0.04 \end{bmatrix}$$

7. Utilizing Eq. (C-13), the estimated standard deviation of unit weight is

$$\mathbf{V}^T\mathbf{PV} = \begin{bmatrix} 0.00 & -0.04 & -0.01 & -0.02 & 0.04 \end{bmatrix}$$

$$\times \begin{bmatrix} 1 & 0 & 0 & 0 & 0 \\ 0 & 1 & 0 & 0 & 0 \\ 0 & 0 & 4 & 0 & 0 \\ 0 & 0 & 0 & 2 & 0 \\ 0 & 0 & 0 & 0 & 2 \end{bmatrix} \begin{bmatrix} 0.00 \\ -0.04 \\ -0.01 \\ -0.02 \\ 0.04 \end{bmatrix} = 0.0060$$

$$\sigma_0 = \sqrt{\frac{0.0060}{5 - 2}} = \pm 0.045 \text{ ft}$$

8. By Eq. (C-14), the estimated standard deviation of adjusted bench-mark elevations of A and B are

$$S_A = \sigma_0 \sqrt{Q_{AA}} = \pm(0.045)\sqrt{0.250} = \pm 0.023 \text{ ft}$$
$$S_B = \sigma_0 \sqrt{Q_{BB}} = \pm(0.045)\sqrt{0.188} = \pm 0.019 \text{ ft}$$

Note in these calculations that terms in the radicals are diagonal (1, 1 and 2, 2) elements of the $(\mathbf{A}^T\mathbf{PA})^{-1}$ matrix. Note also that B has a lower standard deviation than A, indicating its precision is better. A check of Figure C-2 reveals that this should be expected, because the bench marks closest to A are 2 mi away, and those closest to B are only 1 mi away. Thus the weights of level circuits coming into B are higher than those coming into A, giving B a higher precision.

C-8 LEAST-SQUARES ADJUSTMENT OF STATIC DIFFERENTIAL GPS OBSERVATIONS

It was previously noted that least squares is essential in adjusting GPS observations. Its principal application in this work is for adjusting the massive quantities of redundant data that result after several receivers have made repeated observations on multiple satellites. The development of these observation equations and their reduction are beyond the scope of this text. Another valuable application of least squares in reducing GPS data, however, which is within the scope of this text, involves adjusting redundant measurements of ΔX, ΔY, and ΔZ components of base lines in static differential GPS surveying. The procedure is illustrated with an example.

Figure C-3 depicts the geometry of a recent static differential GPS survey performed by the NGS near the Oregon–California border. In the survey two High-Accuracy Reference Network (HARN) points (see Section 19-7) were held fixed, station *Altamont*, which is in the Oregon HARN, and station *0204* from the California HARN. Two new points, *Median 2* and *K–785*, were set on the Oregon Institute of Technology campus. For convenience, these stations are labeled A, B, C, and D, as shown in Figure C-3. The fixed geodetic coordinates of the two HARN stations, as well as their known X, Y, Z three-dimensional coordinates (see Section 20-8) are given in the figure. Table C-1 lists the observed ΔX, ΔY, and ΔZ base line components together with their standard deviations.

From the known X, Y, and Z coordinates of stations *Altamont* and *0204*, and the measured ΔX, ΔY, and ΔZ components, coordinates of new stations *Median 2* and *K–785* can be calculated. An adjustment is necessary, however, due to the existence of redundant measurements. In applying least squares to this problem, observation equations are written which relate the unknown adjusted coordinates of new stations C and D to the observed ΔX, ΔY, and ΔZ values and their residual errors. As shown in Figure C-3, there are five different base line measurements involved, consisting of AC, BC, AD, BD, and CD. The adjustment can be performed in three separate parts, one each for the ΔX, ΔY, and ΔZ measurements. In adjusting the five ΔX measurements, the following observation equations are written:

$$X_C = X_A + \Delta X_{AC} + v_1$$
$$X_C = X_B + \Delta X_{BC} + v_2$$
$$X_D = X_A + \Delta X_{AD} + v_3$$
$$X_D = X_B + \Delta X_{BD} + v_4$$
$$X_C - X_D = \Delta X_{CD} + v_5 \tag{C-17}$$

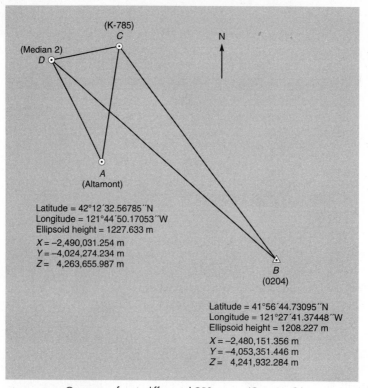

Figure C-3 Geometry of static differential GPS survey. (Courtesy Oregon Institute of Technology and Prof. Earl F. Burkholder.)

TABLE C-1 OBSERVED BASE LINE VECTORS FROM DIFFERENTIAL GPS SURVEY

VECTOR COMPONENT	STA TO STA	OBSERVED LENGTH (m)	STANDARD DEVIATION (m)
ΔX	$A - C$	−945.793	0.011
ΔX	$B - C$	−10,825.707	0.066
ΔX	$A - D$	−1,281.910	0.013
ΔX	$B - D$	−11,161.821	0.090
ΔX	$C - D$	−336.115	0.001
ΔY	$A - C$	4,536.057	0.015
ΔY	$B - C$	33,613.264	0.011
ΔY	$A - D$	4,717.555	0.016
ΔY	$B - D$	33,794.735	0.103
ΔY	$C - D$	181.506	0.003
ΔZ	$A - C$	3,804.397	0.016
ΔZ	$B - C$	25,528.141	0.099
ΔZ	$A - D$	3,767.438	0.016
ΔZ	$B - D$	25,491.165	0.104
ΔZ	$C - D$	−36.964	0.002

From Eqs. (C-17) the **A, X, L**, and **V** matrices for performing a weighted least-squares adjustment according to Eq. (C-10) are

$$A = \begin{bmatrix} 1 & 0 \\ 1 & 0 \\ 0 & 1 \\ 0 & 1 \\ 1 & -1 \end{bmatrix} \quad X = \begin{bmatrix} X_C \\ X_D \end{bmatrix} \quad L = \begin{bmatrix} X_A + \Delta X_{AC} \\ X_B + \Delta X_{BC} \\ X_A + \Delta X_{AD} \\ X_B + \Delta X_{BD} \\ \Delta X_{CD} \end{bmatrix} \quad V = \begin{bmatrix} v_1 \\ v_2 \\ v_3 \\ v_4 \\ v_5 \end{bmatrix}$$

Numerical values for the **L** matrix and weights in the **P** matrix (inverses of variances, or inverses of the squares of standard deviations from Table C-1) are

$$L = \begin{bmatrix} -2,490,977.047 \\ -2,490,977.063 \\ -2,491,313.164 \\ -2,491,313.177 \\ -336.115 \end{bmatrix}$$

$$P = \begin{bmatrix} \left(\dfrac{1}{0.011}\right)^2 & 0 & 0 & 0 & 0 \\ 0 & \left(\dfrac{1}{0.066}\right)^2 & 0 & 0 & 0 \\ 0 & 0 & \left(\dfrac{1}{0.013}\right)^2 & 0 & 0 \\ 0 & 0 & 0 & \left(\dfrac{1}{0.090}\right)^2 & 0 \\ 0 & 0 & 0 & 0 & \left(\dfrac{1}{0.001}\right)^2 \end{bmatrix}$$

After processing the above matrices through weighted least-squares algorithm Eq. (C-10), the adjusted values (**X** matrix) were obtained. These are $X_C = -2,490,977.048$ m and $X_D = -2,491,313.163$ m.

In similar fashion, observation equations for adjusting the ΔY measurements are

$$Y_C = Y_A + \Delta Y_{AC} + v_6$$
$$Y_C = Y_B + \Delta Y_{BC} + v_7$$
$$Y_D = Y_A + \Delta Y_{AD} + v_8$$
$$Y_D = Y_B + \Delta Y_{BD} + v_9$$
$$Y_C - Y_D = \Delta Y_{CD} + v_{10} \qquad\qquad \text{(C-18)}$$

After forming the matrices for Eqs. (C-18) and computing numerical values for them, the ΔY adjustment was performed, again using Eq. (C-10). As before, weights were based on the inverses of corresponding variances. The adjusted Y coordinates of stations C and D that resulted were $-4,019,738.181$ m and $-4,019,556.675$ m, respectively.

Following the same procedure, observation equations for adjusting the ΔZ measurements were written, and the adjustment was performed. They yielded adjusted Z coordinates of 4,267,460.387 m and 4,267,423.423 m, respectively, for stations C and D.

From the adjusted X, Y, and Z three-dimensional coordinates of stations *Median 2* and *K-785* a transformation (beyond the scope of this text)[1] was performed to obtain

---

[1]The transformation procedure and equations are given in Burkholder, E. F. 1993. (See Bibliography at the end of this appendix.)

the geodetic latitudes and longitudes, and ellipsoid heights of these stations. Their values are now available for use by students and local surveyors.

C-9 LEAST-SQUARES ADJUSTMENT OF TRAVERSES, TRILATERA-TION, AND TRIANGULATION

Traversing (described in Chapters 12 and 13) and trilateration and triangulation (discussed in Chapter 19) are traditional ground surveying methods for conducting horizontal surveys. They establish either the X and Y coordinates (usually in state plane systems) or geodetic latitudes and longitudes of points. The basic measurements made in these types of surveys are horizontal angles and horizontal distances. As with other types of surveys, they are most appropriately adjusted by the method of least squares.

Unlike rules of thumb such as the compass rule for traverse adjustment, least squares adjusts all distance and angle measurements simultaneously while also satisfying all geometrical conditions. Furthermore, the condition that the sum of the squares of the residuals times their corresponding weights be a minimum is also enforced, which produces adjusted quantities having the highest probability. An important aspect of adjusting horizontal surveys by least squares is that weights, appropriate to the quality of the measured angles and distances, can be assigned.

To adjust horizontal surveys by the method of least squares, it is necessary to write observation equations for both horizontal distances and horizontal angles. The observation equations are nonlinear, so they must first be linearized using Taylor's theorem and then solved iteratively. These procedures are beyond the scope of this text, but they can be found in references cited in the bibliography at the end of this appendix.

BIBLIOGRAPHY

Burkholder, E. F. 1993. "Using GPS Results in True 3-D Coordinate System." *Journal of Surveying Engineering* 119 (No. 1): 1.

Curry, S. and R. Sawyer. 1990. "Getting the Most Out of Least Squares." *Point of Beginning* 16 (No. 3):46.

Ghilani, C. D. 1990. "A Surveyor's Guide to Practical Least Squares Adjustments." *Surveying and Land Information Systems.* 50(No. 4):287.

Hintz, R. J., et al. 1988. "Least Squares Analysis in Temporal Coordinate and Measurement Management." *Surveying and Land Information Systems* 48 (No. 3):173.

Mikhail, E. M. 1976. *Observations and Least Squares.* New York: Harper & Row.

Mikhail, E. M. and G. Gracie. 1981. *Analysis and Adjustment of Survey Measurements.* New York, Van Nostrand.

Smith, W. K. and D. J. Jarnes. 1978. "Least Squares Adjustment of Triangles and Quadrilaterals in Which All Angles and Distances Are Observed." *Surveying and Mapping* 47(No. 2):125.

Wolf, P. R. 1980. *Adjustment Computations: Practical Least Squares for Surveyors,* 2nd ed. Rancho Cordova, Calif., Landmark.

APPENDIX D
EXAMPLE NOTEFORMS

MEASURING DISTANCES

Sta.	Fwd. Dist.	Back Dist.	Mean	Error	Ratio
A					
	321.18	321.22	321.20	.04 = 321.20	1 8,000
B					
	276.54	276.60	276.57	.06 = 276.57	1 4,600
C					
	100.30	100.29	100.30	.01 = 100.30	1 10,000
D					
	306.77	306.81	306.79	.04 = 306.79	1 7,700
E					
	255.47	255.50	255.48	.03 = 255.48	1 8,500
A					

WITH A STEEL TAPE

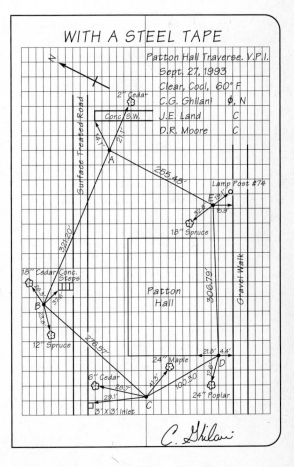

Patton Hall Traverse. V.P.I.
Sept. 27, 1993
Clear, Cool, 60° F
C.G. Ghilani   Ø, N
J.E. Land   C
D.R. Moore   C

PLATE D-1

DIFFERENTIAL LEVELS

Sta.	+ B.S.	H.I.	− F.S.	Elev.	Adj. Elev.
BM Mil.	1.33			2053.18	2053.18
		2054.51			
TP1	0.22		8.37	2046.14	2046.14
		2046.36	7.91		
TP2	0.96		8.91	2038.45	2038.44
		2039.41			
TP3	0.46		11.72	2027.69	2027.68
		2028.15			
BM Oak	11.95		8.71	2019.44	2019.42
		2031.39			
TP4	12.55		2.61	2028.78	2028.76
		2041.33			
TP5	12.77		0.68	2040.65	2040.62
		2053.42			
BM Mil.			0.21	2053.21	2053.18
Σ =	+40.24	Σ =	−40.21		
	Page Check:				
	2053.18				
	+ 40.24				
	2093.42				
	− 40.21				
	2053.21	Check			

GRAND LAKES UNIV. CAMPUS

BM Mil. to BM Oak

BM Mil. on GLU Campus   29 Sept. 1993
SW of Engineering Bldg.   Clear, Warm 70° F
9.4 ft. north of sidewalk   T.E. Henderson N
to instrument room and   J.F. King   ⌀
1.6 ft. from Bldg. Bronze   D.R. Moore   ⊼
disk in concrete flush   Lietz Level   #6
with ground, stamped "Mil"

BM Oak is a temporary
project bench mark located
at corner of Cherry and
Pine Sts., 14 ft. West of
computer laboratory. Twenty
penny spike in 18" Oak
tree, 1 ft. above ground.

Loop Misclosure = 2053.21 − 2053.18 = 0.03

Permissible Misclosure = $0.02 \sqrt{n} = 0.02 \sqrt{7}$
$= 0.05$ ft.

Adjustment = $\dfrac{0.03}{7} = 0.004'$ per H.I.

J.E. Henderson

PLATE D-2

Station	+ Sight	HI	− Sight	Int. Sight	Elev.
BM Road	10.15	(370.62) 370.63			360.48
0+00				9.36	361.26
0+20				9.8	360.8
1+00				6.5	364.1
2+00				4.3	366.3
2+60				3.7	366.9
3+00				7.1	363.5
3+90				11.7	358.9
4+00				11.2	359.4
4+35				9.5	361.1
TP 1	7.34	(366.48) 366.50	11.47		359.16
5+00				8.4	358.1
5+54				11.08	355.40
5+74				10.66	355.82
5+94				11.06	355.42
6+00				10.5	356.0
7+00				4.4	362.1
TP 2	2.56	(363.77) 363.80	5.26		361.24
8+00				1.2	362.6
9+00				3.9	359.9
9+25.2				3.4	360.4
9+25.3				4.6	359.2
9+43.2				2.2	361.6
BM Store			0.76		363.04
Σ	20.05		17.49		(363.01)

PROFILE LEVELS

BM ROAD to BM STORE

BM Road 3 miles SW of Mpls. 200 yrds. N of Pine St. over pass 40 ft. E of ₵ Hwy. 169 Top of RW conc post No. 268.

SW Minneapolis on Hwy 169

6 Oct. 1993

₵ Hwy. 169, painted "X" — Cool, Sunny, 50° F

West drainage ditch — R.J. Hintz — N

N.R. Olson — ∅

R.C. Perry — ⟂

Summit — Wild Level #3

Sag

Summit — COPY

E gutter, Maple St. — +20.05
₵ Maple St. — −17.49
W gutter, Maple St. — + 2.56
360.48
363.04

Summit — 363.04 − 363.01 = Misclosure = 0.03

Top of E curb, Elm St.
Bottom of E curb, Elm St.
₵ Elm St.
BM Store. NE corner Elm St. & 4th Ave. SE corner
Store foundation wall. 3″ brass disc set in grout.
BM store elev. = 363.01

R.J. Hintz

BORROW-PIT LEVELING

Point	+ Sight	HI	− Sight	Elev.	Cut
BM Road	4.22	364.70		360.48	
A,0			5.2	359.5	1.5
B,0			5.4	359.3	1.3
C,0			5.7	359.0	1.0
D,0			5.9	358.8	0.8
E,0			6.2	358.5	0.5
A,1			4.7	360.0	2.0
B,1			4.8	359.9	1.9
C,1			5.2	359.5	1.5
D,1			5.5	359.2	1.2
E,1			5.8	358.9	0.9
A,2			4.2	360.5	2.5
B,2			4.7	360.0	2.0
C,2			4.8	359.9	1.9
D,2			5.0	359.7	1.7
A,3			3.8	360.9	2.9
B,3			4.0	360.7	2.7
C,3			4.6	360.1	2.1
D,3			4.6	360.1	2.1
A,4			3.4	361.3	3.3
B,4			3.7	361.0	3.0
C,4			4.2	360.5	2.5
BM Road			4.22	360.48	

SECOND & OAK STREETS

hn	
BM Road-Description p.5	Madison, WI
1.5	Cool, Cloudy, 60° F
2.6	B.A. Dewitt N
2.0	B.K. Harris φ
1.6	E.A. Custer ⅄
0.5	11 Oct. 1993
4.0	Kern Level #13
7.6	
6.0	
3.6	
0.9	
5.0	
8.0	
7.6	
3.4	
5.8	
10.8	
6.3	
2.1	Grade elevation 358.0′
3.3	
6.0	Volume = area of base × $\frac{h_1+h_2+h_3+h_4}{4}$
2.5	
91.1	4
22.8 × 27 = 337 cu.yd.	400

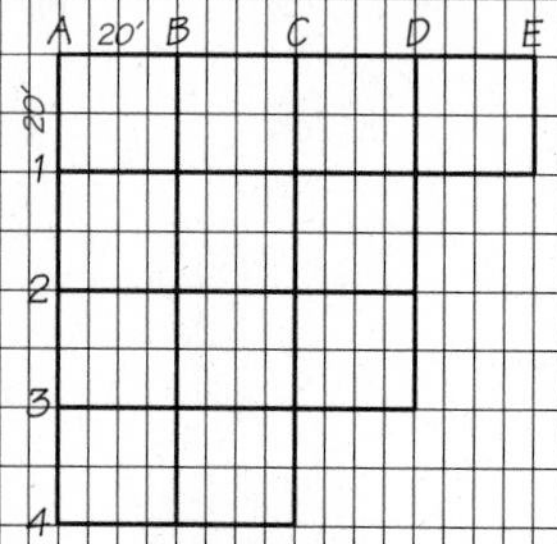

$B.A. Dewitt$

PLATE D-4

TRAVERSING WITH A

Instrument at sta 101

Instrument at sta 101

h_e=5.3 h_r=5.3

Sta. Sighted	FL/FR	Horiz. Circle	Zenith Angle	Horiz. Dist.	Elev. Diff.
104	L	0°00'00"	86°30'01"	324.38	+19.84
102	L	82°18'19"	92°48'17"	216.02	−10.58
104	R	180°00'03"	273°30'00"		
102	R	262°18'18"	267°11'41"		

Instrument at sta 102

h_e=5.5 h_r=5.5

101	L	0°00'00"	87°11'19"	261.05	+10.61
103	L	95°32'10"	85°19'08"	371.65	+30.43
101	R	180°00'02"	272°48'43"		
103	R	275°32'08"	274°40'50"		

Instrument at sta 103

h_e=5.4 h_r=5.4

102	L	0°00'00"	94°40'48"	371.63	−30.42
104	L	49°33'46"	90°01'54"	145.03	− 0.08
102	R	180°00'00"	265°19'14"		
104	R	229°33'47"	269°58'00"		

TOTAL STATION INSTRUMENT

Topo Control Survey

19 Oct. 1993

Cool, Sunny, 48° F

Pressure 29.5 in.

Total Station #7

Reflector #7A

M.R. Duckett—⊼

N. Dahman —∅

T. Ruhren —N

Sketch

M. R. Duckett

DOUBLE DIRECT ANGLES

Hub	Dist.	Single ∠	Double ∠	Avg. ∠
A		38°58.0′	77°56.8′	38°58.4′
	321.31′			
B		148°53.6′	297°47.0′	148°53.5′
	276.57′			
C		84°28.1′	168°56.2′	84°28.1′
	100.30′			
D		114°40.3′	229°20.9′	114°40.4′
	306.83′			
E		152°59.4′	305°58.6′	152°59.3′
	255.48′			
A				
Σ	1260.49′			539°59.7′
		Misclosure		0°00.3′

Σ interior ∠s = (N-2) 180°
= (5-2) 180°
= 540°00′

PATTON HALL TRAVERSE

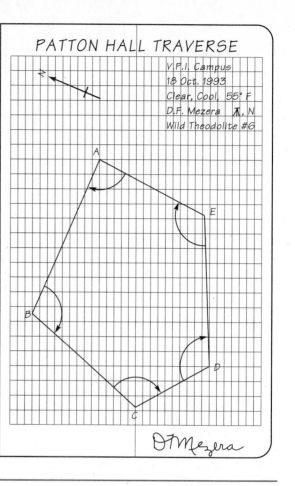

V.P.I. Campus
18 Oct. 1993
Clear, Cool, 55° F
D.F. Mezera ∠, N
Wild Theodolite #6

DFMezera

PLATE D-6

STADIA SURVEY

Station	Stadia Interval	Azimuth	Vert. ∡ or Rod R.	Horiz. Dist.	Elev.
⊼ @ ▱B,		Elev. 177.42,		h.i. = 5.0	
▱A	6.74	148°04′	−0°34′	675	170.7
▱C	4.21	60°12′	−1°35′	422	165.8
1	0.91	90°45′	8.3	92	174.1
2	1.66	120°20′	−2°12′	167	171.0
3	3.15	126°30′	−2°06′	316	165.9
4	4.60	143°45′	−1°23′	461	166.0
5	7.85	141°30′	−0°38′	786	168.7
6	2.47	167°20′	8.6	248	173.8
7	1.97	172°20′	9.6	198	172.8
8	4.99	181°15′	3.2	500	179.2
9	5.79	221°45′	+1°02′	580	187.8
10	3.47	256°00′	+1°50′	348	188.5
11	1.17	342°05′	2.4	118	180.0
12	1.71	350°15′	2.4	172	180.0
▱D	4.49	6°10′	5.2	450	177.2
▱A	6.74	148°04′	−0°34′	675	170.7
⊼ @ ▱C,		Elev. 165.77,		h.i. = 5.2	
▱B	4.20	240°12′	+1°34′	421	177.3
13	3.21	286°00′	+2°01′	322	177.1
14	2.36	32°05′	8.5	237	162.5
15	2.60	41°50′	10.0	261	161.0
16	4.59	68°30′	−1°22′	460	154.8

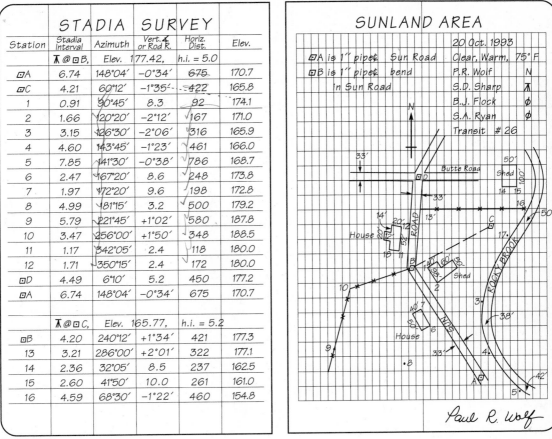

SUNLAND AREA

20 Oct. 1993

▱A is 1″ pipe in Sun Road — Clear, Warm, 75° F

▱B is 1″ pipe, bend — P.R. Wolf

in Sun Road — S.D. Sharp

B.J. Flock

S.A. Ryan

Transit #26

Paul R. Wolf

CROSS-SECTION LEVELING

Sta.	Sight +	H.I.	Sight −	Elev.	
5+00			9.5		
4+00			12.6		
TP 1	10.25	106.61	1.87	96.36	
3+00			2.1		
2+50			5.8		
2+00			7.4		
1+35			9.7		
1+00			5.6		
0+50			7.6		
0+00			8.5		
BM Pod	8.51	98.23		89.72	

PLATE D-8

HONOLULU-KAILUA HIGHWAY

Diamond Highway
25 Oct. 1993
Warm, Sunny 70°
A.C. Chun
R.E. Neilan   N
W.E. Grube   Ø,C
M.L. Hagawa   C
Lietz level #10

99.2	101.5	97.4	97.1	95.8	97.0	103.8
7.4	5.1	9.2	9.5	10.8	9.6	2.8
52	30	10		12	28	45
102.3	99.9	98.4	94.0	100.1	101.5	98.7
4.3	6.7	8.2	12.6	6.5	5.1	7.9
48	24	8		10	25	50
	95.2	95.8	96.6	96.1	94.4	91.1	95.7
	3.0	2.4	1.6	2.1	3.8	7.1	2.5
	50	25	10		8	31	48
95.1	92.8	89.5	93.8	92.4	90.7	93.4	96.6
3.1	5.4	8.7	4.4	5.8	7.5	4.8	1.6
48	32	15	8		10	25	50
	92.3	90.0	90.8	90.8	91.3	93.2	95.9
	5.9	8.2	7.4	7.4	6.9	5.0	2.3
	54	30	10		9	25	40
	85.4	88.9	85.7	88.5	89.8	91.7	94.1
	12.8	9.3	12.5	9.7	8.4	6.5	4.1
	48	25	10		8	15	45
	88.6	97.2	92.2	92.6	95.8	93.6	95.4
	9.6	1.0	6.0	5.6	2.4	4.6	2.8
	52	28	12		10	28	50
	90.0	97.0	92.7	90.6	94.4	95.4	85.5
	8.2	1.2	5.5	7.6	3.8	2.8	12.7
	50	25	8		9	24	42
	88.6	96.1	92.0	89.7	93.5	97.0	91.5
	9.6	2.1	6.2	8.5	4.7	1.2	6.7
	50	25	10		8	25	50

BM Pod—Kalini Valley, Oahu, Ewa-makai corner
Hibiscus and Kiawe Drives. Spike in 30"monkey pod
tree, 2 ft. above ground.

Ruth E. Neilan

8" SEWER STAKEOUT

(1) Station	(2) +Sight	(3) HI	(4) -Sight	(5) Ground Elev.	(6) Pipe Flow Line
BM 16	2.11	102.76		100.65	
0+00			6.21	96.55	96.55
+00₵			3.20	99.56	96.55
+50₵			3.91	98.85	95.95
1+00₵			4.07	98.69	95.34
+31₵			8.22	94.54	94.97
+50₵			4.01	98.75	94.74
2+00₵			4.52	98.24	94.14
+33.7₵			5.03	97.73	93.73
+33.7			9.03	93.73	93.73
BM 16			2.11√	100.65	

Flowline Calculations

Line drops 50′ (1.206%) = 0.60′ per 50′

Example
Sta. 0+50 = 96.55 − 0.60 = 95.95
1+00 = 96.55 − 1.21 = 95.34
1+31 = 96.55 − 131 (1.206%) = 94.97

(7) Cut/Fill	
	Third Street, Statesboro, GA
	1 Nov. 1993
See page 23, Book 67 for	Cool, Clear, 65° F
description.	F.P. Barnes
Floor of Existing Catch Basin	P.A. Hartzheim N
C 3.01	J.C. Storey ∅
C 2.90	Level #14
C 3.35	
F 0.43	Existing Catch Basin
C 4.02	New 8" Sewer
C 4.09	Existing Manhole
C 4.00	
Floor of Existing Manhole	

Existing 8" Clay Pipe
Sink Hole
Existing 10" Cast Iron

Grade = Fall = (96.55 − 93.73)(100) = 1.206%
Dist.   233.7

Paul Hartzheim

ALIGNMENT OF

Station	Chord	Total Def.	Calc. Bearing	Mag. Bearing	Curve Data
68 P.O.T.	100.00				
67	62.92				
			N24°42'E	N24°45'E	
P.T. 66+37.08	37.08	4°12'00"			
66	99.99	3°49'48" 45"			Δ=8°24'
					R=2864.79'
65 P.O.C.	99.99	2°49'45"			D=2°00'
					L=420.00'
64	99.99	1°49'45"			T=210.38'
					E=7.71'
63	82.92	0°49'45"			M=7.69'
					Defl./ft. =0.6 min.
P.C. 62+17.08	17.08	0°00'00"			
			N16°18'E	N16°30'E	
62	100.00				
61	100.00				
60	100.00				
59 P.O.T.					

PLATE D-10

LAFAYETTE HIGHWAY

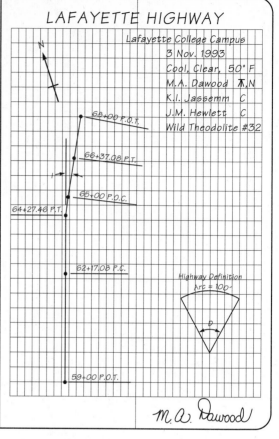

Lafayette College Campus
3 Nov. 1993
Cool, Clear, 50° F
M.A. Dawood ⚒,N
K.I. Jassemm C
J.M. Hewlett C
Wild Theodolite #32

68+00 P.O.T.
66+37.08 P.T.
65+00 P.O.C.
64+27.46 P.T.
62+17.08 P.C.
59+00 P.O.T.

Highway Definition
Arc = 100'

M. a. Dawood

APPENDIX E
TRIGONOMETRIC FORMULAS FOR SOLVING TRIANGLES

TABLE E-1 TRIGONOMETRIC FORMULAS FOR THE SOLUTION OF RIGHT TRIANGLES

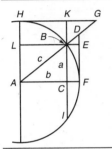

Let A = angle BAC = arc BF, and let radius $AF = AB = AH = 1$. Then,

$$\sin A = BC \qquad\qquad \csc A = AG$$
$$\cos A = AC \qquad\qquad \sec A = AD$$
$$\tan A = DF \qquad\qquad \cot A = HG$$
$$\text{vers } A = CF = BE \qquad \text{covers } A = BK = LH$$
$$\text{exsec } A = BD \qquad\qquad \text{coexsec } A = BG$$
$$\text{chord } A = BF \qquad\qquad \text{chord } 2A = BI = 2\,BC$$

In the right-angled triangle ABC, let $AB = c$, $BC = a$, $CA = b$. Then

1. $\sin A = \dfrac{a}{c}$

2. $\cos A = \dfrac{b}{c}$

3. $\tan A = \dfrac{a}{b}$

4. $\cot A = \dfrac{b}{a}$

5. $\sec A = \dfrac{c}{b}$

6. $\csc A = \dfrac{c}{a}$

7. $\text{vers } A = 1 - \cos A = \dfrac{c - b}{c}$
$$= \text{covers } B$$

8. $\text{exsec } A = \sec A - 1 = \dfrac{c - b}{b} = \text{coexsec } B$

9. $\text{covers } A = \dfrac{c - a}{c} = \text{vers } B$

10. $\text{coexsec } A = \dfrac{c - a}{a} = \text{exsec } B$

11. $a = c \sin A = b \tan A$

12. $b = c \cos A = a \cot A$

13. $c = \dfrac{a}{\sin A} = \dfrac{b}{\cos A}$

14. $a = c \cos B = b \cot B$

15. $b = c \sin B = a \tan B$

16. $c = \dfrac{a}{\cos B} = \dfrac{b}{\sin B}$

17. $a = \sqrt{c^2 - b^2}$
$$= \sqrt{(c - b)(c + b)}$$

18. $b = \sqrt{c^2 - a^2}$
$$= \sqrt{(c - a)(c + a)}$$

19. $c = \sqrt{a^2 + b^2}$

20. $C = 90° = A + B$

21. $\text{area} = \frac{1}{2}ab$

TABLE E-2 TRIGONOMETRIC FORMULAS FOR THE SOLUTION OF OBLIQUE TRIANGLES

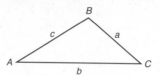

NO.	GIVEN	SOUGHT	FORMULA
22	A, B, a	C, b, c	$C = 180° - (A + B)$
			$b = \dfrac{a}{\sin A} \times \sin B$
			$c = \dfrac{a}{\sin A} \times \sin (A + B) = \dfrac{a}{\sin A} \times \sin C$
		Area	$\text{Area} = \frac{1}{2}ab \sin C = \dfrac{a^2 \sin B \sin C}{2 \sin A}$
23	A, a, b	B, C, c	$\sin B = \dfrac{\sin A}{a} \times b$
			$C = 180° - (A + B)$
			$C = \dfrac{a}{\sin A} \times \sin C$
		Area	$\text{Area} = \frac{1}{2}ab \sin C$
24	C, a, b	c	$c = \sqrt{a^2 + b^2 - 2ab \cos C}$
25		$\frac{1}{2}(A + B)$	$\frac{1}{2}(A + B) = 90° - \frac{1}{2}C$
26		$\frac{1}{2}(A - B)$	$\tan \frac{1}{2}(A - B) = \dfrac{a - b}{a + b} \times \tan \frac{1}{2}(A + B)$
27		A, B	$A = \frac{1}{2}(A + B) + \frac{1}{2}(A - B)$
			$B = \frac{1}{2}(A + B) - \frac{1}{2}(A - B)$
28		c	$c = (a + b) \times \dfrac{\cos \frac{1}{2}(A + B)}{\cos \frac{1}{2}(A - B)} = (a - b) \times \dfrac{\sin \frac{1}{2}(A + B)}{\sin \frac{1}{2}(A - B)}$
29		Area	$\text{Area} = \frac{1}{2}ab \sin C$
30	a, b, c	A	$\text{Let } s = \dfrac{a + b + c}{2}$
31			$\sin \frac{1}{2}A = \sqrt{\dfrac{(s - b)(s - c)}{bc}}$
			$\cos \frac{1}{2}A = \sqrt{\dfrac{s(s - a)}{bc}}$
			$\tan \frac{1}{2}A = \sqrt{\dfrac{(s - b)(s - c)}{s(s - a)}}$
32			$\sin A = \dfrac{2\sqrt{s(s - a)(s - b)(s - c)}}{bc}$
			$\cos A = \dfrac{b^2 + c^2 - a^2}{2bc}$
33		Area	$\text{Area} = \sqrt{s(s - a)(s - b)(s - c)}$

APPENDIX F
COMPUTER PROGRAMS

F-1 INTRODUCTION

This appendix contains listings of two computer programs, together with examples of output illustrating their use. The programs are written in BASIC. Keyboard entry of these listings, if made exactly as shown, will produce the results in the examples and can be used to solve similar problems.

The programs, described in more detail in succeeding sections, perform (1) traverse computations and (2) azimuth calculations from astronomical observations. In each program, "prompts" tell the user to enter data required for the solution.

It should be noted that these programs have been kept short intentionally and thus do not contain certain options that could be incorporated to make them more convenient. Permission is granted for their unrestricted use. However, the publisher and authors assume no responsibility for problems that may arise as a result of employing them.

F-2 PROGRAM FOR TRAVERSE COMPUTATION

The following program performs traverse computations for both types of closed traverses—that is, the "polygon" type that returns to its starting station, and the "link" type that closes on another point of known position. The computations include determining departures and latitudes, balancing them by the compass rule, and calculating linear misclosure and relative precision. From given coordinates of the starting station, coordinates of all traverse points are determined, and if the traverse is a polygon, its area is calculated by the coordinate method.

Prior to using the program, traverse stations must be identified by consecutive numbers beginning with 1. Input to the program consists of (1) the number of courses in the traverse; (2) entry of the number 1 if the traverse is a polygon, 2 if it is a link type; (3) length (in feet) and azimuth (in degrees, minutes, and seconds) of each successive course beginning with line 1–2; and (4) the coordinates of station 1 for a polygon traverse, or the first and last stations for the link type. In numbering traverse stations, 1 must be assigned to a station whose coordinates are known. The program can accept lengths in meters; however, feet are assumed, and based on that assumption, the area is calculated and listed in acres. If meters are used, lengths and coordinates will be correctly listed in the output in meters. However, the listed area must be multiplied by 4.3560 to convert it to hectares.

The following are the program listing and example output for solving Examples 13-4 and 13-7 of the text. Note that, except for roundoff, the results agree with those given in Tables 13-4 and 13-6.

F-2.1 Program Listing

```
100 '                ****  TRAVERSE COMPUTATION  ****
110 '
120 DEFDBL A-H, O-Z
130 RD = 3.14159265# / 180
140 KEY OFF
150 COLOR 7, 1, 1
160 CLS
170 LOCATE 1, 26: PRINT "-- Traverse Computation --"
180 LOCATE 3, 6: INPUT "Results to: Screen(1), Printer(2), or File(3)"; ANS1
190 IF ANS1 <> 1 AND ANS1 <> 2 AND ANS1 <> 3 THEN 180
200 IF ANS1 = 1 THEN OPEN "SCRN:" FOR OUTPUT AS #1: GOTO 250
210 IF ANS1 = 2 THEN OPEN "LPT1:" FOR OUTPUT AS #1: GOTO 240
220 LOCATE 7, 6: LINE INPUT "Output file name ... ", FILENAME$
230 OPEN FILENAME$ FOR OUTPUT AS #1: GOTO 250
240 LOCATE 5, 25: PRINT "Make sure your printer is on"
250 LOCATE 9, 6: LINE INPUT "Problem number ... ", PROBNO$
260 LOCATE 11, 6: INPUT "Number of courses in traverse ... ", NC
270 DIM DS(NC), DG(NC), MN(NC), SC(NC), RA(NC), XU(NC + 1), YU(NC + 1), XB(NC +
1), YB(NC + 1), X(NC + 1), Y(NC + 1), XP(NC + 1), YP(NC + 1)
280 LOCATE 13, 6: PRINT "Traverse type :"
290 LOCATE 14, 9: PRINT "1 - Polygon (closes on itself)"
300 LOCATE 15, 9: PRINT "2 - Link (starts and ends on different known points)"
310 LOCATE 16, 7: PRINT "Select :        "
320 LOCATE 16, 16: INPUT "", TT
330 IF TT <> 1 AND TT <> 2 THEN SOUND 2000, 3: GOTO 310
340 IF TT = 1 THEN TRAV$ = "Polygon" ELSE TRAV$ = "Link"
350 CLS
360 DX = 0
370 DY = 0
380 PE = 0
390 LOCATE 12, 21: PRINT "input azimuth as deg, min, sec"
400 LOCATE 17, 15: PRINT "eg: 248° 14' 34.5'' is entered   248,14,34.5"
410 FOR I = 1 TO NC
420 LOCATE 5, 9: PRINT "Course #"; I
430 LOCATE 6, 12: PRINT "Distance ..." + SPACE$(15)
440 LOCATE 7, 12: PRINT "Azimuth .... " + SPACE$(15)
450 LOCATE 6, 25: INPUT "", DS(I)
460 LOCATE 7, 25: INPUT "", DG(I), MN(I), SC(I)
470 LOCATE 9, 9: PRINT "Change this reading, yes(1) no(2)..." + SPACE$(5)
480 LOCATE 9, 45: INPUT "", ANS2: IF ANS2 = 1 THEN 420
490 PE = PE + DS(I)
500 RA(I) = RD * (DG(I) + MN(I) / 60 + SC(I) / 3600)
510 ACOS(I) = COS(RA(I))
520 ASIN(I) = SIN(RA(I))
530 XU(I) = DS(I) * SIN(RA(I))
540 YU(I) = DS(I) * COS(RA(I))
550 DX = DX + XU(I)
560 DY = DY + YU(I)
570 NEXT I
580 CLS
590 LOCATE 16, 20: PRINT SPACE$(50)
600 LOCATE 17, 14: PRINT SPACE$(50)
610 LOCATE 15, 6: PRINT "Coordinates of start point:"
620 LOCATE 16, 9: INPUT "X ... ", X(1)
630 LOCATE 17, 9: INPUT "Y ... ", Y(1)
640 IF TT = 1 THEN 710
650 LOCATE 18, 43: PRINT "Coordinates of end point:"
660 LOCATE 19, 46: INPUT "X ... ", X(NC + 1)
670 LOCATE 20, 46: INPUT "Y ... ", Y(NC + 1)
680 CX = DX - (X(NC + 1) - X(1))
690 CY = DY - (Y(NC + 1) - Y(1))
700 GOTO 750
710 X(NC + 1) = X(1)
720 Y(NC + 1) = Y(1)
730 CX = DX
740 CY = DY
750 CL = SQR(CX * CX + CY * CY)
760 PC = INT(PE /(100*CL))*100
770 PRINT #1, "-- Traverse Computation " + STRING$(56, "-"): PRINT #1,
780 PRINT #1, "    Problem Number "; PROBNO$; "  :  "; TRAV$; " traverse.": PRIN
T #1,
790 PRINT #1, " Course"; TAB(13); "Length"; TAB(26); "Azimuth"
800 FOR I = 1 TO NC
810 J = 0
820 IF TT = 2 THEN 850
830 IF I <> NC THEN 850
840 J = NC
```

```
850 PRINT #1, I; "-"; I + 1 - J; TAB(9);
860 PRINT #1, USING "######,.###    ### ## ##.#"; DS(I); DG(I); MN(I); SC(I)
870 XB(I) = XU(I) - CX * DS(I) / PE
880 YB(I) = YU(I) - CY * DS(I) / PE
890 X(I + 1) = X(I) + XB(I)
900 Y(I + 1) = Y(I) + YB(I)
910 NEXT I
920 PRINT #1, "              ----------"
930 PRINT #1, USING "         ######,.###  = Total Length "; PE
940 PRINT #1,
950 PRINT #1,
960 PRINT #1, TAB(9); "Unbalanced"; TAB(34); "Balanced"
970 PRINT #1, TAB(7); "Dep"; TAB(18); "Lat"; TAB(31); "Dep"; TAB(42); "Lat"
980 XX = 0
990 FOR I = 1 TO NC
1000 XU(I) = INT(XU(I) * 1000 + .5) / 1000
1010 YU(I) = INT(YU(I) * 1000 + .5) / 1000
1020 XB(I) = INT(XB(I) * 1000 + .5) / 1000
1030 YB(I) = INT(YB(I) * 1000 + .5) / 1000
1040 X(I) = INT(X(I) * 1000 + .5) / 1000
1050 Y(I) = INT(Y(I) * 1000 + .5) / 1000
1060 PRINT #1, USING " +####.###  +####.###    +####.###  +####.###"; XU(I); YU
(I); XB(I); YB(I)
1070 IF TT = 2 THEN 1090
1080 XX = XX + X(I) * Y(I + 1) - Y(I) * X(I + 1)
1090 NEXT I
1100 DX = INT(DX * 1000 + .5) / 1000
1110 DY = INT(DY * 1000 + .5) / 1000
1120 PRINT #1, "   --------  ---------"
1130 PRINT #1, USING " +####.###  +####.###"; DX; DY
1140 PRINT #1,
1150 PRINT #1, TAB(20); "Coordinates"
1160 PRINT #1, TAB(4); "Point"; TAB(18); "X"; TAB(31); "Y"
1170 IF TT = 2 GOTO 1220
1180 FOR I = 1 TO NC
1190 PRINT #1, USING "   ##  #########,.###  #####,.###"; I; X(I); Y(I)
1200 NEXT I
1210 GOTO 1250
1220 FOR I = 1 TO NC + 1
1230 PRINT #1, USING "   ##  #########,.###  #####,.###"; I; X(I); Y(I)
1240 NEXT I
1250 PRINT #1,
1260 PRINT #1, "     Linear Misclosure ="; INT(1000 * CL) / 1000; "ft."
1270 PRINT #1, "     Relative Precision = 1 in"; PC
1280 IF TT = 2 THEN 1300
1290 PRINT #1, "     Area = "; INT(10000 * ABS(XX / (2 * 43560!))) / 10000; "acr
es"
1300 PRINT #1, : PRINT #1, STRING$(80, "-"): PRINT #1, : PRINT #1,
1310 CLOSE
1320 LOCATE 22, 6: INPUT "Run another traverse (Y/N) ... ", GO$
1330 IF GO$ <> "Y" AND GO$ <> "y" THEN GOTO 1360
1340 ERASE DS, DG, MN, SC, RA, XU, YU, XB, YB, X, Y
1350 GOTO 150
1360 LOCATE 22, 1: PRINT SPACE$(25) + " Computations completed "
1370 KEY ON
1380 END
```

F-2.2 Listing of Output for Examples 13-4 and 13-7

TRAVERSE COMPUTATION
PROBLEM 13-4 : POLYGON TRAVERSE

COURSE	LENGTH	AZIMUTH	UNBALANCED		BALANCED	
			DEP	LAT	DEP	LAT
1–2	285.100	26 10 0.0	+ 125.724	+ 255.882	+ 125.661	+ 255.964
2–3	610.450	104 35 12.0	+ 590.774	− 153.738	+ 590.639	− 153.561
3–4	720.480	195 30 6.0	− 192.560	− 694.271	− 192.719	− 694.061
4–5	203.000	358 18 30.0	− 5.993	+ 202.912	− 6.038	+ 202.971
5–1	647.020	306 54 6.0	− 517.401	+ 388.499	− 517.544	+ 388.687
	2,466.050	= Total Length	+ 0.545	− 0.717		

COORDINATES

POINT	X	Y
1	10,000.000	10,000.000
2	10,125.661	10,255.964
3	10,716.301	10,102.404
4	10,523.581	9,408.342
5	10,517.544	9,611.313

Linear Misclosure = 0.9 ft
Relative Precision = 1 in 2700
Area = 6.2579 acres

TRAVERSE COMPUTATION
PROBLEM 13-7 : LINK TRAVERSE

			UNBALANCED		BALANCED	
COURSE	LENGTH	AZIMUTH	DEP	LAT	DEP	LAT
1–2	1,045.500	62 55 47.0	+930.964	+475.789	+930.913	+475.849
2–3	1,007.380	139 12 57.0	+658.032	−762.764	+657.982	−762.706
3–4	897.810	57 25 30.0	+756.573	+483.384	+756.529	+483.435
4–5	960.360	340 55 56.0	−313.737	+907.668	−313.784	+907.722
	3,911.050	= Total Length	+2031.833	+1104.077		

COORDINATES

POINT	X	Y
1	12,765.480	43,280.210
2	13,696.393	43,756.059
3	14,354.375	42,993.353
4	15,110.904	43,476.788
5	14,797.120	44,384.510

Linear Misclosure = 0.294 ft
Relative Precision = 1 in 13200

F-3 PROGRAM FOR AZIMUTH FROM ASTRONOMICAL OBSERVATIONS

The following program solves Eq.(18-3) for the azimuth of Polaris, other stars, or the sun by the hour-angle method. The horizontal angle from the mark to the body must then be applied manually to obtain the line's azimuth. The program will accept any number of direct plus reversed observations, and if more than one is entered, it performs the calculations independently for each observation and lists the azimuth for each observation and the mean azimuth.

Input consists of (1) the total number of observations, direct plus reversed; (2) the latitude of the observation station (in degrees, minutes, and seconds); (3) the longitude of the station (in degrees, minutes, and seconds); (4) the body's declination (in degrees, minutes, and seconds) at 0 hours universal time (UT) on the day of observation; **(5)** the body's declination (in degrees, minutes, and seconds) at 0 hours UT on the day after observation; **(6)** the Greenwich hour angle (in degrees, minutes and seconds) of the body at 0 hours UT on the day of observation; **(7)** the Greenwich hour angle (in degrees, minutes, and seconds) at 0 hours UT on the day after observation; and **(8)** the universal times (in hours, minutes, and seconds) of each observation.

The program listing is presented, followed by output for solving Examples 18-2 and 18-4 of the text. In Example 18-2 for a Polaris observation, all four sets of data are entered and the star's mean azimuth (359°31′04″) is computed. To obtain the line's azimuth, the mean of the four horizontal angles (49°38′40″) is subtracted, giving a line azimuth of 309°52′24″.

In Example 18-4 for a sun observation, the mean azimuth of the sun's left edge is 266°30′44″, and after subtracting the mean semidiameter (19°55″) and horizontal angle (212°25′34″), an azimuth of 53°45′15″ is obtained.

F-3.1 Program Listing

```
100 '      ****  AZIMUTH FROM SUN OR POLARIS : HOUR ANGLE METHOD  ****
110 '
120 PI = 3.14159265#
130 RO = PI / 180
140 KEY OFF
150 COLOR 7, 1, 1: CLS
160 LOCATE 1, 17: PRINT "-- Celestial Reduction : Hour Angle Method --"
170 LOCATE 3, 6: INPUT "Results to: Screen(1), Printer(2), or File(3)"; ANS1
180 IF ANS1 <> 1 AND ANS1 <> 2 AND ANS1 <> 3 THEN 170
190 IF ANS1 = 1 THEN OPEN "SCRN:" FOR OUTPUT AS #1: GOTO 270
200 IF ANS1 = 2 THEN OPEN "LPT1:" FOR OUTPUT AS #1: GOTO 230
210 LOCATE 7, 6: LINE INPUT "Output file name ... ", FILENAME$
220 OPEN FILENAME$ FOR OUTPUT AS #1: GOTO 270
230 LOCATE 5, 25: PRINT "Make sure your printer is on"
240 LOCATE 10, 25: PRINT "Make sure your printer is on then"
250 LOCATE 11, 24, 0: PRINT "press any key to continue"
260 WHILE INKEY$ = "": WEND
270 CLS
280 LOCATE 10, 25: PRINT SPACE$(29)
290 LOCATE 11, 24: PRINT SPACE$(31)
300 LOCATE 3, 5: INPUT "Observer's Latitude  (deg,min,sec) ... ", LATD, LATM, LA
TS
310 LOCATE 4, 16: INPUT "Longitude (deg,min,sec) ... ", LONGD, LONGM, LONGS
320 OBSLAT = RO * (LATD + LATM / 60 + LATS / 3600)
330 OBSLONG = RO * (LONGD + LONGM / 60 + LONGS / 3600)
340 LOCATE 6, 5: PRINT "Celestial body : "
350 LOCATE 7, 8: PRINT "1 - Polaris (or other star)"
360 LOCATE 8, 8: PRINT "2 - Sun"
370 LOCATE 9, 6: PRINT "Select :        "
380 LOCATE 9, 15: INPUT "", BODY
390 IF BODY <> 1 AND BODY <> 2 THEN SOUND 2000, 3: GOTO 370
400 IF BODY = 1 THEN BODY$ = "Polaris (or other star)" ELSE BODY$ = "Sun"
410 LOCATE 10, 22: PRINT BODY$
420 FOR I = 7 TO 9: LOCATE I, 5: PRINT SPACE$(35): NEXT I
430 LOCATE 12, 5: INPUT "Number of observations ... ", N
440 LOCATE 13, 5: PRINT "Declination (deg,min,sec) at 00 hrs UT on:"
450 LOCATE 14, 12: INPUT "day of observation ...... ", D00D$, D00M, D00S
460 LOCATE 15, 12: INPUT "day after observation ... ", D24D$, D24M, D24S
470 D00 = RO * (ABS(VAL(D00D$)) + D00M / 60 + D00S / 3600)
480 IF VAL(D00D$) < 0 THEN D00 = -D00
490 D24 = RO * (ABS(VAL(D24D$)) + D24M / 60 + D24S / 3600)
500 IF VAL(D24D$) < 0 THEN D24 = -D24
510 LOCATE 16, 5: PRINT "GHA (deg,min,sec) at 00 hrs UT on :"
520 LOCATE 17, 12: INPUT "day of observation ...... ", GHA00D, GHA00M, GHA00S
530 LOCATE 18, 12: INPUT "day after observation ... ", GHA24D, GHA24M, GHA24S
540 GHA00 = RO * (GHA00D + GHA00M / 60 + GHA00S / 3600)
550 GHA24 = RO * (GHA24D + GHA24M / 60 + GHA24S / 3600)
560 PRINT #1, "-- Celestial Reduction : Hour Angle Method "; STRING$(32, "-"): P
RINT #1,
570 PRINT #1,
580 PRINT #1, SPACE$(5); "Observer's  Latitude :";
```

```
590 PRINT #1, USING "  ## deg ## min ##.## sec"; LATD; LATM; LATS
600 PRINT #1, SPACE$(16); "Longitude :";
610 PRINT #1, USING " ### deg  ## min  ##.## sec"; LONGD; LONGM; LONGS
620 PRINT #1, : PRINT #1, SPACE$(5); N; "observations made to "; BODY$
630 PRINT #1, : PRINT #1, SPACE$(5); "Declination at 00 hours on"
640 PRINT #1, SPACE$(10); "day of observation      : "; D00D$;
650 PRINT #1, USING " deg  ## min  ##.# sec"; D00M; D00S
660 PRINT #1, SPACE$(10); "day after observation  : "; D24D$;
670 PRINT #1, USING " deg  ## min  ##.# sec"; D24M; D24S
680 PRINT #1, SPACE$(5); "GHA at 00 hours on"
690 PRINT #1, SPACE$(10); "day of observation      :";
700 PRINT #1, USING " ### deg  ## min  ##.## sec"; GHA00D; GHA00M; GHA00S
710 PRINT #1, SPACE$(10); "day after observation  :";
720 PRINT #1, USING " ### deg  ## min  ##.## sec"; GHA24D; GHA24M; GHA24S
730 PRINT #1, : PRINT #1, SPACE$(7); STRING$(47, "-")
740 IF ANS1 = 1 THEN PRINT #1, " Entered Information"
750 IF ANS1 = 1 THEN PRINT #1, " Press any key to continue": WHILE INKEY$ = "":
WEND
760 ZSUM = 0
770 FOR I = 1 TO N
780 CLS
790 LOCATE 5, 5: PRINT BODY$; " Observation #"; I
800 LOCATE 6, 8: PRINT "UT (hrs,min,sec) ..... "; SPACE$(20)
810 LOCATE 6, 32: INPUT "", UTH, UTM, UTS
820 UT = (UTH + UTM / 60 + UTS / 3600)
830 GHA = GHA00 + (GHA24 + (2 * PI) - GHA00) * (UT / 24)
840 IF GHA > 2 * PI THEN GHAT = (GHA - 2 * PI) * 180 / PI ELSE GHAT = GHA * 180
/ PI
850 GHAD = INT(GHAT)
860 GHAM = INT((GHAT - GHAD) * 60)
870 GHAS = (((GHAT - GHAD) * 60) - GHAM) * 60
880 DOBS = D00 + (D24 - D00) * (UT / 24)
890 IF BODY = 2 THEN DOBS = DOBS + ((.0000395#) * D00 * SIN(7.5 * UT * PI / 180#
))
900 IF DOBS < 0 THEN DS$ = "-" ELSE DS$ = "+"
910 DOBST = DOBS * 180 / PI
920 DOBSD = INT(ABS(DOBST))
930 DOBSM = INT((ABS(DOBST) - DOBSD) * 60)
940 DOBSS = ((ABS(DOBST) - DOBSD) * 60 - DOBSM) * 60
950 PRINT #1, : PRINT #1, SPACE$(10); "Observation #"; I
960 PRINT #1, SPACE$(15);
970 PRINT #1, USING "Time    : ## hrs ## min ##.#  sec"; UTH; UTM; UTS
980 PRINT #1, SPACE$(15); "Decl    : "; DS$;
990 PRINT #1, USING "## deg  ## min  ##.## sec"; DOBSD; DOBSM; DOBSS
1000 PRINT #1, SPACE$(15);
1010 PRINT #1, USING "GHA     : ### deg  ## min  ##.## sec"; GHAD; GHAM; GHAS
1020 LHA = GHA - OBSLONG
1030 IF LHA > 2 * PI THEN LHA = LHA - 2 * PI
1040 IF LHA < 0 THEN LHA = LHA + 2 * PI
1050 LHAT = LHA * 180 / PI
1060 LHAD = INT(LHAT)
1070 LHAM = INT((LHAT - LHAD) * 60)
1080 LHAS = (((LHAT - LHAD) * 60) - LHAM) * 60
1090 PRINT #1, SPACE$(15);
1100 PRINT #1, USING "LHA     : ### deg  ## min  ##.## sec"; LHAD; LHAM; LHAS
1110 Z = ATN(-SIN(LHA) / (COS(OBSLAT) * TAN(DOBS) - SIN(OBSLAT) * COS(LHA)))
1120 IF BODY = 1 THEN GOTO 1190
1130 IF LHA > PI THEN GOTO 1170
1140 IF Z > 0 THEN Z = Z + PI
1150 IF Z < 0 THEN Z = Z + 2 * PI
1160 GOTO 1200
1170 IF Z < 0 THEN Z = Z + PI
1180 GOTO 1200
1190 IF Z < 0 THEN Z = Z + 2 * PI
1200 Z = Z * 180 / PI
1210 ZD = INT(Z)
1220 ZM = INT((Z - ZD) * 60)
1230 ZS = (((Z - ZD) * 60) - ZM) * 60
1240 PRINT #1, SPACE$(15);
1250 PRINT #1, USING "Azimuth : ### deg ## min  ##.## sec"; ZD; ZM; ZS
1260 ZSUM = ZSUM + Z
1270 NEXT I
1280 MEANAZ = ZSUM / N
1290 MEANAZD = INT(MEANAZ)
1300 MEANAZM = INT(60 * (MEANAZ - MEANAZD))
1310 MEANAZS = 60 * (60 * (MEANAZ - MEANAZD) - MEANAZM)
1320 MEANAZS = INT(10 * MEANAZS) / 10
```

```
1330 LOCATE 18, 1: PRINT SPACE$(5) + STRING$(52, "-")
1340 LOCATE 19, 1
1350 PRINT SPACE$(10) + "Mean Azimuth of"; BODY$; TAB(34);
1360 PRINT USING " :   ###° ##' ##.#''"; MEANAZD; MEANAZM; MEANAZS
1370 LOCATE 20, 1: PRINT SPACE$(5) + STRING$(52, "-")
1380 PRINT #1, : PRINT #1, SPACE$(5) + STRING$(62, "-")
1390 PRINT #1, SPACE$(10) + "Mean Azimuth of "; BODY$; TAB(34);
1400 PRINT #1, USING " :   ### deg  ## min ##.# sec"; MEANAZD; MEANAZM; MEANAZS
1410 PRINT #1, SPACE$(5) + STRING$(62, "."): PRINT #1, : PRINT #1,
1420 CLOSE
1430 LOCATE 22, 5: INPUT "Reduce another set (Y/N) ... ", GO$
1440 IF GO$ = "Y" OR GO$ = "y" THEN GOTO150
1450 LOCATE 22, 1: PRINT SPACE$(25); " Computations completed "
1460 KEY ON
1470 END
```

F-3.2 Listing of Output for Example 18-2

CELESTIAL REDUCTION : HOUR-ANGLE METHOD

Observer's latitude : 43° 05' 24.00"
       longitude : 89° 26' 00.00"

4 observations made to Polaris (or other star)

Declination at 00 hours on
   day of observation : 89° 13' 56.3"
   day after observation : 89° 13' 56.1"

GHA at 00 hours on
   day of observation : 216° 27' 36.00"
   day after observation : 217° 26' 23.00"

Observation # 1
Time   :   1 h   30 min   49.0 sec
Decl   : +89°   13'   56.28"
GHA   : 239°   13'   33.37"
LHA   : 149°   47'   33.42"
Azimuth : 359°   28'   36.17"

Observation # 2
Time   :   1 h   39 min   00.0 sec
Decl   : +89°   13'   56.28"
GHA   : 241°   16'   38.49"
LHA   : 151°   50'   38.49"
Azimuth : 359°   30'   33.62"

Observation # 3
Time   :   1 h   44 min   33.0 sec
Decl   : +89°   13'   56.28"
GHA   : 242°   40'   7.05"
LHA: : 153°   14'   6.99"
Azimuth : 359°   31'   54.37"

Observation # 4
Time   :   1 h   49 min   46.0 sec
Decl   : +89°   13'   56.28"
GHA   : 243°   58'   34.75"
LHA   : 154°   32'   34.74"
Azimuth : 359°   33'   11.38"
Mean azimuth of Polaris: 359° 31' 03.9"

F-3.3 Listing of Output for Example 18-4

CELESTIAL REDUCTION : HOUR-ANGLE METHOD

Observer's latitude :   42°   45'   10.00"
            longitude :   73°   56'   30.00"

4 observations made to sun

Declination at 00 hours on
      day of observation   :   22°   17'   10.0"
      day after observation   :   22°   24'   26.0"

GHA at 00 hours on
      day of observation   : 180°   29'   38.00"
      day after observation   : 180°   27'   09.00"

Observation # 1
Time   :   20 h   46 min   10.0 sec
Decl   : +22°   23'   28.62"
GHA   : 131°   59'   58.96"
LHA   :   58°   03'   28.99"
Azimuth : 266°   10'   56.98"

Observation # 2
Time   :   20 h   46 min   56.0 sec
Decl   : +22°   23'   28.85"
GHA   : 132°   11'   29.01"
LHA   :   58°   14'   59.05"
Azimuth : 266°   19'   11.59"

Observation # 3
Time   :   20 h   49 min   05.0 sec
Decl   : +22°   23'   29.48"
GHA   : 132°   43'   43.65"
LHA: :   58°   47'   13.69"
Azimuth : 266°   42'   12.68"

Observation # 4
Time   :   20 h   49 min   52.0 sec
Decl   : +22°   23'   29.71"
GHA   : 132°   55'   28.69"
LHA   :   58°   58'   58.70"
Azimuth : 266°   50'   33.98"
Mean azimuth of sun: 266° 30' 43.8"

APPENDIX G
ANSWERS TO SELECTED NUMERICAL PROBLEMS

CHAPTER 2

2-1(c)	4125.00 ft
2-2(c)	404.87 m
2-3(c)	1262 ft
2-4(c)	75.2388 sq Gunter ch
2-6(b)	50.23 ft
2-8(b)	3.69 acres
2-11(d)	1370
2-12(c)	3.5205×10^8
2-13(b)	0.50534, 1.12661, and 1.50964 radians; sum = pi radians
2-16	(a) 728.575 ft; (b) $\pm$ 0.065 ft; (c) $\pm$ 0.021 ft
2-21	(a) 728.531 $-$ 728.619, 60%; (b) 728.468 $-$ 728.682, 80%
2-26	(a) 46°30′02″; (b) $\pm$ 14.1″; (c) $\pm$ 5.8″
2-31	$\pm$ 0.01 ft
2-34(a)	$\pm$ 0.15 ft
2-38(a)	Area = 528,160 ft^2; $\sigma = \pm$ 140 ft^2
2-41(a)	Elev. diff. = $-$ 66.08 ft; $\sigma = \pm$ 0.20 ft

CHAPTER 4

4-2	(a) 2.67 ft; (b) 564 ft
4-6(c)	650.88 ft
4-8	536.99 ft
4-12	966.17 ft
4-15	576.06 ft
4-17	199.97 ft
4-20	97.04 ft
4-22	200.15 ft
4-26	382.391 m
4-29	266.075 m
4-32	653.67 ft

4-37	29.7 lb
4-38(d)	0.0037 m
4-40	$M = 486.27$ ft; $\sigma = \pm 0.033$ ft; $E_{90} = \pm 0.054$ ft
4-44	± 0.05 ft

CHAPTER 5

5-4	(a) 20.000 m
5-8	(a) 8.8 mm; (b) 10.00 mm; (c) 85.5 mm
5-9	(a) $\frac{1}{6000}$; (b) $\frac{1}{52,000}$; (c) $\frac{1}{216,000}$
5-14	(a) 5.5 ppm; (b) 4.5 ppm [(a) is greater]
5-16	(a) 0.7 in; (b) 0.4 in
5-18	(a) 4°C; (b) 10°C
5-19	1429.29 ft
5-22	516.95 ft
5-24	61.65 m

CHAPTER 6

6-3	0.000051 ft; 0.00021 ft; 0.00082 ft; 0.0052 ft; 0.021 ft
6-6	19.8 km
6-8	86 m
6-11	(a) 0.129 ft; (b) 0.515 ft; (c) 1.292 ft
6-15	2543.74 ft
6-21	0.048 ft
6-23	6.12 ft
6-26	HI = 976.51 ft; elev X = 968.65 ft

CHAPTER 7

7-6	0.45 ft
7-8	0.020 ft
7-11	(Partial answer): Adjusted elevs: TP1 = 1239.34 ft; BMX = 1236.72 ft; TP2 = 1235.58 ft
7-14	Elev. diff. = −1.62 ft; Y reading = 6.36 ft
7-17	Misclosure = 0.02 ft; elev A = 1105.785 ft
7-19	(Partial answer): Elev. TP1 = 228.800 m; Elev. BMY = 230.037 m
7-21	(Partial answer): Adjusted elevs.: $1+00 = 1276.8$; $2+00 = 1270.4$; $3+00 = 1275.4$; $4+00 = 1273.5$; $5+00 = 1270.2$
7-24	+3.90%
7-32	(a) 0.141 ft; (b) 0.088 ft; (c) 0.083 ft

CHAPTER 8

8-4	(a) 0.65392 rad; (b) 52°26'12"; (c) 28.2922°
8-6	N32°05'E; S50°42'35"E; S57°48'W; N65°W
8-9	Azimuths: 37°15'; 128°28'; 204°31'; 303°46' Angles: 91°13'; 76°03'; 99°15'
8-12	60°23'
8-16	$A = B = C = 90°00'$; $D = 143°12'$; $E = 126°48'$
8-20	Bearings: BC = N58°52'W; CD = S44°53'W; DE = S35°38'E; EA = S77°18'E
8-24	BC = S72°42'E; CD = S0°42'E; DE = S71°18'W; EA = S36°42'W

8-27   23°17′25″; 32°02′57″; 19°54′41″; 34°48′27″; 22°14′17″
8-30 $BC = 42°31′$; $CD = 40°07′$; $DE = 319°43′$; $EF = 287°01′$; $FG = 251°36′$;
          $GH = 235°06′$; $HI = 175°56′$; $IJ = 97°13′$; $JA = 104°31′$
8-33 $BC = 133°20′$; $CD = 36°32′$; $DE = 346°33′$; $EF = 301°00′$; $FA = 339°22′$

CHAPTER 9

9-6   (a) N6°35′W; (b) S6°35′E; S83°25′W
9-10   N41°13′E
9-12   2°30′ West
9-20 $AB = $ N53°55′W; $BC = $ S54°00′W
9-24   S85°20′W
9-27   27°25′ East
9-31 $B = 1°15′$ East; $BC = $ N83°25′E; $C = 3°05′$ West; $CD = $ N78°50′W
9-33 $A = 107°45′$; $B = 128°15′$; $C = 107°15′$; $D = 44°15′$; $E = 152°30′$

CHAPTER 10

10-7(c)   1′

CHAPTER 11

11-1(c)   26″
11-2(d)   1′09″
11-10   64°13.7′
11-13 $x = 49°36.5′$; $y = 310°23.0$; misclosure $= -0.5′$
11-15 $BAC = 63°25′23″$; $CAD = 163°11′41″$; $DAB = 133°22′56″$
11-18   ± 16″
11-20   0.019 ft
11-22   (a) 10″; (b) 0.073 ft
11-29   Index error $= -03′$; Angle $= -7°11′$
11-33(c) ± 1.4″
11-36(c) 08′36″; 03′26″; 01′23″; 00′52″; 00′10″

CHAPTER 12

12-7 $BC = $ N61°03′E; $CD = $ S61°15′E; $DE = $ S31°00′E; $EF = $ S49°50′E; $FG = $
          N40°08′E
12-10   134°59′
12-13   86°30′ Left
12-18   ± 24″
12-25 D
12-29   (Partial answer): Misclosure $= -06′$; $AB = 275.46$ ft; $CD = 191.44$ ft; $EF = $
          426.39 ft

CHAPTER 13

13-2   Misclosure $= 05′$; after rounding, first, third, fifth, sixth, and eighth angles receive
          a correction of $-01′$.
13-7   (Partial answer): Bearings: $BC = $ S5°07′E; $CD = $ S73°25′W; $DE = $ N82°48′W;
          $EF = $ S80°29′W; $FA = $ N4°27′W

13-8   (Partial answer): Departure $CD = -420.72$ ft; latitude $CD = -125.29$ ft; linear misclosure = 0.36 ft; relative precision = $\frac{1}{13,000}$

13-9   (Partial answer): Departure $CD = -420.75$ ft; latitude $CD = -125.27$ ft; $X_c = 11,790.93$ ft; $Y_c = 9535.93$ ft; length $AD = 1491.55$ ft; bearing $EB = $ N62°00′02″E

13-19   (Partial answer): Unbalanced dep $BC = 621.30$ ft; unbalanced lat $BC = 32.20$ ft; linear misclosure = 0.69 ft; relative precision = $\frac{1}{3200}$; balanced dep $BC = 621.44$ ft; balanced lat $BC = 32.07$ ft; final coordinates: $X_c = 1679.74$ ft; $Y_c = 1431.80$ ft

13-23   (Partial answer): Unbalanced dep $CD = -737.07$ ft; unbalanced lat $CD = -1041.39$ ft; linear misclosure = 0.98 ft; relative precision = $\frac{1}{4100}$; balanced dep $CD = -737.29$ ft; balanced lat $CD = -1041.17$ ft; adjusted length $CD = 1275.79$ ft; adjusted bearing $CD = $ S35°18.2′W

13-24   (Partial answer): Preliminary azimuth $CD = 37°27′19″$; angle misclosure = 05″; unbalanced dep $CD = 747.86$ ft; unbalanced lat $CD = 976.24$ ft; misclosure in dep = 0.26 ft; misclosure in lat = −0.21 ft; relative precision = $\frac{1}{16,000}$; balanced dep $CD = 747.80$ ft; balanced lat $CD = 976.29$ ft; final coordinates $X_D = 38,061.25$ ft; $Y_D = 27,612.44$ ft; final length $BC = 1379.76$ ft; final azimuth BC = 130°24′46″

13-29 CB is probably too short by 10 ft.

13-32 $AB = 748.18$ ft, 355°46′14″; $BC = 523.21$ ft, 61°05′18″; $CD = 586.68$ ft, 149°18′56″; $DE = 858.88$ ft, 234°50′38″

13-33   True bearing $AB = $ N46°20.9′E; length $AB = 2706.2$ ft

13-35 $A = 32°29′06″$; $B = 116°50′27″$; $C = 80°01′53″$; $D = 94°17′28″$; $E = 216°21′06″$

CHAPTER 14

14-1   4.815 acres
14-5   2.1053 acres
14-8   9.6293 acres
14-9   6.9331 acres
14-15   94.8101 acres
14-16   64.7005 acres
14-20   9.2191 acres
14-24   (a) GA: E dep = 10, N lat = 2; (b) 1019.80 ft; (c) 31.4509 acres
14-28   Length $DA = 446.96$ ft; bearing $DA = $ S85°17.9′E; area = 1.5715 acres
14-30   2.3230 acres
14-33 $AE = BF = 165.51$ ft
14-35   21.76 acres

CHAPTER 15

15-8 $K = 100.5$
15-13 $H = 339.1$ ft; $V = -28.8$ ft
15-15 $H = 184.4$ ft; $V = +21.1$ ft
15-17   977.8 ft
15-20 $H = 324.0$; $\text{elev}_B = 471.9$ ft
15-22   5′44″
15-24   Dist = 298.2 ft; elev = 990.3 ft

CHAPTER 16

16-5   6.7%
16-7   10%

16-13   63
16-26   0.47 in

CHAPTER 17

17-5   6.7 ft
17-8   Each elevation that is evenly divisible by 10 from 1290 through 1400. Heavier contours: 1300, 1350 and 1400.
17-10   2 ft; 1 ft or 0.5 m; 20 ft or 10 m
17-17   0.3 in

CHAPTER 18

18-5   39°47′ North
18-8   10:20:38.7
18-9   20°, 2′28″; 25°, 1′55″; 30°, 1′33″; 35°, 1′17″; 40°, 1′04″; 45°, 0′54″
18-10   20°, 8.3″; 25°, 8.0″; 30°, 7.6″; 35°, 7.2″; 40°, 6.8″; 45°, 6.2″
18-12(c)   4:00 AM PST
18-13(d)   1:06:48.8
18-17   Azimuth = 0°00′00″; altitude = 43°05′N
18-19(b)   3:44 PM PST
18-23   59°22′
18-26(a)   N51°27′09″W
18-28   37°42′36″
18-36   42°54′N
18-39   38°06′33″N

CHAPTER 19

19-3   436.05 m
19-10   2.74 mm; 7.12 mm
19-19 $\epsilon'' = 15.9''$; angular misclosure = −00.6″; $A = 51°36′24.2''$; $B = 71°52′26.5''$; $C = 56°31′25.2''$
19-21 $X = 23,235.44$ ft; $Y = 22,381.98$ ft
19-24   263°27′26.1″
19-27   19,838.37 m
19-32 $C = 0.0188$ ft/ft of interval
19-35   First-Order, Class II

CHAPTER 20

20-24   (Partial answer):

TIME (SEC)	ACCEL. (FT/SEC)	VEL. (FT/SEC)	INCR. DIST. (FT)	CUM. DIST. (FT)
5	5	25	22.5	62.5
10	5	50	47.5	295.0
13	−10	20	25	355.0
17	−5	0	2.5	395.0

20-26(a)   22°30′
20-27(a)   9°38′44″

CHAPTER 21

21-7   (Partial answer): For elev = 3000 ft; factor = 0.99985649
21-9   (Partial answer): For 3000 m, factor = 0.99953404
21-13   6614.01 ft and 118°15′36″
21-15(a) 895.22 ft
21-16(b) 1643.343 m
21-17 AB = 2187.32 ft; BC = 2991.62 ft; CA = 3923.07 ft
21-18 AB = 2187.46 ft; BC = 2991.82 ft; CA = 3923.33 ft
21-19   Grid azimuths: AB = 268°54′07″; BC = 186°08′14″; CA = 39°43′08″
21-20   (Partial answer): Dep BC = −319.86 ft; lat BC = −2974.67 ft; linear misclosure
          = 1.20 ft; relative precision = 1/7600
21-21   (Partial answer): Dep BC = −319.91 ft; lat BC = −2975.06 ft; X_C =
          2,303,787.79 ft; Y_C = 560,223.19 ft

CHAPTER 22

22-7   Linear misclosure = 0.00 ft
22-17   Linear misclosure = 2.00 ft; no
22-21   Linear misclosure = 0.06 ft; area = 11,690 ft^2
22-22 xy = 254.88 ft; By = 368.50 ft
22-30   Linear misclosure = 0.00 ft; area = 24,130 ft^2

CHAPTER 23

23-4(a)   68.5 ft
23-5(a)   186.2″ (03′06.2″)
23-6(b)   16.6 miles
23-10   (Partial answer): Area = 40 acres
23-12(a) ¼ mi square; 40 acres
23-13(b) 3840 rods
23-17   479.31 ch
23-23   Single

CHAPTER 24

24-8   (Partial answer): 11+00, elev = 851.50 ft; 12+50, elev = 850.30 ft
24-10   (Partial answer): 2+00, cut = 6.49 ft; 5+00, cut = 5.60 ft
24-11   (Partial answer): 3+00, rod = 4.48 ft; 6+00, rod = 7.36 ft
24-15   (Partial answer): AC = 107.70 ft; FB = 141.42 ft
24-17   79,600 ft^2
24-18   5.30 ft
24-29   (Partial answer): 12+00, 11.04 ft; 14+00, 7.64 ft; 16+00, 4.24 ft

CHAPTER 25

25-1(a) $D_a = 7°38'22''$; $D_c = 7°38'42''$

25-4 $T = 336.90$ ft; $L = 663.64$ ft; $E = 35.91$ ft; $M = 35.10$ ft; $R = 1562.61$ ft; $PC = 47 + 25.58$; $PT = 53 + 89.22$

25-6 $T = 401.55$ ft; $L = 796.25$ ft; $E = 32.04$ ft; $M = 31.64$ ft; $D = 2°17'31''$; $PC = 119 + 23.25$; $PT = 127 + 19.50$

25-7   (Partial answer): $PC = 24 + 36.83$; $PT = 34 + 63.14$; $c_a = 63.17$ ft; $c = 99.99$ ft; $c_b = 63.14$ ft; $\delta\ 26 + 00 = 2°35'01''$; $\delta\ 30 + 00 = 8°55'01''$

25-12   (Partial answer): $PC = 34 + 48.57$; $PT = 44 + 48.57$; $c_a = 51.43$ ft; $c = 100.00$ ft; $c_b = 48.57$ ft; $\delta\ 37 + 00 = 2°18'17''$; $\delta = 42 + 00 = 6°53'17''$

25-14   (Partial answer): $I = 12°05'44''$; $PC = 104 + 60.12$; $PT = 109 + 28.33$; $c_a = 39.88$ ft; $c = 100.00$ ft; $c_b = 28.34$ ft; $\delta\ 105 + 00 = 0°30'54''$; $\delta\ 108 + 00 = 4°23'24''$

25-17   (Partial answer): $PC = 129 + 16.20$; $PT = 136 + 42.26$; $130 + 50$, $TC = 133.78$ ft, $\delta = 1°28'27''$; $133 + 00$, $TC = 383.45$ ft, $\delta = 4°13'44''$; $135 + 00$, $TC = 582.57$ ft, $\delta = 6°25'57''$

25-22   (Partial answer): $I = 38°38'$; $PI = 56 + 58.85$; $PC = 51 + 86.30$; $PT = 60 + 95.32$; $c_a = 13.70$ ft; $c = 50.00$ ft; $c_b = 45.32$ ft; $\delta\ 55 + 00 = 6°39'58''$; $\delta\ 60 + 00 = 17°17'28''$

25-24   1392.03 ft

25-30   (Partial answer): $PC = 54 + 39.49$; $PT = 62 + 84.10$; $56 + 50$, $TC = 210.46$, $\delta = 2°16'50''$; $59 + 00$, $TC = 509.72$, $\delta = 5°31'50''$; $61 + 50$, $TC = 708.38$, $\delta = 7°41'50''$

25-35 $X_p = 1132.94$ ft; $Y_p = 1593.26$ ft

25-38   490 ft

25-41   (Partial answer): $\Delta_s = 3°15'$; $X = 299.90$ ft; $Y = 5.67$ ft; $o = 1.42$ ft; $TS = 52 + 89.26$; $SC = 55 + 89.26$; $c_a = 10.74$ ft; $c = 100.00$ ft; $c_b = 89.26$ ft; $53 + 00$, $\delta = 0°00'05''$; $55 + 00$, $\delta = 0°32'04''$

CHAPTER 26

26-3   (Partial answer): $BVC = 45 + 50$, 961.75 ft; $EVC = 55 + 50$, 964.25 ft; $48 + 00$, 967.06 ft; $53 + 00$, 968.31 ft

26-5   (Partial answer): $BVC = 86 + 55$, 137.98 ft; $EVC = 92 + 05$, 133.44 ft; $87 + 50$, 135.56 ft; $90 + 00$, 132.46 ft

26-11   (Partial answer): $L = 10.00$ sta; $BVC = 31 + 00$, 595.00 ft; $EVC = 41 + 00$, 627.50 ft; $34 + 00$, 608.42 ft; $39 + 00$, 623.80 ft

26-13   (Partial answer): $g = +1.40\%$; $BVC_1 = 64 + 00$, 1236.00 ft; $EVC_1 = BVC_2 = 70 + 00$, 1252.20 ft; $68 + 00$, 1248.53 ft; $EVC_2 = 74 + 00$, 1250.00 ft; $73 + 00$, 1252.01 ft

26-19   712.45 ft

26-21 $51 + 00$; 969.31 ft

26-30   22.5436 sta

26-33   7.53 ft

CHAPTER 27

27-2   (Partial answer): For fill $= 6$ ft, $A = 228$ ft^2; For fill $= 18$ ft, $A = 1512$ ft^2

27-5 $V_e = 1740$ cy

27-7 $V_e = 293$ cy

27-10   (Partial answer): Total cumulative volume $= -705$ cy

27-13 $A_1 = 66.7$ ft^2; $A_2 = 91.6$ ft^2

27-15	$c_p = 0.5$ cy; $V_p = 292.5$ cy; no
27-18	Left = 33.0 ft at elev 963.1 ft; right = 42.6 ft at elev 960.4 ft; $A = 391$ ft^2
27-21	$V_e = 802$ cy; $V_p = 794$ cy
27-24	$V_e = 458$ cy; $c_p = -7$ cy; $V_p = 465$ cy
27-26	$V = 5.07$ acre-ft
27-30	Discharge = 382 ft^3/sec

CHAPTER 28

28-4(c)	1:11,960
28-5(b)	$H = 10,000$ ft
28-7(c)	$s = 1{:}29{,}100 = 1''/2420$ ft
28-9	55.92 acres
28-12	$AB = 7990$ ft
28-14	338 acres
28-15(a)	$h = 360$ ft
28-16	$H = 5650$ ft
28-18	$AB = 3380$ ft
28-21	$h_A = 2664$ ft; $h_B = 2435$ ft
28-25	16
28-30	24 flight lines; 45 photos per line; total photos = 1080

CHAPTER 29

29-7(a)	9,000,000
29-21(a)	5 ft

INDEX

ABBREVIATIONS

CONSTRUCTION SURVEYS

b.b.	batterboards
B.L.	building line
C.B.	catch basin
C.G.	center line of grade
$\mathcal{C}$L	center line
const.	construction
C	cut
esmt.	easement
F	fill
F.G.	finish grade, Fin. Gr.
F.H.	fire hydrant
FL	fence line
F.L.	flow line (invert)
F.S.	finished surface
G.C.	grade change
G.P.	grade point
G.R.	grade rod (s.s. notes)
L	left (x-sect. notes)
M.H.	manhole
PL	property line
P.P.	power pole
pvmt.	pavement
R	right (x-sect. notes)
R/W	right-of-way
S.D.	storm drain
S.G.	subgrade
S.L.	spring line
spec.	specifications
Sq.	square
s.s.	slope stake, side slope
Std.	standard
Str. Gr.	straight grade
X-sect.	cross-section

H & T	hub and tack
H.C.	house connection sewer
I.B.	iron bolt (bar)
I.P.	iron pipe; iron pin
L & T	lead and tack
max.	maximum
min.	minimum
M.H.W.	mean high water
M.L.L.W.	mean low low water
M.L.W.	mean low water
Mon.	monument
No.	number
P	pipe; pin
Rec.	record
St.	street
Std. Surv. Mon.	standard survey monument
Std. Trav. Mon.	standard traverse monument
2″ × 2″	two- × two-inch stake
X	cross cut in stone
yd	yard

PUBLIC LANDS SURVEYS

ac	acres
AMC	auxiliary meander corner
bdy., bdys.	boundary, boundaries
BT	bearing tree
CC	closing corner
ch, chs	chain, chains
cor., cors.	corner, corners
corr.	correction
decl.	declination
dist.	distance
frac.	fractional (sec., etc.)
Gr.	Greenwich
G.M.	guide meridian
lk, lks	link, links
meas.	measurement
mer.	meridian
mkd.	marked
Mi. Cor.	mile corner
MC	meander corner
M.S.	mineral survey

PROPERTY SURVEYS

A	area
C.F.	curb face
ch "X"	chiseled X cross
C.I.	cast iron
diam.	diameter
Dr.	drive
ER	end of return
Ex.	existing